Foundations of
DISCRETE MATHEMATICS

K.D. Joshi

Department of Mathematics
Indian Institute of Technology
Powai, Bombay
India

JOHN WILEY & SONS
New York   Chichester   Brisbane   Toronto   Singapore

First published in 1989 by
WILEY EASTERN LIMITED
4835/24, Ansari Road, Daryaganj
New Delhi 110 002, India

Distributors:

Australia and New Zealand:
JACARANDA WILEY LTD.
GPO Box 859, Brisbane, Queensland 400₁, Australia

Canada:
JOHN WILEY & SONS CANADA LIMITED
22 Worcester Road, Rexdale, Ontario, Canada

Europe and Africa:
JOHN WILEY & SONS LIMITED
Baffins Lane, Chichester, West Sussex, England

South East Asia:
JOHN WILEY & SONS, INC.
05-04, Block B, Union Industrial Building
37 Jalan Pemimpin, Singapore 2057

Africa and South Asia:
WILEY EXPORTS LIMITED
4835/24, Ansari Road, Daryaganj
New Delhi 110 002, India

North and South America and rest of the world:
JOHN WILEY & SONS, INC.
605 Third Avenue, New York, NY 10158, USA

Library of Congress Cataloging in Publication Data

Joshi, K.D.
    Foundations of discrete mathematics/K.D. Joshi.
        p.   cm.
    Includes index
       I. Mathematics – 1961–   2. Electronic data processing—
Mathematics.   3. Combinatorial analysis.   I. Title.
QA39.2.J67   1989   88–14811
510—dc19   CIP

ISBN 0-470-21152-0 John Wiley & Sons, Inc.
ISBN 81-224-0120-1 Wiley Eastern Limited

Printed in India at Prabhat Press, Meerut,

To my discreet wife
SWARADA
for her continuous support

List of Standard Symbols

Symbol	Meaning
ϕ	the empty set
$\mathbb{N}$	the set of natural numbers ($=$ positive integers)
$\mathbb{Z}$	the set of all integers
$\mathbb{Z}_m$	the set of residue classes modulo a positive integer m
$\mathbb{Q}$	the set of rational numbers
$\mathbb{R}$	the set of real numbers
$\mathbb{C}$	the set of complex numbers
$\mathbb{R}^n$	the set of ordered n-tuples of real numbers
S^1	the set of complex numbers of length 1
$\lvert X \rvert$	cardinality of a set X
$P(X)$	the power set of a set X
$P_r(X)$	the set of all r-subsets of X, i.e.
	$$\{A \subset X : \lvert A \rvert = r\}$$
$n!$	'n factorial' ($= 1, 2, 3 \ldots (n-1)n$).
$\binom{n}{m}$	'n choose m' $\left(= \dfrac{n(n-1)\ldots(n-m+1)}{m!} \right)$
D_n	dihedral group of order $2n$
S_n	symmetric group on n symbols
A_n	alternating group on n symbols
$p(n)$	number of partitions of an integer n
P_n	number of partitions of a set with n elements. ($= n$th Bell number)
F_n	nth Fibonacci number
C_n	nth Catalan number
H_n	nth harmonic number
$o(G)$	order of group G
$\subset$	contained in (and possibly equal to)
$\subsetneq$	contained in but not equal to
$\mid$	divides

Suggested Course Coverage

This book is intended to be covered in two one-semester courses with 14 weeks of instruction per semester. Because of the diversity of preferences on the part of the instructors and the diversity of the backgrounds, the calibres and the needs on the part of students, it is impossible to prescribe a uniform style or schedule of coverage. However, the following is a pattern which may be applied under 'average' conditions.

The core of the text lies in Chapters 2 to 7. The level of the material and the style of presentation is kept elementary. Also the answers to most exercises are given at the end of the book. Because of these features, coupled with some initiation from the instructor, it is expected that an average sincere student should be able to read and understand most of the material largely on his own. In that case each section can be covered generally in one week (assuming 3 to 4 hours of instruction per week). The instructor is strongly urged not to duplicate the proofs given here but instead to supplement them with numerical and diagrammatic illustrations (which are somewhat lacking in this book), with comments about the subtle points in the proofs and with alternate proofs wherever possible. Depending on the level of the class, the instructor may also wish to skip some of the exercises and/or to supplement them with simpler exercises designed to give computational drill.

The suggested schedule of coverage is as follows :

First Course : Spend about 2 weeks on Chapter 1. Then spend about one week per section in Chapters 2, 3 and 4.

Second Course : Spend about one week per section in Chapters 5, 6 and 7. Allow a little extra time for Sections 6.2, 6.3 and 6.4.

Preface

This book is intended to be more than just a textbook of discrete mathematics. Its ultimate goal is to make a strong case for the incorporation of discrete mathematics into the basic core curriculum of undergraduate mathematics. This is a rather ambitious task and deserves some elaboration.

Mathematics can be broadly divided into two parts; the continuous mathematics and the discrete mathematics depending upon the presence or absence of the limiting process. (The distinction is brought out more fully in Chapter 1.) Discrete mathematics is conceptually easier and more akin to human experience than continuous mathematics. Ironically, it is these very qualities which give the impression that discrete mathematics is elementary, indeed trivial, and does not deserve to be studied at the collegiate level. So it is the continuous mathematics which has long dominated the scene. This is also consistent with the history of applications of mathematics. Apart from the age-old applications of arithmetic, algebra, geometry and trigonometry, nearly all real-life applications of mathematics till the end of the nineteenth century were through physics. Even today, what is generally understood by 'applied mathematics' consists of topics such as mechanics of solids and fluids, heat transfer, electromagnetic theory etc. In the view of the nineteenth century physics, energy and the other physical variables were assumed to be continuous. This explains why the concept of a limit acquired such a paramount position, a natural corollary being the standardisation of the undergraduate mathematics curriculum so as to include a sequence of two or three calculus courses, followed by one or two courses in differential equations, numerical analysis, complex analysis and so on. There is hardly any room for discrete mathematics in this programme. Whatever little discrete mathematics is needed is considered mostly a matter of common sense or something which one just picks up along the way.

This picture began to change in the twentieth century. Applications of mathematics to physics continued to flourish, especially during the first half where they got a big boost because of the theory of relativity. But applications to other fields, such as statistics, operations research, economics, design of circuits, logic, computer science etc. also got in full swing. An even more important development was the change that took place in the conception of mathematics. Its focal point changed from the concept of a number to the concept of a set (more on this point in Chapter 3). By its very nature this new non-numerical mathematics was more amenable to the methods of discrete mathematics than to those of continuous mathematics. This resulted in a tremendous increase in the real-life applications of discrete mathematics. Things such as finite fields, which at one time were considered to be too abstract to have any practical relevance were shown to have down-to-earth applications (e.g. in designing economical codes).

As a result, the attitude towards discrete mathematics is changing slowly but surely. The skepticism that once prevailed about its relevance is waning. Rich research contributions are being made to it. At the pedagogical level, specialised elective courses catering to various branches of discrete mathematics such as applied algebra, combinatorics, graph theory, linear programming have been started. Many books are being written on these subjects.

Welcome as all these signs are, the undergraduate curricula are yet to accord discrete mathematics its due place in the mainstream of mathematics. The 'core' courses continue to be dominated, almost exclusively, by continuous mathematics. Where specialised courses in discrete mathematics such as those mentioned above do exist, they are generally treated as peripheral, of interest only to certain classes of students (mostly majoring outside mathematics). They are rarely considered to be an integral part of 'basic' mathematics. As a result, we have the paradoxical situation that although in a compulsory calculus course a student is thoroughly drilled into finding maxima and minima, these methods are rarely used in an actual problem, whereas the methods that *are* actually used in practice (e.g. the simplex method) have to be picked up from specialised, 'elective' courses!

Ironically, again, the very strengths of discrete mathematics have come in the way of its entry into the mathematics core. One of the strong points of discrete mathematics is its powerful applications to fields like computer science, engineering and operations research. Indeed, a good deal of discrete mathematics owes its development to problems in some of these areas. In this respect the discrete mathematics behaves no differently from continuous mathematics. Many concepts of the classical continuous mathematics also originated from applications to physical sciences. However, in the case of continuous mathematics, the process of isolating the underlying mathematical thought from a particular application began a long time ago. As a result, it is possible today to give coherent courses in continuous mathematics (already named above), in which, although there are sufficient hints about the applicability of the various concepts, the emphasis is on the mathematical aspects of those concepts and not on any particular applications.

In the case of discrete mathematics, there is an inherent difficulty in separating the underlying mathematics from its applications. It is a fact that many results of discrete mathematics, when stripped of their applications, appear either too trivial or too abstract. In either case, their introduction into the core courses in mathematics is not looked at favourably by the pedagogists of mathematics. A good case in point is the well-known pigeon-hole principle or the double counting argument or some results in graph theory. When ingeneously applied, they work wonders. But when seen all by themselves, they appear so trivial that one wonders if they deserve even to be stated explicitly. As examples of the other kind, we have the theories of groups and fields. Both are replete with profound results. But without some down-to-earth applications (such as Polya's theory of counting or coding theory), they are likely to be disposed of as sterile intellectual exercises. Many mathematicians are either unaware of these applications or tend to dispose them of as too esoteric. Consequently, although both the group theory and the field theory are highly respected branches of graduate mathematics, they are still not considered to be parts of 'everybody's mathematics' the way double integrals and differential equations are.

Because discrete mathematics is so wedded to its own applications, it is not surprising that most of the currently available books on discrete mathematics are applications-oriented. As a rule, they are written for non-mathematicians such as computer scientists or engineers. In such books mathematics is only a means and not the goal and often gets a treatment which is at best utilitarian. As a result, even though some of these books are widely acclaimed outside the mathematical circles, they often fail to appeal to a traditional mathematician. He is not readily convinced that their mathematical contents deserve to be studied mathematically regardless of their particular applications.

Another strong point of discrete mathematics is its infinite variety of interesting problems. Recently a number of books of the 'problems and solutions' type have appeared on discrete mathematics, especially on combinatorics and graph theory. Such books are an intellectual treat for those who love to solve problems. They are also excellent sources of references. In books of this kind, it is possible to pack within a few pages a huge amount of information which, in a conventionally written book, would probably occupy several times as much space. A mature reader, who is in a position to supply the missing details, can learn a lot from such books. Naturally, these books lack coherence and cannot be used as texts in regular courses. There is very little motivation or elaboration of the ideas involved. If not taken in the proper spirit, these books are liable to give the impression that discrete mathematics is a scattered bunch of intellectual puzzles rather than a coherent subject capable of a systematic study.

In short, the dilemma of discrete mathematics is that on one hand, in order to firmly establish it as a fundamental branch of mathematics, it must be delinked from problems and applications of a particular kind. But on the other hand, in doing so there is a danger that the very heart of the subject may be lost.

In the present book, I have made an attempt to find a way out of this dilemma. Keeping in mind the role of problems in discrete mathematics, in the first chapter I have given a fairly long list of some typical problems and a few comments about them. Their solutions are, however, intentionally postponed. Instead, these problems are used from time to time to provide motivation for the various stages of development of the theory. The theory itself is developed systematically, in the traditional definition-theorem-proof pattern so characteristic of mathematics. Following the practice of modern mathematics I have taken a set as a starting point. The general theme is to show how a suitable set empowered with a suitable additional structure provides a convenient mathematical model for a given problem. I am aware that the average reader of this book may not be mathematically mature. I have therefore made a conscious effort to develop his mathematical maturity. This is done through comments about the nature of discrete mathematics, its place in mathematics as a whole and its relationship with continuous mathematics, a review of mathematical logic, and through general comments about the process of abstraction. A special emphasis is laid on motivating the definitions and results.

The applications of discrete mathematics have not been entirely ignored. In fact, I have given quite a few of them. However, as indicated above, the emphasis is more on the *applicability* of discrete structures rather than on any particular applications. So I have generally avoided applications of a technical nature which would demand a knowledge of some other fields, such as computer science or economics. In the few places where technical applications do occur (e.g. the applications of Boolean algebra to switching circuits), the relevant background has been developed. For all other applications, I have chosen real-life problems which can be understood and appreciated even by a layman. (Admittedly, some of these problems appear rather contrived.) One can of course garb a problem, at least superficially, so that it looks like a problem in some other field. For example, in a problem of putting balls into boxes we can think of a ball as a piece of data and a box as a memory register and the problem now becomes a problem in computer science! Whatever be the selling value of such tricks, I have generally refrained from them.

Inasmuch as the introduction of discrete mathematics into the under-graduate curricula is still in its infancy, the contents and the degree of coverage of the topics in a book like this are open to debate. Although there are dozens of different texts on calculus. they more or less cover the same topics and to the same degree of depth. A similar standardisation is also slowly taking place in textbooks covering particular aspects of discrete mathematics such as combinatorics, graph theory or applied algebra. The present book, however, is meant to give a unified rather than a piece-meal treatment of discrete mathematics. Naturally, I could not go as deep as a book specialising in any one area. Still I have attempted to reach a reasonable degree of depth.

The original plan was to include all the meterial in a single book 'Introduction to Discrete Structures'. But this proved impracticable in view of the size. So it was necessary to split it into two separate books, 'Foundations of Discrete Mathematics' and 'Applied Discrete Structures'. The first book, the present one, gives the fundamental concepts and techniques of discrete mathematics and a fairly thorough exposure to algebra. The second book is more applications oriented. (See the Epilogue for a detailed preview of 'Applied Discrete Structures'.) It contains a review of the first book, with the help of which it can be read independently of the first book.

The present book has seven chapters. The first chapter introduces the subject matter. It also contains the list of problems mentioned earlier which are nicknamed for a ready reference in future. The results of the second chapter are elementary. Still, they are treated rather formally, so as to familiarise the reader with the style of presentation to be encountered in the later chapters. The third chapter introduces the process of abstraction, studies two elementary structures on sets and gives the generalities about algebraic structures. The next three chapters deal with specific algebraic structures. The last chapter presents advanced counting techniques based on generating functions and recurrence relations.

Each chapter is divided into four sections. Each section contains exercises followed by 'Notes and Guide to Literature'. The latter are generally intended to direct the interested reader to appropriate references for further reading or to acknowledge credit to the sources from where I have borrowed something. A few historical remarks are also made occasionally. However, such remarks and references are only indicative. No claim is made about thier being complete or most up-to-date.

There are virtually no prerequisites for reading this book. Some facts about power series and differential equations will be referred to in Chapter 7. But that is more by way of relating discrete mathematics with continuous mathematics rather than a strict pre-requisite. Although this book is avowedly written to make a case for discrete mathematics, the idea is definitely not to be little continuous mathematics. A great majority of the readers of this book will have already studied calculus or at least be studying it concurrently. Consequently, wherever possible, I have indulged into comments about continuous mathematics as well. While a mature reader may find them platitudinous, I hope they will help the average reader gain new insights into the nature of continuous mathematics and thereby appreciate discrete mathematics even more.

Like most authors of textbooks, I had to strike a balance between pairs of mutually conflicting virtues such as expanse versus depth, clarity versus brevity and abstract versus concrete. The objectives of including almost all standard topics, of reaching a reasonable depth and of building up the mathematical maturity of the reader so that he can handle abstract concepts soon proved to be somewhat incompatible with each other and threatened to blow the size of the book out of proportion even after it was split into two. As a result, I was forced to make a few unpleasant choices. Some of the standard results had to be relegated to the exercises. It is hoped that with the generous hints given, the reader can work them out. (Answers to most of the exercises are given are the end of the book.) Also the diagrammatic and the numerical illustrations have been kept to a minimum. The emphasis is more on thoughts and less on numerical dexterity. I am of course, not unaware of the pedagogical importance of diagrams and worked-out examples. Had space permitted, I would have loved to include more of them. Another reason is that there already exist books which cater to these aspects very nicely. For the algebraic part we recommend William J. Gilbert's *'Modern Algebra with Applications'* and for the combinatorial part, Alan Tucker's *'Applied Combinatorics'*.

Exercises form an integral part of the book. The results of many of them are used freely in the text. There are virtually no exercises whose sole purpose is to provide numerical drill. Nearly all exercises require some thinking for their solution, the degree and the quality of which obviously vary considerably, Some of the exercises merely ask the reader to supply parts of a proof (occasionally an entire proof). A few require the application of the results proved in the text while a few others are meant to prove some standard results which could not be incorporated in the text. Hints for solution and comments about the significance are given liberally. The degree of difficulty of a problem, especially one where some thought is involved, is always a matter of personal opinion. Many challenging problems look deceptively simple once you know their solutions. It is therefore, very difficult to rank the exercises quantitatively in terms of their difficulty and to give the estimated time for working them out as is done by Donald Knuth in his pioneering volumes of *'The Art of Computer Programming'*. My own experience with such a time scale is, in fact, that when I could not do a problem within the stipulated time, it unnecessarily created an inferiority complex. I have therefore refrained from giving any quantitative assessment of the difficulty of a problem. Still, some qualitative indication seems to be in order. So I have put a star (*) over those problems, which, in my opinion, require a little originality of thought. Unusually demanding exercises are doubly starred (**). These include some standard theorems whose proofs are far from simple and also a few problems which, to my knowledge, are unsolved. These exercises are not really meant to be solved, even by the highly gifted student. He is merely expected to appreciate their difficulty. Here I recall with full agreement, a comment by I.N. Herstein in his *'Topics in Algebra'* that the value of a problem is not so much in coming up with the answer as in the ideas and the attempted ideas it forces on the would-be solver.

The material is so arranged that it can be covered in two courses of one semester each. (See the 'Suggested course coverage' for more details about this.) In fact, it is hoped that the present book, along with *'Applied Discrete Structures'* would eventually be used as texts for a sequence of four one-semester courses in discrete mathematics. Such a sequence, coupled with the traditional sequence in continuous mathematics, would provide a college student with a solid mathematical background, regardless of what he specialises in later on.

It was mostly by accident that I was prompted to write this book. My own mathematical training did not include discrete mathematics and like many others in my position I thought of it as something which is possibly useful but inherently trivial. Then I happened to read C.L. Liu's fine book *'Introduction to Combinatorial Mathematics'* and used it as a text for a course to students of computer science. In doing so, I realised that the mathematics in it was important enough to deserve a place in the mainstream of mathematics. Other books which have influenced me profoundly are those of Knuth and Herstein, mentioned above. The latter's influence extends not only to the coverage of algebra in this book, but also to my style in general. I am deeply indebted to these three authors.

My own expertise in discrete structures being limited to algebra, I frequently had to consult others on many points. I am grateful to many colleagues of mine who helped me by informal discussions and by pointing out appropriate references. I must especially mention prof. G.A. Patwardhan and Prof. M.N. Vartak. If despite their best help, any errors have occurred, I am entirely responsible.

Financial support for the preparation of the manuscript of this book was given by the Curriculum Development Cell at the Indian Institute of Technology, Bombay and is hereby thankfully acknowledged. I also thank Mr. Parameswaran for his sincerity and patience in typing the manuscript. The printers too have done a neat job. The rather large number of misprints that still remain is due mostly to my own negligence in proof-reading and is a sad testimony to the observation that the author himself is generally a poor proof-reader of his own work. A few misprints have been corrected in the 'Errata'.

The introduction of discrete mathematics in the core programme is still in the experimental stage and not yet fully implemented. I shall, therefore, be most interested in the views, comments and suggestions, not only from mathematicians who may be teaching it but from others as well. I hope that through such a dialogue a standard syllabus of discrete mathematics will evolve in near future. The twentieth century is coming to an end and the talk of orienting ourselves for the twentyfirst century has become a fashion of the day. Let us hope that before the dawn of the twentyfirst century, the undergraduate mathematics curriculum is freed, at least partially, from the tight grip of the nineteenth century physics.

Bombay,   K.D. JOSHI
February 20, 1989

Contents

One

Introduction and Preliminaries

This chapter is meant to give an idea of the kind of things that would be discussed in this book and also to review the prerequisites needed for their understanding. The spirit of discrete mathematics, especially in comparison to the continuous mathematics is discussed in Section 1. A rather long list of problems is given in Section 2. The comments about them in Section 3 guide the reader to the chapters in which the machinery needed to solve the problems will be developed. Section 4 gives a warm-up in logic.

1. What is Discrete Mathematics?

Discrete mathematics began at least as early as man (or perhaps some other animals) learnt to count. The fundamental idea behind counting is to establish a one-to-one correspondence (or a bijection, as it is technically called) between two sets of objects and even today this continues to be one of the most widely used devices in discrete mathematics. Nearly all the mathematics that we pick up in early school comes under discrete mathematics. This includes, the addition, multiplication and other arithmetical operations we do with integers and rational numbers. The fairly interesting topics of permutations and combinations and related problems in probability are an important part of discrete mathematics.

Thus 'discrete mathematics' is far from a new innovation in the history of mathematics. Perhaps the only new thing about it is its name. The dictionary meaning of the word 'discrete' is 'separate and distinct, unrelated, made up of distinct parts'. But this only remotely reflects the way the word is used in mathematics. In fact, the best way to understand the spirit of discrete mathematics is by camparing it with non-discrete mathematics! The latter is more popularly known as the **continuous or continuum mathematics.** Let us, therefore, take a look at what continuous mathematics is and how it came into being.

The earliest (and, till fairly recently, all) applications of mathematics to real life involved the process of assigning numbers to describe the sizes of various sets. The various acts to which these sets are subjected often naturally correspond to various mathematical operations on the numbers representing them. For example, suppose two bags of apples are to be emptied into a third, larger, bag (which is originally empty) and we want to find how many apples the third bag would contain. We can do so by simply adding the sizes (that is, the numbers of apples) of the two bags. Similarly cutting a piece of a wire into two equal parts amounts to dividing its length (or rather, the number which represents its length) by two.

The crucial question now is, which numbers should be assigned to the sets under study? Should they be natural numbers, integers (including negative integers as well as zero), rational numbers, real numbers or complex numbers? Moreover, how do we assign them? The answer depends upon the nature of the objects of study and also upon the type of study. In the two examples given above, we regard each apple as an individual unit. We ignore the differences in the sizes of the apples. The numbers assigned to the two bags containing them are whole numbers, that is, positive integers. On the other hand, in the case of a piece of a wire there is no natural unit. We arbitrarily choose some unit of length (such as an inch, a centimeter etc.) and assign to the wire the number by which this unit must be multiplied so as to get a segment of the same length as the given piece. This number need not be an integer and even when it is so, the number that comes out as the answer (namely, the length of each part of the wire after it is divided equally into two) need not be an integer. Thus, integers are adequate to handle the first problem but not the second. It is tempting to try to salvage the situation in the second problem by choosing the unit of length so as to make the answer come out as a whole number. But this trick may not work when more than one line segments are simultaneously involved. For instance, if one of the segments represents the side of a square and the other its diagonal, then it can be shown that it is impossible to find a common unit of length which would make the lengths of both as integers, or even rational numbers. Technically, this is expressed by saying that the side and the diagonal of a square are not *commensurate*. As a more poignant example, the diameter and the circumference of a circle are not commensurate with each other (although this fact is extremely non-trivial to prove).

The difference between discrete and continuous mathematics is basically the difference between a bag of apples and a piece of wire*! In the former, the apples sit apart discretely from each other while in the latter, the

---

*Other similies are also possible. An analog computer corresponds to continuous mathematics while a digital one to the discrete mathematics. An expert in Company Law would probably regard the stock of a company as a continuous variable and its share capital as a discrete variable!

points on a wire spread themselves continuously from one end to the other. Integers are adequate to handle the first but not the second. Given any positive real nnmber α, we could conceive of a piece of wire whose length is α times a fixed unit. Thus, in continuous mathematics, all real numbers must be allowed as possible representatives of some real-life objects (hence the name 'real' numbers). Therefore the length of an object is called a *continuous variable*. What applies to length applies equally well to many other quantities such as time and mass and hence also for all quantities derived from them such as area, volume, density, speed, momentum, energy and so on. All these are continuous variables. At least that was the view that dominated mechanics till fairly recently; that is, till the quantum theory came on the scene. The term 'classical mechanics' is now synonymous with 'continuum mechanics', 'continuum' being the name given to the real number system.

The problems as well as the methods of solution in continuous mathematics differ fundamentally from those in discrete mathematics. The difference, in fact, starts right from the terminology used. In discrete mathematics we 'count' the number of objects while in continuous mathematics we 'measure' their sizes. Many concepts in either branch become meaningless or at least inapplicable when carried to the other. There is, nevertheless, an interplay between the two. A problem in continuous mathematics can often be thought of as the limiting case of a similar problem in discrete mathematics. Put another way, methods of discrete mathematics can often be used to provide approximate solutions to problems in continuous mathematics.

We proceed to illustrate these remarks with a simple problem, which will be referred to as the '**House Problem**'. Let us suppose there is a straight road, one kilometer long, on which there is a house at every one-tenth of a kilometer. There would thus be eleven houses on the road (including the houses at the two ends of it). Suppose we pick two (distinct) houses at random. What is the probability that the distance between them will not exceed, say, one half kilometer? Because the houses are discretely located, this is a typical problem in discrete mathematics. Let us number the houses along the road by integers from 0 to 10. If x and y are two houses (with $y \geqslant x$) then the distance between them is simply $1/10 \, (y - x)$ kilometers. It is tempting to think that the answer to our problem is $1/2$. But this is clearly wrong because there are more pairs of houses that are closely located than those that are remotely located. It is easy to see that there are in all $\frac{1}{2}(11 \cdot 10) = 55$ pairs of distinct houses. This is, therefore, the total number of cases. As with all probability problems, we have to proceed on the assumption that each of these 55 cases is equally likely. (This is, in fact, the interpretation given to the phrase 'at random'.) The next step in our solution is to count the so-called 'number of favourable cases'. For convenience, whenever we consider a pair of houses, say, (x, y) let ua always suppose that $x < y$. We then want the number of such pairs with $x \lessdot y$ (since the two houses are distinct) and for which $1/10\ (y - x) \leqslant \frac{1}{2}$. Here x can take any of the eleven values from 0 to 10. When x is 0, y has to be between 1 and 5 (both included). When x is 1, y can vary from 2 to 6. This will go on till x is 5. Thereafter the variation of y will be restricted in its upper bound. For example when x is 7 y has to be either 8, 9 or 10 (see Figure 1.1). Keeping track of the number of possible values y can take for various values of x and adding, we get the number of favourable cases as $5+5+5+5+5+5+4+3+2+1+0$, that is, 40. Thus the answer to our problem, namely the probability that the distance between two randomly selected distinct houses be less than or equal to 1/2 kilometer, is 40/55 or 8/11.

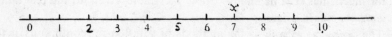

Figure 1.1 : The discrete House Problem

Let us now try the continuous version of this problem. Let us suppose that our one kilometer long road is literally packed with houses in the extreme; that is, there is a house at every point of it. (This may sound like an unrealistic problem, but if the road happens to be in a city like Bombay, it is awfully close to reality!) We now ask the same question, namely, if two distinct houses are picked at random, what is the probability that the distance between them does not exceed 1/2 kilometer? The method of solution to the earlier problem breaks down completely, because the total number of cases and the number of favourable cases are both infinite now and so the answer would be ∞/∞ which is meaningless. How do we tackle the problem then?

There are two methods for doing this. One is to regard the continuous version as a limiting case of the discrete version. Let us, therefore, revisit the discrete house problem, assuming this time that there is a house at every $1/n$th kilometer where n is a positive integer. (In the problem that we just solved n was 10.) Let p_n be the probability that the distance between two distinct, randomly selected houses is at most 1/2. Now as n tends to infinity the distance between consecutive houses tends to zero, the number of houses tends to infinity and the problem becomes more and more like the problem we are trying to solve. It is therefore reasonable to expect that if we solve the problem for each n, that is, if we compute p_n for every n and then take the limit of the sequence $\{p_n\}$ as n tends to infinity, then this limit would be the answer to our problem. Even if we are not able to actually evaluate this limit (a fairly common situation with sequences), still, for sufficiently large n. p_n would at least give us an approximate answer.

If we take this approach, then, in order to compute p_n we have to make two cases, depending upon whether n is even or odd. (It will be a good exercise for the reader to pin-point the reason for this.) If n is even, say $n = 2k$ where k is a positive integer, then it is easy to show, by reasoning similar to above, that $p_n = \dfrac{3k+1}{4k+2}$; while for n odd, say, $n = 2k + 1$, p_n comes out to be $\dfrac{3k}{4k+2}$. As $k \to \infty$ both these expression tend to a common limit, namely 3/4. Thus p_n converges to 3/4 and this is the required probability.

There is also another method to solve the continuous house problem which does not involve its approximation by the discrete version. We begin by taking a new look at the solution given above for our original problem, with eleven houses on the road. In Figure 1.1, we pictured them with dots on a straight line. Let us now picture the ordered pairs of houses by dots in a square as in Figure 1.2. We need not explain in detail how this is done, because the idea is precisely that of cartesian coordinates. For example, the (3/10, 8/10) (the circled point in the figure) represents the pair consisting of the third and the eighth house. Because of our convention that the house with a smaller number will be listed first, we have drawn only the upper triangular half of the full square. There are in all 66 dots in this triangle. Since we want pairs of distinct houses, we ignore the dots on the diagonal. This leaves 55 dots, the same as the total number of cases. The favourable cases correspond to the dots on and below the line $y = x + \frac{1}{2}$.

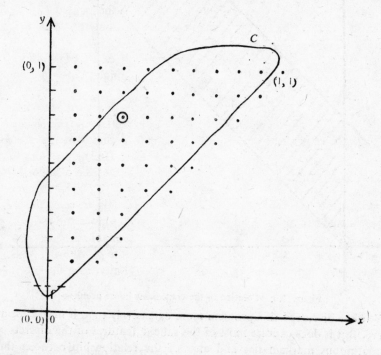

Figure 1.2: Re-interpretation of the solution to the discrete house problem.

They are shown by enclosing them with a curve C. Their count comes out to be 40, exactly as before.

So far we have not done anything new. We have merely given a geometric interpretation to our earlier solution. But in this new formulation, the solution can be easily adapted for the continuous house problem. Once again, we picture pairs of houses by points in a square and consider only the upper triangular half of it as in Figure 1.3. This corresponds to the set of all posible cases. The set of favourable cases corresponds to the trapezoidal region between the parallel lines $y = x$ and $y = x + \frac{1}{2}$. Both the sets contain infinitely many points. So we cannot compare them on that ground. But there is another way to compare them. We simply take their areas! The area of the upper triangle is 1/2 square kilometers while that of the trapezeum is 3/8 square kilometers. By taking the ratio, we get 3/4 as the required probability. This is of course the same as the answer obtained by the first method. Note incidentally, that in the continuous version it does not matter whether the two houses picked are distinct or not. Points corresponding to pairs of identical houses lie on the line $y = x$. The line itself has no area and so the area of the trapezeum is unaffected whether points of the boundary are included in it or not.

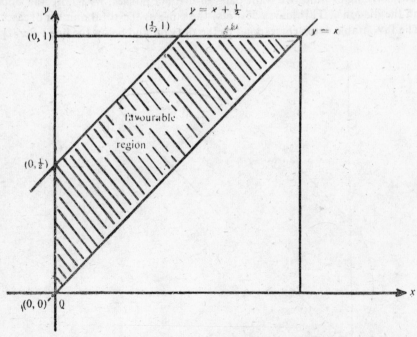

Figure 1.3: **Solution to the continuous house problem**

We have discussed this problem more extensively than it may appear to deserve. But it does capture most of the salient features of the discrete and the continuous mathematics and also of the relationship between them. Some sort of a limiting process is essential for continuous mathematics. (It is tempting to think that the second solution to the continuous house problem was free of any limiting process. But it is not so. A rigorous definition of area for planar figures does involve limits.) Actually, even the precise definition of a real number involves the limiting process in some form or the other. In discrete mathematics, on the other hand, there is hardly any scope for the limiting process. The sets involved are generally finite. (In fact the phrase 'finite mathematics' is often taken to mean the same as 'discrete mathematics'.) There is no way that a discrete variable can come 'infinitesimally close' to something. That is why, concepts like instantaneous speed have no meaning in discrete mathematics although they occur so frequently in continuous mathematics. We may even write a symbolic equation:

Discrete mathematics $+$ Limiting process $=$ Continuous mathematics

This equation is suggestive in many ways. It shows that continuous mathematics arises as a limiting case of discrete mathematics. An approximate solution to a continuous problem can be found using discrete mathematics. Sometimes the tables can be turned around. When a discrete variable assumes values which are very close (although not infinitesimally close) to each other, it may be thought of as a continuous variable for many purposes and so the methods of continuous mathematics become applicable. For example, in the house problem with n houses, for large values of n, the problem is very nearly the continuous house problem. Since we have an independent method of solving the latter, the answer (namely 3/4) can be taken as a good approximation to the answer for the discrete problem for all large n.

Limiting process is undoubtedly an important milestone in the history of human thought. Because of this, for quite some time, discrete mathematics was considered to belong to the infancy of man and continuous mathematics to his manhood. In recent times, however, discrete mathematicians are no longer considered as second class citizens in the mathematical world. In fact, today discrete mathematics is one of the most flourishing branches of mathematics. There are four good reasons for the revival of interest in discrete mathematics. We discuss them briefly.

The first is a very practical one. Although in theory a continuous variable can assume for its value any real number in some interval, in practice, even with the best measuring techniques, we can measure its value only upto a certain degree of accuracy. Thus, even when continuous mathematics gives us the exact answer to a problem as a nice real number, to put it into practice we have to resort to approximations, the commonest example of this being the approximation of the ubiquitous π as 22/7 or as 3.1416. In many problems, the things are even worse. In such problems, continuous mathematics gives the answer as a certain limit. But this limit cannot be evaluated concretely in a closed form. A case in point is the perimeter of an ellipse with semi-major axis a and semi-minor axis b. Using calculus it is easy to write the perimeter as a definite integral (which, by definition, is the limit of a sum), namely, as

$$\int_0^{2\pi} \sqrt{a^2 \sin^2 t + b^2 \cos^2 t} \; dt.$$

But there is no way to evaluate this integral. (If $a = b$ then the ellipse reduces to a circle and the integral can be evaluated to give the familiar answer $2\pi a$; but in general it cannot be so evaluated.) In other words, even if we know the values of a and b there is no handy formula in which we can plug them and the answer would pop out. The best we can do is to evaluate this integral approximately. This is essentially a discrete process and the branch of mathematics which handles such problems is called *numerical analysis.*

Secondly, the very premise on which most of the applications of continuous mathematics to real life are based, has been seriously questioned recently. As remarked earlier, classical machanics held that physical quantities such as mass and energy are continuous variables. The quantum theory, on the other hand, holds that energy comes only in quanta, that is, in whole multiples of some fixed unit, much like discrete objects like apples or balls. Another example is that of the electrical charge. A charge which is smaller in magnitude than the charge on an electron is not known to exist. Any electrical charge must, therefore, be an exact multiple of the charge on an electron. In other words, electrical charge is not a continuous but a discrete variable. Of course, the charge on the electron is so small, that for many practical purposes we may very well regard the electrical charge as a continuous variable, the same way as we could think of the discrete house problem as the continuous house problem when the houses were very densely packed.* This is, in fact, actually done. Consequently, continuous mathematics has not been entirely out-dated. But, at least in principle, its role as the only 'true' representative of the real world (with discrete mathematics playing a subordinate role) has been shaken. The balance has now definitely tilted in favour of discrete mathematics.

The third, and probably the most important, reason for the recent upsurge in the interest in discrete mathematics is the change that has taken place in mathematics itself! Till about a couple of centuries ago, number was the soul of mathematics. Whenever mathematics was to be applied to any problem, the first step was to assign suitable numbers to the various entities in the problem. (Euclidean geometry did proceed without numbers for quite some time; but with the invention of cartesian co-ordinates, it too was subsumed completely by numbers.) Things have changed considerably

---

*Another, common example is that the population of a country is often taken as continuous variable, even though it is, most certainy, a discrete variable.

during the last century. The focal point of today's mathematics is not numbers but sets, or more precisely, sets with additional structures (which we shall study later). Of course, numerical problems continue to have their importance. But the growth of non-numerical mathematics has also been considerable. As an example of such a problem, consider the following problem, which we call as the '**Dance Problem**" "Suppose at a dance party, every boy dances with at least one girl and no girl dances with all the boys. Prove that there must exist boys, b, b' and girls g, g' in the party such that b dances with g and b' with g', but neither b dances with g' nor b' with g."

This problem involves no numbers whatsoever. Neither the continuous variables (such as the heights of the boys or the dimensions of the girls) nor the discrete variables (such as the number of boys or the number of girls) are of any relevance in this problem! We shall present the solution in the third chapter. It will depend on a certain 'additional structure' on the set of girls.

Of course, not all problems of modern mathematics are so completely divorced from numbers. Many of them do involve numbers. But their real crux is not these numbers but rather certain additional structures on some sets involved in the problem. This will be more fully discussed in the third chapter. Suffice it to say here that the concept of additional structures on sets has tremendously increased the applicability of methematics. It has enriched both the discrete and the continuous mathematics. But from the point of view of practical applications, it has benefited discrete mathematics more than continuous mathematics, because the sets that we come across in practice are usually finite sets. Discrete mathematics, empowered with these additional structures constitutes *discrete structures* which is the subject matter of this book.

Finally, the advances in computer technology have also given discrete mathematics a big boost. On one hand they have made discrete mathematics a workable proposition; while on the other hand they have also posed certain problems the solution of which again requires discrete structures. As mentioned earlier, we often have to content ourselves with an approximate solution to a problem. The techniques for doing this were known for a long time. But their practical utility was marred by the fact that even to get a moderately good approximation, the computations involved were horrendous, if attempted by hand. The high speed computers have overcome this difficulty. Therefore numerical methods (which belong to discrete mathematics) are being increasingly used. If we want to draw the graph of a function, then unless the function is of a particularly simple type (say, a linear one), we can only plot a few points on it (and that, too, only approximately) so as to get a rough idea of the exact graph. This is a discrete process. But with modern advances in computer graphics, it can be done to such a level of fineness that the naked eye simply cannot tell the difference between the exact graph and its discrete, computerised version. The recent trend is, in fact, to go for discrete approximations even when the exact solution to a continuous problem can be found with just a little hard work! (It is rumoured that solving differential equations would soon become an obsolete art because of the ease with which series solutions can be obtained to a high degree of accuracy using a computer; much the same way as cameras have rendered obsolete the art of drawing objects exactly as they are.)

We hope the reader has by now grasped the spirit and the importance of discrete mathematics. In the exercises below we give a few more examples to stress the difference and interplay between discrete and continuous mathematics.

Exercises

1.1 Prove that two line segments are commensurate if and only if the ratio of their lengths is a rational number.

1.2 Prove that the side of a square is not commensurate with its diagonal.

1.3 Verify that the solution to the discrete house problem with $n = 2k$ is $\dfrac{3k + 1}{4k + 2}$ and with $n = 2k + 1$ is $\dfrac{3k}{4k + 2}$. Why is it necessary to make two separate cases depending upon whether n is even or odd?

1.4 Let α be a real number between 0 and 1. Generalise the continuous house problem by finding the probability that the distance between two distinct, randomly picked houses is at most α. (If $\alpha = 1/2$, this reduces to the problem we discussed.)

1.5 Similarly generalise and solve the discrete house problem with houses at every $1/n$th kilometer, n being a positive integer. (You will find that depending upon α, several cases may have to be made.)

1.6 In the discrete house problem with 11 houses, assume that the number of persons living in the ith house is f_i for $i = 0, 1, 2, ..., 10$. What is the average number of persons per house? What is the average population per unit length? Which spot on the road is most densely populated? (These are, of course, trivial questions. They are given here because in the next exercise we shall study their continuous versions and also for future reference).

1.7 When we consider the continuous version of the last exercise, we encounter several difficulties. First of all, we have to let the number of persons be a real variable, even though it may sound ridiculous to think of a fractional human being. A more serious hinderance is that we cannot simply let $f(x)$ be the number of persons living at the point x. If we do so, the total number of persons would be $\sum\limits_{0 \leqslant x \leqslant 1} f(x)$. There is no way to define an infinite sum like this. However, we can take the help of definite integrals. We would like the function $f(x)$ to have the property that for any interval $[a, b]$, the number of persons living in it should equal the integral

$$\int_a^b f(x)\, dx.$$

Such a function is called the *density function* of the corresponding variable (in this case, the number of persons). Prove that for every x_0 in the interval $[0, 1]$, $f(x_0)$ equals

$$\lim_{\Delta x \to 0} \frac{\text{Number of persons living in the interval } [x_0, x_0 + \Delta x]}{\Delta x}$$

(This may, in fact, be taken as the definition of the density function, or more precisely, the linear density function. Thus we see that even a rigorous definition of it involves a limiting process, similar to the definition of the instantaneous speed.) Now answer all the three questions of the last exercise. (The dimensions of the quantity $f(x)$ would be persons per unit length.)

1.8   Study the two-dimensional version of the last problem. That is, replace the straight line road by a planar region, say, D (which may be thought of as a town or a city). For a point (x, y) in D, define the laminar or the areal population density at (x, y) as a certain limit. Use double integrals instead of ordinary definite integrals.

1.9   If a discrete variable assumes values $x_1, x_2, \ldots, x_n$ (say) with probabilities $p_1, p_2, \ldots, p_n$ respectively (so that $\sum_{i=1}^{n} p_i = 1$), then its *expected value* (or average) is defined as $\sum_{i=1}^{n} p_i x_i$. In our original discrete house problem, with 11 houses, find the expected distance between two randomly picked houses.

1.10 When we attempt the continuous version of the last exercise we come across difficulties analogous to those in Exercise (1.7). If x is a continuous variable taking values in an interval, say, $[a, b]$, we can no longer let $f(x_0)$ be the probability that x equals x_0. Instead we have to let $p(x_0)$ be the *probability density* at x_0, defined as a certain derivative as before. Now, in the continuous house problem, find the expected distance between two randomly chosen houses.

1.11 Suppose in Exercise (1.6), there is a shopping centre on the road at a distance α from the house at 0. What is its average distance from the houses on the road? What is the average distance which the persons living on the road would have to traverse to go from their houses to the shopping centre? (Caution: The two averages need not be the same.)

1.12 Study the continuous version of the last problem.

1.13 Suppose a savings bank allows an interest at the rate of 10 percent per annum on deposits. If the interest is compounded n times every year (with rests at every $1/n$th year) where n is positive integer, and a depositor invests one rupee, how much money will be at his credit after one year? (Assume, for the purpose of this problem that a rupee can be divided into as many equal parts as we like; although strictly speaking, such an assumption is contrary to the spirit of discrete mathematics.)

1.14 In the continuous version of the last problem, the interest is said to be, quite appropriately, *continuously compounded*. Obtain the answer in this case both as a limit of the answer in the discrete case and also directly by solving a suitable differential equation.

Notes and Guide to Literature

Although we shall primarily deal with discrete structures, some acquaintance with continuous mathematics will be needed, at least to appreciate discrete mathematics, if for nothing else. This may be had through any standard course in calculus (with applications) in which the role of the limiting process has been duly emphasised. There are many standard text books on calculus, for example, Thomas and Finney [1] or Kreyszing [1].

We shall only need some very elementary concepts from probability and statistics. An exposition through elementary college level courses should be adequate. Two good references are Meyer [1] or Parzen [1]. There are numerous treatises on numerical analasis. As a treatise which gives the theoretical justifications for the methods used see, for example, Isaacson and Keller [1]. For a more recent treatment which gives not only the methods but computer programmes which can be implemented on personal computers, see Chapra and Canale [1].

Non-commensurability of the diameter and the circumference of a circle amounts to saying that π is an irrational number. For a proof, see Niven [1].

Continnously compounded interest is not such a hypothetical thing. some banks do allow it (although, of course, in the pass-books they can enter only approximate amounts)! It is also the nature's law of growth (or decay) that the rate of growth be always proportional to the quantity present. It is obeyed by radioactive substances.

2. Typical Problems

In this section we discuss the general types of problems that will be studied in this book.* The very genesis of discrete mathematics is its

---

*Actually, some of them will be covered in a different book. See the Epilogue.

application to real life. Problems would therefore play a dominant role in this book. Accordingly, we give in this section a list of some typical, concrete problems. The reader is invited to try them 'by hand'. (A few of them can indeed be so tackled.) Where he can solve them, he is encouraged to modify them, generalise them or to prove stronger results. On the other hand, even if you cannot get a problem, at least try some special cases of it; see if you can solve the problem with some additional hypothesis. In fact, this is a general piece of advice applicable to all the problems to come. In order not to water the things down, we give only the hard-core problems here. Comments will follow in the next section along with guidance to the chapters where they will be solved fully. But first a few generalities about the problems of discrete mathematics.

It may appear at first sight that the problems of discrete mathematics are simple, at least as compared to those of continuous mathematics because the latter involve a limiting process while the former do not. To some extent this view is substantiated if we compare the discrete and the continuous versions of various problems given in the exercises of the last section. The former generally involve little more than arithmetical computation while the latter require elaborate concepts such as derivatives and integrals (even double integrals). As a concrete example, in Exercise (1.6), we asked which spot on the road was most densely populated. This simply amounts to comparing the eleven numbers $f_0, f_1, ..., f_{10}$ with each other and finding which of them is the largest. In the continuous version in Exercise (1.7), the solution amounts to maximising the density function $f(x)$ over the interval $[0, 1]$. Because there are infinitely many points in the interval $[0, 1]$, the simple-minded method applicable to the discrete version no longer works. The difficulty, in fact, starts right from the existence of a maximum. As a simple example, suppose $f(x)$ is defined by

$$f(x) = \begin{cases} x & \text{for } 0 \leqslant x < 1/2 \\ 0 & \text{for } x = 1/2 \\ 1-x & \text{for } 1/2 < x \leqslant 1 \end{cases}$$

This function is bounded above (with $1/2$ as the supremum or the least upper bound). Yet, there is no point x_0 in $[0, 1]$ at which $f(x_0) = 1/2$. In other words, this function has a supremum but no maximum! Such an apparently paradoxical situation cannot arise in the discrete case. Note that the function $f(x)$ here is not continuous at $1/2$. If $f(x)$ is continuous at every point of $[0, 1]$, then there is a theorem which asserts the existence of a maximum for $f(x)$, that is, a point x_0 in $[0, 1]$ such that $f(x_0) \geqslant f(x)$ for all x in $[0, 1]$. This is a fairly non-trival result whose proof requires some very basic properties of real numbers. Unfortunately, even with such a deep result at our disposal, we are still a long way from the solution. The trouble is that although the theorem gurantees the existence of a point x_0 as above, it gives no construction for finding it in finitely many steps. There is, in fact, no such method available. If $f(x)$ satisfies some additional conditions (such as, having a continuous derivative) then there are methods for finding approximate answers. But they are fairly involved.

Thus we see that at least in this problem, the discrete version of it admits a very easy, almost mechanical solution. It is completely free from the complications that arise in the continuous version brcause of the subtle differences between a supremum and a maximum or between proving the existence of something and actually finding it (the kind of differences which a calculus teacher tries to emphasise to his students with all the zeal of a missionary talking to cannibals)! It is because of reasons like this that discrete mathematics was considered to belong to the infancy of man and continuous mathematics to his manhood. As remarkcd earlier, this view no longer holds good today. Nevertheless, there is still a class of persons who lose interest in a problem the moment they see that it can be solved in a finite number of steps by a mechanical procedure. 'Give it to a computer and forget about it' is their reaction!

But, what might be the end for such persons is the very beginning for some others! This category of persons includes computer programmers (or at least the brains behind them). All the problems they deal with are finitistic in the sense that they can be solved in a finite number of steps by a systematic, mechanical procedure (an *algorithm*, as it is technically called). Do these persons then merely translate these algorithms into computer programs and feed them to a machine? No. What an intelligent programmer looks for is not just some algorithm that will work but one that will work efficiently in terms of time and other factors, such as the amount of memory used. The design and analysis of algorithms is a highly challenging branch which we shall discuss briefly in a later chapter.* For the mement, we can illustrate the difference between an inefficient and an efficient algorithm with reference to the problem we have been discussing, namely, finding the maximum out of the eleven numbers $f_0, f_1, \ldots f_{10}$. The first method is to go on scanning these numbers one by one and comparing each one of them with all its predecessors. If it is less than at least one of them, obviously it cannot be the largest. If, on the other hand, it is greater than or equal to all its predecessors, then we call it the tentative maximum. The tentative maximum that survives after all the numbers upto and including f_{10} have been exhausted, is obviously the maximum, the answer to our problem. It is easy to see that this method requires 55 comparisons. (Remember this was also the number of pairs of distinct houses in the discrete house problem in the last section.)

Now, are these 55 comparisons absolutely necessary? No. One way to

*See the Epilogue.

reduce then is to note that in the procedure above, the moment we found some number (say f_6) less than some predecessor (say f_2) of it, it is absolutely futile to go on comparing it with subsequent predecessors of it (in this case f_3, f_4 and f_5). If we omit such redundant comparisons, how much saving do we effect? The answer depends upon the original sequence, $f_0, f_1 \ldots,$ f_{10}. If it is monotonically increasing (that is, $f_{i-1} \leqslant f_i$ for all $i = 1, 2, \ldots, 10$), we would still need all 55 comparisons. But if it is strictly monotonically decreasing, only 10 comparisons would do. In general, the answer would of course lie between these two extremes. The best we can do is to give the expected, or the average number of comparisons that would be needed by this method. Finding such averages is itself a challenging problem.

In the present problem, though, the analysis of the last algorithm is only of academic interest because there is another way to reduce the number of comparisons still further. We start with f_0 as the tentative maximum. (This was also the case in the last two methods). Now, every time we come across a new number f_i, we simply compare it with the tentative maximum at that time. If it exceeds the tentative maximum then we set the tentative maximum to be f_i and consider the next number, f_{i+1}. Otherwise we go directly to f_{i+1}. This way we can find the maximum of the eleven numbers $f_0, f_1, \ldots, f_{10}$ in just ten comparisons. Can we do better than this? No ! It is a good exercise for the reader to show that to find the maximum out of 11 numbers, no matter which method we follow, at least ten comparisons will be needed. In other words, from the point of view of number of comparisons, the algorithm we have just considered is the best possible. We hasten to add here that this is not a very common situation. In other words, in many problems, it is generally far from easy to prove that a particular algorithm is the best possible one. This is, therefore, yet another challenge to the designer of an algorithm!

In summary, we see that the simple (and, in fact, trivial from the point of view of continuous mathematics) problem of finding the maximum of eleven numbers leads to many interesting improvements and challenging problems. On this ground alone, discrete mathematics is far from a trivial subject, even though the sets involved in it are generally finite. But there is more to come. Even in finite mathematics, an element of the infinite creeps in because of the fact that although each integer is finite, the set of all positive integers is infinite! In any actual problem of real life, we would be dealing with a set whose size is a fixed positive integer. But naturally we would like our method to be applicable to similar problems of a general, undetermined size. For example, the problem we discussed above was to find the maximum of 11 numbers. But there is nothing particularly sacrosanct about 11. We might as well be faced with the problem of finding the maximum of 12 or of 100 numbers, say. We would obviously not like to duplicate all our work every time. We, therefore, want a solution that will work for *every* positive integer n. In the problem that we are discussing, there are no difficulties involved in generalising our comments from the special case of 11 numbers to the general case of *n* numbers. But this is not always so. For many problems, although the special cases can be worked out 'by hand', finding the answer in the general case can be a formidable task. We shall see many instances of this later on.

Secondly, even when we are dealing with a very particular case, so that, theoretically, the solution can be obtained by mechanically going through a finite number of cases, their number can be so enormously large that even the best computers available today (or in near future) would take ages to find the answer ! Since we would like to have the answer in our life-time, it is no solution to say 'Let a computer do it'. One then has to look for the solution by some very ingeneous methods. Although we shall not study very many problems of this type, we mention one such problem here just to illustrate the point. We nickname it as the **'Religious Conference Problem'**.

'Suppose there are 10 recognized religions in India. At a religious conference, delegations fröm 10 states in India are invited. Each state sends a delegation of 10 persons, one each of the ten religions. Thus, in all, there are 100 delegates, 10 from each state and also 10 of each religion. Can these 100 delegates be arranged in a 10×10 square so that in each row (and column) every state and every religion will be represented?

The answer to this problem is of course, a simple 'Yes' or 'No', with justification, that is, showing an actual arrangement if it exists or else proving that it does not exist. We would of course get it eventually if we examine all possible arrangements of the 100 delegates in a sqaure. But the number of such arrangements is 100!, that is, the product $1 \times 2 \times 3 \times 4 \times \ldots$ $\ldots \times 98 \times 99 \times 100$. This number is larger than a trillion ($= 10^{18}$). True, we can weed out many arrangements easily (for example, those in which the first row contains some religion twice). But even then the number of cases left is too large to allow a case by case study. It is therefore not very surprising that this problem remained unsolved for about two centuries.

Finally, we must remember that our subject matter is not discrete mathematics in the traditional sense but rather discrete structures, that is, discrete mathematics empowered with study of additional structures on finite sets. When we study additional structures on sets, we find that some theorems hold regardless of whether the underlying sets are finite or not. But there are also many interesting and non-trivial results which hold only for finite sets. What is needed to prove them is not some routine, brute-force computation but rather some cool, deductive reasoning (similar to the reasoning required to prove theorems in geometry without the use of coordinates, vectors or trigonometry). Indeed, the arguments would be so full of variety and ingenuity that nobody can call discrete structures a dull or a trivial subject.

With these generalities about the problems to be discussed in this book, we now give a list of sample problems. Following our earlier practice we nickname each one of them for ready reference in future.

1. **The Tournaments Problem :** In a knock-out tennis tournament, the players at each round are paired off and a match is played between the players of each pair. (In case of an odd number of players at any round, one of them gets a bye). The winners enter the next round. This process continues till a champion is found. If 1,000 players enter the tournament, how many matches would be needed to decide the champion?

2. **The Locks Problem :** A box contains secret documents and only 5 persons are privileged to have access to them. As a further security measure, it is desired that when any three but no fewer of these 5 persons come together they should be able to open the box. Design a system of locks and keys which will achieve this.

3. **The Envelopes Problem :** 10 letters and 10 envelopes carry matching addresses. If one letter is put in each envelope at random, what is the probability that all the letters are in the wrong envelopes?

4. **The Test Problem :** 20 students appear for a test with 10 questions each of which is to be answered 'True' or 'False'. Prove that there exist two (distinct) students who answer at least six questions identically.

5. **The Postage Problem:** In how many ways can a postage of two rupees be affixed on an envelope, using stamps of denominations 10 paise, 20 paise and 30 paise?

6. **The Division Problem:** Is the number $2 \cdot 3 \cdot 5 \cdot 7 \cdot 11 \cdot 13 \cdot 17 + 1$ divisible by 19?

7. **The Landlord Problem:** A landlord shares a house with a tenant. There is an electric lamp in the common passage with two switches, one with the landlord and one with the tenant. In addition, the landlord has a 'hidden' switch. Design a circuit in which, when the hidden switch is closed, either of the two other switches can control the state of the lamp independently but when the hidden switch is open, the landlord's switch has exclusive control of the state of the lamp.

8. **The Business Problem:** A certain socialist country puts the following restrictions upon businesses:

(i) All businesses having import licenses as well as those not manu-facturing essential commodities must employ local personnel and must not employ any skilled personnel.

(ii) All businesses having import licenses as well as all those not employing local personnel must employ skilled personnel and must not manufacture essential commodities.

(iii) No business shall employ local personnel without obtaining an import license.

Prove that it is impossible to do any business in that country.

9. **The Stone Problem:** You are given a stone weighing 40 kilograms. Cut it into four parts using which it will be possible to weigh any integral multiple of 1 kg upto 40 kilograms with only one use of a balance, the parts of the stone being allowed to be placed in either pan.

10. **The Regions Problem:** Into how many regions is a plane divided by n straight lines no two of which are parallel and no three of which are concurrent?

11. **The Shares Problem:**   A certain company makes a fresh allotment of shares. Thereafter, at the end of every year it issues one bonus share per every share which has matured for two years or more. (The bonus shares, after their maturity, also keep on acquiring further bonus shares every year.) A shareholder buys one share at fresh allotment and thereafter neither buys nor sells any shares, but simply keeps on collecting the bonus shares due to him. Find how many shares he will have during the nth year.

12. **The Vendor Problem:** A vendor sells tickets costing 1 rupee each. 100 customers approach him one by one. 50 customers tender 1 rupee coin each while 50 tender 2 rupees coin each and claim a change of 1 rupee each. The vendor has no money to start with. What is the probability that he will not run out of change when needed?

13. **The Casino Problem:** The management of a casino installs a gambl-ing machine in its lobby. On this machine, a player tosses a fair coin at each round. Each toss costs him a rupee. The game is over when three heads show in a row (called a 'win') or at the end of the tenth round whichever is earlier. Assume that a gambler continues to play a game till there is some possibility of a win but not afterwards. The machine is meant only to attract customers and the management wants to run it on a 'no profit no loss' basis. What amount of a reward for a win will ensure this in the long run?

14. **The Capital Problem:** There are three major cities A, B and C in a newly born state. A capital is to be built at a place from which the sum of the straight line distances to the three cities will be minimum. At what point in the plane (containing A, B and C) should it be built?

15. **The Head Office Problem:** A Company has branch offices at the cities A, B, C, D, E, F, G which are joined by roads as shown in Figure 1.4, where the numbers indicate the lengths of the roads in some fixed unit. The company wants to build a head office from which the sum of the distances to the branch offices (along the shortest possible paths available) is minimised. Where should it be built?

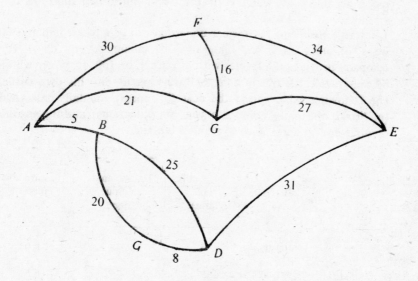

Figure 1.4: The Head Office Problem

16. **The Little Travelling Salesman Problem:** Suppose there are n cities and there is a direct path between every two of them. Construct a circular tour of these cities in which every city is visited exactly once. How many such tours are there? Prove or disprove that the shortest such tour must visit the closest pair of cities consecutively.

17. **The Diet Problem:** The staple diet of a person consists of a combination of two cereals, A and B. The protein and fat contents and the costs (all expressed in suitable units per kilogram) of these cereals are as shown in the following table:

Cereal	Protein	Fat	Cost
A	40	80	10
B	60	30	3

The person is advised to have a protein intake of at least 300 units and a fat intake of at most 500 units per month. However, because of personal tastes, he insists that at least 30 per cent of his staple diet should consist of cereal A. Can he have a diet which meets these requirements? If so, which is the most economical such diet?

18. **The Cattle Problem:** A farmer undertakes to supply a company 300 litres of milk and 450 kilograms of manure every day. The company provides the farmer with a pasture of 100 acres on which he can graze cattle free of charge. But he has to buy his own cattle. The following table shows the cost (in rupees), the daily milk yield (in litres), the daily manure output (in kilograms) and the pasture requirement (in acres) of a cow and a buffalo.

Animal	Cost (Rs.)	Daily milk yield (litres)	Daily manure output (kgs)	Pasture area (acres)
Cow	2,200.00	3	3	1.0
Buffalo	3,500.00	5	7	1.5

What is the minimum investment on the cattle the farmer has to make to fulfill the contract?

Exercises

2.1   Prove that it is impossible to find the maximum out of 11 numbers in less than ten comparisons.

2.2   Explain the relationship between the Tournaments problem and the problem of finding the maximum out of 1,000 numbers.

2.3   If x, y are real numbers, prove that their maximum equals

$$\tfrac{1}{2}(x + y + |x - y|),$$

Obtain a similar formula for their minimum.

2.4 Generalise the last exercise so as to get an inductive expression for the maximum of *n* real numbers.

2.5 Does the last exercise provide a method for finding the maximum of *n* real numbers which avoids comparisons of real numbers?

2.6 The Religious Conference Problem can obviously be stated for the case of *n* religions and *n* states, where *n* is positive integer, prove that for $n = 2$ it has no solution while for $n = 3$ it has a solution.

**2.7 Prove that the Religious Conference Problem also has a solution for $n = 4, 5, 7, 8$ and 9 but no solution for $n = 6$.

2.8 According to a popular story about a famous mathematician, in his student days at Cambridge, he was so confident that he would top the merit list at the coveted tripos examination, that on the day the results were put up he asked his servant to go and find out who was the second in rank. The obedient servant returned and said 'It is you, Sir'! Poor fellow then had to send him again to find out who was the first. In contrast with this, prove that any algorithm for finding the second largest from a given set of real numbers must necessarily also find the largest one. (In other words, although the knowledge of the second largest element does not imply knowledge of the largest element, any method for finding out the second largest element will necessarily give, as a by-product, the largest element as well. Thus there is a subtle difference between knowing something and finding it out.)

2.9 Prove that any algorithm for finding the third largest from a given set of real numbers must necessarily also give, collectively, which two numbers are among the top two, although it may not necessarily tell which is the largest and which is the second largest.

2.10 Prove that the second largest (and hence also the largest) element of a set of 11 (distinct) real numbers can be determined with 13 comparisons.
[Hint: Use your answer to Exercise (2.2).]

2.11 Generalise the result of the last problem for the case of *n* real numbers. (Note, incidentally, that the problem of finding the second best maximum hardly arises in continuous mathematics. Why?)

*2.12 Prove that, in general, less than 13 comparisons will not suffice to find the second largest element of a set of 11 numbers.

Notes and Guide to Literature

A proof of the theorem that every continuous real-valued function on a closed and bounded interval has a maximum may be found in almost any book on real analysis. Such theorems are called **existence theorems** because they assert the existence of something without giving any method for finding it. This intriguing fact has led to two lines of development. One is to look for methods for actually finding the solution (in this case the maximum), or at least an approximate solution. This has been going on for a long time. But recently, some mathematicians have started questioning the very foundations of mathematics which produce such a beautiful but 'useless' result. It turns out that the root cause lies in our classification of statements either as 'true' or 'false' even if we have no way of finding out to which category a particular statement belongs. This has led to a new school of thought, the so-called **constructivist mathematics**, in which a third category is kept for statements which are not known to be either true or false at present. A definitive reference on this type of mathematics is Bishop [1].

The Religious Conference Problem is a paraphrase of what is more commonly known as the problem of finding mutually orthogonal Latin squares. Euler proved that no solution exists for $n = 6$ and conjectured that the same is true for all n of the form $n = 4k + 2$ (the case $n = 10$ being the first unsettled one). Bose and Shrikhande [1] disproved Euler's conjecture by actually giving a construction of the desired type of arrangement. This led to an even more challenging problem of whether there exists what is known as a projective plane of order 10. It is still open. More details (both on Latin squares and finite projective planes) may be found in the book of M. Hall [1].

The anecdote about the mathematician in Exercise (2.8) is taken from an article by Roth [1].

The simile of a missionary talking to cannibals was used originally by Littlewood to comment upon the somewhat prolix style of Hardy in his classic textbook 'A Course of Pure Mathematics' [1].

Results about the minimum numbers of comparisons needed to find out the rth largest element from a given set of real numbers have been summarised in Knuth [1] (volume 3, Section 5.3.3).

3. Comments about Typical Problems

We hope the reader has tried and solved a few (but not all, as otherwise there would be little need to go further!) of the typical problems in the last section. We now make comments about them one by one and indicate what would be needed to solve them. In subsequent chapters these problems will often be used to provide motivation for certain concepts to be studied. We already considered the Dance Problem in Section 1. We now go to the problems in the last section in the order they are listed.

1.   *The Tournament Problem*:   The problem, as it is, is very straightforward and the answer is obtained simply by summing up the number of matches in each round. Thus $500 + 250 + 125 + 62 + 31 + 16 + 8 + 4 + 2 + 1$ or 999 matches will be needed for finding the champion. You might wonder if all problems in discrete mathematics are so trivial (after all, this is a

'typical' problem!). But if you are of the perceptive type, you will not fail to notice that the answer (999) is awfully close to the number of participants (1000). If further, you are scientifically minded, you would try the problem with different numbers of participants say, 100; 10,000, or some random figure such as 12,597,341. You will find that the number of matches to decide the champion is, respectively, 99; 9,999 and 12,597,340. You are now convinced that the number of matches is always one less then the number of participants. But your own personal conviction and verification does not constitute a mathematical proof! There are infinitely many integers and you will need to prove the assertion for all of them, something which you can never do by finitely many experiments. This only illustrates an earlier remark that even in finite mathematics, an element of the infinite creeps in.

So we need a proof that will show that for every positive integer n, $n-1$ matches will be needed to decide the champion among n players by the knock-out tournament method. One way to prove this is by induction (or more precisely, the second principle of mathematical induction). We leave this as an exercise. The real point of this problem is that there is an extremely short, ingeneous proof which uses only some elementary facts from set theory. This will be presented in Chapter 2.

2. *The Locks Problem*: There is, of course, nothing special about the figures 5 and 3. If m and n are any positive integers with $m \leqslant n$ then we can ask the same problem with n persons of which any m but no fewer should be able to open the box. The simplest interesting case arises for $m = n = 2$. This corresponds to the safety locker arrangement commonly provided at the banks. There the solution is to have two locks, one of which can be opened only by the bank and the other only by the customer. It is not difficult to generalise this solution. But it would be more natural to conceive it using set theory. So again we defer it to Chapter 2.

3. *The Envelopes Problem*: This is what is more formally called the problem of finding all derangements of 10 symbols. (A derangement is defined as an arrangement of objects in which no object is in its own position.) It would be horrendous to do this by hand, and in any case we would like to find the answer for any n (not just for $n = 10$). The technique adopted for doing this is called the principle of inclusion and exclusion. This will be studied in Chapter 2.

4. *The Test Problem*: It is instructive to reformulate this problem. A **binary sequence of length** n (where n is some positive integer) is defined as a sequence of n terms each of which is either 0 or 1. Now if we let '1' correspond to 'True' and '0' to 'False' then the answers given by a student can be represented by a binary sequence of length 10. The problem now amounts

to showing that given any 20 binary sequences of length 10, there are at least two among them which differ from each other in at most 4 terms. The solution is simple once we quantify the 'difference' between two binary sequence of length 10. This amounts to putting an additional structure on the set of such sequences. Consequently, this problem, like the Dance Problem, will be relegated to the chapter on sets with additional structures (Chapter 3).

5. *The Postage Problem*: We assume that the stamps of each denomination are indistinguishable from each other, so that all that matters is *how many* stamps of each kind are used and not *which particular ones* are used. Also, from the statement of the problem, it must be understood that once the stamps are selected, the manner in which they are affixed on the envelope is irrelevant. For otherwise, the answer will be infinite since even a single stamp can be placed on an envelope in infinitely many ways: (Problems in which certain laminas are to be 'tiled' with given patterns are also important but we shall not discuss them.)

   With these simplifications the problem amounts to finding the number of ways a total of two rupee (i.e. 200 paise) can be made up from the three kinds of stamps. We might as well take 10 paise as a unit. The problem then reduces to splitting a collection of 20 units into classes of sizes 1, 2 or 3 units. The technical name for such a splitting is a *partition*. We shall formally define them in Chapter 2. Theoretically, the problem could be done by hand by simply listing down all possible partitions of 20, in which no part has size exceeding 3. If for example, instead of a postage of 2 rupees we want to affix only a postage of 50 paise, then we have to partition 5. This can be done in 5 ways: (i) $1 + 1 + 1 + 1 + 1$ (corresponding to 5 stamps of 10 paise) (ii) $2 + 1 + 1 + 1$ (corresponding to one 20 paise stamp and three 10 paise stamps) (iii) $2 + 2 + 1$ (iv) $3 + 1 + 1$ and (v) $3 + 2$. Similarly for the given problem it can be shown that the answer comes out as 44. But there are many disadvantages in such straightforward methods of counting. First, in listing the various possibilities (in this case, partitions with part sizes at most 3), there is a danger that while searching for them we may miss some of them and thereby get a wrong count. This difficulty can be overcome by doing the search systematically and exhaustively. But the time it takes is prohibitive. Secondly, we would obviously like a method which will work for the partitions of any positive integer n, not just for $n = 20$ as in this problem, because there is nothing particularly great about a postage of 2 rupees; it could just as well be some other amount. So let a_n denote the number of partitions of n into parts of sizes 1, 2 or 3. We computed a_5 as 5 and mentioned (without proof) that a_{20} (which is the answer to our problem) is 44. But what is a_n for a 'general' value of n?

   To answer questions like this we need advanced counting techniques, to be studied in Chapter 7. The essential idea is very simple. A partition of n

amounts to a factorisation of x^n into powers of x. For example, consider the factorisation of x^{20} as $(x^3)^2(x^2)^5(x)^4$. This corresponds to taking two 30 paise stamps, five 20 paise stamps and four 10 paise stamps. So the problem amounts to finding the number of possible factorisations of x^{20} into powers of x^3, x^2 and x. Although such a paraphrase is not by itself a solution, it suggests that the machinery of algebra can be used and we shall exploit it.

The reader may wonder why on earth would anybody want to count all possible ways of affixing the postage when in practice any one way would suffice. As a somewhat contrived answer, suppose a spy wants to send some secret messages which are to be coded in terms of the combinations of stamps affixed on the envelopes. Then under the conditions of the problem, he can send upto 44 distinct messages. More importantly, the present problem is only a sample of problems involving counting. Counting the number of all ways a certain thing can happen is important in various connections, for example in finding out the probability of some event or in predicting how long an algorithm may take to do a given problem and so on.

6. *The Division Problem*: Like the Tournaments Problem, this problem also has at least one straightforward way for solution. By sheer computation we see that the number $2 \cdot 3 \cdot 5 \cdot 7 \cdot 11 \cdot 13 \cdot 17 + 1$ is 510,511 which is indeed divisible by 19 (with quotient 26869). But brute force method is not always the best. In the present problem we are not interested in the quotient, but only in seeing whether $2 \cdot 3 \cdot 5 \cdot 7 \cdot 11 \cdot 13 \cdot 17 + 1$ is divisible by 19. This can be done, without carrying out the given multiplication, much more easily by the use of the so-called *residue* or *modular arithmetic*. Although by way of examples, we shall mention this in Chapter 3, to put it in proper perspective we need ring theory from Chapter 6.

By the way, there is a reason for considering the particular number $2 \cdot 3 \cdot 5.7 \cdot 11 \cdot 13 \cdot 17 + 1$ (although the method would apply to any product). Note that this number is the product of the first seven prime numbers plus one. Using an analogous construction, Euclid showed that there are infinitely many prime numbers.

7. *The Landlord Problem*: If anybody thinks that discrete structures have no practical uses, this problem should disprove it! If nothing else, it helps landlords harrass unwanted tenants! The circuit involved here is a combination of two familiar circuits. The first, popularly known as the 'staircase circuit' is a circuit in which either of the two switches can control the state of the lamp independently. The second circuit is even simpler, because a single switch controls the state of the lamp. Consequently, anybody with a little experience about designing circuits, can do the problem by hand. But a systematic solution requires the use of Boolean algebras. They will be studied in Chapter 4. They are applicable to two-state devices

such as switches and lamps. (A switch is either open or closed; a lamp is either on or off).

8. *The Business Problem*: We are, of course, not commenting about socialism! Our interest is only in showing how Boolean algebras can help streamline a jungle of clumsily written laws or statements. In fact Boole (after whom Boolean algebras are named) originally called his theorems as laws of thought. We shall study this problem in applications of Boolean algebras to logic. It can be argued that this problem can be done by common sense. But actually, the logic that we shall study will not be significantly different from common sense. In complicated problems, common sense may bog down by itself, but not when it is aided by the machinery of Boolean algebras.

9. *The Stone Problem*: Boolean algebras provide an appropriate tool for handling two-state systems. In this problem, though, we have three possible states for each part of the stone. It may be placed in the same pan as the object to be weighed, or in the opposite pan or it may not be used at all! Such devices are called ternary devices. Although Boolean algebras are not the natural means for handling them, we shall club them together because some of the ideas involved are the same.

10. *The Regions Problem*: Let a_n be the number of regions into which the plane is decomposed by n straight lines, no two of which are parallel and no three of which are concurrent. Without these restrictions, the number of regions would go down and the problem would become much more complicated, depending upon how many of the lines are parallel to a given direction and how many sets of concurrent lines there are. The lines satisfying the given restrictions are said to be in **general position,** because, although we shall not do it, can be shown that given a finite collection of straight lines at random, the probability that two of them will be parallel or three of them will be concurrent is zero. (The argument resembles the comment made in the continuous house problem in Section 1, that the case of two houses being the same corresponds to points on a line and a line has no area and so does not contribute to probability.)

For small values of n, a_n can be found by actually drawing diagrams. Thus we see $a_1 = 2$, $a_2 = 4$, $a_4 = 7$, $a_4 = 11$ and so on. We may also set $a_0 = 1$ because with no line in the plane the whole plane is a single region. If you are good at guessing (a highly rewarding ability in mathematics), you would probably be able to guess, from these few values, a formula for a_n in general. If your guess comes out to be correct for $n = 5$ and $n = 6$ (say) you have reason to be convinced. But, once again, that is not a mathematical proof. You will have to prove your guess for all n, say by induction.

But there is more to this. Guessing answer and proving that it is in

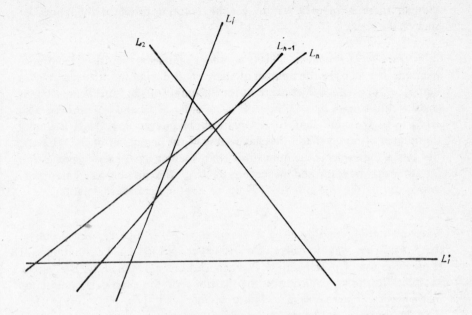

Figure 1.5: The Regions Problem

fact correct is one thing. But finding it out is quite another. The latter is always more desirable, whenever it is feasible. In the present problem it is so. To do this, instead of trying to obtain a formula for a_n directly in terms of n, we express a_n in terms of a_{n-1}. Suppose the n lines are L_1, L_2, ..., L_{n-1}, L_n (see Figure 1.5). Of these, the first $n-1$ lines, namely, L_1, L_2,..., L_{n-1} have already decomposed the plane into a_{n-1} regions. Some of these regions will be divided into 2 each by the nth line, L_n. The difference $a_n - a_{n-1}$ is therefore precisely the number of regions (out of a_{n-1}) through which the line L_n passes. Because of our assumption, L_n intersects each one of the lines L_1, L_2, ..., L_{n-1} in distinct points (which need not, however, lie in the same order on the line L_n). These $n-1$ points of intersection cut the line L_n into n parts; 2 unbounded and the remaining $n-2$ bounded. Each of these parts obviously lies in a different region formed by the first $n-1$ lines. So the line L_n passes through precisely n regions out of the a_{n-1} regions formed by L_1, L_2, ..., L_{n-1}. We thus get an important relationship,

$$a_n = a_{n-1} + n \text{ for all } n = 1, 2, 3, \dots .$$

(This is probably what you had guessed too.)

A relation like this is called a **recurrence relation** or a **difference equation**. To solve a recurrence relation means to find a formula for a_n (as a function of n) which will satisfy the relation subject to some initial conditions (in this case, $a_0 = 1$ is the initial condition.) Systematic methods for solving recurrence relations will be studied in Chapter 7. In the present problem, though, there is an easy way to get the solution provided you know the sum of the series $1 + 2 + \ldots + n$.

11. *The Shares Problem:* This is another example on the use of recurrence relations. We let b_n be the number of shares had during the nth year. We see that $b_1 = 1, b_2 = 1, b_3 = 2$ (because at the beginning of the third year, a bonus share would come), $b_4 = 3, b_5 = 5, b_6 = 8, b_7 = 13$ and so on. We also set $b_0 = 0$. We note that for every positive integer n, $b_n - b_{n-1}$ is simply the number of bonus shares that have come at the beginning of the nth year. But this is the same as the number of shares which have been in possession for at least two years and hence equals b_{n-2}, the number of shares possessed during the $(n-2)$th year. Thus we get the recurrence relation,

$$b_n = b_{n-1} + b_{n-2}$$

with the initial conditions $b_0 = 0$ and $b_1 = 1$. Unlike the last problem, there is no easy way to solve this recurrence relation. The solution will require the use of power series! Power series are associated with complex analytic functions, which are the bastions of the classical, continuous mathematics. The reader may rightly wonder what possible business they have in discrete mathematics, in which there is no limiting process and hence no room for concepts like differentiability. But there is really no reason for surprise if we recall our earlier comment that even in finite mathematics, an element of the infinite creeps in because of the fact that the set of integers is infinite. And once infinity comes in, limiting process is not far away. We shall see, in fact, in the seventh chapter that power series are a powerful tool even for the discrete mathematician.

12. *The Vendor Problem:* What matters in this problem is the order in which the customers approach the vendor. Each such order corresponds to a sequence of 100 coins, of which 50 are of 1 rupee and 50 are of 2 rupees each. In the next chapter we shall see that there are in all

$$\frac{100!}{50!\ 50!}$$

such sequences. This is, therefore, the total number of cases.

The crucial part now is to find the number of favourable cases, that is, the number of those sequences in which at no time the number of 2 rupee coins exceeds the number 1 rupee coins. (If this happens, the vendor runs out of change.) An unusually simple and elegant argument for this is due to Andre and will be given in Chapter 2. But the standard method is to use recurrence relations. For this we paraphrase the problem slightly. Think of each 1 rupee coin as a left parenthesis, (, and each 2 rupee coin as a right parenthesis,). We denote by a_n the number of ways to put n pairs of parentheses in a balanced manner. It is easily seen that number of favourable cases in our original problem is a_{50}. It is not easy to find a_{50} directly (except by Andre's method). But we can write a recurrence relation for a_n, solve it for all n and then substitute $n = 50$. This is in fact the beauty of the method of recurrence relations. It is easier to obtain the answer for all n than for a particular n!

Now for the recurrence relation itself. We are letting a_n be the number of balanced arrangements of n pairs of parentheses. In any such arrangement the very first parenthesis must be a left one. There is a unique r $(1 \leqslant r \leqslant n)$ such that this first left parenthesis gets 'cancelled' after r pairs of parentheses, as in Figure 1.6, where we indicate this mutually cancelling pair by putting a sign X on their top. Now inside this pair of parentheses, there are $r-1$ pairs of parentheses which must balance themselves. This can be done in a_{r-1} ways for $r > 1$. Moreover, after the r pairs of parentheses in our original arrangement, the remaining $n-r$ pairs must balance themselves.

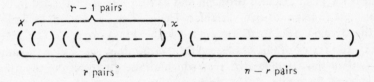

Figure 1.6 :   **Paraphrase of the Vendor Problem**

This can be done in a_{n-r} ways for $r < n$. From the elementary counting techniques to be studied in the next chapter, we shall get

$$a_n = a_{n-1} + a_1 a_{n-2} + a_2 a_{n-3} + \ldots + a_{n-3} a_2 + a_{n-2} a_1 + a_{n-1}$$

For the solution, we must wait till **Chapter 7.**

13.   *The Casino Problem*:   In all problems involving tosses of coins, a fair coin (also called an unbiassed coin) means one for which the probability of a head showing is $\frac{1}{2}$. In the present problem, the outcome of the game depends solely on the 'luck' of the player. The luckiest player can get a reward having spent as little as 3 rupees, while the most unlucky player can spend 10 rupees and still get no win. Good luck of a player of course spells bad luck for the management! The problem here is to match the amount of the reward with what an 'averagely' lucky player would spend. The trouble is how to define average luck. In the present case we have to consider all possible outcomes of the game, count the probability of each of them, add up the total 'revenue' they generate and divide this amount equally among the winners. Even then, there is no guarantee that in a particular day or a particular week the management will wind up even. All we can say is that this will be the case 'in the long run'.

The problem of counting the number of winners in this problem comes under what is called 'pattern recognition' and will be studied in Chapter 7 as an application of recurrence relations. Whatever be the ethical objections to applying mathematics to gambling, it is a fact that many developments in probability theory owe their origin to gambling. Moreover, the ideas underlying gambling also figure in many serious applications where there is an element of uncertainty. For example, the insurance companies are faced with a similar problem when they have to decide how much premium should be charged to cover a given risk.

14. *The Capital Problem*: Let the three cities be at the points A, B, C in a plane. If A, B, C are collinear, the problem is trivial, because obviously the capital should be built at the city which is in between the other two. So suppose A, B, C are not collinear. We then look for a point P in the plane spanned by A, B, C for which the function $| PA | + | PB | + | PC |$ is minimum, where $| PA |$ denotes the length of the straight line segment from A to P. This is a typical calculus problem and can be solved by the methods for minimising a function of two variables. The answer comes out as follows: (i) If the triangle ABC is obtuse angled with obtuse angle $\geqslant 120$ degrees, then P is the vertex of this obtuse angle (ii) in all other cases, P is the unique point (within the triangle ABC) at which all the three sides of the triangle ABC subtend an angle of 120 degrees. Actually, we shall not be studying problems of this type in this book. This one is given only so as to compare it with next problem.

15. *The Head Office Problem*: This problem differs from the last one in that no new roads are to be built. We are looking for a point P from which the sum of the distances to the cities along the shortest *available* paths is minimum. Obviously the point P has to be either in one of the cities or on some path between them. The latter possibility can be eliminated by a simple argument which we leave as an exercise. This leaves only finitely many possible candidates for P. Thus this problem belongs to discrete mathematics unlike the last one which belongs to continuous mathematics. As with many other problems in discrete mathematics, the crucial question here is not the existence of a solution, but rather a method for finding it. Embodied in this problem is the problem of finding the shortest distance between two points in a map. This problem is of independent interest in itself and several algorithms for it are known. Problems like this are studied in what is called *network analysis* and in the chapter on graphs we shall make a brief reference to it (see the Epilogue).

16. *The Little Travelling Salesman Problem*: From the elementary counting techniques we shall develop in the next chapter, the number of all possible circular tours in which each city is visited exactly once will be $n!$. But because the starting point on a tour could be any of the n cities, the number of distinct tours is only $(n-1)!$. Further, it does not matter in which sense the tour is completed. So the number of distinct tours comes down to $\dfrac{(n-1)!}{2}$ which is still a fabulously large number even for relatively small values of n, say $n = 15$ or 20.

The real problem now is to determine efficiently which of these tours is the shortest. This is called the 'Travelling Salesman Problem' and is today one of the most talked about problems in discrete mathematics. Once again, the difficulty is not in showing the existence of the shortest tour. Because there are only finitely many tours, one can always list them all down, compute the length of each one of them and then take their minimum. But because of the large number of tours, this is an exceedingly long procedure. What people have been struggling for is an *efficient* algorithm, or else a proof that no such algorithm exists. Now exactly, what is meant by 'efficient'? This is an important question and will be discussed in the chapter on analysis of alogrithms*. A rough idea can be given as follows. The time taken by algorithm to find the shortest tour of n cities will obviously be a function of n and will increase as n increases. It is the qualitative rate of growth of this function that will be used as a yardstick of efficiency of the algorithm. Rate of growth of a function is a concept from continuous mathematics. So once again, in the chapter on analysis of algorithms it will not be surprising if methods from continuous mathematics come in.

The Little Travelling Salesman Problem given here is meant only to give a little idea of the difficulties involved. A layman is most apt to start the shortest tour with a pair of cities which are closest to each other. But even for $n = 4$ this may fail as shown in Figure 1.7 where the triangle ABC is isosceles and right angled at A and P lies on the altitude through A and very close to the foot of the perpendicular. If the cities are P, A, B and C then $|PA|$ is clearly the shortest inter-city distance but the shortest tour is $P-C-A-B-P$.

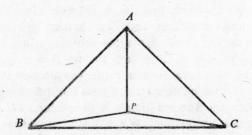

Figure 1.7: The Little Travelling Salesman Problem for $n = 4$.

17. *The Diet Problem*: This is yet another minimisation problem. Let x, y denote the amounts (in kilograms) of the cereals A, B in a monthly

*See the Epilogue.

diet. Then the cost of the diet is $10x + 3y = f(x, y)$ (say). The problem is to minimise f as a function of x and y. The variables x, y are continuous at least in theory (although, in practice, you can't ask a grocer to give you exactly, say $e + \sqrt{\pi}$ kilograms of rice!). The function f has continuous partial derivatives of all orders. So this looks like a typical calculus problem of finding the minimum of a function of several variables. However, if we apply the calculus method we see that $\dfrac{\partial f}{\partial x} = 10$ and $\dfrac{\partial f}{\partial y} = 3$ at all points. So the function f has no critical points. The reason for failure is that the conditions of the problem put certain constraints on the variables x and y. First of all, x and y have to be non-negative, because a person cannot eat a negative amount of cereal. The requirement about the protein intake amounts to the inequality $40x + 60y \geqslant 300$ or equivalently, $2x + 3y \geqslant 15$. Similarly the fat requirement is $8x + 3y \leqslant 50$. As for the last requirement, we have $\dfrac{x}{x + y} \geqslant \dfrac{30}{100}$, or in other words, $3y - 7x \leqslant 0$. The problem is therefore to minimise the function $f(x, y) = 10x + 3y$ subject to the constraints:

  (i) $x \geqslant 0$, (ii) $y \geqslant 0$, (iii) $2x + 3y \geqslant 15$, (iv) $8x + 3y \leqslant 50$, and
  (v) $3y - 7x \leqslant 0$.

Let D be the set of all points (x, y) in the plane satisfying all these five inequalities. It is easy to sketch D. From elementary coordinate geometry we know that a line with equation of the form $ax + by = c$ decomposes the plane into two half-planes of which it is the common boundary. In one of the half-planes the inequality $ax + by \geqslant c$ holds while in the other, the inequality $ax + by \leqslant c$ holds. So to sketch D, we draw the lines $2x + 3y = 15$, $8x + 3y = 50$, and $3y - 7x = 0$ and see by inspection that D is the triangle shown in Figure 1.8. D is called the **feasible region** because its points correspond to situations in which all the constraints are satisfied. In this case, since D is non-empty it follows that the person can indeed have at least one possible diet. To find the cheapest such diet amounts to minimising f over the domain D in the x-y plane. Again the well-known calculus method of Lagrange multipliers fails because here the set D is not the set of points where some function g takes a constant value. The function f to be minimised here is actually a very simple function called a linear function, a concept we shall study in Chapter 6. The constraints in the problem are of the form of what are called linear inequalities. We shall study problems of this kind in a chapter* called linear programming. It turns out that many real-life problems of optimisation can be paraphrased as linear programming problems. Calculus methods are, as a rule, useless in solving them. The solution has to be found by solving certain systems of linear equations.

18.   *The Cattle Problem*:   This problem is of the same spirit as the last

*See the Epilogue

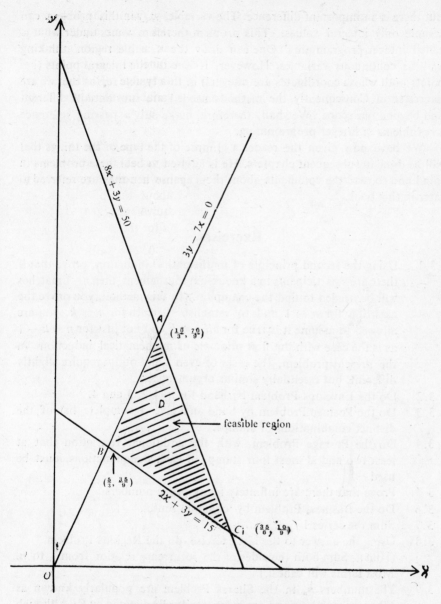

Figure 1.8: The Diet Problem

one. If we denote by x and y the numbers of cows and buffalos respectively, then the problem is to minimise the function $2200x + 3500y$ subject to the constraints

(i) $x \geqslant 0$, (ii) $y \geqslant 0$, (iii) $3x + 5y \geqslant 300$, (iv) $5x + 7y \geqslant 450$, and

(v) $x + (1 \cdot 5) y \leqslant 100$.

Here too the function to be minimised as well as all inequalities are linear.

But there is an important difference. The variables x, y in this problem can assume only integral values. This problem therefore comes under what is called 'integer programming'. One can draw the 'feasible region' thinking x, y as continuous variables. However, it is only the integral points (i.e. points both whose coordinates are integers) in the feasible region that we are interested in. Consequently the methods needed are substantially different and beyond our scope. We shall, therefore, make only a passing reference to problems of integer programming.

We have now given the reader a glimpse of the type of the things that will be done in subsequent chapters. He is advised to bear these problems in mind and to read the comments about them again when they are referred to later in this book.

Exercises

3.1   Using the second principle of mathematical induction, prove that if there are n participants in a knock-out tournament, then $n-1$ matches will be needed to find the champion. (In this method, you prove the assertion for $n = 1$ and to establish its truth for $n = k$, you are allowed to assume it is true for *all $n < k$* and not just for $n = k - 1$ as is the case with the first principle of mathematical induction. In the present problem, the cases of even n and odd n require slightly different, but essentially similar arguments.)

3.2   Do the Envelops Problem by hand for $n = 2$, 3 and 4.

3.3   Do the Postage Problem by hand and give a complete list of the distinct combinations of stamps.

3.4   Do the Postage Problem, with the additional restriction that at least two and at most four stamps of each denominations must be used.

3.5   Prove that there are infinitely many prime numbers.

3.6   Do the Business Problem by 'common sense'.

3.7   Sum the series $1 + 2 + ... + n$.

3.8   Using the answer to the last exercise, do the Regions Problem.
      (Hint: Sum both the sides of the recurrence relation from 1 to n. Most terms will cancel.)

3.9   The numbers b_n in the Shares Problem are popularly known as **Fibonacci numbers** and are more standardly denoted by F_n. Although we have not yet found a closed expression for F_n a number of properties about them can be proved without it by induction, simply from the recurrence relation $F_n + F_{n+1} = F_{n+2}$. For example, prove the following results, in which n, m, p, q are positive integers.

(a) $F_{n+m} = F_m F_{n+1} + F_{m-1} F_n$.
     (Hint: Use induction on m).

(b) $F_{n+1} F_{n-1} - F_n^2 = (-1)^n$.

(c) If p divides q then F_p divides F_q.

(Hint: Use (a) and induction on the ratio q/p.)

(d) $F_n^2 + F_{n+1}^2$ is also a Fibonacci number.

*3.10 Do the Capital Problem.

3.11 Prove that in the Head Office Problem, we may as well suppose that the head office is located at one of the cities. In other words, if Q is a point on a road joining city X to city Y (say) such that the sum of the shortest distances from Q to all the cities is the minimum possible, then either Q is X or Y or else it does not matter where on the road Q lies so that we might set $Q = X$ or $Q = Y$ arbitrarily.

(Hint: The shortest possible path from Q to a given city must pass either through X or through Y. Suppose m such paths pass through X and n through Y. Consider the cases $m > n$, $m < n$ and $m = n$.)

3.12 Verify that in Figure 1.8, the shortest tour is in fact

$$P - C - A - B - P.$$

More generally, prove that this holds as long as P lies on the altitude through A and its distance from the mid-point of BC does not exceed $\dfrac{9 - 4\sqrt{2}}{14} |BC|$.

3.13 Sketch the 'feasible region' for the Cattle Problem.

Note and Guide to Literature

In this section we have touched many topics. References to them will be cited when they are more fully discussed later in this book. As a general reference we cite the various volumes, and especially the first volume, of Knuth's book, 'The Art of Computer Programming', where the reader will find an encyclopaedic collection of known results on almost anything even remotely related to discrete structures and also numerous historical and bibliographical references. For topics in algebra we recommend Herstein's book of the same title, with its highly readable style and collection of problems. The book of Dornhoff and Hohn is a recent one and treats algebra from the point of view of applications. Standard books on combinatorics include Tucker [1], Liu [1], Krishnamurthy [1] M. Hall [1], Riordan [1]. Although titled 'Combinatiorial Mathematics for Recreation', Vilenkin's [1] book goes fairly deep into the theory and has a charming variety of problems.

Euclid (300 B C.) is, of course, more famous for his treatise on geometry. His proof of the fact that there are infinitely many primes is to this date considered one of the best mathematical arguments.

The Fibonacci numbers appear in many apparently unrelated contexts. In nature, certain lower organisms grow much like the shares in the Shares Problem (if we think of a bonus share as an offspring). So, some of the lower Fibonacci numbers appear frequently in nature, as the number of petals in a flower or the number of scales on a snake's skin. There is a huge literature on the Fibonacci numbers. There is even a journal, the *Fibonacci Quarterly* which is devoted to research about them.

4.   Review of Logic

The prerequisites needed for most of this book will be minimal. All the chapters upto the sixth will be independent of calculus, except perhaps for some examples (which may be omitted without loss of continuity). Of course, as noted before, calculus will help the discerning reader compare the various concepts in discrete mathematics with their analogues (if any) in continuous mathematics. Serious use of calculus will be made only in the seventh chapter. There we shall need some facts about sequences and series, about complex analytic functions and about some special functions (such as logarithm and exponential functions). These will be reviewed as and when they are needed. An exposure to vector algebra and matrices will help in Chapters 6 and 7. But these too are not stringent prerequisites as the treatment will be largely self-contained.

What is needed more crucially is not a bundle of theorems in mathematics but rather a certain frame of mind. This can be developed through a working knowledge of mathematical logic and of set theory. We proceed to review logic here. (Set theory will be discussed in the next chapter.) Before doing so, we emphasise that the purpose here is not to give a bunch of definitions and theorems but rather to inculcate a certain discipline of thought. What counts is the spirit and not the technicalities.

The kind of logic that is used in mathematics is called **deductive logic** as opposed to the **inductive logic** which is used in experimental science or in everyday life. In the latter we generalise from a part to the whole, a commonly cited example being that since every dog around us barks, we conclude that all dogs bark. As was made clear in the comments about the Tournaments Problem, this type of reasoning cannot be used in mathematical proofs (although, undoubtedly, it has some illustrative value). Mathematical arguments must be strictly deductive in nature. This means, the truth of the statement to be proved must be established assuming the truth of some other statements. For example, in geometry we deduce the statement that the sum of the three angles of a triangle is 180 degree from the statement that an external angle of a triangle equals the sum of the other two angles of the triangle. Of course, this latter statement has to be deduced from some other statements. Eventually a stage will arise where the truth of the statement to be proved cannot be deduced from that of some others. Such statements are called *axioms* or *postulates*. Their truth is an article of faith. As long as there is no inherent contradiction among themselves, any system of statements may be taken as axioms and theorems may be deduced from them. If the theory so developed is to have any 'practical' applications, the axioms must obviously conform to our experience (although, as noted earlier, experience cannot provide a proof for them). As our experience and the purpose of applications vary, we may modify the axioms and develop new theories. For example, one of the axioms in the classical euclidean geometry is that through any pair of distinct points there passes one and only one line. When the surface of the earth was regarded as a planar one, this was certainly consistent with experience, with the usual interpretation of line as the shortest distance path. But with a spherical earth this is no longer true for points which are diametrically opposite. This has led to development of new geometries called **Riemannian geometries** whose axioms are different from those of the euclidean geometry.

The point is that whatever be the axiom scheme, the rules of deducing theorems from them do not change. In the chapter on Boolean algebras we shall study these rules under 'valid arguments'. Suffice it to say here they all conform to our common sense and we in fact use them almost instinctively in everyday life. By way of illustration we state two such rules here. In stating each of them, first we list a few statements, called the **premises**, then draw a line and write another statement called the **conclusion**. Each rule asserts that whenever the premises hold true so does the conclusion.

Rule 1:   If it rains the streets get wet.
                It rains.

                The streets get wet.

Rule 2:   All men are mortal.
                Socrates is a man.

                Socrates is mortal.

Neither of these sounds very bright. But actually, the proofs of even the most profound theorems* in mathematics consist of chains of such tiny bits of

---

*By the way, not all results proved in mathematics are called 'theorems'. Many of them are called 'propositions', 'lemmas' or 'corollaries'. From a strict deductive point of view, there is no difference among them. The distinction rests on some extrinsic aspects such as utility, depth and beauty. A lemma is useful in a limited context (often only as a preparatory step for some theorem) and is too technical to have an aesthetic appeal. A proposition is like a mini-theorem. A 'true' theorem carries with it some depth and a certain succinctness of form and often represents the culmination of some coherent piece of work, while a corollary is like an outgrowth of a theorem.

reasoning. Genius is needed not for these individual bits but for combining them suitably. Indeed, as we remarked in the last section, the logic we shall study will not be significantly different from common sense. We therefore, conclude our preliminaries about logic by emphasising only those points where a layman (or a person familiar only with the conventional, numerical mathematics) may experience a little difficulty.

1.   *Bivalued Logic*:   The kind of logic we shall use will be bi-valued in that, for every statement there will be only two possibilities, either 'true' or 'false'. Thus every statement is either true, or false but not both, regardless of whether or not we have a way of knowing which way it is. (It is this very point on which constructivists differ. But we shall not follow them). There are many statements which at present are not known to be either true or false. A well-known example of this is the *Goldbach's conjecture* which states that every even integer greater than 2 can be expressed as a sum of two prime numbers. Although this has been verified for an impressive range of cases, nobody has proved it for all cases. Nor has it been disproved, that is, nobody has so far discovered even a single even integer greater than 2 which cannot be expressed as a sum of two primes. Still, even today, the statement is either true or false, even though it may take years to find out which way it is; much the same way as a missing person is either alive or dead at a particular time even though at that time there may be no way to find out which possibility holds.

Another point to note about bi-valuedness is that there is nothing in between 'true' and 'false'. 'Truth' means complete and absolute truth, without qualification. There is no such thing as 'very true', 'almost true', 'substantially true', 'partially true', or 'having an element of truth', although we commonly use such phrases in practice. The reason we use such expressions is that the statement involves, directly, or indirectly, some quantity and the degree of the truth of the statement is measured by how close the actual quantity is to the quantity implicit in the statement. For example, take a simple statement, 'John is tall'. In this statement, some standard of tallness is implicit, which may, of course, depend upon the context. Let us suppose, for instance, that this standard is a minimum height of 180 centimeters. Now if John's height is, say, 150, 160, 170, 175, 179, 180, 183, 186 and 190 centimeters then we would probably describe the statement respectively as, grossly false, false, with a shade of truth, substantially true, almost true, true, quite true, very true and an understatement! From a mathematical point of view, however, the statement is false ('equally false') in the first five cases and true ('equally true') in the remaining four.

2.   *Statements about a Class*:   The remarks about quantification of truth also apply to statements made about a class. The statement 'All rich men are happy' is about the class of all rich men. Goldbach's conjecture above

is a statement about the class of all even integers greater than 2. A layman is apt to regard these statements as true (or at least as 'nearly true') when they hold in a large number of cases. Even if there are a few exceptions, he is likely to ignore them and say 'The exception proves the rule!'. In mathematics, this is not so. Even a single exceptional case, (a **counter-example**, as it is called) renders false a statement about a class. Thus even one unhappy rich man makes the statement 'All rich men are happy' as false as millions of such men would do. In other words, in mathematics we interpret the words 'all' and 'every' quite literally, not allowing even a single exception. If we want to make a true statement after taking the exceptional cases into account, we would have to make a different statement such as, 'All rich men other than Mr. X are happy' or 'At least ninety per cent rich men are happy' and so on. But loose expressions such as 'almost all', 'all except a few', 'a great many' cannot be used in mathematical statements, unless, of course, they have been precisely defined earlier. (There is, indeed, one such common interpretation given to the phrase 'almost all'. When used in connection with statements about an infinite class, 'almost all' means 'all with the exception of finitely many'. For example the statement 'Almost all prime numbers are odd', is true because there is only one even prime (namely 2) while the statement 'Almost all positive integers can be expressed as a sum of three perfect squares' is false because one can construct infinitely many positive integers which are not so expressible. Phrases like 'almost everywhere' are also used with a very precise meaning in certain contexts. But we shall not study them. We only semark that the expression 'lines in general position' used in the comments on the Regions Problem is of this type.)

There is another type of statements made about a class. They do not assert that something holds for all elements of the class but that it holds for some, or at least one element. Take for example the statements, 'There is (or exists) a man who is eight feet tall'. or '200 is divisible by some power of 2'. These statements refer respectively to the class of all men and to the class of all powers of 2. In each case, the statements says that there is at least one member of the class having a certain property. It does not say how many such members are there. Nor does it say which ones they are. Thus the first statement tells us nothing by way of the name and the address of the eight-feeter and the second one does not say which power of 2 divides 200. These statements are, therefore, not as strong as, respectively, the statements, say, 'Mr. X in Bombay is eight feet tall' or '2^3 divides 200', which are very specific. A statement which merely asserts the existence of something without naming it or without giving any method for finding it is called an **existence statement**. In Section 2 we already saw an existence theorem. Because of our commitment to bi-valued logic we have to accept the possibility of an existence statement being true even when it is not specific. (Once again, this is a point of difference with constructivist logic.

Note, incidentally, that we make no distinction between the statements

'All men are mortal' and 'Every man is mortal.' They convey exactly the same meaning and hence **are logically equivalent** in the sense that either both are true together or else both are false. In general, we shall not distinguish between statements which are paraphrases of each other. Thus, 'I own this house' will be taken the same as 'This house belongs to me'. Whenever there is a subtle mathematical difference between two statements which are likely to be taken to mean the same in practice, it will be pointed out.

3. *Negation of a Statement*: The bi-valuedness of logic also has some repurcussions on negations of statements. Formally, **the negation** of a statement is a statement which is true precisely when the original statement is false and vice-versa. The simplest way to negate a statement is to precede it with the phrase 'It is not the case that ...'. Thus, the negation of 'Ram is rich' is 'It is not the case that Ram is rich'. But this is too mechanical and is not very useful either. We therefore paraphrase this as 'Ram is not rich' or even as 'Ram is poor' provided we agree that 'rich' and 'poor' are antonyms, that is, words of opposite meaning. If a statement is denoted by some symbol p, its negation is denoted by p', $-p$ or $\sim p$ and read as 'not p'. Where symbols are used in writing a statement, the negation is written by putting a slash (/) over the symbol which incorporates the principal verb of the statement. Thus, the negation of '$x = y$' is '$x \neq y$'.

Now, if 'poor' is the opposite of 'rich', the opposite of 'very rich' should be 'very poor'. Thus we are apt to negate a statement, 'Ram is very rich' as 'Ram is very poor.' But this is incorrect. There are various degrees of richness ranging from the very rich to the very poor. The original statement is about Ram's degree of richness and its logical negation simply says that he lacks the very high degree. But this does not mean that he is necessarily at the other end. Perhaps he is just average. So the correct negation of 'Ram is very rich' is 'Ram is not very rich'. In other words, negation should not be confused with antithesis, when there is a whole spectrum of other possibilities. The logical negation of 'This ball is black' is 'This ball is not black' and not 'This ball is white'. Similarly, the negation of 'The book is on the table' is 'The book is not on the table' and not 'The book is below the table'.

Similar considerations apply when we deal with statements like 'All men are mortal'. Its negation is not 'All men are immortal'. In view of the comments we made about the truth of statements about a class, it is false, even when it fails to hold just in one case, that is, when there is even one man who is not mortal. So the correct logical negation is 'There is (or exists) an immortal man'. Not surprisingly, the negation of an existence statement is a statement asserting that every member of the class (to which the existence statement refers) fails to have the property asserted by the existence statement. Thus, the negation of 'There exists a rich man' is 'No man is rich' or equivalently, 'Every man is poor'. If we keep in mind these simple facts, we can almost mechanically write down the negation of any complicated statement. Note also that the double negation (that is, the negation of the negation) of a statement is logically equivalent to the original statement. In symbols $(p')'$ is equivalent to p.

4. *Vacuous Truth*: An interesting (and often confusing to a beginner) point arises while dealing with statements about a class. A class which contains no elements at all is called a **vacuous** or **empty** or **null** class. For example the class of all four-legged men is empty because no man has four legs. In the House Problem, if there is only one house then the class of pairs of distinct houses is empty. We can continue this list much longer. But now consider a statement, 'Every four-legged man is happy'. Is it true or false? We cannot call it meaningless. It has as definite a meaning as the statement 'Every rich man is happy'. We may call the statement useless, but that does not debar it from being true or false. Which way is it then? Here the reasoning goes as follows. Because of bi-valued logic, the statement 'Every four-legged man is happy' has to be either true or false (but not both). If it is false then its negation is true. But the negation is the statement 'There exists a four-legged man who is not happy'. But this statement can never be true because there exists no four-legged man whatsoever (the question of his being happy or unhappy not arising at all). So the negation has to be false and hence the original statement is true! A layman may hesitate in accepting this reasoning and we give some recognition to his hesitation by calling such statements as **vacuously true**, meaning thereby that they are true because there cannot be anything to render them false. Of course, from a logical point of view truth has no further qualification and so a vacuously true statement is just as true as a statement whose truth has been established by a long, hard work. Note, by the way, the statement 'Every four-legged man is unhappy' is also true (albeit vacuously). There is no contradiction here because the statements 'Every four-legged man is is happy' and 'Every four-legged man is unhappy' are *not* the negations of each other.

What is the use of vacuously true statements? Certainly, no mathematician goes on proving theorems which are known to be vacuously true. But such statements sometimes arise as special cases of a more general problem. For example, suppose we want to show that under certain conditions, given n lines in a plane are in general position (which means that no two of them are parallel and no three are concurrent). If we try to do this by induction on n, we see that the starting step, namely the case $n = 1$ is vacuously true.

5. *Conjunction and Disjunction*: The **conjunction** of two statements is obtained by putting the word 'and' between them. It is true when both the statements are true and false otherwise. There is absolutely no restriction on

the two statements. One can even form the conjunction of a statement and its negation. Of course, such a conjunction will be always false.

The **disjuunction** of two statements is obtained by putting 'or' between them. It is true when at least one of them is true and false otherwise. The only point to stress about it is the meaning of 'or'. In practice we often use it to mean either one but not both of the possiblities. This is called the **exclusive use of 'or'**. For example, 'I shall spend my vacation in Bombay or in Pune', is often taken to mean that the vacation is to be spent either in Bombay or in Pune but not in both. Indeed, sometimes the very nature of the two possibilities is that they cannot hold simultaneously. For example, 'Either the book is on this shelf or else it is stolen'. If we want to indicate that both the possibilities can hold simultaneously we add the words 'or both' in practice. For example, 'A person with such handwriting must be a doctor or a crook or both'. In mathematics, however, it is unnecessary to add 'or both', because the word 'or' is always used in the **inclusive sense,** that is, so as to include the possibility of both the statements holding true. This is, of course, consistent with the practice of the disjunction being true when at least one (perhaps both) of the statements is true. If we want to use 'or' in the exclusive sense in mathematics, we can do so only by specifying 'but not both'. Thus, 'Either x divides y or y divides x but not both.'

6. *Implication Statements*:   These are the most frequently occurring statements in mathematics and so deserve a careful study. If p and q are two statements then by $p \to q$ (or $p \Rightarrow q$) we denote the implication statement 'p implies q' or 'If p then q'. The layman's interpretation of this is 'Whenever p holds q holds' or 'the truth of p forces the truth of q'. For example, if p is the statement 'It rains' and q the statement 'The streets get wet' then $p \to q$ reads 'If it rains then the streets get wet' and means that whenever it rains the streets must get wet. (Even when it does not rain the streets may get wet for some other reasons but the implication statement is not saying anything as to what happens when it does not rain.) Many statements about a class can be expressed in an implication form so as to convey the same meaning. For example, the statement 'All rich men are happy' is equivalent to 'If a man is rich then he is happy'. In an implication statement '$p \to q$' the statements p and q are called, respectively, the **hypothesis** and the **conclusion.**

It is interesting to note that the statement '$p \to q$' is logically equivalent to the implication statement '$q' \to p$'', called its **contrapositive**. We use this equivalence frequently in the *reductio-ad-absurdum argument*. Instead of showing directly that the truth of p implies that of q, we show that if q fails then p cannot hold. In law, the defence of *alibi* is actually an instance of this.

The mathematical interpretation of an implication statement is the same as a layman's. But a few points need to be stressed. First, as noted above, the implication statement '$p \rightarrow q$' is completely silent as to what happens if p does not hold. In particular it does not say that if p fails then q must fail, although, in practice we often attach this extra meaning to it. For example if a person says 'if Monday is a holiday, I shall come', we normally take this to mean that he would come if Monday is a holiday and also that he would not come if Monday is not a holiday. In mathematics, this extra meaning is never attached. If it is also intended, it has to be expressed by a separate implication statement, namely, '$p' \rightarrow q'$', or equivalently by its contrapositive '$q \rightarrow p$'. The familiar name for this new statement is the **converse** of the original implication statement '$p \rightarrow q$'. Note that the hypothesis of the original statement is the conclusion of the converse and vice-versa. The truth of an implication statement should not be confused with that of its converse. The two are quite independent of each other. Numerous examples can be given where an implication statement is true but the converse is false and vice-versa.

Sometimes, in the statement $p \rightarrow q$ the hypothesis, that is, p, is itself the negation of some statement, as, for example, the statement 'If it does not rain the crops will die'. In such a case it is customary to replace the phrase 'if not' by the single word 'unless'. With this change, the present statement would become 'Unless it rains, the crops will die', We warn once again that this statement says nothing whatsoever about the survival of the crops in the event it does rain. Here again, a logician differs from a layman who would interpret this present statement to mean that if it rains crops will be saved. The safest way to correctly interpret statements involving 'unless' is to substitute for it 'if not'.

In view of the immense importance of implication statements in mathematics, let us consider some other ways of paraphrasing them. Suppose p and q are any statements. Then $p \rightarrow q$ can be read in any of the following ways:

(i) p implies q.
(ii) q follows from p.
(iii) q is a (logical) consequence of p.
(iv) If p is true then q is true.
(v)   If q is false then p is false.
(vi) p is false unless q holds.
(vii) p is a sufficient condition for q.
(viii) q is a necessary condition for p.
(ix) p is true only if q is true.

Item (i) is just the definition, while (ii), (iii) and (iv) are its paraphrases. As we have seen before, (v) is the contrapositive of (i) and (vi) a rephrasing of (v). The last three are the only versions which call for a comment. Of these, (vii) is fairly straightforward. For example to say 'If it rains the streets get wet' clearly amounts to saying that 'Raining is a sufficient condition for the streets to get wet', or that 'In order that the streets get wet, it suffices if it rains'. Thus, the use of the word 'sufficient' here conforms to its ordinary meaning.

It is a little confusing to use version (vii) in the case of some statements. For example, in the example just given, the statement would read 'Wetting of streets is a necessary condition for it to rain'. This sounds absurd. The trouble is with the word 'condition'. In practice, it has the connotation of a prerequisite, that is, something which is to exist prior to the happening of some event. In the present case the question of streets getting wet arises only after the rain and that is why it is hard to swallow that wetting of streets is a necessary condition for it to rain. Perhaps, another example would clarify the situation. Consider the statement, 'If two triangles are congruent then they are similar'. This means that in order that two triangles be congruent they must at least be similar to each other. Congruency can never occur if similarity does not hold. In other words, similarity of the triangles is a necessary condition for them to be congruent. Whether it is sufficient or not is not the concern of the statement, it is the business of the converse statment. Necessity and sufficiency should never be confused with each other. In a sense, they are converse to each other.

About the last version, 'p is true only if q is true' it is once again necessary to distinguish a layman from a logician. When a layman says 'I shall come only if I am free', he generally means that he will come if he is free but not otherwise. A logician, however, makes no such commitment when he makes the same statement. All he is saying is that his being free is a necessary condition for his coming, that is, his coming will be impossible if he is not free. He is saying nothing at all as to what he will do if he is free. Here, too, it is vital to distinguish between 'if' and 'only if'.

There is one exception to the preceding remarks. When something is defined in terms of a condition, it is customary to cite this condition as sufficient, even though it is in fact sufficient as well as necessary. Thus, when we say 'A triangle is called **equilateral** if all its sides are equal', it also means that a triangle all whose sides are not equal will not be called equilateral. In other words, here 'if' means 'if and only if'. This usage is unfortunate but very standard. Fortunately, it appears exclusively in definitions and nowhere else.

In mathematics it often happens that we combine together an implication statement along with its converse. For example take the well-known theorem, 'The sum of opposite angles in a cyclic quadrilateral is 180 degrees and conversely'. If we let p be the statement' $ABCD$ is a cyclic quadrilateral' and q the statement '$\angle A + \angle C = 180$ degrees' then the statement of the theorem is the conjunction of $p \rightarrow q$ and $q \rightarrow p$. These types of statements come up so frequently that it is convenient to have a shorter notation for them. The most natural choice is to use arrows in both direction, that is, to use $p \leftrightarrow q$ or $p \Leftrightarrow q$. Here again it is convenient to list down a number of versions of this statement.

(i) p and q imply each other.

(ii) p and q are equivalent to each other.

(iii) p holds if and only if q holds.

(iv) q is a characterisation of p. (This version is generally used only when p, q express some properties of the same object).

(v) q holds if p does and conversely.

(vi) q holds if p does, but not otherwise,

(vii)   if p is true then q is true and if p is false so is q.

(viii) q is a necessary as well as a sufficient condition for p.

Of course many other formulations are possible in view of the symmetry of p and q. Such statements are called 'if and only if' statements. The expression 'if and only if' appears so often in mathematics that it is customary to abbreviate it to 'iff'. Thus, the geometric theorem quoted above can be stated as 'A quadrilateral is cyclic iff the sum of its opposite angles is 180 degrees.'

A theorem of this sort is really equivalent to two separate theorems which are converses of each other. If we write the statement symbolically as $p \leftrightarrow q$ (or as $p \Leftrightarrow q$) then the implication $p \rightarrow q$ is called the **direct implication** or the **'only if' part** of the theorem while the other way implication, $q \rightarrow p$ is called **converse implication** or the **'if' part** of the theorem. In general, separate proofs are needed for both the parts. Occasionally, it so happens that the steps used in the proofs of the direct implication are all reversible. In such a case, the converse is said to follow by reversing the proof of the direct implication. It is by no means the case that both the implications are of the same degree of difficulty. There are many theorems in which one of the implications is simple, almost to the point of being trivial, but the other way implication if fairly involved. As an example take the well-known remainder theorem which states, 'Let $f(x)$ be a polynomial in the variable x. Then a real number b is a root of f (i.e. $f(b) = 0$) iff $(x-b)$ is a factor of $f(x)$'. In this case, the 'if' part is trivial, but the 'only if' part is not so immediate.

The concept of implication leads naturally to that of comparison of relative strengths of statements. In practice, we say that a certain statement or piece of information is stronger than another if the knowledge of the former subsumes knowledge of the other. For example we say it is stronger to say that a certain person lives in Kerala than to say that he lives in India. This is so because anyone can infer the latter from the former by sheer commonsense, provided of course, that he knows that Kerala is a part of India.

Mathematically, we say that a statement p is **stronger** than a statement

q (or that q is **weaker** than p) if the implication statement $p \rightarrow q$ is true. A few comments are in order. First of all, 'stronger' does not necessarily mean 'strictly stronger'. Note for example that every statement is stronger than itself. The apparent paradox here is purely linguistical. If we want to avoid it we should replace the word 'stronger' by the phrase 'stronger than or possibly as strong as'. However, the use of the word 'stronger' in this context is fairly standard. If $p \rightarrow q$ is true but its converse is false, then we say that p is **strictly stronger than** q (or that q is **strictly weaker** than p). For example it is strictly stronger to say that a given quadrilateral is a rhombus than to say it is a parallelogram. Second, given two statements p and q it may happen that neither is stronger than the other. Indeed, the two statements may not be related at all. In such a case we say that their strengths are not comparable to each other. For example, the statement '$ABCD$ is a rectangle' and the statement '$ABCD$ is a rhombus'. The word 'sharper' is used sometimes for 'stronger'. This usage is common when the two statements deal with estimates or approximation of something.

What happens if out of two statements, each is stronger than the other? As we have already noted, in such a case we say that the two statements have the **same (or equal) strength** or that they are (mutually) **equivalent**. For example the statement, '$ABCD$ is a cyclic quadrilateral' and the statement, '$ABCD$ is a quadrilateral in which $\angle A + \angle C = 180$ degrees' are equivalent to each other. Earlier in this section we defined the logical equivalence of two statements to mean their simultaneous truth or falsehood. It is obvious that in such a case they are of equal strength.

A large part of mathematics is concerned with the determination of relative strengths of statements, that is, with the comparison of strengths of statements. This is a task of varying degree of difficulty. In some cases the comparison of strengths or the equivalence of two statements is a matter of common sense or of using synonymous expressions. For example the statement 'I own this house' is equivalent to the statement 'This house belongs to me'. In some cases, on the other hand, equivalence may not be obvious and needs to be established by some proof (as in the case of the statement '$ABCD$ is a cyclic quadrilateral' and the statement '$ABCD$ is a quadrilateral in which $\angle A + \angle C = 180$ degrees').

7. *Logical Precision in Mathematics*: The importance of logic as a discipline of thought, in mathematics, cannot be over-emphasised. Logical reasoning being the soul of mathematics, even a single flaw of reasoning can thwart an entire piece of a research work. We already pointed out that in mathematics every theorem has to be deduced from the axioms in a strictly deductive manner. (The term 'mathematical induction' is somewhat misleading because it is actually a case of logical deduction. Logical induction has no place in mathematics). Of course, one can always use theorems proved earlier by oneself or by others. But every step has to be

justified, This is the rule for all mathematics, whether pure or applied, classical or modern. But it deserves to be emphasised here because in the classical mathematics (except perhaps for euclidean geometry) the concern is usually with numbers. Consequently the required justifications are based upon some very basic properties of numbers. These are few in number and are used so frequently that their specific mention is rarely made. For example, if we get an equation like $(x+3)\,3(x-3)=30$ in some problem, we mechanically solve it in the following steps:

$$(x + 3)\,(x - 3) = 10$$
$$x^2 - 9 = 10$$
$$x^2 = 19$$
$$x = \sqrt{19} \text{ or } -\sqrt{19}$$

Although no justification is given for these steps, they require various properties of real numbers such as associative, commutative and cancellation laws for multiplication and addition, distributivity of addition over implication and finally, the existence of square roots of real numbers, Because of over-familiarity we tend to ignore them. But in this book we shall consider 'abstract' algebraic systems where some of these laws do not hold. Then the justification for each step will have to be given carefully, starting from the axioms. Considerations of space (and of the level of this book) would, of course, not permit us to be always so complete, and may have to borrow some theorems without proof. Whether a proof is actually given or not, the crucial point is to appreciate that a proof is needed even when the statement may seem 'obvious'.

The kind of logical perfection sought in insisting that every statement be proved from axioms by a deductive reasoning, has another manifestation in mathematics, which is relatively of recent origin. It is seen in definitions. Just as every mathematical statement has to be proved no matter how obvious it appears, every mathematical concept has to be rigorously defined no matter how intuitively clear it may be. Just as every statement is deduced from some others proved earlier, every concept is to be defined in terms of other concepts defined earlier. For example, everybody understands what a triangle is. But as a mathematical term it has to be defined, say, as 'a figure bounded by three line segments'. This definition, of course, requires that the mathematical terms appearing in it, namely, 'figure', 'bounded', 'three' and 'line segment' be defined earlier. Just as in deducing theorems we ultimately reach axioms whose truth has to be taken for granted, in trying to give definitions for mathematical concepts, ultimately we reach some terms which cannot be further defined. Such terms are called **primitive terms**. No formal meaning is assigned to these terms and one is free to interpret them as he likes as long as such interpretation is consistent with the axioms involving them. Indeed such interpretations enlarge the applicability of mathematices. For example, the primitive concepts of geometry are 'point', 'line' and the 'incidence' relation (that is, the relationship between a line and a point 'lying' upon it). We generally interpret a point as a dot (having no dimensions), a line as a set of points having only length but no breadth and a point incident upon a line as a dot belonging to the line. But one is equally entitled to think of a point as a lock, a line as a key and to think a lock as incident upon a key if it can be opened by that key. If such a system of locks and keys satisfies the axioms of geometry (for example, that every two distinct points are incident upon a unique common line) then all the theorems proved from these axioms will be applicable to it.

Although the perfection achieved through the practice of starting from a few primitive terms and then defining everything else unabiguously is aethetically appealing (and also practically useful as indicated above), we shall not always adhere rigidly to this discipline in giving definitions because it has a pedagogical disadvantage that often, some very intuitive concepts require highly clumsy and unappealing definitions. What is frequently done in the interest of logical-precision is as follows. Suppose we want to define something, say, A, which we understand intuitively but cannot describe rigorously. We then look for something, say, B, which can be rigorously described and which is so inexorably related to A that knowing A is as good as knowing B. We then define A as B. For convenience, we call this as the 'definition trick'. A simple instance in which this trick is applied is in the definition of an infinite sequence of real numbers. People have been handling such sequences for centuries and calculus books contain a large number of theorems about them. But what *is* a sequence of real numbers? We may try to answer this something like, 'A sequence of real numbers is an infinite succession of real numbers'. But this is at the most a description of a sequence. It cannot be taken as a *definition* of a sequence because it immediately raises the question, 'What is a succession?'. We may try to dodge it by taking 'succession' as a primitive term. But then we might as well have started with 'sequence' as a primitive term in the first place! Actually, there is another way out. Let N and R denote, respectively, the sets of positive integers and of real numbers. Now given a sequence, say, $\{a_n\}$, or real numbers we define a function f from N into R by $f(n) = a_n$ for all $n \in N$. (We trust the reader is familiar with functions. Anyway, we shall study them in the next chapter.) It is clear that knowing the sequence $\{a_n\}$ is as good as knowing the function f. So we simply define an infinite sequence of real numbers as a function from N into the set of real numbers. More generally, we define a sequence of complex numbers as a function from N into the set of complex numbers, a sequence of monkeys as a function from N into the set of monkeys and so on. These are perfectly rigorous definitions, provided the terms 'function' and 'positive integers' have been defined earlier. This can actually be done, taking a 'set' as a primitive term.

In fact, it turns out that if we take the concept of a set as a primitive one, then almost every concept we come across in highly diverse branches of mathematics can be defined quite unambiguously in terms of it. Everything that can be done using real numbers can be done using sets, because real numbers themselves can be defined as certain special sets. The converse is not true. There are many problems where real numbers are useless but which can be handled through sets. One such problem is the Dance Problem given in Section 1. It is for this reason that the focal point of today's mathematics is not numbers but sets. It is no exaggeration to say that sets are the alphabets of the language in which mathematics is expressed today. We shall briefly study them in the next chapter.

As indicated above, we shall not be very fussy about rigorous definitions of concepts which are intuitively clear and whose formal definitions only serve to make them precise. The purpose of this rather lengthy review of logic was merely to develop a certain maturity on the part of the reader so that when he later comes across 'abstract' concepts, he will not be baffled. We remark once again that basically all that is needed is common sense along with a habit of carefully weighing the statements one sees or makes. Expertise in logic cannot by itself be a substitute for mathematical acumen. But it at least prevents you from going astray.

Exercises

4.1 Write the logical negations of the following statements :

    (i) For every man there exists a woman who loves him.

    (ii) Gopal is intelligent and rich.

    (iii) Gopal is intelligent but not rich.

    (iv) Gopal is either intelligent or rich.

    (v) If it rains the streets get wet.

    (vi) For every $\varepsilon > 0$, there exists $\delta > 0$ such that for all x, y in X, $|x-y| < \delta$ implies $|f(x)-f(y)| < \varepsilon$. (Readers familiar with calculus will recognise this as the definition of uniform continuity of the function f on the set X. But that is not important here.)

4.2 Why is it that the correct negation of 'If it rains the streets get wet' is not 'If it rains the streets do not get wet' ?

4.3 Take a few theorems you know and cast them in the various forms of an implication statement. Also, state their converses. Give examples of 'iff' theorems in various versions.

4.4 Give an example of an 'iff' theorem where the converse implication is proved merely by reversing the proof of the direct implication.

4.5   Give examples of theorems which are true but whose converses are false.

4.6   Give examples of theorems whose converses are also true but where the two differ considerably in their depth.

4.7   Let p, q, r be statements and suppose q is stronger than r. Show that the implication statement $p \to q$ is stronger than the implication statement $p \to r$ (i.e. $(p \to q) \to (p \to r)$ is true given that $q \to r$ is true). In other words, among two implication statements with the same hypotheses, the one with the stronger conclusion is stronger. What can be said about the relative strengths of two implication statements having the same conclusion but different hypotheses ?

4.8   In each of the following pairs of statements determine which of the two statements is stronger. If the two statements are not comparable, or are equivalent, justify why?

   (i)   Statement p :   For every man there exists a woman who loves him.
         Statement q :   There exists a woman who loves every man.

   (ii) p: The diagonals of a parallelogram bisect each other.
          q: The diagonals of a rhombus bisect each other.

   (iii) p: The diagonals of a rhombus bisect each other.
           q: The diagonals of a rhombus bisect each other at right angles.

   (iv) p: One of my friends is an actor and one of my friends is a cricketer.
          q: One of my friends is an actor and a cricketer.

   (v) p: If a man is rich he is also intelligent.
         q: If all men are rich then all men are also intelligent.

   (vi) p: If Monday is a holiday I shall come.
          q: If Monday is not a holiday I shall not come.

   (vii) p: π is an irrational number.
           q: There exists an irrational number.

   (viii) p: This glass is half filled.
            q: This glass is half empty.

4.9   Decide which of the following arguments are logically valid. In each case you are given some statements as premises and some statement as a conclusion. Your problem is to decide whether the truth of the conclusion necessarily follows from the truth of all the premises simultaneously.

   (i)   Premises: No man is rich unless he is intelligent.
                        John is a rich man.
            Conclusion: John is intelligent.

   (ii)   Premises: Every man is mortal.
            Conclusion: John is mortal.

(iii)   Premises: If it rains the streets get wet.
                      It does not rain.
         Conclusion: The streets do not get wet.

(iv)   Premises: If it rains the streets get wet.
                      If the streets get wet, accidents happen.
                      It rains.
         Conclusion: Accidents happen.

(v)   Premises: Same as in (iv), except the third premise.
         Conclusion: If it rains, accidents happen.

(vi)   Premises: The streets get wet only if it rains.
                      It rains.
         Conclusion: The streets get wet.

(vii)   Premises: If it rains the streets get wet.
                      If it rains, accidents happen.
                      The streets get wet.
         Conclusion: Accidents happens.

(viii)   Premises: Every human being is either a man or a woman.
         Conclusion: Either every human being is a man or every human being is a woman.

4.10 Point out the logical fallacy in the following proof which shows that every cyclic quadrilateral is a rectangle.

Proof: Let $ABCD$ be a cyclic quadrilateral. Draw a circle with the diagonal AC as a diameter. Then B, D lie on this circle as $ABCD$ is given to be cyclic. But then, $\angle B$, $\angle D$ are both angles in a semi–circle and so each is a right angle. Similarly, drawing a circle with BD as a diameter we see that $\angle A$ and $\angle C$ are also right angles. So $ABCD$ is a rectangle.

4.11 A **vicious circle** in logic consists of a finite sequence $p_1, p_2, \ldots, p_n$ of terms in which the definition of p_1 involves p_2, that of p_2 involves $p_3 \ldots$, that of p_{n-1} involves p_n and the definition of p_n involves p_1. (For example if we define a fool as an idiot, an idiot as a lunatic and a lunatic as a fool we get a vicious circle.) Prove that because of primitive terms, there are no vicious circles in mathematics.

4.12 Take an English-into-English dictionary (any other language will also do). Start with any word and note down any word occurring in its definition as given in the dictionary. Take this new word and note down any word appearing in its definition. Repeat the process with this new word until a vicious circle is formed. Prove that a vicious circle is unavoidable no matter which word one starts with.

(Caution: The vicious circle may not always involve the original word).

Notes and Guide to Literature

The review of logic here is intended only to build a certain discipline of thought. Many of these topics will be dealt with more formally when we study Boolean algebras. But even there we shall not go very deep into mathematical logic as such. A good reference on it is Mendelson [1].

The material here overlaps considerably with the first chapter of Joshi [1].

Euclid's geometry is even today considered as a monument of axiomatic deduction. It may be noted, however, that in the middle ages, proofs as we understand today were note always given. It is said that as great a mathematician as Euler (1707–1783) derived numerous formulas about series but did not prove the convergence of even one of them. It is believed that Weierstrass (1815–1897) was the first one to insist upon rigorous proofs. The insistence upon rigorous definitions is even more recent. The need for it was pointed out most acutely by the Russell's paradox (which we shall mention in the next chapter.)

Although, because of bivalued logic, truth cannot be quantified, we can nevertheless talk of the probability of a statement being true. For example suppose A and B are classes of 100 students each. Let in class A there be 30 intelligent students and 70 dumb students and in class B let the respective figures be, say, 98 and 2. Then the probability that a 'typical' student is intelligent is .30 in class A and .98 in class B. However, for each particular student the statement that he is intelligent is either true or false, there being no other possibility.

Precise formulations of terms like 'almost all' and 'almost everywhere' require the concept of measure. Standard references on measure theory are Halmos [2], Royden [1].

Riemannian geometry, named after Riemann (1826–66) is the mathematical base of Einstein's theory of relativity.

Two

Elementary Counting
Techniques

As remarked at the end of the last chapter, set theory is the alphabet of modern mathematics. In this chapter we acquaint the reader with this alphabet in the first section. In the second section we study some elementary methods for counting the number of elements in a finite set. The third section deals with the applications of these methods to some problems of counting. The fourth section deals with the so-called principle of inclusion and exclusion. More advanced counting techniques will be studied in chapter 7.

1. Sets and Functions

Since we are going to take a set as a primitive concept, we do not have to define it. It is, nevertheless, instructive to see what difficulties would arise in an attempt to define it.

Cantor defined a set as a plurality conceived as a unity. In other words, the concept of a set involves mentally putting together a number of things (or 'objects' as they are technically called) and assigning to the things so put together a collective identity as a whole, that is, an identity separate from that of each of these things*. The things are called 'elements' or 'members' of the set obtained by conceiving them together. Thus, a set

---

*A reader familiar with the elements of Company law may recall that a company has its own legal personality, apart from that of each of its shareholders (that is why, a company may go bankrupt even when all its shareholders thrive in wealth). A partnership, on the other hand, has no legal personality. This is in fact, the basic difference between a company and a partnership.

consists of or is comprised of its members but is itself a different entity from any of its members. For example, a team (i.e. a set) of cricket players is not equal to any one of its players (not even the captain). This distinction is to be scrupulously observed even when a set consists of just one element, thus a one–man team is to be distinguished from the lone player in it. Actually, a team is a concept which has no material existence at all. Of course we may assign to a set certain attributes in terms of the corresponding attributes of its members, either by making a convention or by common sense implications. For example we may define 'A team of players is going from Madras to Calcutta' to mean that each player in it is going from Madras to Calcutta. On the other hand a set has certain attributes which cannot be described in terms of any attributes of individual members. For example 'a large flock of birds' is not the same as 'a flock of large birds'. Similarly the length of a line–segment is a property of the segment as a whole and not of the individual points which constitute it.

It is obvious that sets of material objects are as old as human thought. But we often have to consider sets of abstract objects, such as integers, real numbers, lines etc. Indeed, we frequently come across sets whose objects are themselves sets of some other objects, for example the set of all teams participating in some tournament. However, a little care is necessary in handling such sets. For example, a set of sets of lions is not a set of lions much the same way as a set of lions is not a lion.

If we can form sets whose elements are themselves sets, a question naturally arises whether we can form the set of all sets, that is, all possible sets at all. Note that this set will be extraordinary in the sense that it will, unlike a set of lions or a set of real numbers, be a member of itself! Let us agree to call a set as ordinary if it is not a member of itself, and extraordinary otherwise. Most of the sets that we come across are ordinary, but some (e.g. the set of sets just considered) are extraordinary. Now let S be the set of all ordinary sets. A deceptively innocuous question is whether the set S is ordinary. We are doomed to get a contradiction whether the answer is affirmative or negative. Such a situation is known as a **paradox**. This particular paradox is due to the philosopher-mathematician **Russel**. It has revolutionised the approach to set theory for it clearly shows that a set cannot admit such a simplistic definition as 'a collection of objects'. Either we must put some restriction on these objects (e.g. by requiring that they be material objects) or else we have to admit that most, but not all collections of objects are sets. The second alternative is the lesser of the evils.

The approach to set theory in which an attempt is made to define a set and postulate a number of axioms about sets so as to avoid paradoxes is known as the **axiomatic set theory**. Although of considerable interest in its own rights, we shall not follow this approach, because our interest in sets is primarily as building blocks from which we can construct everything else. It turns out that most of the concepts in mathematics can be expressed very succinctly and precisely in terms of sets. Even the integers (and hence the real numbers as well as the complex numbers) as well as their addition, multiplication, etc. can be defined in terms of sets. Although we shall not go this far, we shall have ample evidence to show the utility of sets in giving precise expressions to what we can describe only intuitively otherwise. It is no exaggeration to say that sets are the alphabets of modern mathematics.

Because of our interest in sets as a means rather than as an end, we follow the so-called naive approach to set theory. In this approach, the paradoxes are avoided by confining ourselves only to certain sets. The most common choice is to fix some set, say U (to be called the universal set or the universe) and to agree that all the sets under consideration would consist only of elements of U. The choice of the universe need not be specified, but of course must be large enough to include all the symbols, expressions, and concepts that we deal with. The existence of a universal set with sufficiently nice properties is a matter of axiomatic set theory.

After these general remarks about set theory, let us now list; specifically, the preliminaries we shall need about sets. For our purpose, 'set' will be a primitive term and the terms 'set', 'family', 'aggregate' and 'collection' will be synonymous. The term 'class' will be of a wider import, in that it will be used not only for sets but also when we put together too many things to form a set. For example we shall speak of 'the class of all sets' rather than 'the set of all sets'. A 'member' or an 'element' of a set will be another primitive term, and the relation of being a member will be denoted by ε. The expression '$a \in S$' will be read variously as 'a is a member (or an element) of S' or as 'a belongs to S' or 'a is contained in S' or 'S contains a'. The last two versions, however, are not particularly recommended to a beginner as they are likely to be confused with set inclusions, which we are going to define. Sets will generally be denoted by capital letters and their elements by small letters. To denote sets whose elements are themselves sets, we shall generally (but not always) use script letters, such as $\mathscr{B}, \mathscr{C}, \mathscr{D}, \mathscr{G}, \mathscr{H}, \mathscr{L}, \mathscr{S}$ etc. and we shall describe such sets as 'families' of or 'collections' of sets rather than as sets of sets. Occasionally the word 'point' is also used instead of the word 'element' or 'member' and, in the same vein, '$a \in S$' and '$a \notin S$' are sometimes expressed, respectively, by 'a lies or is in (or inside or within) S' and 'a is or lies outside S'.

A set is completely determined by its elements. This means that two sets having exactly the same elements must be identical. There are two ways of specifying a set directly in terms of its elements. The first is to list all the elements together within curly brackets. Thus $\{1, 2, \text{dog}\}$ is the (unique) set whose elements are 1, 2 and dog. In so doing, neither the order of appearance of the elements nor the repetition of any elements makes any difference. For example the sets $\{1, 2, \text{dog}\}$, $\{2, 1, 2, \text{dog}\}$, $\{\text{dog}, 2, 2, 1\}$ are all identical. It is, of course, not necessary to write down all the elements if it is possible (either from the context or because of some convention) to infer all the elements from those that are actually listed. Thus, $\{1, 2, 3, ..., 19, 20\}$ is clearly the set of all integers from 1 to 20. Similarly, $\{2, 4, 6, 8, 10, 12, 14, ..., 2n, ...\}$ is the set of all positive even integers.

Another method of specifying a set by specifying its elements, is by specifying what is known as a characteristic property of the set. **A characteristic property** of a set is a property which is satisfied by each member of that set and by nothing else. The same set may have more than one characteristic property. Each set has a trivial characteristic property, namely, the property of belonging to that set. This property is, of course, useless if we want to describe the set in the first place. If, however, we know some other characteristic property (and usually such a property is present when we conceive a set) of a set, then the set can be described in terms of it. The standard notation for a set so described is $\{x: \ \}$ or $\{x | \ \}$. Here x is a dummy symbol and could have been replaced by any other dummy symbol. The space between : (or |) and} is to be filled by a statement to the effect that x has the property in question. For example, the set of cows owned by a farmer F can be denoted by $\{x: x$ is a cow and is owned by $F\}$ or $\{x \mid x$ is a cow and is owned by $F\}$. Similary, $\{y:$ there exists a positive integer x such that $y = x^2\}$ is the set whose elements are $1, 4, 9, 16, 25, ...,$ etc. In such cases it is customary to abbreviate the notation as $\{y: y = x^2$ for some positive integer $x\}$. As in the present case, it often happens that all the elements of a set can be expressed in a certain common form in terms of one (or more) variable. In such a case it is customary to denote the set by writing such a form in curly brackets and specifying what values the variable can take. For example the set just described could have been written as $\{x^2: x$ a positive integer$\}$ or as $\{x^2: x = 1, 2, 3, ...\}$. Note that here too x is dummy symbol.

When a set is described by specifying some characteristic property and the statement expressing this property is the conjunction of several statements, the word 'and' is often dropped and only commas are used to denote the conjunction. For example $\{x: x$ real, $x > 0, x^2 < 2\}$ is the set of those real numbers which are greater than 0 and whose squares are less than 2. As in the present case, it often happens that the elements of the set in question are required to come from some other set. In such cases it is customary to write this requirement before the colon (:) rather than after the colon. For example if we denote by **R** the set of real numbers and **R**+ the set of positive real numbers, then the present set could have been denoted by $\{x \in \mathbf{R}: x > 0, x^2 < 2\}$ or by $\{x \in \mathbf{R}^+: x^2 < 2\}$.

In the same vein, when a set is characterised by a number of conditions which are of a similar form and can be written in terms of a dummy variable, it is customary to write only the values assumed by this dummy variable and to omit a specific mention of the requirement that all the conditions obtained by assigning the various values to the dummy variable are to be simultaneously satisfied. For example let S be the set of those integers which are divisible by every integer from 1 to 20. Then S can be written as $S = \{x: x$ an integer, x is divisible by i for all $i = 1, 2, ..., 20\}$ or as $S = \{x: x$ an integer, x is divisible by $i, i = 1, 2, ..., 20\}$. Note that in the latter expression the words 'for all' are omitted and are to be understood. A beginner is warned against interpreting such omission as 'for some'. The set $\{x: x$ an integer, x is divisible by some $i, i = 1, 2, ..., 20\}$ will include many elements not present in the set S. The distinction between the quantifiers 'for all' and 'for some' is vital and confusing the two with each other may be disastrous. The situation is admittedly difficult for a beginner since some authors tend to omit both the quantifiers and leave it to the reader to infer from the context which of them is intended. To avoid confusion, in this book the existential quantifier ('for some' or 'for at least one') will always be mentioned specifically.

It is possible to conceive a set with no elements at all. Such a set is variously known as an **empty** set or a **void** or a **vacuous** or a **null** set. Examples of such sets are the set $\{x: x$ an integer and $x^2 = 2\}$ or the set of all six-legged men. One can of course give many other examples. Note, however, that all these sets are equal because they consist of identical elements (viz. no elements at all). To say that two empty sets are unequal would mean that at least one of them contains an element which is not present in the other. Since such an element does not exist, we have to agree that they are equal. By the same logic, any statement to the effect that a certain property holds for every member of the empty set is true, albeit vacuously so. The unique empty set is denoted by 0 or by ϕ. These notations are so standard that they are used even when the same symbols may also represent something else. No confusion results because of such a double use.

Given two sets S and T we say that S is a **subset** of T (or T is a **super-set** of S, or S is **contained** in T or T **contains** S) if every element of S is also an element of T. When this is the case we write $S \subset T$ or $T \supset S$. If $S \subset T$ but $S \neq T$, then we say that S is a **proper subset** of T and writs $S \subsetneq T$. The reader is cautioned that some authors use the notations $\subseteq$ and $\subset$, where we use $\subset$ and $\subsetneq$, respectivelly. Obviously, two sets are equal if and only if each is a subset of the other. This is indeed the most straightforward way of proving that two sets are equal. The words 'sub-family' and 'subcollection' are synonymous with subset. Words, such as 'subaggregate', 'superfamily', could be defined but are rarely used. The term 'subclass' will not be defined formally, but has the same relationship with a class as a subset has with a set. This relationship is known as **inclusion**. We shall define set inclusion formally when we define a relation.

Given two sets S and T, the **complement** of S in T (or with respect to T) denoted by $T - S$ (or by $T \sim S$ or $T \backslash S$) is the set of such elements which are in T but not in S. Thus $T - S = \{x: x \in T \text{ and } x \notin S\}$ or $T - S = \{x \in T : x \notin S\}$. Note that we are not requiring here that S be a subset of T, although the complement $T - S$ is always a subset of T. When the set with respect to which complements are considered is understood, the complement of S is denoted by S', $\sim S$ or by $c(S)$.

If S is a set then the set of all subsets of S is called the **power** set of S and will be denoted by $P(S)$. Note that the emtpy set and the set S itself are always members of the power set $P(S)$. In particular, a power set is never empty. It is easy to show that if S has n (distinct) elements then $P(S)$ has 2^n elements and this is the reason for the name 'power set'. Elements of the power set are in general quite different from those of the original set and the two should not be confused with each other. If x is an element of the original set S then $\{x\}$ will be an element of the power set, that is, a subset of the original set; whereas x itself may or may not be an element of $P(S)$.

Given two sets S and T we define their **union** (sometimes called **join**) to be the set $\{x: x \in S \text{ or } x \in T\}$. It is denoted by $S \cup T$ or by $S + T$. It consists of elements which belong to at least one of the two sets S and T. we can similarly define the union of three or more sets, as the set of all elements which belong to at least one of them. The **intersectiou** (also called the **meet** of two sets S and T is defined as the set $\{x : \in S \text{ and } x \in T\}$. It is denoted by $S \cap T$ or $S \cdot T$ or even ST. It consists of those elements which belong to both S and T. Similary, the intersection of three or more sets is the set of elements which belong to all of them. It may happen that ST is empty even though neither S nor T is empty. If $S \cap T = \phi$, then S and T are said to be (mutually) **disjoint**. If $S \cap T \neq \phi$, the two sets S and T are said to **intersect** (or **meet**) each other at points of $S \cap T$ simply said to **intersect**.

If the sets are denoted by some indices, say, $A_1, A_2, ..., A_n$ then their union and intersection is denoted, respectiveiy, by $\bigcup_{i=1}^{n} A_i$ and $\bigcap_{i=1}^{n} A_i$, This is analogous to the Σ notation for summation. Here i is a 'dummy' indexing variable and could be replaced by any other symbol.

Because the concepts of complementation, union and intersection appear so frequently, it is helpful to form some graphic intuition about them. This is done using what are calld **Venn diagrams**. In such diagrams, some universal set is pictured as some planar region and various subsets of it as suitable subregions of it. One such Venn diagram is shown in Figure 2.1. Here X is the universal set, A and B are subsets of X (shaded by horizontal and vertical lines respectively.) The various subsets that can be formed from A and B

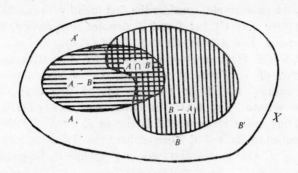

Figure. 21 : Venn Diagram

are shown in the diagram. The advantage of such diagrams is that they suggest some results which then can be verified by reasoning using definitions. For example, this Venn diagram clearly shows, that three the sets $A-B$, $B-A$ and $A \cap B$ are pairwise disjoint (i.e. every two of them are mutually disjoint) and that their union is precisely $A \cup B$. As a slightly more non-trivial example, we see that $A' \cap B'$ equals the complement of $A \cup B$, or in other words, $(A \cap B)' = A' \cup B'$. An analytical proof, of course, has to be given as we do in the following proposition which lists a number of elementary properties of the concepts introduced so far. Most of them are known by certain names, which are also given against them. The full significance of these names will come only later.

(1.1) Proposition: Let X be any set. Then for any three subsets A, B, C of X the following properties hold:

(i) $A \cup B = B \cup A$ and $A \cap B = B \cap A$ (*commutative laws*)

(ii) $A \cap (B \cup C) = (A \cap B) \cup (A \cap C)$ and $A \cup (B \cap C) = (A \cup B) \cap (A \cup C)$ (*Distributive Laws*)

(iii) $A \cup \phi = A$ and $A \cap X = A$ (Indentities for $\cup$ and $\cap$)

(iv) $A \cap A' = \phi$ and $A \cup A' = X$, where ' indicates complementation *w.r.t.* X. (*Properties of complements.*)

(v) $A \cup A = A$ and $A \cap A =$ (*Laws of Tautology*)

(vi) $A \cup X = X$ and $A \cap \phi = \phi$ (*Laws of Absorption*)

(vii) $A \cup B \cup C = A \cup (B \cup C)$ and $(A \cap B) \cap C = A \cap (B \cap C)$ (*Associative Laws*)

(viii) If $A \cup B = X$ and $A \cap B = \phi$ then $B = A'$ (*Uniqueness of complements*)

(ix) $(A')' = A$ (*Law of Double Complementation*)

(x) $(A \cup B)' = A' \cap B'$ and $(A \cap B)' = A' \cup B'$ (*De Morgan's Laws*).

Proof: Most of these properties are simple, almost to the point being trivial. They can all be established in the most straightforward manner of proving that two sets are equal, namely, by showing that each is a subset of the other. As an example, consider the first identity in (ii), namely, $A \cap (B \cup C) = (A \cap B) \cup (A \cap C)$. Let S and T denote, respectively, the sets on the two sides; i.e. $S = A \cap (B \cup C)$ and $T = (A \cap B) \cup (A \cap C)$. We have to show that $S = T$. First we show $S \subset T$. So we start with an arbitrary element, say x, of S. Then by definition, $x \in A$ and also $x \in B \cup C$. The latter means $x \in B$ or $x \in C$. In the first case, $x \in A \cap B$ while in the second, $x \in A \cap C$. In either case, $x \in (A \cap B) \cup (A \cap C)$, i.e. $x \in T$. Thus every element of S is also in T, i.e. $S \subset T$. Now for the other way inclusion let $x \in T$. Then either $x \in A \cap B$ or $x \in A \cap C$. In either case $x \in A$. Also in the first case, $x \in B$ while in the second, $x \in C$. In any case $x \in B \cup C$. So $x \in A \cap (B \cap C)$, i.e. $x \in S$. This proves that $T \subset S$. Putting it all together, $S = T$ as was to be shown. The proof of the second identity in (ii) is similar.

We have given this proof in more detail than it deserves because it is th first proof of this type. As another illustration we prove the first of the two De Morgan's laws, a little tersely this time. Let $x \in (A \cup B)'$. Then $x \notin A \cup B$. Then $x \notin A$ since otherwise $x \in A \cup B$. So $x \in A'$. Similarly $x \in B'$. Hence $x \in A' \cap B'$. Conversely, let $x \in A' \cap B'$. Then $x \in A'$ and $x \in B'$. Now if $x \in A \cup B$ then either $x \in A$ or $x \in B$. The first case contradicts that $x \in A'$ while the second case is ruled out becaure $x \in B'$. so $x \notin A \cup B$. Therefore $x \in (A \cup B)'$.

We leave the proofs of the remaining assertions as exercises. ◼

The interesting point to observe is that once the first four properties are established, the remaining can be derived in a purely formal manner, without actually looking at the elements of the sets. For example consider the second law of tautology, namely, $A \cap A = A$. We prove this as follows. Because of (iii), we have $A = A \cup \phi$. By (iv) $\phi = A \cap A'$ and so $A = A \cup (A \cap A')$. Now, the second identity in (ii) is true for any *any* three subsets A, B and C of X. In particular, it is true if we take B as A and C as A'. This gives $A = A \cup (A \cap A') = (A \cup A) \cap (A \cup A')$. Now once again, $A \cup A'$ is X by (iv) and so we get $A = (A \cap A) \cap X$. Finally, (iii) is true if we replace A by any subset of X. We replace A by $A \cap A$ and get $(A \cap A) \cup X = A \cap A$. Putting it all together, we get $A = A \cap A$ as was to be proved.

The reader may wonder what possible advantage this proof has when a direct proof of the fact that $A = A \cap A$ is so immediate right from the definition of the intersection of two sets as the set of points common to both of them. This is a very pertinent question and its full answer will come only when we study 'abstract' Boolean algebras. For the moment, we leave it as an intructive exercise to the reader to prove the other assertions above by this approach.

As another result about the basic concepts involving sets, we characterise set inclusion in the following proposition. The reader is also urged to verify its truth by drawing appropriate Venn diagrams.

1.2 Proposition : Let X be a set and let A, B be any subsets of X. Then the following statements are equivalent:

(i) $A \subset B$, i.e. A is a subset of B

(ii) $A \cap B' = \phi$

(iii) $A \cap B = A$

(iv) $A \cup B = B$

(v) $B' \subset A'$.

Proof: Once again, the proof itself is trivial. But let us first see what the proposition says. It does *not* say that any of the five assertion listed is true. It merely says that the five assertions are mutually equivalent. That is, when any one of them holds, all must hold. We shall see many propositions of this type, where what is to be established is the relationship between certain statements and not the truth of any one of them.

In proving a proposition like this, we would have to take each of the given statements and prove that its truth implies that of every other. This procedure can sometimes be instructive but is often too lengthy, (in the present instance, we would have to prove 20 implication statements). Moreover, frequently, some of the statements do not directly imply some others. In such propositions, therefore, we resort to a simple fact from logic. Let p, q, r be any statements. If the implications $p \to q$ and $q \to r$ are true then the implication $p \to r$ is also obviously true. This is called the **law of Syllogism**. Because of this law, it suffices to prove the equivalence in a 'cyclic' manner. For this, we arrange the statements in a convenient order, prove that each one implies the next and complete the cycle by showing that the last statement implies the first. Sometimes it is more convenient to have two or more cycles. Of course, then one has to further prove that some member of one cycle is equivalent to some member of another. If some statement is common to the two cycles, then obviously this verification need not be done.

With these generalities about propositions of this form, let us now get down to the actual proof. We shall prove the equivalence of the statements in two cycles (i) $\Rightarrow$ (ii) $\Rightarrow$ (iii) $\Rightarrow$ (iv) $\Rightarrow$ (i) and (i) $\Rightarrow$ (v) $\Rightarrow$ (i). (Many other choices are also possible). We prove each of these implications separately.

(i) $\Rightarrow$ (ii). Here we are assuming (i) as true. That means we are given that A is a subset of B. We have to prove that $A \cap B'$ is the empty set. For this, suppose there is some element, say x, in $A \cap B'$. Then by definition $x \in A$ and $x \in B'$. But since $A \subset B$, $x \in A$ implies that $x \in B$. We thus get a contradiction to $x \in B'$. Thus there cannot be any element in $A \cap B'$, i.e. $A \cap B' = \phi$. So (ii) holds whenever (i) does.

(ii) ⇒ (iii) Here we are given $A \cap B' = \phi$ and we have to prove that $A \cap B = A$. clearly $A \cap B$ is always a subset of A and so we always have $A \cap B \subset A$. For the other way inclusion, suppose $x \in A$. If $x \notin B$ then $x \in B'$, whence $x \in A \cap B'$ which is given to be empty; a contradiction. So $x \in B$ and thus $x \in A \cap B$. Therefore $A \subset A \cap B$, and hence $A = A \cap B$.

(iii) ⇒ (iv). This also could be done by taking elements of the sets $A \cup B$ and B. But we give an alternate argument, based on the properties proved in the last proposition. We have,

$B = B \cap X$ (by property (iii) in the last proposition)

$\quad = B \cap (A \cup A')$ (by property (iv))

$\quad = (B \cap A) \cup (B \cap A')$ (by Property (ii))

$\quad = (A \cap B) \cup (A' \cap B)$ (by Property (i))

$\quad = A \cup (A' \cap B)$ (as we are given $A \cap B = A$)

$\quad = (A \cup A') \cap (A \cup B)$ (Property (ii) again)

$\quad = X \cap (A \cup B)$ (Property (iv))

$\quad = A \cup B$ (Property (iii)).

(iv) ⇒ (i). A is always a subset of $A \cup B$, regardless of what B is. So $A \subset A \cup B$. We now merely replace $A \cup B$ by B as we are assuming (iv).

(i) ⇒ (v). Suppose $A \subset B$. We have to show $B' \subset A'$. Let $x \in B'$. Then $x \notin B$. If $x \in A$ then $x \in B$, a contradiction. So $x \in A'$ as was to be proved.

(v) ⇒ (i). This is similar to the last implication. ∎

Besides union and intersection, there is one more way to generate new sets from two (or more) sets. We often have to simultaneously consider two variables, say, x and y of which x ranges over a set A and y over a set B (say). (The sets, A, B need not be distinct.) It is then convenient to represent this situation by an equivalent single variable z which ranges over a suitable set constructed from A and B. We already saw an instance of this in the House Problem in Section 1 of the last chapter, where we remarked that picking two houses on the road was equivalent to picking a single point from the square constructed there. More generally, given any two sets A and B, we define their **cartesian product** (or simple **product**) denoted by $A \times B$ to be the set of all ordered pairs (x, y) such that x is from A and y is from B, in symbols $A \times B = (x, y): x \in A, y \in B$. We do not define an ordered pair*, but remark that unless $x = y$, (x, y) is not the same as (y, x). The word 'cartesian' here comes from cartesian coordinates, because of which a point in plane is represented by an ordered pair of real numbers (its co-ordinates w.r.t. some fixed frame of reference). The word 'product' is justified because if the sets A and B have m and n elements respectively, then it can be shown that $A \times B$ has mn elements. Note however, that as sets, $A \times B$ and $B \times A$ are not equal even though they have the same number of elements. Similarly we can define the cartesian product of three sets A, B, C as the set of all ordered triples (x, y, z) in which $x \in A$ and, $y \in B$ $z \in C$. More generally, for any positive integer n one can consider ordered n-tuples of elements and use them to define the cartesian product of n sets, say, A_1, A_2,..., A_n. It is not necessary that these sets be all distinct. In fact, it may even happen that all of them are equal to some set, say A. Then it is customary to call $A_1 \times A_2 \times ... \times A_n$ the product of n copies of A or the nth **power** of A. Clearly, $A^2 = A \times A$.

We now turn to the concept of functions which is of paramount importance in mathematics. Intuitively, if X and Y are sets then a function, say f, from X to Y is a rule of correspondence which assigns to every element of X, a unique element of Y. If x stands for an arbitrary element of X then the unique element of Y which is assigned to x under f is denoted by $f(x)$. Common notations for a function f from a set X to a set Y are $f : X \to Y$ or $\xrightarrow{f} Y$. The sets X and Y are called, respectively, the **domain** and the **codomain** of f. The set G_f defined by $\{(x, f(x)) : x \in X\}$ is called the **graph** of f. The term is obviously of geometric origin, because if both X and Y are subsets of the real line then $X \times Y$ is a subset of the cartesian plane and for a function $f : X \to Y$, the set G_f indeed represents the graph of f in the usual sense. Note that for every x in X there is one and only one y in Y such that $(x, y) \in G_f$. (Geometrically this means that every 'vertical' line through a point in X meets the graph at precisely one point.) Note further that a function is completely determined by its graph. This fact, along with the 'definition trick' (mentioned in Section 4 of the last chapter) gives the formal definition of a function. Formally, a **function** from a set X to a set Y is defined as a subset G of $X \times Y$ having the property that for each $x \in X$, there is one and only one $y \in Y$ such that $(x, y) \in G$. (If $X = \phi$, then we may take $G = \phi$. Thus there exists a function from the empty set to any set.)

We can give numerous examples of functions. But before doing so, some points need to be emphasised because the practice adopted by various authors about them is not uniform. First, we require that every function must be defined on the entire domain set. For example, if X and Y are each the set of real numbers then the formula $f(x) = 1/x$ does not define a func-

---

* The formal definition is slightly clumsy. The ordered pair (x, y) is defined as the set $\{\{x, y\}, x\}$. Note that mere $\{x, y\}$ would not do because $\{x, y\}$ is the same as $\{y, x\}$, while we do not want $(x. y)$ to be the same as (y, x).

tion from X to Y unless $f(0)$ is also specified. Some authors do allow such functions and then define the domain of a function to be the set of those points where it is defined. We shall not adopt this practice. Similarly, we require a function to be single-valued, that is, $f(x)$ is uniquely defined for every $x \in X$. For example, if X is the set of positive real numbers and Y the set of all real numbers than we cannot define a function $f : X \to Y$ simply by $f(x) =$ square root of x. It is necessary to specify which of the two square roots is to be taken. The choice may be arbitrary. For example, we may define $f : X \to Y$ by

$$f(x) = \begin{cases} \text{non-negative square root of } x \text{ if } x \text{ is rational} \\ \text{negative square root of } x \text{ if } x \text{ is irrational.} \end{cases}$$

Note that there are many real numbers (for example, $e + \pi$) which are not known to be rational or irrational. Even for such numbers the function above is well-defined, because it has a unique value, the only trouble being that we do not know this value today. (Once again such a situation is not allowed in constructivist mathematics.)

Another point to note about functions is their notation. In calculus books it is customary to denote a function by a symbol like $f(x)$ where x is a variable which ranges over the domain set. Since we shall have occasions to consider sets whose elements are themselves some functions, it is necessary not to confuse a function either with a formula for it or with any value assumed by it. We therefore denote functions by single symbols like $f, g, h, \phi, \theta, \lambda$ etc. An expression like $f(x)$ denotes the **value** of f at x, that is, the element in the codomain set which is associated to x under the function f. In this notation, x is sometimes called an **argument of** f.

As examples of 'non-mathematical' functions, let H be the set of all human beings and M the set of all men. We can then define a function $f : H \to M$ by $f(x) =$ father of x. As another example, in the Tournaments Problem, let M and P denote, respectively, the set of all matches played and the set of all participants. We define two functions f and g from M to P, to be called, respectively, the winner function and the loser function. If $m \in M$ then m is a match played and we let $f(m)$ be the player who wins it and $g(m)$ be the player who loses it. The loser function considered here is very useful as we shall see later.

The words **'transformation'**, **'operator'**, **'map'** and **'mapping'** are really synonyms of 'function' although by convention they are used only in some specific contexts. We shall use 'function' as the general term and reserve the terms 'transformation, and 'operator' for certain special types of functions.

Suppose f and g are functions such that the codomain of f coincides with the domain of g; say $f : X \to Y$ and $g : Y \to Z$. Then we define their **composition** or **composite**, denoted by $g \circ f$ (or sometimes by gf) to be the function from X to Z given by $(g \circ f)(x) = g(f(x))$ for $x \in X$. For example let X, Y, Z each be the set of real numbers. Let f, g be defined by $f(x) = x^2$ for $x \in X$ and $g(y) = \sin y$ for $y \in Y$. Then $(g \circ f)(x) = \sin(x^2)$. Note that in this case the composite $f \circ g$ is also defined but is not equal to $g \circ f$, because $f \circ g(x) = \sin^2 x$ which is in general different from $\sin(x^2)$. We can also give examples where even though $g \circ f$ is defined, $f \circ g$ is not defined. If f, g, h are three functions for which $g \circ f$ and $h \circ g$ are both defined then it is clear that $h \circ (g \circ f)$ and $(h \circ g) \circ f$ are both defined and are easily seen to be equal because if x is any element of the domain of f then $[h \circ (g \circ f)](x)$ and $[(h \circ g) \circ f](x)$ both equal $h(g(f(x)))$. We denote this function by $h \circ g \circ f$.

The simplest functions are the so-called **constant** functions. They assume the same value for all values of the argument and are often denoted by this common value. For any set X, the function $1_X : X \to X$ (or $id_X : X \to X$) defined by $1_X(x) = x$ for all $x \in X$ is called the **identity function** on X. More generally, if Y is a superset of X then the function $i : X \to Y$ defined by $i(x) = x$ for $x \in X$ is called the **inclusion function** of X into Y. If $f : X \to Y$ is a function and $A \subset X$ then the **restriction** of f to A, denoted by $f|A : A \to Y$ is the function defined by $(f|A)(x) = f(x)$ for all $x \in A$. Equivalently, it is the composite of f and the inclusion of A into X. A function $f : X \to Y$ is said to be **injective** (or **one-to-one**) if for all x_1, $x_2 \in X$, $f(x_1) = f(x_2)$ implies $x_1 = x_2$. In other words, a function is injective if it takes distinct points of the domain to distinct points of the codomain. A function $f : X \to Y$ is said to be **surjective** (or **onto**) if for each $y \in Y$ there is some $x \in X$ such that $f(x) = y$. A function which is both injective and surjective is called a **bijective** function or a **bijection**. A bijection of a set onto itself is called a **permutation** of that set. It is easy to show that a function $f : X \to Y$ is bijective iff there exists a function $g : Y \to X$ such that $g \circ f = id_X$ and $f \circ g = id_Y$. When such a function exists it is unique and called the **inverse** function of f. It is denoted by f^{-1}. Note that the inverse function is also a bijection. The term 'one-to-one correspondence' is sometimes used for 'bijection' and is thus different from 'one-to-one function.

Let $f : X \to Y$ be a function and suppose A, B are subsets of X, Y, respectively. Then the **direct image** (or simply, **image**) of A under f, denoted by $f(A)$ is defined as the set $\{f(x) : x \in A\}$. The set A is said to be **taken** (or **mapped**) by (or under) f onto $f(A)$. The **inverse image** (or **pre-image**) of B under f, denoted by $f^{-1}(B)$ is defined as the set $\{x \in X : f(x) \in B\}$. The inverse image is defined even where f is not a bijection. In case f is a bijection, it coincides with the direct image under the inverse function f^{-1} and so the notation causes no ambiguity. The set $f(X)$ is called the **range** of f. It is evident that a function is onto iff its range is the whole codomain.

Just as Venn diagrams provide graphic intuition for sets, arrows between such diagrams graphically represent functions. One such diagram is shown in Figure 2.2. Arrows are also drawn from several points in the domain to

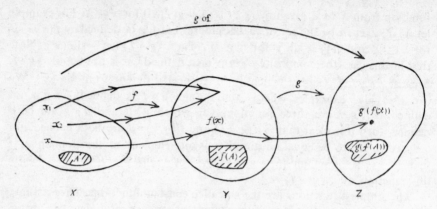

Figure 2.2:   Functions and Composites

the respective values of the function at these points. From the figure we see at once that f is not injective. Similarly for a subset A of X, its image under f is shown as a subset of Y and an arrow is put from A to $f(A)$ We see some simple properties of direct images, namely,

$$f(A_1 \cup A_2) = f(A_1) \cup f(A_2) \text{ and } (g \circ f)(A) = g(f(A)).$$

The proof of these and other similar properties is left as an exercise.

We conclude this section by illustrating how the basic concepts of a set and a function can be used to provide precise formulations of various concepts and problems. In the last chapter we already saw how a precise definition of a sequence can be given as a function with domain $\mathbb{N}$, the set of positive integers.* Similarly an $m \times n$ matrix of real numbers can be defined rigorously as a function from $A \times B$ into $\mathbb{R}$, the set of real numbers, where A and B are respectively the sets $\{1, 2,..., m\}$ and $\{1, 2,..., n\}$, instead of describing it loosely as a rectangular array with m rows and n columns. An arrangement of distinct balls into distinct boxes amounts to defining a function from the set of balls into the set of boxes. A requirement such as no box be empty clearly amounts to saying that the corresponding function is onto.

To illustrate the use of sets in representing problems, we consider two problems, the Locks Problem and the Dance Problem. We shall only paraphrase them set-theoretically. The solutions will be given later on.

In the Locks Problem let the five persons be p_1, p_2, p_3, p_4 and p_5. Let us denote the set of locks to be put by L. Now for each $i = 1, 2,..., 5$ we let L_i be the set of those locks in L which can be opened by the person p_i. Then $L_1, L_2,..., L_5$ are subsets of L. The requirement in the problem amounts to saying that the union of any three of these five subsets be the whole set L while the union of any two should not be the whole set L. Using De Morgan laws, this can be translated further as follows. For each i, let M_i

*A finite sequence of length n is a function from $\{1,..., n\}$.

be the complement of L_i in L. Then the problem amounts to finding a suitable set L and some five subsets M_1, M_2,..., M_5 of L such that (i) for any i and j, $M_i \cap M_j \neq \phi$ and (ii) for any three distinct i, j, k,

$$M_i \cap M_j \cap M_k = \phi.$$

In the Dance Problem, let us denote the girls present by g_1, g_2,..., g_n. Let B be the set of boys at the party. For each $i = 1, 2,..., n$ we let B_i be the set of boys who dance with the girl g_i. Each B_i is a subset of B. The information that every boy danced with at least one girl amounts to saying that $\bigcup_{i=1}^{n} B_i = B$, while the statement that no girl danced with all the boys means that no B_i equals the whole set B. Now, for a paraphrase of the assertion of the problem, note first that for any i and j, $B_i \subset B_j$ is equivalent to saying that every boy who dances with g_i also dances with g_j. Therefore to say that there is some boy who dances with g_i but not with g_j is equivalent to saying that B_i is not a subset of B_j. The problem now amounts to showing that there exist some two girls, say, g_i and g_j such that neither $B_i \subset B_j$ nor $B_j \subset B_i$.

Of course we have not yet solved either the Locks Problem or the Dance Problem. But we have made a good start. We remark that reducing a problem in terms of sets is not, by itself, a guarantee for its solution. The solution may require a good deal of more work, depending on the type and the degree of difficulty of the problem. But such a reduction helps us crystallise our thoughts and gain some insight as to the direction in which to proceed for solution. It also helps tremendously in presenting the solution concisely and precisely. Experience with the two problems just considered shows that clever laymen (not familiar with set theory) can also sometimes hit the right ideas. But they are awfully clumsy in expressing them, using unintelligible, vague expressions all the time and constantly interspersing, 'No, this is not what I meant'. If nothing else, set theory at least provides us a means to say exactly what we mean and for this reason alone deserves to be studied.

Exercises

1.1   Complete the proof of Proposition (1.1).

*1.2   In Proposition (1.1), for properties from (v) onwards, give alternate proofs, using the first four properties.

1.3   In Proposition (1.2), give a direct proof for each of the 20 possible implication statements. (Our proof already covers 6 of these implications).

1.4   Let X be a set and A, B be any two subsets of X.

    (a)   Prove that $A \cup B$ is the smallest subset (in the sense of inclusion) of X which contains both A and B. In other words, prove that (i) $A \cup B$ contains both A and B, and (ii) if C is a subset of X which contnins A as well as B then C contains $A \cup B$.

(b)   Obtain a similar characterisation of $A \cap B$.

1.5   Let X and Y be any sets. Suppose $A \subset X$ and $B \subset Y$. Prove that $A \times B \subset X \times Y$. (For obvious geometric reasons, $A \times B$ is called a **box** with **sides** A and B.) Generalize this to the case of the cartesian product of n sets.

1.6   In the last exercise, prove that the intersection of two boxes is again a box (perhaps the empty box) but that the union of two boxes need not be a box. If $A = A_1 \cup A_2$ and $B = B_1 \cup B_2$, express $A \times B$ as a union of 'smaller' boxes.

1.7   Which of the following formulas define functions from the set of real numbers, to itself? Which of these functions are injective, surjective, bijective?

(i) $f(x) = x^2$ for all x

(ii) $f(x) = x^3 - x$ for all x

(iii) $f(x) = \sqrt{1-4x^2}$

(iv) $f(x) = e^x$ for all x

(v) $f(x) = \begin{cases} \text{any prime which divides } x \text{ if } x \text{ is a positive integer} \\ 0 \qquad \text{otherwise} \end{cases}$

(vi) $f(x) = \begin{cases} 1 \text{ if Goldbach's conjecture is true} \\ 0 \qquad \text{otherwise.} \end{cases}$

1.8   For any three sets X, Y, Z prove that there exists a bijection between $(X \times Y) \times Z$ and $X \times (Y \times Z)$ and also a bijection between $X \times Y$ and $Y \times X$.

1.9   Extend the definition of the composite $g \circ f$ of two functions f and g to the case where the range of f is contained in the domain of g. Give an example where $g \circ f$ is defined but $f \circ g$ is not.

1.10   Let $f: X \to Y$ and $g: Y \to Z$ be functions. Prove that

(i)   for any $A \subset X$, $(g \circ f)(A) = g(f(A))$.

(ii)   for any $B \subset Z$, $(g \circ f)^{-1}(B) = f^{-1}(g^{-1}(B))$.

(iii) for any $A \subset X, f^{-1}(f(A)) \supset A$ and for any $B \subset Y, f(f^{-1}(B)) \subset B$ and $f^{-1}(Y-B) = X - f^{-1}(B)$.

(iv)   for any two subsets A_1, A_2 of X and B_1, B_2 of Y,

$$f(A_1 \cup A_2) = f(A_1) \cup f(A_2)$$

and

$$f(A_1 \cap A_2) \subset f(A_1) \cap f(A_2),$$
$$f^{-1}(B_1 \cup B_2) = f^{-1}(B_1) \cap f^{-1}(B_2)$$

and

$$f^{-1}(B_1 \cap B_2) = f^{-1}(B) \cap f^{-1}(B_2),$$
$$A_1 \subset A_2 \Rightarrow f(A_1) \subset f(A_2)$$

and

$$B_1 \subset B_2 \Rightarrow f^{-1}(B_1) \subset f^{-1}(B_2).$$

1.11 (a) Let $f: X \to Y$ be a function. Prove that the following statements are equivalent to each other:
   (i) f is injective

   (ii) for every $A \subset X, f^{-1}(f(A)) = A$

   (iii) for every $A_1, A_2 \subset X, f(A_1 \cap A_2) = f(A_1) \cap f(A_2)$

   (iv) for any set Z and for any two functions $g, h: Z \to X$, $f \circ g = f \circ h$ implies $g = h$ (in other words, f can be cancelled from the left).

   (v) either $X = \phi$ or there exists a function $g: Y \to X$ such that $g \circ f = id_X$ (such a function g is called a **left inverse** of f).

   (b) Obtain similar characterisations of surjective functions.
   (c) Find necessary and sufficient conditions for the composite of two functions to be injective, surjective, bijective.

1.12 Let $f: X \to X$ be a function. The functions $f \circ f, f \circ f \circ f, \ldots\ldots$ are denoted respectively by $f^2, f^3, \ldots$, etc. By convention, f^0 denotes 1_X. A point $x_0 \in X$ is called a **fixed point** of f if $f(x_0) = x_0$. Prove that a fixed point of f is also a fixed point of f^2. Does the converse hold?

1.13 Let $\mathbb{Z}_2$ be the set $\{0, 1\}$ and X any set. For $A \subset X$, the function $f_A: X \to \mathbb{Z}_2$ defined by $f_A(x) = 1$ if $x \in A$ and $f_A(x) = 0$ if $x \notin A$ is called the **characteristic function** of A, so called because the subset A is completely characterised by the function f_A. If $A, B \subset X$, express $f_{A \cap B}, f_{A \cup B}$ and f_{X-A} in terms of f_A and f_B. What are f_ϕ, f_X?

1.14 Let M be the set of all men in some town and let H, R and I be, respectively, the subsets of all happy men, rich men and intelligent men. Express the following subsets in terms of H, R, I and their complements
   (i) the set of all men who are happy and rich but not intelligent.
   (ii) the set of all men who are not intelligent but are either happy or rich.

(iii) the set of all men who are exceptions to the statement 'every rich man is happy'.

(iv) the set of all men who conform to the statement 'every rich man is happy'.

1.15   Suppose that in the town in the last exercise, there is monogamy, all men are married and live with their wives. Let W be the set of all women in the town and let A, B, C be respectively the subsets of all rich women, beautiful women and happy women. Define $f: M \to W$ by $f(x) =$ wife of x for $x \in M$. Express the following sets in terms of R, H, I, A, B, C, their complements and f.

(i)   the set of all married women (in the town)

(ii) the set of all unmarried women (including widows)

(iii) the set of all rich men who are married to beautiful women

(iv) the set of women whose husbands (if any) are intelligent but neither happy nor rich

(v)   the set of men who are exceptions to the rule 'a happy man has a happy wife'.

*1.16   Paraphrase the Business Problem in terms of sets.

1.17   Prove that the loser function in the Tournaments Problem is one-to-one. What is its range?

*1.18   This exercise shows how functions of two variables can be looked upon in a certain way. The result itself is simple and will not be frequently needed in the sequel. But we give it as a drill in 'abstract' reasoning. If A and B are any two sets then the set of all functions from A to B will be denoted by B^A. The reason for this notation will be explained in the next section. Now let X, Y, Z be any sets and f be a function from the cartesian product of $X \times Y$ into Z. Fix some $y \in Y$. Define $f_y: X \to Z$ by $f_y(x) = (x, y)$ for $x \in X$. The function f is of two variables while f_y is a function of only one variable. It is said to be obtained from f by fixing the second variable.

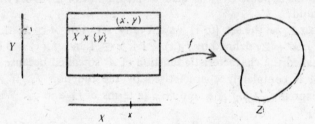

Figure 2.3:   Function of Two Variables.

f_y is also suggestively denoted by $f(-, y)$. If we fix the first variable, f would give rise to a function from Y to Z.

(a)   Prove that f_y is effectively the restriction of f to the box $X \times \{y\}$ contained in $X \times Y$ (see Figure 2.3).

(b) Define a function $\hat{f}$ (read 'f hat') from Y to Z^X by $\hat{f}(y) = f_y$ for $y \in Y$. Prove that $\hat{f}$ determines f uniquely, that is, if we know $\hat{f}$, we can recover the original function f from it.

(c) Prove that there is a bijection between $Z^{X \times Y}$ and $(Z^X)^Y$. By analogy with exponentiation for real numbers, this is called the **exponential law for functions.**)

Notes and Guide to Literature

By far, a most reccommendable reference on set theory is the book of Halmos [1]. The axiomatic approach to set theory along with the construction of natural numbers as certain sets may be found in the Appendix to Kelley [1]. Once natural numbers are constructed, then step-by-step one can construct, integers, rational numbers, real numbers and finally complex numbers. These constructions are fairly standard and may be found, for example, in Joshi [1]. This elaborates our earlier remark that numbers can be constructed from sets.

If $f : X \to Y$ is a function and $x \in X$, the value of f at x is denoted by some authors by xf rather than by $f(x)$. With this convention, the composite of functions is denoted the opposite way. One advantage of this notation is that the functions appear in the same order in which they act. It is probably for this reason that the new notation is increasingly followed. Nevertheless, we stick to the old notation. Note also that some authors call 'codomain' as 'range'.

We remarked that in specifying a set by its elements, it makes no difference how many times a particular element, is repeated. Sometimes, however, such repetitions do matter. For example, when we consider the set of roots of a polynomial, we would like to count each root according to its multiplicity. Similarly when we consider the set of share-holders of a company, we would like to count each share holder as many times as the number of shares he holds. The appropriate mathematical concept to handle such situations is a 'multi-set', which we shall study in the next chapter.

The perceptive reader must have sensed some resemblance between complements and negation, between union and disjunction and between intersection and conjunction. Such resemblance is not fortuitous and will be fully brought out when we study Boolean algebras.

2. Cardinalities of Sets

The discussion in the last section was independent of the sizes of the sets. The concept of the number of elements in a set never appeared in it (except in the remark we made after defining cartesian products to justify the term 'product'). In practice, the size of a set is obviously a very important concept. The technical name for it is the **cardinality** or the **cardinal number** of the set. For finite sets it coincides with the notion of the number of elements in a set. Since we shall be interested mostly in finite sets, we shall omit the definition of cardinality, which is rather technical. Thus, for our purpose, the cardinal number of a set S is simply another name for the number of elements in S. It is commonly denoted by $|S|$ or $n(S)$ of $\#(S)$. For brevity, a set of cardinality n will be called an *n-set* and a subset of cardinality m an *m-subset*. Note that a 1-subset (or **singleton**) corresponds to a point of a set although conceptually the two are different.

Now, what is a finite set ? The word 'finite' comes from the Latin verb 'finire' which means 'to end'. This is consistent with the intuitive meaning of a finite set. If we start counting its elements one by one, this process will end eventually. That brings another question. What is counting ? When we count, we go on assigning successively larger positive integers to the elements of a set, making sure that no element gets counted more than once. This gives a one-to-one function from the set S into the set of positive integers. If the set is finite, the range of this function will be a set of the form $\{1, 2,..., n\}$ for some positive integer n. We then get a bijection between S and $\{1, 2,..., n\}$. Therefore, formally we define:

2.1 **Definition:** A set S for which there exists a bijection between S and the set $\{1, 2,..., n\}$ for some positive integer n, is called **finite**.

Note that this definition presupposes that positive integers are already defined. As remarked in the 'Notes and Guide to Literature' of the last section, positive integers can indeed be defined as certain sets and their basic properties (popularly known as *Peano axioms*, one of which is the principle of induction) can actually be derived as theorems, But we shall make no attempt to achieve this degree of completeness. So we shall take these properties for granted. Using them it can be shown that the integer n that appears in the definition above is unique for a given set. This unique integer n is formally called the cardinality of S. By convention we also take the empty set to be finite and assign it the cardinal number 0.

Most of the sets that we come across in practice are finite. This includes the set of students in a college, the set of persons living in a country, the set of all sand particles on a beach and the set of all atoms in some physical object. These are all finite sets, even though some of them are so large that in practice we tend to think of them as infinite. Whenever we deal with 'practical' problems, there will be a tacit assumption that the sets involved are finite, unless otherwise stated or implied by the context. For example, in the Dance Problem, we assume that the set of boys and the set of girls are both finite. (Actually, if this assumption does not hold, the result can be shown to be false.)

The fundamental problem of combinations is to find the cardinality of a given set which is expressed in terms of some other sets of known cardinalities*. Often we want to do something more than mere counting. For example, we may want to list the elements of a set in some order. But in general we are lucky if we can at least count how many elements there are in a given set, because with the infinite variations in the problems that arise in practice, there is no golden method which will work in all problems. Some of the more elaborate methods will be studied in later chapters. Here we study a few elementary counting techniques. They will all be based upon the following simple property which we take as an axiom, popularly called the **principle of addition**.

2.2 Axiom: If S and T are any two (finite) sets and $S \cap T = \phi$, then $|S \cup T| = |S| + |T|$.

We remark that although we are taking this property as an axiom (and therefore not giving any proof for it), it can actually be proved if we go to the definition of positive integers as certain sets. As a matter of fact, the very definition of addition involves the process of taking the union of two mutually disjoint sets. So the property above is actually an easy consequence of certain definitions. However, as remarked before, we are not after such degree of completeness. Therefore, we shall take it as an axiom. It is certainly consistent with our experience. The condition $S \cap T = \phi$, is necessary to ensure that no element in the union $S \cup T$ is counted twice. Before we derive further consequences of this axiom, it is convenient to have a slight extension of it to the case of the union of any number of sets.

2.3 Proposition: Let $A_1, A_2, ..., A_n$ be pairwise disjoint, finite sets. Then

$$\left| \bigcup_{i=1}^{n} A_i \right| = \sum_{i=1}^{n} |A_i|.$$

Proof: We prove this by induction on n. There is nothing to prove for $n = 1$, while the case $n = 2$ is covered by the axiom above. Assume the result holds for $n = k$. We want to prove it for $n = k + 1$. For this, let $S = \bigcup_{i=1}^{k} A_i$ and $T = A_{k+1}$. Then

$$S \cap T = \left(\bigcup_{i=1}^{k} A_i \right) \cap A_{k+1}$$

which is easily seen to be equal to

$$\bigcup_{i=1}^{k} (A_i \cap A_{k+1})$$

(cf. property (ii) in Proposition (1.1).) Since we are assuming that $A_i \cap A_j = \phi$

---

*The basic question in combinatories is often put as 'In how many ways can a certain thing be done? We can paraphrase this using sets. Let S be the set of all possible ways the particular thing can be done. The question now is equivalent to asking the cardinality of S.

for all $i \neq j$, it follows that $S \cap T = \phi$. Therefore, from our axiom,

$$|S \cup T| = |S| + |T|$$

or in other words,

$$\left| \bigcup_{i=1}^{k+1} A_i \right| = \left| \bigcup_{i=1}^{k} A_i \right| + |A_{k+1}|.$$

By induction hypothesis, the first term on the right hand side equals $\sum_{i=1}^{k} |A_i|$. So the right hand side equals $\sum_{i=1}^{k+1} |A_i|$ completing the proof. ▆

2.4 Corollary: If in the proportion above, all sets A_i are of the same cardinality (say m) then $\left| \bigcup_{i=1}^{n} A_i \right| = mn$.

Proof: This is obvious, since $m + m + \ldots + m$ (n times) is precisely mn. (Actually this is the definition of multiplication of integers, but we are not doing it that way.) ▆

As another easy consequence of Axiom (2.2) we prove;

2.5 Proposition: Let X be any (finite) set and A any subset of it. Then

$$|X - A| = |X| - |A|.$$

Proof: We let $S = A$ and $T = X - A$ and apply the axiom. We get,

$$|X| = |S \cup T| = |S| + |T| = |A| + |X - A|.$$

The result now follows by subtracting $|A|$ from both the sides. ▆

This simple proposition is useful in finding the probability of an event by first computing what is known as its **complementary probability**, that is, the probability that the event will not occur. It so happens sometimes that it is easier to find the number of unfavourable cases than the number of favourable cases. To find the latter, we apply the last proposition (provided, of course, that we know the total number of cases). We shall see instances of this later. The following corollary of the last proposition conforms to our intuition that a part cannot be equal to the whole.

2.6 Corollary: If X, A are as above then $|A| \leqslant |X|$. Also if A is a proper subset of X then $|A| < |X|$.

Proof: We have, $|X| = |A| + |X - A|$. Since $|X - A| \geqslant 0$, the first assertion is clear. Also, if A is a proper subset of X then $X - A$ is non-empty and so $|X - A| > 0$, proving the second assertion. ▆

In Axiom (2.2) we required that the sets S and T be disjoint. When if $S \cap T \neq \phi$? We can then still find $|S \cup T|$ by summing up $|S|$ and $|T|$ but we have to make up for the fact that the elements of $S \cup T$ have been counted twice. The situation is pictured in Figure 2.4. Although later on we shall consider a much more general formula for the union of any number of sets, this special case is worth mentioning separately because of its intuitive nature.

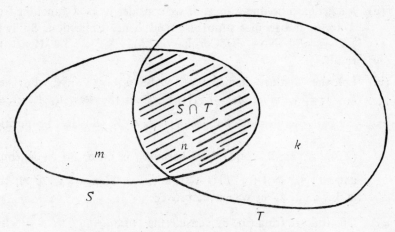

Figure 2.4: Cardinality of the Union of Two Sets.

2.7 Proposition: For any two finite sets S and T,

$$|S \cup T| = |S| + |T| - |S \cap T|.$$

Proof: Let m, n and k denote respectively the cardinalities of the sets $S-T$, $S \cup T$ and $T-S$. It is easy to see that these three sets are pairwise disjoint and that their union is $S \cup T$. So by Proposition (2.3), $|S \cup T| = m + n + k$. It is also clear that S is the union of the two disjoint subsets $S-T$ and $S \cap T$. Hence, $|S| = |S-T| + |S \cap T| = m + n$. Similarly $|T| = k + n$. The desired result now follows by substitution. ▪

Let us now see how the existence of functions with certain properties implies some relationship between the cardinalities of their domains, codomains and ranges. In the following proposition we list a number of such results.

2.8 Proposition: Let $f: X \to Y$ be a function with range R, where X, Y are finite sets. Then,

(i)   if f is a bijection, then $|X| = |Y|$
(ii) if f is one-to-one, then $|X| \leqslant |Y|$
(iii) if f is onto, then $|Y| \leqslant |X|$
(iv) $|R| \leqslant |X|$ (without any restriction on f).

Proof: If X is empty then all the results are trivially true. So let us suppose $X \neq \phi$. Note that Y must also be non-empty because there cannot exist any function from a non-empty set to the empty set. Let $m = |Y|$.

(i) By definition, there exists a bijection, say, $g: Y \to \{1, 2, ..., m\}$. The composite of two bijections is easily seen to be a bijection (cf. Exercise (1.11)). So $g \circ f: X \to \{1, 2, ..., m\}$ is a bijection. But this means that $m = |X|$. So $|X| = |Y|$.

(ii) f is given to be one-to-one. If we consider it as a function from X to R, it becomes onto also and hence a bijection. So by (i), $|X| = |R|$. Now, by Corollary (2.6), $|R| \leqslant |Y|$ Hence the result.

(iii) Let the distinct elements of y be y_1, y_2, ..., y_m. For each $i = 1, 2, ..., m$ let $A_i = f^{-1}(\{y_i\})$. Then the sets $A_1, A_2, ..., A_m$ are pairwise disjoint and also $\bigcup_{i=1}^{m} A_i = X$. So by Proposition (2.3), $|X| = \sum_{i=1}^{m} |A_i|$. Now because f is onto, no A_i is empty and so $|A_i| \geqslant 1$ for all $i = 1, 2, ..., m$. Hence $|X| \geqslant m$; that is, $|X| \geqslant |Y|$ as was to be proved.

(iv) This follows from (iii) by considering once again f as a function from X onto R. ∎

Note that the argument used for proving (iii), also gives the slightly more general result that if every point of the codomain has at least k preimages for some positive integer k then $|X| \geqslant k|Y|$. Similarly (i) can be generalized to say that if every $y \in Y$ has exactly k preimages then $|X| = k|Y|$. This generalisation will be frequently used in the sequel.

The preceding proposition is hardly profound. But it has some very interesting consequences. The ingenuity lies, of course, in defining a suitable function. As an illustration let us do the Tournaments Problem with n participants. Let P be the set of participants and M the set of matches played. In the last section we defined the loser function $f: M \to P$ by $f(m) =$ the player who loses the match m; for $m \in M$. It is easy to show that this function is one-to-one (cf. Exercise (1.17)). Let R be its range. Then by (ii) in the proposition above, $|M| = |R|$. But now what does R consist of? Clearly it contains those (and only those) players who lose at least one match. The rules of the game are such that the champion never loses a match but every other player does. So R is the entire set P except the champion. Hence by Proposition (2.5), $|R| = |P| - 1 = n-1$. This means, $|M| = n-1$, that is, $n-1$ matches are played to find the champion from n players. See how effortlessly the result falls out!

For other applications, it is convenient to reformulate the proposition above slightly. We state the result without proof because basically the proof simply amounts to taking the contrapositive of the assertions proved above.

2.9   Theorem: Let $f: X \to Y$ be a function where X, Y are finite sets. Then,

(i)   if $|X| = |Y|$ then f is one-to-one if and only if it is onto.
(ii) if $|X| > |Y|$ then f cannot be one-to-one. More generally, if k is a positive integer such that $|X| > k |Y|$ then there exist at least $k + 1$ distinct points, say, $x_1, x_2, ..., x_{k+1}$ in X such that $f(x_1) = f(x_2) = \cdots = f(x_{k+1})$. ∎

The reason for calling this result a 'theorem' is that the second statement in it is very famous by a different name. It is called the **pigeon-hole principle.** In essence it says that when we put letters into pigeonholes, if there are more letters than there are pigeonholes, then at least one pigeonhole must contain more than one letter. This seems to say little more than common sense. But once again, although the pigeonhole principle itself is trivial, when cleverly applied, it can yield nontrivial results. We illustrate this with two problems.

2.10   Problem: Suppose 14 students in a class appear at a university examination. Prove that there exist at least two among them whose seat numbers differ by a multiple of 13. (There is nothing very special about the number 13 here. In Western countries there is a superstition that it signifies evil.)

Solution: Let S be the set of 14 students. The trick is to define a suitable function f on S in such a way that $f(x) = f(y)$ would imply that the seat numbers of the students x and y differ by a multiple of 13. For this we note that given positive integers p and q, we can divide p by q and get a quotient a and a remainder r; that is we can find integers a and r such that $p = aq + r$ where r, being the remainder, can only have the values $0, 1, ..., q-1$. (This is formally known as the *Euclidean Algorithm* and we shall study its consequences in the chapter on rings. Actually, p can be a negative integer, but we always require the remainder to be non-negative.) So we define $f: S \to \{0, 1, ..., 12\}$ by $f(x) =$ remainder left when the seat number of x is divided by 13. Here the domain has cardinality 14 and the codomain has cardinality 13. So the pigeon-hole principle applies and we get that there exist distinct students x and y such that $f(x) = f(y)$. Let the seat numbers of x and y be m and n. Then by definition, there exists integers a and b, respectively, such that $m = 13a + f(x)$ and $n = 13b + f(y)$. Since $f(x) = f(y)$, we get that $m - n = 13(a - b)$ which is clearly a multiple of 13 since $a - b$ is an integer. ∎

For the next problem, we urge the reader to try it first on his own. (It can actually be done without the use of the pigeon-hole principle. But the argument becomes clumsy.) Many problems in combinatories (and more generally in mathematics) look deceptively simple once you know their solutions. In order to truly appreciate the solution, it is therefore necessary to give the problem a serious try.

2.11   Problem: If 101 integers are selected from the set $\{1, 2, ..., 200\}$, prove that among the selected integers there exist two integers such that one of them is a multiple of the other.

Solution: Before giving the solution let us see what is so special about 101. When one integer is a multiple of some other, it is at least twice as big as the other. Therefore if two integers selected from $\{1, 2, ..., 200\}$ are both 'large', neither could be a multiple of the other. In particular if we select the 100 integers from 101 to 200, the result does not hold. Now if we select just one more integer, say x, then it has to be from 1 to 100 and so either $2x$ or $4x$ or $8x$ ... etc. is among 101 to 200 and so the result holds. Of course, this does not solve the problem (because we are not given that the selected integers must include all integers from 101 to 200), but it suggests that twofolds may have some special role to play in the solution.

Now, for the solution itself, we let S be once again the set of selected integers and then construct a suitable function f with domain S. Let $x \in S$. Then either x is odd or else it is a twofold of some integer y, i.e. $x = 2y$. Again y is itself odd or else it is a twofold of some z, $y = 2z$. We go on repeating this until we get an odd integer (which may be equal to 1). Formally, we write x as $2^r t$ where r is a non-negative integer and t is odd. We now define $f(x)$ to be this odd integer t. Because x is from 1 to 200, $f(x)$ can be from 1 to 199; thus f takes values only in the set $\{1, 3, 5, 7, ..., 197, 199\}$. This set has cardinality 100, less than that of the domain S. So by the pigeon-hole principle there exist distinct $x, y \in S$ such that $f(x) = f(y)$. By definition, $x = 2^r f(x)$ and $y = 2^s f(y)$ for some non-negative integers r and s. Since $f(x) = f(y)$ we see that $x/y = 2^{r-s}$ which is an integer if $r \geqslant s$ while if $s \geqslant r$ then y/x is an integer. In either case the assertion of the problem holds.   ▨

We have proved something stronger than the problem asks. We have found x and y among the selected integers such that x/y (or y/x) is not only an integer but a power of 2. If the problem was stated like this, perhaps we would have got some hint of the proof. This is the case with some problems. A stronger assertion is easier to prove because it carries with it a built-in clue for the solution.

The propositions we have proved so far also allow us to find the cardinality of the cartesian product of sets (as remarked in the last section.) The result which follows is popularly called the **principle of multiplication**.

2.12   Proposition: Let X, Y be finite sets. Then $|X \times Y| = |X| \times |Y|$.

Proof: Let $m = |X|$ and $n = |Y|$. If $n = 0$, then Y and $X \times Y$ are both empty and the result holds trivially. Assume $n > 0$ and let the distinct elements of Y be $y_1, y_2, ..., y_n$. For each $i = 1, 2, ..., n$ let A_i be the set $X \times \{y_i\}$. Then A_i is a subset of $X \times Y$ (cf. Exercise (1.5)) and $X \times Y = \bigcup\limits_{i=1}^{n} A_i$. Also for $i \neq j$, $y_i \neq y_j$ and so $A_i \cap A_j = \phi$. Further, for every i, the function $f_i : X \to A_i$ defined by $f_i(x) = (x, y_i)$ is clearly a bijection. So by Proposition (2.8), Part (i), $|A_i| = m$ for every i. The result now follows from Corollary (2.4). (A pictorial representation of the proof is given in Figure 2.5.) ∎

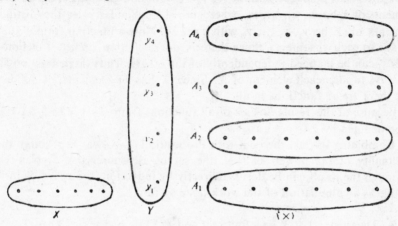

Figure 2.5: Cardinality of a Cartesian Product.

The preceding result can be readily extended, by induction, to the case of the product of n sets, the same way as Proposition (2.3) is an extension of Axiom (2.2). The proof is left as an exercise.

2.13 Proposition: For any finite sets $A_1, A_2, ..., A_n$ we have
$$|A_1 \times A_2 \times ... \times A_n| = |A_1| \times |A_2| \times ... \times |A_n|.$$ ∎

The result of Proposition (2.12) is often expressed by saying that if something can be done in m ways and when it is done in any one of these ways, some other thing can be done in n ways then the two things can together be done in mn ways. (In Proposition (2.12), the first 'thing' is to choose an element of X and the second is to pick an element of Y.) Although this formulation is somewhat loose, it helps in many problems. In fact, the number of ways to do the second thing need not be independent of the way the first thing is done. Thus if the first thing can be done in m ways and when it is done in the ith way the second thing can be done in n_i ways, for $i = 1, 2, ..., m$ then the two things can be done together in N ways where $N = n_1 + n_2 + ... + n_m$ The reasoning is analogous to the proof of Proposition (2.12). As an application if there are five men having $2, 3, 1, 0$ and 2 wives respectively then a couple can be invited in $2 + 3 + 1 + 0 + 2$ ways, i e., in 8 ways. The extension to the case of n things (instead of 2) is also obvious.

Using this reasoning we count the number of functions from one set to another. Because of its basic character, we record the result as a theorem.

2.14 Theorem: Let X and Y be finite sets. Then the number of distinct functions from X to Y is $|Y|^{|X|}$.

Proof: If X is empty then there is only one function from X to Y and the assertion holds with the convention that any integer raised to 0 is 1. Assume X is non-empty. If Y is empty, there can be no function from X to Y and again the result holds. So assume X, Y are both non-empty. Let the distinct elements of X be x_1, x_2, ..., x_m where $m = |X|$. Similarly let the distinct elements of Y be y_1, y_2 ..., y_n with $n = |Y|$. Now a function from X to Y assigns to each x_i some y_j. Since there is no restriction on the functions, each x_i can be mapped independently of the others. Thus there are n possible ways to map each element of X. Since X has m elements, it follows that there are n^m functions in all. ∎

Because of this result, the set of all functions from a set X to a set Y is often denoted by Y^X.

Combining the last theorem with Proposition (2.8), we can count the cardinality of the power set (i.e., the set of all subsets) of a given set. Although the result can be derived directly by induction, the proof is instructive as a culmination of the work done so far.

2.15 Theorem: Let X be a finite set and $P(X)$ its power set. Then

$$|P(X)| = 2^{|X|}$$

Proof: Let $\mathbb{Z}_2$ be the set $\{0, 1\}$. Let F be the set of all functions from X to $\mathbb{Z}_2$. By the theorem obove, $|F| = 2^{|X|}$ and so the proof would be completed if we find some bijection between $P(X)$ and F. For this, define $\theta : P(X) \to F$ as follows. If A is a subset of X, we let $f_A : X \to \mathbb{Z}_2$ be the characteristic function of A, defined by $f_A(x) = 1$ if $x \in A$ and 0 if $x \notin A$ (see Exercise (1.13)). We define $\theta(A) = f_A$. (A beginner often finds this definition confusing. But it need not be so. Note that the codomain of the function θ is F which itself consists of functions from X to $\mathbb{Z}_2$. So if we take a typical element, say A, of the domain of θ(viz , $P(X)$), then $\theta(A)$ itself should be a function from X to $\mathbb{Z}_2$. We let this function be the characteristic function of A. Thus we can write $[\theta(A)](x) = 1$ or 0 according as $x \in A$ or $x \notin A$.) Now to show that θ is a bijection, we first show it is one-to-one. Suppose $\theta(A) = \theta(B)$ where $A, B \in P(X)$. We must show $A = B$. Let $x \in A$. Then $\theta(A)(x) = 1$. So $\theta(B)(x) = 1$, which means $x \in B$. Thus $A \subset B$. Similarly $B \subset A$ and so $A = B$. Next, to prove that θ is onto, let $f \in \mathbb{Z}_2^X$. This means f is some function from X into $\mathbb{Z}_2$ and we have to find some $A \in P(X)$ such that $\theta(A) = f$, i.e. such that f equals the characteristic function of A. Obviously we have no choice but to let $A = f^{-1}(\{1\})$. Then $A \subset X$, i.e. $A \in P(X)$. Now if $x \in A$, then $x \in f^{-1}(\{1\})$ and so $f(x) = 1$; while if $x \notin A$ then $f(x) \neq 1$, but then $f(x) = 0$ because f assumes only two possible values, namely, 1 and 0. Therefore $f(x) = 1$ or 0 according as $x \in A$ or $x \notin A$. This is precisely the definition of the characteristic function of A. So $f = f_A = \theta(A)$, proving that θ is onto. Since θ is already injective, it is a bijection and, as noted earlier, this completes the proof. ∎

In Theorem (2.14), there was no restriction on the functions (other than those imposed by the very definition of a function, namely that it be single-valued and defined over the entire domain set). If we want to count only functions satisfying some additional requirements, the reasoning has to be tailored to the type of requirements and in many cases is far from easy. For example, we count below the number of injective functions from one set to another. But there is no easy formula for the number of surjective functions. Before proving the result, we paraphrase the problem slightly, because it is the paraphrased version that is more well-known. Let n and r be positive integers. By an r-**permutation** of n objects (or symbols) we mean an ordered arrangement of r of these objects, or in other words a sequence of length r in which no object appears more than once. We can make this precise using the language of functions. Let Y denote the set of n objects. A sequence of length r of these objects obviously corresponds to a function f from the set $\{1, 2, ..., r\}$ to the set Y. The requirement that no object appears more than once is equivalent to saying that the function f is injective. We can therefore state our results in the form of r-permutation of n objects.

2.16 Theorem: The number of r-permutations of n objects (or equivalently, the number of injective functions from a set of cardinality r to a set of cardinality n) is the product $n \cdot (n-1) \cdot (n-2) \ ... \ (n-r+1)$.

Proof: We simply have to modify the proof of Theorem (2.14) slightly. Let $X = \{x_1, x_2, ..., x_r\}$ and $Y = \{y_1, y_2, ... \ y_n\}$ be sets of cardinalities r and n respectively. Let f be an injective function from X to Y. Then $f(x_1)$ can be anything from $\{y_1, y_2, ..., y_n\}$. Once $f(x_1)$ is fixed, say $f(x_1) = y_i$, then, however, $f(x_2)$ cannot equal y_i because f is one-to-one. So, for $f(x_2)$ there are only $n-1$ choices namely, $y_1, y_2, ..., y_{i-1}, y_{i+1}, ..., y_n$. When $f(x_2)$ has been fixed, $f(x_3)$ will have $n-2$ possible choices. Continuing this way, we see that f can be constructed in $n \cdot (n-1) \ (n-2), \ ......(n-r+1)$ ways. This is, therefore, the total number of injective functions from X to Y, or equivalently, the number of r-permutations of n objects. ∎

If $r > n$, then the expression in the last theorem has 0 as a factor and so we see that there can be no r-permutation of n objects for $r > n$. Of course this is also obvious from the pigeon-hole principle. But it is always nice (and comforting) to note how a general formula is consistent with its special cases proved earlier by different reasonings.

The case $r = n$ is especially interesting. An n-permutation of n object is simply called their permutation. We already defined this term in the last section and in view of the comments we made before Theorem (2.16), the two meanings are consistent with each other. From the formula above, the number of permutations of a set with n elements is $n \cdot (n-1) \cdot (n-2) \ldots \ldots 3.2.1$. It is the product of the first n positive integers. This expression appears so frequently in combinatories, that it is convenient to have a shorthand notation for it. It is denoted by $n!$ or by $\lfloor n$ (read as '...n factorial'). By convention we set $0! = 1$. The number of r-permutations of n objects is also denoted often by a special symbol, $_nP_r$ (or ^nP_r). Clearly $_nP_r = \dfrac{n!}{(n-r)!}$ for $0 < r \leqslant n$.

For example, let $n = 3$. Then $n! = 6$. The six permutations of a 3-element set, say $\{a,b,c\}$ are given by abc, acb, bca, bac, cab and cba. There are 12 2-permutations of a set with 4 elements, say a, b, c, d. They are $ab, ac, ad, ba, bc, bd, ca, cb, cd, da, db$ and dc. The function $n!$ grows very rapidly as n grows. Even for relatively small values of n such as $n = 8$ (note that $8!$ exceeds 40,000), if we want to list all permutations of n symbols, we cannot do so haphazardly. If we do, we are very prone to miss some permutation or to repeat some permutation. Systematic methods for listing the permutations of a given set will be studied in the next chapter.

We often have to consider arrangements of objects in which some of the objects are indistinguishable from each other. For example, suppose we have to form strings containing, say, three red beads, 2 blue beads and 1 white beads. Beads of the same colour are to be regarded as indistinguishable and so two strings which result from each other by mere reshufflings within beads of the same colour are to be regarded as identical. (As another example, if we want to form different words by permuting the letters of some given word, say, 'content' we do not distinguish between the two n's and the two ts'. The general problem can be represented as follows. Suppose we have n objects which are classified into k distinct types, so that no two objects of the same type are to be distinguished from each other. Suppose there are n_1 objects of type 1, n_2 objects of 2,..., n_k objects of type k where $n_1, n_2, \ldots, n_k$ are positive integers (it may happen that $n_i = 1$ for some values of i). Clearly $n_1 + n_2 + \ldots + n_k = n$. The question now is how many distinct permutations of these n objects there are. The following theorem provides the answer.

2.17. Theorem: The number of distinct permutations of a set with n_1 objects of type 1, n_2 objects of type 2,..., n_k objects of type k equals

$$\frac{(n_1 + n_2 + \ldots + n_k)!}{n_1! \, n_2! \ldots n_k!}.$$

Proof: The proof will be an application of generalisation of (i) in Proposition (2.8) (see the comment following its proof). Let us denote objects of type 1 by $a_1, a_2, \ldots, a_{n_1}$. Even though they are to be regarded as identical, because of our notation, we distinguish among them for a moment. Similarly denote the objects of type 2 by $b_1, b_2, ..., b_{n_2}$; those of type 3 by $c_1, c_2...,$ c_{n_3} and so on. (It may be argued that if k is large, we may run out of the alphabet. But we can always introduce additional symbols if necessary. Alternatively, we can use double suffixes. Thus, the objects of type 1 will be denoted by $x_{1,1}, x_{1,2}, ..., x_{1,n_1}$, those of type 2 by $x_{2,1}, x_{2,2}...,x_{2,n_2}$ and so on).

Now let $n = n_1 + n_2 + ... + n_k$ and let S be the set of these n 'distinct' objects. Let X be the set of all permutations of S. By Theorem (2.16), $|X| = n!$. Let Y be the set of all permutations of the given objects, that is, where we do not distinguish between two objects of the same type. Our goal is to find the cardinality of Y and we do so in terms of the cardinality of X by defining a suitable function $f : X \rightarrow Y$ as follows.

Take any typical element in the set X. This is some permutation of the n objects. Now we merely remove the subscripts of the objects in this permutation. We then get an element of Y, i.e., a permutation of the original set of objects. We denote this assignment by the function f. For example, $f(b_1 a_2 c_2 a_3 a_2 c_1 d_1 b_2) = bacaacdb$. Let us now examine how many pre-images each point of Y has under f. Note that every element of Y is a permutation in which there are n_1 a's (without suffixes), n_2, b's, n_3 c's,.... and so on. Now we can assign the n_1 suffixes to these n_1 a's (of course we must not assign the same suffix twice), the n_2 suffixes to the n_2 b's and so on. When we do this, we shall get an element of X which will be mapped under f to the element of Y with which we started. (For example if we take *bacaacdb* in Y, then,

$$b_1 a_1 c_1 a_2 a_3 c_2 d_1 b_2, \ b_1 a_1 c_2 a_3 a_2 c_1 d_1 b_2, \ b_2 a_2 c_1 a_1 a_3 c_2 d_1 b_1$$

and many other elements of X are all mapped to it under f). Now, the n_1 suffixes can be put on the $n_1 a$'s in $n_1!$ ways (because each suffix is to be used only once). Similarly the assignment of suffixes to the b's can be made in $n_2!$ ways, to the c's in $n_3!$ ways and so on. Together, for a given element of Y, we can assign the suffixes, in $n_1! \times n_2! \times ... \times n_k!$ ways and thereby get just these many points of X which are all taken to the same element of Y under f.

Thus, every point of Y has exactly $n_1 \times! \times n_2! \times ... n_k!$ pre-images under f. So by the extension of Proposition (2.8),

$$|X| = (n_1! \times n_2! \times ... \times n_k!) |Y|.$$

Since $|X|$ is already known to be $n!$ and $n = n_1 + n_2 + ... + n_k$, it follows that

$$|Y| = \frac{(n_1 + n_2 + ... + n_k)!}{n_1! \ n_2!...n_k!}$$

as was to be proved. ∎

As an application, by rearranging the letters of the word 'content' we get $\frac{7!}{2!2!}$ or 1,260 distinct words. (Of course, not all these words are meaningful, but that is a problem of English rather than of mathematics).

The reasoning used above can also be used to find the number of r-combinations of n objects. Formally, an r-**selection** or r-**combination** of n objects is an unordered arrangement of r distinct objects out of these objects. In the language of sets, if S is the set of n objects then an r-combination of these objects is nothing but a subset of S of cardinality r, or an r-subset of S. To find the number of such r-subsets, let X be the set of all r-permutations and Y the set of all r-combinations of n objects, say $a_1, a_2, ..., a_n$. We define $f: X \to Y$ by ignoring the order. That is, if we take any r-permutations say, $a_{i_1}, a_{i_2} ... a_{i_r}$ then we let $f(a_{i_1}, a_{i_2}, ... a_{i_r})$ be simply the set $\{a_{i_1}, a_{i_2}, ..., a_{i_r}\}$ which is an r-subset. If we permute $a_{i_1}, a_{i_2}, ..., a_{i_r}$ among themselves, we get the same r-subset, i.e. the value of f is unaffected. But there are exactly $r!$ distinct permutations of r objects. So every point of Y has exactly $r!$ preimages. So by the extension of Proposition (2.8) (i), we get $|X| = r! |Y|$. We already know that

$$|X| = {}_nP_r = \frac{n!}{(n-r)!} \ .$$

We therefore get the following important result.

2.18. Theorem: The number of distinct r-combinations of n objects equals $\dfrac{n!}{r!\,(n-r)!}$, (which also equals $\dfrac{n \cdot (n-1)...(n-r+1)}{r!}$). ∎

This number appears so frequently in mathematics that it is convenient to have a short notation for it. It is commonly denoted by ${}_nC_r$, ${}^nC_r$ or by $\begin{pmatrix} n \\ r \end{pmatrix}$. It is also called a **binomial coefficient** because it appears in the well-known binomial theorem, which we shall study later. Note incidentally that

$$\begin{pmatrix} n \\ r \end{pmatrix} = \begin{pmatrix} n \\ n-r \end{pmatrix}, \text{for } 0 \leqslant r \leqslant n.$$

There are numerous identities involving ${}_nC_r$ (and sometimes ${}_nP_r$). They are called **combinatorial identities** because they can generally be proved by a combinational argument which consists of showing that both the sides of the identity represent the cardinality of the same set, counted in two different ways. Of course sometimes it is just as easy to establish the identity purely algebraically. For example in the identity

$$\begin{pmatrix} n \\ r \end{pmatrix} = \begin{pmatrix} n \\ n-r \end{pmatrix}$$

given above each side equals

$$\frac{n!}{r!(n-r)!} \ .$$

For a combinatorial proof, we observe that selecting r objects out of n amounts to rejecting the remaining $(n-r)$ objects. The $n-r$ objects to be rejected can be chosen from n objects in $\binom{n}{n-r}$ ways. As another illustration, we prove one more identity.

2.19 Proposition: For $1 \leqslant r \leqslant n$, we have

$$\binom{n}{r} = \binom{n-1}{r} + \binom{n-1}{r-1}.$$

Proof: An algebraic proof can be given by expanding the terms. For a combinatorial proof, note that the left hand side represents the number of ways to select r objects out of n objects. Let S be the set of these n objects. Fix some $x \in S$ and let $T = S - \{x\}$. Then $|T| = n-1$. The r-subsets of S are of two types; those containing x and those not containing x. An r-subset of the first type will have its remaining $r-1$ elements from T while an r-subset of the second type will have all its r elements from T. Since $|T| = n-1$, T will have $\binom{n-1}{r-1}$ subsets of cardinality $r-1$ and $\binom{n}{r}$ subsets of cardinality r. Together, there will be $\binom{n-1}{r-1} + \binom{n-1}{r}$ subsets of S having r elements each.

Although this argument is reasonably clear, it is instructive to formulate it rigorously so as to bring out exactly which of the results in this section have been used where. We let $P_r(S)$ be the set of all r-subsets of S. Now let $Q = \{A \in P_r(S) : x \in A\}$ and $R = \{A \in P_r(S) : x \notin A\}$. Then $Q \cap R = \phi$ and $Q \cup R = P_r(S)$. So by Axiom 2.2, $|P_r(S)| = |Q| + |R|$. There is evidently a bijection between Q and $P_{r-1}(T)$. Hence $|Q| = |P_{r-1}(T)|$ by Proposition 2.8. Similarly there is a bijection between R and $P_r(T)$ and so $|R| = |P_r(T)|$. The result now follows by substituting the values of $|P_r(S)|$, $|P_{r-1}(T)|$ and $|P_r(T)|$ given by Theorem 2.18. ∎

Combinatorial arguments can also be used to prove some other results. As a somewhat trivial example, we prove:

2.20 Proposition: For every non-negative integer n, $n < 2^n$.

Proof: Let X be a set with $|X| = n$. Let $P(X)$ be, as usual, the power set of X. Define $f: X \to P(X)$ by $f(x) = \{x\}$ for $x \in X$. Then f is one-to-one. So by Proposition 2.8, $|X| = |R|$ where R is the range of f. Note that R is a proper subset of $P(X)$, because ϕ, the empty set is in $P(X)$ but not in R. So by Corollary 2.6, $|R| < |P(X)|$. Moreover, by Theorem 2.15, $|P(X)| = 2^n$. The result now follows. ∎

The fact that the cardinality of a set must always be a whole number implies that whenever it is expressed as a ratio of two integers, the denominator must divide the numerator. As an application of this type of a reasoning we prove one result.

2.21 Theorem: The product of any r consecutive positive integers is divisible by $r!$.

Proof: Let the integers be $n, n + 1, n + 2,..., n + r-1$. We have to show that $n \cdot (n + 1)...(n + r - 1)$ is divisible by $r!$. But $\dfrac{n \cdot (n + 1)...(n + r - 1)}{r!}$ is precisely $\dbinom{n + r - 1}{r}$, which is the number of r-subsets of a set with $n + r - 1$ elements. So $\dfrac{n(n + 1)...(n + r - 1)}{r!}$ is an integer. ∎

These applications, interesting as they are, are somewhat incidental. The real applications of the preceding theorems are in actual counting problems. They will be given in the next section. We conclude the present section with a brief discussion of infinite sets. The results we have proved hold only for finite sets. Most of them are either meaningless or false for infinite sets. That is why the study of infinite sets requires substantially different techniques.

The difficulty begins right from the definition of the cardinal number of an infinite set because the intuitive description of it as the number of distinct elements in the set, fails. It is tempting to set the cardinality of all infinite sets as ∞. But this solution is not satisfactory for certain purposes. For example, let $\mathbb{N}$ be the set of positive integers and $\mathbb{R}$ the set of real numbers. Then both are infinite sets. But it can be shown that there cannot exist any bijection between them. So if we assign them the same cardinality, then it will not be a true measure of their sizes. A bijection from one set to another simply amounts to renaming the elements of the domain set with the corresponding elements of the codomain. We would certainly not like to sacrifice the existence of a bijection as a criterion for equality of cardinality of two sets, whatever be the formal definition of cardinal numbers.

The standard approach to cardinal numbers of sets (including infinite sets) starts by defining two sets to have the same cardinal number if there exists a bijection between them. One then takes certain 'standard' sets and shows that given any 'abstract' set X, there is precisely one standard set Y such that there exists a bijection between X and Y. We then define the cardinal number of X as the set Y. It may appear strange at first sight that we define equality of cardinal numbers even before we define cardinal numbers. But it need not be so. The situation is very much like weighing objects with a balance. A balance alone cannot tell the weight of an object. It merely tells whether two objects are of equal weight. To find the weight of a given object, we balance it with some standard weight, which is then called the weight of the object. For finite sets, the 'standard' sets are, of course, the sets of the form $\{1, 2,..., n\}$ for positive integer n.

Having defined cardinal numbers, one goes on to define their addition and multiplication and study their properties. This branch of mathematics is called **cardinal arithmetic**. We shall not study it but remark that it is not just a routine extension of the arithmetic of finite cardinal numbers. To illustrate what goes wrong let X, Y, Z be respectively the sets $\{n \in \mathbb{N}: n \geqslant 2\}$, $\{n \in \mathbb{N}; n \text{ even}\}$ and $\{n \in \mathbb{N} : n \text{ odd}\}$. These are all proper subsets of $\mathbb{N}$, But there exists a bijection between $\mathbb{N}$ and each of these sets. For example, define $f: \mathbb{N} \to X$ by $f(n)=n+1$ for $n \in \mathbb{N}$; $g; \mathbb{N} \to Y$ by $g(n)=2n$ and $h; \mathbb{N} \to Z$ by $h(n) = 2n-1$. Thus the sets X, Y, Z all have the same cardinality as $\mathbb{N}$. This defies our intuition that a part cannot be equal to the whole (cf. Corollary (2.6)). Note also that Y and Z are complementary subsets whose union is $\mathbb{N}$. Still each of them has the same cardinality as $\mathbb{N}$. Thus when an infinite cardinal number is multiplied by 2, it is not increased!

Actually more bizarre things can happen with infinite cardinals. Let X be a finite set with cardinality n. Then the cartesian product $X \times X$ has cardinality n^2 which is much bigger than n for large n. If, however, X is infinite then $X \times X$ has exactly the same cardinality as X because there exists a bijection between X and $X \times X$. We shall not prove this. But we shall illustrate it for the case $X = \mathbb{N}$, the set of positive integers. To construct a bijection f from $\mathbb{N}$ to $\mathbb{N} \times \mathbb{N}$ amounts to listing the elements of $\mathbb{N} \times \mathbb{N}$ as an infinite sequence in which every element appears exactly once. This is done pictorially in Figure 2.6 where we start from the point $(1, 1)$ and each arrow points to the next term in the sequence.

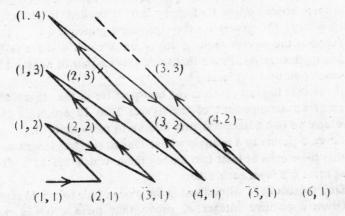

Figure 2.6: Cardinality of $\mathbb{N} \times \mathbb{N}$.

We shall rarely deal with cardinalities of infinite sets. Still, the set of positive integers deserves to be mentioned, for on one hand, it is the limiting case of finite subsets while on the other hand, in a certain sense it is the 'smallest' infinite set. Figuratively the set of positive integers is the bridge between the discrete and the continuous mathematics. Limiting process, in its simplest form, appears as limits of sequences. And sequences, as we know, are nothing but functions defined on the set of positive integers.

The cardinal number of the set of positive integers is denoted by a special symbol $\aleph_0$ (read 'aleph naught' or 'aleph zero'). $\aleph$ is the first letter of the Hebrew alphabet. Obviously any set X for which there exists a bijection between $\mathbb{N}$ and X also has cardinality $\aleph_0$ and conversely. There is a special name for such sets.

2.22   Definition: A set X is called **enumerable** (or **denumerable**) if there exists a bijection $f : \mathbb{N} \to X$ (such a bijection is often called an **enumeration** of the set X). A set which is either finite or denumerable is called **countable**. A set which is not countable is called **uncountable**.

For example, the set of all positive integers, the set of all even positive integers, the set $\mathbb{N} \times \mathbb{N}$ are all denumerable and hence countable. The set of real numbers is uncountable, although a proof of this fact is not easy. and requires a fairly deep property of the real number system (known as completeness).

A few properties of countable sets will be given as exercises.

Exercises

2.1   Prove the generalisation of proposition 2.8, parts (i) and (iii).

2.2   Prove Theorem 2.9.

2.3   Suppose there are k boxes with capacities to hold, says, $n_1, n_2 \ldots, n_k$ objects respectively. Let $n = n_1 + n_2 + \ldots + n_k$. If n objects are put in these boxes, prove that every box is packed to its capacity. How does this generalise the pigeonhole principle ?

2.4   Suppose the figures from 1 to 12 on a clock dial are reshuffled among themselves. Prove that there exists a pair of adjacent figures which add up to at least 14.

2.5   Prove that the last result is the best possible in the sense that there exists an arrangement of the figures 1 to 12 around a clock in which no two adjacent figures add up to more than 14.

2.6   Given 5 points in a triangle whose longest side has length a, prove that there exist at least two among them which are at a distance at most $a/2$ from each other.

2.7   Do Problem 2.11 without using the pigeon-hole principle explicitly.

2.8   Given a positive integer n, prove that there exists a positive integer which is divisible by n and whose decimal representation consists of 0's 1's only. (Hint: Consider a suitable function on the set $\{1, 11, 111, 1111, 11111,...\}$.)

*2.9 Suppose that at a party there are at least 6 persons. Prove that, either there exist 3 persons every two of whom know each other or else there exist 3 persons no two of whom know each other.

2.10 Let $P_r(X)$ denote the set of all r-subsets of a set X. Prove that given positive integers n, r and k, there exists an integer N (depending only on n, r and k) such that whenever X is a set with $|X| \geqslant N$ and $P_r(X)$ is written as a union of n subsets, say, $P_r(X) = S_1 \cup S_2 \cup \ldots \cup S_n$, then there exists a k-subset Y of X such that $P_r(Y) \subset S_i$ for some $i = 1, 2, ..., n$. (This is known as **Ramsey's theorem.)

2.11 Express the pigeonhole principle and the result of Exercise 2.9 as special cases of Ramsey's theorem.

2.12 A k-**ary sequence** is defined as a sequence which takes only k possible values, which are generally denoted by $0, 1, ..., k-1$. For $k = 2, 3, 4, 5$ the resulting sequences are called, respectively, **binary**, **ternary**, **quaternary** and **quintary** sequences. How many k-ary sequences of length n are there?

2.13 Prove Theorem 2.15 by induction on the cardinality of X.

2.14 Suppose we prove theorem 2.15 as follows : 'Let the distinct elements of X be $x_1, x_2, .., x_n$. For each subset A of X we define a binary sequence $a_1, a_2, ..., a_n$ in which $a_i = 1$ or 0 according as $x_i \in A$ or $x_i \notin A$. This gives a bijection between $P(X)$ and the set of all binary sequences of length n. Since there are 2^n such sequences, $|P(X)| = 2^n$. Is this proof significantly different from the proof given the text?

2.15 Find $_nP_2$ using proposition 2.12 and Corollary 2.5.

2.16 Prove the following identities combinatorially :

(i) $\dbinom{n}{0} + \dbinom{n}{1} + \dbinom{n}{2} + ... + \dbinom{n}{n} = 2^n$

(ii) $\dbinom{n}{k}\dbinom{k}{r} = \dbinom{n}{r}\dbinom{n-r}{k-r}, r \leqslant k \leqslant n$

(iii) $\dbinom{2n}{2} = 2\dbinom{n}{2} + n^2$

(iv) $n! = 1 \times 1! + 2 \times 2! + 3 \times 3! + ... + (n-1) \times (n-1)! + 1.$

2.17 Let $S \subset X \times Y$ where X, Y are finite sets. For $x \in X$, let

$$G_x = \{y \in Y : (x, y) \in S\}.$$

Similarly for $y \in Y$, let

$$F_y = \{x \in X : (x, y) \in S\}.$$

    (a) Interpret the stes G_x, F_y geometrically.

    (b) Prove that $\sum_{x \in X} |G_x| = \sum_{y \in Y} |F_y|$. (An argument based on this simple result is called a **double counting argument**. It is useful in proving identities.)

2.18 At a party there are more boys than girls. If each boy dances with exactly 2 girls, prove that there is at least one girl who dances with at least 3 boys.

2.19 Let m, n, p, q be positive integers. Prove that:

    (i) $(mn+pq)!$ is divisible by $(m!)^n (p!)^q$

  *(ii) $(n^2)!$ is divisible by $(n!)^{n+1}$.

2.20 Assume that the probability of a person's birthday falling on a given date is $1/365$ (ignore leap years). In a class with n students, what is the probability that two of them have the same birthday? (Hint: Consider complementary probability. The desired probability is fairly high even for relatively low values of n. For $n = 25$, there is more than 50% chance.)

2.21 Let $f: \mathbb{N} \to \mathbb{N} \times \mathbb{N}$ be the bijection shown in Figure 2.6. Prove that the inverse function $f^{-1}: \mathbb{N} \times \mathbb{N} \to \mathbb{N}$ is given by $f^{-1}(x, y) = \frac{1}{2}(x+y-1)(x+y-2) + y$.

2.22 Prove the following theorems:

    (i) A subset of a countable set is countable.

    (ii) If $f: X \to Y$ is surjective and X is countable, then Y is countable.

    (iii) The cartesian product of two (and hence any finite number of) countable sets is countable.

    (iv) The union of two (and hence any finite number of) countable sets is countable.

    (v) If X is countable, then the set of all finite subsets of X is countable.

2.23 Using the last exercise prove that the set of all integers, as well as the set of all rational numbers is countable.

2.24 Using the decimal expansion of real numbers prove that the set of real numbers is uncountable. Hence show that irrational numbers exist and in fact their set is uncountable. (Hint: If $a_1, a_2, a_3, \ldots, a_n, \ldots$ is a denumeration of real numbers let x be the real number whose decimal expansion is $0.b_1 b_2 \ldots b_n \ldots$ where b_n is a digit between 1 and 8 which is different from the digit in the nth place of decimal expansion of the number a_n. Then $x \neq a_n$ for all n. In constructing x, the digits 0 and 9 are excluded so as to avoid the possibility of a real number having two different expansions such as, $3.141600000000\ldots = 3.1415999999999\ldots$).

2.25 Prove that every infinite set contains a denumerable subset.

2.26 Using the last exercise prove that if X is an infinite set and Y is a finite set then the sets $X \cup Y$ and $X - Y$ have the same cardinality as X. (Figuratively, a finite subset of an infinite set is like a drop in a bucket. See also the meaning given to the expression 'almost' all' in Section 1.4).

*2.27   If X is a set, prove that there cannot be a bijection from X to $P(X)$, the power set of X. (This is easy for finite sets. For infinite sets there is a surprisingly short but tricky proof due to Cantor).

2.28   What is wrong with the following 'proof' that every triangle is equilateral? 'Let ABC be any triangle. Through every point P on the side AB, there is a unique line parallel to BC. Let this line intersect the side AC at a point Q. The function which takes P to Q is a bijection. So there are as many points on the side AB as on AC. Hence the two sides are equal. Similarly every other pair of sides is equal.'

2.29   Find a bijection between the sets of points inside a square and inside a circle.

2.30   Prove that a set X is infinite if and only if there exists a proper subset Y of X and a bijection $f: X \to Y$.

Notes and Guide to Literature

Nearly all the results proved in this section are so obvious that they are generally used without proofs, or sometimes even without formally stating them. We have deliberately given a systematic approach starting from an axiom not so much for the sake of the results, but to acquaint the reader with the discipline of axiomatic deduction.

For cardinal arithmetic involving infinite cardinal numbers see, for example, Halmos [1].

The argument given in the hint to Exercise (2.24) is an example of what is called a *diagonalisation argument*. The name becomes suggestive if we write the decimal expansions of $a_1, a_2, \ldots$ one below the other. The number x is constructed by keeping off the 'diagonal' which consists of the nth place of decimal in the expansion of a_n, $n = 1, 2, 3 \ldots$

It is interesting to note that Exercise (2.24) establishes the existence of irrational numbers without proving any particular number (such as $\sqrt{2}$, e or π) to be irrational. It is an example of proving existence through proving abundance.

Ramsey's theorem is the starting point of a number of interesting developments in mathematics. For a thorough exposition, see Graham, Rothschild and Spencer [1].

While it is obviously impossible to make any sweeping generalisation, a rule of thumb can be laid down that for infinite sets the methods of continuous mathematics are needed. Methods of discrete mathematics apply well for finite sets, especially those of relatively small cardinalities. For finite sets of large cardinalities, the tool is statistics.

3. Applications to Counting Problems

In this section we apply the elementary techinques developed in the last section to some counting problems. There is considerable variety in the types of problems that arise and sometimes ingenuity is needed in selecting the right technique. We can therefore only give a glimpse of the problems. Real expertise has to be developed through practice.

As a simple example, we take the Locks Problem. In Section 1, we paraphrased it in terms of sets. We are looking for a set L and five subsets M_1, M_2, M_3, M_4 and M_5 of L such that (i) $M_i \cup M_j \neq \phi$ for all i, j and (ii) $M_i \cap M_j \cap M_k = \phi$ for all distinct i, j, k. Thus, for every (unordered) pair $\{i, j\}$ of distinct indices there must be at least one element, say, $x_{i, j}$ in $M_i \cap M_j$. There are 10 such pairs by Theorem (2.18). because $\binom{5}{2} = 10$. Now if we take two distinct unordered pairs say $\{i, j\}$ and $\{p, q\}$ then out of the four indices i, j, p and q at least three are distinct. Therefore the element $x_{i, j}$ cannot be equal to $x_{p, q}$, for otherwise it will be common to at least three of the subsets $M_1, ..., M_5$. Thus L contains at least 10 distinct elements, one corresponding to each pair of indices. This puts a lower bound on the number of locks needed. We now show that a solution with 10 locks is in fact possible. Indeed, let $L = \{x_{i, j}: 1 \leqslant i \leqslant 5, 1 \leqslant j \leqslant 5, i \neq j\}$ where we regard $x_{i, j}$ and $X_{j, i}$ as the same element. Now for each $i = 1, 2, ..., 5$ let $M_i = \{x_{i, j}: 1 \leqslant j \leqslant 5; j \neq i\}$. Then for any $i \neq j$, $M_i \cap M_j = \{x_{i, j}\} \neq \phi$ while for any three distinct i, j, k, $M_i \cap M_j \cap M_k = \phi$. To translate the solution back into the language in which the problem was posed, we recall that M_i is the set of locks which the person p_i cannot open. Each M_i has 4 elements. So every person can open 6 locks. The distribution of keys to each lock can be succinctly given as follows. Note that for every two distinct persons p_i and p_j, $x_{i, j}$ is a lock which neither of them can open but which everybody else can open. So every lock has 3 keys. In all there are 10 locks, 3 keys to each lock and every person has 6 keys. (The same answer will be obtained in Chapter 4 by another method.)

As a far more ingeneous application of the elementary counting techniques, we present Andre's solution to the Vendor Problem. In Chapter 1, Section 3, we reduced the problem to counting the number of balanced arrangements of n pairs of parentheses. (In the original problem $n = 50$). Let S be the set of all arrangements of n pairs of parentheses and let B be the subset of S consisting of balanced arrangements. $|S|$ is easily seen to be $\binom{2n}{n}$ of (cf. Exercise (2.17)). To find $|B|$ we find $|S - B|$ and then apply Proposition (2.5). In other words we find out the number of unbalanced arrangements of n pairs of parentheses and subtract it from $\binom{2n}{n}$.

In any unbalanced arrangement of n pairs of parentheses, there will always be a first stage at which the number of right parentheses exceeds the number of left parentheses upto that point. Obviously the parenthesis at this stage will be a right parenthesis. Figure 2.7(a) shows one such unbalanced arrangement, the right parenthesis at which the 'balancing' breaks down is shown with an arrow. Let us call this parenthesis as the critical parenthesis. After the critical parenthesis, there will be an odd number of

$$((( ) ( )) ( ))) ( ) (\qquad\qquad ((( ) ( )) ( ))) ( ) ( )$$

↑

Critical parentheses   Critical parenthesis

Figure 2.7: Andre's solution to the Vendor Problem.

parentheses, the number of left parentheses among them will be one more than the number of right parentheses. Let us interchange them. That is, replace every right parenthesis occurring after the critical parenthesis by a left one and vice-versa. The parentheses upto and including the critical parenthesis are to be left unaffected. Figure 2.7(b) shows the new arrangement obtained from the unbalanced arrangement in Fig. 2.7(a). Note that in the new arrangement, there are $n+1$ right parentheses and $n-1$ left parentheses.

The crucial point now is to observe that the original unablanced arrangement of n pairs of parentheses can be recovered from the new arrangement. In fact given any arrangement of $n+1$ right and $n-1$ left parentheses, we simply go on scanning it from left to right till we come across a point at which the number of right parentheses exceeds the number of left ones. We again call this the critical parenthesis. Then we interchange all the left and the right parentheses occuring after the critical parenthesis. This gives an unbalanced arrangement of n pairs of parentheses. These two operations are clearly inverses of each other. Thus, we have defined a bijection between the set $S-B$, of all unbalanced arrangements of n pairs of parentheses, and the set, X, say, of all arrangements of $n+1$ right and $n-1$ left parentheses. Hence $|S-B| = |X|$ by Proposition (2.8). But from Theorem (2.17), it follows that

$$|X| = \frac{(2n)!}{(n+)!\,(n-1)!}.$$

Hence

$$|S-B| = \frac{(2n)!}{(n+1)!\,(n-1)!}$$

So,

$$|B| = |S| - |S-B| = \frac{(2n)!}{(n!)(n!)} - \frac{(2n)!}{(n+1)!\,(n-1)!} = \frac{(2n)!}{n!(n+1)!}.$$

As noted in Chapter 1, Section 3, this is the number of favourable cases, while $|S|$ is the total number of cases. So the required probability is simply $\frac{1}{n+1}$. In the original Vendor Problem n was 50 and so the probability that the vendor will not run out of change is $\frac{1}{51}$ or slightly less than 2 per cent.

In other words, even though there is an adequate numbers of 1 rupee coins with the customers, if they approach the vendor randomly, he is almost sure to run out of change.

In many problems we have to find the number of ways to do something subject to some restriction. Mere paraphrasing of the problem in terms of sets and functions may not be of much help. The tricky part in such problems is often to grasp the essence of the restriction imposed, that is, to realise that every permissible way of doing the thing under question is equivalent to a permutation or combination of some other objects associated with the problems. These 'associated objects' may not be given in the problem itself, and often some ingenuity is needed to conceive them, analogus to the ingenuity needed in solving problems of euclidean geometry where we construct some additional lines (such as perpendiculars, angle bisectors) not present in the statement of the original problem. We illustrate this technique with two examples.

3.1 Problem: There are n guests at a party. Two of them do not get along well with each other. In how many ways can they be seated in a row so that these two persons do not sit next to each other?

Solution: Let us name the guests as $x_1, x_2, ..., x_n$ with x_1 and x_2 as the quarreling members,. We can do the problem by considering the various positions in which x_1 can be seated and then finding, for each such position, the number of positions for x_2. But there is a better way. The total number of ways to seat n persons is $n!$. We now count the number of forbidden ways, i.e., those arrangements in which x_1 and x_2 do sit next to each other. Such arrangements fall into two mutually disjoint categories; those in which x_1 is immediately followed by x_2 (as viewed from one end of the row) and those in which x_2 is immediately followed by x_1 (as viewed from the same end). By symmetry both these categories have the same cardinality. So we find the cardinality of the first category, i.e. the number of arrangements in which x_1 is immediately followed by x_2. To do this we treat $x_1 x_2$ as a single person. (This is the new 'object' we are introducing). This 'person' along with $x_3, ..., x_n$ gives a total of $n-1$ persons, any permutation of which corresponds to an arrangement of $x_1, x_2, ..., x_n$ in which x_1 is followed immediately by x_2. There are $(n-1)!$ permutations of $n-1$ persons. The other category also has $(n-1)!$ arrangements. Since the two categories are disjoint, the total number of forbidden ways is $2(n-1)!$ and the answer to the problem is $n!-2(n-1)!$. ∎

3.2 Problem: In how many ways can r men and s women be seated in a row so that no two women sit next to each other?

Solution: This is a generalisation of the last problem. (The last problem arises as a special case if we treat the quarreling guests as women and all other guests as men.) But the method of the last problem cannot be applied directly here. Let us call the men as x_1 ..., x_r and the women as y_1, ..., y_s. By the method of the last problem, we can count the number of arrangements in which a particular pair of women is adjacent. But this will have to be done for every pair and there will be many overlaps. For example, an arrangement in which y_1, y_2, y_3 occurs will get discarded at least twice, once because y_1 and y_2 are adjacent and once again because y_2 and y_3 are together. There is a way to handle such overlaps and we shall study it in the next section. But even that method does not work smoothly for this particular problem.

We therefore try a new approach. Let us first arrange the r men. This can be done in $r!$ ways. Now for any one such arrangement of men, there will be $r-1$ gaps between adjacent pairs of men. Besides, there will be two gaps, one at each end of the row. These 'gaps' are the extra objects we are introducing into the problem. There are in all $r+1$ of them. Now the arrangement of the men and women can be completed by placing the s women into these $r+1$ gaps. The restriction that no two women be next to each other amounts to saying that no gap should be assigned to more than one woman. By Theorem (2.16) the number of ways to do this in

$$_{(r+1)}P_s \text{ or } \frac{(r+1)!}{(r-s+1)!}.$$

This is the number of permissible arrangements for every arrangement of men. But there are $r!$ ways to arrange the men. So the answer to the problem is $\dfrac{r!\,(r+1)!}{(r-s+1)!}$. ∎

We now turn to problems about k-ary sequences (cf. Exercise (2.12)). A k-ary sequence of length n can be formally defined as a function from the set $\{1, 2, ..., n\}$ into a set of cardinality k (which is generally taken to be the set $\{0, 1, ..., k-1\}$. This is also often called the **alphabet**, its elements being called **letters.** In this context, a sequence of length n is also called a **word** or a **string** of length n. The study of such strings is quite important in the formation of compilers for computer programs. Binary sequences are especially important because they can be used to represent many things in many contexts (see, for example, Exercise (2.14)). Also since only two symbols are needed for every entry, mechanical implementation of binary sequences becomes very easy with binary devices such as a switch or a ferromagnetic bit.

The number of all k-ary sequences of length n is k^n, by Theorem (2.14). We often want to count the number of such sequences satisfying some condition. We prove one such result.

3.3 Proposition: The number of binary sequences of length n in which the digit 1 occurs an even number of times* is 2^{n-1}. This is also the number of sequences of length n in which 1 occurs an odd number of times.

Proof: Let S be the set of all binary sequences of length n. Let A be the set of those in which 1 occurs an even number of times and B be the set of those in which 1 occurs an odd number of times. Clearly A and B are disjoint and $A \cup B = S$. So $|A| + |B| = |S|$. Also we know $|S| = 2^n$ for all n. Our interest is in showing that $|A| = 2^{n-1}$ and $|B| = 2^{n-1}$. Now define $f : S \to S$ by $f(a_1 a_2 \ldots a_n) = a_1' a_2' \ldots a_n'$ where $a_i' = 0$ or 1 according as $a_i = 1$ or 0 for $i = 1, 2, \ldots, n$. Verbally, f interchanges the 0's and 1's occurring in a sequence. It is easy to see that if n is odd then f maps A into B and B into A. Also f is clearly a bijection. So $|A| = |B|$. Hence

$$|A| = \tfrac{1}{2} \cdot 2^n = 2^{n-1}.$$

Thus the result holds for odd n.

For even n, the function f maps A into itself and B into itself and so $|A|$ and $|B|$ cannot be related to each other directly. However, we can utilise the fact that the result has already been proved for sequences of odd length. Note that if n is even then $n - 1$ is odd. Now, let C be the subset of S consisting of those sequences whose last digit is 1 and D be the set of those whose last digit is 0. Obviously $|C| = |D| = 2^{n-1}$, because a sequence in C (or in D) is completely determined by its first $n - 1$ entries. Also $C \cap D = \phi$ and $C \cup D = S$. Because of this, A is the disjoint union of $A \cap C$ and $A \cap D$. So $|A| = |A \cap C| + |A \cap D|$. Now note that a sequence in $A \cap C$ is obtained by taking a sequence of length $n - 1$ containing an odd number of 1's and appending a 1 at the end. Since $n-1$ is odd, the number of sequences of length $n - 1$ with an odd number of 1's is 2^{n-2}. It follows that $|A \cap C| = 2^{n-2}$. It can similarly be shown that $|A \cap D| = 2^{n-2}$. Hence $|A| = 2^{n-2} + 2^{n-2} = 2^{n-1}$. Since $|B| = |S| - |A|$, it also follows that $|B| = 2^{n-1}$. Thus we have established the result for even n also. ■

Using the last proposition we can count the number of k-ary sequences in which some digit, say 1, occurs an even number of times, for any $k \geqslant 2$. We could give the extension right here but we prefer to defer it to the next chapter where we will be in a position to present it avoiding unnecessary clumsiness.

We conclude the section with a discussion of problems involving distribution of objects into boxes. Such problems are important because many other problems can be reduced to them. Depending upon whether we treat the objects and/or the boxes as distinct or non-distinct there are four possible cases. We consider them separately.

---

* A sequence in which 1 does not appear at all is also to be included here.

(i) *Distinct Objects and Distinct Boxes*: Every placement of distinct objects into distinct boxes amounts to defining a function from the set of objects into the set of boxes. The number of all such functions was found in Theorem (2.14). A restriction that no box should contain more than one object amounts to saying that the corresponding function be injective. The number of such functions was obtained in Theorem (2.16). A requirement that no box be empty is equivalent to saying that the corresponding function be surjective. As remarked earlier, in this case there is no easy formula for the number of onto functions. We shall revert to this problem later in this section.

When we put objects into boxes, sometimes the order of objects in each box matters. In such cases, we have to distinguish between two ways of putting the objects into the boxes even when each box contains the same set of objects, but ordered in different ways. We illustrate such a situation in the following problem.

3.4. Problem: A noveau riche wants to wear four distinct rings on the five fingers of his right hand. In how many ways can he do so? (Ignore the differences in the sizes of the rings and the fingers. Also assume each finger to be long enough to hold all the rings).

Solution: Here the order in which rings appear on each finger is important. Let us name the rings as r_1, r_2, r_3 and r_4. Now r_1 can be placed on any of the 5 fingers. So there are 5 ways to place the ring r_1. Suppose it has been placed in one of these ways. Then the number of ways to place the ring r_2 is not 5 but 6. Because when it is placed on the same finger as r_1, it can be either above or below r_1. It is as if the ring r_1 cuts the finger on which it appears into two fingers and so there are six different fingers on which r_2 can be placed. By the same reasoning for any one arrangement of r_1 and r_2, there are 7 ways to place r_3. Therefore the total number of ways to wear the 4 rings on 5 fingers is $5 \times 6 \times 7 \times 8$ or 1,680.

There is another way to arrive at the answer. Let us name the fingers as f_1, f_2, f_3, f_4 and f_5. After wearing the rings in any one manner, let us cut the fingers (hypothetically) and arrange them one after the other gluing adjacent fingers. This gives a linear arrangement of four rings and four junction points where adjacent fingers are glued together. Conversely every linear arrangement of four rings and four junction points determines a way of wearing the rings on the five fingers. This corrrespondence gives a bijection between the set of all ways of wearing the rings on five fingers and the set of all linear arrangements of 4 rings and 4 junction points. So our problem reduces to count the latter. This is preceisely the number of arrangements of 8 objects (4 rings and 4 junction points) of which 4 (namely the junction points) are indistinguishable from each other. By Theorem (2.17), this number equals $\dfrac{8!}{4!}$ which is $5 \times 6 \times 7 \times 8$. ▨

Note that the second solution is an instance of the technique of introducing new objects into the problem and interpreting the problem in terms of them. The generalisation to the case of r objects and n boxes is immediate and left to the reader.

3.5. Proposition: The number of ways to put r objects into n boxes so that the order in each box is important is $n(n+1)(n+2)...(n+r-1)$. ∎

(*ii*) *Distinct objects into Non-distinct Boxes*: Suppose we have a set S distinct objects which are placed in each boxes. The set of objects placed in each box is a subset of S. This gives us a collection of subsets of S. Obviously every two members of this collection are mutually disjoint and the union of all of them is the whole set S. Therefore, a placement of objects of a set S into non-distinct boxes is equivalent to a collection of mutually disjoint subsets of S whose union is S. This gives us a paraphrase of the problem intrinsically in terms of S, without involving the boxes.

When we consider a collection of subsets of a set S whose union is S, there is obviously little point in including the empty subset in the collection. There is a special name for such collections.

3.6. Definition: A **decomposition** (or **partition**) of a set S is a collection of mutually disjoint, non-empty subsets of S whose union is S. These subsets are called the **parts (or members)** of the decomposition.

For example, $\{\{1, 3\}, \{2, 5\}, \{4\}\}$ is a partition of the set $\{1, 2, 3, 4, 5\}$ into three parts. The problem of finding all partitions of a set S of cardinality n into m parts is equivalent to the problem of placing n distinct objects into m non-distinct boxes so that no box is empty. To find the number of all such partitions is an important problem because many other problems can be reduced to it. Unfortunately, there is no easy formula in terms of n and m which will give us the number of partitions of a set S with n elements into m parts, simply by substituting the values of n and m. But the problem is too important to be given up. What do we do now?

The way out is quite ingeneous and is frequently adopted in similar dilemmas in other branches of mathematics as well. Whenever we cannot find a formula for something, simply give it a name and a symbol! That is exactly what we do in the following definition.

3.7. Definition: The number of ways to put n distinct objects into m non-distinct boxes so that no box is empty (or equivalently the number of partitions of a set with n elements into m parts), is denoted by

$$S_{n,m} \left(\text{ or by } \left\{ \begin{matrix} n \\ m \end{matrix} \right\} \right)$$

and is called a **Stirling number of the second kind***.

What have we achieved? Actually, nothing; because simply assigning a name to the solution does not solve the problem. The real use of such names or symbols is not in the problems from which they arise. Their importance lies in the fact that the solutions to many other problems can be expressed in terms of these symbols. As a very familiar example of this type of a situation, recall from elementary calculus that the integral $\int_1^x \dfrac{dt}{t}$ cannot be expressed as a familiar function of x. (Here by 'familiar' we mean a combination of algebraic and trigonometric functions.) So we give it a new name, the natural logarithm** of x, denoted by $\ln x$. Of course, this new function is useless in evaluating the integral $\int_1^x \dfrac{dt}{t}$ unless we have some other way of computing $\ln x$ (in which case it would amount to evaluating the integral in the first place). But, as it turns out, many other integrals such as $\int \tan x \, dx$ or $\int (1 + x^2)^{1/2} \, dx$ can be expressed in terms of the natural logarithm. So it is worthwhile to study the properties of the logarithms and also to find methods for evaluating them at least approximately.

So, that is what we shall do with Stirling numbers of the second kind. First, we shall illustrate how some other problems can be reduced to them and then we shall see a way to evaluate them. We begin with a problem that was left open earlier in this section.

3.8. Proposition: The number of ways to put n distinct objects into m distinct boxes so that no box is empty (or equivalently, the number of surjective functions from a set with n elements to a set with m elements) is $m! \, S_{n,m}$.

Proof: Let us denote the objects by $x_1, x_2, ..., x_n$ and their set by X and the boxes by $B_1, B_2, ..., B_m$ and their set by B. We are assuming the boxes to be distinct. Let S be the set of all arrangements of the x_i's into the boxes B_j's so that no box is empty. Let T be the set of all arrangements of the

---

*The usual definition is different but equivalent to ours. Also, as the name indicates, there are Stirling numbers of the first kind as well denoted by

$$ s_{n,m} \quad \text{or by} \quad \begin{bmatrix} n \\ m \end{bmatrix}. $$

But we shall not define them here.

**$\ln x$ can be defined differently. But the present definition is very standard.

objects into the boxes so that no box is empty, if we do not distinguish between the boxes. Then $|T| = S_{n,m}$ by definition. Our goal is to prove that $|S| = m! \, S_{n,m}$ and to do this it would suffice to find a function $f: S \to T$ such that the inverse image of each element of T has precisely $m!$ elements (c.f. Proposition (2.8)). Construction of this function is quite simple. Given an element of S, i.e. given an arrangement of the x_i's into the B_j's, we simply view it as an arrangement into non-distinct boxes. It is as if the boxes are carrying labels from 1 to m initially and we are removing these labels. These labels can be put back into the boxes in $m!$ ways. And when we do so we would be getting different arrangements as viewed into distinct boxes but the same arrangement as viewed into non-distinct boxes, (see Figure 2.8, where $n = 5$ and $m = 3$).

Note that the requirement that no box be empty is crucial. If some of the boxes are empty then an interchange of the labels of two empty boxes will not give rise to a new arrangement as viewed into distinct boxes. Thus $m!$ distinct elements of the set S correspond to the same element of T. As noted before, this completes the proof.

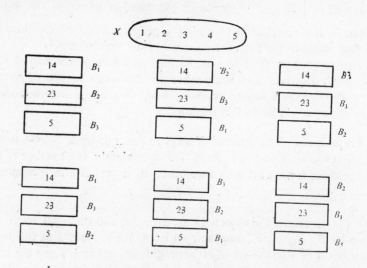

Figure 2.8:   Placement of objects into boxes

Next, we count the number of ways to put n distinct objects into r non-distinct boxes, without any restriction about the boxes being non-empty.

3.9. Proposition: The number of ways to put n distinct objects into r non-distinct boxes is $\displaystyle\sum_{m=0}^{r} S_{n,m}$.

Proof: In every arrangement there will be m non-empty boxes for a unique m, $0 \leqslant m \leqslant r$. (The case $m = 0$ can occur only if $n = 0$ and we may set $S_{0,0} = 1$ by convention.) For each m, the number of arrangements is, $S_{n,m}$ by definition. The result follows by summation. ∎

The Stirling numbers also appear in other problems but we cannot discuss them here. We now consider the question of computing $S_{n,m}$ where m, n are positive integers. Clearly, $S_{n,m} = 0$ for $m > n$ and $S_{n,n} = 1$ because there is only one way to put n objects into n non-distinct boxes so that no box is empty, namely, to put one object in each box. $S_{n,1}$ is also easily seen to be 1 for all n. As a less trivial result, the following proposition computes $S_{n,2}$.

3.10 Proposition: For all $n \geqslant 2$, $S_{n,2} = 2^{n-1} - 1$.

Proof: Let X be a set with n elements. By definition, $S_{n,2}$ is the number of partitions of X into two mutually disjoint non-empty subsets whose union is X. These subsets must therefore be of the form A and $X - A$ for some subset of X. Note that $A \neq \phi$, and also $A \neq X$ (otherwise $X - A$ would be ϕ). Every other subset of X gives rise to a partition of X of the desired type. Note, however, that the partition arising out of a non-empty proper subset A of X is the same as the partition arising out of its complement. So every partition gets counted twice. Formally. we have defined a function f from the set $P(X) - \{\phi, X\}$ into the set, say S, of all partitions of X into two parts by $f(A) = \{A, X - A\}$. Then $f(A) = f(X - A)$. So every point of S has exactly two pre-images. Hence $|P(X) - \{\phi, X\}| = 2|S|$. But $|P(X)| = 2^n$. So $|P(X) - \{\phi, X\}| = 2^n - 2$. The result now follows. ∎

For $m > 2$, there is no easy formula for $S_{n,m}$. For small values of n and m one can find $S_{n,m}$ by actually considering all partitions of a set with n elements. For higher values this becomes impracticable and also unreliable (because some possible partitions are likely to slip us). There is, however, an important recurrence relation which allows us to compute a Stirling number by first computing the lower Stirling numbers.

3.11. Theorem: For all positive integers m and n,

$$S_{n,m} = m S_{n-1,m} + S_{n-1, m-1}.$$

Proof: As usual let X be the set $\{x_1, x_2, \ldots, x_n\}$ and S be the set of all partitions of X into m parts. Then $|S| = S_{n,m}$. We divide S into two subsets as follows. In any partition of X there will be a unique part to which x_n belongs. This part may be simply $\{x_n\}$ or else it will contain some other elements of X. Now let P be the set of those partitions in which $\{x_n\}$ is a part and Q be the set of those in which the part containing x_n also contains some other elements of X. Clearly $|S| = |P| + |Q|$.

Computing the cardinality of P presents no problem because every partition in P obviously corresponds to a partition of the set of remaining elements, $\{x_1, x_2,...,x_{n-1}\}$ into $m-1$ parts. So $|P| = S_{n-1,\ m-1}$.

However, computing the cardinality of Q is not so immediate. Paraphrasing the problem in terms of boxes, Q is the set of all arrangements of the $n-1$ objects $x_1, x_2,...,x_{n-1}$ into m boxes so that no box is empty. But these boxes are *not* all identical! The box containing the object x_n stands out as different from the remaining $m-1$ boxes which are indistinguishable from each other. We now proceed as in the proof of Proposition 3.8. We let R be the set of all arrangements of the objects $x_1, x_2,...,x_{n-1}$ into the m boxes $B_1, B_2,...,B_m$ where we assume that B_m is the box containing x_n, and where we regard the boxes as distinct. By Proposition 3.8, $|R| = m!\ S_{n-1,m}$. Now every arrangement in R gives rise to an arrangement of Q if we treat the boxes as identical, subject to the restriction that B_m is not to be identified with any other box. Then we can reshuffle the labels of only $m-1$ boxes among themselves and still get the same element of Q. This can be done in $(m-1)!$ ways. So, by a reasoning we have used a number of times so far, $|R| = (m-1)!\,|Q|$. But $|R| = m!\ S_{n-1,m}$. So $|Q| = m\ S_{n-1,m}$.

Since $|S| = |P| + |Q|$ and we have expressed all the three terms in terms of Stirling numbers, the result follows.   ∎

To see how this theorem can be applied, we calculate

$$S_{5,3} = 3\,S_{4,3} + S_{4,2}$$

$$= 3\,(3\,S_{3,3} + S_{3,2}) + S_{4,2}$$

$$= 3(3 + 4 - 1) + 8 - 1 \text{ (using Proposition 3.10}$$
$$\text{and the fact that } S_{n,n} = 1)$$

$$= 25.$$

Theoretically we can compute, by repeated applications of theorem 3.11, $S_{n,m}$ for any m, n. But this is a time-consuming process. In view of the applications of the Stirling numbers, readymade tables, listing their values upto a fairly large value of n are available. There are also numerous identities about Stirling numbers. A few of them will be given as exercises.

(iii)   *Non-distinct Objects and Distinct Boxes:*   In this case it is convenient to think of the objects as balls, beads, or some tags, mechanically produced so that you cannot tell one apart from the other. Suppose we have r such objects and we want to put them into n distinct boxes. The number of ways to do so can be obtained using Proposition 3.5. Call the boxes as $B_1, B_2,...,B_n$. Even though the objects are non-distinct, let us temporarily call them $x_1, x_2,..., x_r$. Let S be the set of ways to put these r objects (temporarily regarded as distinct) into the distinct boxes so that the order in which the objects appear in each box matters. Now arrange the boxes in a row in the order $B_1, B_2, \ldots B_n$. This amounts to arranging the r objects in a row, separated by the $n-1$ 'walls' between adjacent boxes as in Figure 2.9 where $r = 10$ and $n = 5$. Let T be the set of all arrangements of the x_i's into the B_j's where we do not distinguish between the x_i's. As usual we get a function from S to T obtained by forgetting the difference between two objects. Under this function, $r!$ distinct elements of S (obtained by permuting the x_i's among themselves) go to the same element of T. So $|S| = r! \, |T|$. But by Proposition 3.5.

$$|S| = n(n+1)\ldots(n+r-1).$$

Therefore

$$|T| = \frac{n(n+1\ldots(n+r-1)}{r!} = \binom{n+r-1}{r}$$

It is also instructive to derive the result directly, by an argument analogous to the second proof of Proposition 3.5. If we drop the suffixes on the x_i's in Figure 2.9, we see that every distribution of r identical objects into n distinct boxes is equivalent to a permutation of the r objects along with the $n-1$ interbox walls (denoted by w's). By Theorem 2.17, the number of such permutations is

$$\frac{(n+r-1)!}{(n-1)!\,r!} \text{ or } \binom{n+r-1}{r}.$$

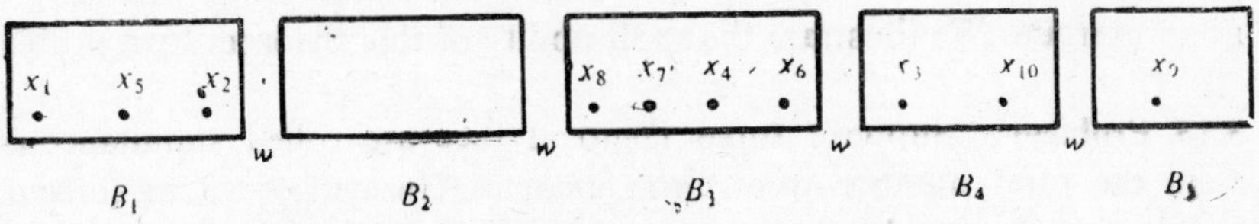

Figure 2.9: Non-distinct Objects into Distinct Boxes

As an example, 8 coins (of the same denomination) can be given to 3 children in $\binom{10}{8}$ or 45 ways. It is interesting to derive from the result above, a formula for the number of selections with repetitions allowed. We often have a situation where there are n piles of objects, the objects in each pile being indistinguishable from each other. Suppose we want to pick r objects from this collection of n types of objects with repetitions allowed freely, which means that we can pick as many objects of each type as we like as long as the total number of objects picked is r. Of course, in order that this be feasible there must be an adequate supply of each type of objects. Specifically, each of the *n* piles must contain at least *r* objects. To play it safe, we take each pile as infinite although literally it need not be so. The following theorem computes the number of such *r*-selections with repetitions allowed.

3.12 Theorem: The number of *r*-selections from *n* types of objects with free repetitions allowed equals the number of ways to put *r* non-distinct objects into *n* distinct boxes and hence equals $\binom{n+r-1}{r}$.

Proof. Let there be *n* piles. The members of each pile will be called 'things' rather than 'objects' because we shall call something else as objects. Suppose the things of each type are stored in a box, so that there are *n* distinct boxes. We have to choose a total of *r* things from these *n* boxes. Let us indicate our selection of a thing by putting some label over it. These labels will be indistinguishable from each other and in all there will be *r* of them. Because the things in each box are identical, what matters is not which ones of them are selected but rather simply how many of them are selected, or equivalently how many labels are put into that box. Thus the problem is equivalent to distributing *r* identical labels into *n* distinct boxes. These labels will be our objects. By the formula above, the number of ways to put *r* identical labels into *n* distinct boxes is $\binom{n+r-1}{r}$. This completes the proof. ∎

There is yet another way to derive the same result which will be given as an exercise. We illustrate the applicability of this theorem by a problem.

3.13. Problem: Suppose three identical dice are rolled simultaneously. Find the total number of distinct outcomes. (Instead of rolling three dice simultaneously, essentially the same problem could be asked by rolling the same dice thrice in succession and ignoring the order of the three scores.)

Solution: In all problems involving dice, we assume, unless otherwise specified, that each die has six faces marked with figures from 1 to 6. If we have three such dice, it is as good as having six different boxes, the first containing three replicas of the figure 1, the second, three replicas of the figure 2, and so on, Each possible outcome on rolling three dice simultaneously is equivalent to selecting 3 figures from these 6 boxes with repetitions allowed. So by the last theorem, the number of distinct outcomes is

$$\binom{6+3-1}{3} = \binom{8}{3} = 56. \quad ∎$$

(iv) *Non-distinct objects and Non-distinct Boxes*: This case is analogous to (ii) which was reduced to the problem of partitioning a set. However, in the present case even the objects are non-distinct. Suppose we have n identical objects (say balls) which are to be put into m identical boxes. Once again, let us first consider the case where every box is to be nonempty. Let $n_1, n_2, ..., n_m$ be the numbers of balls in these. The indexing here is purely arbitrary because since the boxes are identical we cannot call one of them as the first, one as the second. Each n_i is positive integer and obviously $n_1 + n_2 + ... + n_m = n$. Note that the integers n_i's (each counted with its multiplicity if any) completely determine the arrangement of the balls into the boxes. Thus the problem is reduced to partitioning the integer n, i.e. expressing it as a sum of m positive integers (with repetitions allowed). In order to systematise them, we reshuffle the indices if necessary and suppose that $n_1 \geqslant n_2 \geqslant n_3 \geqslant ... \geqslant n_m$ (an ascending order would do as well). This leads to the following definition.

3.14 Definition: A **partition** of a positive integer n into m *parts* is a sequence of positive integers $n_1, n_2, ..., n_m$ such that $n_1 \geqslant n_2 \geqslant ... \geqslant n_m$ and $n_1 + n_2 + ... + n_m = n$. The number of such partitions is denoted by $P_{n,m}$.

For example 5 can be partitioned into three parts in two ways, either as $3 + 1 + 1$ or as $2 + 2 + 1$. So we see $P_{5,3} = 2$. A recurrence relation for $P_{n,m}$ (analogous to the one for Stiriling numbers in Theorem (3.11)) is easily obtained and will be given as an exercise. Analogous to proposition 3.9, we have the following result whose proof is also left to the reader.

3.15 Proposition : The number of ways to put n non-distinct objects into r non-distinct boxes is $\sum_{m=1}^{n} P_{n,m}$. ∎

Sometimes the number of parts is not important. We want to consider all partitions of an integer n, into whatever number of parts. Obviously, in any such partition, the number of parts will be some integer between 1 and n. Consequently, the number of all partitions of n equals $\sum_{m=1}^{n} P_{n,m}$. This number is quite important in applications and is commonly denoted by $p(n)$. For small values of n, $p(n)$ can be found by explicitly writing down all possible partitions of n. For example we see that $5 = 5 = 4 + 1 = 3 + 2 = 3 + 1 + 1 = 2 + 2 + 1 = 2 + 1 + 1 + 1 = 1 + 1 + 1 + 1 + 1$ and this gives $p(5) = 7$. However, there is no easy formula for $p(n)$. But the answers to many other problems (appearing in many diverse branches of mathematics) can be expressed in terms of the partition function $p(n)$. So it has become one of the standard functions in mathematics and tables for its values are available. (See the comments made after Definition (3.7)).

Numerous results are known about partitions. Some of them require the use of generating functions and will therefore be deferred to Chapter 7.

However, a few results can be proved by some elementary but rather tricky arguments. A most ingeneous device to handle a partition is to look at it graphically. If $n = n_1 + n_2 + \ldots + n_m$ is a partition of n (where $n_1 \geqslant n_2 \geqslant n_3 \geqslant \ldots \geqslant n_m \geqslant 1$) then we draw a block of dots in which the m rows from top to bottom have $n_1, n_2, \ldots, n_m$ dots and the dots in each row appear below those of the upper row (if any), aligning from the left-most dot. Such a block of dots is called the **Ferrer's graph** of the given partition. Figure 2.10 shows the Ferrer's graph of the partition $7 + 4 + 4 + 2 + 1$ of 18.

Figure 2.10.   Ferrer's graph of a partition.

If we take the Ferrer's graph of a partition $n = n_1 + n_2 + \ldots + n_m$ then clearly it has n_1 columns and their lengths decrease as we go from the left to the right. The left-most column has m dots. If we take the numbers of dots in each column from left to the right, we get another partition of n, into n_1 parts and in which the largest part has size m. This partition is called the **dual partition** of the original partition. For example, the dual partition of the partition $7 + 4 + 4 + 2 + 1$ of 18 whose Ferrer's graph is shown in figure 2.10 is $5 + 4 + 3 + 3 + 1 + 1 + 1$.

Many results about partitions can be visualised easily in terms of their Ferrer graphs. As an example, the concept of the dual partition just introduced leads to the following result, which would otherwise appear somewhat awkward.

3.16 Proposition: $P_{n,m}$, that is, the number of partitions of an integer n into m parts also equals the number of partitions of n in which the largest part is of size m. Similarly the number of partitions of n into at most m parts is the same as the number of partitions of n in which the part sizes do not exceed m. ∎

Let us apply this proposition to the Postage Problem. We noted earlier that this problem reduces to finding the number of partitions of 20 into parts of sizes 1, 2 and 3. We denoted this number by a_{20}. In view of proposition 3.16, a_{20} equals the number of partitions of 20 into at most 3 parts. In other words $a_{20} = P_{20,1} + P_{20,2} + P_{20,3}$. Evidently $P_{20,1} = 1$.

As for $P_{20,2}$, any partition of 20 into 2 parts is uniquely determined by the size of the smaller part, which can be any integer from 1 to 10. So $P_{20,2} =$ 10. (More generally, for any positive integer n, $P_{n,2} = n/2$ if n is even and $P_{n,2} = (n-1)/2$ if n is odd). However, computing $P_{20,3}$ is not so easy. Here the smallest part size, say x, can be any integer from 1 to 6. For a given x we have to consider the number of partitions of $20-x$ in two parts both of which are $\geqslant x$. It is easily seen that for $x = 1, 2, 3, 4, 5$ and 6, these numbers are, respectively, 9, 8, 6, 5, 3 and 2. Summing together $P_{20,3} = 33$. Hence $a_{20,3} = 1 + 10 + 33 = 44$.

Note, however, that this method would not work if the denominations of the stamps were, say, 10, 30 and 40 for in that case we would have to consider partitions of 20 into parts of sizes 1, 3 or 4 and proposition 3.16 would not apply. In Chapter 7 we shall do the Postage Problem again by a method with a wider applicability.

Exercises

3.1   Do the generalised Locks Problem with n persons out of which any m but no fewer are to be able to open the box.

3.2   Vary the Locks Problem by calling one of the five persons as the leader. The leader is to be able to open the box with the concurrence of at least one other person out of the remaining four. For the remaining four members, the requirement is the same, namely, any three but no fewer of them should be able to open the box. Design a system of locks and keys.

3.3   Vary the Vendor Problem by assuming that there are n persons with one 1 rupee coin each and k with 2 rupee coins. Find the probability that the Vendor will not run out of change. (Obviously, the answer is 0 if $n < k$.)

3.4   As a further variation, suppose that the Vendor has q one rupee coins to start with. If in the line there are n persons with 1 rupee coins and k with 2 rupees coins, where $q < k \leqslant q + n$ and they approach the vendor randomly, prove that the probability of his not running out of change is

$$\left[\binom{n+k}{n} - \binom{n+k}{n+q+1}\right] \bigg/ \binom{n+k}{n}.$$

3.5   In how many ways can r integers be selected from set $\{1, 2, \ldots, n\}$ so that no two consecutive integers are selected?

3.6   Do Problem (3.1) by the first method (that is, by considering the various positions in which x_1 may be placed).

*3.7   Let ABC be an isosceles triangle in which $AB = AC$ and $\angle A = 20$

degrees. Let D and E be points on the sides AC, AB respectively such that $\angle DBC = 60$ degrees and $\angle ECB = 50$ degrees. Prove that $\angle EDB = 30$ degrees. (Hint: Consider a point P on AC such that $\angle PBC = 20$ degrees. This problem has little to do with combinatorics. It is meant just to illustrate the remark regarding some ingeneous constructions needed in problems in geometry.)

3.8   Suppose a city has m parallel roads going East-West and n parallel roads going North-South. How many rectangles are formed with their sides along these roads? If the distance between every consecutive pair of parallel roads is the same, how many shortest possible paths are there to go from one corner of the city to its diagonally opposite corner?

3.9   Take an 8×8 chess-board and remove two diagonally opposite corner squares. Prove that it is impossible to pair off the remaining 62 squares in such a way that every pair contains two adjacent squares, i.e. squares having one side in common. Do the problem both without and with colouring the squares. This illustrates the facility gained by introduction of an additional device, in this case, the colouring of squares.

3.10 Suppose there are 20 players of different heights. These are to be divided into two teams, A and B, of 10 players each so that for each $i = 1, 2,..., 10$, the ith tallest player in team A will be taller than the ith tallest player in B. In how many ways can this be done?

3.11 Suppose in Exercise (3.8), 2^k persons start from the southwest corner of the city where $k \leqslant \min \{m-1, n-1\}$. Half of these persons proceed eastaward and half northward. They travel with the same speed and reach the next junction at the end of one unit of time. Persons arriving at each junction again bifurcate, half of them going eastward and half northward and then continue to travel with the same speed. If $0 \leqslant j \leqslant k$ find how many persons there will be at each junction, at the end of j units of time.

3.12 Suppose a person walks along a straight line starting at a point 0. He walks at a uniform speed of 100 meters per minute. But at the end of every minute he is likely to reverse his direction with probability 1/2. Find the probability that he will be at a spot $100 k$ meters from 0 at the end of t minutes, where k, t are integers and t is positive.

*3.13 In the last problem suppose there is a dangerous zone on the line beginning at a distance of 1 km. from 0. The person may hit the boundary of this zone and go back. But once inside this zone, his motion stops. Find the probability that the person is still walking at the end of the tth minute where t is a positive integer.

3.14 How many paths are there from point A to point B in Figure 2.11

if no portion of a path is to be traversed more than once (whether in the forward or in the backward sense)?

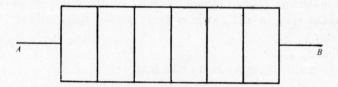

Figure 2.11: Diagram for Exercise (3.14)

3.15 Into how many parts are the diagonals of a convex octagon decomposed, given that no three of these diagonals are concurrent except at a vertex? Generalise to a convex n-gon.

3.16 Given an arithmetic progression of length n, how many subprogressions of length 3 may be formed from it?

*3.17 Find the number of regions into which a convex n-gon is split by its diagonals if no three of them are concurrent, except at a vertex.

3.18 Prove that for every positive integer n,

$$\binom{n}{0} + \binom{n}{2} + \binom{n}{4} + \cdots + \binom{n}{2q} + \cdots$$

$$= \binom{n}{1} + \binom{n}{3} + \binom{n}{5} + \cdots + \binom{n}{2q+1} + \cdots = 2^{n-1}.$$

Deduce that $\sum_{j=0}^{\infty} (-1)^j \binom{n}{j} = 0$.

(Although each sum is superficially infinite, it is actually finite, in view of the fact that $\binom{n}{r} = 0$ for $r > n$).

3.19 A *palindrome* is a word which reads the same backward or forward (e.g., 'MADAM', 'ANNA'). Find how many palindromes of length n can be formed from an alphabet of k letters.

3.20 How many k-ary sequences of length n are there in which no two consecutive entries are the same?

3.21 How many ternary sequences of length n are there which either start with 012 or end with 012?

3.22 Prove Proposition (3.5).

3.23 A conference, attended by 100 delegates, is held in a hall. The hall has 3 doors, marked *A*, *B*, *C*. At each door, an entry book is kept and the delegates entering through that door sign in it in the order in which they enter. If each delegate is free to enter anytime and through any door he likes, how many different sets of three lists would arise in all? (Assume every person signs only at his first entry.)

3.24 Find $S_{n, n-1}$.

3.25 For positive integers *m*, *n* prove that,

$$S_{n+1, m+1} = \sum_{k=m}^{n} \binom{n}{k} S_{k, m}$$

$$= \sum_{k=m}^{n} (m+1)^{n-k} S_{k, m}.$$

3.26 In how many ways can 3 blue, 4 white and 2 red balls be distributed into 4 distinct boxes?

3.27 A person has three sons. He owns 101 shares of a company. He wants to give these to his sons so that no son should have more shares than the combined total of the other two. In how many ways can he do so?

3.28 Prove that the number of ways to put *r* identical objects into *n* distinct boxes so that every box is non-empty is $\binom{r-1}{r-n}$. Interpret this in terms of *r*-selection of *n* types of objects with repetitions allowed.

3.29 10 balls are picked from a large pile of red, blue and white balls. How many such selections contain less than 5 red balls?

3.30 A sequence of real numbers is said to be **monotonically increasing** if every term is greater than or equal to its predecessor, if any. A monotonically decreasing sequence is defined similarly. If every term is greater than its prelecessor (i.e. if strict inequality holds) the sequence is called **strictly monotonically increasing**. Prove that a monotonically increasing sequence of length *r* taking values in the set $\{1, 2, \ldots, n\}$ corresponds uniquely to an *r*-selection of *n* types of objects with repetitions allowed.

3.31 Let *S* be the set of all monotonically increasing sequences of length *r* taking values in the set $\{1, 2, \ldots, n\}$ and let *T* be the set of all strictly monotonically increasing sequences of length *r* taking values in the set $\{1, 2, \ldots, n+r-1\}$. Prove that the function $f: S \to T$ defined by
$$f(a_1, a_2, \ldots, a_r) = (a_1, a_2+1, a_3+2, \ldots, a_i+i-1, \ldots, a_r+r-1)$$
is a bijecction. Use this to give an alternate proof of Theorem (3.12).

3 32   For positive integers m, n prove that

$$P_{n,m} = P_{n-1,m-1} + P_{n-m,m}.$$

Also show that for every n,

$$p(n) = P_{2n,n}.$$

3.33   Prove proposition (3.15).

3.34   What is wrong with the following argument which attempts to show that

$$S_{n,m} = n! \, P_{n,m}?$$

'Let X be a set with n elements, say,

$$X = \{x_1, \ldots, x_n\}.$$

Given any partition of X into m parts, we get a partition of the integer n into m parts if we count the number of elements in each part. This gives us a function f from the set of all partitions of the set X into the set of all partitions of n, each into m parts. If we reshuffle the indices of the x's, we get the same partition of the integer n. But the indices can be reshuffled in $n!$ ways. So under f, every point has $n!$ preimages. Hence the result.'

3.35   A **triangular partition** of an integer n is defined as a partition of n into three parts such that the sum of any two exceeds the third. Clearly such a partition corresponds to a triangle with integer sides and perimeter n. Let t_n be the number of triangular partitions of n. Prove that, for all $n \geqslant 4$,

$$t_n = \begin{cases} t_{n-3} & \text{if } n \text{ is even} \\ t_{n-3} + k & \text{if } n = 4k + 1 \text{ for some positive integer } k \\ t_{n-3} + k & \text{if } n = 4k - 1 \text{ for some positive integer } k \end{cases}$$

3.36   A partition is called **self-dual** if it coincides with its dual partition. Prove that the number of self-dual partitions of n equals the number of partitions of n into parts of different and odd sizes.

*3.37   Suppose there are 10 houses in a row. The occupants of these houses go on a vacation, one by one, in a random order. They being good neighbours, every occupant, while leaving, checks that things are in order in the two houses immediately neighbouring his (one house in case of a house at an end). What is the probability that every house, except that of the occupant leaving last, has been checked by its neighbour after its occupant has left?

Notes and Guide to Literature

The problems in this section are standard. For more numerical problems, see Vilenkin [1]. For more identities on Stirling numbers and for tables of them see Knuth [1], Vol. 1. There is a huge literature about partitions. On the basis of a few observations, Ramanujam conjectured and proved a number of interesting congruence relationships about the partition numbers, a few of them are given in M. Hall [1].

The movements in Exercises (3.11) and (3.12) are examples of what are called random walks. They are important in statistical mechanics, where the number of particles is huge and it is assumed that they move randomly. A good reference is Parzen [1].

4.   Principle of Inclusion and Exclusion

Suppose a set S is expressed as the union of a finite number of its subsets, say,

$$S = S_1 \cup S_2 \cup \ldots \cup S_n.$$

If $S_i \cap S_j = \phi$ for all distinct i and j, then by Proposition (2.3).

$$|S| = |S_1| + |S_2| + \ldots + |S_n|.$$

But if the sets S_i's over-lap this no longer holds, because if some element is common to, say, two subsets S_i and S_j then in $|S|$ it will be counted only once while in the right hand side it is counted more than once. Ignoring such overlaps is in fact one of the most common pitfalls that occur in solving combinatorial problems.

However, as remarked in the solution of problem (3.2), there is a method available in which we keep a systematic track of such overlaps and get the cardinality of the union of a finite number of sets. The basic idea is to take an element x, to include it in our count as many times as it appears and then to make up for its excessive inclusion (because of overlaps) by excluding it an appropriate number of times. That is why, the result is called the principle of inclusion and exclusion. A special case of this was stated in Proposition (2.7), where we saw that

$$|S \cup T| = |S| + |T| - |S \cap T|.$$

Here on the left hand side, every elememt of $S \cup T$ is counted only once. On the right hand side, elements which are present in only one of S and T count only once each. But elements of $S \cap T$ get counted twice, once in $|S|$ and again in $|T|$. To correct this, we 'exclude' them once, i.e. we subtract $|S \cap T|$.

Although the basic idea is the same, the result looks much more complicated when the set is expressed as a union of n sets with $n > 2$. The difficulty can be illustrated with the case $n = 3$. Suppose

$$S = S_1 \cup S_2 \cup S_3.$$

Then in general,

$$|S| \neq |S_1| + |S_2| + |S_3|$$

because elements which are common to more than one S_i get counted more than once. So, to make up for this, we subtract

$$|S_1 \cap S_2| + |S_2 \cap S_3| + |S_1 \cap S_3|$$

from the right hand side and get

$$|S_1| + |S_2| + |S_3| - |S_1 \cap S_2| - |S_2 \cap S_3| - S_1 \cap S_3|.$$

This may appear to be equal to $|S|$. But it need not be so! The reason is that we have over-corrected the error. Points which are common to all the three sets S_1, S_2, S_3 are counted in each of the first three terms. So they are included thrice each. But these points also appear in each of the term that is subtracted. This means they are also excluded thrice. Effectively, then, points of $S_1 \cap S_2 \cap S_3$ are not included in this expression. To make up for this we must add $|S_1 \cap S_2 \cap S_3|$. Thus

$$|S| = |S_1| + |S_2| + |S_3| - |S_1 \cap S_2| - |S_2 \cap S_3|$$
$$- |S_1 \cap S_3| + |S_1 \cap S_2 \cap S_3|.$$

This is the correct result. It can be graphically visualised through the Venn diagram in Figure 2.12.

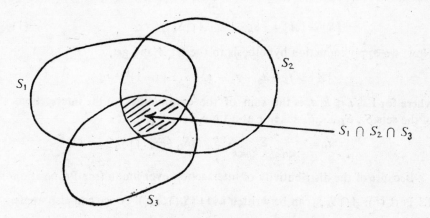

Figure 2.12. Cardinality of the Union of Three Sets.

We now state and prove the general result.

4.1. Theorem (Principle of Inclusion and Exclusion). Let $S_1, S_2, \ldots, S_n$ be finite sets and let

$$S = S_1 \cup S_2 \cup \ldots \cup S_n.$$

Then

$$|S| = s_1 - s_2 + s_3 - s_4 + \ldots + (-1)^{n+1} s_n$$

where $s_1 = |S_1| + |S_2| + \ldots + |S_n| = $ sum of the cardinalities of the sets S_i's.

$s_2 = \sum\limits_{1 \leqslant i < j \leqslant n} |S_i \cap S_j| = $ sum of the cardinalities of intersections of the S_i's taken two at a time.

$s_3 = \sum\limits_{1 \leqslant i < j < k \leqslant n} |S_i \cap S_j \cap S_k| = $ sum of the cardinalites of the intersections of the S_i's taken three at a time and so on.

(Caution: s_2 is often confused with the number of elements belonging to at least two of the S_i's. This interpretation is incorrect. Actually s_2 is a sum of cardinalities of certain subsets. It is not, by itself, the cardinality of any easily identifiable subset. A similar warning holds for $s_1, s_3, s_4 \ldots$).

Proof: We proceed by induction on n. For $n = 1$, there is nothing to prove. The case $n = 2$ is covered by Proposition 2.3 and will be used in the proof of the inductive step.

Suppose the result holds for all finite sets expressed as the unions of k (or less) subsets. We shall show that it holds for sets which are expressed as unions of $k + 1$ subsets. So let

$$S = S_1 \cup S_2 \cup \ldots \cup S_k \cup S_{k+1},$$

where each $|S_i|$ is finite. Now let $A = \bigcup\limits_{i=1}^{k} S_i$. Then by Proposition 2.3,

$$|S| = |A| + |S_{k+1}| - |A \cap S_{k+1}|. \tag{1}$$

Now, we apply induction hypothesis to the set A and get,

$$|A| = t_1 - t_2 + t_3 - t_4 + \ldots + (-1)^{k+1} t_k$$

where for $1 \leqslant r \leqslant k$, t_r is the sum of the cardinalities of the intersections of the sets $S_1, S_2, \ldots, S_k$ taken r at a time, i.e.

$$t_r = \sum\limits_{1 \leqslant i_1 < i_2 \ldots < i_r \leqslant k} |S_{i_1} \cap S_{i_2} \cap \ldots \cap S_{i_r}|.$$

Because of the distributivity of intersection over union (see Proposition 1.1 Part (ii), $A \cap S_{k+1}$ can be written as $\bigcup\limits_{i=1}^{k} (S_i \cap S_{k+1})$, So we can also apply induction hypothesis to it and get

$$|A \cap S_{k+1}| = u_1 - u_2 + u_3 - u_4 + \ldots + (-1)^{k+1} u_k$$

where for $1 \leqslant r \leqslant k$,

$$u_r = \sum\limits_{1 \leqslant i_1 < i_2 < \ldots < i_r \leqslant k.} |(S_{i_1} \cap S_{k+1}) \cap (S_{i_2} \cap S_{k+1}) \cap \ldots \cap (S_{i_r} \cap S_{k+1})|.$$

Therefore we get, by substitution in (1).

$$|S| = |S_{k+1}| + t_1 - (t_2 + u_1) + (t_3 + u_2) - (t_4 + u_3) + \ldots$$
$$+ (-1)^{r+1} (t_r + u_{r-1}) + \ldots + (-1)^{k+1} (t_k + u_{k-1}) + (-1)^{k+1} u_k.$$

To complete the proof we now show,

(i) $s_1 = |S_{k+1}| + t_1$

(ii) $s_r = t_r + u_{r-1}$ for $r = 2, 3, ..., k$ and

(iii) $s_{k+1} = u_k$.

Of course (i) is obvious from the definitions of s_1 and t_1. (iii) also poses no problem because s_{k+1} is simply

$$|S_1 \cap S_2 \cap ... \cap S_k \cap S_{k+1}|.$$

But the set

$$S_1 \cap S_2 \cap ... \cap S_k \cap S_{k+1}$$

equals the set

$$(S_1 \cap S_{k+1}) \cap (S_2 \cap S_{k+1}) \cap ... \cap (S_k \cap S_{k+1})$$

(see again Proposition 1.1 parts (vii) and (v)): so its cardinality equals u_k.

A little argument is needed for (ii). By definition, s_r equals

$$\sum |S_{i_1} \cap S_{i_2} \cap ... \cap S_{i_r}|$$

as the sum ranges over all r-tuples $(i_1, i_2, ..., i_r)$ of indices in which

$$1 \leqslant i_1 < i_2 < ... < i_{r-1} < i_r \leqslant k + 1.$$

We classify such r-tuples into two types: (I) those in which $i_r \leqslant k$ and (II) those in which $i_r = k + 1$. The summation over r-tuples of type (I) obviously gives t_r. So the problem is reduced to proving that the summation over r-tuples of type (II) equals u_{r-1}. A typical r-tuple of type (II) is of the form $(i_1, i_2, ..., i_{r-1}, k + 1)$ where $1 \leqslant i_1 < i_2 < ... < i_{r-1} \leqslant k$. But the set $S_{i_1} \cap S_{i_2} \cap ... \cap S_{i_{r-1}} \cap S_{k+1}$ equals, by the same argument as above, the set

$$(S_{i_1} \cap S_{k+1})(S_{i_2} \cap S_{k+1}) \cap ... \cap (S_{i_{r-1}} \cap S_{k+1}).$$

But u_{r-1} was, by definition, the sum of cardinalities of all such sets, for all possible $(r-1)$-tuples,

$$(i_1, i_2, ..., i_{r-1}) \text{ with } 1 \leqslant i_1 < i_2 < ... < i_{r-1} \leqslant k.$$

So u_{r-1} equals the summation over r-tuples of type (II). As noted before, this completes the proof. ∎

For application, it is convenient to paraphrase the last theorem in a different form. We often have a set S and some n properties of elements of S. We want to find the number of those elements of S which have exactly r of these properties, where $0 < r \leqslant n$. Of course, every property determines a subset of S, namely the set of those elements of S which have that property. So the question can also be framed as 'Given n subsets, say $S_1, S_2, ..., S_n$ of a set S, how many elements of S are common to precisely r of these subsets?'. The case $r = 0$ is an immediate consequence of the theorem above and is itself often called the principle of inclusion and exclusion. The general case will be treated later.

4.2 Theorem: Let S be a finite set and $S_1,\ S_2,\ldots,S_n$ be subsets of S. Let $'$ denote complementation w.r.t. S. Then

$$|\bigcap_{i=1}^{n} S_i'| = s_0 - s_1 + s_2 - s_3 + \ldots + (-1)^n s_n$$

where $s_0 = |S|$ and $s_1,\ldots,s_n$ have the same meaning as in Theorem 4.1.

Proof: By de Morgan's laws (Proposition 1.1 part x),

$$\bigcap_{i=1}^{n} S_i' = S - \bigcup_{i=1}^{n} S_i$$

and so by Corollary 2.6,

$$|\bigcap_{i=1}^{n} S_i'| = |S| - |\bigcup_{i=1}^{n} S_i|.$$

The result now follows immediately from the last theorem. However, we give an alternate argument because the ideas used in it are useful elsewhere. We take a typical element x of S and find how many times it is counted on either side of the equality to be proved. If these two counts tally for every x, the equality is established. Now if $x \in S_i'$ for all $i = 1, 2,\ldots,n$ then it is counted once in the left hand side, viz., $|\bigcap_{i=1}^{n} S_i'|$. Also, in the right hand side, x is counted only in the first term, namely $|S|$. Hence the two counts are equal. Suppose now that x belongs to precisely k of the subsets $S_1, S_2,\ldots,S_n$ for some k, $1 \leqslant k \leqslant n$. Without loss of generality we let $x \in S_i$ for $i = 1, 2,\ldots,k$ and $x \notin S_i$ for $i > k$. Now x contributes nothing to the left hand side. As for the right hand side, x contributes only to the first $k + 1$ terms, namely, $s_0, -s_1, s_2,\ldots,(-1)^k s_k$. The contribution to s_0 is obviously 1. Let us see how often x contributes to a typical term, $(-1)^r s_r$, $1 \leqslant r \leqslant k$. Recall that, s_r is the sum of the cardinalities of the intersection of the set $S_1, S_2,\ldots,S_n$ taken r at a time. The element x will appear in one such intersection, say, $S_{i_1} \cap S_{i_2} \cap \ldots \cap S_{i_r}$ if and only if the indices $i_1, i_2,\ldots,i_r$ are all from 1 to k. Such indices may be chosen in $\binom{k}{r}$ ways. We conclude that x contributes $(-1)^r \binom{k}{r}$ to the term $(-1)^r s_r$ on the right hand side. The proof will, therefore, be complete if we prove the identity, $1 - \binom{k}{1} + \binom{k}{2} - \binom{k}{3} + \binom{k}{4} + \ldots + (-1)^k \binom{k}{k} = 0$. But this is an easy consequence of Proposition 3.3. Note that the positive terms add up to the number of binary sequences of length k with an even number of 1's while the terms with a negative sign add up to the number of those with an odd number of 1's. Each equals 2^{k-1}. (This is in fact the solution to Exercise 3.18.) ▮

This argument may appear a bit elusive at first sight. But actually it grasps the very essence of the principle of inclusion and exclusion. We take an arbitrary element, find how may times it has been included in the count and how many times it has been excluded and finally show that in sum total, it has been counted just the right number of times (which will be either 1 or 0 depending upon whether or not that element belongs to the set whose cardinality is being found). For a reader who is still not convinced, we shall present the same argument in a more rigorous form in an exercise.

The general case can be obtained by a similar argument.

4.3 Theorem: Let S and S_1, S_2,..., S_n be as before. Then the number of elements which belong to exactly m of these subsets equals

$$\sum_{r \geqslant m} (-1)^{r-m} \binom{r}{m} s_r.$$

Proof: Let T be the set of those elements of S which belong to precisely m of the sets S_1, S_2,..., S_n. We have to show that $|T|$ equals the given sum. Now let x be an element of S. Suppose x belongs to exactly k of the subsets S_1, S_2,..., S_n and once again, without loss of generality suppose $x \in S_i$ for $i \leqslant k$ and $x \notin S_i$ for $i > k$. If $k < m$ then $x \notin T$. Also x does not contribute anything to s_r for $r \geqslant m$. So it contributes nothing to the summation. If $k = m$, then x is in T and in the summation it contributes only to one term, namely, to $\binom{m}{m} s_m$. Here s_m is the sum of the cardinalities of the intersections of S_1,..., S_n taken m at a time. Only one of these intersections, namely, $S_1 \cap S_2 \cap ... \cap S_m$ contains x. So x is counted only once in the summation. The remaining case is $k > m$. Here x contributes nothing to $|T|$. The contribution to s_r is $\binom{k}{r}$. If we let $j = r - m$ then the problem reduces to proving the identity,

$$\sum_{j=0}^{k-m} (-1)^j \binom{m+j}{m} \binom{k}{m+j} = 0.$$

The proof of this is left as an exercise. ▮

To apply the principle of inclusion and exclusion, we must first know, either from the data directly or by some computations, the cardinalities of the intersections of all possible sub-families of the given family of sets. This can be a horrendous task, because if there are n sets S_1, S_2,..., S_n

then there are $2^n - 1$ such intersections (Why?). But often, there is some symmetry among these sets which simplifies the computations. Secondly, if some intersection is known to be empty, then many others have to be so. For example, if $S_1 \cap S_2 = \phi$, there is no need to consider intersections of subfamilies of $\{S_1, S_2, ..., S_n\}$ in which both S_1 and S_2 are present.

We do a few illustrative problems.

4.4 Problem: In a language survey of students it is found that 80 students know English, 60 know French, 50 know German, 30 know English and French, 20 know French and German, 15 know English and German and 10 students know all the three languages. How many students know (i) at least one language (ii) English only (iii) French and one but not both out of English and German (iv) at least two languages?

Solution: Let E, F, G denote, respectively, the sets of students knowing English, French and German. Because only three sets are involved, the problem can be done vividly using a Venn diagram as shown in Figure 2.13. The various regions in it represent the various subsets obtained by taking intersections of E, F, G and their complements. For example the shaded region represents $G \cap E' \cap F'$ where ' stands for complementation. Now in each region we go on putting its cardinality. We start with $E \cap F \cap G$ which is given to contain 10 elements. Since $E \cap F$ is the disjoint union of

$$E \cap F \cap G \text{ and } E \cap F \cap G'$$

and

$$|E \cap F| = 30,$$

it follows that

$$|E \cap F \cap G'| = 20.$$

After filling in completely it is easy to answer any of the given questions simply by expressing an appropriate set as the disjoint union of some of these seven regions. This gives the answers, (i) 135, (ii) 45, (iii) 30 and (iv) 45.

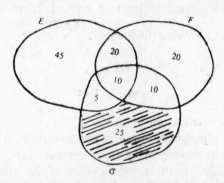

Figure 2.13: Venn Diagram for Problem (4.4)

However, with more than three sets the Venn diagrams tend to be cumbersome and at any rate it is instructive to be able to handle the problem 'abstractly' as well. So we apply the principle of inclusion and exclusion. (i) comes as a direct consequence of it. For (ii) we let

$$S_1 = E \cap F \text{ and } S_2 = E \cap G.$$

Then

$$|\, S_1 \cup S_2 \,| = 30 + 15 - 10 = 35.$$

We want the cardinality of $E - (S_1 \cup S_2)$ which is $80 - 35 = 45$. For (iii) let

$$S_1 = F \cap E, \; S_2 = F \cap G.$$

Then we want

$$|\, (S_1 \cup S_2) - (S_1 \cap S_2)\,|$$

which equals, by Theorem (4.1),

$$|\, S_1 \,| + |\, S_2 \,| - 2 \,|\, S_1 \cap S_2 \,|$$

or $30 + 20 - 2.10$, i.e. 30. For (iv) let

$$S_1 = E \cap F, \; S_2 = F \cap G, \; S_3 = E \cap G.$$

Note that in this case,

$$S_1 \cap S_2, \; S_2 \cap S_3, \; S_1 \cap S_3$$

and

$$S_1 \cap S_2 \cap S_3.$$

are the same set, namely $E \cap F \cap G$. So by theorem (4.1)

$$|\, S_1 \cup S_2 \cup S_3 \,| = |\, S_1 \,| + |\, S_2 \,| + |\, S_3 \,| - 2 \,|\, E \cap F \cap G \,|$$
$$= 30 + 20 + 15 - 20 = 45. \quad \blacksquare$$

4.5 Problem: How many permutations of the integers from 1 to n are there in which at least one integer is left in its own place? (If we think of each permutation as a bijection from the set $\{1, 2,...,n\}$ into itself, then the question amounts to asking how many permutations have at least one fixed point, cf. Exercise (1.12).).

Solution: Let S_i be the set of all those permutations which fix i, $i = 1$, $2,...,n$. We want $|\, S_1 \cup S_2 ... \cup S_n \,|$. Clearly $|\, S_i \,| = (n-1)!$ for each i, because having fixed i, the permutation may reshuffle the remaining $n-1$ elements among themselves and this can be done in $(n-1)!$ ways.* Now

---

*Note that we are not saying that a permutation in S_i has i as its *only* fixed point. It may have others. This point should always be kept in mind in applying the principle of inclusion and exclusion. Whenever a statement is made to the effect that an element has (or lacks) certain properties, it should be interpreted exactly as it is, without making any unwarranted inferences as to whether it lacks (or has) the remaining properties, as we often do in practice. If a student tells about his examination results as 'I failed in chemistry' we tend to think that he cleared all other subjects. In mathematics we must not do so; for the student may have failed in other subjects too!.

for $i \neq j$, $S_i \cap S_j$ consists of those permutations of $\{1, 2, ..., n\}$ in which both i and j are fixed. Since the remaining elements may be permuted among themselves in $(n-2)!$ ways,

$$| S_i \cap S_j | = (n-2)!$$

for every pair $\{i, j\}$ of distinct indices. There are $\binom{n}{2}$ such pairs. Similarly there are $\binom{n}{3}$ triples of distinct indices and for every one such triple,

$$\{i, j, k\}, \ | S_i \cap S_j \cap S_k | = (n-3)!$$

Note the considerable symmetry in this problem. Continuing in this manner, and applying Theorem (4.1), we get

$$| \bigcup_{i=1}^{n} S_i | = \binom{n}{1} (-1)! - \binom{n}{2} (n-2)!$$

$$+ \binom{n}{3} (n-3)! \ldots \ldots + (-1)^{n+1} \binom{n}{n} 0!$$

$$= \frac{n!}{1!\,(n-1)!} \cdot (n-1)! - \frac{n!}{2!\,(n-2)!} \cdot (n-2)!$$

$$+ \ldots + (-1)^{n+1} \frac{n!}{n!}$$

$$= n! \left(\frac{1}{1!} - \frac{1}{2!} + \frac{1}{3!} - \frac{1}{4!} + \ldots + (-1)^{n+1} \frac{1}{n!} \right).$$

As there is no way to sum this series, the answer has to be left in this form.

The last problem gives us a way to find the number of derangements of n symbols. Recall that a derangement is an arrangement in which no element is left in its original position. We considered such derangements in commenting upon the Envelopes Problems. The number of derangements of n symbols is often denoted by D_n.

4.6 Theorem: The number of derangements of n symbols, D_n, is

$$n! \left(1 - \frac{1}{1!} + \frac{1}{2!} - \frac{1}{3!} + \ldots + (-1)^n \frac{1}{n!} \right).$$

Consequently, the probability that a given permutation of n symbols has no fixed points is

$$1 - \frac{1}{1!} + \frac{1}{2!} - \frac{1}{3!} + \cdots + (-1)^n \frac{1}{n!}.$$

Proof: The result follows immediately by subtracting from $n!$, the total number of permutations, the number of those having at least one fixed point, as obtained in the last problem. (Or one could have directly applied Theorem 4.2 instead of 4.1.) ∎

As noted in the solution of Problem 4.5, there is no way to evaluate

$$1 - \frac{1}{1!} + \frac{1}{2!} - \frac{1}{3!} + \cdots + (-1)^n \frac{1}{n!}.$$

of course, the first two terms cancel each other. But there is some reason for retaining them. A reader familiar with calculus would know that for every complex number z,

$$e^z = \sum_{n=0}^{\infty} \frac{z^n}{n!}.$$

In particular,

$$e^{-1} = \frac{1}{e} = \sum_{n=0}^{\infty} \frac{(-1)^n}{n!}.$$

Therefore the probability calculated above is the partial sum of this series for $1/e$. Since $n!$ is very large for even relatively small values of n, the series converges very rapidly. Hence even for small values of n (say, $n = 10$), the probability is very nearly equal to $1/e$ which is approximately 0.3678794418.

As another application of the principle of inclusion and exclusion we return to the problem of finding the number of surjective functions from one set to another.

4.7 Theorem: Let X, Y be finite sets with n and m elements respectively. Then the number of onto functions from X to Y equals

$$\sum_{r=0}^{m} (-1)^r \binom{m}{r} (m-r)^n$$

Proof: Let F be the set of all functions from X to Y. By theorem 2.14, $|F| = m^n$. Now let $Y = \{y_1, y_2, \ldots, y_m\}$. For each $i = 1, 2, \ldots, m$ let F_i be the set of those functions from X to Y whose ranges do not contain the point y_i. Every such function may be thought of as a function from X into the set $Y - \{y_i\}$ whose cardinality is $m-1$. So again by Theorem 2.14,

$$|F_i| = (m-1)^n.$$

For $i \neq j$, $F_i \cap F_j$ is the set of functions which take values in the set $Y - \{y_i, y_j\}$ whose cardinality is $m - 2$.

So $F_i \cap F_j \,| \,= (m-2)^n.$

In general for any r-tuple, $(i_1, i_2, ..., i_r)$ with

$$1 \leqslant i_1 < i_2 < ... < i_r \leqslant m,$$

we have

$$| \, F_{i_1} \cap F_{i_2} \cap ... \cap F_{i_r} \,| = (m-r)^n.$$

There are $\begin{pmatrix} m \\ r \end{pmatrix}$ such r-tuples. The result now follows Theorem 4.2, because a function from X to Y is onto iff it is in none of the F_i's. ∎

As a consequence, we get an expression for Stirling numbers.

4.8 Theorem: For positive integers n and m,

$$S_{n,m} = \sum_{r=0}^{m-1} \frac{(-1)^r (m-r)^{n-1}}{r! \, (m-r-1)!}.$$

Proof: In Proposition (3.8) we counted the number of surjective functions from a set with n elements to a set with m elements as $m! \, S_{n,m}$. Equating this with the count in the last theorem, we get the result. ∎

4.9 Problem: There is is an elevator in a four storeyed building which has a ground floor and four other floors marked 1, 2, 3, 4. Seven persons get in the elevator at the ground floor. In how many ways can they be discharged at the remaining floors, assuming that (i) at every floor at least one person gets out and (ii) the order of persons coming out on the same floor is immaterial?

Solution: Because of the second assumption, every possible way of discharging the persons on the floors corresponds uniquely to a function which assigns to each person, the floor at which he gets out. The first assumption implies that this function is surjective. So the answer to the problem is simply the number of surjective functions from the set of 7 persons to the set of four floors. By Theorem 4.7 this equals,

$$4^7 - \begin{pmatrix} 4 \\ 1 \end{pmatrix} 3^7 + \begin{pmatrix} 4 \\ 2 \end{pmatrix} 2^7 - \begin{pmatrix} 4 \\ 3 \end{pmatrix}. \quad ∎$$

Sometimes the data of a problem is not sufficient to give an exact count of the set in question. In such cases, we can nevertheless find upper and lower bounds for its cardinality. Such a bound is said to be **sharp** if we can demonstrate at least one instance, consistent with the data of the problem, in which it is attained, i.e. in which the cardinality of the set equals the bound. As a general result of this type we have :

4.10 Theorem: Let $S_1, S_2, ..., S_n$ be finite sets. For $r = 1, 2, ..., n$ let s_r be the sum of the cardinalities of the intersections of these sets taken r at a time. Let

$$t_r = \sum_{j=1}^{r} (-1)^{j+1} s_j.$$

Then, for every $r = 1, 2, ..., n-1$, $|\bigcup_{i=1}^{r} S_i|$ lies between t_r and t_{r+1}. More specifically,

$$t_{r+1} \leqslant |\bigcup_{i=1}^{n} S_i| \leqslant t_r \text{ for } r \text{ odd}$$

and

$$t_r \leqslant |\sum_{i=1}^{n} | S_i | \leqslant t_{r+1} \text{ for } r \text{ even.}$$

These bounds are sharp.

Proof: In computing $|\sum_{i=1}^{n} S_i|$, we take s_1; then subtract s_2 so as to exclude the elements which have been included too often because of their being common to at least two of the S_i's; then again add s_3 so as to include the elements that have been excluded too often because of their being common to at least three of the S_i's and so on. If in this process we go upto s_r, then we have correctly counted all elements of $\bigcup_{i=1}^{n} S_i$ which belong to exactly r or less of the S_i's. But a correction is due for those which belong to more than r of the S_i's. This correction is positive or negative depending upon whether r is even or odd. But in any case it is an over-correction because an element which belongs to more than $r + 1$ subsets gets counted too often in this correction. So it follows that the correct value of $|\bigcup_{i=1}^{n} S_i|$ lies somewhere between t_r and $t_r +$ (the next correction); i.e. between t_r and t_{r+1}. Keeping in mind the signs of these corrections (depending upon whether r is even or odd), we get the desired inequalities. As for their sharpness, consider a situation in which no element is common to more than r subsets, i.e. the intersection of every $r + 1$ subsets is empty. Then $s_{r+1} = 0$ and so $t_r = t_{r+1}$. Therefore equality holds. ∎

The preceding theorem combined with Corollary (2.6) often gives the desired bounds on the cardinalities as we now illustrate.

4.11 Problem: Suppose in an examination with four subjects A, B, C, D the percentages of candidates passing in them are respectively 70, 75, 80 and 85. Find upper and lower bounds for the percentage of candidates passing all the courses. How are these bounds affected (i) if it is known that everybody who clears A also clears C, (ii) if it is known that everybody who clears C clears at least one of A and B?

Solution: Let S be the set of all candidates appearing for the examination. We may suppose $|S| = 100$ for convenience. Let A, B, C, D also denote the sets of those candidates who clear the corresponding subjects. Then we are given that

$$|A| = 70, \ |B| = 75, \ |C| = 80$$

and $|D| = 85$ and we have to find upper and lower bounds for:

$$|A \cap B \cap C \cap D|.$$

Of course 100 and 0 are always two bounds. But obviously we look for sharp bounds. Since

$$A \cap B \cap C \cap D$$

is contained in each A, B, C, D clearly 70 (the lowest of the cardinalities of the sets), is an upper bound. (It is also true that

$$A \cap B \cap C \cap D$$

is contained in

$$A \cap B, B \cap C \cap D$$

etc. but we are not given the cardinalities of these sets). Also this upper bound is sharp, because we can have $A \subset B \subset C \subset D$, in which case $A \cap B \cap C \cap D = A$. As for the lower bound, let $A' = S - A$ etc. Then by Theorem 4.14,

$$|A' \cup B' \cup C' \cup D'| \leqslant |A'| + |B'| + |C'| + |D'| = 90.$$

Since $A \cap B \cap C \cap D = S - (A' \cup B' \cup C' \cup D')$, it follows that $|A \cap B \cap C \cap D| \geqslant 10$. Also we can have a situation in which A', B', C', D', are mutually disjoint and consequently $|A \cap B \cap C \cap D| = 10$. So the lower bound 10 is also sharp.

If it is known that $A \subset C$, then $A \cap B \cap C \cap D$ is simply $A \cap B \cap D$. By the same reasoning as above, the upper and lower bounds for $|A \cap B \cap D|$ are obtained as 70 and 30. Both are easily seen to be sharp. So the lower bound is improved while the upper one is unaffected.

If it is given that $C \subset A \cup B$ then $|C| \leqslant |A \cup B|$. But

$$|A \cap B| = |A| + |B| - |A \cap B|.$$

This gives

$$|A \cap B| \leqslant |A| + |B| - |C| = 70 + 75 - 80 = 65.$$

Since $A \cap B \cap C \cap D \subset AB$ we get $|A \cap B \cap C \cap D| \leqslant 65$. That this upper bound is sharp is seen from the Venn diagram in Figure 2.14, in which $C = A \cup B$ and the shaded area is $A \cap B$. Thus the upper bound is reduced.

But there is no change in 10 as the lower bound for $|A \cap B \cap C \cap D,|$, because in the case where A', B', C', D' are mutually disjoint, $C \subset A \cap B$. ∎

What happens if in numerical examples such as above the cardinality of some subset comes out to be less than some lower bound for it? Then obviously the data is inconsistent! For example in the problem above if it was given that $|A \cap B| = 40$, this contradicts the estimate

$$|A \cap B| = |A| + |B| - |A \cup B| \geqslant |A| + |B| - |S| = 45.$$

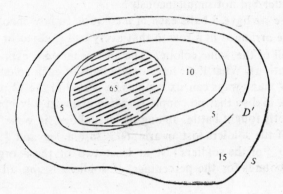

Figure 2.14: Venn Diagram for Problem 4.11.

So the data is inconsistent. Such inconsistency is known as **numerical inconsistency.**

Exercises

4.1   In how many ways can the four walls of a room be painted with three colours so that no two adjacent walls have the same colour?

4 2   Suppose the room has two doors, one on each of a pair of opposite walls. Now, in how many ways can the walls be painted with three colours so that no adjacent walls have the same colour and the walls with the doors in have the same colour?

4.3   Suppose n persons are given one card each. These cards are collected, shuffled and again distributed to the persons, one to each. What is the probability that no person gets the same card again?

4.4   How many permutations of the integers 1 to 8 are there in which no even integer remains in its own place?

4.5   How many permutations of the integers $1, 2, ..., n$ are there in which no two adjacent positions are filled by consecutive integers (in an increasing order)? (For examale, for $n = 4$, the permutation 1432 is allowed but not the permutation 1423).

4.6   The finished products in a factory have to undergo two separate tests for quality control under two machines. On the same product the two tests may be performed in either order but not simultaneously. Also no machine can handle more than one product at the same time. If the time taken for each product for each test is one hour, prove that the number of ways to carry out the testing of n (distinct) products in n hours is $n! \, D_n$ where D_n is the number of derangements of n objects.

**4.7 What will be the answer to the last problem if each product is to undergo three tests of equal duration which may be carried out in any order but not simultaneously?

4.8 Suppose we have 5 balls each of n colours. In how many ways can they be arranged in a row so that every ball is adjacent to at least one ball of the same colour? (The answer may be expressed as a summation). What if we have only 4 balls of each colour?

4.9 In how many ways can six couples be seated in the 12 positions of a clock dial so that no couple is seated diametrically opposite?

4.10 In a hotly fought battle, among the casualities, it was found that 50% of the soldiers lost an arm, 60% lost a leg and 45% lost an eye. 40% of the soldiers lost at least two of these organs. Find sharp bounds for the percentage of soldiers losing all the three organs.

4.11 In how many ways can 10 men and 7 women board a bus with 20 seats of which 5 are reserved only for women?

4.12 10 persons go for a picnic. 3 of them are vegetarians. They carry 20 (distinct) food packets of which 10 are vegetarian and 10 non-vegetarian. In how many ways can these packets be distributed so that each picnicker gets two packets?

4.13 Prove the identity at the end of the proof of Theorem 4.3. (Hint: Use Exercise 2.16 (ii).)

4.14 The purpose of this exericse is to give a rigorous form to the somewhat intuitive argument in the second proof of Theorem 4.2. Let S be a finite set and $P(S)$ its power set. Define a function $f : S \times P(S) \to \mathbf{R}$ (the set of real numbers) by $f(x, A) = 1$ or 0 according as $x \in A$ or $x \notin A$ for $x \in S$ and $A \in P(S)$. Prove the following properties:

   (i) If we fix $A \in P(S)$, and consider f as a function of one variable (see Exercise 1.18) then it is essentially the characteristic function of the subset A(cf. Exercise 1.13).

   (ii) For any $A, B \in P(S)$, and $x \in S$,

   (a) $f(x, S - A) = 1 - f(x, A)$

   (b) $f(x, A \cup B) = f(x, A) + f(x, B) - f(x, A \cap B)$

(c) $f(x, A \cap B) = f(x, A) f(x, B)$.

(iii) For any $A \in P(S)$, $|A| = \sum\limits_{x \in S} f(x, A)$.

(iv) Let $S_1, S_2, \ldots, S_n$ be subsets of S. Let I be the index set $\{1, 2, \ldots, n\}$. For every subset J of I, let $S_J = \bigcap\limits_{i \in J} S_i$. (By convention we let $S_\phi = S$.) For $0 \leqslant r \leqslant n$, let $P_r(I)$ be the set of all r-subsets of I. Then s_r equals $\sum\limits_{J \in P_r(I)} |S_J|$ and hence

$$\sum_{x \in S} \sum_{J \in P_r(I)} f(x, S_J).$$

(v) If an element $x \in S$ belongs to precisely k of the subsets $S_1, S_2 \ldots, S_n$ then $\sum\limits_{J \in P_r(I)} f(x, S_J)$ equals $\dbinom{k}{r}$, with the understanding that $\dbinom{0}{0} = 1$.

(vi) For every $x \in S$,

$$f(x, \bigcap_{i=1}^{n} S_i') = \sum_{r=0}^{n} (-1)^r [\sum_{J \in P_r(I)} f(x, S_J)].$$

Summing over the two sides of (vi) as x varies over S, we get the proof of Theorem 4.2.

4.15 If n is a positive integer then let $\phi(n)$ denote the number of positive integers $\leqslant n$ which are relatively prime to n, i.e., which have no common prime factor with n. (For example, if $n = 20$ then such integers are 1, 3, 7, 9, 11, 13, 17, and 19 and so $\phi(20) = 8$). The function ϕ so defined is called the **Euler ϕ-function.**

(i) Using the principle of inclusion and exclusion find $\phi(200)$, $\phi(300)$ and $\phi(1,030)$.

(ii) If p is a prime prove that $\phi(p^r) = p^r - p^{r-1}$.

(iii) If n has a prime factorisation as $p_1^{r_1} p_2^{r_2} \ldots p_k^{r_k}$ prove that

$$\phi(n) = (p_1^{r_1} - p_1^{r_1-1})(p_2^{r_2} - p_2^{r_2-1}) \cdots (p_k^{r_k} - p_k^{r_k-1}),$$

using the principle of inclusion and exclusion.

(iv) If m and n are positive intergers which are relatively prime to each other then prove that $\phi(mn) = \phi(m) \phi(n)$. Because of this property, the Euler ϕ-function is called **multiplicative.**).

Notes and Guide to Literature

The principle of inclusion and exclusion covers all the counting techniques studied in this chapter. Its application is somewhat limited by the fact that computing the cardinalities of the various intersections can be difficult, and even where it is easy, the answer has often to be left in a summation form. Still its theoretical applications are important. It gives a formula for the Stirling numbers. Also Exercise 4.15 shows an application to number theory. The Euler ϕ-function is very important in algerbra and number theory. We shall have occasions to consider it later. For a more detailed discussion of this and other multiplicative functions, see Hua [1].

Three

Sets with Additional Structures

In this chapter we elaborate the remark made earlier that the focal point of today's mathematics is sets with additional structures. In the first section we study some generalities about such structures and describe how they arise from an attempt at abstraction. The second section deals with binary relations, which are among the simplest additional structures on a set. In the third section we specialise to order relations which are especially important in applications. The last section introduces some basic concepts about algebraic structures. In the next three chapters we shall deal with certain particular types of algebraic structures. The present chapter is a prerequisite for them.

1. Abstraction and Mathematical Structures

Birbal, the legendary wizard in the court of the Mogul Emperor Akbar, was noted for his wit. One of the anecdotes about him relates that he was once asked by the Emperor to give one common answer to three highly unrelated questions. The questions were (1) Why does the horse ail? (2) Why does the earthen pot crack? and (3) Why does the bread* char? The diversity of the questions would be baffling to anyone. But not so for Birbal, who quipped instantaneously, "Because you fail to move it, Jehanpenah." He was right. The meaning of movement would of course change according to the context. For the horse it is the trotting exercise, for the earthen pot it is the spinning motion on a potter's wheel and for the bread it is a quick flipover to ensure uniform baking.

Although Birbal might not have intended it, he had hit one of the major forces behind the development of mathematics, namely to look for

---

*Not the modern bread in the form of a loaf, but the traditional, flat, Indian bread (*roti*) which is baked on a roasting pan or directly on fire.

similarities among apparently diverse things, to isolate what is common to them and then to concentrate on this 'abstracted' portion (often forgetting its origin!). For example, given two problems (1) If Shivaji the Great was born in 1630 A.D. and died in 1680 A.D., for how long did he live? and (2) If there were 1680 birds on a tree and 1630 of them flew away, how many would be left?, a mathematician would hardly treat them as different. He would consider them as the same mathematical problem put in different garbs. This is so because the 'essence' or the 'abstract' of the two problems is the same, namely to subtract 1630 from 1680 and get 50 as the answer. The interpretations of these figures vary according to the problem. But (unfortunately) they are not considered to be the business of a mathematician.

Thus, the process of abstraction is hardly new. It has been there for centuries. But till recently, the abstraction was numerical, that is, numbers were assigned to the various objects involved in the real life problems. A mathematician then would work with these numbers, formulate the problem in terms of these numbers, inventing new concepts (such as a limiting process) if necessary, then look for methods for solving it exactly (or approximately). The final answer would be in the form of some numbers which would be transferred back to the original problem, where they would be interpreted in terms relevant to that problem. This is the gist of the development of mathematics, or at least the applied part of it. Of course, many collateral developments also took place. For example, although differential equations originally arose in connection with problems in mechanics, electricity etc., the theory for solving them became so rich that it inspired many mathematicians to try to solve other equations, even though the latter might not have originated from any 'practical' problems.

While this line of development still continues; as was remarked earlier, in modern times, the doamain of mathematics has been enlarged to include non-numerical problems as well. In Chapter 1, we gave the Dance Problem as an example of such a non-numerical problem. Historically, one of the most celebrated non-numerical problems is the **Konigsberg Bridge Problem**, solved by Euler. Through the city of Konigsberg, flew the Pregel river, with two islands, *A* and *B* in it. The islands were connected to each other and also to the banks by seven bridges as shown in Figure 3.1. The problem is to start from any one of these islands and to return to it after having traversed every bridge exactly once (without using any other means. of course, such as rowing or swimming). It is all right to come to the same island any number of times during the journey, but no bridge may be traversed more than once, whether in the forward or in the backward sense. Many people tried, unsuccessfully, to perform such a round walk and were convinced that it is impossible to do it. But nobody, till Euler, could *prove* rigorously that it cannot be done. (There is obviously a world of difference between your being unable to do some thing and that thing being impossible!)

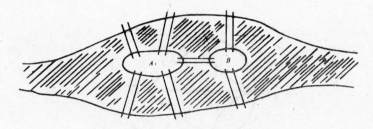

Figure 3.1: **Konigsberg Bridge Problem**.

As we shall see later in this book,* to prove the impossibility of such a walk is really very simple, once the problem is formulated properly. Why did, then, the problem baffle so many? The reason is that people, probably because of their pre-occupation with numbers, simply could not lay their hands on the essence of the problem. Many quantities (such as the lengths of the bridges, the areas of the islands, the speed of walking), which would be important if the problem is looked upon as a problem of motion in mechanics, are simply irrelevant in the present problem! (That is why, they are not given in the statement of the problem.) It follows that if we want to abstract and isolate the essence of the problem then we must look away from these numerical data. This is what Euler did as follows.

Since the shapes and sizes of the islands are irrelevant, we may as well represent them by points. Similarly we represent the two banks of the river by points marked C and D. We thus get four points A, B, C and D. Now each of the seven bridges may be represented by a curve joining two of these points. The lengths and curvatures of these curves are unimportant. Perhaps the only thing we should insist is that these curves should not cross each other except at end points. We then get the configuration of points and curves shown in Figure 3.2, which represents the essence of the problem.

In the new formulation the problem amounts to asking, whether there exists a permutation of the seven curves $C_1, \ldots, C_7$ and choice of orientations for them such that (i) the terminal point of each curve coincides with the initial point of the next curve, if any, and (ii) the terminal point of the last curve coincides with the initial point of the first curve.

What has been achieved because of the new formulation? For one thing, we can now solve the problem. There are 7! possible permutations of the seven curves C_1 to C_7. Also each curve has two possible orientations. Thus, in all there are $7! \, 2^7$ cases and if we examine them one by one we would either come across a case in which the arrangement of the curves

*See the Epilogue.

(along with their chosen orientations) gives an affirmative answer to the problem or else, after exhausting all these $2^7.7!$ case we would be able to say conclusively that a round walk of the desired type does not exist.

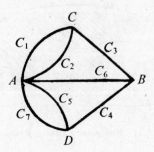

Figure 3.2.   The Essence of the Konigsberg Bridge Problem.

This will be a rigorous, mathematical proof, as opposed to a mere personal conviction, of the impossibility of such a walk based on a few unsuccessful trials. Of course, there are far better ways to arrive at the answer. (Euler's original solution is, in fact, the easiest and will be studied in the chapter* on graphs). But time-consuming as the method may be, it is sure to give us the answer eventually.

But the real advantage of abstracting the essence of the problem is that it tremendously increases the applicability of the solution. In arriving at the abstraction, we stripped the original Konigsberg Bridge problem of the irrelevant details (such as the shapes and sizes of the islands, the lengths of the bridges and so on). It is quite possible that some other problem, when stripped of its inconsequential details may reduce to exactly the same problem as the Konigsberg Bridge problem. As a simple (and somewhat artificial) example, suppose at a tea party, some tea spills over a table and its stream streches across the full length of the table, with two dry spots in it. Suppose somebody lays seven spoons as shown in Figure 3.3. A few sugar particles stick to these spoons. An ant trapped at one of the two dry spots wants to eat all these particles. Can it do so and return to its original position, without getting wet, without leaving the table top and without going over the same spoon more than once? The solution to the Konigsberg Bridge problem provides an answer to this problem as well. But this is not very impressive because in this case the similarity between the two problems is so obvious that most people would refuse to call them as different in the first plate. So we consider another example.

Suppose we have four chemical compounds, say, A, B, C and D. Some of these compounds can be converted to some others by putting them into

*See the Epilogue.

certain processors. Each processor is capable of carrying out two mutually opposing chemical reactions and can be used either way. For example, if compound A is obtained from compound B by oxidation through a processor then B can be obtained from A by reduction through the same processor. Suppose we have seven such processors, P_1, P_2...,P_7 and they act as follows (we list only one way conversions because the other way conversions are possible by 'reversing' the processor). P_1 and P_2 convert A to C; P_5 and P_7 convert A to D; P_3 converts B to C; P_4 converts B to D and P_6 converts A to D. Can we start with a compound, say, A, make it pass through each processor exactly once and get back the same compound?

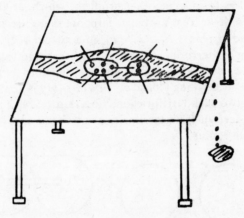

Figure 3.3: Problem of the Spilled Tea.

This problem does look quite different from the Konigsberg Bridge problem. But if we represent the four compounds by four points, and each processor by a curve joining the points corresponding to the two compounds on which that processor acts, then we get exactly the same diagram as in Figure 3.2. So the solution to the Konigsberg Bridge problem could conceivably have applications in chemistry!

We could continue to list further applications (just as we can cite numerous examples in which the equation 1680-1630 = 50 is used). All we have to do is to give suitable interpretations to the four points and to the seven curves joining them. While such applications may be interesting to others, a mathematician will be more concerned with the common underlying structure on which they are built. It consists of a set of 11 elements, four of which are points (often called 'vertices') and the remaining seven are curves (often called 'edges'). There is some inter-relationship among these elements which can be expressed by specifying, for each edge, the pair of vertices which are joined by it. This amounts to defining a function from the set of edges into the set of all 2-subsets of the set of vertices.

Formally, the mathematical structure we obtained consists of three gadgets, (1) a set V (2) a set E and (3) a function f from E into $P_2 (V)$ (i.e. the set of the all subsets of V having two elements each). The technical name for such a structure is a graph*. For the moment our interest is not so much in this particular structure but rather in its genesis. We start with something, forget some of its details which are irrelevant for our purpose, isolate the essence and form an 'abstract' mathematical structure. We prove theorems about this structure in the abstract. Such theorems are applicable not only in the original context but in any other situation in which the 'abstract' structure has been given a 'concrete' interpretation.

As another illustration of this procedure, let us consider the Test problem discussed in Chapter 1. We shall solve it using a mathematical structure which arises from some other problem, which, at first sight, has little resemblance with the Test problem. Suppose there are 10 Kings with their palaces located at points $P_1, P_2,..., P_{10}$ in a planar region. Tired of constantly fighting for territory, they come together and sign a pact. According to this pact, each king will be given a territory bounded by a circle with its centre at the king's palace. This territory is to be exclusively his. Obviously no two such territories may overlap (except perhaps for points on their boundaries). If all territories are to be of equal size, what will be their maximum radius?

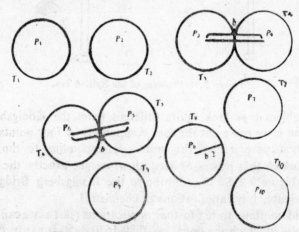

Figure 3.4: Kings with a Territorial pact.

This problem is quite simple. Our common sense tells us that if the territories are the largest possible then at least two of them will touch each other, because if there is some leeway left, we could enlarge the territories slightly without overlapping. In such a case the radius of each territory will be half the distance between the closest pair of palaces. But how do we prove it rigorously that with this choice for the radius the territories will not overlap? The proof is not difficult. Consider all the possible

---

*More precisely, a multigraph.

$$45 \left(= \left(\begin{array}{c} 10 \\ 2 \end{array} \right) \right)$$

pairs of the points $P_1, P_2, ..., P_{10}$. Let b be the minimum of the distance between these 45 pairs. (It may happen that there are more than one pair of palaces at a distance b apart from each other). Denote by $T_1, T_2, ..., T_{10}$ the circular territories of radius $b/2$ each, centred at $P_1, P_2, ..., P_{10}$ respectively as in Figure 3.4. Now suppose, if possible, some point P is common to the interiors of two of these territories, say T_i and T_j. Then the distance of P from P_i as well as P_j is less than $b/2$ each. So the sum of these two distances is less than b. By a well-known geometric property of triangles, the distance between P_i and P_j is less than this sum. (If P_i, P and P_j are collinear then it equal this sum). But this means that P_i and P_j are at a distance less than b from each other, contradicting the definition of b. Thus it follows that the territories cannot have any common interior points, i.e. they cannot overlap except perphaps for boundary points. It is also clear that a radius larger than $b/2$ will not work, because with such a radius, the midpoint of the line segment joining the closest pair of palaces will be a common interior point to the territories centered at these palaces.

Thus we see that of all the many attributes of the plane, the most relevant to this problem is the concept of the distance between various points. Let us isolate this concept, so that it can be applied to other problems. Let X denote the set of all points in the plane. If $P, Q \in X$, let $d(P, Q)$ denote the distance between P and Q. This is a real number depending on both P and Q. Using the language of functions, we have a function d from the cartesian product $X \times X$ into $\mathbf{R}$, the set of real numbers which assigns to each ordered pair (P, Q) of points of the plane, the distance between P and Q. This distance is often called the **euclidean distance**, because it occurs so very frequently in the euclidean geometry.

Now if we want to isolate this concept and put it into an abstract, general form we would have to let X be any 'abstract' set and d an 'abstract' real-valued function on $X \times X$. If x, y are two points of X then the real number $d(x, y)$ will be called the distance between them. The resulting mathematical structure is called a metric space. Examples of metric spaces will be obtained by letting X be some concrete set and letting

$$d: X \times X \rightarrow \mathbf{R}$$

be some concrete function. We already saw one such example where X was the set of points in the plane and d was the euclidean distance function. But we can give many others. We list a few below:

(i) X = set of real nnmbers, $d(x, y) = |x - y|$
(ii) X = set of real numbers, $d(x, y) = x + y^2$
(iii) X = set of all elephants,

$$d(x, y) = \begin{cases} 1 \text{ if the elephant } x \text{ is taller than the elephant } y \\ 0 \text{ otherwise} \end{cases}$$

(iv) $X =$ set of real numbers, $d(x, y) = xy$
(v) $X =$ set of real numbers, $d(x, y) = x^2 + y^2$

(vi) $X =$ any set, $d(x, y) = \begin{cases} 1 & \text{if } x \neq y \\ 0 & \text{if } x = y \end{cases}$

(viii) $X =$ the set of four vertices associated with the Konigsberg Bridge problem (Figure 3.2).
$d(x, y) =$ number of curves joining x and y (Thus $d(A, A) = 0$, $d(A, C) = 2$, $d(C, A) = 2$, $d(A, B) = 0$, $d(C, D) = 0$, etc.)

We could continue this list much longer. In fact since we have not put any restriction on the set X or on the function d (other than it be real-valued), it does not take much work or imagination to give examples of metric spaces. Whatever theorems we prove for metric spaces in the abstract would be applicable to all the examples given above, and also to any other examples of metric spaces. The trouble, though, is that there is not much one can prove, given only an abstract set X (about which nothing is known in general) and an abstract function d: $X \times X \to \mathbf{R}$. In order to prove some non-trivial theorems, we would have to assume as axioms some properties of the function d. Effectively this amounts to changing the definition of a metric space so as to make it more restrictive. Obviously, in doing so we must be careful not to exclude our original example, namely the plane and the euclidean distance function. For example, if we assume as an axiom that the distance between every two distinct points be the same then this axiom is no longer satisfied by the euclidean distance function for the plane (although it is satisfied in example (vi) above). This means that the properties which we wish to assume as axioms for an abstract metric space must be chosen from among those which are true for the euclidean distance function. Let us therefore list some of the properties of the euclidean distance function. We do this in the next proposition.

(1.1) Proposition: Let X be the set of points in a plane and let $d(P, Q)$ denote the euclidean distance for two points P and Q in X. Then,

1. $d(P, Q) \geqslant 0$ and $d(P, Q) = 0$ if and only if $P = Q$. (This property is called **positivity**.)
2. For all $P, Q, d(P, Q) = d(Q, P)$ **(Symmetry)**
3. For all $P, Q, R, d(P, R) \leqslant d(P, Q) + d(Q, R)$ and equality holds if and only if P, Q, R are collinear with Q lying between P and R. **(Triangle inequality)**

4. For all P, Q there exists a unique point M in the plane such that $d(P, M) = d(M, Q) = \frac{1}{2} d(P, Q)$. **(Mid-point property).**

Proof: The proof depends upon how $d(P, Q)$ is defined. In elementary books on geometry, the euclidean distance is taken as a primitive concept. No formal proofs are given for (1) and (2) (which effectively amounts to taking them as axioms). The inequality in (3) is proved as a property of triangles (hence the name), namely that the sum of the lengths of the two sides of any triangle exceeds the length of the third. For (4), the existence of the mid-point is again taken as an axiom, while its uniqueness follows as a consequence of (3).

In an analytic approach to geometry, points of the plane are represented by their cartesian coordinates w.r.t. some fixed frame of reference. If $P \equiv (x_1, y_1)$ and $Q \equiv (x_2, y_2)$ then $d(P, Q)$ is defined as:

$$\sqrt{(x_1 - x_2)^2 + (y_1 - y_2)^2}$$

Then the proofs of all the properties reduce to some simple facts about real numbers (such as, the fact that the square of every non-zero real number is positive). We leave the proofs as exercise because our concern is not proving these properties *per se* but in studying their consequences. ∎

We can list more properties of the euclidean distance function. But first let us take stock as to which of these properties are satisfied by the examples of the 'abstract' distance function given above. Note that the concept of collinearity has no meaning for an abstract set. Consequently the second part of property (3) is meaningless. Therefore, for an abstract distance function $d : X \times X \to \mathbf{R}$, by triangle inequality we shall simply mean, $d(x, z) \leqslant d(x, y) + d(y, z)$ for all $x, y, z \in X$ (the points x, y, z need not all be distinct). We list the results summarily, leaving their verification as exercises.

(i) satisfies all four properties.

(ii) satisfies none of them.

(iii) also satisfies none of the four properties (although it satisfies a part of the first property, namely, if $x = y$ then $d(x, y) = 0$).

(iv) satisfies symmetry but no others.

(v) satisfies symmetry and the triangle inequality. It also satisfies a part of the positivity condition, namely, the distance between any two elements is non-negative. It does not satisfy the mid-point property.

(vi) satisfies the first three properties. The mid-point property is satisfied if and only if the set X is either empty or consists of just one point.

(vii) satisfies the first two properties but not the other two.

Now we come to a basic question. Which of these (or some other) properties should we assume as axioms for an abstract metric space? Questions like this arise every time we want to define a new mathematical structure, starting from some particular example (in this case the euclidean distance) as a model and isolating some of its properties. So some general discussion regarding the selection of axioms is in order. Recall that the very purpose of axioms is to provide foundations for the theorems to be proved. Naturally, the stronger our axiom system, the more powerful theorems we would be able to deduce from it. These profound theorems would make our theory rich in terms of its depth. But there is a price we have to pay for this. As our axiom system becomes more and more powerful, it also becomes more and more restrictive in the sense that the number of examples that satisfy all the axioms goes on decreasing. Consequently, although we can prove more and deeper theorems, the domain where they become applicable will shrink. For example, if we take symmetry and positivity as our axioms for an abstract distance function then among the seven examples given above, our theory will be applicable to (i), (vi) and (vii). If we further include triangle inequality then (vii) will have to go out. And if the mid-point property is also added, then (i) will be the only example to which our theory would apply.

Thus we see that depth and generality are conflicting virtues. So we seek an optimum balance between the two. That is, we want our axiom system to be strong enough so as to enable us to prove non-trivial theorems from it and at the same time should not be so restrictive that we would be forced to throw away interesting applications. To hit upon the right scheme of axioms is not an easy task. There is also some room for a difference of opinion as to which is the optimum balance, because depending upon individual tastes, one can lean more towards generality than depth or vice versa. Ultimately, it is only through the test of time that a particular scheme of axioms gets recognition as one leading to a fairly non-trivial and still sufficiently general theory.

The definition of a metric space which is now accepted in the mathematical world has well stood this test of time. We shall give this definition first, then we shall prove a theorem about metric spaces. Finally, we shall apply this theorem to a suitably constructed metric space and thereby obtain a solution to the Test Problem.

1.2 Definition : A metric space consists of a set X and a function $d : X \times X \to \mathbf{R}$, called the **distance function** or the **metric**, satisfying the following conditions:

1.   For all $x, y \in X$, $d(x, y) \geqslant 0$ and equality holds if and only if $x = y$.
2.   For all $x, y \in X$, $d(x, y) = d(y, x)$
3.   For all $x, y, z \in X$, $d(x, z) \leqslant d(x, y) + d(y, z)$.

The plane with the euclidean distance is of course a foremost example of a metric space. In the list of the seven examples given above, (i) and (vi) are metric spaces. Another important example will be given after proving a general theorem about 'abstract' metric spaces. Note, by the way, that on the same set it may be possible to define more than one distance function. For example, on the set of real numbers, one distance function is given by $d_1(x, y) = |x - y|$ (this is the same as the example (i) above) while another distance of function is given by $d_2(x, y) = |x^3 - y^3|$. Thus although the underlying set is the same, the metric space whose distance function is d_1 is not the same as the metric space whose distance function is d_2. (A beginner may find this a little difficult to swallow, but it need not be so. Different metrics induce different structures on the underlying set and it is the structure that counts. It is somewhat like this. The same mound of clay may be moulded into various forms and then we call it a jug, a plate and so on even though the material is the same.) To take into account this fact, a metric space whose underlying set is X and whose distance function is d is denoted by the ordered pair $(X; d)$. In the example just given, $X = \mathbb{R}$ and the two metric spaces $(\mathbb{R}; d_1)$ and $(\mathbb{R}; d_2)$ are to be distinguished from each other. Of course, we do not have to be so fussy all the time. Where the distance function d is understood, we may denote a metric space $(X; d)$ simply by X.

Now, for the theorem we are going to prove, we first define the abstract analogue of the circular territorial regions in the problem of the ten kings considered above.

1.3 Definition. Let $(X; d)$ be a metric space and let $x \in X$. Let r be a non-negative real number. Then the sets $B_d(x, r)$ and $C_d(x, r)$, defined by,

$$B_d(x, r) = \{y \in X : d(x, y) < r\}$$

and

$$C_d(x, r) = \{y \in X : d(x, y) \leqslant r\}$$

are called, respectively, the **open** and the **closed balls** of radius r and centre x, or for short, open r-ball and closed r-ball with centre x.

If X happens to be the plane and d the euclidean distance then $B_d(x, r)$ consists of a circular region (without the boundary) while $C_d(x, r)$ consists of a circular region with the boundary included. The euclidean distance can also be defined on the three dimensional space so as to give a metric space. In this case, the open and the closed balls in fact turn out to be solid balls (without and with boundary respectively) in space. Hence the name. Of course for other metric spaces, the open and the closed balls may not look like 'balls'. Note also that they depend as much on the metric d as on their centre and their radius. For example, consider the two metric spaces $(\mathbb{R}; d_1)$ and $(\mathbb{R}; d_2)$ defined above. We invite the reader to prove that the sets $B_{d_1}(2, 1)$ and $B_{d_2}(2, 1)$ respectively the open intervals

$(1, 3)$ and $(1, 3^{1/3})$ while the corresponding closed balls are the closed intervals $[1, 3]$ and $[1, 3^{1/3}]$ respectively. (Hence the names 'open' and 'closed'). Once again, where only one distance function d is involved, we may suppress it from notation and denote $B_d(x, r)$ by $B(x, r)$ and $C_d(x, r)$ by $C(x, r)$.

Obviously, in any metric space $C_d(x, r)$ is larger than $B_d(x, r)$, the difference between the two being the set $\{y \in X : d(x, y)\} = r$. In all the examples given above this difference was much smaller as compared to the open ball $B_d(x, r)$ and consequently we tend to think that the open and the closed balls are 'nearly' equal. However, this does not hold in all metric spaces. For example let X be any set and define $d : X \times X \to \mathbb{R}$ by $d(x, y) = 1$ if $x \neq y$ and 0 otherwise. Then $(X; d)$ is a metric space. Verification of the positivity and the symmetry property is trivial. For triangle inequality, let $x, y, z \in X$. We have to show $d(x, z) \leqslant d(x, y) + d(y, z)$. If the right hand side is 0 then $x = y = z$ and so the left hand side is also 0. If the right hand side is not 0 then it is at least 1 and so the inequality holds because the left hand side can be at most 1. Note that in this metric space all distinct points are equidistant from each other. Let x be any point of this space. Then $B_d(x, 1)$ consists only of x but $C_d(x, 1)$ is the whole set X! Thus $C_d(x, 1)$ is considerably larger than $B_d(x, 1)$. This is inconsistent with our intuition but we have to accept it as it is a logical consequence of our definitions. The reason for this behaviour is that, unlike the earlier examples, this metric does not have the mid-point property. In fact no point in this space has any close neighbours, the nearest neighbours being at a distance of 1. So every point is isolated. For this reason, this metric is often called the **discrete** metric. It deserves to be studied, if for no other reason, then at least because it is a namesake of the subject matter of this book, namely 'discrete' structures! In fact our interest will be in finite sets and it is not difficult to show that every metric on a finite set is discrete in the sense that every point of it is isolated.

Another point which defies our intuition is that open balls of the same radii but with different centres need not be of the same 'size'. This is true for the euclidean distance on the plane and also for the metric space $(\mathbb{R}, d_1)$ considered above. But it is false for the metric space $(\mathbb{R}, d_2)$. $B_{d_2}(0, 1)$ is an open interval of length 2 but $B_{d_2}(1, 1)$ is an open interval of length $2^{1/3}$. We invite the reader to pinpoint the reason for this anomalous behaviour.

Instances like this are to be expected because, since we are not making all properties of the euclidean distance as the axioms for an abstract distance function, it is but natural that something that holds for the euclidean distance may fail in some other examples of metric spaces. However, theorems which are based only on the three properties we have chosen (namely positivity, symmetry and the triangle inequality) will hold good in every metric space. We prove one such simple result.

1.4   Theorem.   Let $(X; d)$ be a metric space. Let x, y be distinct points of X and suppose $b = d(x, y)$. If $0 \leqslant r \leqslant b/2$ then the open balls $B_d(x, r)$ and $B_d(y, r)$ are mutually disjoint. If $0 \leqslant r < b/2$ then the closed balls $C_d(x, r)$ and $C_d(y, r)$ are also mutually disjoint.

Proof.   Note first that $b > 0$ because of the positivity condition since $x \neq y$. Now suppose $0 \leqslant r \leqslant b/2$. Let, if possible, z be a common point of $B_d(x, r)$ and $B_d(y, r)$. Then, by definition, $d(x, z) < r$ and $d(y, z) < r$. By symmetry we get that $d(z, y) < r$. Now by triangle inequality,

$$d(x, y) \leqslant d(x, z) + d(z, y) < r + r = 2r \leqslant b.$$

Thus we see that $d(x, y) < b$ contradicting that $d(x, y) = b$. Therefore the sets $B_d(x, r)$ and $B_d(y, r)$ are mutually disjoint. The proof of the second assertion is similar and left to the reader. ∎

This theorem, as applied to the plane with the euclidean distance was at the crux of the solution to the problem of the kings discussed above. We shall now construct a new metric space and apply the theorem to it, so as to get a solution to the Test Problem.

Let X be the set of all binary sequences of length n, where n is some positive integer. If $x_1, x_2, \ldots, x_n$ is such a sequence we shall denote it by $\bar{x}$ for short. Now suppose $\bar{y} = y_1, y_2, \ldots, y_n$ is another such sequence. Then we define $d(\bar{x}, \bar{y})$ to be the sum $\sum_{i=1}^{n} |x_i - y_i|$. Note that each x_i and y_i is either 0 or 1 (because the sequences are binary). So $|x_i - y_i| = 0$ if x_i equals y_i and $|x_i - y_i| = 1$ if x_i is different from y_i. Therefore $d(\bar{x}, \bar{y})$ is also equal to the number of i's for which $x_i \neq y_i$, i.e. to the number of places at which the two sequences $x_1, x_2, \ldots, x_n$ and $y_1, y_2, \ldots, y_n$ do not match with each other. The number $d(\bar{x}, \bar{y})$ is called the **Hamming distance** between the binary sequences $\bar{x} = x_1, \ldots, x_n$ and $\bar{y} = y_1, \ldots, y_n$. For example, if $n = 8$, $\bar{x} = 01010101$ and $\bar{y} = 10010001$ then $d(\bar{x}, \bar{y}) = 3$ because these two sequence differ in the first, second and the sixth term.

The importance of this concept stems from the following result.

1.5   Theorem.   The set of binary sequences of length n along with the Hamming distance is a metric space:

Proof :   We have to show that the function d satisfies the three axioms of a metric space, namely, positivity, symmetry and the triangle inequality. Let

$$\bar{x} = x_1 \, x_2 \ldots \ldots x_n$$

and

$$\bar{y} = y_1 \, y_2 \ldots y_n \in X.$$

Then

$$d(\bar{x}, \bar{y}) = \sum_{i=1}^{n} |x_i - y_i|.$$

Each term in this summation is non-negative. So the sum is also non-negative and will be 0 if and only if

$$| x_i - y_i | = 0, \text{ i.e. } x_i = y_i,$$

for all $i = 1, 2, \ldots, n$. But this is equivalent to saying that $\bar{x} = \bar{y}$. So positivity is established. Symmery is also immediate from the fact that for every i,

$$| y_i - x_i | = | x_i - y_i |.$$

For the triangle inequality, suppose

$$z = z_1 z_2 \ldots z_n \in X.$$

Then

$$d(\bar{x}, \bar{z}) = \sum_{i=1}^{n} | x_i - z_i |.$$

Now for each i,

$$| x_i - z_i | \leqslant | x_i - y_i | + | y_i - z_i |.$$

(This is known as the triangle inequality for the absolute value of real numbers. We hope the reader is familiar with it as it is used innumerably often in calculus.) If we sum both the sides of this inequality for $i = 1$ to n, we get that

$$d(\bar{x}, \bar{z}) \leqslant d(\bar{x}, \bar{y}) + d(\bar{y}, \bar{z}).$$

This completes the proof. We have given the argument using the definition of d as a certain sum. It will be instructive for the reader to paraphrase it, especially for the triangle inequality, in terms of the interpretation of $d(\bar{x}, \bar{y})$ as the number of places where the sequences $\bar{x}$ and $\bar{y}$ differ from each other ▪.

Because of this theorem, all the theorems about metric spaces become applicable to the set of binary sequences. So far we have proved one such theorem. To apply it, let us take a closer look at the open and the closed balls in the metric space we have just constructed. Because the Hamming distance assumes only integral values, it suffices to consider balls whose radii are integers. Moreover, we need consider only the closed balls, because an open ball of radius r is equal precisely to a closed ball (with the same centre) of radius $r - 1$.

1.6 Lemma: Let X be the set of all binary sequences of length n and let d be the Hamming distance. Then in the metric space $(X; d)$, every closed ball of radius r (where r is a positive integer), has precisely

$$\sum_{k=0}^{r} \binom{n}{k} \text{ elements.}$$

Proof: Let

$$\bar{x} = x_1 \, x_2 \ldots x_n \in X.$$

For each k, let

$$S_d \, (\bar{x}, k) = \{\bar{y} \in X : d \, (\bar{x}, \bar{y}) = k\}.$$

Then

$$C_d \, (\bar{x}, r) = \bigcup_{k=0}^{r} S_d \, (\bar{x}, k).$$

Also if $k_1 \neq k_2$ then

$$S_d \, (\bar{x}, k_1) \cap S_d \, (\bar{x}, k_2) = \phi.$$

So by Proposition (2.2.3).

$$| \, C_d(\bar{x}, r) \, | = \sum_{k=0}^{r} | \, S_d \, (\bar{x}, k) \, |.$$

Thus our result will be established if we show that for every k,

$$| \, S_d \, (\bar{x}, k) \, | = \binom{n}{k}.$$

Now $S_d \, (\bar{x}, k)$ consists of those sequences $\bar{y} = y_1 \, y_2 \ldots y_n$ in X whose entries differ from those of $\bar{x}$ in precisely k places (out of n). The choice of k places from n places may be made in $\binom{n}{k}$ ways. For every such choice, there is precisely one sequence whose entries differ from those of $\bar{x}$ in those k places. This is so because the sequences are binary. Hence if we are given that y_i is different from x_i, then x_i uniquely determines y_i. In fact,

$$y_i = 1, \text{ if } x_i = 0$$

and

$$y_i = 0 \text{ if } x_i = 1.$$

Thus we see that there are $\binom{n}{k}$ elements of X which are each at a distance k from $\bar{x}$. As noted before this completes the proof. ∎

Having computed the size of the closed r-balls, we now get an upper bound on how many of them can be packed in the space X without overlapping (like the non-overlapping territories in the Kings problem above). We first introduce a notation. If x is a real number, then by $\lfloor x \rfloor$ we shall denote the greatest integer not exceeding x and by $\lceil x \rceil$ we denote the least integer greater than or equal to x. For example,

$$\lfloor 43.12 \rfloor = 43, \lceil 43.12 \rceil = 44, \lfloor 43 \rfloor = 43 = \lceil 43 \rceil, \lfloor -7.5 \rfloor = -8$$

and $\lceil -7.5 \rceil = -7$. (In the literature, $\lfloor x \rfloor$ is often denoted by $[x]$ and is called **integral part** of x.) These notations appear frequently in discrete mathematics because we often have to convert real numbers to integers. Note that if m is an integer and x is a real number such that $m \leqslant x$, then, by definition, $m \leqslant \lfloor x \rfloor$.

1.7 Theorem: Let $(X; d)$ be as above. Suppose there are m mutually disjoint closed balls of radius r each in X where r is some non-negative integer. Then

$$m \leqslant \left\lfloor \frac{2^n}{\sum\limits_{k=0}^{r} \binom{n}{k}} \right\rfloor$$

Proof: By the lemma above, each closed r-ball has $\sum\limits_{k=0}^{n} \binom{n}{k}$ elements. If there are m such balls and they are mutually disjoint, then by Proposition (2.2.3), their union has cardinality $m \sum\limits_{k=0}^{r} \binom{n}{k}$. But this union is a subset of X and so its cardinality cannot exceed that of X by Corollary (2.2.6). Now, $|X| = 2^n$ by Proposition (2.3.3). So, putting it all together, we get

$$m \sum\limits_{k=0}^{r} \binom{n}{k} \leqslant 2^n \text{ or } m \leqslant \frac{2^n}{\sum\limits_{k=0}^{r} \binom{n}{k}} .$$

This expression need not be an integer. But m is an integer. So

$$m \leqslant \left\lfloor \frac{2^n}{\sum\limits_{k=0}^{r} \binom{n}{k}} \right\rfloor . \blacksquare$$

Combining the last theorem with Theorem (1.4), we now get an upper bound for the number of points in X which are sufficiently far apart from each other. We prefer to word the result in terms of its contrapositive, because of the application to follow.

1.8 Theorem: Let $(X; d)$ be as above. Let t be an odd positive integer and let $r = \dfrac{t-1}{2}$. Then if we take more than

$$\left\lfloor \dfrac{2^n}{\displaystyle\sum_{k=0}^{r} \begin{bmatrix} n \\ k \end{bmatrix}} \right\rfloor$$

points in X, then at least two of these points must be at a distance less than t from each other.

Proof: Suppose the number of distinct points taken is m. Suppose no two of them are at a distance less than t from each other. Then for every two distinct points, say, $\bar{x}$ and $\bar{y}$ among these m points, $d(\bar{x}, \bar{y}) \geqslant t$. So

$$r < \frac{d(\bar{x}, \bar{y})}{2}.$$

Therefore by Theorem (1.4), the closed balls $C_d(\bar{x}, \bar{r})$ and $C_d(\bar{y}\,\bar{r})$ are mutually disjoint, Since this hold for every pair of distinct points from the m points taken, it follows that X contains a collection of m mutually disjoint closed r-balls. Therefore, by the last Theorem,

$$m \leqslant \left\lfloor \dfrac{2^n}{\displaystyle\sum_{k=0}^{r} \begin{pmatrix} n \\ k \end{pmatrix}} \right\rfloor.$$

This contradicts our assumption that more than

$$\left\lfloor \dfrac{2^n}{\displaystyle\sum_{k=0}^{r} \begin{pmatrix} n \\ k \end{pmatrix}} \right\rfloor$$

points have been taken, and establishes the result. ∎

Put differently, the last result says that for every non-negative integer r, the integer

$$\left\lfloor \dfrac{2^n}{\displaystyle\sum_{k=0}^{r} \begin{pmatrix} n \\ k \end{pmatrix}} \right\rfloor$$

is an upper bound on the maximum number of binary sequences of length n, every two of which are at least $2r + 1$ apart. This upper bound is called the **Hamming bound** and is important in coding theory. (See the Epilogue.)

We now have all the machinery needed to solve the Test Problem. In this problem, 20 students appear for a test with 10 questions. As commented in Chapter 1, Section 3, the answer-book of each student corresponds to a binary sequence of length 10. To say that two students answer at least six questions identically is equivalent to saying that their answers to at most four questions are different, i.e. the Hamming distance between their answer-books is 4 or less. We apply Theorem (1.8) with $n = 10$ and $t = 5$. Then $r = 2$ and

$$\left\lfloor \frac{2^n}{\displaystyle\sum_{k=0}^{r}\binom{n}{k}} \right\rfloor = \left\lfloor \frac{1024}{1 + 10 + 45} \right\rfloor = \lfloor 18.3 \rfloor = 18.$$

Since we have more than 18 students, it follows that the answer-books of at least two of them must be at a distance less than 5 apart from each other. Actually the problem can be solved even with 19 students instead of 20.

The reader may ask whether all this lengthy procedure is necessary to arrive at a solution to the Test Problem. After all, the original problem does not involve the Hamming distance. Could we not simply consider, for every student, the set of all binary sequences which differ from the answer-book of that student in at most two places and then show that these sets are mutually disjoint? This question is important because a similar question can be asked whenever we apply a general result about mathematical structures of a certain type (in this case a metric space) to a particular instance of it (in this case the Hamming distance). Strictly speaking the answer to such questions is 'yes'. Whatever can be done through the use of general mathematical structures can, at least in theory, be done directly in a particular case. So if our interest is in only one particular problem, then abstract mathematical structures are not indispensible. Why do we study them then? One answer was already given above, namely, the same structure may be applicable in more than one problem and so studying it in the abstract saves duplication of work. The greater the number of instances where a structure can be applied, the more worthwhile it is to study it in the abstract. A fitting analogy would be that if you want to send just one or two greeting cards to persons in your neighbourhood, you might as well arrange to deliver them personally; but if you want to send them in bulk you better mail them (and thus use the postal structure) even though one of the cards may be addressed to your next-door neighbour !

However, there is one more reason to justify the study of abstract mathematical structures. It is valid even when we have just one or two applications of that structure. True, the abstract structure is not indispensible in the sense that whatever can be done with it can also be done directly without it. Sometimes, it is indeed preferable to do so. But many times the direct solution looks so clumsy and contrived that one would wonder how on earth it was thought of. For example in the direct solution to the Test Problem, it is far from easy to see the significance of the set of all binary sequences which differ from a given binary sequence in at most two places. But with the introduction of the Hamming distance, the same set corresponds to a very natural concept, namely, a closed ball of radius 2.

As another example, take the concept of the dual partition of a given partition of a positive integer. We introduced it at the end of Chapter 2, Section 3 and used it successfully to give a solution to the Postage Problem. Strictly speaking, this is a purely set-theoretic concept and we could have defined it as such. Consider a partition of a positive integer n into k parts of sizes $a_1, \dots, a_k$ (say) with $a_1 \geqslant a_2 \geqslant \dots \geqslant a_k > 0$. Here all the k parts have positive sizes. For each i, let b_i be the number of parts whose size is $\geqslant i$. Clearly $b_1 = k$ and $b_i = 0$ for $i > a_1$. Also

$$b_1 \geqslant b_2 \geqslant \dots \geqslant b_{a_1} > 0.$$

For each $i = 1, 2, \dots$, the number of parts of size i is $b_i - b_{i+1}$. It follows that $\sum_{i=1}^{a_1} i(b_i - b_{i+1}) = n$. But if we actually write this sum out we see, after cancellations, that it is nothing but $b_1 + b_2 + \dots + b_{a_1}$. Thus we have proved that $b_1 + b_2 + \dots + b_{a_1}$ is also a partition of n. It is not hard to show, further, that this is precisely the dual of the original partition of n. We could have therefore defined the dual partition in this manner. But this definition does not appear as elegant as the earlier one, based on the Ferrer's graph of a partition. When we draw the Ferrer's graph of a partition, we are in essence putting an additional structure on the underlying set, namely the structure derived from the geometry of the cartesian plane. The earlier definition of a dual partition is not substantially different from the one given above. But it appears more natural because of our long familiarity with the geometry of the plane. The geometric structure also helps in the proof of a result like that of Exercise (2.3.36), where a direct proof would be quite awkward. As we shall see in Chapter 6, the geometric structure is also inherent in certain concepts involving matrices.

So far we introduced two mathematical structures, a graph and a metric space. Typically, a mathematical structure consists of an 'abstract' set, called the underlying set, (sometimes more than one set as in the case of a graph). On this set, we have an additional structure. Depending upon the type of study we are interested in, the form of this structure will vary. Usually it is in the form of some function or some set associated with the underlying set. A few properties of such functions or sets are often assumed as axioms. We shall study many such structures in this and the next few chapters. We conclude this section with a mathematical structure which is designed to take into account the repetitions in a set. According to our convention, the repetitions of elements in specifying a set by listing its elements do not affect the set. Thus we regard $\{1, 2, 1, 2, 2, 3, 2, 1\}$ the same set as {1, 2, 3}. Many times, however, the frequency with which a particular element appears in a list may be relevant. For example, when we consider the set of marks obtained by the students in a class at some examination, we would like to count each figure as many times as the number of students who score it. Similarly, in considering the set of roots of a polynomial, we would like to count each root according to its multiplicity.

The appropriate concept to handle the repetitions of elements of a set is a multiset. It can be formally defined as an ordinary set along with a certain additional structure on it. This additional structure is in the form of a function which tells us how many times a particular element is to be repeated.

1.9   Definition: A **multi-set** is an ordered pair (S, f) where S is a set and $f: S \to \mathbb{N}$ is a function, called the **frequency** or **multiplicity** or **weight function**.

A convenient way of writing a multi-set is by writing each element as many as its frequency. Thus, the multi-set {1, 1, 2, 3, 3, 3, 3, 6} is the multi-set (S, f) where $S = \{1, 2, 3, 6\}$ and $f: S \to \mathbb{N}$ is defined by $f(1) = 2$, $f(2) = 1, f(3) = 4$ and $f(6) = 1$. The roots of the polynomial $x^4 - x^3$ form the multi-set {0, 0, 0, 1}.

Many concepts for sets can be generalized for multi-sets. For example if (S, f) and (T, g) are two multi-sets then we say (S, f) is a **submulti-set** of (T, g) if first of all, $S \subset T$ and secondly, for every $x \in S, f(x) \leqslant g(x)$. Given any set S we can define $f: S \to \mathbb{N}$ by $f(x) = 1$ for all $x \in S$. Then the multi-set (S, f) can obviously be identified with the set S, because no element is repeated more than once. In this sense, the theory of multisets is more general than the theory of sets. Some of the things that hold for sets may fail for multisets. A few results about multisets will be given as exercises.

Exercises

(Some of these exercises are meant only to develop your intuition. Rigorous proofs may not be possible at this stage. Nevertheless, try to be as precise as you can.)

1.1   Give an example of a problem (other than those given in the text) which reduces to the same problem as the Konigsberg bridge problem.

1.2   For which of the graphs in Figure 3.6 is it possible to have a round tour in which every edge is traversed exactly once?

1.3   Give an example of a graph in which there exists a tour with every edge traversed exactly once but no such round tour exists.

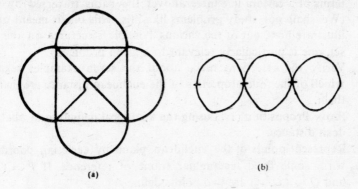

(a)

(b)

Figure 3.5: Graphs for Exercise (1.2)

1.4 Suppose there is a green meadow M with a little pond P in it as shown in Figure 3.6. A herd of cows is originally at the point A.

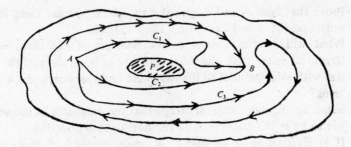

Figure 3.6: Cows in the Meadow with a Pond.

Each cow moves from point A to point B without going through the pond and without going around the pond. Prove that the cows fall into two categories depending upon the position of the pond vis-a-vis the path of the cow. Classify the three cows whose paths are shown in Figure 3.6.

1.5 Which of the following attributes correctly describes the classification of the cows in the last theorem? That is, given two cows traversing paths C_1 and C_2 (say), which of the following quantities tell us whether the two cows are in the same class or not?

  (i) the lengths of C_1 and C_2,

 (ii) the curvatures of C_1 and C_2,

(iii) the area bounded by C_1 and C_2.

1.6 (For those who know a little complex analysis). Prove that the classification in Exercise (1.4) can be characterised rigorously in terms of a certain line integral over the paths traversed by them. (We shall not study problems like this. This one is meant only to illustrate how, out of the various possible structures on the same set, one is particularly relevant to a given problem.)

1.7 Verify the assertions made about the seven examples regarding which of the four properties of the euclidean distance are shared by them.

1.8 Prove Proposition (1.1) using the analytical definition of the euclidean distance.

1.9 Represent points of the euclidean plane by cartesian coordinates w.r.t. some fixed rectangular frame of reference If $P = (x_1, y_1)$ and $Q = (x_2, y_2)$ are two points define

$$d_1(P, Q) = | x_1 - x_2 | + | y_1 - y_2 |$$

and

$$d_2(P, Q) = \max \{ | x_1 - x_2 |, | y_1 - y_2 | \}.$$

Prove that both d_1 and d_2 give rise to metric spaces. Describe the open and the closed balls w.r.t. these metrics.

1.10 What will be the answer to the problem of the kings with a territorial pact if all the territories are to be squares of the same size with sides parallel to the east-west and the north-south directions?

1.11 Same as the last problem except that the diagonals of the squares are to be in the east-west and the north-south directions.

*1.12 If in Exercise 1.10 we allow rectangles instead of squares, then prove that the solution is not unique in general.

**1.13 Do Exercise 1.10, with a further relaxation, namely that the sides of the rectangles need not all be parallel to the same direction.

1.14 Let X be a finite set and d a metric on X. Prove that, there exists a real number $\lambda > 0$, such that for all $0 < r \leqslant \lambda$, and for all $x \in X$, $B_d(x, r)$ consists only of the point x. (In this sense, every metric on a finite set behaves locally like the discrete metric.)

1.15 Let $(X_1: d_1)$ and $(X_2; d_2)$ be metric spaces. Let $X = X_1 \times X_2$. Define $d: X \times X \to \mathbf{R}$ by any one of the following formulas:

(i) $d((x_1, x_2), (y_1, y_2)) = \sqrt{[d_1(x_1, y_1)]^2 + [d_2(x_2, y_2)]^2}$

(ii) $d((x_1, x_2), (y_{,1} y_2)) = d_1(x_1, y_1) + d_2(x_2, y_2)$

(iii) $d(x_1, x_2), (y_1, y_2)) = \max. \{d_1(x_1, y_1), d_2(x_2. y_2)\}$. Prove that each of these is a metric on $X \times X$. Generalize to the case when we have n metric spaces $(X_1; d_1), (X_2, d_2), ..., (X_n, d_n)$. (The first metric is called the **Pythagorean metric**. Can you justify the name?)

1.16 Prove that the Hamming distance can be thought of as a special case of the metric (ii) in the last exercise (generalised to the case of n metric spaces.) (Hint: Note that a binary sequence of length n corresponds uniquely to an element of the product

$$\mathbb{Z}_2 \times \mathbb{Z}_2 \times \ldots \times \mathbb{Z}_2$$

(n-times) where $\mathbb{Z}_2$ is the set $\{0, 1\}$).

1.17 Prove that there do not exist three (distinct) binary sequences of length 10, every two of which are at least at a distance 7 apart from each other; even though the Hamming bound for the number of such sequences is 5. (Thus the Hamming bound is not always sharp.)

1.18 Let n, t be positive integers with $t \leqslant n$.

$$M = \left\lceil \frac{2^n}{\sum\limits_{k=0}^{t-1} \binom{n}{k}} \right\rceil.$$

Prove that there do exist at least M binary sequences of length n every two of which are at least at a distance t apart from each other. (This gives a lower bound for the number of such sequences. It is called the **Gilbert bound**.)

1.19 For $n = 10$ and $t = 6$, prove that the Gilbert bound is also not sharp.

1.20 Generalise the concept of Hamming distance and Theorem 1.7 to k-ary sequences length n.

1.21 Define the intersection and union of two multi-sets in such a way that they come out to be, respectively, the largest common sub-multi-set of the two and the smallest common super-multi-set of the two.

1.22 Suppose (A, f) is a submulti-set of a multiset (X, g). Define

$$B = (X - A) \cup \{x \in A : f(x) < g(x)\}$$

and $h: B \to \mathbb{N}$ by

$$h(x) = \begin{cases} g(x) & \text{if } x \in X - A \\ g(x) - f(x) & \text{if } x \in A. \end{cases}$$

The multiset (B, h) is called the *complement* of (A, f) in (X, g). Prove, however, that $(A, f) \cap (B, h)$ need not be empty.

1.23 The **cardinality** or the **weight** of a multi-set (S, f) is defined as $\sum\limits_{s \in S} f(s)$. Generalise the results about the cardinalities of sets (in Chapter 2, Section 2) to multi-sets.

1.24   Let (S, f) be a multi-set of cardinality n. **A permutation** of (S, f) is defined as a function

$$\theta: \{1, 2, \ldots, n\} \to S$$

such that for every $x \in S$,

$$|\theta^{-1}(\{x\})| = f(x).$$

Show how this generalises the concept of a permutation of a set. Using Theorem 2.2.17 obtain a formula for the number of permutations of a multiset.

1.25   In a survey of 98 persons, each person was asked to indicate his preference for two products by a rating from 0 (total dislike) to 9 (total liking) for each of them. No person showed total dislike for both the products or total liking for both. Also no two persons gave the same ratings to both the products. Prove that it is impossible to pair off these 98 persons in such a way that in each pair the two persons have a minimum difference of taste.

1.26   What would have happened if, Birbal, instead of answering 'Because you fail to move it' had answered 'Because you fail to treat it properly'?

Notes and Guides to Literature

This Section gives the fundamentals of mathematical structures. Actual examples will be encountered in the chapters to come, where additional comments will be made.

There is some variation in the definition of a graph. We shall study them later (see the Epilogue). Standard references include Harary [1] or Deo [1].

The theory of metric spaces is very important in analysis and topology. See for example, Simmons [1].

The problem about the kings with a territorial part is a special case of a more general problem of fitting in a given figure, mutually disjoint replicas of some other figure. The general problem is considerably complex.

The concept of Hamming distance also appears in coding theory (see the Epilogue).

For permutations of multisets, see Knuth [1], vol. 3, Chapter 5, Section 1.

2.   Binary Relations on Sets

Given a set X, one of the simplest and yet one of the most frequently occuring additional structures on X consists of what is called a binary relation on X. As the name implies, a binary relation on a set X describes some type of relationship which may exist between certain pairs of elements of X. It is the relationship and not the two particular elements that matters. For example, let X be the set of human beings in a town. Two particular elements, say, A and B may be related to each other in any one or more of the many possible ways, e.g. (i) A and B work at the same place (ii) A and B are neighbours of each other (iii) A is exactly 7 inches taller than B (iv) A loves B (v) A loves B but B does not care (vi) B is A's wife (vii) A and B live in the same town. There may be many other pairs of persons who also exhibit some of these relationships between them. The significance of these relationships of course varies depending upon our point of view: (i), (ii) and (vi) may be relevant in a census; a dress designer may be concerned only with (ii), (iv) and (v) may be vitally interesting to a gossip. (vii) does not seem to say much because A and B are already known to be from the same town. Still it is a relationship, a relationship which exists between every two members of X, as a matter of fact. At the other extreme, we may have an impossible relationship such as, (viii) A is taller than B and also B is taller than A. There is no pair of persons who are so related. Finally we may also have a relationship like (ix) A is more than 18 years old. At the outset one may refuse to call this as a relationship between A and B because, B is not at all 'involved' in it. Still, it is a perfectly well-defined relationship and may be relevant, for example, in opening a joint bank account if the rules require the first account holder to be an adult.

Having given so many examples of relationships which may exist between two persons at a time, let us now see how we can give a rigorous, mathematical definition of a binary relation. Let us take example (iv) above. So, let X be the set of all persons in a town. Now define

$$R = \{ (x, y) \in X \times X : x \text{ loves } y \}.$$

Then R is a subset of $X \times X$ and consists of all ordered pairs of persons in X in which the first person loves the second. (We are not requiring that x and y be always different; it may happen that some person loves himself!) Thus the relationship (iv) determines the set R, which is a subset of $X \times X$. Conversely the subset R determines the relationship (iv) completely. Given two persons, say, A and B in the town whether A loves B or not amounts to asking whether the ordered pair (A, B) is in R or not. The same holds for all other relationships considered above. Each one of them determines some subset of $X \times X$ and this subset, in turn, determines the relationship. These subsets, of course, vary depending upon the relationship. For the relationships (vii) and (viii) they are respectively, $X \times X$ and ϕ, the empty set.

We now appeal to the 'definition trick' mentioned in Chapter 1, Section 4. When we have two things, inexorably related to each other, we define one of them to be the other. This leads to the following formal definition.

2.1. Definition : A **binary relation** on a set X is defined as any subset of $X \times X$. If $R \subset X \times X$, and $(x, y) \in R$ we say 'x is R-related to y' or 'x is related to y under R' and often write 'xRy'. More generally, for every positive integer n, we can define **an n-ary relation** on X as a subset of $X \times X \times ... \times X$ (n times). We shall rarely consider n-ary relation for $n \neq 2$. So by relation, from now on, we shall mean a binary relation unless otherwise specified.

Despite the motivation given above, it may still appear somewhat arbitrary and smug to define a ralation on a set X as a subset of $X \times X$. But actually it need not be so. Conceiving a relationship through the set of pairs which have it is not artificial, it is in fact natural. We often hint at certain relationships without naming them, by citing some instances of the intended relationships. Consider, for example, the statement, 'New Delhi is to India what London is to England or Paris to France'. We all understand that the implied relationship is that of being the capital of the country. (Strictly speaking it could be different. London and Paris are also the largest cities of England and France respectively and if this was the intended relationship than New Delhi would have to be replaced by Calcutta. The difficulty here is quite understandable. A set cannot be uniquely specified by merely giving two or three of its elements, as there may be other sets containing these elements. That is why, when we identify a relation on a set X with a subset of $X \times X$, we take the set of *all* ordered pairs of elements of X related under that relation.) As a poetic example, take the statement, 'without you I am like a fish out of water or like a night without a moon', suggestion of a particular relationship through some instance of it is, in fact, the very essence of a simile.

It may, of course, happen that two apparently different relationships determine the same relation on a set X. The chances of this are more, the smaller the set X. Suppose, for instance that X consists of three persons, say, A, B, C. Let the heights of these persons be, respectively, 190, 180 and 170 cms. and let their weights be 80 kg., 70 kg and 60 kg respectively. Then the two relations, 'x is taller than y' and 'x is heavier than y' are the same for the set X, each corresponds to the set

$$\{(A, B), (B, C), (A, C)\}.$$

There is nothing wrong in this, because as far as the set X is concerned, higher weight is indeed equivalent to greater height.

Another point which a beginner sometimes finds confusing is that we are allowing *every* subset of $X \times X$ to be a binary relation on X. As we saw above, certain subsets of $X \times X$ arise from some relationship for elements of X. Is this true of all subsets of $X \times X$? For example, let $X = \mathbb{N}$, the set of positive integers. Let $R = \{(1, 1), (2, 2) (2, 4), (3, 3), (3, 9), (4, 4), (4, 16), (5, 5), (5, 25),..., (n, n), (n, n^2 ...)\}$. It is clear that R corresponds to the relation 'y equals either x or x^2'. But what if we take

$$S = \{1, 39), (2, 105), (3, 87)\}?$$

Does S correspond to some relationship for elements of $\mathbb{N}$? The answer is 'yes'. But this relationship may not be as 'natural' as the one for *R*. Put differently, what we want is a characteristic property of the set *S* (see Chapter 2, Section 1). Theoretically, there is always one such property, namely, the property of belonging to *S*. This may sound evasive but philosophically it is not so. The very fact that out of all possible pairs of positive integers, we chose to pick only the three pairs, (1, 39), (2, 105) and (3, 87) and no others, itself serves to distinguish them. Admittedly the choice was arbitrary. Somebody else may make a different choice. But he too would get a perfectly well-defined binary relation.

With so much preamble about what a binary relation on a set is, let us now get down to their properties. Since every subset, without restriction, of $X \times X$ is a binary relation on X, it is clear that the number of binary relations on a set with n elements is 2^{n^2}. There are two ways of visualising a binary relation on a set X. One is the **cartesian representation,** where we think of $X \times X$ as the real line (or some subset of it). Then a relation on X determines a subset of the cartesian plane. One advantage of this representation is that the points on the line whose equation (in cartesian co-ordinates) is $y = x$ correspond to ordered pairs both whose elements are equal. Formally, for any set X, the **diagonal** on X, denoted by $\triangle X$, is defined as the set

$$\{(x, x) \in X \times X : x \in X\}.$$

This is a subset of $X \times X$ and plays a crucial role in certain concepts associated with binary relations. Figure 3.7 shows the relation R defined by xRy iff $x \leqslant y$ on the interval [0, 1]. Note that points on the diagonal are included in R. If the relation were $x < y$ then they would not be included.

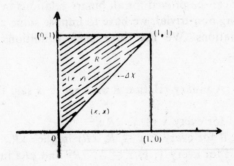

Figure 3.7: Cartesian Representation of a Relation.

Another method for visualising a binary relation R on a set X is known as the **graphic representation**. It is convenient when the cardinality of the set R (i.e. the number of pairs related under R) is relatively small. In this method we picture elements of X by points in a plane. Whenever xRy, we praw a directed arrow (which may be curved) from x to y. These arrows may cross each other, but we ignore the crossings. (If X is a finite set, the crossings can be eliminated by allowing the curves to lie in space instead of just in the plane.) Note that if a point x is related to itself, then there will be a closed curve or a 'loop' at x. Figure 3.8 shows the graphic representation of the relation $x \leqslant y$ on the set $\{1, 2, 3, 6\}$.

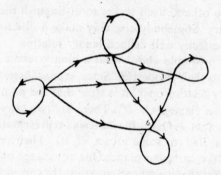

Figure 3.8:   Graphic Representation of a Binary Relation.

The reader will notice that this representation looks very much like a graph, if we take points of X as vertices and the arrows as edges. This is indeed so, except that the edges are now directed in a particular sense. Note also the presence of loops. The structure that we get here is formally called a digraph (a short form of directed graph) and we shall study it later. (See the Epilogue).

Since we are allowing every subset of $X \times X$ to be a relation, obviously there is little that can be proved for all binary relations in general. In order to prove something non-trivial, we have to impose some additional conditions on the relations. We list three such conditions in the following definition.

2.2.   Definition: A binary relation R on a set X is said to be

   (i) **reflexive** if for every $x \in X$, xRx,
  (ii) **symmetric** if for every $x, y \in R$, xRy implies yRx,
 (iii) **transitive** if for every $x, y, z \in X$, xRy and yRz imply xRz.

Before giving examples of relations having these properties, it is instructive to interpret them in terms of the two representations of relations we have studied. Let us first take the graphic representation. Evidently, the first condition, reflexivity, is equivalent to saying that at every point there is a loop. The second condition says that whenever there is an arrow joining one point to another, there is also an opposite arrow joining the second point to the first. (For loops, we do not draw opposite loop because the symmetry condition is always satisfied when $y = x$.) The transitivity condition, in graphic representation, means that whenever you can 'transit' from x to y by an arrow and also from y to z by an arrow then you can also transit directly from x to z by an arrow. (Hence the name).

For the cartesian representation, reflexivity of a relation R on a set X is equivalent to saying that all points on the diagonal are included in R, i.e. $\Delta X \subset R$. The diagonal also plays a crucial role in the interpretation of symmetry. From elementary co-ordinate geometry it follows that for any real numbers x_0, and y_0 the point (y_0, x_0), in the mirror image of the point (x_0, y_0), in the line $x = y$ (i.e. the line segment joining (x_0, y_0) and (y_0, x_0) is bisected at right angles by the line $x = y$). It is then clear that a relation R is symmetric if and only if it is symmetrically situated about the diagonal line. Hence the name. Unfortunately, there is no natural interpretation of a transitivity of a relation in terms of its cartesian representation.

As a simple application of these interpretations let us count the number of reflexive and symmetric relations on a set.

2.3. Proposition: On a set with n elements, there are 2^{n^2-n} reflexive relations, $2^{(n^2+n)/2}$ symmetric relations and $2^{(n^2-n)/2}$ relations which are both reflexive and symmetric.

Proof: Let X be a set with n elements. The diagonal ΔX contains n elements and its complement, $X \times X - \Delta X$ contains $n^2 - n$ elements. Now a reflexive relation on R contains ΔX and the remainder $R - \Delta X$ can be any subset of $X \times X - \Delta X$. So the number of reflexive relations on X is the same as the number of subsets of $X \times X - \Delta X$, which by theorem (2.2.15) is 2^{n^2-n}. Now, $X \times X - \Delta X$, consists of all pairs (x, y) with $x \neq y$ with $x, y \in X$. Group together (x, y). with (y, x). Then $X \times X - \Delta X$ is decomposed into $(n^2 - n)/2$ such groups, each containing two elements which are mirror images of each other in the diagonal. Let Y be a set formed by picking any one element from each of these duplets. (If X is a set of real numbers then the choice can be made uniformly by taking only those pairs (x, y) for which $x < y$. This gives Y as the 'upper' triangular half of the square $X \times X$, with the diagonal removed.) Now, a symmetric relation R on X is determined completely by what points of ΔX are in it and what points of Y are in it (because for all $x \neq y$, R will either contain both (x, y) and (y, x) or neither of the two and precisely one of these elements is in Y). Thus a symmetric relation corresponds to a subset of $\Delta X \cup Y$ while a relation which is both symmetric and reflexive must contain ΔX and so corresponds to a subset of Y. Since

$$|Y| = \frac{n^2 - n}{2}$$

and

$$| \Delta X \cup Y | = | \Delta X | + | Y | = n + \frac{n^2 - n}{2} = \frac{n^2 + n}{2} ,$$

the result follows, again using Theorem (2.2.15). ∎

The proposition above also tells us how to give examples of reflexive and symmetric relations. We take any set X and choose suitable subsets of $X \times X$ as in the proof. But let us give some 'natural' examples. The relation $x \leqslant y$ on the set of real numbers is reflexive, transitive but not symmetric. However, if on the same set we define xRy by $x \leqslant y^2$ then R has none of the three properties. If we define S to be the union of the last two relations, we get a relation on $\mathbf{R}$, which is reflexive but neither symmetric nor transitive. Note that xSy iff $x \leqslant y$ or $x \leqslant y^2$ Then 1S2 but 2$\cancel{S}$1. Also 50S10 and 10S7 but 50$\cancel{S}$7. The relation $x < y$ on $\mathbf{R}$ is transitive but neither reflexive nor symmetric. On any non-empty set X, the empty relation is symmetric and transitive but not reflexive. The conditions for symmetry and transitivity are satisfied vacuously in this case. If $(X; d)$ is a metric space and r is a positive real number then the relation R defined by xRy iff $d(x, y) \leqslant r$ is reflexive and symmetric and but not transitive in general.

As other examples, let X be a set of statements. If $p, q \in X$, define pRq iff the implication statement $p \to q$ is true. Clearly R is reflexive and transitive but not symmetric in general. Let X be the set of all straight lines in plane. If we define parallel lines as those not having any point in common then parallelism is a symmetric relation which is neither reflexive nor transitive. But if we define two lines as parallel whenever they are both perpendicular to a common line, then parallelism is reflexive, symmetric as well as transitive. Relations which enjoy all these three properties are very important and are given a special name.

1.4   Definition: A binary relation which is reflexive, symmetric and transitive is called an **equivalence relation**. A few standard notations for equivalence relations are $\equiv$, $\sim$ or $\cong$.

As remarked just before the definition, parallelism (in the second sense) is an equivalence relation on the set of all straight lines in a plane. But these relations arise very frequently and so we give a few more examples.

1.   Let X be a set of statements. For $p, q \in X$ define pRq iff $p \Leftrightarrow q$ is true (i e. iff both the implication statements $p \Rightarrow q$ and $q \Rightarrow p$ are true). Then R is easily seen to be an equivalence relation. In Chapter 1, Section 4 we called two statements p and q as logically equivalent iff the statement $p \Leftrightarrow q$ is true. In other words, logical equivalence is indeed an equivalence relation. More generally, this is the case with many other types of equivalence. Whenever two

things are considered equivalent in some context, it turns out to be an equivalence relation.

2. Suppose we want to count the number of distinct ways to seat n distinct guests on n (indistinguishable) chairs placed evenly around a circular table. In problems like this the crucial question is to decide which arrangements are to be regarded as not distinct. In the present problem, let S be the set of all possible ways to seat the n guests on the n chairs. Then $|S| = n!$. But in this problem what matters is the relative position of the guests. So we regard two arrangements as equivalent to each other if one can be obtained from the other by asking each guest to move the same number of places to the right, because doing so does not affect the relative position of the guests. We leave it to the reader to verify that this is indeed an equivalence relation on the set S.

3. Let $X = \mathbb{Z}$, the set of all integers. Let n be fixed positive integer. Define a relation R on X by aRb iff the integer $a - b$ is divisible by n, i.e. iff there exists an integer p such that $a - b = np$. Then aRa for all $a \in X$, because $a - a = 0 = n \cdot 0$. If aRb then $a - b = np$ for some integer p. But then $b - a = n \cdot (-p)$ and since $-p$ is also an integer, it follows that bRa. For transitivity, suppose $a - b = np$ and $b - c = nq$, where p, q are integers. Then

$$a - c = (a - b) + (b - c) = n(p + q)$$

and since $p + q$ is an integer we get aRc. Putting it all together, we see that R is an equivalence relation. There is a special name and notation for this relation. If aRb, we write $a \equiv b$ (modulo n) or $a \equiv b$ (mod n) or sometimes simply $a \underset{n}{\equiv} b$ and read 'a is congruent to b modulo n'. For $n = 2$, this is also called the **parity** relation. Instead of saying $a \equiv b$ (mod 2), we also say a and b have the same parity. Note that this happens iff a, b are both even or both odd.

4. Let $\mathcal{F}$ be a collection of sets. If X, Y are two elements of $\mathcal{F}$, then they are themselves sets. We say X and Y are **equipollent** if there exists a bijection from X to Y. Then equipollency is an equivalence relation: reflexivity follows from the fact that the identity function on every set is a bijection, symmetry from the fact that the inverse function of a bijection is a bijection and transitivity from the fact that the composite of two bijections is a bijection. Note that two sets have the same cardinality if and only if they are equipollent.

5. Let $(X; d_1)$ and $(Y; d_2)$ be two metric spaces. We say $(X; d_1)$ is **isometric** (or **congruent**) to $(Y; d_2)$ if there exists a bijection $f: X \to Y$ which preserves distance in the sense that for all $x, y \in X$,

$$d_1(x, y) = d_2(f(x), f(y)).$$

Such a bijection is called an **isometry** or **congruence.** By an argument similar to that in the last example, it follows that congruency is an equivalence relation. In school geometry, congruency is studied for triangles only but as our definition shows, it is a concept applicable for any two geometric figures, or, indeed, two metric spaces. Obviously congruent objects would have the same geometric properties such as areas and volumes.

6.   In example (4) we dealt with sets and defined an equivalence relation for them in the form of a bijection. In example (5), on the other hand, we dealt with sets with some additional structure (namely a distance function). To define the equivalence of two such sets, we required the bijection to preserve this additional structure. This is a typical situation Two mathematical structures of the same type are said to be equivalent if there is a bijection between their underlying sets which preserves (or is compatible with) their respective additional structures. The name for such equivalence can change. In geometric context it is called congruence, for algebraic structures (which we shall study later) it is called an isomorphism, and so on. In all cases it is an equivalence relation. We leave it to the reader to define the equivalence of multi-sets.

7.   As one more example let us consider the set, say X, of all quarternary sequence of length n. If $\bar{x} = x_1 \ldots x_n$ and $\bar{y} = y_1 \ldots y_n \in X$, define $\bar{x} \sim \bar{y}$ iff the 2's and the 3's occurring in them are the same. For example, $012311220330 \sim 112301220331$. It is clear that $\sim$ is an equivalence relation. This relation will be used later.

8.   As the final example, let X, Y be any sets and $f : X \to Y$ any function. Define a relation R on Y by xRy iff $f(x) = f(y)$. Then R is an eqaivalence relation. By varying the sets X and Y and the function f, we can get many examples of equivalence relations this way. For example, the relations of being of equal age, having the same hometown etc. on the set of human beings.

Having given examples of equivalence relations, let us now see what they do to the set on which they are defined. We begin with an important definition.

2.5 Definition: Let R be an equivalence relation on a set X. Let $x \in X$. Then the set $\{y \in X : xRy\}$ is called the **equivalence class** of R determined by x or R-equivalence class determined by x and is denoted by $R[x]$ or simply by $[x]$ when the relation R is understood.

Of course, a similar concept could have been defined for any relation (not just an equivalence relation) on X. But, for equivalence relations it has certain important properties which we now prove.

2.6 Proposition: Let R be an equivalence relation on a set X. Then for every $x \in X$, $x \in [x]$. Also for any x, $y \in X$, either $[x] = [y]$ or $[x] \cap [y] = \phi$.

Proof: The first assertion follows from reflexivity of R. For the second assertion, suppose x, $y \in X$ and $[x] \cap [y] \neq \phi$. Then we claim that $[x] = [y]$. Since $[x] \cap [y] \neq \phi$, there exists $z \in X$ such that $z \in [x]$ and $z \in [y]$. By definition, this means that xRz and yRz. By symmetry of R, yRz implies zRy. By transitivity of R, xRz and zRy give xRy. By symmetry of R, yRx. Now suppose $w \in [x]$. Then xRw. This, coupled with yRx gives yRw by transitivity. This means $w \in [y]$. Hence $[x] \subset [y]$. Similarly $[y] \subset [x]$. So $[x] = [y]$ as was to be shown. ▮

Because of this proposition, it follows that every element of X belongs to precisely one equivalence class of R. Let now $\mathscr{D}$ be the collection of all R-equivalence classes. Then every member of $\mathscr{D}$ is non-empty; every two members are mutually disjoint and the set X is the union of members of $\mathscr{D}$. From Definition (2.3.6), $\mathscr{D}$ is a decomposition of X. Conversely given any decomposition $\mathscr{D}$ we define a relation R on X by letting xRy iff x and y belong to the same member of $\mathscr{D}$. Then R is easily seen to be an equivalence relation on X and the equivalence classes of R are precisely the members of $\mathscr{D}$. We have thus proved the following theorem.

2.7 Theorem: There is a one-to-one correspondence between the set of equivalence relations on a set X and the set of all decompositions of X. This correspondence is obtained by assigning to each equivalence relation, the collection of its equivalence classes. ▮

As an illustration of the decomposition induced by an equivalence relation, let X be the set of all students in a college. If we define two students to be related iff they have the same height then we get an equivalence relation (see Example (8) above) on X. In the corresponding decomposition of X, all students of the same height will be grouped together. Students of different heights will be in different groups. In other words, we have classified the students according to their heights. Similarly the equivalence relation of parallelism classifies the lines in a plane according to their directions (we do not distinguish between opposite directions). Generally, whenever we classify the elements of a set according to some criterion it amounts to defining a suitable equivalence relation on it. In Example (7) above, the equivalence relation classifies the quarternary sequences according to the pattern of 2's and 3's appearing in them.

The concept of an equivalence relation and the classification it induces is useful in various ways. First, it serves to clarify ideas when we wish to regard certain elements of a set as indistinguishable from each other. For example, the problem considered in Example (2) above, really amounts to counting the numbers of equivalence classes under the equivalence relation defined there, because the relative position of the guests in two arrangements is the same iff they are equivalent. Obviously each class has n elements and so the answer is $n!/n = (n-1)!$.

As another application, let us calculate the number of k-ary sequences of length n in which 1 occurs an even number of times. In proposition (2.3.3) we answered a similar question for binary sequences and obtained 2^{n-1} as the answer, that is half the total number of binary sequences of length n for $n \geqslant 1$. (For $n = 0$, there is only the null or the empty sequence and by convention we regard that 1 occurs in it an even number of times. This exceptional case will also figure in the proof below.)

2.8. Theorem. The number of k-ary sequences of length n in which 1 occurs an even number of times is $[k^n + (k-2)^n]/2$.

Proof: We have already proved the result for $k = 2$. Interestingly it also holds for $k = 1$. For $k = 1$, the sequence will consist only of 1's and then depending upon whether n is even or odd, this sequence will be included or not included.

Assume now $k > 2$. The exceptional case $n = 0$ is covered by the fact that the null sequence is to be regarded as having an even number of 1's. So assume now $n > 0$. Now let X be the set all k-ary sequences of length n. We divide X into two subsets first. Let Y consist of those sequences which do not contain any 0 or 1 and let Z consist of the remaining sequences. Then a sequence in Y is a sequence of the remaining $k-2$ symbols. So $|Y| = (k-2)^n$ and hence $|Z| = k^n - (k-2)^n$. Now every sequence in Y is to be included in our count because it contains 1 an even number of times (namely 0 times). This is also like the exceptional case. Let us now count how many sequences in Z have an even number of 1's. For this, we classify them according to the pattern of the symbols $2, 3, ..., k-1$, (cf. Example (7) above where the case $k = 4$ was considered). This gives an equivalence relation on Z. Let the corresponding equivalence classes be $Z_1, Z_2, ..., Z_p$. Then $\sum\limits_{i=1}^{p} |Z_i| = |Z|$. We do not know p nor what each $|Z_i|$ is. But we can do without finding it out. Let us find how many sequences in a typical equivalence class, say Z_l, have an even number of 1's. By definition, Z_l consist of all k-ary sequences which have a common pattern of the symbols 2, 3, ..., $k-1$. Let r be the number of places filled by these symbols. Then $r < n$, for if $r = n$ then the sequence would be in Y and not in Z. Thus all sequences in Z_l have these r terms in common. The remaining $n-r$ terms have to be either 0 or 1. Hence an element of Z_l corresponds uniquely to a binary sequence of length $n-r$. So $|Z_l| = 2^{n-r}$. (For example let $k = 4$ and $n = 10$ and Z_l consist of all sequences of the form $-23--2-33-$ where the blanks can be filled with 0's and 1's. Then $r = 5$ and $|Z_l| = 2^{10-5} = 2^5$.) Now clearly half of these sequences have an even number of 1's, by Proposition

(2.3.3) This holds for all $i = 1, 2, \ldots, p$. So the number of sequences in Z having an even number of 1's is

$$\sum_{i=1}^{p} \tfrac{1}{2} \mid Z_i \mid = \tfrac{1}{2} \sum_{i=1}^{p} \mid Z_i \mid = \tfrac{1}{2} \mid Z \mid = \tfrac{1}{2} [k^n - (k-2)^n].$$

Since all sequences in Y are also to be included, we get the total number of sequences with an even number of 1's as $[k^n + (k-2)^n]/2$. ∎

The equivalence classes of Example (3) above deserve to be considered in detail. There, the relation was that aRb iff $a - b$ is divisible by n. It follows that for any $a \in \mathbb{Z}$, the equivalence class $[a]$ consists of all integers of the form $a + kn$ where k is any integer. For example, for $n = 12$, the equivalence class $[17]$ is the set $\{\ldots, -31, -19, -7, 5, 17, 29, 41, \ldots\}$. Note that the difference between every two distinct elements of an equivalence class is at least n. It follows that no equivalence class can contain more than one integer from 0 to $n-1$. On the other hand, by the euclidean algorithm (see problem (2.2.10)), given any integer a we can find an integer r from 0 to $n-1$ such that $a \equiv r \pmod{n}$, i.e. $[a] = [r]$. Thus we see that every equivalence class contains precisely one integer from 0 to $n-1$. So in all there are n equivalence classes. They are called the **congruence classes modulo** n or **residue classes modulo** n or simply the **residues modulo** n, the last two names coming from the fact that the integer r above is called the residue left upon division by n. The set of residue classes modulo n is often denoted by $\mathbb{Z}_n$. We shall have many occasions to consider it later.

The concept of an equivalence class is sometimes put to a theoretical use, namely to give precise definitions of certain terms. Take once again the example where X is the set of all students in a college and R is the relation defined by xRy iff x and y have the same height. This relation classifies the students according to their heights. Now suppose we are given this classification beforehand without knowing that it has been obtained on the basis of height. If an outsider takes a close look at these classes, he would notice that all students in the same class have the same height and no two students in different groups have the same height. In other words, the height of a student is characterised by the equivalence class to which he belongs.

Now we come to the crucial points. Can we *define* the height of a student as the equivalence class to which he belongs? According to the 'definition trick' we often do this sort of a thing in mathematics. It does sound extremely artificial to tell the height of a student not as so many centimeters but as a certain equivalence class! But actually it is not so far removed from practice. When we say 'John is 180 centimeters tall', we have in mind some object of height 180 cms (which may be a ruler, a tree, or, most likely, some other person of this height) and what we really convey is that John is in the same equivalence class as this object. Indeed, this is how a layman often tells height. So, after all, it is not all that arbitrary to define height as an equivalence class under a suitable equivalence relation. As another example, we may define the direction of a straight line in a plane as its equivalence class under the relation of parallelism. We shall not give many definitions of this type because we shall not be very particular to give precise definitions where they tend to be clumsy. We have discussed this point only to advise the reader not to get perplexed when he sees such definitions in reading other books on mathematics. For example, he will often find the cardinal number of a set X, defined formally as the class of all sets which are equipollent to X.

The correspondence between an equivalence relation and a decomposition of a set, given by Theorem 2.7, allows us to visualise an equivalence relation. Also any concept about decomposition can be translated to a corresponding concept about equivalence relations and vice versa. We discuss one such concept.

2.9 Definition: Let $\mathcal{D}$ and $\mathcal{E}$ be two decompositions of a set X. Then $\mathcal{D}$ is said to be **coarser** than $\mathcal{E}$ (or $\mathcal{E}$ is said to be **finer than** $\mathcal{D}$, or a **refinement** of $\mathcal{D}$) if every member of $\mathcal{E}$ is contained in some member of $\mathcal{D}$.

Let us denote the decompositions $\mathcal{D}$ and $\mathcal{E}$ respectively by

$$\mathcal{D} = \{D_1, D_2,..., D_m\}$$

and

$$\mathcal{E} = \{E_1, E_2,..., E_n\}.$$

If $\mathcal{E}$ is a refinement of $\mathcal{D}$ then for every E_i there exists some D_j such that $E_i \subset D_j$. This D_j is unique since the D's are mutually disjoint. It is clear that each D_j is the union of those E_i's which are contained in it. In other words, $\mathcal{E}$ may be thought of as obtained from $\mathcal{D}$ by further decomposing each member of $\mathcal{D}$. A simple example is to let X be some country, the D_j's be its states and the E_i's the districts in the various states. We illustrate this for a hypothetical country in Figure 3.9, where the state boundaries are shown by thick lines and the district boundaries by dotted lines.

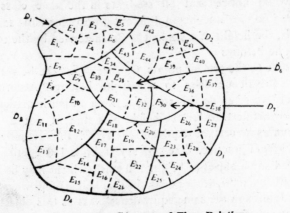

Figure 3.9: Coarser and Finer Relations.

(We may ignore points on these boundaries. We may make some arbitrary convention by which such points are considered to lie in one of the regions on whose boundaries they lie.)

Now let R and S be the equivalence relations corresponding to the decompositions $\mathscr{D}$ and $\mathscr{E}$ respectively. What relationship between R and S corresponds to the fact that $\mathscr{D}$ is coarser than $\mathscr{E}$? The name is provided in the following definition.

2.10 Definition: Let R and S be two binary relations on a set X. Then S is said to be **stronger** than R (or R is said to be **weaker** than S) if for all $x, y \in X$, xSy implies xRy. (In the terminology of Chapter 1, Section 4, this says that the statement 'xSy' is stronger than the statement 'xRy'. Hence the name. As subsets of $X \times X$, the condition simply means that $S \subset R$.)

We now state the relationship between the concepts defined by the last two definitions. The proof is extremely simple and is left as an exercise, for it will give the reader a chance to review many of the concepts defined earlier.

2.11 Proposition: Let R and S be two equivalence relations on a set X and let $\mathscr{D}$ and $\mathscr{E}$ be, respectively, the corresponding decompositions of X. Then S is stronger than R if and only if $\mathscr{E}$ is finer than $\mathscr{D}$. ∎

Because of this proposition, a stronger relation is sometimes called a finer relation. If we go back to the example of the states and the districts above then the truth of the proposition above is obvious, because given two persons, it is certainly stronger to say that they are from the same district than to say that they are from the same state.

An especially instructive reformulation of this proposition arises when the relation R is induced by some function on X in the manner of Example 8 above. The result is often expressed in a different terminology. So first we introduce this terminology.

2.12 Definition: Let S be an equivalence relation on a set X and let $\mathscr{E}$ be the corresponding decomposition of the set X. Then $\mathscr{E}$ is often denoted by X/S and is called the **quotient set** of X by the relation S. The function p: $X \to X/S$ defined by $p(x) = [x]$, the equivalence class of S containing x, is called the **quotient function** or the **projection function**.

2.13 Definition: Let S be an equivalence relation on a set X and suppose Y is some other set. Then a function $f: X \to Y$ is said to be **compatible** with S or to **respect** S if for all $x, y \in X$, xSy implies $f(x) = f(y)$.

For example, let X be the set of all persons in a country and let S be the relation, xSy iff x and y live in the same district. (The equivalence classes induced by S are precisely the sets of persons living in the various districts). If we define $f(x)$ to be the state the person x lives in then f respects the relation S because two persons living in the same district obviously live in the same state. But if we define $f(x)$ = height of x then f does not respect S (except in the unlikely event that all the persons in each district are of equal height). As another example, let X be the set of all straight lines in the plane which are not parallel to some fixed line, say, the x-axis. Let S be the equivalence relation of parallelism, discussed above. For a line L, let $f(L)$ be the acute angle between L and the x-axis. Then f is compatible with parallelism. But if we let $f(L)$ be the point at which L meets the x-axis then f is not compatible with parallelism.

Let us now consider the reformulation of proposition 2.11.

2.14 Proposition: Let S be an equivalence relation on a set X and p: $X \to X/S$ the corresponding quotient function. Also let $f : X \to Y$ be any function where Y is some set and let R be the equivalence relation on X induced by f, i.e. xRy iff $f(x) = f(y)$.

Then the following conditions are equivalent:

(i) S is stronger than R

(ii)   the function f respects the relation S

(iii) there exists a function $g : X/S \to Y$ such that $g \circ p = f$ (this is also expressed by saying that f **factors through** p). Moreover, such g is unique.

Proof : We prove the equivalence in a cyclic manner. (i) $\Rightarrow$ (ii). Assume S is stronger than R. To show that f respects S, let $x, y \in X$ with xSy. Then xRy because S is stronger than R. But then $f(x) = f(y)$ by the definition of R. (ii) $\Rightarrow$ (iii). Assume that f respects S. Let us try to define a function g from X/S to Y. A typical element of X/S is of the form $[x]$, where $x \in X$ and $[x]$ is the unique equivalence class of S containing x. It is tempting to define $g([x])$ simply as $f(x)$, because $f(x)$ is an element of y. But there is a catch here. It is quite possible that S-equivalence class $[x]$ is the same as the S-equivalence class $[y]$ for some $y \neq x$. Now, $g([y]) = f(y)$. But since $[x] = [y]$, it is necessary to have $g([x]) = g([y])$, or equivalently, $f(x) = f(y)$. How do we ensure it? A question like this is quite important and has to be scrupulously answered every time it arises. When we denote an equivalence class by $[x]$, and define g in terms of the value of f at x, we are effectively choosing x as a representative of the equivalence class. If the value of g would change just because we choose a different representative for the same equivalence class then the function g will not be well-defined because it will be really speaking a function of the particular representative and not of the equivalence class as such.

The situation is analogous to a spokesman of a political party, giving the party's reaction to some event. Different parties may have different reactions. But if two spokesmen of the same party issue two different reactions, they can hardly be said to be giving the party's reaction, they are only giving their personal reactions. Coming back to our original problem, if x and y are in the same equivalence class under S, then by (ii), $f(x) = f(y)$. So g is well-defined in the sence that its value on a particular class depends only on that class and not on a particular representative. It remains to verify that the composite $g \circ p$ equals f. But this is immediate from the very definition of g, because for every $x \in X$, $p(x) = [x]$ and $g([x]) = f(x)$. For uniqueness, suppose $h : X/S \to Y$ is such that $h \circ p = f$. Then $h[x] = h(p(x)) = f(x) = g([x])$, for all $[x] \in X/S$, showing that $g = h$.

(iii) $\Rightarrow$ (i) Assume that f factors through p, i.e. there exists a function $g : X/S \to Y$ such that $g \circ p = f$. We want to show that S is stronger than R. So let $x, y \in X$ and suppose xSy. Then by very definition of p,

$$p(x) = p(y)$$

since both equal the same equivalence class under S. But then

$$g(p(x)) = g(p(y)),$$

giving

$$f(x) = f(y).$$

Recalling the definition of R, this gives xRy as was to be proved. ∎

This proposition is really not profound. But the reader may find it a bit too abstract. To make it more visual, we paraphrase the statement (iii) in it diagrammatically in Figure 3.10. In (a), we show a triangle at whose vertices are the sets X, X/S and Y. Two of the 'sides' of this triangle correspond to the functions p and f.

    (a)   (b)

Figure 3.10: Factoring through a Projection Function.

To complete the triangle we need a function g from X/S to Y as in (b). But merely having some function g would not do. We want the function g in such a way that in (b), when we go from X to Y we must get the same result whether we go do directly (by f) or through X/S (i.e. by the composite function $g \circ p$). A triangle with this property is said to be a **commutative triangle** and this property is expressed by putting a circular arrow inside it. A commutative square of functions is defined analogously. The reader will find it very helpful to visualise equalities involving composites of functions in term of appropriate commutative diagrams.

To illustrate Proposition 2.14 with concrete example, let X be the set of all students in a residential school and let Y be the set of all hostels in which these students are accommodated. Define $f: X \to Y$ by $f(x) =$ the hostel in which X is accommodated, for $x \in X$. Define a relation S on X by xSy iff x and y are studying in the same year. Then S in an equivalence relation and each equivalence class consists of all students in a particular year. The quotient set X/S consists of all the batches of students. Now to say that the function f respects the relation S amounts to saying that all students studying in the same year are accommodated in he same hostel. When this is the case, the concept of 'the hostel of a particular batch' is well-defined and correspondents to the function g.

We conclude this section with a discussion of the concept of the equivalence relation generated by a given relation R on a set X. Let S be the smallest equivalence relation on X which contains R. The relation R need not be reflexive, symmetric or transitive. But S, being an equivalence relation has to have all these three properties. So S has to be obtained from R by adding some pairs of points. The extension from R to S can be done very systematically in three steps. They can be conveniently visualised in terms of the graphic representation of binary relations. First we let

$$R_1 = R \cup \Delta X;$$

i.e. R_1 is obtained by adding to R all pairs of the form (x, x) for $x \in X$, if necessary. Clearly R_1 is a reflexive relation. However R_1 need not be symmetric. So we let

$$R_2 = R_1 \cup R_1^{-1}$$

where R_1^{-1} is called the **inverse relation** of R_1, and is defined by

$$R_1^{-1} = \{(y, x) \in X \times X : (x, y) \in R_1\}.$$

The name is suggestive because, in the graphic representation, R_1^{-1} is obtained from R_1 by reversing the arrows. Clearly R_2 is symmetric. It is also reflexive because it contains ΔX. As the last step, we now extend R_2 to an equivalence relation. If R_2 is transitive, then it is itself the desired extension. If R_2 is not transitive then there exist x, y, z such that

$$(x, y) \in R_2, (y, z) \in R_2$$

but $(x, z) \notin R_2$. So we must add (x, z) to R_2. Graphically, whenever there are two arrows in succession, we put a direct arrow. But even if we do this for all pairs of arrows in succession, the resulting relation may not still be transitive. There might be four elements x, y, z, w such that (x, y), (y, z) and (z, w) are in R_2. Then because of our construction, (x, z) will be added. But in order to have transitivity we would also have to add (x, w). More generally, whenever $x_1, x_2, \ldots x_n$ are elements of X such that

$$(x_1, x_2), (x_2, x_3), \ldots, (x_{n-1}, x_n) \in R_2$$

then we would have to include (x_1, x_n) in the extension. The following proposition shows that if we do this for all possible finite sequences, then we do get an equivalence relation.

2.15 Proposition: Let R_2 be a reflexive and symmetric relation on a set X. Let $S = R_2 \cup \{(x_1, x_n)$: there exist $x_2, \ldots, x_{n-1} \in X$ such that $(x_i, x_{i+1}) \in R_2$ for $i = 1, 2, \ldots, n-1\}$. Then S is an equivalence relation on X. Moreover, it is the smallest equivalence relation on X containing R_2 in the sense that if T is an equivalence relation on X containing R_2 then $S \subset T$.

Proof: Since R_2 is reflexive and S contains R_2, S is reflexive. For symmetry suppose $(x, y) \in S$. If $(x, y) \in R_2$ then $(y, x) \in R_2$ by symmetry of R_2 and hence $(y, x) \in S$. If $(x, y) \notin R_2$ then (x, y) is of the form (x_1, x_n) where n is some positive integer and there exist $x_2, x_3, \ldots, x_{n-1} \in X$ such that for each $i = 1, 2, \ldots, n-1$,

$$(x_1, x_{i+1}) \in R.$$

But then, by symmetry of R_2, (x_{i+1}, x_i) also belongs to R_2. Considering the sequence $x_n, x_{n-1}, \ldots, x_2, x_1$ it now follows that $(x_n, x_1) \in S$, i.e. $(y, x) \in S$. Thus we see that S is symmetric.

It only remains to prove that S is transitive. For this let (x, y) and $(y, z) \in S$. Then there exist integers m and n ($\geqslant 2$) and

$$x_1, \ldots, x_m, y_1, \ldots, y_n \in X$$

such that

$$x_1 = x, x_m = y = y_1, y_n = z, (x_i, x_{i+1}) \in R_2$$

for all $i = 1, 2, \ldots, m-1$ and $(y_j, y_{j+1}) \in R_2$ for all $j = 1, 2, \ldots, n-1$. Now let $z_i = x_i$ for all $i = 1, \ldots, m$ and $z_i = y_{i-m}$ for $i = m+1, \ldots, m+n$. Then $(z_i, z_{i+1}) \in R_2$ for $i = 1, 2, \ldots, m+n-1$. Also $z_1 = x$ and

$$z_{m+n} = y_n = z. \text{ So } (x, z) \in S.$$

(Graphically we have concatenated the chain of arrows from x to y with the one from y to z.) Thus S is transitive and hence an equivalence relation. If T is any other equivalence relation on X containing R_2 then T must also contain, by repeated applications of transitivity, all pairs of the form (x_1, x_n) where there exist $x_2, \ldots, x_{n-1} \in X$ such that $(x_i, x_{i+1}) \in R_2$ for $i = 1, 2, \ldots, n-1$. But this means T must contain S. Thus S is the smallest equivalence relation on X containing R_2. ∎

2.16   Definition: The equivalence relation S obtained from R_2 (and hence ultimately from R) is said to be **generated** by R.

In Figure 3.11 we show graphically the three steps of the transition from R to S. For convenience, instead of drawing separate curves for two opposite arrows, they are indicated by putting two arrowheads in opposite directions on the same curve. (For loops it is unnecessary to do so as

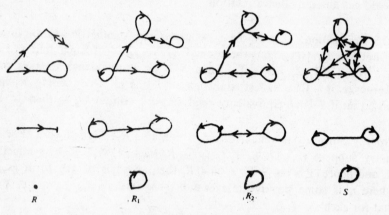

Fig. 3.11:   **Generating an Equivalence Relation.**

observed earlier.) The process described above is reminiscent of the spread of epidemics. As soon as something gets contaminated, so does everything that is in touch with it and this process continues till we get well-isolated 'islands' with no more room left for spreading. Generating from an arbitrary subset, a subset of a particular type is a general process and we shall see many instances of it. For the moment we remark that the equivalence relation in Example (2) above is actually generated by a much smaller relation. Let us call one arrangement of guests as adjacent to another if the second arrangement is obtained from the first by asking each guest to move to the seat immediately on his right. The adjacency relation so defined on the set S is neither reflexive, nor symmetric nor transitive. But it generates the equivalence relation in (2).

Exercises

2.1   For the relationships (i) to (ix) for the set of persons in a town, see which relations are reflexive, symmetric and transitive.

2.2   Prove that the set $S = \{(1,39), (2,105), (3,87)\}$ is the same as the set

$$\{(x, y) \in \mathbb{N} \times \mathbb{N}: 1 \leqslant x \leqslant 3, y = -42x^2 + 192x - 111\}.$$

(In this formulation the relation S appears to be less arbitrary, because it is defined 'according to some formula'. But this is really an illusion for two reasons. For one, the formula itself is very arbitrary. Secondly, given any arbitrary numbers $a_1, a_2, ..., a_r$ we can always find a polynomial $p(x)$ such that $p(1) = a_1$, $p(2) = a_2, ..., p(r) = a_r$ and thereby give the impression that the numbers $a_1, a_2, ..., a_r$ follow some regular pattern. The method for finding this polynomial is due to Lagrange.)

2·3　Let R be a binary relation on a set X and let R^{-1} be its inverse relation, i.e.

$$R^{-1} = \{(y, x) \in X \times X, x, y \in R\}.$$

What is the interpretation of R^{-1} in terms of the cartesian representation of R? Prove that the relations $R \cup R^{-1}$ and $R \cup R^{-1}$ are symmetric for any R. Prove that R is symmetric if and only if $R = R^{-1}$. Show also that $(R^{-1})^{-1} = R$.

2.4　Let X be a set of cardinality n. How many relations on X are neither reflexive nor symmetric? Also express the number of equivalence relations on X as a sum of Stirling numbers.

2.5　Let R be a binary relation on a set X and let Y be a subset of X. Then $Y \times Y$ is a subset of $X \times X$ and $R \cap (Y \times Y)$ is a subset of $Y \times Y$ and hence a binary relation on the set Y. It is called the **restriction** of R to Y and is denoted by R/Y. Prove that if R is reflexive, symmetric or transitive then R/Y also has the corresponding properties. (The relation R/Y is also said to be **induced** on Y by R. This is a general construction. An additional structure on a set induces a structure of a similar type on a subset of the original set. Sometimes, the subset has to satisfy certain restrictions. Some of the properties of the original structure hold good for such 'substructures' also. Such properties are called **hereditary**. Thus the properties of reflexivity, symmetry, and transitivity are hereditary. But not all concepts pass so nicely to substructures as the following exercise shows.)

2.6　Show by an example that if R is a relation on a set X, S is the equivalence relation on X generated by R and Y is subset of X then S/Y need not be the same as the equivalence relation on Y generated by R/Y.

2.7　What are the coarsest and the finest decompositions of a given set X? What are the corresponding equivalence relations?

2.8　Let $X = \mathbb{Z}$, the set of all integers, n some fixed positive integer and let

$$R = \{(x, y) \in X \times X : y = x + n\}.$$

Prove that the equivalence relation on X generated by R is precisely the relation of congruence modulo n.

2.9 Let R be a transitive relation on a set X. By induction, prove the following statement for all $n \geqslant 2$:

$$\text{'For any } x_1, x_2,..., x_n \in X, \text{ if}$$

$$(x_i, x_{i+1}) \in R \text{ for all } i = 1, 2..., n-1$$

then

$$(x_1, x_n) \in X.'$$

(This result was actually used in the proof of Proposition 2.15.)

2.10 Let R_1 and R_2 be two equivalence relations on a set X. Prove that $R_1 \cap R_2$ is also an equivalence relation on X. Generalise to the intersection of more than two equivalence relations. Show by an example, however, that the union of two equivalence relations need not be an equivalence relation.

*2.11 Let R be a relation on a set X. Prove that the equivalence relation, S, on X generated by R coincides with the intersection of all equivalence relations on X containing R. (This gives an 'external' view of S. However, as a method of constructing S, it is not very useful because it is impracticable to look at all possible equivalence relations containing R. The construction we have given proceeds 'internally' from R to S.)

2.12 If R_1 and R_2 are equivalence relations on a set X, how is the decomposition corresponding to $R_1 \cap R_2$ related to the decompositions correspoding to R_1 and R_2?

2.13 Two decompositions $\mathscr{D}$ and $\mathscr{E}$ of a set X are said to be **mutually orthogonal** if for all $D \in \mathscr{D}$ and $E \in \mathscr{E}$, $D \cap E$ contains at most one point. (For example, if X is a rectangular array of dots then the decomposition of X into its rows is orthogonal to the decomposition of X into its columns.) How does this concept translate for the corresponding equivalence relations?

2.14 Let R be a symmetric and transitive relation on a set X. Suppose for every $x \in X$ there is some $y \in X$ such that xRy. Prove that R is also reflexive and hence an equivalence relation on X. (In other words every symmetric and transitive relation is an equivalence relation, provided it 'touches' all points.)

2.15 Let R be a symmetric and transitive relation on a set X. Prove that there exists a subset Y of X such that $R \subset Y \times Y$ and R, regarded as a relation on Y, is an equivalence relation.

2.16 Let $f : X \to Y$ be a function and T a relation on the set Y. Define a relation R on X by

$$xRy \text{ iff } (f(x), f(y)) \in T \text{ for } x, y \in X.$$

Prove that if T is reflexive, symmetric or transitive then R also has the corresponding properties. How does this generalise Example 8 of equivalence relations?

2.17   (a) Prove that for positive integers n, k with $k \geqslant 2$,

$$\sum_{r=0}^{\infty} \binom{n}{2r} (k-1)^{n-2r} = \frac{k^n + (k-2)^n}{2}$$

(b) Prove that the number of k-ary sequences of length n in which both 0 and 1 occur an even number of times each is

$$\frac{1}{4}\left[k^n + 2(k-2)^n + (k-4)^n \right] \text{ for } k \geqslant 2.$$

2.18   Find the number of k-ary sequence length n in which 1 appears an even number of times and 0 appears an odd number of times.

1.19   Suppose S and T are equivalence relations on sets X and Y respectively and $f : X \to Y$ is a function which is compatible with them in the sense that for all $x, y \in X$, xSy implies

$$f(x) \ T \ f(y).$$

Let

$$p : X \to X/S \text{ and } q : Y \to Y/T$$

be the projection functions. Prove that there exists a unique function $g : X/S \to Y/T$ such that $g \circ p = q \circ f$. Express this result in terms of commutative diagrams. Prove that the converse also holds. How does this generalise Proposition (2.14)?

2.20   Let R and S be equivalence relations on the sets X and Y respectively. Let $Z = X \times Y$. Define a relation R on Z by letting

$$(x_1, y_1) \ T \ (x_2, y_2)$$

iff $x_1 R x_2$ and $y_1 S y_2$. Prove that T is an equivalence relation on Z. What do the equivalence classes look like?

2.21   In Exercise 2.16 above, suppose f is a bijection and T is an equivalence relation. Then R is an equivalence relation on X. How are the equivalence classes of R are related to those of T?

Notes and Guide to literature

The material in this section is elementary and standard. Some authors consider, more generally, a relation from one set X to another set Y as a subset of $X \times Y$. But any such relation is also a subset of $Z \times Z$, where $Z = X \cup Y$. Therefore, there is really no loss to confine to the case of a relation on a set.

3. Order Relations

An equivalence relation on a set serves to classify its elements into mutually disjoint subsets according to some common property. In the present section we consider a relation which ranks them according to some criterion for comparison. Such relations arise frequently in practice. Words are listed in a dictionary in the alphabetical order, the winners of a competition are ranked in the order of merit, the guests at a dinner are seated according to the order of their importance, the letters in a file are arranged in a chronological order and so on. In all these examples, the objects are put in a sequential manner from the first to the last. But this may not always be possible. For example, in a competition, several qualities may matter and a prize may have to be divided between two contestants not because they are equal but because one is superior to the other in some respects and inferior in some other respects. Similarly, if some of the letters in the file are undated then we may not know where to place them. From their contents we may be able to tell, in some cases, that they came before or after certain other letters. But such a comparison may not always be possible, with the result that certain pairs of letters will have to be left as incomparable.

We now formally define an order relation on a set X.

3.1 Definition: A binary relation R on a set X is called a **partial order** (or **partial ordering**) if,

(i) R is reflexive,
(ii) R is transitive and
(iii) R is **anti-symmetric**, i.e. for all $x, y \in X$ xRy and yRx implies $x = y$.

Standard notations for partial order relations are $\leqslant$, $\propto$ (which are generally read as 'is less than or equal to' or 'precedes') or $\geqslant$, ∞ (which are read as 'is greater than or equal to' or 'follows'). Note that the inverse relation of an order relation (see Exercise 2.3) is also an order relation and is called the **reverse order**. If a partial order is denoted by $\leqslant$, $\propto$ etc. then its reverse order is denoted almost exclusively by $\geqslant$, ∞ etc.

We already cited some examples of partial orders. Before giving other examples, it is worthwhile to comment on the definition, especially because different authors tend to adopt different conventions. Certainly, transitivity is the very essence of ordering and any definition of it ought to include it. Whether reflexivity should be included or not is largely a matter of taste, depending upon whether in an inequality one wants to include a possible equality or not. For those who do not, there is the following definition of what is known as strict order.

3.2 Definition: A binary relation S on a set X is called a **strict partial order**, if,

(i) S is **irreflexive**, i.e. every $x \in X$, $(x, x) \notin S$.
(ii) S is transitive
(iii) S is **asymmetric**, i.e for every $x, y \in X$, $(x, y) \in S$ implies $(y, x) \notin S$.

There is an obvious relationship between the two concepts which is stated in the next proposition. The proof is left to the reader.

3.3 Proposition: If R is a partial order on a set X then $R - \Delta X$ is a strict partial order on X. If S is a strict partial order on a set X then $S \cup \Delta X$ is a partial order. (Here ΔX is the diagonal on X.) ▨

Thus we can convert partial orders to strict partial orders and vice versa. If partial orders are denoted by $\leqslant, \propto, \subseteq$, etc. then the corresponding strict orders are denoted by $<, \propto, \subset$ etc. If $x < y$, we say x is **strictly** (or **properly**) less than y. Sometimes partial orders are themselves denoted by symbols like $<, \propto$ and $\subset$ and then the corresponding strict orders are denoted by $\underset{\neq}{<}, \underset{\neq}{\propto}$ and $\underset{\neq}{\subset}$.

The presence of antisymmetry in the definition of a partial order (or its counterpart, asymmetry, in the definition of a strict partial order) calls for a comment. Some authors do not require it. Let us see what would happen without it. Let X be the set of all students in a class and for $x \in X$ let $h(x)$ be the height of x. Define $\leqslant$ on X by $x \leqslant y$ iff $h(x) \leqslant h(y)$. Clearly $\leqslant$ is reflexive and transitive. But it is not anti-symmetric if there exist two distinct students, say a and b, of equal heights. This does not sound very strange. But look at the corresponding, strict order, which is defined by $x < y$ iff $x \leqslant y$ and $x \neq y$. Then $a < b$ and $b < a$. This does sound strange because we are apt to interpret $a < b$ as a is shorter than b (although this is not the correct interpretation).

Thus we see that if we do not have antisymmetry then the corresponding strict order would not conform to its natural interpretation. We would always have to be on the guard while handling it. On the other hand, if we include antisymmetry in the definition of a partial order, then we would have to sacrifice so many naturally occurring examples. One such example is the set of students compared according to their heights. Another would be the file of letters, to be arranged chronologically, with two letters carrying the same dates.

In other words, although antisymmetry is desirable, its inclusion has a price. Fortunately, the difficulty is not a serious one. In many cases it can be supposed not to arise at all. For example, in the case of the heights of the students, if we measure them very minutely then it is extremely unlikely that two distinct students would have exactly the same height. There will always be some difference from the point of view of probability and even an iota of difference would ensure antisymmetry. However, this solution is not a sound one from our point of view because it is based on the assumption that the height is a continuous variable and such an assumption is contrary to the very spirit of discrete mathematics. So we have to accept the possibility that two distinct students may have the same height. Even then the situation is not so hopeless. After all, we are comparing the heights and not the students per se. Let us group together students of the same heights (which amounts to defining an equivalence relation on the set of the students). Now the concept of the height of a group is well-defined and if we compare the groups in terms of it then antisymmetry does hold, because no two distinct groups can have the same height. More generally, we have the following result.

3.4 Proposition: Let X be a set and $\leqslant$ a binary relation on X which is reflexive and transitive. Define a binary relation on X by xRy if $x \leqslant y$ and $y \leqslant x$. Then R is an equivalence relation on X. Let X/R be the quotient set and $p: X \to X/R$ the projection function. Then there is a partial order $\underline{\propto}$ on X/R such that for all $x, y \in X$, $x \leqslant y$ implies $p(x) \underline{\propto} p(y)$.

Proof: The verification that R is an equivalence relation on X is simple and left as an exercise. Let now X/R consist of the equivalence classes under R. (Note that if $\leqslant$ is itself antisymmetric to begin with then each equivalence class contains only one element and we may as well identify X/R with X.) Elements of X/R are of the form $[x]$ for $x \in X$, where [ ] denotes the R-equivalence class. If $[x]$, $[y]$ are two elements of X/R, we define $[x] \underline{\propto} [y]$ iff $x \leqslant y$ in X. To ensure that this gives a well-defined binary relation on X, we must verify that it is independent of the choice of representatives of the equivalence classes (see the comment in the proof of Proposition 2.14). In other words, suppose $[x] = [z]$ and $[y] = [w]$. Then we must show that $[x] \leqslant [y]$ iff $[\dot{z}] \leqslant [w]$. So suppose $[x] \underline{\propto} [y]$, i.e. $x \leqslant y$. Since $[x] = [z]$ we have $z \leqslant x$ (and also $x \leqslant z$) and similarly $[y] = [w]$ gives $y \leqslant w$ (and $w \leqslant y$). Hence by transitivity of $\leqslant$, we get $z \leqslant w$, i.e. $[z] \underline{\propto} [w]$. Similarly $[z] \underline{\propto} [w]$ implies that $[x] \underline{\propto} [y]$. Thus $\underline{\propto}$ is a well-defined binary relation on X/R. By its very definition, $p(x) \underline{\propto} p(y)$ whenever $x \leqslant y$ in X, because $p(x)$, $p(y)$ are simply $[x]$ and $[y]$ respectively. It only remains to verify that $\underline{\propto}$ is a partial order on X/R. Reflexivity and transitivity of $\underline{\propto}$ follow from the corresponding properties of $\leqslant$. For antisymmetry, suppose $[x] \underline{\propto} [y]$ and $[y] \underline{\propto} [x]$. Then $x \leqslant y$ and $y \leqslant x$ which means xRy and hence $[x] = [y]$ (even though x may not be equal to y). ∎

The significance of this proposition is that whenever we have a reflexive and trans tive relation on a set, we can always assume antisymmetry by passing to a suitable quotient set, if necessary, and in this passage we regard as equivalent such elements which would have been equal, had there been antisymmetry. A construction like this appears frequently in mathematics. Whenever we want two things to be equal but they are not, we define a suitable equivalence relation R under which they are equivalent and pass to the quotient set. Of course, this equivalence relation must not be unnecessarily large, or else we would be treating too many things as equal. This procedure is expressed by saying that the two things under consideration are **equal modulo** R. The proposition above says that every reflexive and transitive relation is also antisymmetric modulo an equivalence relation. So there is no significant loss of generality in retaining antisymmetry in the definition of a partial order. From now onwards, we accept Definition (2.1).

Finally, we answer one more question about the definition, namely, why the word 'partial'?; 'Partial' as used here is the opposite of 'total'. So we must define a total ordering first, as we indeed do.

3.5 Definition: A **total** or **linear** or **simple** order on a set X is a relation $\leqslant$ on X which is reflexive, transitive, antisymmetric and which has the following property, known as the **law of dichotomy**: for every $x, y \in X$ either $x \leqslant y$ or $y \leqslant x$. (verbally, every two elements are **comparable** to each other.)

Put differently, the law of dichotomy says that the relation $\leqslant$, along with its inverse relation $\geqslant$, cover the totality of all ordered pairs (x, y), with $x, y \in X$. Hence the name 'total'. The justification for the term 'linear' will be given a little later. Apparently, there is no simple justification for the term 'simple'. Probably it is used just to indicate that certain theorems about partial orders take a particularly simple form when specialised to simple orders. 'Dichotomy' literally means branching into two. Here the two possibilities ($x \leqslant y$ and $y \leqslant x$) are the 'branches'.

We now give examples partial orders, some of which will be total orders. Note that if $\leqslant$ is a partial order on a set X and $Y \subset X$ then $\leqslant/Y$, i.e., the restriction of $\leqslant$ to Y (see Exercise (2.5)) is also a partial order and this way we can get more examples. Note further that if $\leqslant$ is linear so is $\leqslant/Y$. It may happen, however, that $\leqslant/Y$ is total but $\leqslant$ is not. A **partially ordered set**, abbreviated, as p.o. set or a **poset** is defined as a pair $(X, \leqslant)$ where X is a set and $\leqslant$ is a partial order on X. This is yet another example of a mathematical structure. A linearly ordered subset of a poset is called a **chain**, a **tower** or a **nest**.

Examples:

1. A foremost example is that of the usual ordering for real numbers. This is a total order.

2. On the set of non-negative integers, define $a \mid b$ to mean that 'a divides b', i.e., 'there exists an integer n such that $b = na$'. It is easily seen that $\mid$ is a partial order. It is not total; for example, neither $10 \mid 15$ nor $15 \mid 10$. We could have defined a similar relation on the set of all integers but it would not have been antisymmetric.

3. Let X be any set and $P(X)$ its power set. Then the inclusion defines a partial order on $P(X)$. This is also not a total order because we can have subsets A, B of X such that neither $A \subset B$ nor $B \subset A$.

4. Let D be the set of all partitions of a set X. If $\mathcal{D}, \mathcal{E} \in D$, let $\mathcal{D} \leqslant \mathcal{E}$ mean $\mathcal{E}$ is a refinement of $\mathcal{D}$ (see Definition (2.9)). Then it is easily seen that $(D, \leqslant)$ is a poset. Again, it is not totally ordered.

5. Let $(X_1, \leqslant_1)$ and $(X_2, \leqslant_2)$ be posets. Let $X = X_1 \times X_2$. We define a partial order $\leqslant$ on X by

$$(x_1, x_2) \leqslant (y_1, y_2)$$

iff either $x_1 \leqslant_1 y_1$ or

$$(x_1 = y_1 \text{ and } x_2 \leqslant_2 y_2).$$

In other words, we first compare the first co-ordinates and if they are equal then we compare the second co-ordinates. More generally, this can be done for the cartesian product of any finite number of posets, the order of these posets being crucial. (In other words, although there is an obvious one-to-one correspondence between $X_1 \times X_2$ and $X_2 \times X_1$, it is not compatible with the orderings we get.) When we list words in a dictionary, ths is how they are arranged, first, by their first letters and in case of words having the same first letters, by their second letters and so on. (In an actual dictionary, not all words are of the same length. But this can be tackled by putting blanks at the end of the shorter words and declaring that the blank is to be regarded as the first symbol of the alphabet.) For this reason, the ordering defined here is called **lexicographic** or **dictionary** ordering. Note that if $\leqslant_1, \leqslant_2$ are linear, so is $\leqslant$.

6. Let S be a set of statements. Define $p \leqslant q$ to mean p implies q. Then $\leqslant$ is a reflexive and transitive relation. It is not antisymmetric in general. However, by the construction in Proposition (2.3), we get a partial order on the set of equivalence classes of statements where two logically equivalent statements are regarded as equivalent.

7. If $(X_1, \leqslant_1)$ and $(X_2, \leqslant_2)$ are posets we can also define a partial order on $X_1 \times X_2$ by

$$(x_1, x_2) \leqslant (y_1, y_2)$$

iff $x_1 \leqslant_1 y_1$ and $x_2 \leqslant_2 y_2$. But unlike the lexicographic order, even when $\leqslant_1$ and $\leqslant_2$ are total, $\leqslant$ need not be so. If $x_1 \leqslant_1 y_1$ and $y_2 \leqslant_2 x_2$, then the pairs (x_1, x_2) and (y_1, y_2) are incomparable unless they are equal. (This was the essence of one of the examples at the beginning, where two contestants in a competition are incomparable.)

8.   Let $X = \mathbf{R} \cup \{*\}$ where $*$ is some point not on the real line. Define $\leqslant$ on X as

$$\{(x, y) \in \mathbf{R} \times \mathbf{R} : x \leqslant y \text{ in the usual order}\} \cup \{(*, *)\}.$$

In other words, we retain the usual order on $\mathbf{R}$ and declare that the point $*$ is not comparable with anything except itself. This seems a very artificial way to define a relation, but it satisfies all the conditions of a partial order and is useful as a counterexample.

9.   Let $\mathbf{R}^* = \mathbf{R} \cup \{\infty, -\infty\}$ where ∞ and $-\infty$ are just two symbols not representing any real number. Define $\leqslant$ on $\mathbf{R}^*$ as $\{(x, y) \in \mathbf{R} \times \mathbf{R} : x \leqslant y \text{ in the usual order}\} \cup (\{-\infty\} \times \mathbf{R}^*) \cup (\mathbf{R}^* \times \{\infty\})$. In other words, we once again retain the usual order on $\mathbf{R}$ but extend it to the points ∞ and $-\infty$ by declaring that $-\infty$ is less than everything else and ∞ is greater than everything else. $\mathbf{R}^*$ with this ordering is called the **extended real line**. Note that it is a chain.

The study of partial orders can be taken in three directions. One is the usual one, common to the study of all mathematical structures, namely, to prove theorems about them in the abstract and then apply these theorems to specific examples. But because of the frequent occurence of order relations in real life, two other problems are very important. One is to put on a given set X a suitable ordering subject to certain constraints. The criteria of suitability and the nature of the constraints would, of course, change from problem to problem. Worded differently, the problem is to list the elements of the given set X in a suitable manner. Many times these elements are pieces of information or 'data' as they are called technically. Problems of this type are therefore studied under what is called 'data structures'. Although we shall not study them, we shall do one listing problem here. Another problem commonly encountered in practice is to transform, with a suitable permutation, elements listed in one linear order to the same elements listed in some other order. This is known as the **sorting** problem. We shall discuss it briefly later in this book. (See the Epilogue).

Let us take the first line first. We give below a few definitions. Note that if $\leqslant$ is a partial order on a set then so is the reverse order, $\geqslant$. Whenever a concept is defined for $\leqslant$; the same concept for $\geqslant$, after translating in terms of $\leqslant$, gives what is known as the **dual** concept. For example, the concept of a lower bound is dual to that of an upper bound because a lower bound of a set is also an upper bound of the same set under the reverse order relation. This saves the duplication of work because it is unnecessary to give separate definitions of the dual concepts. Similarly once a result is proved about such concepts, the corresponding 'dual' result follows simply by duality.

3.5.   Definition: Let $(X, \leqslant)$ be a partially ordered set, $A \subset X$ and $x \in X$. Then we say x is **an upper bound** of A if for all $a \in A$, $a \leqslant x$. x is called a **maximum** (or **largest** or **greatest**) element of A if $x \in A$ and x is an upper bound of A, x is called a **maximal** element of A if $x \in A$ and for every $a \in A, x \leqslant a$ implies $x = a$. The dual concepts are a **lower bound**, a **minimum** (or **smallest** or **least**) element and a **minimal** element respectively. x is said to be a **least upper bound** (or **l.u.b.** or **supremum**) of A if x is a least element of the set of all upper bounds of A. The dual concept is a **greatest lower bound** (or **g.l.b.** or **infimum**). A is said to be **bounded above** if it has at least one upper bound. The dual concept is 'bounded below'. Finally, a set which is both bounded above and below is called **bounded**. If X itself is bounded, the partial order $\leqslant$ is said to be bounded.

The reader must have seen most of these concepts defined (and illustrated) for the case of the usual order on the real line. For linear orders, they behave much the same way as for the real line. But, for partial orders which do not satisfy the law of dichotomy, one has to be cautious. Note, for example, the distinction between a maximum and a maximal element. If x is a maximum element of A then clearly it is also a maximal element of A. The converse holds if A is a chain but not in general. The catch is that the maximality of an element simply means that there is no element which properly beats it, i.e., it beats every element with which it is comparable. It does not necessarily mean that it beats every other element. Beating presupposes comparison. So if there is no comparability, automatically there is no beating. Thus, an element can be a maximal element without being a maximum element, because there may not be very many elements with which it is comparable. Fot the same reason, note also that while a set can have at most one maximum element, it may have more than one maximal elements. (If x and y are two maximum elements of a set A then $x \leqslant y$ and $y \leqslant x$, giving $x = y$ by antisymmetry. But if x and y are merely maximal elements of A, then this argument breaks down because x and y may not be comparable.) As an example, in Example (2) above, let $A = \{1, 2, ..., 10\}$. Then A has 6, 7, 8, 9 and 10 as its maximal elements. Even when a maximal element is unique it need not be a maximum element. For example, in Example (8), above, the whole set X has

* as its only maximal element. Still it has no maximum element. Note also that sometimes a set may have no maximal element; for example the set of positive integers in Example (2) above. (Curiously, 0 is the maximum element for the whole set there.)

Things are somewhat better for finite sets as we now show. The result, although not profound, is important because in discrete mathematics we often deal only with finite sets.

3.7. Theorem: Let $(X, \leqslant)$ be a poset and A a non-empty, finite subset of X. Then A has at least one maximal element. Also A has a maximum element iff it has a unique maximal element. (Similar assertions hold for minimal elements.)

Proof: We start with any element, say x_1 of A. If x_1 is not a maximal element of A then, by definition, there exists an element $x_2 \in A$ such that $x_1 < x_2$. If x_2 is not a maximal element of A, then again, there exists $x_3 \in A$ such that $x_2 < x_3$. Note that $x_1 < x_3$ (by transitivity of the strict order) and so $x_2 \neq x_1$. Again, if x_3 is not maximal, we get an element x_4 of A, different from x_1, x_2, x_3 such that $x_3 < x_4$. Continuing in this manner we get a sequence $x_1 < x_2 < x_3 < \ldots$. Since every time we are getting a new element and the set is finite, this process must stop at some stage. But then A would have a maximal element.

For the second assertion, suppose A has a maximum element, say x. Then it is also a maximal element. Further there can be no other maximal element in A, for if y were such an element then $y \leqslant x$ (since x is a maximum element), which would force $y = x$ by maximality of y. Thus if A has a maximum element then A has a unique maximal element. This part does not require that A is finite. However, for the converse implication. we do need it. Suppose A has a unique maximal element, say x. We claim it is also a maximum element of A. If not, then there exists some $x_1 \in A$ such that x_1 is not comparable with x; for if x_1 were at all comparable with x then $x_1 \leqslant x$ (by maximality of x) and if this holds for all $x_1 \in A$, it would mean that x is a maximum element. So x_1 is not comparable with x. From now onwards the argument is similar to the one used in the proof of the first assertion. If x_1 is not a maximal element of A then there exists $x_2 \in A$ such that $x_1 < x_2$. Note that x_2 cannot be comparable with x (otherwise, $x_2 \leqslant x$ and this would give $x_1 < x$, contradicting that x_1 was not comparable with x). Continuing we get a sequence

$$x_1 < x_2 < x_3 < \ldots.$$

of elements of A none of which is comparable with x. By finiteness of A, this sequence must terminate and hence some x_n would be a maximal element of A, different from x, contradicting the assumption. This proves the converse implication. The assertion about minimal elements follows by duality. ∎

As an application of this theorem, we can do the Dance Problem. In Chapter 2, Section 1 we paraphrased it by considering B as the set of boys, and B_i as the set of those boys who danced with the ith girl, g_i, $i=1, 2, ..., n$. The problem reduces to showing that there exist i, j such that neither $B_i \subset B_j$ nor $B_j \subset B_i$. We interpret this in terms of a suitable partial order. Let $P(B)$ be the power set of B, partially ordered by set inclusion $\subset$ (see Example (3) above). Now let $A = \{B_1, B_2, ..., B_n\}$. Then A is a fiinite subset of $P(B)$. (We are assuming here that the number of girls is finite. The assertion of the problem need not hold without this assumption.) We have to show that at least two elements of A are incomparable under the relatlon $\subset$. If this is not the case then A is a chain. By Theorem (2.6), A has a maximal element, say, B_r and since A is a chain, B_r is also the maximum element, i.e., $B_i \subset B_r$ for all $i = 1, 2, .., n$. But then

$$B_r = \bigcup_{i=1}^{n} B_i.$$

The conditions of the problem imply that

$$\bigcup_{i=1}^{n} B_i = B,$$

the set of all boys. But this means $B_r = B$, or in other words, that the rth girl, g_r, danced with all boys, contradicting the data of the problem. So A cannot be a chain, and, as noted before, this completes the proof.

Another application of finiteness is in simplifying the graphic representation of an order relation $\leqslant$ when the underlying set, say X, is finite. Recall that in this representation, we picture element of X as points in a plane and whenever $x \leqslant y$ we draw an arrow from x to y. Such diagrams tend to be messy even for a relatively small set X. For example, if $|X| = 10$ and $\leqslant$ is a total order on X then its graphic representation would contain 55 arrows. (Fortunately, because of antisymmetry, there are no reverse arrows.) It is possible to reduce the number of arrows and still retain the information conveyed by the graphic representation of the partial order. Let us see how this can be done.

First, all loops on elements of X can be removed. Since $\leqslant$ is always reflexive, such loops can be understood even without being drawn. However, this is not a very big saving. A really substantial saving comes because of transitivity. Whenever $x, y, z \in X$ and there is an arrow from x to y and an arrow from y to z, there is no need to draw a direct arrow from x to z because such an arrow can be inferred from transitivity of $\leqslant$. Even the arrow from x to y can be eliminated if we find some element, say w, 'between' x and y, i.e., such that $x < w$ and $w < y$. Again we can look for something in between x and w. Unfortunately, if X is an infinite set then his process may go on forever. For example, if X is the set of real numbers and $\leqslant$ the usual order on it, then between every two real numbers there are infinitely many real numbers and so every arrow that we draw is superfluous. But then, if we do not draw any arrows, we would not get the usual ordering on R. (The ordering that can be inferred from a diagram with no arrows drawn is the trivial ordering in which $x \leqslant y$ iff $x = y$. But this is not the usual ordering on $\mathbb{R}$.)

However, for finite sets, things are better as we now show. (Even for infinite sets, the argument given below applies if the ordering happens to be what is called a well ordering. This is an important concept in mathematics and although we shall not need it, it will be briefly touched at through the exercises.) First we need a definition.

3.8 Definition: Let $(X, \leqslant)$ be a partially ordered set and let $x, y \in X$. Then y is said to **cover** x if $x < y$ and there is no $z \in X$ such that $x < z$ and $z < y$.

In other words, y covers x iff $x < y$ and there is nothing between x and y. As noted above, in the usual ordering on the real line no element covers any element. But if we take the restriction of this order to the set of integers, then an integer n is covered by $n + 1$. As another example, in Example (2) above note that x is covered by y if and only if $x \mid y$ and the ratio y/x is a prime number. This also shows that the same element may be covered by more than one element.

In the next proposition, we characterise the concept of covering in terms of the concepts defined earlier.

3.9 Proposition: Let $(X, \leqslant)$ be a poset and $x \in X$. Let

$$A = \{z \in X : x < z\}.$$

Then an element $y \in X$ covers x if and only if y is a minimal element of A.

Proof: This is a straightforward consequence of the definitions and is left as an exercise. ■

Now, coming back to the problem of reducing the number of arrows in the graphic representation of an order relation, we show that for finite posets, it suffices to draw only those arrows which go from elements of the set to those which cover them Formally, the result can be stated as follows:

3.10 Theorem: Let X be a finite set and $\leqslant$ a partial order on X. Define a binary relation R on X by xRy iff y covers x (w.r.t. $\leqslant$). Then $\leqslant$ is generated by R, i.e., is the smallest order relation on X containing R.

Proof: Clearly, as subsets of $X \times X$, R is contained in $\leqslant$. We have to show that we can get $\leqslant$ by suitably extending R. First we add ΔX for the sake of reflexivity. So let $R_1 = R \cup \Delta X$. Then R_1 is reflexive. Now we come to the critical part of the argument, namely, showing that the order relation generated by R_1 is the original relation $\leqslant$. For this we have to show that given and x, $y \in X$ with $x \leqslant y$, we can find some sequence of elements $x_1, x_2, \ldots, x_n$ in X such that $x = x_1$, $y = x_n$ and $(x_i, x_{i+1}) \in R_1$ for all $i = 1, 2, \ldots, n-1$ (cf. the proof of Proposition (2.15)). If $x = y$, then $(x, y) \in \Delta X$ and hence $(x, y) \in R_1$, so we take $x_1 = x$ and $x_2 = y$. If $x < y$ then we proceed by induction on the number of elements in between x and y (i.e. the number of elements z such that $x \lessdot z$ and $z < y$). Let r be the number of such elements. If $r = 0$, then y covers x and so $(x, y) \in R$ by definition. So again $(x, y) \in R_1$. Suppose $r > 0$. Let A be the set $\{z \in X : x < z \text{ and } z < y\}$. Then, by definition, $|A| = r$. Fix some $z \in A$ and let

$$B = \{w \in X : x < w < z\}$$

and

$$C = \{w \in X : z < w < y\}.$$

Then B and C are proper subsets of A (since $z \notin B$ and also $z \notin C$) and so $|B| < r$ and $|C| < r$. So by induction hypothesis, there exist sequences $x_1, \ldots, x_n$ and $y_1, y_2, \ldots, y_m$ such that $x = x_1$, $z = x_n$, $z = y_1$ and $y = y_m$ and $(x_i, x_{i+1}) \in R_1$ for $i = 1, \ldots, n-1$ and $(y_j, y_{j+1}) \in R_1$ for $j = 1, 2, \ldots, m-1$. Concatenating these two sequences (see again the proof of Proposition (2.15)) we get a sequence of the desired type with first term x and last term y. This completes the induction and proves our assertion that if $x < y$ in X then the pair (x, y) belongs to the smallest transitive relation containing R_1. So it follows that $\leqslant$ is generated by R_1. ∎

Intuitively we may think of the relation R in this theorem as the 'skeleton' of the given relation $\leqslant$. The graphic representation of R is called the **Hasse diagram** of $\leqslant$. It is obtained by picturing elements of X as points and drawing an arrow from x to y whenever y covers x. In Figure 3.12 we show the Hasse diagrams of two posets. The first is the power set of a three element set, say $\{a, b, c\}$, and the second is the set of integers from 1 to 10 partially ordered as in Example (2) above.

The Hasse diagram of a (finite) poset vividly describes most of the concepts associated with the partial order. For example, two elements are comparable iff there is a path from one of them to the other along the arrows. An element is maximal iff there is no arrow issuing from it. Note that if X is linearly ordered then its Hasse diagram can be drawn by picturing points of X on a straight line and drawing arrows between consecutive points. (This justifies the name 'linear'.) Hasse diagrams are also useful in constructing counterexamples. It is often easier to conceive a desired counterexample through its Hasse diagram.

Although we defined the concept of a supremum in Definition (3.5), so far we did not discuss it. This is not because the concept is only of peri-

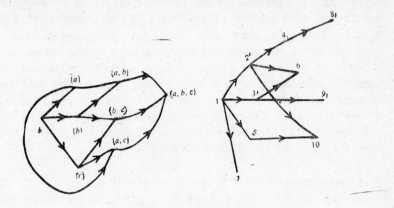

Fig. 3.12: Hasse Diagram of Posets

pheral importance. On the contrary, it is one of the most pivotal concepts in mathematics. Let us see how. Note first of all that if a set A has a supremum then it is unique because it is the least element of some set, namely the set of all upper bounds of A. Moreover, if A has a maximum element then clearly this element is also its supremum. So the concept of a supremum has an independent interest only in the absence of a maximum. For linear orders, we saw in Theorem (2.6), that every finite subset has a maximum. So, to get really interesting examples of suprema in linear orders we must necessarily work with infinite sets. This leads to a limiting process and therefore goes beyond the purview of discrete mathematics. For example, let A be the set $\{x \in \mathbf{R}: 0 \leqslant x < 1\}$. Then w.r.t. the usual order on $\mathbf{R}$, A has 1 as its supremum, but not as the maximum because $1 \notin A$. A contains points which are arbitrarily close to 1 but not the point 1 itself. This is the very essence of a limiting process, coming infinitesimally close without actually touching. Indeed, the very definition of a real number involves the concept of a supremum (or some other, equivalent form of the limiting process). Let $\mathbf{Q}$ be the set of all rational numbers and let

$$A = \{x \in \mathbf{Q}: x > 0 \text{ and } x^2 < 2\}.$$

Then A is bounded above in $\mathbf{Q}$; for example, it is easy to show that 2 is an upper bound of A. But A has no supremum in $\mathbf{Q}$. As a subset of $\mathbf{R}$, A does have a supremum, namely $\sqrt{2}$. But $\sqrt{2}$ is not a rational number. Figuratively, absence of $\sqrt{2}$ amounts to a hole in the set of rational numbers. Real numbers can be constructed by patching up these holes and every real number can be obtained as the supremum of some subset of rational numbers. Interesting as these topics are, they are beyond the scope of this book and so we abandon this line here.

Thus, for total orders, a supremum either coincides with the maximum or else leads to a limiting process. However, for partial orders which are not total orders, the picture is quite different. In such a case, even a finite set need not have a maximum. Still, it may have a supremum. Such suprema (and infima) are very important and are given a special name, as in the following definition. We assume that the set in question has only two elements, because this is the most interesting case. (A set with only one element trivially has its lone element for both its supremum and infimum.) If every set with two elements has a supermum then it is easy to show, by induction on the cardinality, that every finite set does, (cf. the derivation of Proposition (2.2.3) from Axiom (2.2.2)).

3.11 Definition: Let $(X, \leqslant)$ be a poset and let $x, y \in X$. Then the supremum and the infimum of the set $\{x, y\}$ (in case they exist), are called, respectively, **the join** and **the meet** of x and y and are generally denoted by $x \vee y$ (read 'x join y' or 'x wedge y') and '$x \wedge y$' (read 'x meet y'). A poset in which every pair of elements has a meet and a join is called a **lattice**. The Hasse diagram of a lattice is called a **lattice diagram.**

Obviously if x and y are comparable under $\leqslant$, then $x \vee y$ is the greater of the two and $x \wedge y$ is the smaller of the two. Consequently, every totally ordered set is a lattice. As other examples, the power set of any set, partially ordered by inclusion, is a lattice. The intersection of two subsets is their meet and their union is their join (cf. Exercise 2.1.4). Example 2 above of partial orders is also a lattice, with the meet of x and y being their greatest common divisor and their join the least common multiple. Example 6 is also a lattice, provided the set S has the property that whenever two statements, say, p and q are in S, their conjunction and disjunction are also in S. Indeed it is easy to see that the conjunction of two statements is their meet. Clearly the truth of the conjunction implies that of each of the two statements and any statement whose truth implies that of both p and q must also imply the truth of 'p and q', Similarly the disjunction of p and q is their join. For this reason the conjunction and the disjunction of two statements, say p and q, are often denoted by $p \wedge q$ and $p \vee q$ respectively.

The poset in Example 8 is not is a lattice. For any $x \in \mathbf{R}$, the pair x and * has no meet and no join. We can get many other examples by removing certain elements from lattices. For example, the collection of all non-empty subsets of a set S is in general not a lattice, because two subsets which are mutually disjoint would have no meet.

The theory of lattices is quite important and certain special types of lattices will be studied as Boolean Algebras in the next chapter.

We now consider an example of the problem of listing. This amounts to putting a linear order on a given set, which would generally be a finite set for our purpose. Although the characteristics of a 'good' ordering would depend on the particular context, one general, desirable feature is that there should be an easy retrieval. This means that the index (or rank) of an element should be given by an easy formula directly in terms of the element (and without having to go down the list till that particular element is encountered). Secondly the inverse of this function should also be easily computable, i.e., given a positive integer k (not exceeding the cardinality of the set in question), we should be able to tell the kth element of the list directly in terms of k (and without having to traverse the list physically). Actually, this is a very strong requirement because the existence of such an indexing function for the given set, X, establishes an order isomorphism* between X and the set $\{1,2,...,n\}$ where $n = |X|$. The latter set is a very familiar set and consequently, it is easy to answer any question about the ordering on the set X. For example given $x \in X$, if we want to find its immediate successor (i.e. $y \in X$ such that y covers x) we simply compute the index of x, add 1 to it and then find the unique y with this index. Similarly given x and y in X we can tell which way they are related by looking at their indices. In a good ordering, it is desirable to have a mechanism to answer these questions directly, without going through the computation of the index.

As an illustration, let X be the set of all permutations of a set with n elements, say, the set $\{1, 2, ..., n\}$. Then $|X| = n!$. We put the usual order on the set $\{1, 2, ..., n\}$. Now, every element of X can be thought of as a 'word' of length n (with every 'letter' appearing exactly once). We put the lexicographic ordering on X (see Example (5) above). For example, for $n = 3$, the six permutations in X are ordered as

$$123 < 132 < 213 < 232 < 312 < 321.$$

In this order it is very easy to tell which of the two given permutations comes first; we simply 'scan' both of them from left to right and when we first come across a place in which their entries differ, the permutation with the smaller entry in this place is the smaller. It is also easy to find the successor of a given element, say, $x = a_1 a_2 ... a_n$ of X. We scan x from right to left till the a's keep on 'climbing' and first 'fall down', i.e., we find the integer j such that $a_j < a_{j+1}$ and

$$a_{j+1} > a_{j+2} > ... > a_n.$$

(If such j does not exist then

$$a_1 = n, a_2 = n - 1,...., a_{n-1} = 2, a_n = 1$$

and this is the last permutation in X). Now let r be the unique integer such that $a_r > a_j > a_{r+1}$ (set $r = n$ if $a_n > a_j$). Leave $a_1 .... a_{j-1}$ as they are, replace a_j by a_r and arrange the remaining elements in an ascending order. Thus the original permutation,

$$a_1 \, a_2 \, ... \, a_{j-1} \, a_j \, a_{j+1} \, a_{j+2} \, ... \, a_r \, a_{r+1} \, ... \, a_n$$

*For a formal definition, see Exercise 3.12.

changes to,

$$a_1 \, a_2....a_{j-1} \, a_r \, a_n \, a_{n-1}....a_{r+1} \, a_j \, a_{r-1}.... \, a_{j+1}$$

It is easy to show that this is in fact the immediate successor of the original permutation. For example, for $n = 5$, the successor of 23541 is 24135. Here $j = 2$ and $r = 4$.

It remains to find the index function for this ordering. We could do this recursively, by a formula, which expresses the index of a permutation of n symbols in terms of the index of some other permutation of $n - 1$ symbols. Applying this formula again and again $n - 1$ times, we get the index of the original permutation. However, there is another method and we give it here because it involves a concept, which is important elsewhere also.

3.12 Definition: Let $x = x_1 \, x_2.... \, x_n$ be a sequence of real numbers. By an **inversion** in x, we mean a pair (x_i, x_j) such that $i < j$ but $x_i > x_j$. For each i, let $d_i = $ the number of inversions whose first entry is x_i, i.e.

$$d_i = | \, \{x_j : i < j, \, x_i > x_j\} \, | \, .$$

Then the sequence $(d_1, d_2, ..., d_n)$ is called the **inversion table** or **inversion vector** of x.

Intuitively, an inversion is a pair which is in the 'wrong' order. The permutation $1 \, 2....n$ has no inversions and hence has $(0, 0,..., 0)$ as its inversion table. At the other extreme, the permutation $n(n - 1)....21$ has every pair as an inversion and its inversion table is

$$(n - 1, n - 2,..., 2, 1, 0).$$

The permutation 24351 has $(1, 2, 1, 1, 0)$ as its inversion table. It is obvious that for every $i = 1, 2, ... \, n$,

$$0 \leqslant d_i \leqslant n - i.$$

What is not so obvious is that the inversion table of a permutation completely determines it. For example, let us see how we may recover the permutation 24351 from the inversion table $(1, 2, 1, 1, 0)$. We are given that $d_1 = 1$. This means that there is one integer less than the integer in the first place, which follows it. So the first place must be filled by 2. Then the second entry must be from 1, 3, 4, 5. Since $d_2 = 2$, we know that it must be the third smallest number in this set. So the second entry is 4. The next entry must be 1, 3 or 5 and the fact that $d_3 = 1$ fixes it as 3. Continuing we get 24351 as the permutation.

The same procedure can be generalised to give the following theorem due to **M. Hall.**

3.13 Theorem: Given integers $d_1, d_2, ..., d_n$ such that for all $i = 1, 2,..., n,$

$$0 \leqslant d_i \leqslant n - i,$$

there exists a unique permutation $x = x_1 x_2 .... x_n$ of $\{1, 2, ..., n\}$ whose inversion table is $(d_1, d_2, ..., d_n)$.

Proof: Start with d_1. This is the number of integers from $\{1, 2, ..., n\}$ which are smaller than x_1. So $x_1 = d_1 + 1$. Let

$$S_{n-1} = \{1, 2, ..., n\} - \{x_1\}.$$

Then $x_2 \in S_{n-1}$ and d_2 is the number of elements in S_{n-1} which are smaller than x_2. So x_2 is the $(d_2 + 1)$th smallest element in the set S_{n-1}. Next let $S_{n-2} = S_{n-1} - \{x_2\}$. Then, once again, x_3 is the $(d_3 + 1)$th smallest element of S_{n-2}. Continuing in this manner we determine x_4, x_5...., x_{n-1} and x_n in succession. (The construction can be done conveniently by writing $1, 2, ..., n - 1, n$ in a row and scoring off the elements $x_1, x_2, ...$ as they are determined. For example, in the example above we have $(1, 2, 3, 4, 5)$, $(1, 2, 3, 4, 5)$, $(1, 2, 3, 4, 5)$, $(1, 2, 3, 4, 5)$ $(1, 2, 3, 4, 5)$ and $(1, 2, 3, 4, 5)$ giving 24351 as the permutation) Thus the permutation x is determined uniquely. By the very construction of it, its inversion table is $(d_1, d_2, ..., d_n)$. ∎

It is interesting to note that the number of all permutations of n symbols can also be obtained from this theorem. Each d_j can take $(n - j + 1)$ values, independently of the others. So the number of possible inversion tables is $n \times (n - 1) \times ... \times 3 \times 2 \times 1$ or $n!$. This is also the number of all permutations of n symbols.

In terms of inversion numbers we are now ready to give a formula for the indexing function for the lexicographic order on permutations.

3.14 Theorem: In the lexicographic ordering for the set of all permutations of n symbols, the permutation whose inversion table is $(d_1, d_2, ... , d_n)$ has index (or rank)

$$= 1 + d_1 (n-1)! + d_2 (n - 2)! + .. + d_i (n - i)! + ... + d_{n-1} 1! + d_n 0!.$$

(Actually, the last term need not be included because $d_n = 0$).

Proof: We prove this by induction on n. The case $n = 1$ is trivial because then there is only one permutation and its inversion table is (0). Let us denote the index function for the permutations of n symbols (arranged lexicographically) by Ind_n. Let $x = x_1 x_2 .... x_n$ be a permutation of $1, 2, ..., n$. Let $(d_1, d_2, ..., d_n)$ be the inversion table of x. We have to prove that

$$\mathrm{Ind}_n (x) = 1 + \sum_{i=1}^{n} d_i (n - i)!.$$

Now, for each k with $1 \leqslant k < x_1$ let T_k be the set of those permutations of $\{1, 2, ..., n\}$ in which the first entry is k. Clearly $|T_k| = (n - 1)!$. By the definition of the lexicographic ordering, all elements of T_k come before x, for $k = 1, 2, ..., x_1 - 1$. In all there are $(x_1 - 1)(n - 1)!$ such permutations. Besides these, among those permutations whose first entry is x_1, some permutations will come before x. Specifically, a permutation $x_1 y_2 y_3 .... y_n$ will come before $x_1 x_2 .... x_n$ iff the permutation $y' = y_2 y_3 .... y_n$ comes before the permutation $x' = x_2 .... x_n$ in the set of all permutations of the set $\{1, 2, ..., n\} - \{x_1\}$, arranged lexicographically. Thus we have,

$$\text{Ind}_n(x) = (x_1 - 1)(n_1 - 1)! + \text{Ind}_{n-1}(x') \qquad (*)$$

$\text{Ind}_{n-1}(x')$ is known by the induction hypothesis. (Note that x' need not be a permutation of the set $\{1, 2, ..., n-1\}$. Still we can apply the induction hypothesis because we are proving the result for the set of permutations of any symbols, provided there is some linear order on these symbols, which is used to put the lexicographic order on the set of all permutations of these symbols. The actual symbols are obviously immaterial.) For the sake uniformity in notation, let

$$x_i' = x_{i+1} \text{ for } i = \{1, 2, ..., n-1\}.$$

Then x' is a permutation of the set

$$\{x_1', x_2', ..., x'_{n-1}\}$$

Let

$$(d_1', d_2', ..., d'_{n-1})$$

be its inversion table. Clearly

$$d_i' = d_{i+1} \text{ for } i = 1, 2, ..., n-1,$$

Now

$$\text{Ind}_{n-1}(x') = 1 + \sum_{i=1}^{n-1} d_i'(n-1-i)!$$

$$= 1 + \sum_{i=1}^{n-1} d_{i+1}(n-1-i)!$$

$$= 1 + \sum_{i=2}^{n} d_i(n-i)! \text{ (replacing } i \text{ by } i+1$$

as the index variable of summation).

Substituting this in (*) and noting that $d_1 = x_1 - 1$ completes the proof of the inductive step. Thus we have established the index formula for the permutations of the set $\{1, 2, .., n\}$. But obviously the same formula applies for the set of permutations of any n symbols, on which some linear order is given. (As noted above, the generality so gained is vitally needed in the inductive step.) This completes the proof. ■

For example, for $n = 6$, the index of the permutation 531462 is

$$1 + 4 \times 5! + 2 \times 4! + 0 \times 3! + 1 \times 2! + 1 \times 1! + 0 \times 0! = 532.$$

Thus we see just by inspection and a little calculation, that this element ranks 532nd from the beginning in the lexicographic order, without searching down the list till we encounter this element. What about the reverse question? That is, given an integer k, how do we find the permutation whose index is k? For the answer we need the following theorem:

3.15 Theorem: Let k and n be positive integers with $k \leqslant n!$. Then there exist unique integers $d_1, d_2, \ldots d_n$ such that

$$0 \leqslant d_i \leqslant n-i \text{ for } i = 1, 2, \ldots, n$$

and

$$k = 1 + \sum_{i=1}^{n} d_i (n-1)!.$$

Proof: We prove the existence by induction on n. The case $n = 1$ is trivial. If $k = n!$, then we set

$$d_i = n-i \text{ for all } i = 1, 2, \ldots, n.$$

Then

$$\Rightarrow \sum_{i=1}^{n} d_i (n-i)! = \sum_{i=1}^{n} (n-i)(n-i)! = \sum_{j=1}^{n-1} j(j)! = \sum_{i=1}^{n} (j+1-1) j!$$

$$= \sum_{j=1}^{n-1} (j+1)! - \sum_{j=1}^{n-1} j! = n! - 1 = k - 1$$

and this proves the result.

Suppose $k < n!$. Let d_1 be the largest non-negative integer such that

$$d_1 (n-1)! < k. \text{ (If } k < (n-1)! \text{ then } d_1 = 0).$$

Then

$$0 \leqslant d_1 \leqslant n-1.$$

Let

$$r = k - d_1 (n-1)!$$

Then r is a positive integer and $r \leqslant (n-1)!$ (otherwise

$$(d_1 + 1)(n-1)! < k,$$

contradicting the definition of d_1). Now apply induction and express r as

$$1 + \sum_{j=1}^{n-1} b_j (n-1-1)!$$

where

$$0 \leqslant b_j \leqslant n - 1$$

for all

$$j = 1, 2, \ldots, n - 1.$$

Now put $d_i = b_{i-1}$ for $i = 2, \ldots, n$. Then we get

$$k = 1 + \sum_{i=0}^{n} d_i (n-i)!$$

This completes the inductive step and establishes the existence of the d_i's.

For uniqueness too, we can apply a similar argument. But there is a better way out. Let S be the set of all sequences $(d_1, d_2,...,d_n)$ such that

$$0 \leqslant d_i \leqslant n-i \text{ for } i = 1, 2, ..., n.$$

Clearly $|S| = n!$. Let T be the set $\{1, 2,..., n!\}$. Define $f: S \to T$ by

$$f(d_1, d_2,...,d_n) = 1 + \sum_{i=1}^{n} d_i (n-i)!$$

Then by what we just proved, f is onto. But $|S| = |T| = n!$. Hence by Theorem (2.2.9.), f is one-to-one which is equivalent to the uniqueness of the d_i's.

With a slight change of notation, the preceding theorem says that every non-negative integer can be uniquely expressed as $\sum_{j=1}^{\infty} d_j \, j\,!$ where $0 \leqslant d_j \leqslant j$ for all j and $d_j = 0$ for all except finitely many j's (thus the sum is really finite). This expression is called the **factorial representation** of the integer. For example

$$100 = 2\times 2! + 4\times 4!, \, 200 = 1\times 2! + 1\times 3! + 3\times 4! + 1\times 5!.$$

Theorem 2.14 also answers our question about finding the permutation whose index equals a given integer. We obtain the integers $d_1, d_2,...,d_n$ and simply construct the permutation whose inversion table is

$$(d_1, d_2, ..., d_n).$$

We conclude this section with a discussion of a particularly interesting type of listing problems, known as topological sorting. When we put a linear order on a set there is often a constraint that certain elements must precede some others. For example, when an author writes a book, the order in which the chapters may appear is constrained by some of them being prerequsites to some others. When a number of jobs are to be performed on a machine, one at a time, some of the jobs must be carried out prior to some others and this puts a restriction on the order the jobs are to be done. Such situations can be paraphrased easily using a suitable mathematical structure. Let X be set and $\leqslant$ some given partial order on X. The problem then is to find a total order $\propto$ on X which is consistent with $\leqslant$ in the sense that for all $x, y \in X, x\leqslant y$ implies $x \propto y$. As subsets of $X\times X$ this means that $\leqslant$ is a subset of $\propto$. In other words, we want to 'extend' the given partial order to a total order. This process is known as **topological sorting**. The origin of this peculiar name is interesting. 'Sorting' as used here is simply a synonymn for 'ordering'. The adjective 'topological' comes from 'topology'. This is a branch of mathematics in which, among other things, one studies the problem of embedding one figure into another, allowing the original figure to be 'stretched or bent without cutting or gluing'. The graphic representation of the problem of topological sorting is to embed the graph representing the partial order $\leqslant$ into the graph representing a linear order on the same underlying set. Note that the extension need not be unique as seen from the example Figure 3.13.

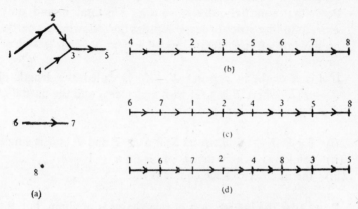

Fig. 3.13: Topological Sorting. **(a)** Hasse diagram of original partial order, **(b)**, **(c)** and **(d)** three possible linear orders containing the given partial order.

From the point of view of discrete mathematics, it is never enough merely to prove the existence of something. We also want a systematic procedure for constructing it. We give below one such simple procedure. (Incidentally, we are tacitly assuming that the set X is finite. If X is infinite, it is still true that every partial order on X can be extended to a linear order. But the proof requires substantially different techniques.)

The procedure consists of outputting minimal elements of subsets of X (w.r.t. the given partial order $\leqslant$ on it), one by one. Let X_1 be X and let x_1 be a minimal element of X_1. Since X is finite, such an element exists by Theorem (2.6). Delete x_1 from X_1, i.e., consider the set $X_2 = X_1 - \{x_1\}$. Let x_2 be a minimal element of X_2. Note that, in the order $\leqslant$, if at all x_1 and x_2 are related, then $x_1 < x_2$ (by minimality of x_1). Next let $X_3 = X_2 - \{x_2\}$ and x_3 be a minimal element of X_3. Once again, if at all x_3 is related to x_1 (or x_2), it is greater than x_1 (or x_2). Continue this process till the set X is exhausted. This gives a listing of the elements of X as $x_1, x_2, ..., x_n$ (where $n = |X|$). In this listing for $i < j$, x_i is either unrelated to x_j, (under $\leqslant$) or else $x_i < x_j$ by minimality of x_i. So if we define $x_i \propto x_j$ simply by $i \leqslant j$, we get a linear order which is an extension of the original partial order $\leqslant$. Note that because minimal elements are not, in general, unique, there is considerable choice involved at every stage and consequently we get many linear orders which are extensions of the same partial order.

Exercises

3.1   Prove Proposition (3.3).

3.2   Verify that the relation R in Proposition (3.4) is indeed an equivalence relation.

3.3   What poset do we get if we apply Proposition (3.4) to the set of all integers on which we define $x \leqslant y$ iff x divides y?

3.4   Prove that a partial order $\leqslant$ on a set X is total if and only if the corresponding strict order $<$ satisfies the following property, known as the **law of trichotomy**: for all $x, y \in X$ either $x < y$ or $x = y$ or $y < x$.

3.5   If A is a chain in a poset X, $|A|$ is called the **length** of A. If $X = P(B)$ where B is a set with n elements and the partial order on X is by set inclusion, prove that

    (i)   for $S, T \in X$, T covers S iff $S \subset T$ and $T - S$ is singleton set

    (ii) the longest chain in X is of length $n + 1$

    (iii) the number of such chainr is $n!$.

3.6   Prove that the intersection of two chains is a chain but that their union need not be a chain.

3.7   Let $(X, \leqslant)$ be a poset and τ be the set of all chains in X. Clearly $\tau \subset P(X)$ and if $\leqslant$ is a linear order then $\tau = P(X)$. Partially order τ by set inclusion (this partial ordering is for certain subsets of X and should not be confused with the original partial order $\leqslant$ which is for elements of X). A maximal element of τ is called a **maximal chain** in X.

    (i)   Prove that a longest chain in X is maximal but the converse need not hold.

    (ii) Prove that every chain is contained in a maximal chain. (Assume that the set X is finite as usual. For infinite posets the existence of maximal chains is a very different matter.)

    (iii) Prove that the set X can be expressed as a union of maximal chains.

3.8   Let m be the largest possible number of mutually incomparable elements of a poset X. Prove that X cannot be expressed as a union of less than m chains.

3.9   In the last exercise prove that X can be expressed as the union of m chains. (This is known as **Dilworth's theorem.)

*3.10   Prove that a sequence of real numbers of length $mn + 1$ (where m and n are positive integers) must contain either a monotonically increasing subsequence of length $m + 1$ or else a monotonically decreasing subsequence of length $n + 1$.

3.11 Prove that the assertion of the Dance Problem may fail if we do not assume the sets of boys and girls to be finite. (Actually, it suffices to assume that at least one of these two sets is finite.)

3.12 Let $(X, \leqslant)$ and $(Y, \leqslant)$ be posets. A bijection $f: X \to Y$ is called an **order isomorphism** (or **order equivalence**) if for all $x, y \in X$, $x \leqslant y$ iff $f(x) \leqslant f(y)$. When such an order isomorphism exists, the posets $(X, \leqslant)$ and $(Y, \leqslant)$ are said to be **order isomorphic** or to be of the same **order type**. Prove that this defines an equivalence relation on any collection of posets. Two posets having the same order type have identical order theoretic properties. For example, if one of them is a lattice, so is the other. If one of them has a chain of certain length, so does the other. Prove that any two finite, linearly ordered posets of the same cardinality are of the same order type. Show that **N** and **Q** each with the usual order are not order isomorphic.

3.13 Let two sets X, Y be equipollent (i.e., suppose there is a bijection between them). Prove that the posets $(P(X), \subset)$ and $(P(Y), \subset)$ are of the same order type.

3.14 Let X be the set of all positive integers which divide 30. Define $\leqslant$ on X by $x \leqslant y$ iff $x \mid y$. Prove that there exists a set Y such that $|Y| = 3$ and $(P(Y), \subset)$ is order isomorphic to $(X, \leqslant)$. However, if $\leqslant$ is the usual order on X, prove that no such set Y exists.

*3.15 Prove that the poset $(P(\mathbf{N}), \subset)$, where **N** is the set of positive integers, contains an uncountable chain.

3.16 A linearly ordered set X is called **dense-in-itself** if for all $x, y \in X$ with $x < y$, there exists $z \in X$ such that $x < z$ and $z < y$. For such a set, prove that:

   (i) If $|X| > 1$, then X is infinite

   *(ii) If $|X| = \aleph_0$ and X has neither a maximum nor a minimum element then X has the same order type as **Q** with the usual ordering.

3.17 Prove that the set of all decompositions of a set, partially ordered by the refinement relation is a lattice. (cf. Example (4) above.)

3.18 Prove that in a lattice, every nonempty, finite subset has a supremum and an infimum.

3.19 Let $(X, \leqslant)$ be a lattice. A subset Y of X is called a **sublattice** of X if for all, $x, y \in Y$, $x \vee y$ and $x \wedge y$ are also in Y. (This condition is commonly expressed by saying that Y is closed under the operations $\vee$ and $\wedge$.)

   (i) Prove that $(Y, \leqslant/Y)$ (i.e., the set Y with the restriction of the ordering $\leqslant$ to Y) is itself a lattice.

   (ii) Prove that the intersection of any two sub-lattices of X is also a sublattice of X. Prove that this need not hold for their union.

  (iii) Let S be a set and T a subset of S, Prove that $P(T)$ is a sublattice of $P(S)$ (partially ordered under $\subset$ as usual).

  (iv) Let S be a set. Pick any two distinct elements, say, x and y of S. Let Y be the set of those subsets of S which do not 'separate x from y', i.e. $Y = \{A \subset S$: either both $x, y \in A$ or neither x, nor y is in $A\}$. Prove that Y is sublattice of $P(S)$.

   (v) Prove that the set of all divisors of a positive integer n as well as the set of all multiples of n are sublattices of the lattice in Example (2),

3.20 Let X be the set $\mathbf{N} \times \mathbf{N}$. For

$$(x_1, y_1), (x_2, y_2) \in X$$

define

$$(x_1, y_1) \leqslant (x_2, y_2)$$

iff $x_1 \leqslant x_2$ and $y_1 \leqslant y_2$ (cf. Example (7)). Prove that $\leqslant$ is a partial order on X and $(X, \leqslant)$ is a lattice. Draw its lattice diagram.

3.21 A poset $(X, \leqslant)$ is said to be **well-ordered** if every non-empty subset of X has a least element.

   (i) Prove that a well-ordered set is totally ordered.

  (ii) Prove that the set of real numbers and the set of all integers (with usual orders) are not well-ordered.

 (iii) Prove that the set of positive integers is well ordered (Hint: By induction on n, prove that every subset of $\mathbf{N}$ containing n has a least element.)

 (iv) Let $(X, \leqslant)$ be a well-ordered set and let ∞ be any symbol not in X. Let $X^* = X \cup \{*\}$. Define $\leqslant$ on X^* by $x \underset{\propto}{} y$ iff either $x, y \in X$ and $x \leqslant y$ or $y = *$

$$\text{(i.e., } \underline{\propto} = \ \leqslant \cup \ \{(x, *): x \in X^*\}).$$

Prove that $(X^*, \leqslant)$ is well-ordered. (This shows how to construct from a given well-ordered set, the 'next larger' well-ordered set.)

   (v) If X is well-ordered, prove that every element of X, except the largest element, if any, is covered by a unique element of X, called its **successor**.

3.22 Prove that the construction given for the next element in the lexicographic ordering on the set of permutations in fact gives the next permutation.

3.23 Let n, r be positive integers with $r \leqslant n$. Denote an r-subset of $\{1, 2, ..., n\}$ by listing its elements in an ascending order. Now put the lexicographic ordering on the set $P_r(n)$, of all r-subsets of $\{1, 2, ..., n\}$. For this ordering find formulas for (i) the next element and (ii) the index of an element.

3.24 Obtain a linear ordering for the set of all subsets of $\{1, 2, ..., n\}$, by first listing the empty set, then all singleton sets, then all subsets with 2 elements and so on in the manner of the last exercise. What is the index function for this ordering?

3.25 There is another way to put a linear order on the power set $P(X)$ where $X = \{1, 2, ..., n\}$. Let Y be the set $\{0, 1, 2, ..., 2^n - 1\}$. Define

$$g: P(X) \to Y \text{ by } g(A) = \sum_{i=1}^{n} f_A(i)2^{i-1}$$

where f_A is the characteristic function of A for $A \subset X$ (see Exercise (2.1.13)). Prove that g is a bijection. Because of this, the usual order on Y can be transferred to an order on $P(X)$. Note that g itself is the index function for this ordering.

3.26 Prove that the Hasse diagram of a poset cannot contain any cycle, i.e., a sequence of consecutive arrows terminating where it starts.

3.27 Let $\leqslant$ be a partial order on a set X. Prove that unless $\leqslant$ is itself linear, there are at least two different linear orders on X which extend $\leqslant$. (Hint: show that there will be at least one i such that the set X_i given in the algorithm for topological sorting has more than one minimal element.)

3.28 Let $X = \{1, 2, ..., n\}$. Which of the linear orders in Exercises (3.24) and (3.25) is an extension of the partial order on $P(X)$ by set inclusion?

3.29 Prove that the method given for topological sorting is exhaustive in the sense that every possible linear extension of the given partial order can be obtained by making an appropriate choice of minimal elements at every stage. Obtain the three extensions of the partial order shown in Figure 3.13 this way.

3.30 A linear order $\leqslant$ on a set X is called **complete** if every non-empty subset of X which is bounded above has a supremum in X (which need not belong to that subset). Prove that:

   (i) $\leqslant$ is complete if and only if every non-empty subset of X which is bounded below has an infimum.

  (ii) every well order is complete.

*(iii) the usual order on $\mathbf{N}$, (the set of all integers) and $\mathbf{R}$ is complete but the usual order on $\mathbf{Q}$ is not complete.

Notes and Guide to Literature

Partial orders arise in highly diverse problems. The study of lattices is especially important. A standard reference on them is Birkhoff [1]. Incidentally, the term 'lattice' is also used in a very different sense in mathematics and physics. The name 'lattice' as used here probably comes because the Hasse diagram of the lattice in Exercise (3.20) resembles the lattice of tetravalent atoms such as carbon or silicon.

Our definition of an inversion differs slightly from that in Knuth [1] Volume 3 where one full chapter (which comprises more than half the third volume) is devoted to the problem of sorting and gives an exhaustive list of sorting algorithms.

A proof of Dilworth's theorem may be found in Hall [1]. It can also be proved using graph theory, see the Epilogue.

The proof of the fact that the usual ordering on **R** is complete requires a careful study of the construction of real numbers. There are two standard ways to construct real numbers from rational numbers, one due to Cantor and one to Dedekind. The second method is designed to make **R** order complete by very construction. Of course, ultimately both the methods yield the same properties of real numbers. For a discussion of both the methods, and for a proof of their equivalence, see, for example, Goffman [1]. Completeness of the continuum (which is the old name for the real number system) may very well be considered to be the corner stone of the continuous mathematics. It would be difficult to think of a really non-trivial concept or theorem of calculus which does not require it.

For infinite posets, the existence of maximal chains requires what is known as the **axiom of choice**. Simply stated, this means that given a collection of mutually disjoint, non-empty sets, it is possible to choose one element from each of them and form a set. This seems very obvious, at least for finitely many sets. But many of its equivalent formulations are not so obvious. One such formulation is, in fact, the existence of a maximal chain in every poset. Another statement equivalent to the axiom of choice is that every set can be well-ordered. For a discussion of this axiom and its use in mathematics, see Halmos [1]. Extending a partial order to a linear order also requires the axiom of choice, in case the underlying set is infinite. Since we shall be dealing mostly with finite sets, we shall rarely need it. We have mentioned it only to stress that the theory of infinite sets needs substantially different techniques from that of finite sets.

4. Algebraic Structures

We have mentioned many times that the limiting process is the essence of continuous mathematics. If the limiting process is taken away, what is left is mostly algebraic manipulations. While the staunch lovers of the continuous mathematics may scoff at this algebraic part as 'raw computation', it is precisely the algebraic aspect that is of interest to us. When properly abstracted, it leads to a powerful species of mathematical structures, called the algebraic structures. It includes Boolean algebras, groups, rings, fields and vector spaces. They will be studied one by one in subsequent chapters. What we present here is some of the most basic concepts about algebraic structures.

Whenever we do some algebraic manipulation such as adding, subtracting, multiplying or dividing on numbers we do so on two numbers at a time. We often encounter expressions like 'the sum of this series' or 'the product of the first n integers'. But they really involve the process of summing or multiplying two numbers at a time and carrying this out repeatedly. (In case of an infinite series, this way we get only the so-called partial sums and then a limiting process is needed.) Two real numbers, subjected to any one of these operations give rise to another real number. It is clear that if we want to define abstract mathematical structures of this type, we must first replace the set of real numbers by an 'abstract' set X. To define an abstract 'addition' on this set X, we must specify for every two elements x and y in X, some element of X to be called their sum and to be denoted by $x + y$. This amounts to defining a function from the cartesian product $X \times X$ into X. Such a function is given a name.

4.1 Definition: A **binary operation** on a set X is a function from $X \times X$ into X. More generally for every positive integer n, an **n-ary operation** on X is a function from the product set $X^n = X \times X \times \ldots \times X$ (n times) into X. (For $n = 1$ and 3 the operation is called **unary** and **ternary** respectively. Since $n = 2$ is the most important case, an 'operation' would mean a binary operation unless otherwise stated.)

Thus addition and multiplication are binary operations on the set of real numbers. When so viewed, the equation $3 + 4 = 7$ should really be expressed as $+ (3, 4) = 7$ (or, more fussily, as $+ ((3, 4)) = 7$), and the equation $3 + 4.2 = 11$ as $+ (3,.(4, 2)) = 11$. This is obviously clumsy. The usual notations are too standard and convenient to be changed. If they are changed, most of the familiar theorems of algebra would appear unintelligible. For example, the result that for all $x, y \in \mathbf{R}$,

$$(x + y) \cdot (x - y) = x^2 - y^2$$

will take the form,

$$\cdot [+ (x, y), - (x, y)] = - [\cdot (x, x), \cdot (y, y)]!$$

True, it is only a matter of habit. But if we want to serve the familiar binary operations for real numbers as guiding lights in the process of defining abstract algebraic structures, we better not alter them, even notationally. On the other hand, we do want uniformity of notation between the usual binary operations on **R** and the abstract binary operations on abstract sets. The only way out is, therefore, to adopt suitable notations for the latter, rather than change the notations for the former. Thus, even though every binary operation on a set X is a function, it is rarely denoted by the standard symbols for a function such as f, g, h etc. The common notations for binary operations are ,·, $+$, $*$ and a few other notations in specific contexts. If $*$ is a binary operation on a set X and $x, y \in X$ then $*(x, y)$ is denoted by $x * y$. Also, to keep conformity with the usual multiplication for real numbers, even for an abstract set, if the binary operation is denoted by ·, $x \cdot y$ is often denoted by xy. A few other such conventions, designed to keep conformity with real numbers, will be pointed out as we proceed.

Note that usual addition and multiplication are also well-defined binary operations on **N**, **Z** and **Q**. However, if we take **N**, then subtraction does not define a binary operation. For, if $x, y \in$ **N** and $x < y$ then $x - y$ is no longer in **N**. (Of course we can define $x - y$ arbitrarily as, say, 25 in all such cases and get a well-defined binary operation on **N** $\times$ **N**; but it can hardly be called subtraction.) On the other hand, on **Z** or **Q**, subtraction does define a binary operation. What is crucially involved here is that whenever one element of **Z** (or **Q**) is subtracted, as a real number, from another element of **Z** (or **Q**) the result stays within **Z** (or **Q**). This is not always true for **N**. When we peform subtraction on two elements of **N**, the result may sneak out of **N**. It is this intuition that lies behind the peculiar name given to this property in the following definition.

4.2 Definition: Let $*$ be a binary operation on a set X. Then a subset Y of X is said to be **closed** under (or w.r.t.) $*$ if for all $x, y \in Y$, $x * y \in Y$.

The term 'closed under' something appears frequently and in many different contexts. But every time the connotation is the same, namely not having to go out of that set, whenever that 'something' is performed to it. For example, a family $\mathscr{F}$ of sets is said to be closed under unions if for every subfamily $\mathscr{G}$ of $\mathscr{F}$, the union of members of $\mathscr{G}$ is also a member of $\mathscr{F}$ (although not necessarily of $\mathscr{G}$). Similarly a set Y of real numbers is said to be closed under suprema if for every subset A of Y, the supremum of A, if at all it exists, is also in Y.

Coming back to Definition 4.2, suppose Y is closed under $*$. Then we get a well-defined binary operation on Y, defined by $*$ itself. This binary operation is said to be **induced on** Y by the binary operation $*$ on the ambient set X. It may be denoted by $*/Y$ but is generally denoted by $*$ itself because as far as its action on any two elements of Y is concerned, it is the same as the original binary operation $*$ on X.

Before giving other examples of binary operations, it is convenient to introduce names for binary operations satisfying certain conditions.

4.3 Definition: Let $*$ be a binary operation on a set X. Two elements

x and y are said to **commute with** each other if $x * y = y * x$. $*$ is called **commutative** if for all x, $y \in X$, $x * y = y * x$. $*$ is called **associative** if for all x, y, $z \in X$, $x * (y * z) = (x * y) * z$. (In other words, $*$ is commutative iff every two elements commute with each other.)

For example, the usual addition on **R** is both commutative and associative, but the binary operation of subtraction is neither commutative nor associative. If we define $*$ on **R** by $x * y = \dfrac{x + y}{2}$ then $*$ is commutative but not associative, while the operation $\odot$ defined by $x \odot y = x$ for all $x, y \in \mathbf{R}$ is associative but not commutative.

Because of commutativity, the order in which the two elements appear is immaterial. The symbol $+$ is generally not used for non-commutative binary operations. In absence of associativity, an expression like $a * b * c$ would be ambiguous since it can be interpreted either as $a * (b * c)$ or as $(a * b) * c$. If $*$ is associative, it is unnecessary to put the parentheses. More generally, if $*$ is an associative binary operation on a set X then for any $a_1, a_2, ..., a_n \in X$ the expression $a_1 * a_2 * ... * a_n$ can have only one meaning. If further, $*$ is also commutative, then the a_i's can be permuted among themselves in any order and all the resulting expressions would be equal. This fact is used innumerably many times for the usual addition and multiplication for real numbers often without being conscious of their commutativity and associativity. In an abstract set, care must be exercised while handling such expressions, making sure that no more properties are used than are assumed as axioms or have already been proved from the axioms. Note, for example, that without associativity, the familiar law of indices which says that $a^{m+n} = a^m . a^n$ for positive integers m and n would be meaningless. Indeed, the difficulty would be to define the power of an element. If $*$ is a binary operation on a set X and $a \in X$ then we may define a^2 as $a * a$. But $(a * a) * a$ need not be the same as $a * (a * a)$ (e.g. let $*$ be subtraction). So we have difficulty in defining a^3. We may try to bypass the difficulty by defining a^3 only as $a^2 * a$ (and more generally, a^n inductively as $a^{n-1} * a$ when $n \geqslant 2$). But then the law of indices would not hold. We leave it to the reader to prove that the law does hold if $*$ is associative (commutativity is not needed). If $*$ is denoted by $+$, the nth power of a, a^n, is also commonly denoted by na. Note that as yet we have given no meaning to powers where the exponent is not a positive integer.

We now list a number of additional examples of binary operations.

1. Let S be any set and $X = P(S)$, the power set of S. Then $\cup$ and $\cap$ are two binary operations on X. Both are commutative and associative.
2. More generally, if $(X, \leqslant)$ is a lattice, then the meet $\wedge$, and the join $\vee$ both define binary operations on X. Both these operations

are also commutative and associative. Every sub-lattice of X is closed under these operations.

3.   Let S be a set and X the set of all functions from S to itself. For $f, g \in X$, let $f \circ g$ be the composite of the functions f and g. Then $f \circ g \in X$. Thus $\circ$ defines a binary operation on X. This operation is associative but not commutative in general. Note that the set of all permutations of S is a subset of X and it is closed under $\circ$, because the composite of two permutations of S is again a permutation of S.

4.   Suppose $*_1$ and $*_2$ are binary operations on two sets X_1 and X_2 respectively. Let $X = X_1 \times X_2$. Then we can define a binary operation $*$ on X by

$$(x_1, x_2) * (y_1, y_2) = (x_1 *_1 y_1, x_2 *_2 y_2).$$

This operation is said to have been obtained by **co-ordinatewise application** of $*_1$ and $*_2$. The construction can obviously be generalised for the product of any number of sets with binary operations. This gives a way to construct new examples of binary operations. (We assume the sets $X_1, X_2 \ldots$ to be non-empty, or else their product would be empty). Clearly $*$ is commutative if and only if $*_1$ and $*_2$ are so. A similar statement holds for associativity. Often the sets $X_1, X_2, \ldots$ and the binary operations $*_1, *_2, \ldots$ are all equal. Then we get a binary operation on the power X^n. A common illustration of this is the 'co-ordinatewise addition' of elements of the n-dimensional euclidean space $\mathbf{R}^n$. (Caution: For $n = 2$, we may identify $\mathbf{R}^2$ with the set of complex numbers by letting an element (x, y) of $\mathbf{R}^2$ correspond to the complex number $x + iy$. Then the co-ordinatewise addition corresponds to the addition of complex numbers. But the co-ordinatewise multiplication of elements of $\mathbf{R}^2$ does not correspond to the usual multiplication of complex numbers.)

5.   On $\mathbf{R}^3$, there is one more binary operation. We identify points of $\mathbf{R}^3$ with vectors in three dimensional space. Specifically to (x_1, x_2, x_3) in $\mathbf{R}^3$ we associate the vector $x_1\mathbf{i} + x_2\mathbf{j} + x_3\mathbf{k}$ where $\mathbf{i}, \mathbf{j}, \mathbf{k}$ is a right handed orthonormal system. Then the cross product of vectors defines a binary operation $\times$ on $\mathbf{R}^3$. This operation is neither commutative nor associative. Note incidentally, that the dot product (also called the scalar product) of vectors does not define a binary operation because the dot product of two vectors is not a vector, it is only a scalar.

6.   Let X, Y be sets and let F be the set of all functions from X to Y. (In earlier notations, $F = Y^X$). Suppose $*$ is a binary operation on Y. Then $*$ induces a binary operation on F by what is known as the **pointwise application** of $*$. This is done as follows. Let $f, g \in F$. Then both f and g are functions from X to Y. Let $x \in X$. Then $f(x)$ and $g(x)$ are elements of Y. So we can apply $*$ to them and

get $f(x) * g(x)$ which is another element of Y. This element depends on x (f, g and $*$ being fixed) and thus we get a function from X to Y which associates to $x \in X$, the element $f(x) * g(x)$ of Y. This new function is denoted by $f \circledast g$. Then $f \circledast g \in F$ whenever $(f, g) \in F \times F$ and thus we get a binary operation on F. For example, if $X = Y = \mathbf{R}$ and $*$ is the usual addition, then for $f(x) = e^x$ and $g(x) = \sin x$ we have $f \oplus g$ defined by $(f \oplus g)(x) = e^x + \sin x$. $f \circledast g$ is often denoted by $f * g$ but a beginner occasionally finds the double role of $*$ confusing; on one hand it is a binary operation on Y and the same symbol is used for a binary operation on F, whose elements are functions into the set Y. It is clear that if $*$ is commutative or associative then $\circledast$ also has the corresponding properties. There are many other properties which pass similarly from $*$ to $\circledast$. Because of this, the construction here is very useful in providing new examples of algebraic structures of a given type. Often, instead of taking the full set F, some of its subsets which are closed under $\circledast$ are more interesting. For example, if $X = Y = \mathbf{R}$ and $*$ is the usual $+$ or $\cdot$ for $\mathbf{R}$, then the set of all continuous function from $\mathbf{R}$ to $\mathbf{R}$ is closed under the binary operations $\oplus$ and $\odot$. (In simple language this means that sum and the product of two continuous functions are continuous.)

7. As an especially instructive example, let n be a positive integer and let $\mathbf{Z}_n$ be the set of residue classes of integers modulo n. Recall from Section 2, that these are the equivalence classes of the relation R on Z defined by xRy iff $x - y$ is divisible by n, for $x, y \in \mathbf{Z}$. We define the addition and the multiplication of these residue classes as follows. Take any two residue classes in $\mathbf{Z}_n$. Then they will be of the form $[a]$ and $[b]$ for some $a, b \in \mathbf{Z}$. Since a, b are integers we can add and multiply them under the usual binary operations on $\mathbf{Z}$. Thus we get $a + b$ and $a \cdot b$ as elements of $\mathbf{Z}$. So, $[a + b]$ and $[a \cdot b]$ are elements of $\mathbf{Z}_n$. It is tempting to define $[a] + [b]$ as $[a + b]$ and $[a] \cdot [b]$ as $[a \cdot b]$. But to make these definitions valid, we must check that they are well-defined, i.e., independent of the choice of the representatives of the equivalence classes (see the comments in the proof of Proposition (2.14)). Specifically, we have to show that if a, b, c, d are integers such that $[a] = [c]$ (i.e., $a \equiv c \bmod n$) and $[b] = [d]$ (i.e. $b \equiv d \bmod n$) then $[a + b] = [c + d]$ and $[a \cdot b] = [c.d]$. These verifications are simple. Since $a \equiv c$ and $b \equiv d$ mod n, there exist integers p and q such that $a - c = np$ and $b - d = nq$. Now $(a + b) - (c + d) = (a - c) + (b - d) = np + nq = n(p + q)$. Since $p + q$ is an integer, it follows that $a + b \equiv c + d \bmod n$. For multiplication, note that $ab - cd = ab - bc + bc - cd = b(a - c) + c(b - d) = bnp + cnq = n(bp + cq)$. Since p, q, b and c are integers, so is $bp + cq$. It follows that $[ab] = [cd]$. We, there-

fore have two well-defined binary operations, $+$ and $\cdot$ on Z_n, called respectively **residue** addition (or **mod n addition**) and **residue multiplication** (or **mod n multiplication**). It is clear that both these operations are commutative and associatives because of the commutativity and associativity of usual $+$ and $\cdot$ on Z.

Using the residue addition and multiplication, we can do the Division Problem elegantly, that is, without the brute force computation. Let $x = 2.3.5.7.11.13.17 + 1$. We have to check whether x is divisible by 19. Let us consider the residue classes modulo 19. Then the problem reduces to asking whether $[x]$ and $[0]$ are the same elements of Z_{19}. Now because of commutativity and associativity of mod 19 multiplication we have,

$$[x] = [2] \cdot [3] \cdot [5] \cdot [7] \cdot [11] \cdot [13] \cdot [17] + [1]$$

$$= [2.11] \cdot [3.13] \cdot [5.7] \cdot [17] + [1]$$

$$= [22] \cdot [39] \cdot [35] \cdot [17] + [1]$$

$$= [3] \cdot [1] \cdot [-3] \cdot [-2] + [1]$$

$$\text{(since } 22 \equiv 3 \text{ mod 19 etc.)}$$

$$= [3.1. \ (-3) \cdot (-2)] + [1]$$

$$= [18] + [1]$$

$$= [19]$$

$$= [0].$$

Thus we see effortlessly that x is divisible by 19. The study of the residue addition and multiplication is called the **residue arithmetic**. The example here is an illustration of its use.

8. Let X be the set of all finite sequences whose entries come from some set S. (We may think of S as an alphabet and then elements of X as 'words'). Given two such sequences, say, $(s_1, s_2, ..., s_m)$ and $(t_1, t_2, ..., t_n)$ we form a third sequence of length $m + n$, namely, $(s_1, s_2, ..., s_m, t_1, t_2, ..., t_n)$. This gives a binary operation on X, called **concatenation** or **juxtaposition**. This operation is associative, but not commutative. (The idea of concatenation was also used in the proofs of Proposition (2.15) and Theorem (3.10)).

9. Given any binary operation $*$ on a set X, we define its **opposite operation** $*'$ by $x *' y = y * x$ for all $x, y \in X$. If $*$ is commutative then $*'$ coincides with $*$. Note that $(*')' = *$. There is an obvious duality between properties of $*$ and $*'$ (analogous to the duality between the properties of a partial order and its reverse order.)

10. When the set X is finite, there is a very handy way to define and represent a binary operation on X, by drawing, what is called the

table of the operation. If $|X| = n$, this table has n rows and n columns, one for each element of X. If $X = \{x_1, x_2, \ldots, x_n\}$ and the binary operation is $*$, then we put $x_i * x_j$ in the ith row and jth column. For example, in Figure 3.14, we show the table for the binary operation of residue multiplication on the set $\mathbf{Z}_6$ (see Example (8) above).

·	[0]	[1]	[2]	[3]	[4]	[5]
[0]	[0]	[0]	[0]	[0]	[0]	[0]
[1]	[0]	[1]	[2]	[3]	[4]	[5]
[2]	[0]	[2]	[4]	[0]	[2]	[4]
[3]	[0]	[3]	[0]	[3]	[0]	[3]
[4]	[0]	[4]	[2]	[0]	[4]	[2]
[5]	[0]	[5]	[4]	[3]	[2]	[1]

Fig. 3.14.   Residue Multiplication for $\mathbf{Z}_6$.

Conversely, we may fill the n^2 places in the table by elements of X in any manner we like and get a well-defined binary operation on X. This way we do not have to describe it by any formula; we are defining it by exhaustively listing its values for all possible ordered pairs of elements of X (that is why, this method requires X to be finite). Analogous to the cartesian representation of a binary relation, certain properties of a binary operation have a vivid interpretation in terms of its table. For example, commutativity is equivalent to symmetry about the diagonal. (For associativity there is no such obvious interpretation.)

We now define two more concepts pertaining to binary operations.

4.4   Definition: Let $*$ be a binary operation on a set X. Then an element e of X is said to be a **left identity** for $*$ if for all $x \in X$, $e * x = x$. A **right identity** is defined similarly. An element which is both a right and a left identity is called a **two-sided identity** or simply **identity** for $*$.

Obviously a left identity for an operation $*$ is right identity for its opposite operation $*'$ (Example (9) above), and vice-versa. So it suffices to give examples or to prove theorems for one of these concepts. Also for a commutative operation, the distinction between left and right identities disappears. For non-commutative operations, several things can happen. For example, let X be any set and define $*$ on X by $x * y = x$ for all $x, y \in X$. Then every element of X is a right identity but there is no left identity for $*$ unless X is a singleton set. The operation of subtraction for real numbers has 0 as the only right identity and no left identity. The following simple but useful result shows that thing are smooth if both types of identities are present.

4.5 Proposition: Let $*$ be a binary operation on a set X and suppose it has both a left identity and a right identity. Then they are equal. Also there are no other left or right identities.

Proof: It suffices to show that any left identity of $*$ equals any right identity for $*$. Let e and f be respectively such identities. Then the element $e * f$ is on one hand equal to f (since e is a left identity) and on the other hand equal to e (since f is a right identity). So $e = f$. ∎

For real numbers, 0 and 1 are, respectively, the identities for addition and multiplication. Generally, even for an abstract set X, when a binary operation on X is denoted by $+$ (as noted before this is rarely done when it is not commutative) its identity (if any) is denoted by 0. When the operation is denoted by $\cdot$, its identity (if any) is denoted by 1 or by e or by some other symbol. Note that the identity function on any set, $1_S: S \to S$ is also the identity element for the binary operation of composition on the set of all functions from S to S (Example (3) above). Hence probably the name. We leave it to the reader to find which operations given above have left or right identities. If e is an identity for a binary operation $*$ on a set X and Y is a subset of X which is closed w.r.t. $*$ then three things can happen: (i) $e \in Y$, in which case e is also the identity of the induced binary operation on Y, (ii) $e \notin Y$, but the induced binary operation on Y has its own identity, e.g. let $X = \mathbf{R}$, $* = \cdot$ and $Y = \{0\}$ and (iii) $e \notin Y$ and the induced operation on Y has no identity, e.g. $X = \mathbf{R}$, $* = \cdot$ and

$$Y = \{x \in \mathbf{R}: x > 2\}.$$

We invite the reader to give more examples of these three possibilities.

Having discussed identities, let us discuss one more important concept, that of an inverse of an element. We often make somewhat loose statements such as 'subtraction is the opposite (or inverse or reverse) of addition' or 'division is the opposite of multiplication'. What do they really mean? Let us take the case of subtraction. When we take a real number a, subtract some other real number b from it, and write the answer as $c(= a - b)$, we are really answering the question, 'which real number, when added to b

will give us a?' In other words, we are solving the equation $b + x = a$ where x is an unknown. The solution can be written as $a + (-b)$ where $-b$ is the negative of b. Thus subtraction can be expressed in terms of addition and the concept of the negative of a real number. Similarly division can be expressed in terms of multiplication and the concept of the reciprocal of a (non-zero) real number. There is somthing common to these two concepts. When a real number and its negative are added, we always get 0, the identity of addition. Similarly when a (non-zero) real number and its reciprocal are multiplied, we always get 1, the identity of multiplication.

Let us now isolate this common feature and put in the wider setting of an 'abstract' binary operation on an abstract set, keeping in line with the spirit of abstraction discussed in Section 1. We have to be a little careful because while both the operations above (namely $+$ and $\cdot$ for **R**) are commutative, our abstract binary operation need not be so.

4.6 Definition: Let e be an identity for a binary operation $*$ on a set X. Let $x \in X$. Then x is called **right invertible** if there exists $y \in X$ such that $x * y = e$. Any such element y is called a **right inverse** of x. The concepts of **left invertibility**, **left inverse**, two **sided invertibility** (also simply called **invertibility**) and **two sides inverse** (simply called **inverse**) are defined similarly.

We could define these concepts when e is just a right identity or a left identity (instead of being two-sided identity). In that case, since right identities need not be unique we would have to define these concepts w.r.t. a particular right identity. The generality so gained is worthwhile sometimes. But for most purposes, the definition above is adequate. As foremost examples, we see that for the usual addition on **R**, every element x is invertible and $-x$ is the unique inverse of x. Similarly for the usual multiplication on **R**, every non-zero x is invertible with its reciprocal as the unique inverse. For the example (3) above, we see from Exercise (2.1.11), that a function f is left-invertible if and only if it is injective (assuming $S \neq \phi$). However, simple examples show that the same function may have more than one left inverse. For example, let $S = \mathbf{N}$ and define $f: S \to S$ by $f(x) = x + 1$. Define $g: S \to S$ by $g(y) = y - 1$ for $y > 1$. Then $g(1)$ may be defined arbitrarily and any such g would be a left inverse of f. Similarly f has a right inverse iff f is onto, but the right inverse may not be unique. However, f is invertible iff f is a bijection and in this case the inverse is unique; it is the inverse function $f^{-1}: S \to S$. This notation is also used in the case of an abstract binary operation, provided the inverse is unique. If the binary operation is denoted by $+$, then instead of x^{-1} we frequently write $-x$. If every element of X has a unique inverse then we get a well-defined function, called **inversion** from X into itself, which assigns to each $x \in X$, its (unique) inverse, x^{-1}. Note that in such a case $(x^{-1})^{-1} = x$ for all $x \in X$. In other words, the inversion function composed with itself gives the identity function. Also note that inversion is a unary operation on the set X.

Unfortunately, without associative law, the inverses do not behave the expected way. The analogue of Proposition 4.5 does not hold, as shown by the operation $*$ on a set $\{a, b, c, d, e\}$ whose table is drawn in Figure 3.15. Here a is the identity. The element b has c as the unique right inverse and d as the unique left inverse. Still it is not invertible. The element e has two inverses, b and c.

$*$	a	b	c	d	e
a	a	b	c	d	e
b	b	d	a	d	a
c	c	d	d	d	a
d	d	a	d	d	b
e	e	a	a	b	c

Fig. 3.15: **Pathological Behaviour of Inverses.**

Things are substantially improved if we have associative law as we now show.

4.7 Proposition: Let $*$ be an associative binary operation on a set X, with identity e. Suppose an element $a \in X$ is right invertible. Then,

   (i) a can be **cancelled from the right**, i.e., whenever x, $y \in X$ and $x * a = y * a$, we have $x = y$.
   (ii) if further, a is also left invertible, then its right and left inverses are unique and equal. Hence a is invertible and has a unique inverse.

Further, if a, b are any invertible elements, so is $a * b$ and

$$(a * b)^{-1} = b^{-1} * a^{-1}.$$

and for every positive integer n, a^n is invertible with inverse $(a^{-1})^n$. Finally if a is invertible, then for every $c \in X$, the equations $x * a = c$ and $a * x = c$ have unique solutions. (However, these solutions need not be equal.)

Ptoof:　Suppose b is a right inverse of a. Then,

(i)　$x * a = y * a$ implies $(x * a) * b = (y * a) * b$, which by associativity gives $x * (a * b) = y * (a * b)$, or, $x * e = y * e$. But this means $x = y$ as desired.

(ii)　Let c be a left inverse of a. By associativity,

$$(c * a) * b = c * (a * b).$$

Now, the left hand side equals $e * b$, i.e. b. Similarly the right hand side equals c. We have thus shown that any left inverse of a and any right inverse of a are equal to each other. This proves that there is only one inverse.

Now let a, b have inverses a^{-1}, b^{-1} respectively. Then $(a * b) * (b^{-1} * a^{-1})$ equals, by associativity, $a * [(b * b^{-1}) * a^{-1}]$ which reduces to e. Similarly $(b^{-1} * a^{-1}) * (a * b) = e$. Thus $a * b$ is invertible with inverse $b^{-1} * a^{-1}$. In particular, if we take $b = a$ then $(a^{-1})^2$ is the inverse of a^2. The assertion that a^n is invertible with inverse $(a^{-1})^n$ can easily be proved by induction on n. The case $n = 1$ is trivial. For $n > 1$, we simply write a^n as $a * (a)^{n-1}$ (note again that associativity of $*$ is used). Then set $b = (a)^{n-1}$. Then by induction hypothesis, $b^{-1} = (a^{-1})^{n-1}$ and so $b^{-1} * a^{-1} = (a^{-1})^n$. For the last assertion, the equation $x * a = c$ has $x = c * a^{-1}$ as a solution as seen by direct calculation. Uniqueness of the solution follows by (i). The equation $a * x = c$ similarly has $x = a^{-1} * c$ as its unique solution. ▟

The fact that the inverse of a product equals the product of the inverses in the reverse order resembles our everyday experience that when we dress up we first put on a shirt and then a coat but when we undress, we first take off the coat and then the shirt! The resemblance will be brought out more fully when we discuss actions of algebraic structures†. Of course, if $*$ is commutative (in addition to being associative), then $(a * b)^{-1} = a^{-1} * b^{-1}$. Using the concept of an inverse, we now define the negative powers of an invertible element $a \in X$. If n is any positive integer, we define $a^{-n} = (a^{-1})^n$ which also equals $(a^n)^{-1}$ by the proposition above. By convention we set a^0 to be the identity element. We then have the following 'laws of indices' whose proof is left as an exercise.

4.8. Proposition: Let a be an invertible element of a set X, with an associative binary operation $*$, having identity e. Then for all integers m and n we have

(i)　$a^{m+n} = a^m * a^n$

(ii)　$(a^m)^n = a^{mn}$. ▟

The fact that an invertible element can be cancelled from the two sides

† See the Epilogue.

of an equation is frequently used in simplification of equations. This property is given a name.

4.9   Definition: A binary operation $*$ on a set X is said to satisfy the **left cancellation law** if for all $x, y, z \in X$, $x * y = x * z$ implies $y = z$. **Right cancellation** law is defined similarly.

Because of (i), in Proposition (4.7), if $*$ is associative and every element is right invertible then $*$ satisfies the right cancellation law. The converse is false. The usual multiplication for the set of positive integers is associative, has an identity and also obeys both the cancellation laws. Still, no element except 1 is invertible in **N**. (As real numbers, all elements of **N** are invertible; but we want the inverses to be elements of **N**.) Interestingly, for finite sets the converse holds as we now prove.

4.10.   Theorem: Suppose $*$ is an associative binary operation on a finite set X. If $*$ satisfies both the cancellation laws, then $*$ has an identity and every element of X is invertible.

Proof: We first show that for any $a, b \in X$, the equation $a * x = b$ has a unique solution for x. Given such a, b, define a function $f_a : X \to X$ by $f_a(x) = a * x$ for $x \in X$. By left cancellation law, f_a is one-to-one. But since X is finite, by Theorem (2.2.9), f_a is onto. In particular b is in the range of f_a. This says precisely that the equation $a * x = b$ has at least one solution. Uniqueness of the solution follows from the left cancellation law. Similarly, using the right cancellation law, we see that for any $a, b \in X$, the equation $x * a = b$ has a unique solution for x. (This solution may differ from that of the equation $a * x = b$.)

Now fix any $a \in X$. Taking $b = a$, there exists a unique $e \in X$ such that $a * e = a$. We contend that e is a right identity for $*$. For this, suppose $c \in X$. We have to show $c * e = c$. Now, by what we showed above, there exists $x \in X$ such that $x * a = c$. But then $c * e = (x * a) * e = x * (a * e)$ (by associativity) $= x * a = c$ as was to be proved. So $*$ has a right identity, namely e. By similar reasoning, $*$ has a left identity. But then, by Proposition (4.5), $*$ has a two sided identity, which we denote by e.

It remains to show that every element of X is invertible. Let $a \in X$. By the existence of solutions proved above, there exist $b, c \in X$ such that $a * b = e$ and $c * a = e$. Then b is a right inverse of a and c is a left inverse of a. But $*$ is associative. So by Proposition (4.7), a is invertible.   ∎

Note that in this proposition it is not enough to have just one cancellation law. For example, if X is any set and we define $x * y = x$ for all $x, y \in X$ then $*$ is associative and satisfies the right cancellation law. But the left cancellation law does not hold and no element has a left inverse (unless X is a singleton set). Note that in this example, $*$ has no identity. If we are given beforehand that $*$ has an identity then the preceding proposition can be proved using just one cancellation law. We leave this as an exercise.

So far we discussed only one binary operation on a set at a time. We frequently have situations where there are two (or more) binary operations on the same set, and they are inter-related in some way. We define the most important inter-relationship of this type.

4.11. Definition: Suppose $\cdot$ and $*$ are two binary operations on a set X. Then $\cdot$ is said to be **left distributive** over $*$ if for all $x, y, z \in X$,

$$x \cdot (y * z) = (x \cdot y) * (x \cdot z).$$

Right distributivity and **two-sided distributivity** (or simply **distributivity**) are defined analogously. [If $\cdot$ is commutative either distributivity implies the other. But commutativity of $*$ has no role to play.]

As a foremost example, the usual multiplication for real numbers is distributive over the usual addition. Here, the addition is not distributive over multiplication. However, sometimes we have two operations, each of which is distributive over the other. For example, in Example (1) above, $\cup$ and $\cap$ are distributive over each other. As an example where only one-sided distributivity holds, let X be any set and $*$ be any binary operation on a set X. Define another binary operation $\cdot$ on X by $x \cdot y = y$ for all $x, y \in X$. Then for all $x, y, z \in X$, both the expressions $x \cdot (y * z)$ and $(x \cdot y) * (x \cdot z)$ equal $y * z$ and hence $\cdot$ is left distributive over $*$. But in general the equality $(y * z) \cdot x = (y \cdot x) * (z \cdot x)$ does not hold unless the operation $*$ is such that $x * x = x$ for all $x \in X$. This example also shows how an inter-relationship (in this case right distributivity) between two operations imposes certain restrictions on one (or both) of them. We shall see more instances of this in later chapters. For the moment we prove one result of this type.

4.12 Proposition: Let $\cdot$ and $*$ be two binary operations on a set X with $\cdot$ distributive over $*$. Suppose $\cdot$ has an identity. Assume $*$ is associative and satisfies both the cancellation laws. Then $*$ is commutative.

Proof: Let 1 be the identity for $\cdot$. Let $a, b \in X$. We have to show $a * b = b * a$. Let $c = 1 * 1$. Then,

$$c \cdot (a * b) = (c \cdot a) * (c \cdot b) \quad \text{(by left-distributivity of } \cdot \text{ over } *)$$
$$= [(1 * 1) \cdot a] * [(1 * 1) \cdot b]$$
$$= [(1 \cdot a) * (1 \cdot a)] * [(1 \cdot b) * (1 \cdot b)] \quad \text{(by right distributivity)}$$
$$= (a * a) * (b * b).$$

But on other hand

$$c \cdot (a * b) = (1 * 1) \cdot (a * b)$$

$$= [1 \cdot (a * b)] * [1 \cdot (a * b)] \text{ (by distributivity again)}$$

$$= (a * b) * (a * b).$$

Hence

$$(a * a) * (b * b) = (a * b) * (a * b)$$

Since $*$ is associative we can remove the parentheses. Also because of cancellation laws, we can cancel a from the left and b from the right. This gives $a * b = b * a$ as desired. ∎

By now we have discussed a fair number of general concepts about binary operations in general. We close this section with a brief discussion of abstract algebraic structures. The reader may possibly find it a little intricate. In subsequent chapters we shall study many 'concrete' algebraic structures. Each such 'concrete' structure will itself be abstract in the sense that it will be defined on an abstract set and it is only after specifying a particular choice of the underlying set that a concrete example of it will be obtained. What we now propose to do is, therefore, an abstraction of the second order. A beginner may find it useful to return to it again after having seen several particular examples of algebraic structures.

4.13 Definition: An **algebraic structure** is an ordered triple, $(X, \mathscr{S}, \mathscr{C})$ where X is a set, $\mathscr{S}$ is a finite sequence of operations on X say

$$\mathscr{S} = (*_1, *_2, ..., *_k),$$

where k is some positive integer and for each $i = 1, 2, ..., k$, $*_i$ is an n_i-ary operation on X, $(n_1, n_2, ..., n_k$ being positive integers) and $\mathscr{C}$ is a set of conditions, each involving one (or more) of these operations on X. The set X is called the **underlying set** of the algebraic structure.

For example, a **monoid** is a triple, $(X, (*), \mathscr{C})$ where X is a set, $*$ is a binary operation X and the set $\mathscr{C}$ consists of two conditions, (i) $*$ be associative and (ii) $*$ have a two-sided identity. There are, of course, many other types of algebraic structures such as Boolean algebras, groups, rings, fields etc. We classify them according to the following rule.

4.14 Definition: Two algebraic structures $(X, \mathscr{S}, \mathscr{C})$ and $(X', \mathscr{S}', \mathscr{C}')$ are said to be of the **same type** (or of the **same category**) if there exists a bijection between $\mathscr{S}$ and $\mathscr{S}'$ and a bijection between $\mathscr{C}$ ane $\mathscr{C}'$ and these two bijections are compatible with each other.

Note that we are not requiring that there be a bijection between X and X'. The definition simply means that if $\mathscr{S} = (*_1, *_2, ..., *_k)$, with $*_i$ an n_i-ary operation X, then we can express $\mathscr{S}'$ as $(*_1', *_2', ..., *_k')$ with $*_i'$ an n_i-ary operation X' and for every condition in $\mathscr{C}$, if we replace each $*_i$ occurring in it by $*_i'$, we get a condition in $\mathscr{C}'$ and vice versa. For example, any two monoids, say, $(X, (*), \mathscr{C})$ and $(X', (*'), \mathscr{C}')$ are algebraic structures of the same type. On the other hand a monoid cannot be of the same type as an algebraic structure in which two binary operations are involved, or one in which only one binary operation is involved but the conditions involving it are different.

While dealing with algebraic structures of the same type, we may omit the sets of conditions, $\mathscr{C}$ and $\mathscr{C}$ from notation.

Obviously, at any one time we would be proving theorems about algebraic structures of the same category. In general, there is very little that would appply to all possible algebraic structures. Nevertheless a few basic concepts can be defined in such generality.

4.14 Definition: Let $(X, \mathscr{S}, \mathscr{C})$ be an algebraic structure. Then an algebraic structure $(Y, \mathscr{S}', \mathscr{C}')$ is called a **substructure** of $(X, \mathscr{S}, \mathscr{C})$ if the following conditions hold:

(i) Y is a subset of X and it is closed under each of the operations in $\mathscr{S}$,

(ii) $\mathscr{S}'$ is the sequence of the operations on Y induced by the operations in $\mathscr{S}$ and

(iii) $\mathscr{C}'$ consists of the conditions in $\mathscr{C}$, with each binary operation in $\mathscr{C}$ replaced by the corresponding induced operation on Y.

In this context the structure $(X, \mathscr{S}, \mathscr{C})$ is called the **ambient** structure.

Clearly, a substructure of an algebraic structure is of the same type as the original structure. For example, a submonoid of a monoid is itself a monoid.

A homomorphism from one algebraic structure to another of the same type is a function between the underlying sets which preserves, or is compatible with the operations. A formal definition is as follows:

4.15 Definition: Let $(X, (*_1, *_2, ..., *_k), \mathscr{C})$ and $(X', (*_1', *_2', ..., *_k'), \mathscr{C}')$ be algebraic structures of the same type. Then a function $f: X \to X'$ is called a **homomorphism**, if for every $i = 1, ..., k$ and for all $x_1, x_2, .., x_{n_i} \in X$ (where n_i is the integer such that both $*_i$ and $*_1'$ are n_i-ary operations on X and X' respectively), we have

$$f(*_i(x_1, ..., x_{n_i})) = *_i'(f(x_1), f(x_2), ..., f(x_{n_i})).$$

For example, if $(X, (*), \mathscr{C})$ and $(X', (*'), \mathscr{C}')$ are monoids then a function $f: X \to X'$ will be called a homomorphism if for all $x, y \in X$,

$$f(x * y) = f(x) *' f(y).$$

In informal terms, the image of a product is the product of the images, or in other words, whether you first multiply and then apply f or whether you first apply f and then multiply, you get the same result.

Homomorphism between two algebraic structures is a fundamental concept in the study of algebraic structures of a given type. Note that we are not requiring the function f to be a bijection. If it is so, then we call f an **isomorphism**, and say that the structures $(X, \mathscr{S}, \mathscr{C})$ and $(X', \mathscr{S}', \mathscr{C}')$ are isomorphic to each other. Isomorphic structures may be considered as replicas of each other. From an abstract point of view they are indistinguishable from each other. In Section 1, we compared two different structures on the same underlying set as two earthenwares of different types formed from the same mound of clay. As a similar analogy, we may regard isomorphic structures as two flasks of exactly the same shape and size but made of possibly different metals.

Suppose $(X, \mathscr{S}, \mathscr{C})$ is an algebraic structure. Let R be an equivalence relation on the underlying set X. Let X/R be the quotient set (See Definition (2.12)). We want to make X/R into an algebraic structure of the same type as the original structure. To do this, we have to define corresponding operations for the set of equivalence classes and the usual question of their being well-defined will arise. The following definition is tailored to meet precisely this difficulty.

4.16. Definition: An equivalence relation R on the underlying set X of an algebraic structure $(X, (*_1, ..., *_k), \mathscr{C})$ is called a **congruence relation** if for all $i = 1, 2, ..., k$, and for all $x_1, ..., x_{n_i}, y_1, ..., y_{n_i} \in X$, $x_j R y_j$ for all $j = 1, 2, ..., n_i$ implies

$$[*_i(x_1, ..., x_{n_i})] \; R \, [*_i(y_1, ..., y_{n_i})].$$

For example, let $X = \mathbf{Z}$ and let $*_1 = +$, the usual addition and $*_2 = \cdot$, the usual multiplication. Then for any positive integer n, the equivalence relation of congruency modulo n is a congruence relation (see Example (7) above). Hence the name. The following basic theorem is now easily proved.

4.17. Theorem: Let R be a congruence relation on the underlying set of an algebraic structure $(X, \mathscr{S}, \mathscr{C})$. Then the quotient set X/R can be made into an algebraic structure in such a way that the quotient function $p: X \to X/R$ is a homomorphism. (The resulting algebraic structure is called a **quotient structure** of the original structure.)

Proof: Suppose $\mathscr{S} = (*_1, *_2, ..., *_k)$, where $*_i$ s an n_i-ary operation on X. Let $[x_1], [x_2], ..., [x_{n_i}] \in X/R$. Define

$$*_i'([x_1], ..., [x_{n_i}]) = [*_i(x_1, ..., x_{n_i})].$$

This is well-defined since R is a congruence relation. This gives the algebraic structure on X/R. Also, since $p(x) = [x]$ for all $x \in X$, it is clear that p is a homomorphism. ▨

If we want to apply this theorem, we must have a way of finding conguence relations on the underlying set. It turns out that if the structure

$(X, \mathscr{S}, \mathscr{C})$ is sufficiently strong (i,e. the conditions in $\mathscr{C}$ are sufficiently power-ful) then certain substructures of $(X, \mathscr{S}, \mathscr{C})$ induce congruence relations on X. We shall see this in particular cases in the chapters to come.

A comment about the notations used for particular types of algebraic structures is in order. First, when we are dealing with structures of the same type, it is customary to suppress $\mathscr{C}$ from notation and to list the con-ditions in $\mathscr{C}$ as the postulates or axioms of that particular type of structures. Secondly, the structures we shall deal with will have only a few operations (often one or two and never more than three). Consequently, it is unneces-sarily fussy to denote them by a sequence. For example, the reader will often find a monoid defined simply as an ordered pair, $(X, *)$ where X is a set and $*$ is an associative binary operation on X having an identity. Also when the operation $*$ is understood, we simply say X is a monoid. Another common simplification of language is that attributes of the underlying set are assigned, by a transfer of epithet, to the algebraic structure in question. (This is also a common practice for other mathematical structures.) Thus when we say that a monoid is finite, it simply means that its underlying set is finite.

Exercises

4.1   How many binary operations are there on a set with n elements? How many of them are commutative? How many have an identity?

4.2   Let $*$ be a binary operation on a set X. Fix $a \in X$. Define $f_a : X \to X$ by $f_a(x) = a * x$ for $x \in X$. This function f_a is called the **left trans-lation** by a. Similarly we define the **right translation** by a. (These functions were already used in the proof of Theorem (4.10).) Prove that a is a left identity for $*$ if and only if the function f_a is the identity function 1_X. (This is probably another justification for the name.)

4.3   It is sometimes helpful to paraphrase the various conditions about binary operations in terms of commutative diagrams. Prove, for example, that a binary operation $*$ on a set X is commutative if and only if the following triangle is commutative where the func-tion $\theta : X \times X \to X \times X$ is defined by $\theta(x, y) = (y, x)$ for all $x, y \in X$.

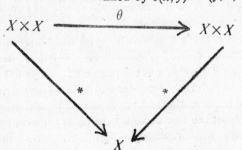

Similarly characterise associativity, distributivity and presence of identities.

4.4 Prove that the left cancellation law for a binary operation ∗ on a set X is equivalent to the assertion that for the function

$$g : X \times X \times X \to X \times X$$

defined by

$$g(a, x, y) = (a * x, a * y) \text{ for } a, x, y \in X, g^{-1}(\Delta X) = X \times \Delta X$$

where ΔX is the diagonal on X (see Section 2.)

4.5 In the examples (1) to (8), find which operations have identities and which elements are invertible.

4.6 Prove Proposition (4.8).

4.7 A lattice is called **distributive** if the binary operations $\wedge$ and $\vee$ are distributive over each other. Prove, in fact, that for any lattice distributivity of $\vee$ over $\wedge$ implies that of $\vee$ over $\wedge$ and vice versa. (Hint: Use absorption laws, namely

$$a \wedge (a \vee b) = a \text{ and } a \vee (a \wedge b) = a$$

for all a, b in a lattice, besides other properties of $\wedge$ and $\vee$.)

4.8 Prove that the power set lattice, the lattice of all positive integers with the partial order defined by divisibility, the lattice of statements with partial order defined by implication and the lattice in Exercise (3.20) are distributive.

4.9 Prove that the lattice of all partitions of a set, with the partial order defined by refinement is not distributive. (Hint: Let the set be the cartesian plane. Consider three partitions of it whose members are, respectivey, all horizontal lines, all vertical lines and all lines of slope 1. More generally, any lattice which contains 3 elements every two of which are non-comparable and every two of which have the same meet and the same join, is non-distributive.)

4.10 Prove that the composition of two homomorphisms is a homomorphism and the composition of two isomorphisms is an isomorphism. Prove also that the inverse of an isomorphism is an isomorphism.

4.11 Prove that $(\mathbf{R}, +)$ and $(\mathbf{R}, \cdot)$ are both monoids. Prove also that the function $f : \mathbf{R} \to \mathbf{R}$ defined by $f(x) = e^x$ is a monoid homomorphism. Is it an isomorphism ?

4.12 Given two algebraic structures with the same underlying set, say $(X, \mathscr{S}, \mathscr{C})$ and $(X, \mathscr{S}', \mathscr{C}')$ we say $(\mathscr{S}', \mathscr{C}')$ is a **stronger** structure than $(\mathscr{S}, \mathscr{C})$ if (i) $\mathscr{S}$ is a subsequence of $\mathscr{S}'$ and (ii) every condition in $\mathscr{C}$ follows from $\mathscr{C}'$. In other words, a weaker structure has fewer and less restrictive algebraic operations. Thus a monoid is a weaker structure than the one in which the binary operation obeys, besides associativity and presence of identities, some other laws such as cancellation laws. Prove that strictly speaking, the structure on the quotient set X/R, defined in Theorem (4.17) may be of a weaker type than the original structure on X. For example, the cancellation law may hold for X but not for X/R.

4.13 We often have situations where the underlying set of an algebraic structure also carries some other structure. In such cases the question of compatibility of the two structures is very important. For example, let $(X, *)$ be a monoid and $\leqslant$ a partial order on X. Then $\leqslant$ is said to be **compatible** with (or **invariant under**) $*$ if for all $a, b, c \in X, a \leqslant b$ implies $a * c \leqslant b * c$ and $c * a \leqslant c * b$. Prove that if $\leqslant$ is compatible with $*$ then for all $a, b, c, d \in X, a \leqslant b$ and $c \leqslant d$ implies $a * c \leqslant b * d$. Prove that the usual order on $\mathbf{R}$ is compatible with the usual addition but not with the usual multiplication on $\mathbf{R}$.

4.14 Let $(X, *)$ be a monoid and d a metric on X. Then d is called **translation invariant** if for all $a, b, c \in X$,

$$d(a * c, b * c) = d(a, b) = d(c * a, c * b).$$

Prove that the usual metrics on $\mathbf{R}$ and on the euclidean plane $\mathbf{R}^2$ are translation invariant if the binary operations are the usual addition. Prove that the metric d on $\mathbf{R}$ defined by $d(x, y) = |x^3 - y^3|$ is not translation invariant. If d is a translation invariant metric on $\mathbf{R}$ (or on $\mathbf{R}^2$) then the open (or closed) balls of the same radius but with different centres are of the same 'size'.

4.15 Let $*$ be a binary operation on a set X. We then define a binary operation on the power set $P(X)$, which may be denoted by $\circledast$ (or by $*$ itself). Given $A, B \subset X$, we let $A \circledast B$ be the set

$$\{x * y; x \in A, y \in B\}.$$

Study which properties pass over from $*$ to $\circledast$.

4.16 Using modulo 3 arithmetic, prove the 'rule of three' which says that a positive integer is divisible by 3 if and only if the sum of its digits, written in decimal expansion, is divisible by 3. (Hint: Note that $10 \equiv 1 \pmod 3$. Hence prove that all powers of 10 are congruent to 1 modulo 3.)

4.17 Obtain similar criteria for divisibility of a positive integer by 9 and by 11.

4.18 According to a Western superstition 13 is an evil figure and when the 13th of a month falls on a Friday, such 'Friday the 13th' is considered an especially evil day. Using modulo 7 arithmetic, prove that in any calendar year there is at least one and at most three Friday the thirteenths.

4.19 If three distinct integers are picked from 1 to 100, find the probability that their sum is divisible by 3. (Hint; First see, in $\mathbf{Z_3}$, in how many ways the residue class [0] can be expressed as a sum of three elements.)

4.20   A seminar consists of 60 lectures, spread over a period of five weeks. There is to be at least one lecture per day. Prove that no matter how the lectures are scheduled, there will be a block of consecutive days in which exactly 13 lectures will be given. Prove also that this can be avoided if the seminar is to run only for 34 days.

4.21   Using residue classes modulo 8, prove that no integer of the form $8n + 7$, (n an integer) can be expressed as a sum of three perfect squares.

4.22   If $*$ is an associative binary operation on a set X and $a_1, ..., a_n \in X$ then the expression $a_1 * a_2 * ... * a_n$ has only one meaning. Prove that in absence of associativity it could have $\dfrac{1}{n}\begin{pmatrix} 2n - 2 \\ n - 1 \end{pmatrix}$ possible meanings. [Hint: Each such meaning is given completely by putting one pair of brackets around the result every time the operation $*$ is carried out. For example, the two meanings of $a_1 * a_2 * a_3 *$ are $[[a_1 * a_2] * a_3]$ and $[a_1 * [a_2 * a_3]]$. Now ignore the left brackets and the $*$'s. Change all right brackets to right parentheses. Also ignore a_1 and change $a_2, a_3, ..., a_n$ to left parentheses. Thus $[[a_1 * a_2] * a_3]$ changes to ( ) ( ) while $[a_1 * [a_2 * a_3]]$ changes to (( )). The trick is to recover from these, the original interpretations. This can be done by changing the left parentheses to $a_2, ..., a_n$, the right parentheses to right brackets and then noting that as we scan from left to right, each right bracket indicates that $*$ has been performed on the two elements of X immediately preceding it (one or both these elements may be the results of earlier applications of $*$.) Show that this gives a bijection between the set of all possible meanings of $a_1 * a_2 * ... * a_n$ and the set of all balanced arrangements of $n - 1$ pairs of parentheses. The latter was counted in the solution to the Vendor Problem in Chapter 2, Section 3.]

4,23   Given a binary operation $*$ on a set X, an element $x \in X$ is called **idempotent** if $x * x = x$. Prove that every right and every left identity is an idempotent. If X is finite and $*$ is associative, prove that X contains at least one idempotent. [Hint: consider a sequence of suitable powers of any $x \in X$.]

4.24   In this exercise you may assume that every positive integer has a unique factorisation into prime powers. This fact will be formally proved in **Chapter 6, Section 2**.

Let S be the set of all functions from **N** to **R**. Let δ, c, i and μ be the functions defined by

$$\delta(n) = \begin{cases} 1 & \text{if } n = 1 \\ 0 & \text{if } n > 0 \end{cases}$$

$$c(n) = 1 \text{ for all } n$$

$$i(n) = n \text{ for all } n$$

$$\mu(n) = \begin{cases} 1 \text{ for } n = 1 \\ 0 \text{ if } p^2 \mid n \text{ for some prime } p \\ (-1)^r \text{ if } n \text{ is the product of } r \text{ distinct primes} \end{cases}$$

For $f, g \in S$, define their **convolute** $f * g$ to be the function whose value at $n \in \mathbf{N}$ is $\sum_{k \mid n} f(k) g\left(\dfrac{n}{k}\right)$ where the sum ranges over all positive integers dividing n (including 1 and n). Prove that:

(i) the functions δ, c, i and μ are all multiplicative (see Exercise (2.4.15) for definition).

(ii) convolution defines a commutative and an associative binary operation on S with δ as the identity element.

(iii) if f, g are multiplicative, so is $f * g$. (Hint; If m, n are relatively prime and $k \mid mn$ then k can be uniquely expressed as uv where $u \mid m$ and $v \mid n$).

(iv) $f \in S$ is invertible (w.r.t. $*$) iff $f(i) \neq 0$. (Hint: In the converse implication, define the value of the inverse function at n by induction on n, assuming it is already defined on all proper divisors of n.)

(v) the functions c and u are inverses of each other.

(vi) the function $c * c$ gives the number of divisors of a positive integer.

(vii the function $i * c$ gives the sum of divisors of a positive integer. ($i * c$ is often denoted by σ).

(viii) if ϕ is the Euler function in Exercise (2.4.15) then $\phi * c = i$. (Hint: Classify the integers in $\{1, ..., n\}$ according to their greatest common divisors with n. If $k \mid n$, note that there are precisely $\phi\left(\dfrac{n}{k}\right)$ integers whose g.c.d. with n is k.)

(ix) $\phi = i * \mu$.

(x) ϕ is a multiplicative function. (Another proof was given in Exercise (2.4.15).

[The function μ is called the **Mobius inversion**. For $f \in S$, $f * c$ and $f * \mu$ are called, respectively, the **Mobius transform** and the **inverse**

Mobius transform of f. The terms 'transform', 'convolution' as well as the peculiar notation δ come from analogous concepts in continuous mathematics for functions from $\mathbf{R}$ to $\mathbf{R}$.]

4.25   The concept of a Mobius function can be extended to certain posets. Let $\leqslant$ be a partial order on a set X. Then $\leqslant$ is a subset of $X \times X$ and we let S be the set of all functions from $\leqslant$ to $\mathbf{R}$. For $(x, y) \in \leqslant$ (i.e. for $x, y \in X$ with $x \leqslant y$) we assume there are only finitely many $z \in X$ with $x \leqslant z$ and $z \leqslant y$. (A poset satisfying this property is said to be *locally finite*. Clearly every finite poset is locally finite.) For $f, g \in S$, define their convolute

$$f * g: \leqslant \to \mathbf{R}$$

by

$$(f * g)(x, y) = \sum_{x \leqslant z \leqslant y} f(x, z)\, g(z, y).$$

Show that $*$ is associative (although not necessarily commutative) and has an identity. Prove that $f \in S$ is invertible w.r.t. $*$ iff $f(x, x) \neq 0$ for every $x \in X$. Define $\mu : \leqslant \to \mathbf{R}$ to be the inverse of the constant function with value 1. (In case $X = \mathbf{N}$ and $\leqslant$ means divisibility, prove that $\mu(x, y)$ is the same as $\mu\left(\dfrac{y}{x}\right)$ where μ is the Mobius function of the last exercise.) Deduce that if

$$g(x, y) = \sum_{x \leqslant z \leqslant y} f(x, z)$$

for all $x \leqslant y$ then

$$f(x, y) = \sum_{x \leqslant z \leqslant y} g(x, y)\, \mu(y, z).$$

(This is called **Mobius inversion formula**).

Notes and Guide to Literature

The material in this section is basic and we shall elaborate more on it in the chapters to come. For more on the residue arithmetic and its use in computer science see Tremblay and Manohar [1].

Regarding Exercise (4.21), it can be shown that every positive integer can be expressed as a sum of four perfect squares. This famous theorem is called Lagrange's four square theorem. For a proof, see Herstein [1].

For more on Mobius transforms see Hua [1]. From Exercise (4.25), a number of interesting results can be derived, by choosing the poset appropriately. See Bender and Goldman [1].

Four

Boolean Algebras

Having discussed the generalities about algebraic structures in the fourth section of the last chapter, we now proceed to study particular algebraic structures. The logical starting point would, perhaps, be an algebraic structure of a simple type, having only one binary operation and with as few axioms as possible (keeping in mind the general remarks made in Section 1 of the last chapter regarding depth versus generality). However, we prefer to start with a relatively rich structure, called a Boolean algebra. The reason for this is twofold. On one hand we already have with us certain particular Boolean algebras (although we have not called them so). Secondly, among all possible 'abstract' algebraic structures (other than those modelled after the real number system), the Boolean algebras have the most down-to-earth applications. Hopefully, this will convince the reader of the need for abstraction.

In Section 1 we define Boolean algebras and study some of their properties. Section 2 is devoted to the study of Boolean functions. In the third and the fourth sections we consider applications to the electric circuitry and to logic respectively.

1. Definition and Properties

In practice we often come across things which have two natural states and they exist, at any one time, in one and only one of them. For example, a statement is either true or false, an electric switch is either closed or open, a car is either at rest or in motion and so on. In some cases, depending upon the needs of the problem, we may divide these states into further categories. For example, we may classify moving cars according to their speeds. But in many problems, this may be irrelevant. This is especially true of electrical circuits. While in some problems, the magnitude of the current flowing in a circuit may be crucial, in some other types of problems, the only thing that matters is whether current is flowing in it or not. In case of statements, we have already agreed not to consider the various degrees of truth but to recognise only two categories, 'true' or 'false'.

Devices which occur in two states are called **binary** or **two-state** devices. These two states are called opposite or complementary states. We arbitrarily assign the symbol 1 to one of these states and the symbol 0 to the other. Two devices which are always in the opposite states are said to be **complementary** to each other. We also define the **conjunction** of two binary devices as another binary device which is in the state 1 iff both of them are in state 1. The concept of **disjunction** of two binary devices is defined similarly, as a device which is in state 1 iff at least one of them is in state 1. The interpretation of these three concepts, namely, complement, conjunction and disjunction would of course change depending upon the nature of the devices.

The structure of a Boolean algebra is meant to isolate these three basic concepts about two state devices. The actual definition is as follows:

1.1 Definition: A **Boolean algebra** is an ordered quadruple $(X, +, \cdot, ')$ where X is a non-empty set, $+$ and $\cdot$ are two binary operations on X and $'$ is a unary operation on X (i.e., $'$ is a function from X to X and if $x \in X$, we denote, $'(x)$ by x') satisfying the following conditions:

    (B1)   both $+$ and $\cdot$ are commutative,

    (B2)   both $+$ and $\cdot$ have identities denoted by 0 and 1 respectively,

    (B3)   both $+$ and $\cdot$ are distributive over each other and,

    (B4)   for all $x \in X$, $x + x' = 1$ and $x \cdot x' = 0$.

If $x \in X$, then x' is called the **complement** of x. If $x, y \in X$ then $x + y$ and $x \cdot y$ are sometimes called the disjunction and the conjunction of x and y, respectively.

A foremost example of a Boolean algebra is provided by taking X as the power set of some set, say S, defining $+$ as $\cup$, $\cdot$ as $\cap$ and $'$ as complementation w.r.t. the set S. Then the empty set ϕ is the identity for $+$ while the whole set S is the identity for $\cdot$. For this reason, ϕ and S are often denoted by 0 and 1 respectively. We must, of course, verify that the four axioms (B1) to (B4) are satisfied. This is precisely the content of Proposition (2.1.1), parts (i) to (iv). The reader may wonder which is the two state system associated with this Boolean algebra. The answer is that for every fixed element x of S, every subset of S is a two state system, because for any such subset, say, A of S there are only two possibilities, either $x \in A$ or $x \notin A$ (in which case $x \in A'$, the complement of A).

If we take $S = \phi$, then the Boolean algebra $P(S)$ consists of only one element. This is a **trivial Boolean algebra**. In this Boolean algebra, the elements 0 and 1 coincide. As we shall see below, the elements 0 and 1 are always distinct, provided the Boolean algebra has at least two elements. The simplest non-trivial Boolean algebra has 0 and 1 as its only elements. An example of such a Boolean algebra is obtained by taking the power set of a singleton set. This Boolean algebra is very important for certain purposes and is often denoted by Z_2. (Recall that the same symbol is used for the set of residue classes modulo 2. The underlying set is the same for both, namely, $\{0, 1\}$. The operation of $\cdot$ also coincides with the modulo 2 multiplication. But the operation of $+$, as Boolean algebra, does not coincide with that of the modulo 2 addition, because $1 + 1 = 1$ in a Boolean algebra, as we shall soon prove, but in the modulo 2 addition. $1 + 1 = 0$. This double usage of the same symbol is somewhat unfortunate. But the context will always make it clear, which algebraic structure is meant.)

Other examples of Boolean algebras will be given later on. Following the general constructions given in Section 4 of the last chapter, we can generate new Boolean algebras from old ones. For example we can take the cartesian product of two (or more) Boolean algebras, and make it into a Boolean algebra by defining $+$, $\cdot$ and $'$ coordinatewise. Similarly given a Boolean algebra $(B, +, \cdot, ')$ and any set S, the set of all functions from S to B is a Boolean algebra under pointwise operations. A third method is to take subalgebras of a Boolean algebra. A non-empty subset Y of a Boolean algebra $(X, +, \cdot, ')$ is called a **subalgebra** if it is closed under the operations $+$, $\cdot$, $'$. We prove that Y itself is a Boolean algebra.

1.2. Proposition: Let Y be a subalgebra of a Boolean algebra $(X, +, \cdot, ')$. Then Y, with the induced operations (which we continue to denote by the same symbols, $(+, \cdot$ and $')$ is a Boolean algebra with the same identity elements as X.

Proof: We first show that 0 and 1 are in Y. Since Y is non-empty, there exists some $x \in Y$. But then $x' \in Y$, since Y is closed under complementation. Since $x, x' \in Y$ and Y is closed under $+$ and $\cdot$, we get $x + x' \in Y$ and $x \cdot x' \in Y$. Thus both 0 and 1 are in Y. Now all the axioms of a Boolean algebra are satisfied for Y. These follow from the corresponding properties for X. ∎

An an instructive example of a subalgebra, let X be the power set Boolean algebra of a set S. Let $\mathscr{D}$ be a decomposition of S whose members are, say, $S_1, S_2,..., S_k$ Let $Y = \{A \subset S : \text{for each } i = 1, 2, ..., k, \text{ either } S_i \subset A \text{ or } S_i \cap A = \phi\}$. In other words, Y consists of those subsets of S which completely contain those members of $\mathscr{D}$ which they intersect. Y cannot contain any subset of S which intersects some S_i only partly. It is easy to show that Y is a subalgebra of $P(S)$. As a Boolean algebra, it can be shown that Y is isomorphic to the power set Boolean algebra $P(\mathscr{D})$.

We could go on looking for more examples of Boolean algebras. But there is an interesting theorem which stops us by saying that, from an abstract point of view, our search will not yield any more variety. Specifically the theorem says that every Boolean algebra is isomorphic to some sub-algebra of a suitable power set Boolean algebra. This famous theorem is called the **Stone Representation Theorem**. We shall prove it for finite Boolean algebras (the infinite case requires the axiom of choice). In studying abstract algebraic structures we generally identify two structures that are isomorphic to each other. Modulo this identification, the Stone representation theorem says that if we study all power set Boolean algebras and their subalgebras, then we have exhausted the world of Boolean algebras. Whatever theorems hold for the subalgebras of power set Boolean algebras also hold for *all* Boolean algebras.

Such representation theorems represent landmarks in the study of mathematical structures because they represent abstract mathematical structures as some concrete, familiar structures (hence the name 'representation theorems'). They serve to measure the true generality arising out of abstraction. We shall see a few other representation theorems later. However, despite their theoretic achievements, such theorems generally do not provide a direct simplification of the study of the respective abstract structures. For example, in the case of the Stone representation theorem, to prove it requires considerable spadework on abstract Boolean algebras. Some representation theorems are surprisingly easy to prove (as we shall see for groups). But then, their utility is correspondingly limited. Although they reduce the study of the abstract structures of a particular type to that of certain particular, concrete structures, the latter itself is a formidable task, not any less difficult than the former.

Therefore, despite the Stone representation theorem, (which, any way, has not been proved yet), we continue to work in terms of abstract Boolean algebras. We begin with a few basic properties, which are derived directly from the axioms (see the comments made after the proof of Proposition (2.1.1)). The proofs provide an excellent example of axiomatic deduction.

1.3. Theorem: Let $(X, +, \cdot,)$ be a Boolean algebra. Then the following properties hold for all elements x, y, z of X,

(i) $x + x = x$ and $x \cdot x = x$. (Laws of **Tautology** or **Idempotency**)

(ii) $x + 1 = 1$ and $x \cdot 0 = 0$

(iii) $x + x \cdot y = x$ and $x \cdot (x + y) = x$ (Laws of **Absorption**)

(iv) $x + (y + z) = (x + y) + z$ and $x \cdot (y \cdot z) = (x \cdot y) \cdot z$

(**Associative Laws**)

(v) If $x + y = 1$ and $x \cdot y = 0$ then $y = x'$

(**Uniqueness of Complements**)

(vi) $(x')' = x$ (**Law of Double Complementation**)

(vii) $(x + y)' = x' \cdot y'$ and $(x \cdot y)' = x' + y'$

(De Morgan's Laws)

(viii) $0' = 1$ and $1' = 0$

(ix) $0 \neq 1$ unless X has only one element.

Proof: Before actually giving the proofs, we make an important observation. There is an absolute symmetry between the operation $+$ and $\cdot$ in the definition of a Boolean algebra. Consequently, whenever any identity holds in a Boolean algebra, we can replace all occurrences of $+$ with $\cdot$ and vice versa (with a corresponding interchange of 0 and 1) and get a new identity, called the **dual** of the original identity. Its proof can be given dualising each step in the proof of the original identity. This is known as the **principle of duality**. Formally, it states that if the ordered quadruple $(X, +, \cdot, ')$ is a Boolean algebra, so is the ordered quadruple $(X, \cdot, +, ')$. We see that in most of the assertions to be proved, there are two statements which are the duals of each other. Because of the principle of duality, it suffices to prove either one of the two statements.

We now prove the assertions one by one.

(i)   We have, $x = x \cdot 1$ (by property $B2$ in the definition)

$$= x \cdot (x + x') \qquad \text{(by } B4)$$
$$= (x \cdot x) + (x \cdot x') \qquad \text{(by } B3)$$
$$= (x \cdot x) + 0 \qquad \text{(by } B4 \text{ again)}$$
$$= x \cdot x \qquad \text{(by } B2 \text{ again)}$$

Hence $x \cdot x = x$. The other assertion, $x + x = x$ follows by duality. The reason for the name 'Laws of tautology' (and also for the name 'Laws of Absorption') will be given in Section 4.

(ii) $x + 1 = (x + 1) \cdot 1 \qquad \text{(by } B2)$

$$= (x + 1) \cdot (x + x') \qquad \text{(by } B4)$$
$$= x + (1 \cdot x') \qquad \text{(by } B3)$$
$$= x + x' \qquad \text{(by } B2)$$
$$= 1 \qquad \text{(by } B4)$$

(iii) From (ii) we have, $y + 1 = 1$. Multiplying both sides by x we get $(y + 1) \cdot x = 1 \cdot x$. By $B3$, $B2$ and $B1$ the left hand side reduces to $x + (x \cdot y)$ while the right hand side equals x by $B2$. Hence $x + (x \cdot y) = x$ as was to be proved.

(iv) This is really a remarkable property, because in most algebraic structures, it is customary to assume associativity as an axiom. In the case of Boolean algebras, this property can be proved as a consequence of other axioms and hence it is redundant to include it as an axiom. Of course, till we have proved it, we have to be careful to insert parentheses wherever necessary to avoid ambiguity.

Let us prove $x \cdot (y \cdot z) = (x \cdot y) \cdot z$. Let the left and right hand sides be denoted by a and b respectively. Instead of proving $a = b$ directly, we shall first prove, separately, that $a + x = b + x$ and $a + x' = b + x'$. Then multiplying the two we shall get $(a + x) \cdot (a + x') = (b + x) \cdot (b + x')$. By $B3$, $B4$ and $B2$, the left hand side reduces to a while the right hand side reduces to b. This would prove the result.

$$\text{Now,} \quad a + x = [x \cdot (y \cdot z)] + (x \cdot 1) \qquad \text{(by } B2\text{)}$$

$$= x \cdot [(y \cdot z) + 1] \qquad \text{(by } B3\text{)}$$

$$= x \cdot [1] \qquad \text{(by (ii) proved above)}$$

$$= x$$

$$\text{while} \quad b + x = [(x \cdot y) \cdot z] + x$$

$$= [(x \cdot y) + x] \cdot [z + x] \qquad \text{(by } B3\text{)}$$

$$= [x + (x \cdot y)] \cdot [x + z] \qquad \text{(by } B1\text{)}$$

$$= x \cdot (x + z) \qquad \text{(by (iii) proved above)}$$

$$= x \qquad \text{(again by (iii))}.$$

Hence $a + x = b + x$.

$$\text{Next,} \quad a + x' = [x \cdot (y \cdot z)] + x'$$

$$= [x + x'] \cdot [(y \cdot z) + x'] \qquad \text{(by } B3\text{)}$$

$$= 1 \cdot [(y \cdot z) + x'] \qquad \text{(by } B4\text{)}$$

$$= (y \cdot z) + x' \qquad \text{(by } B2\text{)}$$

$$= (y + x') \cdot (z + x') \qquad \text{(by } B3\text{)}$$

$$\text{while} \quad b + x' = [(x \cdot y) \cdot z] + x'$$

$$= [(x \cdot y) + x'] \cdot [z + x'] \qquad \text{(by } B3\text{)}$$

$$= [(x + x') \cdot (y + x')] \cdot (z + x') \qquad \text{(by } B3 \text{ again)}$$

$$= [1 \cdot (y + x')] \cdot (z + x') \qquad \text{(by } B4\text{)}$$

$$= (y + x') \cdot (z(+ x') \qquad \text{(by } B2\text{)}$$

So $a + x' = b + x'$ and as noted before this completes the proof.

(v) Suppose $x + y = 1$ and $x \cdot y = 0$. Then

$$x' = x' \cdot 1 = x' \cdot (x + y) = (x' \cdot x) + (x' \cdot y) = 0 + (x' \cdot y) = x' \cdot y.$$

Similarly $y = 1 \cdot y = (x' + x) \cdot y = x' \cdot y$ since $x \cdot y = 0$. Thus both x' and y equal $x' \cdot y$ whence $x' = y$. We have omitted the justifications this time. By now the reader should be in a position to supply them. This shows that the complement of an element x is characterised by the property that when added to x it gives 1 and when multiplied with x it gives 0.

We caution the reader against attempting to prove the uniqueness of complements from the equation $x + x' = x + y$ by cancelling x from both sides. This is so because cancellation law has not been established. Actually, from (ii) we have $x + 1 = y + 1 (= 1)$ for all $x, y \in x$. Thus we see that the cancellation law does not hold for $+$. Similarly it fails for $\cdot$.

(vi) We have $x + x' = 1$ and $x \cdot x' = 0$. By $B1$, this gives, $x' + x = 1$ and $x' \cdot x = 0$. But because of the uniqueness of complements just proved, this means x is the complement of x', i.e., $x = (x')'$.

(vii) Let $a = x + y$ and $b = x' \cdot y'$. We have to show $a' = b$. By uniqueness of complements, it suffices to prove that $a + b = 1$ and $a \cdot b = 0$. Now, $a + b = x + y + (x' \cdot y')$ (because of associativity we need not put parentheses)

$$= x + [(y + x') \cdot (y + y')]$$
$$= x + [(y + x') \cdot 1]$$
$$= x + y + x'$$
$$= y + x + x'$$
$$= y + 1$$
$$= 1 \qquad \text{(by (ii))}$$

Also $\quad a \cdot b = (x + y) \cdot (x' \cdot y')$

$$= (x \cdot x' \cdot y') + (y \cdot x' \cdot y')$$
$$= (0 \cdot y') + (0 \cdot x')$$
$$= 0 + 0 \quad \text{(by (ii))}$$
$$= 0.$$

This proves $(x + y)' = x' \cdot y'$. The other assertion follows either by duality or by replacing x' by x, y' by y and taking complements.

(viii) Since $0 + 1 = 1$ and $0 \cdot 1 = 0$ (from (ii)), it follows, by (v) again that $1 = 0'$ and $0 = 1'$.

(ix) Suppose $0 = 1$. We have to show that X has only one element. Let $x \in X$. Then $x = x \cdot 1 = x \cdot 0 = 0$. So 0 (or 1) is the only element of X. ∎

Using the various laws proved so far (or assumed as axioms) we can simplify an algebraic expression involving elements of a Boolean algebra, much the same way as we simplify expressions involving real variables. Because of commutativity and associativity, the sum or the product of any number of terms is defined unambiguously, without having to insert parentheses. As a further notational simplification, we make the convention that, as with real numbers, $+$ will rank higher than $\cdot$. This means that an expression like $x \cdot y + z \cdot w$ will be interpreted always as $(x \cdot y) + (z \cdot w)$ and not as $x \cdot [y + (z \cdot w)]$ or as $x \cdot (y + z) \cdot w$ or as $[(x \cdot y) + z] \cdot w$ etc. In view of the symmetry between $+$ and $\cdot$, we could have as well decided to let $\cdot$ rank higher than $+$. But we prefer our convention because it is adopted for the real number system with which we are so familiar. To further keep up with real numbers, we shall often suppress $\cdot$ from notation. Of course, analogy with expressions involving real numbers should not be relied upon blindly. Certain laws, such as cancellation laws, which hold good for real numbers no longer apply in Boolean algebras. On the other hand, there are certain features of Boolean algebras such as complementation and distributivity of $+$ over $\cdot$ which render possible certain simplifications which would not hold far real numbers. As a commonly used example, $x + x'y$ simplifies to $x + y$ since $x + x'y = (x + x')(x + y) = 1 \cdot (x + y) = x + y$. Another point to note is that because of tautology, the same term (or factor) may be used any number of times. For example, the expression $xyz + xyz' + x'yz$ equals $xyz + xyz + xyz + x'yz$ which further reduces to $xy + yz$ or to $y(x + z)$. Yet another simplifying feature is provided by the absorption law. In any expression we may ignore all terms which are multiples of some other term and any factor which is obtained by adding some terms to some other factors.

As an application of these methods we now present a systematic, algebraic solution to the Business Problem. (Another solution will be presented in the fourth section.) We let B be the set of all businesses in the country and consider the power set Boolean algebra, $(P(B), \cup, \cap, ')$. We denote by I, E, L and S respectively, the subsets of B consisting of all businesses having import licenses, manufacturing essential commodities, employing local personnel and employing skilled personnel. We let V_1, V_2, V_3 be the subsets of those businesses which violate the three given rules respectively. The problem amounts to showing that $V_1 + V_2 + V_3 = 1 = B$. For this, we first express each V_i as a Boolean expression involving I, E, L and S (which are elements of our Boolean algebra), cf. Exercise (2.1.16). Now, the first rule is violated by two types of businesses, (a) those which have import licences but either do not employ local personnel or do employ skilled personnel and (b) those businesses which do not manufacture essential commodities and either do not employ local personnel or do employ skilled personnel. The set of businesses of type (a), is precisely $I \cap (L' \cup S)$ or $I(L' + S)$, in our notation. Similarly the set of businesses of type (b) is $E'(L' + S)$. So, V_1, the set of violators of the first rule, is the union of these two, i.e.,

$$I(L' + S) + E'(L' + S)$$

or $(I + E')(L' + S)$. By similar reasoning it follows that

$$V_2 = (I + L')(S' + E)$$

and

$$V_3 = LI'.$$

Now

$$
\begin{aligned}
V_1 + V_2 + V_3 &= (I + E')(L' + S) + (I + L')(S' + E) + LI' \\
&= IL' + E'L' + IS + E'S + IS' + L'S' + IE + L'E + LI' \\
&= L'(I + E' + S' + E) + I(S + S' + E) + E'S + LI' \\
&= L'(1 + I + S') + I(1 + E) + E'S + LI' \\
&= L' \cdot 1 + I \cdot 1 + E'S + LI' \\
&= L' + I + E'S + LI' \\
&= L' + I' + I + E'S \quad \text{(since } L' + LI' = L' + I') \\
&= 1 + L' + E'S \\
&= 1.
\end{aligned}
$$

This shows that every business violates at least one rule and so it is impossible to do any business in the country. Sometimes things are not so bad. If the rules are different, the expression $V_1 + V_2 + V_3$ may not reduce to 1. Still, it is worthwhile to simplify it, because then taking its complement we get a handy expression for $V_1'V_2'V_3'$ which is precisely the set of all lawful businesses. Paraphrasing it gives a simplified, but equivalent version of the original rules. We illustrate this in the following problem.

1.4. Problem: Suppose that in the Business Problem, the third rule is changed, by an amendment, to 'No business shall employ skilled personnel without obtaining an import license', the first two rules being unaffected. Simplify the rules to an equivalent set of rules.

Solution: We proceed as above. The sets V_1 and V_2 remain the same. But V_3 changes to SI'. So $V_1 + V_2 + V_3$ would now reduce, instead of 1, to $L' + I + E'S + SI'$ which in turn becomes $L' + E'S + I + S$ and finally $L' + I + S$ (since $E'S$ is absorbed in S). Therefore, the set of lawful businesses, is $(L' + I + S)'$ which equals $LI'S'$ by De Morgan's laws. Hence the system of rules is equivalent to the following simple system of rules; (i') every business must employ local personnel (ii') no business shall have an import license and (iii') no business shall employ skilled personnel. ■

In this problem, the simplified version of the rules shows something which is not obvious from the given rules, namely, that the original system is independent of whether the business manufactures essential commodities or not. The first two rules do ostensibly make references to essential commodities. But our work shows that, if we consider the system of the three rules as a whole then businesses manufacturing essential commodities do not have any more advantage or disadvantage as compared to those businesses which do not manufacture essential commodities.

Theorem (1.3) was a generalisation of Proposition (2.1.1) in the sense that when Theorem (1.3) is applied to a power set Boolean algebra, we get precisely the results of Proposition (2.1.1) (except, of course, the results (i) to (iv) which correspond to verification of the axioms of a Boolean algebra). Direct proofs, using elements of the set in question, were possible for Proposition (2.1.1). Such arguments would not work for Theorem (1.3), because the elements of an abstract Boolean algebra need not be subsets of some set; they could be real numbers, some functions, some animals, in fact anything at all. We know nothing about them save what is implied by the axioms. Therefore, whatever, theorems we prove about abstract Boolean algebras, must be deduced strictly from the axioms. (Things would be different if we have the representation theorem at our hand. Such a theorem allows us to regard an abstract structure as a concrete one.)

This situation is fairly typical in the study of abstract mathematical structures. Some particular example serves as a model. We take concepts which are originally defined for this model and see if they can be suitably paraphrased so that they can be defined for the general, abstract context. Similarly, we inquire what theorems can be carried over from the concrete to the general. In the present case, the power set Boolean algebras serve as models for the abstract Boolean algebras. Let $P(S)$ be the power set Boolean algebra of some set S. Elements of $P(S)$ are subsets of S and hence consist of elements of S. We are used to do things in terms of these elements. Let us see which of these things can be carried over to an abstract Boolean algebra. Let $A \in P(S)$. Then $|A|$, the cardinality of A is the number of elements in A. There is no easy, direct way to generalise this concept for elements of an abstract Boolean algebra. But let us take some other concept, say, that of inclusion. Let $A, B \in P(S)$. Then we say $A \subset B$ if every element of A is also an element of B, and this gives a partial order on $P(S)$. As it is, this concept cannot be generalised for an abstract Boolean algebra, X. If $x, y, \in X$, we cannot define $x \subset y$ to mean that every element of x is also an element of y, because x may not be a set. It may be an elephant, a flower and so on. So it is meaningless to talk of an element of x. (The reader may ask whether it is not meaningless to talk of the sum of two elephants or the complement of an elephant. Such a question is baseless. We are starting with the assumption that $(X, +, \cdot, ')$ is a Boolean algebra. If elements of X happen to be elephants then it means that we are given some rule to define the sum of two such elephants. Whether such a definition is 'meaningful' is an extraneous question.)

Fortunately, in Proposition (2.1.2) we have characterised set inclusion in various ways which involve only the operations $\cap$, $\cup$ and $'$ and do not directly involve the elements of the set. Since these three operations correspond to $+$, $\cdot$ and $'$ in an abstract Boolean algebra, the concept of set inclusion can be generalised for an abstract Boolean algebra, as we now do.

1.5 Definition: Let $(X, , \cdot, ')$ be a Boolean algebra. If $x, y \in X$, we say $x \leqslant y$ if $x \cdot y' = 0$.

There is no harm in denoting $x \leqslant y$ by $x \subset y$ and reading it as 'x is contained in y' instead of 'x is less than or equal to y', as long as we treat the notation $\subset$ and the phrase 'contained in' like proper nouns. This is another point, albeit a minor one, to be noted about abstraction. When the inspiration for a concept comes from a particular example, often the same notation and terminology is used even in the general context. Their etymology may be interesting and instructive but should not be stretched unduly. For example, in the present case, although for power set Boolean algebras $x \leqslant y$ is the same as $x \subset y$, for other Boolean algebras, $x \leqslant y$ may have a very different interpretation. Similarly two element x and y in any Boolean algebra are called disjoint if $x \cdot y = 0$, even though $\cdot$ may not always stand for intersection.

The following theorem captures the important properties of the binary relation just defined.

1.6 Theorem: The relation $\leqslant$ defined above makes the underlying set of a Boolean algebra into a lattice. Moreover, 0 and 1 are the minimum and the maximum elements of this lattice.

Proof: Let $(X, +, \cdot, ')$ be a Boolean algebra. For $x, y \in X$, we have defined $x \leqslant y$ as $x \cdot y' = 0$. We want so show that $(X, \leqslant)$ is a lattice. First we verify that $\leqslant$ is a partial order on X. Reflexivity follows from the fact that $x \cdot x' = 0$ for all $x \in X$. For transitivity suppose $x \leqslant y$ and $y \leqslant z$. Then $xy' = 0$ and $yz' = 0$. Now

$$xz' = x \cdot 1 \cdot z' = x(y + y')z' = xyz' + xy'z = x \cdot 0 + 0 \cdot z = 0 + 0 = 0,$$

proving that $x \leqslant z$. It remains to prove that $\leqslant$ is anti-symmetric. For this, let $x, y \in X$ with $x \leqslant y$ and $y \leqslant x$. Then $xy' = 0$ and $yx' = 0$. Now

$$x = x \cdot 1 = x(y + y') = xy + xy' = xy + 0 = xy.$$

By symmetry y also equals xy. Hence $x = y$.

For the verification of the lattice properties, let $x, y \in X$. We claim that the join of x and y is simply $x + y$ and that their meet is $x \cdot y$. For the first assertion, note first that $x \cdot (x + y)' = x \cdot x'y' = 0 \cdot y' = 0$. So $x \leqslant x + y$. Similarly $y \leqslant x + y$. Hence $x + y$ is an upper bound of the set $\{x, y\}$. To show that it is the least upper bound, suppose z is some other upper bound of $\{x, y\}$, i.e. $x \leqslant z$ and $y \leqslant z$. Then $xz' = 0$ and $yz' = 0$. So

$$(x + y)z' = xz' + yz' = 0 + 0 = 0.$$

This means $x + y \leqslant z$ and hence $x + y$ is the least upper bound of the set $\{x, y\}$, which, by definition, is the join of x and y. The proof that the meet of x and y is $x \cdot y$ is obtained by duality.

Finally, if $x \in X$ then $0 \cdot x' = 0$ and hence $0 \leqslant x$, showing that 0 is the minimum element of X. Similarly, $x \cdot 1' = x \cdot 0 = x$ for all $x \in X$, shows that 1 is the maximum element of X. This completes the proof. ∎

The preceding theorem shows that every Boolean algebra gives rise to a lattice. An interesting question arises, whether every lattice can be obtained from a Boolean algebra. The answer is obviously in the negative. As we just saw, the lattice obtained from a Boolean algebra is always bounded (i.e. has a smallest and a largest element). It turns out that a bounded lattice satisfying a couple of other properties indeed arises from a Boolean algebra. We already defined a distributive lattice as one in which the binary operations of join and meet are distributive over each other (see Exercise (3.4.7)). We make one more definition.

1.7 Definition: Let $(X, \leqslant)$ be a bounded lattice with 0 and 1 as its minimum and maximum elements respectively. Then X is called **complemented**, if for every $x \in X$, there exists some $y \in X$ such that $x \vee y = 1$ and $x \wedge y = 0$. Any such y is called a complement of x.

Note that we are not requiring complements to be unique. For example, in the lattice of Exercise (3.4.9), any two of the three given partitions of the plane are complementary to each other. This particular lattice is not distributive. As may be expected, things are much better for distributive lattices. In fact, such lattices come very close to Boolean algebras as the next theorem shows.

1.8 Theorem: Let $(X, +, \cdot, ')$ be a Boolean algebra. Then the corresponding lattice $(X, \leqslant)$ is complemented and distributive. Conversely if $(X, \leqslant)$ is a bounded, complemented and distributive lattice then there exists a Boolean algebra structure on X, $(X, +, \cdot, ')$ such that the partial order relation defined by this structure coincides with the given relation $\leqslant$.

Proof: The first part needs no proof, because since $\vee$ and $\wedge$ are precisely $+$ and $\cdot$ respectively, the assertion following right from the axioms of a Boolean algebra along with the fact that 0 and 1 are the minimum and the maximum elements of X respectively. It is the converse that is more interesting. Suppose $(X, \leqslant)$ is a bounded, complemented, distributive lattice with 0 and 1 as the smallest and the largest elements. For $x, y \in X$ define

$$x + y = x \vee y$$

and

$$x \cdot y = x \wedge y.$$

Then the binary operations $+$ and $\cdot$ are commutative with 0 and 1 as their respective identities. Their distributivity over each other follows from the very definition of a distributive lattice. Thus the axioms $B1$ to $B3$ in Definition (1.1) are verified. Now for each $x \in X$, we select any one complement of x and denote it, by x'. (Actually, using distributivity it is not difficult to show that complements are unique. But this fact is not needed. In order to have a Boolean algebra, all we need to have is at least one complement for every element.) This gives a function $' : X \to X$ such that the quadruple $(X, +, \cdot, ')$ is a Boolean algebra.

Now let $\propto$ be the partial order on X induced by this Boolean algebra structure, i.e., for $x, y \in X$, $x \propto y$ iff $xy' = 0$. We have to show that $\propto$ coincides with $\leqslant$, i.e., for $x, y \in X$, $x \propto y$ iff $x \leqslant y$. Suppose $x \propto y$. Then $x \cdot y' = 0$. So $x = x \cdot 1 = x \cdot (y + \overline{y'}) = xy + xy' = xy + \overline{0} = x \cdot y$. This means $x = x \wedge y$. But $x \wedge y \leqslant y$. So $x \leqslant y$. Conversely suppose $x \leqslant y$. Then $x = x \wedge y = x \cdot y$. Hence $xy' = xyy' = x \cdot 0 = 0$. So $x \propto y$. This completes the proof. ∎

Many authors define a Boolean algebra as a bounded, complemented, distributive lattice. The preceding theorem shows that this definition is equivalent to ours. It really does not matter which definition is adopted, bacause both approaches will ultimately yield the same results. Still, certain concepts may appear more natural in one approach than in the other and certian results may be easier to prove in one approach than in the other. For example associativity is immediate for the binary operations $\wedge$ and $\vee$ in a lattice. But to prove associativity for a Boolean algebra as we have defined, requires a little work. On the other hand, in some examples such as the algebra of circuits (which will be studied in Section 3), the partial order $\leqslant$ is not a very natural one and so it is a little artificial to conceive them as lattices. Ultimately, however, it is more a matter of taste than of real convenience as to which approach one adopts. Note that the partial orders induced by a Boolean algebra $(X, +, \cdot, ')$ and its dual Boolean algebra $(X, \cdot, +, ')$ are dual to each other.

The partial order structure induced on the underlying set of a Boolean algebra $(X, +, \cdot, ')$ also enables us to prove the representation theorem for it, at least when the set X is finite. We proceed as follows. Suppose, to start with, that X is indeed the power set Boolean algebra of some set, say S. Let $S = \{x_1, x_2, \ldots, x_n\}$ and for each $i = 1, 2, \ldots, n$ let $A_i = \{x_i\}$. (Note again that $x_i \neq \{x_i\}$, $x_i \in S$ while $\{x_i\} \subset S$, i.e. $\{x_i\} \in P(S) \cdot$) Now suppose $A \in P(S)$, i.e. $A \subset S$. Then either $A = \phi$ or else A is of the form

$$\{x_{i_1}, x_{i_2}, \ldots, x_{i_r}\}$$

for some positive integers r and

$$1 \leqslant i_1 < i_2 < \ldots < i_r < n.$$

Also

$$A = \bigcup_{k=1}^{k=r} A_{i_k}$$

or,

$$A = A_{i_1} + A_{i_2} + \ldots + A_{i_r}.$$

Thus we see that every non-zero element of the power set Boolean algebra $P(S)$ can be expressed as a sum of singleton sets. Since two distinct singleton sets are disjoint, this expression is unique.

The key to the representation of an abstract Boolean algebra as a power set Boolean algebra is to obtain a similar expression for an element of an abstract Boolean algebra as a sum of certain 'basic' elements. Which elements should play the role of these 'building blocks'? To get an answer we take another look at singleton subsets. If the power set of any set is partially ordered under inclusion, then the empty set is obviously the minimum element and singleton subsets are characterised as minimal elements in the set of all non-empty subsets. We are now in a position to define their counterparts for an abstract Boolean algebra.

1.9 Definition: Let $(X, +, \cdot, ')$ be a Boolean algera. Then a minimal element of the set $X - \{0\}$ is called an **atom of** X.

Atoms of a power set Boolean algebra $P(S)$, are precisely singleton subsets of S. As we saw above, every element of $P(S)$ can be expressed as a union of atoms much the same way as in chemistry every molecule is obtained by combining atoms. (Hence probably the name 'atom'.) Actually, the concept of an atom can be defined for any poset having a least element. In the terminology of Definition (3.3.8), atoms are precisely those elements which cover the unique least element. For example, if on the set $\mathbf{N}$, we define $x \leqslant y$ if x divides y for $x, y \in \mathbf{N}$, then 1 is the minimum element and prime numbers are the atoms.

In the following proposition we prove a few simple properties about atoms in a finite Boolean algebra and show how they serve as building blocks.

1.10 Proposition: Let $(X, +, \cdot, ')$ be a finite Boolean algebra. Then

(i)   every non-zero element of X contains at least one atom,
(ii) every two distinct atoms of X are mutually disjoint, and
(iii) every element of X can be uniquely expressed as a sum of atoms, specifically if $x \in X$, then x is the sum of all atoms contained in x (with the understanding that an empty sum is 0).

Proof: (i) Let $x \in X$, $x \neq 0$. Then either x is itself an atom (i.e., a minimal element of $X - \{0\}$) or there is some $x_1 \in X$, $x_2 \neq 0$ such that $x_1 < x$. If x_1 is an atom we are done. Otherwise there is some $x_2 \in X$ such that $x_2 \neq 0$ and $x_2 < x_1$. Continuing in this manner, since X is a finite set, there will be some n such that x_n is an atom and $x_n \leqslant x$. (The same idea was used in the proof of Theorem (3.3.7).)

(ii) Let a, b be two atoms of x. If $a \cdot b \neq 0$, then by (i), there exists an atom c such that $c \leqslant a \cdot b$. Since $a \cdot b \leqslant a$, we have $c \leqslant a$. Since a itself is a minimal non-zero element of X, it follows that $a = c$. Similarly $b = c$. Hence $a = b$. In other words, if a, b are distinct atoms then $a \cdot b = 0$.

(iii) Let $x \in X$. If $x = 0$, then x contains no atoms and the assertion holds because of the understanding that an empty sum is to be regarded as 0. So suppose $x \neq 0$. Let $a_1, a_2, \ldots a_k$ be the distinct atoms of X contained in x. We assert that $x = a_1 + a_2 + \ldots + a_k$. At any rate, since $a_i \leqslant x$ for all $i = 1, 2, \ldots, k$ and $a_1 + a_2 + \ldots + a_k$ is the supremum of the set $\{a_1, a_2, \ldots, a_k\}$, we already have $a_1 + a_2 + \ldots + a_k \leqslant x$. To show $x \leqslant a_1 + a_2 + \ldots + a_k$, we must show, by De Morgan's laws, that $x a_1' a_2' \ldots a_k' = 0$. If not, then by (i) there exists an atom b such that $b \leqslant x a_1' a_2' \ldots a_k'$. But $x a_1' a_2' \ldots a_k' \leqslant x$. So $b \leqslant x$, i.e. b is an atom contained in x. So $b = a_i$ for some i. But this means $a_i \leqslant x a_1' a_2' \ldots a_k' \leqslant a_i'$ giving $a_i a_i = 0$, i.e. $a_i = 0$ by the law of tautology. Since no atom can be 0, we get a contradiction, proving that $x \leqslant a_1 + a_2 + \ldots + a_k$ and hence that $x = a_1 + \ldots + a_k$. As for uniqueness, suppose $x = b_1 + b_2 + \ldots + b_r$ where each b_i is an atom and $b_i \neq b_j$ for $i \neq j$. (If $b_i = b_j$ for some $i \neq j$, then, by the law of taultology we may replace $b_i + b_j$ by b_i itself). Then $b_i \leqslant x$ for all $i = 1, 2, \ldots, r$. But $a_1, a_2, \ldots, a_k$ are all the atoms contained in x and so $b_i = a_j$ for some j. Since all the b's are distinct, it follows that $r \leqslant k$. Suppose $r < k$. Then there is some a_j which does not equal any b_i. Without loss of generality suppose $a_1 \neq b_1, b_2, \ldots, b_r$. We shall derive a contradiction as follows. $b_i + b_i' = 1$ for all $i = 1, \ldots, r$. Hence

$$a_1 = a_1 (b_1 + b_1') (b_2 + b_2') \ldots (b_r + b_r').$$

When this product is expanded, every term except $a_1 b_1' b_2' \ldots b_r'$ will contain the product of a_1 and at least one of the atoms $b_1, \ldots, b_r$. By (ii), all such products are 0. Hence $a_1 = a_1 b_1' b_2' \ldots b_r'$. But $a_1 \leqslant x$ gives $a_1 x' = 0$ which gives $a_1 b_1' b_2' \ldots b_r' = 0$ since $x = b_1 + b_2 + \ldots + b_r$. So we get $a_1 = 0$, a contradiction. This shows that every a_j equals some b_i and hence that the expression of x as a sum of atoms is unique. ∎

As an important special case of (iii), the sum of all atoms equals 1. We now have all the machinery needed to prove the Stone representation theorem for finite Boolean algebras.

1.11 Theorem: Every finite Boolean algebra is isomorphic to a power set Boolean algebra, specifically, to the power set Boolean algebra of the set of all its atoms.

Proof: Let $(X, +, \cdot, ')$ be a Boolean algebra. If $|X| = 1$, then X is isomorphic to the power set Boolean algebra of the empty set. Assume $|X| > 1$. Then $0 \neq 1$ (by (ix) in Theorem (1.3)) and so X has at least one atom. Let $a_1, \ldots, a_n$ be the distinct atoms of X. By the last proposition,

$$a_1 + a_2 + \ldots + a_n = 1.$$

Now let S be the set $\{1, 2, \ldots, n\}$. We assert that the power set Boolean algebra $(P(S), \cup, \cap, ')$ is isomorphic to $(X, +, \cdot, ')$. Define $f: P(S) \to X$ as follows. A typical element, say I of $P(S)$ is some subset of $\{1, 2, ..., n\}$. We let $f(I) = \sum_{i \in I} a_i$. In other words, $f(I)$ is the sum of those atoms whose indices are in I. Clearly $f(\phi) = 0$ and $f(S) = 1$, Because of part (iii) of the last proposition, the function, f is a bijection. In order to show that it is an isomorphism of two Boolean algebras, we must show that it preserves the corresponding operations. Specifically, for any two subsets I and J of S, we must show: (i) $f(I \cup J) = f(I) + f(J)$ (ii) $f(I \cap J) = f(I) \cdot f(J)$ and (iii) $f(S - I) = [f(I)]'$. We verify these conditions one-by-one. For notational convenience, let $I = \{k_1, ..., k_r, i_1, ..., i_p\}$ and $J = \{k_1, ..., k_r, j_1, ..., j_q\}$ where p, q, r are non-negative integers and $i_1, ..., i_p \notin J$ and $j_1, ..., j_q \notin I$. Then $I \cup J = \{k_1, ..., k_r, i_1, ..., i_p, j_1, ..., j_q\}$ and $I \cap J = \{k_1, ..., k_r\}$.

Now for (i) $f(I \cup J) = \sum_{s=1}^{r} a_{k_s} + \sum_{t=1}^{p} a_{i_t} + \sum_{u=1}^{q} a_{ju}$

$$= \sum_{s=1}^{r} a_{k_s} + \sum_{t=1}^{p} a_{i_t} + \sum_{u=1}^{q} a_{ju} + \sum_{s=1}^{t} a_{k_s}$$

(by law of tautology)

$$= f(I) + f(J).$$

For (ii), $f(I) \cdot f(J) = (\sum_{s=1}^{r} a_{k_s} + \sum_{t=1}^{p} a_{i_t}) \cdot (\sum_{s=1}^{r} a_{k_s} + \sum_{u=1}^{q} a_{j_u})$

$$= \sum_{s=1}^{r} a_{k_s} \cdot a_{k_s} \quad \text{by (ii) of the last proposition}$$

$$= \sum_{s=1}^{r} a_{k_s} \quad \text{by the law of tautology}$$

$$= f(I \cap J).$$

And finally,

$$f(I) + f(S - I) = f(I \cup (S - I)) = f(S) = 1$$

and

$$f(I) \cdot f(S - I) = f[I \cap (S - I)] = f(\phi) = 0$$

by what is proved earlier. So by uniqueness of complements $f(S - I) = [f(I)]'$. This proves (iii) and completes the proof that f is an isomorphism. So X is isomorphic to $P(S)$. But obviously $P(S)$ is isomorphic to the power set Boolean algebra of the set $\{a_1, ..., a_n\}$. ∎

Note that although the theorem merely asserts the existence of an isomorphism between a given 'abstract' Boolean algebra and some concrete, power set Boolean algebra, the proof does more than that, because it gives an explicit isomorphism. In fact, the proof shows that the structure of an abstract, finite Boolean algebra is completely determined by the set of its atoms. For this reason, this theorem (or rather its proof) is an example of what are known as **structure theorems** in mathematics. Such thorems are very important because they express an abstract mathematical structure in terms of some concrete structure of the same type. As an application of the last theorem (or rather, Proposition (1.10)), we can define the concept of cardinality or of weight for elements of any finite, abstract Boolean algebra. If x is an element of a finite abstract Boolean algebra X, we let $| x |$ be simply the number of atoms of X contained in x. This is consistent with the definition of cardinality in case X is a power set Boolean algebra. The results proved in Chapter 2, Section 2, about cardinalities of finite sets can be generalised to any finite Boolean algebras.

As another application of the last theorem, we have the following result, which is not so easy to establish directly.

1.12 Corollary: If X is a finite Boolean algebra then $| X | = 2^n$ for some non-negative integer n.

Proof: We simply note that the power set of a set with n elements has cardinality 2^n. The result now follows by letting n be the number of atoms in X. ▮

The Stone representation theorem actually deals with all Boolean algebras, not just the finite ones. The idea of the proof is basically the same, namely to consider the atoms and to express every element in terms of the atoms. But there are two difficulties. The first is that we have to consider infinite sums of atoms. This is not a very serious difficulty and can be circumvented by considering suitable functions from the set of atoms into Z_2 (similar to the characteristic functions). The real hurdle is to prove part (i) of Proposition (1.10), namely that every non-zero element contains at least one atom. The proof for the finite case no longer works because in an infinite poset it is possible to have an infinite strictly descending sequence,

$$x_1 > x_2 > x_3 > ... > x_n > x_{n+1} \, ... $$

To get the existence of a minimal non-zero element we have to appeal to the axiom of choice, see the notes at the end of Chapter 3, Section 3. Even after using it, the result we get is that every Boolean algebra is isomorphic to a *subalgebra* of the power set Boolean algebra of the set of its atoms. There do exist Boolean algebras which are not isomorphic to the entire power set Boolean algebras of any sets. An example of such a Boolean algebra will be given in the exercises.

Exercises

1.1 Let S be any set and Z_2 a Boolean algebra with two elements 0 and 1. Prove that the Boolean algebra of all functions from S to Z_2 (under pointwise operations) is isomorphic to the power set

Boolean algebra of S. (Hint: Prove that the bijection constructed in the proof of Theorem (2.2.15) is an isomorphism).

1.2 Suppose S and T are disjoint sets. Prove that the product Boolean algebra $P(S) \times P(T)$ is isomorphic to the Boolean algebra

$$P(S \cup T).$$

1.3 A positive integer is called **square-free** if it is not divisible by the square of any prime number. For example 30 is square-free but 45, 120 are not. Suppose n is a square-free positive integer. Let X be the set of all positive integers dividing n, i.e., $X = \{x \in \mathbf{N} : x|n\}$. For $x, y \in X$ define $x + y$ to be the least common multiple of x and y, $x \cdot y$ to be the greatest common divisor of x and y and x' to be n/x. Prove that $(X, +, \cdot, ')$ is a Boolean algebra. Find a power set Boolean algebra isomorphic to X. Why is it necessary to assume that n is square-free?

1.4 Let $X = \{a, b, c, d\}$. Define two binary operations $+$ and $\cdot$ on X by the following tables:

$+$	a	b	c	d
a	a	b	c	d
b		b	d	d
c	c	d	c	d
d	d	d	d	d

$\cdot$	a	b	c	d
a	a	a	a	a
b	a	b	a	b
c	a	a	c	c
d	a	b	c	d

Prove that there exists a function $' : X \to X$ such that $(X, +, \cdot, ')$ is a Boolean algebra.

(Hint: Instead of verifying the axioms directly, it is much easier to find some known Boolean algebra Y and a bijection $f : X \to Y$ which is compatible with addition and multiplication.)

1.5 In the example given after Proposition (1.2), verify that the sub-algebra Y is indeed isomorphic to $P(\mathscr{D})$.

1.6 (a) Prove that it is impossible to prove the laws of tautology using only $B1$, $B2$ and one of the distributive laws in Definition (1.1).

(b) What is wrong if we attempt to prove part (ii) of Theorem (1.3) as follows?

$$x + 1 = x + x + x' \qquad (\text{since } x + x' = 1 \text{ by } B4)$$

$$\begin{aligned} &\doteq x + x' && \text{(since } x + x = x \text{ by (i))} \\ &= 1 && \text{(by } B4). \end{aligned}$$

1.7 Prove that in a distributive, bounded lattice, the complements are unique, whenever they exist.

1.8 Although neither $+$ nor $\cdot$ in a Boolean algebra $(X, +, \cdot, ')$ satisfies the cancellation law individually, prove that the 'simultaneous cancellation law' holds, i.e. for any $x, y, z \in X$, $x + y = x + z$ and $x \cdot y = x \cdot z$ together imply $y = z$.

1.9 If $x_1, x_2, ..., x_n$ are elements of a Boolean algebra prove that

$$x_1 + x_2 + ... + x_n = 0$$

iff $x_i = 0$ for all $i = 1, 2, ..., n$ and $x_1 \cdot x_2 \cdot .... \cdot x_n = 1$ iff $x_i = 1$ for all $i = 1, 2, ..., n$.

1.10 If x, y are elements of a Boolean algebra, prove that $x = y$ iff $xy' + x'y = 0$. (This and the last result help in reducing any system of equations to a single equation.)

1.11 Prove that the following system of statements:
(i) All happy, intelligent men are rich
(ii) The only way for a poor man to be happy is to be intelligent
(iii) Every rich man is either happy or married
(iv) No married man is unhappy unless he is poor
is equivalent to the single statement that a man is happy iff he is rich.

1.12 A medical officer reported the following observations in a health survey of a certain population:
(i) 40% of the population smoked cigarettes, 30% suffered from cancer and 20% suffered from heart diseases.
(ii) All smokers suffering from heart diseases also suffered from cancer.
(iii) Every person not suffering from heart diseases either refrained from smoking or else had cancer.
Prove that the survey is a fake one.

1.13 A certain university offers four courses A, B, C and D and prescribes the following rules for registration:
(i) Every student must register for at least two courses.
(ii) A student registering for course A must also register for one of the courses B and C but not both.
(iii) Every student registering for course C must also register for at least one of the courses B and D.
(iv) No student may register for courses A and C simultaneously.
(v) No student may register for courses C and D simultaneously without registering for course B
Simplify the rules and prove that one of them is redundant (that is, it is implied by the remaining four rules).

1.14   Suppose $(X, \leqslant)$ is a lattice with a maximum element 1. Then a maximal element of $X - \{1\}$ is called a **co-atom**. Prove that an element a of a Boolean algebra is a co-atom iff its complement a' is an atom. (Atoms and co-atoms are dual concepts. Generally, the suffix co- is used to name the dual concepts. Another such usage occurs in the next exercise.) Prove that every element of a finite Boolean algebra can be uniquely expressed as a product of co-atoms.

1.15   A subset A of a set S is called **cofinite**, if its complement, $S - A$ is finite. Prove that:
   (i)   the union and the intersection of two cofinite sets is cofinite.
   (ii) if S is infinite then every cofinite subset of S is infinite but not every infinite subset is necessarily cofinite.
   (iii) if S is countable then the set of all cofinite subsets of S is countable. (cf. Exercise (2.2.22) part (v).)

1.16   Let Y be the set of all subsets of N which are either finite or cofinite. Prove that Y is a subalgebra of the power set Boolean algebra $P(\mathbf{N})$.

1.17   Prove that Y, regarded as a Boolean algebra by itself, is not isomorphic to the full power set Boolean algebra of any set.
   (Hint: Prove that there exists no set S such that $|P(S)|$ equals $\aleph_0$.)

1.18   Prove directly (without using Corollary (1.12)), that the number of elements in a finite, nontrivial Boolean algabra is even. (Hint: Pair off every element with its complement.)

1.19   Let $(X, +, \cdot, ')$ be a Boolean algebra. For $x, y \in X$ define $x \Delta y$ as $xy' + x'y$. The binary operation Δ so defined is called **symmetric difference** or the **exclusive or**. Δ is also often denoted by $\oplus$ and called the **ring sum**. Prove that:
   (i) Δ is commutative, associate and has 0 as an identity.
   (ii)   Every element of X is invertible w.r.t. Δ.
   (iii) If x, y are two-state devices then $x\Delta y$ is 1 when exactly one of x and y equals 1. (Hence the name 'exclusive or'.)
   (iv) $\cdot$ is distributive over Δ.
   (v)   For all $x \in X$, $x' = x\Delta 1$.
   (vi)   For all $x, y \in X$, $x + y = x\Delta y\Delta(x \cdot y)$.
          (Thus, $+$ can be recovered from Δ).

1.20   Let S be any set and X a Boolean algebra. Let Y be the set of all functions from S to X. Then Y is a Boolean algebra under pointwise operations. If $f, g \in Y$, prove that $f \leqslant g$ iff $f(s) \leqslant g(s)$ in X for all $s \in S$. (In other words the partial order induced by pointwise operations is the same as the pointwise partial order. Fix some $s_0 \in S$ and define $f: S \to X$ by

$$f(s) = \begin{cases} 0 \text{ for all } s \in S, \text{ except for } s = s_0 \\ \text{some atom of } X, \text{ for } s = s_0. \end{cases}$$

Prove that f is an atom of Y.

1.21 Similar to the last exercise, obtain a description of the partial order and of the atoms of the product of two Boolean algebras.

**1.22 Suppose X is a non-empty set, $\cdot$ is a binary operation on X and $': X \to X$ is a function such that (i) $\cdot$ is commutative and associative, (ii) there exists an element 0 in X such that $x \cdot x' = 0$ for all $x \in X$, (iii) for all $x, y \in X$, $x \cdot y = x$ iff $x \cdot y' = 0$. Prove that a binary operation $+$ on X can be so defined that $(X, +, \cdot, ')$ is a Boolean algebra. (This gives an alternate and a highly compact definition of a Boolean algebra.)

Notes and Guide to Literature

The definition of a Boolean algebra, as given here is due to Huntington. The lattice approach is equally popular, especially in logic. For more on the definition in Exercise (1 22), see Dornhoff and Hohn [1]. A proof of the Stone representation theorem for Boolean algebras may be found in Kelley [1].

In the literature on Boolean algebras (and other algebraic structures) the reader may often find additional axioms such as 'if $a, b \in X$ then $a + b \in X$' and 'if $a = b$ then $a + c = b + c$', called respectively as laws of closure and substitution (or uniqueness). This practice comes from the days when a rigorous definition of a binary operation was not formulated. Now that we define a binary operation on a set X as a function from $X \times X$ to X and require that a function always assume values in its codomain and be single valued, it is redundant for us to include such axioms.

2. Boolean Functions

It was remarked in the last section that Boolean algebras provide the appropriate mathematical tool for handling two state devices. It often happens that the state of some two state device depends upon the states of some other two state devices. For example whether current flows through an electrical circuit containing various switches depends upon the states of these switches. The truth or falsehood of a complicated statement depends upon that of several simpler statements. Such dependence can be described by a function called a Boolean function. In this section we study the general theory of such functions. Their applications to circuits and logic will be considered in the next two sections.

We begin by giving a formal, mathematical name for a two-state device.

2.1   Definition: A **Boolean variable** is a variable which assumes only two possible values, 0 and 1.

In other words, a Boolean variable takes values in the set Z_2. Since Z_2 is a Boolean algebra, we can define the addition, multiplication and complements of Boolean variables. For example, if x, y are two Boolean variables, $x + y$ is the Boolean variable which has value 1 when at least one of x and y has value 1, and the value 0 otherwise, (because in Z_2, $0 + 1 = 1 + 0 = 1 + 1 = 1$ and $0 + 0 = 0$).

Conceptually, Boolean variables are similar to real variables with which we are all too familiar. The obvious difference, of course, is that while a real variable assumes infinitely many possible values, a Boolean variable assumes only two possible values. Thus, a Boolean variable is the simplest possible variable, because a variable assuming only one possible value is not a variable at all, it is a constant. A real variable is a continuous variable while a Boolean variable is discrete.

Two Boolean variables x and y are called **independent** if each can assume values independently of the other. This means any of the four possibilities can occur, namely, $x = 0$, $y = 0$; $x = 1$, $y = 1$; $x = 1$, $y = 0$ and $x = 1$, $y = 1$. More generally, if $x_1, x_2, ..., x_n$ are n independent Boolean variables, then together they can be assigned values in 2^n possible ways. Note that two complementary variables x and x' can never be independent.

For real variables, we often replace two real variables, say x and y, by a single variable (x, y) which takes values in the cartesian plane $R \times R$, or R^2. More generally, if we have n real variables, we can replace them by a single variable ranging over the euclidean space R^n. A similar construction is possible for Boolean variables. Let Z_2^n denoted the product set $Z_2 \times Z_2 \times ... \times Z_2$ (n times). (Note that Z_2^n is also the set of all binary sequences of length n.) If $x_1, x_2, ..., x_n$ are Boolean variables, we can replace them by a single sequential variable $(x_1, x_2, ..., x_n)$ taking values in Z_2^n. Mutual independence of $x_1, x_2, ..., x_n$ is equivalent to saying that this new variable $(x_1, x_2, ..., x_n)$ assumes all the 2^n values in Z_2^n. Analogous to the concept of a real-valued function of several real variables, we have the following definition.

2.2   Definition: A **Boolean function** of n variables is any function from Z_2^n to Z_2.

For example suppose $n = 3$ and $f(x_1, x_2, x_3) = x_1 x_2 + x_2' x_3$. Then f is a Boolean function of three variables. $f(1, 0, 1)$ is the value assumed by f when $x_1 = 1$, $x_2 = 0$ and $x_3 = 1$. Since in this case, $x_2' = 1$, we get

$$f(1, 0, 1) = 1.0 + 1.1 = 0 + 1 = 1.$$

Similarly $f(0, 1, 0) = 0$. Notice that computation of the values of a Boolean function is considerably simple because 0 and 1 are the only two possible values and also because the binary operations in the Boolean algebra Z_2 are so simple. For example, if we want to evaluate some function f which is given by writing $f(x_1, x_2, ..., x_n)$ as a sum of terms, the moment we see that at least one of these terms equals 1 for a particular choice of values of the variables $x_1, ..., x_n$, it is unnecessary to evaluate the other terms because no matter what they are, f will have value 1 for that particular choice of values of the variables.

There are two ways to denote a Boolean function f of n variables, $x_1, x_2, ..., x_n$. The first method is to write the table of values of f, in which we actually list down the values of f for all possible elements of the domain set, namely Z_2^n. This is a workable proposition because the domain set is finite. (Even in the case of a function of a real variable, such tables of values are frequently prepared while drawing its graph. But such tables can never be complete since the domain set is infinite.) In such a table there are $n + 1$ columns, one for each x_i and one for f. Sometimes some auxiliary columns may be added as an aid to fill the column below f. There are 2^n rows. In each row, we indicate one possible assignment of the values 0 and 1 to the x_i's by writing the value given to x_i in the column under x_i. In that row, the place in the column below f is filled by the value of f for these values of x_1's. Although the order of the rows can be arbitrary, one standard practice is to write them, starting from $(1, 1, ..., 1)$ and then making changes from the right to the left, so that x_n changes most frequently and x_1 changes only once. In Figure 4.1 (a), we show the table of values of the function f of three variables considered above, namely

$$f(x_1, x_2, x_3) = x_1 x_2 + x_2' x_3.$$

Row No.	x_1	x_2	x_3	f
1	1	1	1	1
2	1	1	0	1
3	1	0	1	1
4	1	0	0	0
5	0	1	1	0
6	0	1	0	0
7	0	0	1	1
8	0	0	0	0

Figure 4.1 (a) : Table of Values of a Boolean Function

There is another way to write the table of values of a Boolean function, which is a little more concise. The idea is to replace several (say k) Boolean variables by a single variable taking 2^k possible values. For example, instead of the two Boolean variables x_1 and x_2, we treat (x_1, x_2) as a single variable, taking 4 possible values, namely $(1, 1)$, $(1, 0)$, $(0, 1)$ and $(0, 0)$. Given a function f of n Boolean variables $x_1, ..., x_n$, we take a suitable k (usually an integer close to $n/2$). We consider $(x_1, ..., x_k)$ as a single variable taking 2^k values and $(x_{k+1}, ..., x_n)$ as another single variable taking 2^{n-k} values. The function f may then be thought of as a function of these two variables and consequently its values can be listed in a table with 2^k rows and 2^{n-k} columns, similar to the table of a binary operation. For example, consider the function $f(x_1, x_2, x_3) = x_1 x_2 + x_2' x_3$ for which a table was constructed above. Here $n = 3$. We take $k = 2$. Then another table for f can be drawn as shown in Fig. 4.1 (b). Both the tables, of course, convey the same information.

(x_1, x_2) ＼ x_3	1	0
$(1, 1)$	1	1
$(1, 0)$	1	0
$(0, 1)$	0	0
$(0, 0)$	1	0

Figure 4.1 (b): Concise Table for a Boolean Function

Even with the concise version, it becomes impracticable to draw the table of values of a Boolean function of n variables even for relatively small values of n. So we look for another, compact method. Such a method is provided by giving an algebraic formula for the function. In Chapter 2, Section 1 we emphasised that a function should not be confused with a formula for it. Indeed for functions of real variables, such closed formulas may not always exist. As a well-known example take the function $f: \mathbf{R} \to \mathbf{R}$ defined by $f(x) =$ the smallest prime greater than x. This is a perfectly well-defined function but there is no known formula expressing $f(x)$ as something like $\exp(-\sin^2 x + x)$.

However, things are much better for Boolean functions as we prove in the following theorem. First we introduce a shorthand notation. Let $x_1, x_2, ..., x_n$ be Boolean variables which are mutually independent. For $i = 1, 2, ..., n$ let $f_i: \mathbf{Z}_2^n \to \mathbf{Z}_2$ be the Boolean function defined by

$$f_i(x_1, x_2, \ldots, x_n) = \begin{cases} 1 \text{ if } x_i = 1 \\ 0 \text{ if } x_i = 0 \end{cases}$$

In other words, $f_i(x_1, x_2, \ldots, x_n) = x_i$. For this reason we shall denote f_i by x_i itself. This double role of x_i (once as a Boolean variable and again as a Boolean function) should cause no confusion. (The function f_i is often called the **projection** on the ith factor and denoted by π_i.)

We now prove what may be called the structure theorems for Boolean functions.

2.3 Theorem: Let

$$x_1, x_2, \ldots, x_n$$

be mutually independent Boolean variables. Then there are 2^{2^n} Boolean functions of these n variables. The totality of such functions constitutes a Boolean algebra. The atoms of this algebra are the 2^n functions of the form

$$x_1^{\varepsilon_1} x_2^{\varepsilon_2} \ldots x_n^{\varepsilon_n}$$

where each $x_i^{\varepsilon_i}$ is either x_i or x_i' (with x_i interpreted as above).

Proof: Let X be the set of all Boolean functions of the n variables. Every element of X is a function from Z_2^n to Z_2. Since $\lfloor Z_2^n \rfloor = 2^n$, it follows from Theorem (2.2.14) that $|X| = 2^{2^n}$. Z_2 is a Boolean algebra. So under point-wise operations (which we denote by $+$, $\cdot$ and $'$) X is a Boolean algebra. We have to find the atoms, i.e., the minimal non-zero elements in X. First we show that all functions of the form

$$x_1^{\varepsilon_1} x_2^{\varepsilon_2} \ldots x_n^{\varepsilon_n}$$

are atoms. Let g be one such function. For notational simplicity we suppose

$$g = x_1 x_2 \ldots x_k x_{k+1}' \ldots x_n'$$

for some k. (In all other cases the argument is the same except for notations.) Note that as a function from Z_2^n to Z_2, g vanishes at all points of Z_2^n except the point $(1, \ldots, 1, 0, \ldots, 0)$ where the first k entries are 1 and the remaining $n - k$ entries are 0 (cf. Exercise (1.9)). For brevity let us call this point e_k. Then $g(e_k) = 1$ but $g(x) = 0$ for all $x \in Z_2^n - \{e_k\}$. So g is not an identically zero function. Hence g is a non-zero element of X. To show it is a minimal non-zero element, suppose $f \in X$ and $f \leqslant g$. Then $f(x) \leqslant g(x)$ for all $x \in Z_2^n$ (cf. Exercise (1,20)). But $g(x) = 0$ for all $x \neq e_k$. So $f(x) = 0$ for all $x \neq e_k$. If $f(e_k) = 0$, then f is the zero element of X and if $f(e_k) = 1$ then $f = g$. Thus g is not properly larger than any non-zero element of X. In other words g is a minimal non-zero element of X, i.e., an atom of X. Thus every element of the form

$$x_1^{\varepsilon_1} x_2^{\varepsilon_2} \dots x_n^{\varepsilon_n}$$

where each $x_i^{\varepsilon_i}$ is either x_i or x_i' is an atom of X. Since there are two possibilities for each i, we get 2^n such elements. But by Corollary (1.12), the number of atoms in a Boolean algebra with 2^{2^n} elements is precisely 2^n. Since we have already constructed 2^n atoms in X, it follows that there can be no other atoms. ∎

Now that we know the atoms of X, we can apply the theory proved in the last section. We record the result as a theorem.

2.4　Theorem: Every Boolean function of n variables x, x_2, ..., x_n can be uniquely expressed as a sum of terms of the form

$$x_1^{\varepsilon_1} x_2^{\varepsilon_2} \dots x_n^{\varepsilon_n}$$

where each $x_i^{\varepsilon_i}$ is either x_i or x_i'. Specifically, for every element, say,

$$\bar{y} = (y_1, y_2, \dots, y_n)$$

of Z_2^n, let $f_{\bar{y}}$ be the Boolean function from $Z_2^n \to Z_2$ defined by $f_{\bar{y}}(\bar{y}) = 1$ and $f_{\bar{y}}(\bar{z}) = 0$ for all $\bar{z} \in Z_2^n - \{\bar{y}\}$, and let $x_{\bar{y}}$ be the term

$$x_1^{\varepsilon_1} x_2^{\varepsilon_2} \dots x_n^{\varepsilon_n}$$

where $x_i^{\varepsilon_i} = x_i$ if $y_i = 1$ and $x_i^{\varepsilon_i} = x_i'$ if $y_i = 0$. (For example, for $n = 5$, and

$$\bar{y} = (1, 0, 1, 0, 0), \ x_{\bar{y}} = x_1 x_2' x_3 x_4' x_5').$$

Then every function

$$f : Z_2^n \to Z_2$$

equals $\Sigma x_{\bar{y}}$ where the sum extends over all $\bar{y}$ such that $f(\bar{y}) = 1$.

Proof:　As proved above, elements of the form $f_{\bar{y}}$ are precisely the atoms of X, the Boolean algebra of all Boolean functions of $x_1, x_2, \dots, x_n$. Note that a function $f : Z_2^n \to Z_2$ contains the atom $f_{\bar{y}}$ iff $f(\bar{y}) = 1$ (see again Exercise (1.19)). Hence by Proposition (1.10), f equals $\Sigma f_{\bar{y}}$ where the sum extends over all $\bar{y} \in Z_2^n$ such that $f(\bar{y}) = 1$. But, because of our understanding to let x_i denote the function which takes $(x_1, x_2, \dots, x_n)$ to x_i, it is easily seen that $f_{\bar{y}}$ is nothing but $x_{\bar{y}}$. Hence $f = \Sigma x_{\bar{y}}$. ∎

As a concrete example, consider the function f whose table of values is shown in Figure 4.1(a). This function takes the value 1 at four points of Z_2^3, namely,

$$(1, 1, 1), \ (1, 1, 0), \ (1, 0, 1) \text{ and } (0, 0, 1).$$

The corresponding atoms are

$$x_1 x_2 x_3, \ x_1 x_2 x_3', \ x_1 x_2' x_3 \text{ and } x_1' x_2' x_3,$$

Hence f equals the sum of these four atom, i.e.,

$$f = x_1x_2x_3 + x_1x_2x_3' + x_1x_2'x_3 + x_1'x_2'x_3.$$

Upon simplification, this reduces to

$$x_1x_2 + x_2'x_3.$$

Thus

$$f(x_1, x_2, x_2) = x_1x_2 + x_2'x_3$$

for all

$$x_1, x_2, x_3 \in Z_2.$$

This was, of course, the original definition of f. What we have shown is that even if we do not know the original formula for f, but simply know (either from the table of values or by some other method) the points where f takes the value 1 then we can always get an algebraic expression for f by adding the corresponding atoms. Of course, the expression so obtained may not be the simplest possible. As in the example above, we can often reduce it by combining suitable terms and using the laws proved for Boolean algebras. The point is that, unlike functions of real variables, Boolean functions always have an algebraic formula. Indeed a few authors *define* a Boolean function of n Boolean variables $x_1, ..., x_n$ as an algebraic expression of these variables involving the operations $+, \cdot$ and $'$. We refrain from doing so because the concept of a function is a general one and although in a particular instance it coincides with a formula, this is no reason to change the definition.

Although the algebraic expression for a Boolean function, as a sum of certain atomic functions is generally not the shortest possible, it has certain advantages. First, it can be written down just by inspection from the table of values. Secondly, it has a certain regularity of form. It consists of a sum of terms, each term has exactly the same number of factors (equal to the number of variables), the order of these factors is regulated and each factor is very simple, either some x_i or x_i'. Because of this regularity of form, the expression is given a special name.

2.5 Definition: The algebraic expression for a Boolean function of n variables as a sum of atoms (as given by Theorem (2.4)) is called the **disjunctive normal form** (abbreviated **D.N.F.**) of that function. The disjunctive normal form of the function which identically equals 1 is called the **complete disjunctive normal form** in n variables.

The word 'disjunctive' here refers to summation, which is more formally called disjunction. The name comes because the function is expressed as a sum of terms each having a certain 'normal' form.

As noted above, the disjunction normal form of a function can be written down by an inspection of its table of values. Conversely, given the disjunctive normal form of a function of n variables we know immediately at what points of Z_2^n it assumes the value 1. Since at all other points the function must vanish, we can construct the table of values of the function. Because of this, we may regard the disjunctive normal form of a function as a compact version of its table of values. For example, if

$$f(x_1, x_2, x_3, x_4) = x_1 x_2 x_3' x_4 + x_1' x_2 x_3' x_4 + x_1 x_2' x_3 x_4'$$
$$+ x_1' x_2' x_3' x_4 + x_1' x_2' x_3' x_4'$$

then f is in its disjunctive normal form and we see that f assumes the value 1 for 5 points in Z_2^4, namely, (1, 1, 0, 1), (0, 1, 0, 1), (1, 0, 1, 0), (0, 0, 0, 1) and (0, 0, 0, 0). Therefore, in the table of values of f, in the column under f, 1 will occur at 5 places, that is, in the rows corresponding to these 5 points. All the remaining 11 places will be filled by 0's.

Sometimes we are already given some function f in an algebraic form (not necessarily in the disjunctive normal form) and we want to cast it into its D.N.F. One method is to write the table of values of f. But this is often too tedious. So we look for other methods. We first write f as a sum of monomials, where by a **monomial** we mean a product in which every factor is some variable or its complement. For example $x_1 x_2' x_4$ is a monomial but $x_1(x_2' + x_3)x_4$ is not a monomial by itself; it is a sum of two monomials, $x_1 x_2' x_4 + x_1 x_3 x_4$. Because of $x_i \cdot x_i' = 0$ we ignore those monomials in which some variable and its complement both occur. Also because $x_i \cdot x_i = x_i$ (by law of tautology) if the same variable occurs more than once in a monomial, we retain only its first occurrence. Of course, not every variable need appear in a given monomial. If neither x_i nor x_i' appears in a monomial, we multiply the monomial by $x_i + x_i'$ (which equals 1) and split it into two monomials, one of which contains x_i and the other contains x_i' and both of which contain all other variables occurring in the original monomial. We repeat this process till every monomial in the original expression of f is expressed as a sum of monomials in which every variable occurs. Now once again apply the law of tautology to weed out the repetitions of monomials. The resulting expression is the disjunctive normal form of the given function. As an illustration we do the following problem.

2.6   Problem: Write the following Boolean functions in their disjunctive normal forms:

(i) $f(x_1, x_2, x_3) = (x_1 + x_2')x_3' + x_2 x_1'(x_2 + x_1' x_3)$

(ii) $g(a, b, c) = (a + b + c)(a' + b + c')(a + b' + c')$
$$(a' + b' + c')(a + b + c')$$

Solution:

(i) $f(x_1, x_2, x_3) = x_1 x_3' + x_2' x_3' + x_2 x_1' + x_2 x_1' x_1' x_3$
$$= x_1(x_2 + x_2')x_3' + (x_1 + x_1')x_2' x_3' + x_1' x_2(x_3 + x_3')$$
$$+ x_1' x_2 x_3$$

$$= x_1x_2x_3' + x_1x_2'x_3' + x_1x_2'x_3' + x_1'x_2'x_3' + x_1'x_2x_3$$
$$+ x_1'x_2x_3' + x_1'x_2x_3$$

$$= x_1x_2x_3' + x_1x_2'x_3' + x_1'x_2x_3 + x_1'x_2x_3' + x_1'x_2'x_3'$$

This is the distinctive normal form.

(ii) It would be too cumbersome to multiply the 5 factors out using distributivity of $\cdot$ over $+$, because this would initially give a sum of 3^5, i.e. 243 terms. But we can use distributivity of $+$ over $\cdot$. Then the product of the first and the fifth factor is $a + b$ while that of the second and the fourth is $a' + c'$. Because of tautology, the same factor may be used again. So the product of the third and the fourth factor is $b' + c'$.

Hence $g(a, b, c) = (a + b)(a' + c')(b' + c')$

$$= (a + b)(a'b' + c')$$

$$= aa'b' + ba'b' + ac' + bc'$$

$$= ac' + bc'$$

$$= a(b + b')c' + (a + a')bc'$$

$$= abc' + ab'c' + abc' + a'bc'$$

$$= abc' + ab'c' + a'bc'. \quad \blacksquare$$

We recall once again that the disjunctive normal form arises by applying the theory of atoms developed in the last section to the Boolean algebra of Boolean functions of Boolean variables. In the proof of Theorem (1.11) we saw how to add, multiply and take complements of elements expressed as sums of atoms. This leads to the proof of the following result.

2.7 Proposition: Let two functions f and g of the same n Boolean variables be expressed in their respective disjunctive normal forms. Then the D.N.F. of $f + g$ is obtained by taking the sum of those terms which appear in the D.N.F. of at least one of f and g; the D.N.F. of $f \cdot g$ is obtained by summing terms which are common to the D.N.F.'s of both f and g and the D.N.F. of f' is obtained by omitting from the complete D.N.F. the terms which appear in the D.N.F. of f.

Proof: As already noted, this is merely a special case of the proof of Theorem (1.11). For the last statement, note that the function which identically equals 1 is precisely the element 1 (i.e., the identity of multiplication) of the Boolean algebra of Boolean functions of n variables. By (iii) in Proposition (1.10), this function is the sum of all atoms, which is precisely the complete disjunctive normal form. $\quad \blacksquare$

The dual concept of a disjunctive normal form is the conjunctive normal form. By Exercise (1.14), every element of X can be uniquely expressed as a product of co-atoms of X. Since co-atoms are precisely the complements of atoms, it follows that the co-atoms of X are all functions of the form $x_1^{\varepsilon_1} + x_2^{\varepsilon_2} + \cdots + x_n^{\varepsilon_n}$ where each $x_i^{\varepsilon_i}$ is either x_i or x_i'. When a Boolean function of $x_1, x_2, \ldots, x_n$ is expressed as a product of factors of this form the function is said to be expressed in its **conjunctive normal form**, abbreviated C.N.F. Each factor in the C.N.F. of a function f corresponds to one point of $\mathbb{Z}_n^2$ where f vanishes. Therefore, like the D.N.F., the C.N.F., of a function can be written down by an inspection of its table of values. For example, for the function $f(x_1, x_2, x_3) = x_1 x_2 + x_2' x_3$, we see from its table in Figure 4.1(a) that f vanishes at four points in $\mathbb{Z}_2^3$, namely, $(1, 0, 0), (0, 1, 1), (0, 1, 0)$ and $(0, 0, 0)$. Take the point $(1, 0, 0)$. It corresponds to the factor $x_1' + x_2 + x_3$, because this is the only co-atom which vanishes when $x_1 = 1$, $x_2 = 0$ and $x_3 = 0$. Similarly, we determine other factors and get

$$f = (x_1' + x_2 + x_3)(x_1 + x_2' + x_3')(x_1 + x_2' + x_3)(x_1 + x_2 + x_3).$$

We could also get this directly from

$$f(x_1, x_2, x_3) = x_1 x_2 + x_2' x_3$$

by dualising the procedure for obtaining the D.N.F. Thus,

$$f(x_1, x_2, x_3) = x_1 x_2 + x_2' x_3$$
$$= (x_1 + x_2' x_3)(x_2 + x_2' x_3) \qquad \text{(by distributive laws)}$$
$$= (x_1 + x_2')(x_1 + x_3)(x_2 + x_3)$$
$$= (x_1 + x_2' + x_3)(x_1 + x_2' + x_3')(x_1 + x_2 + x_3)$$
$$(x_1 + x_2' + x_3) \cdot (x_1 + x_2 + x_3)(x_1' + x_2 + x_3)$$
$$= (x_1' + x_2 + x_3)(x_1 + x_2' + x_3')(x_1 + x_2' + x_3)$$
$$(x_1 + x_2 + x_3).$$

Properties of the conjunctive normal form are dual to those of the disjunctive normal form and hence will not be stated separately. Because of the principle of duality for Boolean algebras, the two forms are really equivalent. But in a particular context, it may be more advantageous to express a given function in its C.N.F. than in its D.N.F. or vice versa. Note that for a function of n variables, if its D.N.F. has r terms then its C.N.F. will have $2^n - r$ factors. Here r also equals the number of points in $\mathbb{Z}_2^n$ where the function takes the value 1. So, if a function vanishes at all except a few points, it is more economical to take its D.N.F. than its C.N.F., but if it vanishes at only a few points, the situation is reversed.

There is also a way to convert either of the two forms to the other using double complementation. We illustrate it for the function f in problem 2.6. There we obtained

$$f = x_1 x_2 x_3' + x_1 x_2' x_3' + x_1' x_2 x_3 + x_1' x_2 x_3' + x_1' x_2' x_3'.$$

Then by proposition (2.7), f' is obtained by omitting these five terms from the complete D.N.F. in the three variables (which has 8 terms). This gives $f' = x_1 x_2 x_3 + x_1 x_2' x_3 + x_1' x_2' x_3$. We now again take complements, but apply the De Morgan's laws to the right hand side. Thus

$$(f')' = (x_1' + x_2' + x_3')(x_1' + x_2 + x_3') \cdot (x_1 + x_2 + x_3').$$

But $(f')' = f$ and thus we have obtained the C.N.F. of f.

Having discussed the generalities about Boolean functions, let us now illustrate how they arise in real life problems. More examples would come in the next two sections. But some initiation can be done right now.

Let us take the Locks Problem. As before, denote the five persons by p_1, p_2, p_3, p_4 and p_5. Each person p_i can either give his assent to open the box (which means he hands over all the keys in his possession) or he may refuse to do so. These being the only two possibilities, we associate a Boolean variable x_i to represent them. We set $x_i = 1$ if p_i agrees to open the box and $x_i = 0$ if he does not. The box also has two states namely open or closed, and we use the Boolean variables b to denote the state of the box, $b = 1$ if the box is open and $b = 0$ if not. The problem now is to express b as a function of the variables $x_1, ..., x_5$ in such a way that b will be 1 iff at least three of the x_i's are equal to 1. From this we can mentally prepare the table of values of f and write f in its disjunctive normal form, giving $b = b_1 + b_2 + b_3$ where b_1 is the sum of ten terms of the form $x_1 x_2 x_3 x_4' x_5'$ (with three variables without $'$ and 2 with $'$), b_2 is the sum of 5 terms of the form $x_1 x_2 x_3 x_4 x_5'$ and $b_3 = x_1 x_2 x_3 x_4 x_5$.

We have not yet said anything about the locks. Indeed, the discussion so far has been completely independent of the manner by which each person signifies his assent or dissent to open the box. In the next section, we shall consider the case where the assent is expressed by pressing an electric switch and the outcome (that is, the state of the box) will be indicated by the lighting of some lamp. For the moment, let us do the problem of designing a system of locks and keys. Each lock L is also a two state device and hence can be represented by a Boolean variable, say y, which we set to 1 if L is open and 0 if L is closed. Now if the keys of a particular lock are with persons $p_{i_1} ..., p_{i_r}$ (say), then the corresponding variable y equals 1 iff at least one of $x_{i_1} ..., x_{i_r}$ equals 1 (regardless of the states of the other x_i's). So $y = x_{i_1} + x_{i_2} + ... + x_{i_r}$. The box will open iff all the locks are open. Hence, b (that is, the Boolean variable representing the state of the box) must be the product of Boolean functions representing the states of the locks. So the problem will be solved if we factor b (for which an expression was obtained above) into factors of the form $(x_{i_1} + x_{i_2} + ... + x_{i_r})$. Corresponding to each such factor there will be a lock whose keys will be given to $p_{i_1}, p_{i_2}, ..., p_{i_r}$. If b cannot be factorised into factors of this form it means the problem has no solution.

To see whether b has a factorisation of the desired form, we first express b in its conjunctive normal form. We convert the D.N.F. of b obtained above as $b_1 + b_2 + b_3$ to the C.N.F. using double complementation as illustrated earlier. We omit the details of the computation but write the final answer as $b = c_1 \cdot c_2 \cdot c_3$ where c_1 is the product of ten factors of the form $(x_1 + x_2 + x_3 + x_4' + x_5')$, c_2 is the product of five factors of the form $(x_1 + x_2 + x_3 + x_4 + x_5')$ and $c_3 = x_1 + x_2 + x_3 + x_4 + x_5$. Now we group together these 16 factors using distributivity of $+$ over $\cdot$ and also the law of tautology (which allows us to use the same factor again and again). As a sample, the product of the four factors

$$(x_1 + x_2 + x_3 + x_4' + x_5'), (x_1 + x_2 + x_3 + x_4' + x_5),$$
$$(x_1 + x_2 + x_3 + x_4 + x_5') \text{ and } (x_1 + x_2 + x_3 + x_4 + x_5)$$

comes out as $(x_1 + x_2 + x_3 + x_4')(x_1 + x_2 + x_3 + x_4)$ which equals $x_1 + x_2 + x_3$. Similarly for every distinct i, j, k we can get $x_i + x_j + x_k$ as the product of four factors in the C.N.F. of b. Thus b ultimately comes out to be equal to the product of 10 factors of the form $x_i + x_j + x_k$ where i, j, k run through triples of distinct indices from 1 to 5. Consequently there will be ten locks and each lock will have 3 keys. This is, of course, exactly the same answer as obtained in Chapter 2, Section 3. But the present method makes a systematic use of Boolean functions and therefore can be applied to more general problems.

As in this problem, the data often has certain symmetry. In the Locks Problem, whether the box can be opened or not depends only on *how many* of the five persons want to open it and not on *which particular persons* want to open it. (We can, of course, change the problem, say, by giving some special powers to one of the persons. It would then no longer be symmetric in all the five persons.) The Boolean functions arising in such problems have a particularly simple form whose advantage will be more apparent in the next section. In the present section we study this symmetry condition. The motivation for the following definition comes from the fact that in the Locks Problem, if any two persons exchange their sets of keys, the solution to the problem remains unaffected.

2.8 Definition: A Boolean function f of n variables $x_1, x_2, .., x_n$ is said to be **symmetric** w.r.t. a pair of variables x_i and x_j if the value of f is unaffected by interchanging x_i and x_j, that is, (assuming $i < j$ without loss of generality) if for all values of $x_1, x_2, ..., x_n$ we have

$$f(x_1, .., x_{i-1}, x_i, x_{i+1}, ..., x_{j-1}, x_j, x_{j+1}, ..., x_n)$$
$$= f(x_1, ..., x_{i-1}, x_j, x_{i+1}, ..., x_{j-1}, x_i, x_{j+1}, ..., x_n).$$

If f is symmetric w.r.t. every pair of variables out of $x_1, ..., x_n$ then f is called **symmetric** among these variables (or simply symmetric).

For example, the function $x_1 x_2 + x_3$ is symmetric w.r.t. x_1 and x_2 but not w.r.t. x_1 and x_3 nor w.r.t. x_2 and x_3. The function $x_1 x_2 x_3 + x_4' (x_1' + x_2' + x_3')$ is symmetric among x_1, x_2 and x_3 but not among all the four variables. The function b in the Locks Problem is a symmetric function of all the five variables $x_1, ..., x_5$. The function $x_1 x_2' + x_2 x_3' + x_3 x_1'$ is also symmetric in the three variables. This is not immediately obvious. But if we rewrite $x_1 x_2' + x_2 x_3' + x_3 x_1'$ in its disjunctive normal form as

$$x_1 x_2' x_3 + x_2 x_3' x_1 + x_3 x_1' x_2 + x_1 x_2' x_3' + x_1' x_2' x_3 + x_1' x_2 x_3'$$

we see that it is symmetric.

More generally, whether a function is symmetric or not can be easily decided by writing it in its D.N.F. (C.N.F. would do equally well). Suppose f is written in its D.N.F. and $x_1 .... x_r x'_{r+1} .... x_n'$ is one of the terms, where r is some integer, $0 \leqslant r \leqslant n$. Let $e_r = (1, ..., 1, 0, ..., 0)$ be the binary sequence whose first r entries are all 1 and the remaining $n - r$ entries are all 0. Then $f(1, ..., 1, 0, ..., 0) = f(e_r) = 1$. Now because of symmetry, we can interchange any two terms in the sequence e_r and f will still assume the same value, namely 1 at this new sequence. By performing a series of such interchanges of two terms at a time, we can transform e_r to any binary sequence of length n in which exactly r terms are 1 and the remaining $n - r$ terms are 0. (A formal proof of this fact will be given in a later chapter. For the moment the reader can convince himself by trying a few cases. For example $(1,1,0,0,0)$ can be transformed into $(0,1,0,1,0)$ by letting it go through $(1,0,1,0,0)$, $(0,1,1,0,0)$ and $(0,1,0,1,0)$ there being only one interchange at any stage. Alternatively, we can modify the definition of a symmetric function so as to requires (reduntantly), that it be invariant under any permutation of the variables.) The number of sequences with r 1's and $(n - r)$ 0's is $\binom{n}{r}$. By symmetry, f takes the value 1 at all of them. Accordingly, the D.N.F. of f will contain all the $\binom{n}{r}$ terms of the form $x_1^{\varepsilon_1} .... x_n^{\varepsilon_n}$ where $x_i^{\varepsilon_i}$ equals x_i for exactly r values of i and equals x_i' for the remaining $(n - r)$ values of i. On the other hand if $f(e_r) = 0$ then the D.N.F. of f cannot contain any of these $\binom{n}{r}$ terms.

We use this reasoning to obtain a handy representation of symmetric functions. First we need a definition.

2.9 Definition: Let f be a symmetric function of n Boolean variables. Then an integer $r (0 \leqslant r \leqslant n)$ is called a **characteristic number** of f if $f(e_r) = 1$ where e_r is the binary sequence of length n whose first r entries are 1 and the remaining $(n - r)$ entries are 0.

For example the symmetric function $f(x_1, x_2, x_3) = x_1x_2 + x_2x_3 + x_3x_1$ has 2 and 3 as its characteristic numbers as we see by actual computation, namely $f(e_0) = f(0,0,0) = 0, f(e_1) = 0, f(e_2) = 1$ and $f(e_3) = 1$. With a little practice, the characteristic numbers can be found by inspection. The characteristic numbers of $x_1x_2' + x_2x_3' + x_3x_1' + x_1'x_2'x_3'$ are 0, 1 and 2. The identically zero function has no characteristic numbers while the function which is identically 1, has every integer between 0 and n as a characteristic number.

Although Definition (2.9) could as well have been made for any Boolean function of n variables (not just for a symmetric function), it is only for symmetric functions that the concept of characteristic numbers becomes powerful. In fact, a symmetric function is completely characterised by the set of its characteristic numbers as we now show. This also justifies the name.

2.10 Theorem: The characteristic numbers of a symmetric Boolean function completely determine it. In other words, given integers

$$0 \leqslant r_1 < r_2 < ... < r_k \leqslant n$$

there exists one and only one symmetric function of n Boolean variables whose characteristics numbers are $r_1, r_2, ..., r_k$.

Proof: For an interger $r, 0 \leqslant r \leqslant n$ let S_r be the set of all $\binom{n}{r}$ terms of the form $x_1^{\varepsilon_1} .... x_n^{\varepsilon_n}$ where exactly r of the n variables occur without primes and the remaining $n - r$ variables occur with prime ($'$). Let f be a symmetric Boolean function of $x_1, x_2, ..., x_n$ with characteristic numbers $r_1, r_2, ..., r_k$. Then by the argument made before Definition (2.9), f contains all the terms in the sets $S_{r_1}, S_{r_2}, ..., S_{r_k}$ and none of the terms in S_r for $r \neq r_1, r_2, ..., r_k$. Therefore, the D.N.F. of f must be the sum of the terms in $S_{r_1}, S_{r_2}, ..., S_{r_k}$. Since two distinct functions cannot have the same D.N.F., f is uniquely determined by this expression. This also show that given any subset, say C, of $\{0, 1, ..., n\}$ we can construct a symmetric function of n variables whose characteristic numbers are precisely the elements of C. All we have to do is to set $S = \bigcup_{i \in C} S_i$ and take the sum of the terms in S. ∎

The reader may note that in the Locks Problem, we had b (the state of the box) as a symmetric function of $x_1, ..., x_5$ with characteristic numbers 3, 4 and 5. We expressed b as $b_1 + b_2 + b_3$. These summands were nothing but the sums of the terms in S_3, S_4 and S_5 respectively.

The concept of characteristic numbers also provides a structure theorem for the set of all symmetric functions of given variables.

2.11 Theorem: The set of all symmetric Boolean functions of n Boolean variables $x_1, x_2, ..., x_n$ is a subalgebra of the Boolean algebra of all Boolean functions of these variables. As a Boolean algebra, it is isomorphic to the power set Boolean algebra of the set $\{0, 1, ..., n\}$.

Proof: Let X denote, as before, the Boolean algebra of all Boolean functions of $x_1, x_2, ..., x_n$ under pointwise operations and let Y be the set of all symmetric Boolean functions of $x_1, ..., x_n$. We show Y is closed under the three operations $+$, $\cdot$ and $'$ $\cdot$ Let $f, g \in Y$. Then f, g are symmetric and hence each is unaffected under any interchange of two variable, Obviously the same is true of $f + g$. Hence $f + g$ is symmetric, i.e., $f + g \in Y$. So Y is closed under $+$ $\cdot$ Similarly, Y is closed under $\cdot$ and $'$ $\cdot$ Since Y is obviously non-empty (at least the constant functions 0 and 1 are always in Y), it follows that Y is a subalgebra of X.

Now let T be the set $\{0, 1, ..., n\}$. We have to show that, Y, as a Boolean algebra by itself, is isomorphic to the power set Boolean algebra $P(T)$. An explicit isomorphism can be defined as follows. For a symmetric function f of $x_1, ..., x_n$ let $C(f)$ denote the set of characteristic numbers of f. Then $C(f) \subset T$, i.e., $C(f) \in P(T)$. Define a function $\theta : Y \to P(T)$ by $\theta(f) = C(f)$ for $f \in Y$. In view of the last theorem, θ is a bijection. To show that θ is an isomorphism, we have to show that θ is compatible with the operations. This amounts to showing that for all $f, g \in Y$, (i) $C(f + g) = C(f) \cup C(g)$, (ii) $C(f \cdot g) = C(f) \cap C(g)$ and $C(f') = T - C(f)$. For (i), suppose r is a characteristic number of $f + g$. Then $(f + g)(e_r) = 1$, i.e, $f(e_r) + g(e_r) = 1$. Since $f(e_r)$, $g(e_r)$ are elements of Z_2, this can happen iff at least one of $f(e_r)$ and $g(e_r)$ equals 1. Therefore, r is either in $C(f)$ or in $C(g)$. This shows $C(f + g) \subset C(f) \cup (g)$. Conversely if $r \in C(f) \cup C(g)$ then either $f(e_r) = 1$ or $g(e_r) = 1$. In either case, $(f + g)(e_r) = 1$. So $r \in C(f + g)$. Hence $C(f + g) = C(f) \cup C(g)$. Similarly we prove (ii) and (iii). Thus θ is an isomorphism between Y and $P(T)$. ∎

In particular, since $|T| = n + 1$ we see that there are 2^{n+1} symmetric Boolean functions of n variables. The importance of symmetric functions comes from the fact that they arise frequently in applications and their characteristic numbers can often be determined directly from the data (as in the Locks Problem). Moreover, in electrical circuits (to be studied in the next section), some particularly simple circuits can be designed for such functions.

Boolean variables provide the appropriate mathematical tool for handling two-state devices, such as statements, switches, boxes, locks, persons (having only a 'yes' or 'no' opinion). However, there are a few entities which have three natural states. For example, a member of a committee may vote 'yes' or 'no' on a resolution or he may abstain. The resolution itself may be either carried (by majority vote) or rejected or the votes may be divided equally upon it. There are also physical devices having three states, e.g. a balance, an iron bar (which may be magnetised in two ways or else not magnetised). Although we shall not develop any mathematical structure suitable for handling ternary devices, some of the reasoning used for Boolean functions may be adapted for three-state devices. We illustrate this by doing the Stone Problem. In this problem we are given a balance and we are allowed to put a weight in either pan (or else not use that weight). So we have a ternary device. Let us agree to place the object to be weighed in the left pan. With this convention, a weight of m kilograms acts as $+m$ when placed in the right pan, 0 when not used and as $-m$ when placed in the left pan. (For example, if the balance is even with a 10 kg. weight in the right pan and 2 kg. weight in the left pan then the object has weight $10-2=8$.) The problem now is to cut the stone weighing 40 kgs. into 4 parts so that every integral multiple of 1 kg. between 1 to 40 kg. can be weighed with only one use of the balance. By interchanging pans and allowing negative weights, this means that every integeral weight between -40 kg to 40 kg. (both included) is to be weighed with only one use of the balance. Note that this gives 81 distinct possible weights and 81 is a power of 3.

Let us now formulate the problem mathematically. Let us call the four parts of the stone as p_1, p_2, p_3, p_4 and let their weights be m_1, m_2, m_3, m_4 (all weights are in kilograms) where m_1, m_2, m_3, m_4 are positive real numbers with $m_1 \leqslant m_2 \leqslant m_3 \leqslant m_4$ and $m_1 + m_2 + m_3 + m_4 = 40$. Now let T denote the set $\{-1, 0, 1\}$, and let S be the set of all sequences of length 4 with values in T. There are $3^4 = 81$ such sequences. Now every element, say (x_1, x_2, x_3, x_4), of S corresponds to a placement of the parts of the stone, if we make the convention that for $i = 1, 2, 3, 4$ the part p_i will be in the left pan if $x_i = -1$, in the right pan if $x_i = 1$ and not be used if $x_i = 0$. (For example $(1, -1, 0, -1)$ indicates the placement where the left pan contains p_2 and p_4 and the right pan contains p_1.) Now define

$$f : S \to \mathbf{R} \text{ by } f(x_1, x_2, x_3, x_4) = m_1 x_1 + m_2 x_2 + m_3 x_3 + m_4 x_4$$

for $(x_1, x_2, x_3, x_4) \in S$. The function f may be called the **weight function** because $f(x_1, x_2, x_3, x_4)$ gives the weight that can be weighed by the placement of the p_i's corresponding to (x_1, x_2, x_3, x_4) (Note that the values of f may be negative also as we are allowing negative weights.) The problem now reduces to determining the real numbers m_1, m_2, m_3, m_4 in such a way that the range of f contains the set A, where $A = \{0, \pm 1, \pm 2, ..., \pm 40\}$. One such solution is provided by taking $m_1 = 1$, $m_2 = 3$, $m_3 = 9$ and $m_4 = 27$ (the verification that every integer between -40 and 40 can be expressed as the weight of some arrangement is left to the reader; for example, $16 = 27 - 9 - 3 + 1$ gives that a weight of 16 kg, is obtained by putting p_1 and p_4 in the right pan and p_2 and p_3 in the left pan.)

Although this answers the Stone Problem, it can hardly be said that we have *obtained* the solution. It is as if we pulled the numbers 1, 3, 9 and 27 out of the blue and they luckily worked. This is all right for a puzzle but not for a genuine mathematical problem. True, these figures are not entirely at random. They are precisely the first four powers of 3 and 3 is a vital number in the present problem. Still, we would like to arrive at the answer by some reasoning. We would also like to know if there is some other solution. Both these will be done through exercises. The crucial point is that the domain of the weight function f above has cardinality 81. So its range, say R, can have cardinality at most 81 by Proposition 2.2.8, part (iv). Since $|A|$ is also 81, it follows that if R contains some element not in A then R cannot contain A (or else $|R|$ would be at least 82, a contradiction). In particular, this forces all the weights m_1, m_2, m_3 and m_4 to be integers since each m_i is obviously in the range of f.

Exercises

2.1 Prepare the tables of values of the following functions. Obtain their disjunctive normal forms. Also obtain their conjunctive normal forms both directly and by conversion from the disjunctive normal forms.

(i) $x_1' x_2 (x_1' + x_2 + x_1 x_3)$

(ii) $a + b + c'$

(iii) $(xy + x'y + x'y')' (x + y)$.

2.2 Let f, g be two Boolean functions of n Boolean variables. Prove that f is a summand of g (i.e., there exists a Boolean function h such that $f + h = g$) if and only if for all $\bar{y} = (y_1, y_2, ..., y_n) \in \mathbb{Z}_2^n$, $f(\bar{y}) = 1$ implies $g(\bar{y}) = 1$.

2.3 Obtain a similar characterisation for f to be a factor of g.

2.4 Let f be a Boolean function. Prove that the conjunctive normal form of f provides the ultimate factorisation of f in the sense that if $f = f_1 \cdot f_2 \cdot \ldots \cdot f_k$ is a factorisation of f, then every factor f_i is the product of some factors in the C.N.F. of f. (This fact was implicitly used in the solution to the Locks Problem.)

2.5 Suppose in the Locks Problem that the person p_1 has a special veto power which he can exercise, in addition to and independently of his decision to open the box on par with other persons (which means even if he has agreed to open the box, as an ordinary member, he can still exercise his veto). Now design a system of locks and keys, using Boolean variables.

2.6   At an examination there are seven subjects A, B, C, D, E, F and G. A candidate is given a grade of 'pass' or 'fail' in each subject. The rules for passing the examination are as follows:

(i)   A candidate must pass in at least five subjects.

(ii)   A candidate must pass in subject A and in at least two out of B, C and F.

(iii) For candidates passing in C, D and G, requirement (ii) shall be waived.

(iv)   For candidates, coming from the scheduled castes, requirement (i) shall be waived.

Express the pass/fail state of a candidate as a Boolean function of 8 Boolean variables (one for each subject and one for indicating whether he comes from a scheduled caste).

2.7   Theree are five industries A, B, C, D and E applying for import licenses. Their qualifications are as follows:
A is a large scale, urban industry employing skilled personnel.
B is a large scale, urban industry employing unskilled personnel.
C is a small scale, rural industry employing unskilled personnel.
D is a small scale rural industry employing skilled personnel.
E is a large scale, rural industry employing unskilled personnel.
The licensing authority is biased in favour of A, B, C and against D, E. However, to dismiss the implications of favouritism, it wants to frame rules so that only A, B, C would get the licenses. Design a simple system of rules for this.

2.8   Decide which of the following functions are symmetric. In case of symmetric functions find their characteristic numbers.

(i) $(a + b'c)(b + c'a) + (c + a'b)$

(ii) $x_1 x_2 x_3 x_4' + x_2 x_3 x_4 x_1' + x_3 x_4 x_2 x_1' + x_4 x_1 x_2 x_3'$

(iii) $x_1 x_2' x_3 x_4' + x_2 x_3' x_4 x_1'$

2.9   Let X be the Boolean algebra of all Boolean functions of n Boolean variables. Let $\leqslant$ be the corresponding order structure on X. (cf. Exercise (1.20)). Prove that for every $f \in X$, there exists a smallest symmetric function g such that $f \leqslant g$ and a largest symmetric function h such that $h \leqslant f$.

2.10   A Boolean function $f(x_1, x_2,...,x_n)$ is said to be **cyclically symmetric** if for all $(x_1...,x_n) \in \mathbb{Z}_2^n$,

$$f(x_1, x_2,..., x_n) = f(x_2, x_3,..., x_n, x_1).$$

Prove that if f is cyclically symmetric then $f(x_1, x_2, ..., x_n)$

depends only on the relative placement of $x_1,..., x_n$ around a circle. (Hence the name). In other words, prove that

$$f(x_1, ..., x_n) = f(x_2, ..., x_n, x_1)$$

$$= f(x_3, x_4, ..., x_n, x_1, x_2) = ... = f(x_n, x_1, ..., x_{n-1}).$$

2.11 Prove that every symmetric function is cyclically symmetric and that the converse is not true in general.

2.12 Prove that for $n = 1, 2, 3$ every cyclically symmetric Boolean function it symmetric.

2.13 The notions of symmetry and cyclic symmetry may be defined for functions of n real variables also. Prove that for such functions, even for $n = 3$, cyclic symmetry does not imply symmetry.

2.14 Suppose f is a cyclically symmetric Boolean function of $x_1, ..., x_n$ where n is a prime. Let $x_1^{\varepsilon_1} x_2^{\varepsilon_2} ... x_n^{\varepsilon_n}$ be an atom other than $x_1 x_2 ... x_n$ and $x_1' x_2' ... x_n'$. Prove that if the D.N.F. of f contains

$$x_1^{\varepsilon_1} x_2^{\varepsilon_2} ... x_n^{\varepsilon_n}$$

then it must also contain certain $n-1$ other atoms obtained from $x_1^{\varepsilon_1} x_2^{\varepsilon_2} ... x_n^{\varepsilon_n}$. What happens if n is not a prime?

*2.15 Using the last exercise show that the number of cyclically symmetric functions of n variables where n is a prime is 2^m where

$$m = \frac{2^n - 2}{n} + 2.$$

2.16 In the Stone Problem prove that at least one of the four parts must weigh 1 kg. (Hint: If not, show that it would be impossible to weigh 39 kg.)

2.17 Let $A = \{0, \pm 1, \pm 2, ..., \pm 40\}$. Let A_{-1}, A_0 and A_1 be three mutually disjoint subsets of A, each having 27 elements such that (i) for all $x \in A_0$, $x + 1 \in A_1$ and $x - 1 \in A_{-1}$,

(ii) for all $x \in A_1$, $x - 1 \in A_0$,

(iii) for all $x \in A_{-1}$, $x + 1 \in A_0$ and (iv) $0 \in A_0$. Prove that

$$A_0 = \{0, \pm 3, \pm 6, ..., \pm 39)\}.$$

2.18 In the Stone problem, if we exclude the part weighing 1 kg. (given by Exercise 2.16), then prove that all multiples of 3 kg, (upto 39 kg.) can be weighed using the remaining three parts. (Hint: Use the last exercise. Let S, f be as in the discussion of the problem and suppose p_1 has weight 1 kg., i e., $m_1 = 1$. Let

$$S_{-1} = \{(x_1, x_2, x_3, x_4) \in S : x_1 = -1\}.$$

Similarly define

$$S_0 = \{(x_1, x_2, x_3, x_4) \in : S : x_1 = 0\}$$

and

$$S_1 = \{(x_1, x_2, x_3, x_4) \in S : x_1 = 1\}.$$

Let

$$A_i = f(S_i), \ i = -1, 0, 1.)$$

2.19 Prove that the only solution to the Stone Problem is

$$m_1 = 1, \ m_2 = 3, \ m_3 = 9 \ \text{and} \ m_4 = 27.$$

(Hint: Let Q_2, Q_3, Q_4 be stones with weights

$$\frac{m_2}{3}, \ \frac{m_3}{3} \ \text{and} \ \frac{m_4}{3}.$$

Then by the last exercise, every multiple of 1 kg. upto 13 kg. can be weighed using Q_2, Q_3 and Q_4. Now apply induction.)

2.20 Let f be a Boolean function of n variables. Prove that for all

$$(x_1, x_2, ..., x_n) \in \mathbf{Z}_2^n,$$

$$f(x_1, x_2, ..., x_n) = f(x_1, ..., x_{n-1}, 1) \ x_n + f(x_1, ..., x_{n-1}, 0) \ x_n'.$$

What is the significance of this result?

2.21 For certain applications, it is more convenient to consider the ring sum, $x \oplus y$ of two Boolean variables x and y than the ordinary sum (the disjunction) $x + y$ (cf. Exercise 1.19). If $x_1, x_2, ..., x_n$ are Boolean variables, prove that $x_1 \oplus x_2 \oplus ... \oplus x_n$ (which is well-defined, since $\oplus$ is associative) equals 1 iff the number of x_i's having value 1 is odd.

*2.22 Prove that every Boolean function f of n Boolean variables can be uniquely expressed as the ring sum of terms of the form

$$x_{i_1} \ x_{i_2} \ ... \ x_{i_r}$$

where r is an integer $0 \leqslant r \leqslant n$, (if $r = 0$, the term is an empty product which is interpreted as 1, just as an empty sum is interpreted as 0). (Hint: Note that $x_i' = x_i \oplus 1$. Apply properties of the ring sum to convert the D.N.F. of f to the desired form). The form of the function given by this exercise is called the **ring normal form**.

Notes and Guide to Literature

This section is preparatory to the next one. For more on the disjunctive, conjunctive and the ring normal forms, see Dornhoff and Hohn[1].

3. Applications to Switching Networks

In this section we apply the theory of Boolean functions of Boolean variables to a very special context, where the variables represent the states of electrical switches and the function represents the state of an electric circuit in which these switches occur. For most of the time we shall deal with what are called **combinational circuits.** In these circuits, the state of the circuit (i.e. whether current is flowing through it or not) at any time depends only on the combination of the states of the switches at that time (i.e. on which of them are closed and which are open). If at two different times, every switch is in the same state, so will be the circuit. On the other hand, there are some circuits for which this is not necessarily true. In such circuits, the state of the circuit is a function not only of the states of the switches but also of time. Such circuits are called **sequential circuits** and we shall mention them briefly.

The mechanical construction of switches will be unimportant for our purpose. Our discussion will also be independent of the manner in which the switches are operated, whether manually or automatically by some sort of a feedback arrangement. That is why the theory has survived the transitions in the mechanism of switches caused by advances in electronics. For fixation of ideas we shall take a switch as a device with two pieces of wire (called **leads**). The switch is said to be **open** when the two wires are electrically insulated from each other and **closed** when current can flow from one into the other. Admittedly this terminology is confusing to a beginner, because these terms are used precisely with opposite meanings in other contexts. For example, when a door is closed it prevents the passage through it and when it is open, such passage is allowed. But the usage is too standard to be changed. A symbolic representation of a switch is shown in Figure 4.2 (a). Two switches which are always in the same state are to be regarded as equal. This means that their leads are joined by some non-conducting device in such a way that the current will flow or stop flowing simultaneously in every pair of leads, (Figure 4.2 (b)).

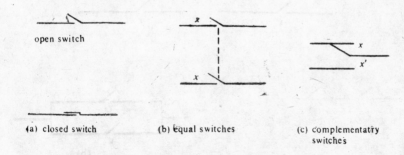

(a) closed switch   (b) equal switches   (c) complementatry switches

Figure 4.2:   Symbolic Representation of Switches.

On the other hand, two switches which are always in the opposite states are said to be **complementary**. If one of them is denoted by x, the other is denoted by x'. A symbolic representation for a pair of complementary switches is given in Figure 4.2 (c).

Since a switch is a two state device it can be represented by a Boolean variable. We assign the elements 1 and 0 of Z_2 respectively to the closed and the open states of a switch, i.e. when a switch is closed, the Boolean variable representing it has value 1 and when it is open, the value 0. In general, the same symbol will be used to denote both a switch and the Boolean variable representing it. In diagrams, for notational brevity we shall treat the two leads of a switch as parts of the same wire separated by a gap which will be filled by the Boolean variable representing the switch, for example——— x ———— or ———— y' ————.

Besides complementation, there are two basic operations, $+$ and $\cdot$ for Boolean variables. Let us see which physical devices serve to represent them. Let x, y be two switches. Then we want $x \cdot y$ to be a switch which will be closed when both x and y are closed and open when at least one of them is open. The simplest way to achieve this is to take one lead from each x and y and join them together. The current will flow from the other lead of x to the other lead of y iff both x and y are closed. This is called the **series arrangement** of the switches x and y and is shown in Figure 4.3 (a). On the other hand, if we join together one lead of x with one lead of y and the other two leads together, then current will flow between these junction points iff at least one of x and y is closed. Therefore, this arrangement, called the **parallel arrangement** of the switches x and y and pictured in Figure 4.3 (b), serves to represent $x + y$. Incidentally, 'parallel' here simply means that the wires on which the switches x and y operate have no electrical contact in between the two junction points. It does not literally mean that they are parallel lines in the geometric sense. Indeed, the wires need not be straight at all! They could be any arcs. There is no harm if they cross each other, as long as they are insulated. Such crossings can be

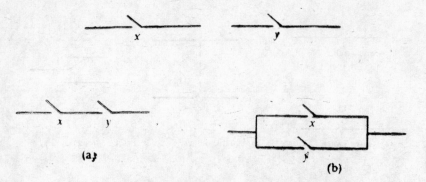

Figure 4.3: Series and Parallel Arrangements of Switches

avoided in space (by a result from graph theory). Since we shall be drawing only planar diagrams for circuits, sometimes crossings are inevitable. In such cases, we show the crossing by curving one of the wires slightly as indicating that there is no electrical contact at the point of intersection.

We now turn to circuits. A complete electric circuit has three parts, namely, a source of power, an output (such as a lamp, a bell etc.) and a third part called the control part, in which we have a combination of switches. For our purpose, the third part will be the most important. We may not always draw the other two parts, when we draw them we shall show the source by some symbol like ▋, and the output by some symbol like L. In more complicated circuits, that is, in the sequential circuits, some of the switches in the control path themselves appear as output or are controlled by the output.* However, for the time being, we shall restrict ourselves to the cases where the control part of a circuit is independent of the other two parts and controls the flow of the current between two points on the circuit (called terminals) depending upon the states of the various switches in it. We shall call such circuits as **two terminal circuits** even though they are, really speaking, only the control parts of some complete circuits. In Figure 4.4 (a) we show a complete circuit and in Figure 4.4 (b), its control part as a two-terminal circuit.

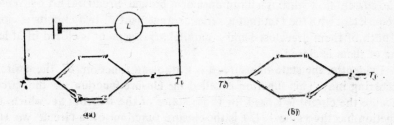

Figure 4.4: Electrical Circuit

Even among two terminal circuits, we shall first consider only the so-called series parallel circuits which are obtained by repeated series and parallel combinations of the simplest circuits, namely, those in which there is a single wire with only one switch on it. (Later we shall see how any two terminal circuit, can be replaced by an equivalent series-parallel circuit.) For example, the circuit in Figure 4.4 (b) is obtained as follows:

  (i) take the series combination of the switch x and the switch w
 (ii) take the series combination of the switch y and the switch z
(iii) take the parallel combination of the circuits (i) and (ii)
       and

---

* This is the essence of what is popularly called a **feedback** arrangement.

(iv)   take the series combination of the circuit obtained in (iii) and the switch z'.

Note that the same switch may appear any number of times in a circuit. In actually constructing the physical circuit the multiple occurrences of a switch would have to be handled by switches capable of simultaneous operation on various pairs of wires. Such multiple switches are costly and are, as far as possible, avoided. This leads to the notion of simplification of a circuit, that is, replacing it by an equivalent circuit in which the total number of switches (counting multiplicities) is as small as possible.

Informally, two circuits are equivalent if, whenever every switch common to them is in the same state in both of them, the two circuits are in the same state, i.e. current flows either through both of them or through neither of them. This idea can be neatly expressed if we use the language of Boolean functions. We already remarked that every switch is a Boolean variable. Now every circuit also has only two states, either some current flows through it or else no current flows through it. (The magnitude of the current flowing through various parts of a circuit may be relevant in some problems, but not for our purpose.) In analogy with switches, we associate a Boolean variable with a circuit and assign it the value 1 if current flows through it (a **closed** or a **completed** circuit) and the value 0 if no current flows through it (an **open** or a **broken** circuit). This convention is consistent with the fact that a series arrangement of two circuits is closed iff both of them are closed and a parallel arrangement is closed iff at least one of them is closed.

Evidently, the state of a circuit is a Boolean function of the switches occurring in it. This function is called the **closure function** of the circuit, because the circuit is closed for those states of the switches at which this function has the value 1. If f is the closure function of a circuit we say f **represents** the circuit or is **realised** by the circuit. These definitions are applicable for all two terminal circuits, not just for series parallel circuits. However, in the case of series parallel circuits, closure functions can be written down mechanically simply by inspection keeping in mind that a series arrangement corresponds to the Boolean multiplication and a parallel arrangement to the Boolean addition. For example, the circuit in Figrue 4.4 (b) realises the Boolean function

$$f(x, y, z, w) = z' (xw + yz).$$

Conversely, given any Boolean function, we can mechanically construct at least one circuit to realise it, even though there may be many others, and some of them may be more economical.

Now, simplification of a circuit can be done at least partly by simplifying the function representing it. (If we allow circuits which are not series parallel, then further simplification may be possible as we shall see later.) For example, the circuit of Figure 4.4 (b) can be simplified by simplifying its closure function

$$f(x, y, z, w) = z'(xw + yz)$$

as

$$f(x, y, z, w) = z'xw.$$

Thus an equivalent circuit can be drawn simply as a series combination of three switches x, w and z', between the terminals T_0 and T_1. We could also see this by inspection. In order that current can flow from T_0 to T_1, it must either pass through x, w and z' or else through y, z and z'. The second possibility can never hold because, by definition, the switches z and z' can never be closed simultaneously. Hence, there is only one path for the current to flow namely through x, w and z', giving xwz' as the closure function. More generally, if we trace all possible simple paths (i.e., paths not having any loops in them) for the current to flow from one terminal to the other, then each path gives rise to a monomial involving the switches and summing these monomial gives the closure function. In small circuits this method is easy to apply and has the additional advantage that it works for any two terminal circuits, not just for the series parallel ones. However, one has to be careful that all possible paths have been traced. An algorithm for tracing all possible paths will be given when we study graph theory*. As an application of this method, we see that the closure function of a 'bridge circuit' in Figure 4.5 (a) (which is not a series parallel circuit) is $f = ab + cd + acd + ceb$ because there are only 4 possible simple paths from T_0 to T_1, namely

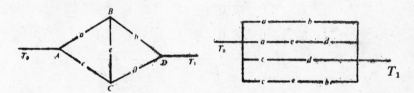

Figure 4.5:   Bridge Circuit and Its Series Parallel Equivalent

$$T_0 - A - B - D - T_1,\ T_0 - A - C - D - T_1,$$
$$T_0 - A - B - C - D - T_1\ \text{and}\ T_0 - A - C - B - D - T_1$$

An equivalent series parallel circuit is shown in Figure 4.5 (b).

In practical problems, we have to construct a circuit with given switches and some output which is to be on for certain combinations of the states of the switches and off for the remaining combinations. We can do this by putting the output (and the power source) in series with a circuit of these switches whose closure function is first to be determined from the data of the problem. As an illustration, let us do the **Landlord Problem**. Let x and y denote the switches controlled by the landlord and the tenant respectively

*See the Epilogue.

and let z be the hidden switch (with the landlord). When z is closed, each of x and y is to control the state of the lamp independently of the other, which means that the change in the state of either one of them (with the other switch remaining as it is) must cause a change in the state of the lamp. We arbitrarily set the lamp on when x and y are both closed. We then get the table of values for the closure function, say, f, of the lamp (or rather, its circuit), shown in Figure 4.6 (a). (Note that when we go from the second to the third row, both x and y change their states and hence the lamp remains in the same state.) The last four rows of the table indicate that when z is open, x has exclusive control of the lamp, i.e., a

Row	z	x	y	$f(x, y, z)$
1	1	1	1	1
2	1	1	0	0
3	1	0	1	0
4	1	0	0	1
5	0	1	1	1
6	0	1	0	1
7	0	0	1	0
8	0	0	0	0

(a)

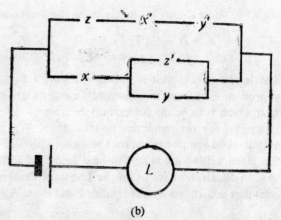

(b)

Figure 4.6:   Solution to the Landlord Problem

change in x changes the state of the lamp, regardless of whether y changes or not.

From this table, we get the D.N.F. of f as $zxy + zx'y' + z'xy + z'xy'$ which upon simplification gives

$$f(x, y, z) = z(xy + x'y') + z'x = zx'y' + x(z' + y).$$

A circuit realising this function is shown in Figure 4.6 (b), where L denotes the lamp.

It is a common practice for several circuits to share some common parts. As a commonest example, the circuits of various appliances in a house share the same source of power and also the main switch. Sometimes the functions representing these circuits have common factors and if so, portions representing these common factors may be shared. In such a case we represent all the circuits with a common initial terminal, say, T_0. If there are n circuits, we put n terminals $T_1, T_2, ..., T_n$ and then between T_0 and T_i put the circuit whose closure function is given (or calculated from the data of the problem), sharing common factors wherever possible. Each T_i is then connected to the ith output and then to T_0 through the power source. Sometimes, in order to share the switches so as to effect an economy, it may be necessary to manipulate the closure functions using the various laws in a Boolean algebra, as is illustrated in the following problem.

3.1 Problem: An aircraft has three engines. Each engine is provided with a switch which closes as soon as there is any mechanical fault in that engine. Although the aircraft can run with just one engine operating, it is desired to have a red lamp appear when there is a fault in any one of the engines and an alarm to ring when there is a fault in any two of them. Design a three terminal circuit for this, sharing switches wherever possible.

Solution: Denote the three switches by x, y, z. Let f, g denote the closure functions for the red lamp (R) and the alarm (A) respectively. A simple calculation shows that $f(x, y, z) = x + y + z$ and $g(x, y, z) = xy + yz + zx$. If we draw separate circuits, we in all need 8 occurrences of switches (because g can be simplified as $x(y + z) + yz$). To see if any economy is possible with sharing of switches, let us first see if f and g have any common factors. For this we take their conjunctive normal forms (cf. Exercise (2.4)). f is already in its C.N.F. while

$$g = (x + y + z)(x + y + z')(x + y' + z)(x' + y + z).$$

Here although $x + y + z$ is a common factor, it will not result in any saving to share it because the remaining three factors of g, even upon simplification would require at least five occurrences of switches anyway. So we write f as $f_1 + f_2$ where $f_1 = x$ and $f_2 = y + z$. Then $g = (y + z)$

$(x + yz)$ and we may try to use the factor $y + z$ common to f_2 and g. It is tempting to try to do this by a circuit as in Figure 4.7 (a). But in this circuit the alarm can ring even when only x is closed, which is not desired. A path by which a current can flow when not desired is called a **sneak path**. It can be avoided by inserting the switch x' as in Figure 4.4 (b), which is a correct solution. We could have drawn the second circuit without the first if instead of writing f as a sum of x and $y + z$ we write

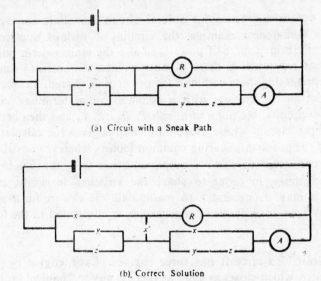

(a) Circuit with a Sneak Path

(b). Correct Solution

Figure 4.7 : Solution to Problem (3.1)

f as $x + x'(y + z)$. Here the two summands are mutually disjoint (i.e. there product is 0) and hence their is no possibility of a sneak path. Note that there is a saving of switches because only 7 occurrences are needed.

We remark that in this problem both f and g are symmetric functions with characteristic numbers 1, 2, 3 and 2, 3 respectively. Later on we shall give an alternate circuit to realise symmetric functions of switches.

Let us now see how an arbitrary two terminal circuit can be replaced by an equivalent series parallel circuit. One method for doing this was discussed already, namely to look for all possible paths for current to flow from one terminal to another. Instead of applying this method in a heuristic manner (which makes it likely that some possible paths may be missed), we now show how it can be applied in a systematric, step-by-step manner to reduce a non-series-parallel circuit to a series parallel circuit.

Obviously, in a non-series parallel circuit, there must be at least one point, other than a terminal, which is joined to more than two points by wires and each of these wires carries some combination of switches. Such a point is called a **star**, and the number of wires meeting at it is called its degree. For example, in the bridge circuit of Figure 4.5, the points B and C

are stars of degree 3 each. (The points A and D are not regarded as starts because they are effectively the terminals.) A star of degree 3 is called a wye because of its resemblance with the capital letter Y.

The key step in reducing a non-series-parallel circuit, circuit is to go on replacing its stars by equivalent arrangements of switches between every pair of points which are joined to the star points. The simplest case is that of a wye as shown in Figure 4.8 (a). The star point P is joined to A, B and C by wires whose closure functions are f, g, h respectively. (Here f, g, h may themselves be switches or some combinations of switches.) Since the

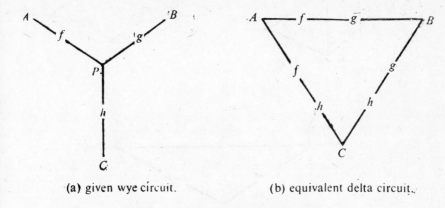

(a) given wye circuit.   (b) equivalent delta circuit.

Figure 4.8: Wye-to-delta Transformation

point P is not a terminal, we eliminate it by putting wires between A and B, B and C and between A and C. These wires are in addition to whatever other wires that may be already present. The resulting new circuit is called a **delta circuit**, 'delta' being another name for a 'triangle'. The transformation is called a wye-to-delta transformation. The new circuit will be equivalent to the old one if and only if for every pair of vertices, the conditions for the current to flow from one of them to the other are identical in the two circuits. Let us take A and B. Then in the wye circuit current will flow from A to B iff both f and g are 1. (We are, of course, not counting here the other paths, if any, available to go from A to B.) This is equivalent to saying that $fg = 1$. So the closure function of the side AB in the delta circuit is fg. Similarly the other two closure functions are determined. Note that the path $A - C - B$ does not provide an additional path for the current to flow from A to B because current can flow through it iff fh and gh are both 1; but in that case is $fg = 1$ any way.

Thus we have eliminated a star of degree 3. Similar constructions apply for eliminating stars of higher degrees. Using this method we can reduce the bridge circuit of Figure 4.5, redrawn in Figure 4.9 (a). We eliminate the star at B and show the resulting circuit in Figure 4.9 (b). Note that the wires between A and C and between C and D present in the original circuit are to be retained. Note also that there is no star in the circuit in (b). Although four wires meet at C, the point C is not a star because it is joined to only two points, namely to A and D, by two wires each. We replace each pair of wires by a single wire whose closure function is the sum of their closure functions. This gives the circuit in Figure 4.9 (c). This is a series-parallel circuit with closure function $ab + (c + ae)(d + eb)$ which upon simplification reduces to $ab + cd + aed + ceb$, which was also obtai-

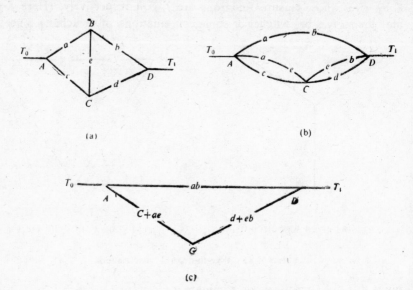

(a)   (b)

(c)

Figure 4.9:   Reduction of a Bridge Circuit

ned earlier by inspecting all possible paths for the current to flow between T_0 and T_1.

It should be noted that although the series-parallel circuit obtained from the bridge circuit is easier to analyse, it is the bridge circuit that is more economical as far as the number of occurrences of the switches is concerned. The wire containing the switch e is like a 'bridge' joining the paths $A - B - D$ and $A - C - D$ (hence the name). By putting more such bridges one can construct non-series parallel circuits which are considerably cheaper than their series parallel equivalents. However, it takes some ingenuity to fortell which Boolean functions can be realised by a bridge circuit. There is no easy method for doing this in general. However, in case of symmetric functions there is a well-known circuit which allows us to realise a symmetric function of n switches say $x_1, x_2, ..., x_n$ as soon as we know its characteristic numbers. Such a circuit is shown in Figure 4.10. The initial terminal is marked as T. We have a triangular grid of wires with horizontal wires carrying the switches $x_1', x_2', ..., x_n'$ and slanting wires carrying the switches $x_1, x_2, ..., x_n$. Note that the switches appear with

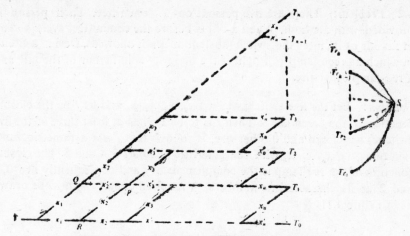

Figure 4.10: Realisation of a Symmetric Function

different multiplicities, x_i occurs $2i$ times (including complementary occurrences), for $i = 1, 2, \ldots, n$. The terminals on the right are numbered $T_0, T_1, \ldots, T_n$. The key to understand the working of this circuit is the observation that current can flow from T to T_i if and only if precisely i of the switches are closed and the remaining $n - i$ are open. To see this, note first that the arrangements of the complementary switches is such that when the current comes at any junction point (such as the point P in the figure) it cannot go back towards T. For example, if the current is at P, then it has come through Q or through R. If it came through Q it cannot go back to Q (because no loops are allowed in the path of a current) and it cannot go to R because x_2 is open, x_2' being closed. So the current at any junction point has only two alternatives, either to go horizontally to the right or else to 'climb up' along the slanting line by one unit, depending upon whether the next switch is open or closed. Since every path from T to T_i involves i climb-ups, current can flow from T to T_i precisely when exactly i of the switches are closed. So if we consider a two terminal circuit with T and T_i as the terminals, its closure function will be the sum of $\binom{n}{i}$ terms of the form $x_1 \cdot x_2 \ldots x_i \cdot x_{i+1}' \ldots x_n'$. This is a symmetric function with i as its only characteristics number. Now if we attach leads to the points $T_{r_1}, T_{r_2}, \ldots, T_{r_k}$ (say) and fuse them together at a point S then current will flow from T to S iff it flows from T to one of $T_{r_1}, T_{r_2}, \ldots, T_{r_k}$. Therefore the closure function of the two terminal circuit between T and S is the symmetric function of $x_1, x_2, \ldots, x_n$ whose characteristic numbers are $r_1, r_2, \ldots, r_k$. Conversely every symmetric function of n variables can be realised by finding its characteristic numbers and joining the corresponding terminals to a common terminal S. We illustrate this in the following problem.

3.2 Problem: There are five persons on a committee. Each person is provided with a switch. When a bill is before the committee, every person in favour of it closes the switch (abstention is not allowed). Design a circuit in which a green lamp will light if the majority is in favour of the bill and a red lamp otherwise.

Solution: Let the five switches be x_1, x_2, x_3, x_4, x_5 and let f be the closure function of the green lamp. Then f is to be 1 when at least three of the five switches are closed and 0 otherwise. It follows that f is a symmetric function of x_1, x_2, x_3, x_4, x_5 with characteristic numbers 3, 4 and 5. The closure function of the red lamp is the complement of f and consequently has 0, 1 and 2 as its characteristic numbers. A required circuit can now be drawn as in Figure 4.11.

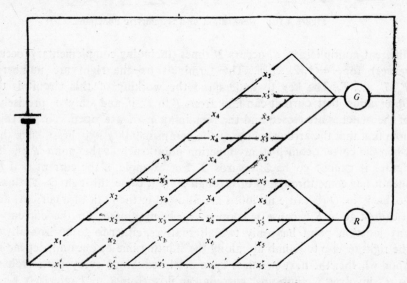

Figure 4.11:   Solution to Problem (3.2)

Note that the disjunctive normal forms of f and f' contain 16 terms each. A series parallel circuit for the problem would be extremely costly.

As another illustration of the circuit for symmetric functions, we can do problem (3.1) by connecting R to the terminals T_1, T_2 and T_3 and the alarm A to the terminals T_2 and T_3. The resulting circuit is not as compact as that in Figure 4.7 (b). But it has a greater adaptability. For example, if we want the red light to go off when the alarm is ringing all we have to do is to cut off the wires joining R to T_2 and T_3.

Sometimes the closure function of a circuit is not symmetric in all the variables. Even in such a case, the circuits for symmetric fuctions are useful if a large factor or summand of the closure function is symmetric among some of the variables. For example we invite the reader to modify the circuit of Figure 4.11 if one of the persons has a special veto power (cf. Exercise (2.5)).

Electrical circuits provide a physical means of expressing a Boolean function of Boolean variables, by realising each Boolean variable as a switch. In electronic data processors, it is convenient to think of a function as a device which produces an output when some inputs are fed to it. This is of course just another way of looking at the definition of a function. Let $f: X \to Y$ be any function, expressed by $y = f(x)$. Here the argument x ranges over the set X. The elements of X are, therefore, the possible inputs and when any one of them say x_1 is fed to the function f, it produces as output the value of f at x_1, i.e., the element $f(x_1)$ of Y. For functions of several variables, say, $x_1, x_2, ..., x_n$ there are n inputs. In the case of Boolean functions each input has only two possible values 1 and 0 and the output is also either 1 or 0.

In signal processing, a binary input is usually represented by the presence or absence of a certain voltage along a wire, called an input lead. The magnitude of the voltage is immaterial for our purpose. If f is a Boolean function of n Boolean variables, say, $y = f(x_1, x_2, ..., x_n)$ then we want to construct a 'black box' with n input leads, one for each x_i and one output lead such that a voltage will appear on the output lead precisely for such values of the inputs at which the function f has value 1. A symbolic representation of such a black box is shown in Figure 4.12.

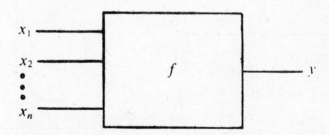

Figure 4.12:   Black Box Representation of Boolean Function

The black box representing a Boolean function can be constructed by suitably combining some very elementary black boxes, called **gates**. These gates are also called **logic elements** for reasons which will be explained in the next section. The internal mechanism of these gates is not our concern. There are three such basic gates corresponding to the three basic operations of a Boolean algebra namely $+$, $\cdot$ and $'$. They are called respectively the OR-gate, the AND-gate and the NOT-gate. These names again come from logic and will be explained in the next section. The OR-gate has two input leads and one output lead. A voltage appears on the output lead when a voltage appears on at least one of the inputs. An AND-gate also has two input leads and an output lead on which a voltage appears iff a voltage appears on both the input leads. The NOT-gate has one input and one output lead and there is a voltage on the output lead if there is no voltage on the input lead and vice-versa. The symbolic representations for these basic logic elements are shown in Figure 4.13. The inputs are marked with Boolean variables and the function represented by the output lead is shown along the output lead.

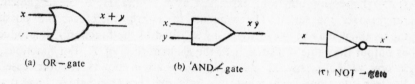

(a) OR – gate   (b) 'AND⊄ gate   (ᴄ) NOT → gate

Figure 4.13:   Basic Logic Elements (= Gates)

It is now a simple matter to construct a black box to realise a given Boolean function in terms of these basic gates. For example, the arrangement of gates shown in Figure 4.14 realises the function $x_1 x_2 + x_3' (x_1' + x_2') + x_1' x_2'$. Note that all crossings of wires are insulated except those which are circled. The 'black box' is shown by a dotted boundary.

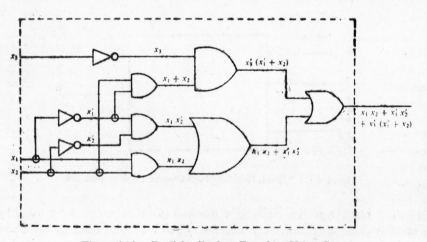

Figure 4.14:   Realising Boolean Functions Using Gates

Because of the various laws such as distributivity and associativity, which hold in a Boolean algebra, the same function may often be represented by more than one black box. Naturally, between two arrangements for the same functions, the one with a smaller number of gates is preferable. For example, to realise the function $xy + xz$ as it is, it takes three gates; two AND-gates and one OR-gate. But writing the same function as $x(y + z)$

(using distributive law), saves one AND-gate. Even between two black boxes requiring the same number of gates of each type, one arrangement may be better than the other on some other ground. For example, take the function $x + y + z + w$ of four Boolean variables. This can be realised in two ways, each using three OR-gates as shown in Figure 4.15. But the arrangement in (b) is superior to that in (a) because it is quicker. It takes some time for each gate to operate. If we denote this time by T, then the time taken by (a) is $3T$ because the gates must operate one after the other (we are not counting here the time it takes for the signal to pass from one end of a wire to another, this time is negligible as compared to the time T). On the other hand, in (b), two OR-gates can function simultaneously and hence the time taken to process the function $x + y + z + w$ is only $2T$. Simultaneous functioning of several gates is popularly known as **parallel processing**.

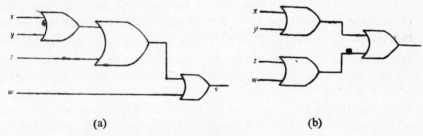

(a)   (b)

Figure 4.15:   Two ways to realise x + y + z + w

We close the section with a brief discussion of sequential circuits. As remarked at the beginning of this section, the state of such a circuit does not depend exclusively on the states of the switches appearing in it. Obviously in such circuits there must be some device which changes its state under a change of certain conditions but which does not return to its original state even when the initial conditions are restored. A simplest device of this type is provided by what is known as a **relay**. There are various types of relays such as magnetic or thermal relays. Every relay, regardless of its type, is a binary device with a circuit called its **control path**. When no electric current flows through the control path, the relay is in its natural or **release** state. When current flows in the control path, the relay is activated and is said to be in its **operate** state. It will continue to be in this state as long as current flows in its control path. When the control path is broken, the relay is deactivated, or released and returns to its release state. The mechanism of change of states depends on the type of the relay. For magnetic relays it is the magnetisation and demagnetisation of an iron bar, for thermal relays it is the expansion and contraction upon a change in temperature. But this is unimportant for our purpose because our interest is not in the construction of relays but rather in their use.

In order to use a relay, it is necessary to attach to it an additional feature called a **contact**. As the name suggests, a contact consists of two thin metallic plates. One of these plates is steady and is symbolically denoted by a vertical arrow ↑ or ↓. The other plate is capable of a slight motion when the relay is activated and returns to its original position (by a springlike arrangement) when the relay is released. Symbolically this movable plate is shown by a horizontal line (——) in its natural state and a slightly slanting line (——) in its activated state. The gap between these plates is small. When the plates touch each other the contact is said to be **made** and when they do not, it is said to be **broken**. There are two types of contacts. In a **make** contact, the plates do not touch in their natural position but when the relay is activated the movable plate moves and touches the stationary plate. Symbolically, a make contact is represented by ——↑ when it is broken and by ——↑ when it is made. On the other hand, in the other type of a contact, called a **break** contact, the plates touch each other in the released state and a gap is created when the relay is in its operate state. A symbolic representation for a break contact is ___↓ in its release state and ——↓ in its operate state. There can be several contacts on the same relay. As a rule, relays are denoted by capital letters such as A, B, X etc. The make contacts on them are denoted by corresponding small case letters a, b, x etc. and the break contacts with small case letters with primes, a', b', x' etc. This notation is consistent with our earlier practice of using ′ for complementation because a contact has only two states, one where it is made (i.e., the plates touch each other) and one where it is broken (i.e., the plates do not touch each other). It is obvious that a make contact and a break contact on the same relay will always be in opposite states, whether the relay is operated or released. Symbolic representations of relays and the contacts on them are shown in Figure 4.16. The standard practice is to write the contacts directly above the relays and to show the link between a relay and a contact on it by a dotted line. (Such a line is only symbolic.) Unlike the circuits considered earlier in this section, it is customary to often show the control circuits of relays in full, that is, with the power source (▓) included in them. The completion of a circuit may be symbolised by earthing the two ends (shown by ═). As a student of electricity knows, the earthing of a lead simply means maintenance of a constant potential at its terminus and not necessarily a physical burying into the earth.

If two pieces of a wire are joined to the two plates of a contact on a relay, then the contact behaves exactly like a switch on that wire. The only difference, perhaps, is that while a switch is generally operated manually, a contact is operated through a relay whose control path may be operated manually. Such a relay therefore provides an indirect way to control the current in some other circuit and is useful when a direct manual control of the current in the other circuit is undesirable for some reason, such as

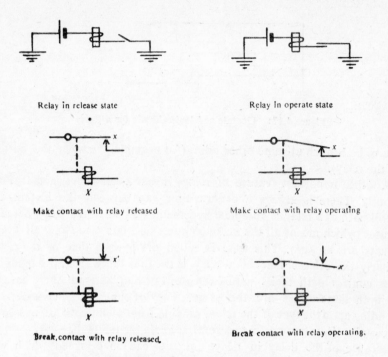

Figure 4.16: Symbolic Representation of Relays and Contacts

safety. But conceptually this use of a relay is not very interesting. From the point of view of Boolean algebra there is little difference between a circuit consisting of ordinary switches and one consisting of contacts on relays whose control paths contain switches, as long as these control paths do not mingle with the contacts. However, things are quite different when the control path of a relay passes through a contact on that relay itself or, more generally, when the control paths of several relays contain contacts on each other. It is this feature of a relay that truly distinguishes it from a switch although both are binary devices. We study the simplest circuit of this type shown in Figure 4.17. Here the relay X is provided with a make contact. When the switch a is initially open no current flows through the control circuit of X and so X is in the released state. (Normally, in all diagrams involving relays, they are shown in the released state. What happens when any of them are activated has often to be visualised mentally.) Now if a is closed, current flows through the control circuit of X. So X is activated and consequently the make contact x is made. A portion of the current will flow through x also. Now even if the switch a is opened, current will continue to flow through x and will hold the relay in the operate state. (Hence the name 'operate and hold' circuit.) Such a circuit is the basic memory device because the operate state of the relay 'remembers' that the switch had been closed. If the relay is to be released, the control path will

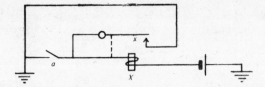

Figure 4.17:   Operate and Hold Circuit for a Relay

have to be broken at some other point (for example, between the battery
and the relay).

Another important feature of relays is that a relay never acts instantaneously. There is always a definite time gap between the instant the
control path of a relay is complete and the instant when the relay actually
operates (which means all the make contacts on it are made and all break
contacts are broken). This time is called the **operate time** of the relay.
Similarly the **release time** of a relay is the time taken from the breaking
of its control path to its actual release. For simplicity we shall assume
that both these times are the same. They of course vary considerably
depending upon the type of the relay, ranging from a millionth of a second
for certain electronic relays to several seconds for thermal relays.

Because of the delay in relays, circuits can be designed which will
repeat some pattern of states periodically. The simplest such circuit, shown
in Figure 4.18, appears in a door bell.

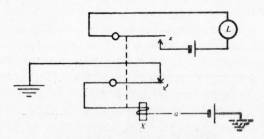

Figure 4.18:   A Door Bell Circuit

In this circuit, there is a break contact on a relay and the control path
of the relay passes through this contact. As soon as the switch a is closed,
the control circuit is complete. After the lapse of the operate time the
break contact is broken. This breaks the control circuit. However, another
time interval, equal to the release time of X has to pass before the relay
is released i.e. the break contact is made again. Now the control circuit is
complete again. This process repeats cyclically till the switch a is opened,
the contact being made and broken in alternate intervals of time. This
make-and-break arrangement is used in a door bell. It may be used, with
a relay of a delay time of several seconds, to have a lamp go on and off

periodically. All we have to do is to put into the circuit of the lamp, a contact on the relay X (other than the break contact mentioned earlier), as shown in Figure 4.18. Such a circuit is called an **output circuit** attached to the relay circuit.

As another illustration, consider the circuit of Figure 4.19 (a) in which there are two relays X and Y and the control path of each passes through a contact on the other (**interlocked relays**). For simplicity let us suppose both the relays have the same delay time, T. The state of the circuit has to be described for various intervals of time T, starting from the instant the switch a is closed. During the first time interval both the relays are at rest (even though the control circuit of Y is complete). At the end of the first interval, the make contact on Y is made and the control path of X is completed. However X is operated only at the end of the second interval. In the third interval both X and Y are in the operate state but the control path of Y is broken. Consequently, at the end of the third interval, Y will come to rest. This will break the control path of X. So the relay X will be released at the end of the fourth interval. We are now back to square one, so to speak, and the states of the relays would repeat cyclically in a sequence till the switch a is opened (hence the name sequential circuits) as shown in the table in Figure 4.19 (b). By attaching suitable output circuits, we can utilise this circuit to provide four consecutive recurring actions of frequency $\frac{1}{4T}$ each.

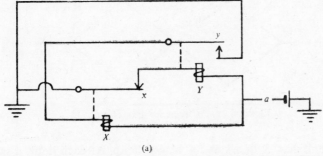

(a)

Time Interval	1	2	3	4	5	6	7	8	9
X	0	0	1	1	0	0	1	1	0
Y	0	1	1	0	0	1	1	0	0

(b)

Figure 4.19: Interlocked Relays of Equal Time Delays

Further problems on the use of relays, both as memory devices and in sequential circuits will be given as exercises.

Exercises

3.1 Find the Boolean functions representing the circuits of Figure 4.20. Simplify the circuits.

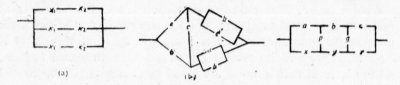

Figure 4.20. Circuits for Exercise (3.1)

3.2 Find the Boolean function realised by the circuit in Figure 4.21 both by finding all possible paths for the current and also by replacing the star of degree 4 in it. (Caution: Such replacement will result in the production of other stars of degree 4 which will then have to be replaced.)

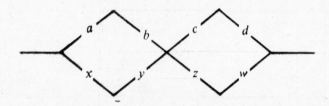

Figure 4.21: Circuit for Exercise (3.2)

3.3 A hall has 3 doors and a central lamp. At each door, a switch is provided. Design a circuit in which each of these three switches can control the lamp independently of the others.

3.4 Generalise the last problem by assuming that the hall has n doors. Prove that the closure function of the lamp is a symmetric function whose characteristic numbers are in an arithmetic progression with common difference 2.

3.5 Prove that the circuit of Figure 4.22 can be used to realise the function in the last problem. (This circuit is said to be obtained from that in Figure 4.10 by the **shifting down** method, because of the wires slanting forward that go downwards (from level 1 to level 0) instead of upwards as in Figure 4.10.)

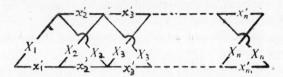

Figure 4.22: Shifting down Method (Exercise (3.5))

3.6 At a party there are several tables. At each table there is a switch which closes if the number of persons sitting at that table is odd. Design a circuit with these switches which will indicate whether the total number of persons at the party is even or odd.

3.7 For Exercise (2.6), design a circuit with 8 switches (one for each subject and one for indicating whether the candidate comes from a scheduled caste) in which a green lamp will light if the candidate passes the examination and a red lamp if he fails.

3.8 A concert is to be recorded on a series of cassettes (the concert being too long to come on one cassette). To avoid gaps that may result because of the time taken to change cassettes, two recording machines are provided. Each machine has a switch. Design a circuit such that when the position of either one of these two switches is altered (the other switch remaining in the same position), the states of both the machines would change (i.e., the machine which is running would stop and the machine at rest would start).

3.9 Three students A, B and C share an apartment and take turns at cooking as per the following schedule:

    (i)   When exactly two of them have a test the third student cooks,

    (ii) When only A has a test, B cooks on even numbered days and C on odd numbered days.

    (iii) When only B has a test, C cooks on even numbered days and A on odd numbered days.

    (iv) When only C has a test, A cooks on even numbered days and B on odd numberd days,

    (v) In all other cases they eat out.

Each student is given a switch which closes iff as he has a test. There is also a fourth switch which closes precisely on even numbered days. Design a circuit with these four switches in whlch a red lamp lights if A is to cook, a blue one if B is to cook, a green one if C is to cook and a white one if they are to eat out.

3.10 An obstacle track has five obstacles. At each obstacle, there is a switch which closes automatically if a fault is committed while clearing that obstacle. Design a circuit in which lamps of different colours would light to indicate the number of faults committed.

3.11   Do the variation of Problem (3.2) in which one of the committee members has a special veto power which he may exercise by pressing an additional switch, independently of his opinion as an ordinary member.

3.12   Suppose that abstention is allowed in Problem 3.2. Then a single switch (having only two states) does not suffice to represent a member's mind. So suppose each person p_i is given two switches x_i and y_i. The person closes x_i if he is in favour and y_i if he is opposed to the bill. If he wishes to abstain, he closes neither. (Closing both x_i and y_i may be either prohibited or taken to mean abstention.) Now design a circuit with these 10 switches in which a green lamp lights if the bill is carried by a majority of those voting, a red lamp if the bill is defeated by a majority of those voting and a white lamp if the votes are equally divided.

3.13   The last exercise illustrated how 2 Boolean variables can be utilised to represent a ternary device (with some wastage.) What is the minimum number of Boolean variables needed to represent a variable taking n possible values? (Any representation of a variable in terms of several Boolean variables is known as a **binary coding** of that variable.)

3.14   Design black box representations, using logic gates, for the following Boolean functions. Use parallel processing wherever possible.

   (i) $xy' + yz' + zx$
   (ii) $x \oplus y$ (that is, the 'exclusive or' or the 'ring sum' of x and y).
   (iii) $x_1 \oplus x_2 \oplus \ldots \oplus x_n$ (this is called the **parity checker**, see Exercise (2.21)).
   (iv) $xyz' + x'y + xy'z$.

3.15   Suppose a designer of black boxes has run out of AND-gates. Show that he can still manage if he has an adequate supply of the other two logical gates, namely the OR-gates and the NOT-gates. (Technically, a set S of gates is called **functionally complete** if every logical gate can be constructed from replicas of the gates in S. Thus {OR, NOT} is a functionally complete set. So is {AND, NOT}.)

3.16   Three other basic logic elements are the XOR-gate, the NAND-gate and NOR-gate. For inputs x and y, their outputs are, respectively, $x \oplus y$, (xy) and $(x + y)'$.

   (a)   Construct these gates.
   (b)   Show that {NAND} is a functionally complete set.
   (c)   Show that {NOR} is also functionally complete.

3.17   Show that any delta in a circuit may be replaced by a wye. In other words, prove that the circuits in (a) and (b) of Figure 4.23

are equivalent if the blanks in (b) are suitably filled. (This is called **delta-to-wye conversion.**

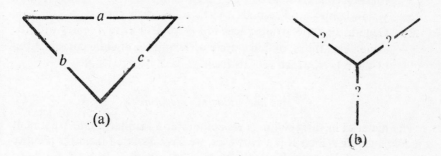

Figure 4.23: Delta-to-wye Conversion (Exercise 3.17)

It is not quite the opposite of the wye-to-delta conversion, because if the two are applied one followed by the other then the circuit that results is not the same as the original one, although it is of course equivalent).

3.18 In Figure 4.24, assume *a* and *b* are pushbutton switches which cannot be operated simultaneously. Prove that if *a* is pressed and released and then *b* is pressed and released then the relay *Y* will remain operated but if first *b* is pressed and released and then *a* is pressed, it will be at rest. (In other words, this circuit can distinguish between the order in which two switches are pressed).

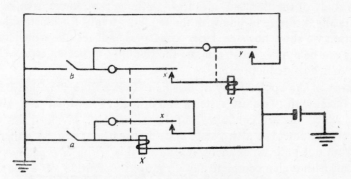

Figure 4.24: Circuit for Exercise 3.18

*3.19 Generalising the construction in the last exercise, design a toy with three pushbuttons marked *A*, *C* and *T* (and using as many 'hidden' relays as needed), in which a bell will ring iff the pushbuttons are pressed in the order *C*—*A*—*T*, and a red lamp will light otherwise, indicating that a fresh start is necessary. (The red lamp should not light at press and release of *C* or at *C*—*A*, but it should light at *C*—*C*, *C*—*A*—*A*, *C*—*T*—*A* etc.).

3.20   By attaching suitable output circuits to the relay diagram in Figure 4.19 (a), design a traffic signal in which a red lamp lights for 30 seconds, followed by a green lamp for 15 seconds and a yellow lamp for 15 seconds and then this cycle repeats.

3.21   Generalising the arrangement in Figure 4.19(a) to n relays with the same delay times, obtain a cycle of length $2n$ of state combinations of the relays, which repeats itself.

Notes and Guide to Literature

The material in this section is elementary and standard. Our treatment is modelled after Whitesitt [1]. However, we have deviated from the practice of considering, directly, Boolean algebra of circuits. Instead we have taken a circuit as a Boolean function of the switches occurring in it and used the theory of Boolean functions. Because of the great importance of the Boolean algebra of functions in switching networks, this algebra is often itself called the **switching algebra**. For more on it, and for methods of simplifying Boolean functions, see Dornhoff and Hohn [1].

Digital processing using logic elements is an important branch of electronics today. See, for example, Strangio [1].

4.   Applications to Logic

As remarked earlier, Boole originally formulated his theorems as laws of thought. In other words, application to logic was the very motivation for the study of the Boolean algebras. Although we have treated them axiomatically, keeping in line with the modern trend for abstraction, in this section we show how the basic concepts and results from Boolean algebras can be applied to streamline the thinking process, especially the process of logical deduction.

The crux of the applicability of Boolean algebra to logic is that Boolean algebras deal with two state systems and in logic (or at least the kind of logic to which we are confining ourselves, namely, bivalued logic) we have every statement as a two state system, because it can be either true or false but not both simultaneously. Indeed this is the *sine-qua-non* of a statement. Although a complete and a rigorous definition of a statement is beyond our scope (for such a definition would cut deep into linguistics and philosophy), we shall take a **statement** (also called **proposition**) as any declarative sentence, conveying a single, definite meaning which is either true or false but not both. Let us examine these three conditions, one by one.

The first requirement, that of being a declarative sentence, rules out all questions, commands and mere phrases from being statements unless, of course, they are used either figuratively or for brevity to mean declarative sentences. For example, 'Who does not want to be rich?' is a statement despite being a question ostensibly. (Whether it is a true statement or a false statement is another matter.) Similarly the phrase 'In the morning' is not a statement by itself. But as an answer to the question 'When did you arrive?' it is a statement, the words 'I arrived' being understood.

The second requirement of a statement, namely, it should have a single definite meaning is more elusive because the meaning is subject to the context. The statement 'Today is Friday' is a different statement every day and it is only when the day is understood that it becomes a statement. Similarly a sentence like 'John is tall' would be a statement only when the particular individual named John and the standard of tallness are expressed or at least understood. To avoid the dependence upon the context, some authors confine their discussion only to what may be called mathematical statements, where, as noted in Chapter 1, every term, other than a primitive term, is given a precise definition. While this chaste practice is mathematically adequate (because all mathematical statements are included anyway), if we adopt it we would have to sacrifice many 'real-life' statements. Secondly, even if we restrict ourselves to only mathematical statements, some imperfection of meaning will always occur because of the primitive terms. Let us, therefore, not be very fussy about the question of meaning, which is really a deep question of philosophy. However, two points deserve to be noted. First, by 'meaningful' we do not necessarily mean 'sensible'. Statements like 'Either Grandma chews gum or missiles are costly' and 'If there is life on Mars then the postman delivers a letter' sound utterly non-sensical and seldom appear in practice (except, of course, in surrealistic literature!). But from the point of view of logic they are as valid examples of statements as 'Either the set A is empty or else it has a least element' and 'If two triangles are congruent then they have equal areas'. Another point about 'meaning' is, as emphasised in Chapter 1, Section 4, we interprete statements to mean exactly what they say. We exclude all connotations and implications.

It is the third requirement, namely that every statement must be either true or false, that really calls for a comment. At first sight it may appear to be superfluous, because every declarative sentence with a definite meaning must be either true or false depending upon whether what it states holds or not. For example the sentence 'The boiling point of water is 100° centigrades' is true if the phenomenon asserted by it in fact holds, that is if water indeed boils at 100°C (but at no lower temperature) and false otherwise. The statement is silent about the conditions under which water is boiled. As it stands, it is false. To make it true, these conditions would have to be specified. But that is immaterial for our purpose. The point is that, every sentence about physical facts is either true or false according as its contents are consistent or inconsistent with observations. The same holds about mathematical statements To be sure, there are many sentences which are not known to be true or false at present. Goldbach's conjecture was cited as an example in Chapter 1, Section 4. As a real life example we can take the statement 'There is life outside our solar system'. What all these examples show is that while it may be a matter of varying degree of difficulty to ascertain the truth or falsehood of a sentence, the fact that it is either true or false is obvious enough. Why do we then include it as an additional requirement of a statement?

The answer lies in our attempt to avoid paradoxes. These paradoxes are of the same spirit as the Russel's paradox in Chapter 2, Section 1 which arises if we let any collection of objects be a set. To avoid it, we had to put some restriction on the collections. Similarly, let us see what happens if we allow every declarative sentence (with a single definite meaning) to be a statement. Consider the sentence 'I am telling you a lie'. This sentence cannot be called true because if so then what it says is true and hence the person uttering it is lying. But this means that what he is saying is not true and hence the statement is false. Similarly if we say the sentence is false then we are forced to conclude that it is true! This example shows that there do exist sentences which cannot be assigned any truth value (i.e., which cannot be called true or false without getting a contradiction). These examples are different from those where we do not *know* the truth value of a statement (such as the Goldbach's conjecture). They also show that the third requirement of a statement, that it should be either true or false but not both is not a superfluous one.

If we look at this example closely we see that the sentence 'I am telling you a lie' differs from some sentence like 'The boiling point of water is 100°C' in one crucial respect, namely it makes a reference to its own truth value. There are many variations of the paradox given above where a sentence makes a self-reference directly or indirectly (for example, two sentences referring to the truth values of each other). Such sentences cannot be assigned truth values and consequently cannot be called statements. Sometimes they are called **metastatements**. They share some of the properties of ordinary statements, but not all. In the same vein terms like **metaknowledge** are used to express knowledge about the presence or absence of knowledge of something. For example, in the statement, 'I know that I know nothing' the second occurrence of the verb 'to know' conveys knowledge but the first occurrence conveys metaknowledge. If we treat metastatements on par with statements (or metaknowledge on par with knowledge) we get many paradoxes (a few of which will be given as exercises).

We shall not pursue the definition of a statement further because as far as the applications of Boolean algebra are concerned, what matters is not the definition of a statement but how the truth or falsehood of a statement depends on that of some other statements much the same way as in the applications of Boolean algebra to electricity, what matters is not the construction of switches but the formation of circuits from these switches. Our primary concern will thus be to express certain statements (or more precisely, their truth values) as Boolean functions of some other statements (or more precisely, of their truth values). We say the **truth value** of a statement is 1 (or T) if it is true and 0 (or F) if it is false. Every statement is, therefore, a Boolean variable, very much like a switch. Just as two switches which are always in the same state are to be regarded as equal, two statements which are always simultaneously true or simultaneously false are to be regarded as equal. (Earlier we called such statements as logically equivalent. Now we are treating them as equal because our concern is only in their truth values. An alternate approach would be to consider, instead of statements *per se* their equivalence classes under the relation of logical equivalence. But we stick to the first approach.) We shall generally denote statement by $p, q, r, \ldots$ etc.

We already defined the basic operations of disjunction, conjunction and negation of statements. If p, q are statements, their disjunction is denoted by $p + q$ (or by $p \lor q$ because the statements form a lattice). The conjunction of p and q is denoted by $p \cdot q$, pq or by $p \land q$. The negation of p is denoted by $\neg p$, $\sim p$ or p'. To justify that the negation indeed corresponds to complementation in a Boolean algebra, we first have to consider what are the identity elements for the operations $+$ and $\cdot$. A statement which is always true is called a **tautology** and is denoted by 1 while a statement which is always false is called a **contradiction** and is denoted by 0. (Although both these terms have been used earlier, in the present context they play the role of constant functions which are identically 1 and 0 respectively.) It is clear that for any statement p, the statement $p + 0$ will be true iff p is true (because 0 can never be true). Hence $p + 0$ has always the same truth value as p. So by our convention, $p + 0 = p$, showing that 0 is an identity for $+$. Similarly 1 is the identity for $\cdot$. Commutativity of $+$ and $\cdot$ is trivial while the distributivity of each over the other can be established directly as follows (see also Exercise (3.4.8)). Let p, q, r be statements. Then $p \cdot (q + r)$ is true iff p is true and at least one of q and r is true. This is equivalent to saying that either p and q are true or else p and r are true. But that is exactly the condition for $pq + pr$ to be true. Thus $p \cdot (q + r) = pq + pr$. Similarly the other distributivity can be proved. As for complementation, it is clear that for any statement p, $p + p'$ is always true and $p \cdot p'$ can never be true. (It is here that bivaluedness of logic is crucially needed.)

Therefore if we let S be the set of all statements and consider the operations $+$, $\cdot$ and $'$ defined respectively by conjunction, disjunction and negation then we have verified all the axioms of a Boolean algebra. However, there is a technical difficulty in calling the set S (along with these operations) as a Boolean algebra because S is not a set at all! We cannot form the set of all statements much the same way as we cannot form the set of all sets. The difficulty involved here is a matter of axiomatic set theory and beyond our scope. Still, to play it safe, we state the results in the following form:

4.1 Theorem: Let S be any non-empty set of statements which is closed under $+$, $\cdot$ and $'$ (i.e. for every $p, q \in S$, $p + q$, $p \cdot q$ and $p' \in S$). Then $(S, +, \cdot, ')$ is a Boolean algebra. ∎

It often happens that the truth value of a statement depends on the truth values of some other statements say $p, q, r,...$ The number of statements $p, q, r,......$ may be finite or infinite. For example, the truth value of $p + q$ depends only on that of p and q. But the truth of a statement like 'every positive integer can be expressed as a sum of four perfect squares' depends on the simultaneous truth of the infinite sequence $p_1, p_2,..., p_n,...$ of statements, where, for each positive integer n, p_n is the statement 'n can be expressed as a sum of four perfect squares'. More generally, the truth of a statement about a class sometimes depends on the truth or falsehood of statements about the individual members. (This is not always the case, because as noted in Chapter 2, Section 1, there are some attributes of a set which cannot be described in terms of any attributes of the individual members).

When the truth value of a function f depends on the truth values of infinitely many statements, there is no easy way to express f in terms of those statements. We can try things like the conjunction of infinitely many statements but that would involve a limiting process. However, if the truth value of f depends upon those of only finitely many statements, say $p_1, p_2,..., p_n$ then f is a Boolean function of the Boolean variables $p_1, p_2, ..., p_n$ and the methods of Section 2 become applicable. The table of values of f is known as the **truth table** of f because the values assumed by f (and also by $p_1,..., p_n$) are the truth values. Knowing the truth table of f we can find the disjunctive normal form of f and hence express f as an algebraic expression involving $p_1, p_2,..., p_n$ and the operations $+$, $\cdot$ and $'$.

We apply this procedure to an implication statement $p \to q$ where p and q are some statements. As defined in Chapter 1, Section 4, $p \to q$ means that whenever p is true so is q. It is a little easier to first write the truth table of the negation of $p \to q$, i.e. of $(p \to q)'$. Note that $(p \to q)'$ holds precisely when $p \to q$ fails. But to say that $p \to q$ fails (i.e. is false) is equivalent to saying that p holds and still q fails, that is, p is true but q is false. In all other cases $p \to q$ holds. Hence we get the truth tables of $(p \to q)'$ and of $p \to q$ as in Figure 4.25.

From this table we see that the disjunctive normal form of

$$p \to q \text{ is } pq + p'q + p'q'$$

which upon simplification gives $p' + q$. In other words, the implication statement $p \to q$ is logically equivalent to 'either p fails or else q holds'. This is consistent with common sense too. In practice we use a sentence like 'That plane is a jet or else my eyes are failing me' to mean 'If my eyes are not failing me then that plane is a jet'. The crucial point in preparing the truth table above was that the implication statement $p \to q$ is completely silent as to the truth of q when p is false. And since it is not saying anything as to what happens if p is false, we have to let it be true regardless of whether q is true or not. The logic adopted here is the same as in

Row	p	q	$(p \to q)'$	$p \to q$
1	1	1	0	1
2	1	0	1	0
3	0	1	0	1
4	0	0	0	1

Figure 4.25:   Truth Table of Implication Statement

vacuous truth. Indeed, vacuously true statements can be expressed as implication statements whose hypthesis is always false. For example 'Every four legged man is happy' can be expressed as 'If a man has four legs then he is happy', which is true because its hypothesis is false.

We also note that equating $p \to q$ with $p' + q$ is consistent with the partial order $\leqslant$ defined on the set of statements by $p \leqslant q$ iff p implies q is true (cf. Chapter 3, Section 3). For, $p' + q = 1$ iff $pq' = 0$ (taking complements) which is precisely the order relation for a Boolean algebra in Definition 1.5.

Having obtained an algebraic expression for an implication statement, we can now handle with ease problems involving logical implications. For example we see at once that an implication statement $p \to q$ is logically equivalent to its contrapositive, $q' \to p'$, because

$$p \to q = p' + q \text{ and } q' \to p' = q + p'$$

(since $(q')' = q$). In real-life problems where a system of rules is given, they are generally in the form of the implication statements. To say that a particular rule is obeyed is equivalent to saying that the corresponding implication statement has truth value 1. Therefore, to say that a system of rules is simultaneously satisfied is equivalent to saying that the conjunction of the corresponding implication statements is true. Factorising this conjunction into simpler factors leads to a simplification of the rules. Similarly if the conjunction comes out to be 0 then it means the rules can never hold simultaneously and hence that the system is inconsistent.

We illustrate this technique for the Business Problem. Take a particular business and let p, q, r, s be respectively the statements that it has an import license, it manufactures essential commodities, it employs local personnel and it employs skilled personnel. Then the three rules are respectively (i) $(p + q') \to (rs')$ (ii) $(p + r') \to (sq')$ and (iii) $p' \to r'$. If the first rule is to be satisfied then $p'q + rs' = 1$. Similarly (ii) holds iff $p'r + sq' = 1$ and (iii) holds iff $p + r' = 1$. Thus the business in question will satisfy all the three rules iff $(p'q + rs')(p'r + sq')(p + r') = 1$. The product of the last two factors is $sq' (p + r')$, which, when multiplied with $(p'q + rs')$ gives 0. So we get $0 = 1$ which is a contradiction. Thus no business can satisfy all the three rules. So it is impossible to do any business in that country.

In Section 1, we tackled the Business Problem by considering the set of all businesses and certain subsets of it. It is instructive to compare the two solutions. There we calculated, for each rule, the set of businesses which violated that rule. It turns out that this set, (or its complement, namely the set of businesses which obey a particular rule) is intimately related to the implication statement which corresponds to that rule. We proceed to study this relationship. First we need a definition which involves the formation of statements.

4.2 Definition: A **predicate** (or more precisely a **unary predicate**) on a set X is a sentence which contains a variable x such that when every occurrence of x is substituted by an element of X we get a statement.

Intuitively, we may think of a predicate as a variable statement which becomes specific when a particular value is assigned to the variable in it. The terminology comes from grammar where every simple sentence has a subject and a predicate. The predicate contains the verb and hence describes the action in the sentence and the subject specifies a particular entity about which the predicate says something.

For example, let X be the set of all positive integers. Consider the sentence 'x is a prime and x^2 can be expressed as a sum of two squares'. Here x is a variable. For convenience let us write this sentence as $p(x)$. When we substitute for x some value, say 5, from X we get the statement '5 is a prime and 25 can be expressed as a sum of two squares'. We denote this statement by $p(5)$. Similarly $p(3)$ is the statement, '3 is a prime and 9 can be expressed as a sum of two squares', $p(10)$ is the statement '10 is a prime and 100 can be expressed as a sum of two squares' and so on. The statement $p(5)$ is true while the statements $p(3)$ and $p(10)$ are false. It is very important to replace every occurrence of the variable by the same element of the underlying set. Otherwise disastrous results are obtained (which often form the gist of some jokes). For example if A says to B, 'I love my wife' and B says 'So do I', then B's statement means that B loves his own wife and not that B loves A's wife! Here there is an implicit predicate $p(x) = $ 'x loves x's wife' on the set of all married men (with only one wife each!) and the statement made by A is $p(A)$. If B wants to make a similar statement, arising from the same predicate, it would be $p(B)$, which reads 'B loves B's wife'. However, if A says 'I love Lucy' and B says 'So do I' then B's statement means 'B loves Lucy'. Here the predicate is different, namely $q(x) = $ 'x loves Lucy'. If Lucy happens to be A's wife then by 'I love Lucy' whether A means $p(A)$ or $q(A)$ is to be inferred from the context in practice. Unintended interpretation would lead to a disaster.

We remark here that the word 'sentence' appearing in Definition (4.2) is technically incorrect. An expression like 'x is intelligent' is not called a sentence because of the variable x occurring in it. It is called just a 'formula' or a 'string'. A sentence is technically defined as a formula with no free variables. This distinction is important in the study of formal languages but we shall ignore it for the moment here.

Note also that the variable x is a dummy variable and could be replaced by any other variable which ranges over the same set as x. Generally, in specifying a predicate, the choice of the variable is so made that the set on which the predicate is defined would be automatically implied. For example, in the predicate 'If a man is rich, he is happy' we can take 'a man' as the variable and obviously it ranges over the set of all men. We can also word the predicate as 'A rich man is happy' where the variable is 'a rich man' and ranges over the set of all rich men.

We can also consider **binary predicates** (also called **2-place predicates**) or more generally n-ary predicates. For example if X is the set of all persons in a town and Y is the set of all houses in it then 'x lives in y' is a binary predicate with two variable x and y ranging over the sets X and Y respectively. If the sets X and Y coincide then a binary predicate over X can be identified with a binary relation on X. Given a binary relation R on X we define a binary predicate on X by 'x is R-related to y', i.e., by '$(x, y) \in R$'. Conversely every binary predicate $p(x, y)$ on X determines a binary relation R on X, if we let $R = \{(x, y): x, y \in X, p(x, y)$ is true$\}$. It is, in fact, more consistent with our intuition to think of a binary relation on a set X as a binary predicate on X rather than as a subset of $X \times X$ as we have defined. But, as just remarked, the two approaches are really equivalent.

Another way to look at a predicate p on a set X is to think of p as a function from X to set of statements. The various operations on statments, such as disjunction etc. can then be defined for predicates by pointwise application. Thus if $p(x)$, $q(x)$ are two predicates (i.e. unary predicates) on a set X then we define their disjunction $(p + q)(x)$, as the predicate $p(x) + q(x)$. Similarly $p \cdot q$ and p' are defined. With these operations, the set of all predicates on a set X forms a Boolean algebra. We now propose to investigate the structure of this Boolean algebra by showing that it is isomorphic to a more familiar Boolean algebra. The key concept on which this isomorphism is based is the following.

4.3 Definition: Let $p(x)$ be a (unary) predicate on a set X. Then the **truth set** of $p(x)$ is defined as the set $\{y \in Y : p(y)$ is true$\}$.

In other words, the truth set of a predicate is the set of those values of the variable for which the statement formed from the predicate is true. For example, the truth set of the predicate $p(x)$ on $\mathbb{R}$ (the set of real numbers) defined by $p(x) =$ '$x^4 - 7 = 9$' is $\{2, -2\}$.

If the same predicate is considered on the set of complex numbers then the truth set would be larger, namely, $\{2, -2, 2i, -2i\}$. Obviously a predicate serves as a characteristic property (see Chapter 2, Section 1) for its truth set, provided the set on which it is defined is understood.

It is quite possible that two different predicates, say, $p(x)$ and $q(x)$ defined on the same set X have the same truth sets. The chances of this are more, the smaller the set X is. For example if X is the set of all students in a small class it may very well happen (by chance) that all intelligent students in it are heavy smokers and vice-versa, even though we expect no correlationship between intelligence and smoking. (A similar remark was made about binary relations, following Definition (3.2.1).) The purpose of predicates is to distinguish among various elements of a set by means of properties which some elements may have and others do not. We therefore regard two predicates on a set as equal if their truth sets are identical. This may sound absurd at the beginning; for example who would agree that 'x is a smoker' is the same as 'x is intelligent'? But if we really look at it from the point of the range of variation of x, it is no longer all that absurd. The extreme case is that of an empty set on which any two predicates are equal. When a set X is not empty but still fairly small, two predicates on X may very well be equal, even though we would hesitate to admit it. Our hesitation is generally due to our knowledge that the two predicates are really the restrictions of two unequal predicates on some superset of X. But if the set X is our whole world, there would be little room for hesitation.

With this understanding about the equality of predicates on a set X, we now determine the structure of the Boolean algebra formed by them.

4.4 Theorem: Let X be any set. Assume that two predicates on X having the same truth sets are equal. Then the Boolean algebra of all predicates on X is isomorphic to the power set Boolean algebra $P(X)$.

Proof: For a predicate $p(x)$, let T_p be its truth set. Let B be the set of all predicates on X. Define $f: B \to P(X)$ by $f(p(x)) = T_p$. We verify easily that $f((p + q)\ (x)) = T_p \cup T_q,\ f((p \cdot q)\ (x)) = T_p \cap T_q$ and $f(p'(x)) = X - T_p$. So once we show that f is a bijection, it would follow that f is an isomorphism. That f is one-to-one follows from the assumption that if $T_p = T_q$ then the predicates $p(x)$ and $q(x)$ are to be regarded as equal. It remains to show that f is onto. Let $A \in P(X)$. Then A is a subset of X. Let $p(x)$ be the predicate '$X \in A$'. Clearly $T_p = A$, that is, $f(p(x)) = A$. So f is onto. ∎

Although this theorem is not very profound, the isomorphism established in it serves to link she two solutions we gave for the Business Problem. Let $p(x)$ and $q(x)$ be two predicates on a set X. Let $r(x)$ be the predicate $p(x) \to q(x)$. Let us compute the truth set T_r of $r(x)$. If $y \in X$ then the statement $p(y) \to q(y)$ is true iff either $p(y)$ is false or else $q(y)$ is true. This shows $T_r = (T_p') \cup T_q$. It follows that the predicate $r(x)$ is identically true (that means $p(x) \to q(x)$ for all $x \in X$ or equivalently, $T_r = X$) iff $((T_p)' \cup T_q)' = \phi$, i.e., iff $T_p \subset T_q$. This is, of course, consistent with our interpretation of

$p(x) \to q(x)$ to mean that whenever $p(x)$ is true so is $q(x)$. Even when $r(x)$ is not identically true, its truth set, $(T_p)' \cup T_q$ tells us in which cases it is true.

Now in the Business Problem, the set X is the set of all businesses in the country. In the second solution, we took some element of X and considered the statements p, q, r, s about it. This actually amounts to considering four predicates, $p(x)$, $q(x)$, $r(x)$, $s(x)$ on X. We expressed the three rules as certain other predicates on X and showed that their conjunction was 0. In the first solution, on the other hand, we considered the truth sets of the predicates $p(x)$, $q(x)$, $r(x)$ and $s(x)$. For each rule, we found its truth set (or rather its complement) and showed that their intersection was null (by showing that the union of their complemets was the whole set). The two solutions are, therefore, just the translations of each other.

The algebraic expression for an implication statement also provides a method for testing the validity of an argument. This was left largely to common sense in Chapter 1, Section 4. But with the aid of the machinery of Boolean algebras we can tackle it quite systematically. Recall that in every argument we have a collection of statements, say, $p_1, p_2, ..., p_n$ which are called **premises** and a statement q called the **conclusion**. The argument is called **valid** if whenever all the premises hold true, so does the conclusion. Now the simultaneous truth of $p_1, p_2, ..., p_n$ is equivalent to the truth of their conjunction $p_1 p_2, ... p_n$. Therefore, an argument is valid or invalid depending upon whether the implication statement $(p_1 p_2, ... p_n) \to q$ is identically true or not. Since $(p_1 p_2 ... p_n) \to q$ equals $p_1' + p_2' + ... + p_n' + q$, the problem of testing the validity of an argument can be solved by simplifying this expression and seeing whether it comes out to be a tautology (i.e., identically equal to 1).

To illustrate this consider the following argument :

Premises :   If it rains, the streets get wet.
If the streets get wet, accident happen.
Accidents do not happen.

Conclusion :   It does not rain.

Let r, s, t denote respectively the statements 'it rains', 'the streets get wet' and 'accidents happen'. Then the three premises are $r \to s$, $s \to t$, and t' or $r' + s$, $s' + t$ and t' and the conclusion is r'. To test the validity, we consider

$$(r' + s)' + (s' + t)' + (t')' + r'$$

which equals $rs' + st' + t + r'$ which reduces to 1 through

$$rs' + s + t + r', r + s + t + r' \text{ and } 1 + s + t.$$

So the argument is valid. We could also establish the validity by drawing a table of values in which we show the truth values of the premises and of the conclusion for every possible combination of truth values of the statements r, s, t. If we find that for every combination in which all premises are 1, the conclusion is also 1 we conclude that the argument is valid. This is a time consuming method. Some saving can be effected by noting that if in a particular row we find that some premise is 0 or the conclusion is 1, we need not do any further computation in that row. Even then, the method is not as efficient as the agebraic method.

If in the example above, the third premise was 'Accidents happen' then to test the validity we would have to consider the statement,

$$rs' + st' + t' + r'$$

which, upon simplification, equals $r' + s' + t'$. This is false when r, s, t are all true. Thus there is at least one situation in which all the premises are true but the conclusion is false. (In terms of the truth table, this means there is at least one row in which the entries in the columns of the premises are all 1 but the entry in the column of the conclusion is 0.) So the argument is invalid.

When the number of premises is large, it is hardly practicable to check the validity of an argument either by the truth table or by a reasoning similar to the above. In such cases we resort to what may be called the **chain rule** about validity. Simply stated, it says that by chaining together two (or more) valid arguments, we get a valid argument. A precise formulation is as follows. Let A_1 be an argument with premises $p_1..., p_k$ and conclusion q_1. Let A_2 be an argument with premises $q_1, p_{k+1}, ..., p_n$ and conclusion q_2. Let A_3 be an argument with premises $p_1, ..., p_k, p_{k+1}, ..., p_n$ and conclusion q_2. Then, if A_1 and A_2 are valid, so is A_3. The proof of the chain rule is easy and left to the reader.

The chain rule is used so frequently in mathematics that an explicit mention is rarely made. Whenever in a proof we cite a theorem previously proved, we are implicitly using the chain rule. By repeated applications of the chain rule a valid argument can be split into a series of some very simple arguments. Among these, probably the most frequently used argument is what is called **modus ponens**. Formally, this is an argument with two premises, one of the form $p \to q$ and the other p and whose conclusion is q (where p, q are any statements whatever.) Verbally, modus ponens is the argument that if an implication statement holds and its hypothesis is true, then so is its conclusion. This is certainly consistent with common sense and it is a triviality to verify the validity of modus ponens. Using modus ponens and the chain rule let us now establish the validity of the argument given above. We follow the notation there and present the reasoning in the form most commonly adopted in mathematical proofs, as a sequence of steps beginning with the premises, ending with the conclusion and defending each step.

(1) $r \to s$ (given as a premise)
(2) $s \to t$ (given as a premise)
(3) $\neg\, t$ (given as a premise)
(4) $\neg\, t \to \neg\, s$ (equivalent to (2))
(5) $\neg\, s$ ((3), (4) and modus ponens)
(6) $\neg\, s \to \neg\, r$ (equivalent to (1))
(7) $\neg\, r$ ((5), (6) and modus ponens).

Hence the argument is valid. This method is generally not very useful in proving that an argument is not valid, because even if we cannot reach the conclusion through one particular sequence of simple arguments, conceivably some other sequence could work. However, for proving validity, it is the most frequently resorted method in mathematics. The desired conclusion appears at the end. But every step along the way, is itself the conclusion of a valid argument. These arguments are often called **minor** or **subordinate arguments.**

We often come across arguments of the following type:

> *Premises*: Every man is mortal
> Socrates is a man.

> ---

> *Conclusion*: Socrates is mortal.

The validity of this argument is obvious. But it is not strictly speaking a case of modus ponens. If the first premise were 'If Socrates is a man, then Socrates is mortal' then this would be a case of modus ponens. As it happens, the first premise is much stronger. Consider the predicate 'If x is a man then x is mortal'. Here the variable x can range over any set. The first premise then says that this predicate is identically true, although we are using it in only a particular instance, namely, when the variable x is given the value 'Socrates'. For this reason, the first premise is called a **major premise** and the second is called a **minor premise**. This form of an argument is called **instantiation** because it consists of taking a particular instance of a statement.

We urge the reader to go back to Exercise 1.4.9 and test the validity of the arguments using the methods of Boolean algebra. The reasoning used for proving validity of arguments can also be used to prove that a system of statements is inconsistent. All we have to do show that 0 can be drawn as valid conclusion of an argument whose premises are the given statements. This provides yet another solution of the Business problem which will be developed through exercises.

Exercises

4.1   Which of the following expressions are statements? Why?

   (1) If it rains the streets get wet.
   (2) If it rains the streets remain dry.
   (3) It rains.
   (4) If it rains.
   (5) John is intelligent and John is not intelligent.
   (6) If John is intelligent then John is intelligent.
   (7) If John is intelligent then John is not intelligent.
   (8) For every man there is a woman who loves him.
   (9) There exists a woman for whom there exists no man who loves her.
   (10) There exists a woman such that no man loves her.
   (11) This sentence is false.
   (12) There is life outside our solar system.

4.2   On both the sides of a piece of paper it is written, 'The sentence on the other side is false'. Are the two sentences so written statements? Why? What if on one side 'The sentence on the other side is true' is written and on the other side 'The sentence on the other side is false'?

4.3   A barber in a village makes an announcement, 'I shave those (and only those) persons in this village who do not shave themselves'. Is this announcement a statement? Why?

4.4   Comment on the usage of the verbs 'to learn' and 'to teach' in the sentence 'The only thing we learn from history is that it teaches us nothing'.

4.5   Suppose God the Almighty can do anything at any time. Can He then make a stone so heavy that He will not be able to lift it? Comment.

4.6   What, if anything, is wrong in the following reasoning?
'An instructor of a course announces to the class that he would give a surprise test next week. The week runs from Monday to Friday and by 'surprise' is meant that on no day prior to the test could the class tell with certitude that the test will fall on the next day. Then the test could not be given on Friday because in that case, by the evening of Thursday, the class would be in a position to predict it. So the test has to be given on a day from Monday to Thursday. But then by the same reasoning, it cannot be given on Thursday, as otherwise, by Wednesday evening the class would come to know about it. Continuing in the same manner, the test would have to be given on Monday. But then it is hardly a surprise test. So it is impossible to give a surprise test, after announcing that there would be one.'

4.7 A clever inspector of police assured his suspects that he would not resort to any third degree methods to extort confessions from them, if they would agree to answer truthfully just two questions, with simple 'yes' or 'no' answers. After a suspect agreed the inspector would ask him the first question and the suspect would reply 'yes' or 'no'. Then this inspector would ask his second question 'Are you guilty?' and to this the suspect had to answer 'yes'. Find out the inspector's first question that would do the trick. Then comment upon this form of questioning. (Forget the legalities involved. Just stick to the logic of it.)

*4.8 A and B are two mathematicians. A third person, C, selects two (not necessarily distinct) integers x and y each less than 100 and greater than 1. He tells A only their sum $x + y$ and tells B only their product xy. A is also told that B is given the product xy but not the sum $x + y$ and B is told that A is given $x + y$ but not $x \cdot y$. The following conversation takes place:

B : I do not know the two numbers x and y.
A : I know you don't.
B : Now I know them.
A : So do I.

Assuming both A and B are telling the truth, find the numbers x and y.

4.9 Verify the following identities for statements p, q, r by writing truth tables for both the sides.

(i) $p \cdot (q + r) = p \cdot q + p \cdot r$

(ii) $p + (q \cdot r) = (p + q) \cdot (p + r)$

(iii) $(p + q) \cdot (pq)' = (p \cdot q') + (p' \cdot q)$

4.10 Do Exercise (1.13) using implication statements.
4.11 Do Exercise (1.11) using implication statements.
4.12 A set X consists of 5 men named Ram, Rahim, Robert, Gopal and Goliath. They have the following properties in regard to their height, financial status, intelligence and age.

Person	Height	Financial status	Intelligence	Age
Ram	Tall	Rich	Dumb	Young
Rahim	Tall	Poor	Dumb	Old
Robert	Short	Poor	Intelligent	Young
Gopal	Tall	Rich	Intelligent	Old
Goliath	Short	Rich	Dumb	Young

Find the truth sets of the following predicates on X

   (i)   A man is rich and intelligent.

   (ii) A man is rich or old but not necessarily tall.

   (iii) A rich man is intelligent.

   (iv) A man whose name begins with an 'R' is rich.

4.13 Prove the validity of modus ponens and of the chain rule.

4.14 Besides modus ponens and instantiation the following elementary forms of valid argument are frequently needed. The names of some of them are given against them. Establish the validity of each one of them. (p, q, r, s are some statements)

   (i) $p \to q$
        $p \to r$
       ———— **(Hypothetical Syllogism)**
        $p \to q$

   (ii) pq
       ————
       p

   (iii) p
      ———— **(Addition)**
      $p + q$

   (iv) q
       r
      ———— **(Conjunction)**
       qr

   (v) $p \to q$
        $s \to r$
      ————
      $ps \to qr$

   (vi) $p \to r$
        $q \to s$
    ————————
    $p + q \to r + s$

4.15 Prove that the rules in the Business Problem are mutually inconsistent by drawing 0 as a valid conclusion from them. (Hint : Use the various arguments in the last exercise.)

4.16 Can the inconsistency of the data in Exercise (1.12) be proved using solely implication statements?

4.17 Prove that the following argument is invalid:

    Premises: If it rains missiles are costly.
                  Missiles are costly.

    Conclusion: It rains.

Can this argument be called invalid because the first premise is absurd, there being no correlation between rain and the cost of missiles?

4.18 If in the last exercise we keep the premises as they are but change the conclusion to 'It does not rain', prove that the resulting argument is still invalid. Why does this not contradict the bivaluedness of logic because of which one of the two statements, 'It rains' and 'It does not rain' must be true?

Notes and Guide to Literature

This section, like the fourth section of Chapter 1, overlaps with the first chapter of Joshi [1]. For a more formal treatment of predicates see, for example, Tremblay and Manohar [1]. The branch of logic dealing with statemeats (or propositions) and their Boolean algebra is called **propositional calculus**. On the other hand, **predicate calculus** deals with the formation of statements in a language.

Five

Group Theory

A group structure on a set is probably the most important algebraic structure that can be induced on it with a single binary operation. The theory of groups is rich both in terms of the profound results in it and in terms of its applications to combinatories, physics and many other fields. In an elementary book like this, we have to be content with only a few glimpses of the theory. In this chapter we present some of the basic results about groups in the first three sections. The last section deals with a special type of groups, called the permutation groups. The properties proved here about them will be applied in the later chapters. Other applications of groups will be given in the chapter on group actions (see the Epilogue).

1. Groups and Subgroups

Suppose S is a set. We often have to consider functions of the set S into itself. These functions are often called operations, transformations or some other names depending upon the context. Let X denote the set of all functions of S into itself. We are rarely interested in the whole set X. The types of functions that are important will of course depend upon the context. In mechanics we are interested in what are called **rigid body motions**, that is, those transformations which preserve the distance between various points (also called **isometries**). In the study of finite state machines, the set S is often called the state space. Each member of S represents some configuration in which a particular system is at a given time. The transformations in this case are the results of giving some instructions to the machine, by which one state is converted into another. As yet another example, which has acquired recent popularity, take a Rubik's cube. Let S be the set of various patterns of squares on the six faces. Then every sequence of motions of the Rubik's cube gives a function from S to S. (The object of the puzzle is to find a sequence of such motions which will transform a given pattern to a standard pattern in which all the squares on each face are of the same colour).

So let S be the set and let G be the set of those functions of S into itself in which we are interested. Then $G \subset X$, the set of all functions from S into itself. The property of G that is most commonly true is that it is closed under composition, that is, if two functions $f: S \to S$ and $g: S \to S$ are in G then so is their composite $f \circ g: S \to S$. If G is not closed under composition, then we usually enlarge it to a subset H of X which is closed under composition. Another property which is often true of transformations is that they are invertible and their inverses are also of the permitted type. In symbols, if $f \in G$ then f^{-1} exists and belongs to G. For example, in the case of a Rubik's cube, if a particular sequence of motions changes one pattern of squares into another we can get back the original pattern by applying, in the reverse order, the inverse motions. Note that if G is a non-empty subset of X which is closed under composition and under inversion then the identify function 1_S is also in G because we take any $f \in G$ and write $1_S = f \circ f^{-1}$.

Sets of functions which are closed under composition and inversion are called groups. The origin of the term is obscure. Possibly, it lies in the fact that functions grouped together so that their composites and inverses are also included have certain nice properties, 'as a group' which an arbitrary set of functions assembled together does not have in general. (We shall study such properties later.) At least historically, the term group was used to mean a group of functions (or a group of *substitutions* as it was called). As at many other places, abstraction set in. The set of functions was replaced by an abstract set G and the composition of functions was replaced by an abstract binary operation on G, satisfying certain conditions. Such abstraction is definitely worth-while because there are many 'naturally occurring' examples of groups other than groups of functions. Nevertheless, the old ties with the groups of functions are important for two reasons. First, in applications, it is these groups that will be important. Secondly, we shall prove in Section 3 that every 'abstract' group is isomorphic to a group of functions. Although this result will not be of much direct use, it will show that, upto isomorphism, the groups of functions are the only possible groups.

Let us now turn to the definition of an abstract group. As noted before it is modelled after groups of functions. So we take an abstract set G, a binary operation $\cdot$ on G and we want to assume as axioms certain properties of $\cdot$ which are true of composition of functions. The foremost such property is associativity. Without associativity, it would not be possible to manipulate algebraic expressions involving $\cdot$ (see Chapter 3, Section 4). Associativity by itself is not a very strong condition. Still there is a very large number of examples where it holds and so there is a name for the algebraic structure where it is the only axiom.

1.1 Definition: A **semigroup** is an ordered pair $(G, \cdot)$ where G is a set and $\cdot$ is an associative binary operation on G.

The set of all functions from a set into itself is a semigroup under the operation of composition. There are of course many other examples. The reader should consult the list of examples of binary operations given in Chapter 3, Section 4. As usual, there are three ways to construct new semigroups from old ones. One is to take a sub-semigroup of a semigroup, that is, a subset closed under the given binary operation. Another is to take the cartesian product of two or more semigroups and define the binary operation coordinatewise. The third method is to take the set of functions from a set into a given semigroup and define the binary operation pointwise. Note that we have not required the presence of an identity element in the definition of a semigroup. If a two sided identity exists (which would then be unique by Proposition (3.4.5)), then the semigroup is called a *monoid* as already defined after Definition (3.4.13). The set of all functions of a set S into itself is actually a monoid. Under the operation of usual addition, the set of non-negative integers is a monoid, while the set of positive integers is a semigroup but not a monoid. An especially interesting example of a monoid is obtained by taking the set of all words in an alphabet S (see again the list of examples in Chapter 3, Section 4) and defining the binary operation by concatenation. The empty word (that is, a sequence of length 0) plays the role of the identity element. This example is quite general in that the choice of the alphabet is arbitrary. By choosing a suitable alphabet we can get a monoid which will be important in a particular context. For example, if we let S be the set of all elementary operations which a machine can do then the monoid of all words in S consists of all series of operations which the machine can do in a finite amount of time.

Thus we see that the theory of monoids would have many applications. Unfortunately, the structure of a monoid is just not strong enough to yield profound results. It turns out that the addition of one more axiom, namely the existence of an inverse for each element, improves the situation dramatically. The inevitable price to be paid is, of course, the sacrifice of those monoids where this condition is not met. For example, in the monoid of words in an alphabet considered above, no element other than the identity element (that is, the empty word) is invertible. It turns out that even after the exclusion of such examples, the list where the theory would be applicable is still impressively large. As a matter of fact, even if we add one more axiom, namely that the operation be commutative, we still have a sizable number of cases where the theory would apply. We therefore make the following two definitions.

1.2 Definition: A **group** is an ordered pair $(G, \cdot)$ where G is a non-empty set and $\cdot$ is a binary operation on G which satisfies the following conditions:

(i) $\cdot$ is associative,

(ii) $\cdot$ has a two-sided identity element and

(iii)   every element of G is invertible.

If further, $\cdot$ is commutative, then the group $(G, \cdot)$ is said to be **abelian** (or **commutative**). Otherwise it is called **non-abelian**.

Before we give examples, a comment is in order regarding the notation. As usual, when the group operation is denoted by $\cdot$ it is often suppressed from notation. Thus we write xy for $x \cdot y$. For abelian groups, the group operation is also often denoted by $+$ and in that case the group is said to be **additively written**. As with Boolean algebras, there is in general no harm in using the same symbol to denote the group operations of two different groups so long as this is not likely to cause any confusion. A confusion would indeed arise if the underlying sets of the two groups are the same. When the group operation is denoted by $+$ (which is almost never done for non-abelian groups), the identity element is generally denoted by 0. In all other cases it is customary to denote it by e or by 1, we shall follow the first notation for 'abstract' groups. The inverse of an element x (which is unique by Proposition (3.4.7)) is denoted by x^{-1} in general and by $-x$ in case of an abelian group which is additively written. Following the comments preceding Proposition (3.4.8), if x is a group element then the power x^n is well-defined for all integers n (positive, zero or negative) and the laws of indices hold. If the group is abelian and additively written, then x^n is denoted by nx and is called a *multiple* rather than a power of x. Unless the group operation has some other suitable name in a particular context, we shall call it 'multiplication' and denote it by $\cdot$.

Now, as for examples of groups, the foremost example is the set G of all bijections of a set X. If for $f, g, \in G$ we define $f \cdot g$ as the composite $f \circ g$ then $(G, \cdot)$ is a group. This group is called the **permutation group** of the set X, and may be denoted by $S(X)$. If X is a finite set with n elements then $S(X)$ has $n!$ elements. For $n = 0, 1$ and 2 this group is abelian. For all other n it is non-abelian. To see this suppose X has at least three distinct elements say x, y and z. Define $f : X \to X$ by $f(x) = y$, $f(y) = x$ and $f(s) = s$ for all $s \in X$, $s \neq x, y$. In words, f interchanges x and y and leaves every other element of X fixed. Similarly, let $g : X \to X$ be $g(y) = z$, $g(z) = y$ and $g(s) = s$ for all $s \in X - \{y, z\}$. Then the composites $g \circ f$ and $f \circ g$ are unequal because $g \circ f(x) = g(y) = z$ while $f \circ g(x) = f(x) = y \cdot$ Note here that $f \circ f = g \circ g = 1_X =$ identity element of $S(X)$. In other words f and g are their own inverses. If we define $h : X \to X$ by $h(x) = y$, $h(y) = z$, $h(z) = x$ and $h(s) = s$ for all $s \in X - \{x, y, z\}$ then $h \circ h \circ h = 1_X$. As examples of abelian groups, one can take the groups of integers, rational numbers, real numbers or complex numbers under usual addition. The set of rational numbers is not a group under usual multiplication because the element 0 has no inverse. However, the set of all non-zero rational numbers is an abelian group under usual multiplication. The same holds for the set of non-zero real numbers and the set of non-zero complex numbers.

Another important class of groups is provided by the sets of residue classes modulo some positive integer. If m is a positive integer then in Chapter 2 Section 2, we defined $\mathbb{Z}_m$ as the set of all equivalence classes of $\mathbb{Z}$ (the set of all integers) under the equivalence relation of congruence modulo m. In Chapter 3, Section 4 we saw that there are two well-defined binary operations $+$ and . on $\mathbb{Z}_m$, namely, the addition and multiplication of residue classes modulo m. Also both these operations are commutative and associative. We claim $(\mathbb{Z}_m, +)$ is a group. Clearly the residue class $[0]$ is an identity for $+$. Also if $[n] \in \mathbb{Z}_m$ then $[-n]$ is the additive inverse of $[n]$.

Under multiplication of residue classes, $\mathbb{Z}_m$ is not a group (except in the trivial case $m = 1$), because the residue class $[0]$ cannot have an inverse. It is tempting to exclude $[0]$ and ask whether $Z_m - \{[0]\}$ is a group under modulo m multiplication. Here the difficulty is that in general $Z_m - \{[0]\}$ is not closed under multiplication. For example, let $m = 10$. The neither $[5]$ nor $[2]$ equals the residue class $[0]$ but $[5] \cdot [2] = [10] = [0]$ modulo 10. So we do not have a well-defined binary operation on $Z_m - \{[10]\}$. Notice here that 10 is not a prime. If m is a prime then things are better as we now show.

1.3 Proposition : Let m be a positive integer greater than 1. Then the set of non-zero residue classes modulo m is closed under residue multiplication modulo m if and only if m is a prime.

Proof : The direct implication is easy. If m is not a prime then $m = a \cdot b$ for some $0 < a < m$ and $0 < b < m$. Then $[a]$ and $[b]$ are both non-zero in Z_m but $[a] [b] = [a \cdot b] = [m] = [0]$. So $Z_m - \{[0]\}$ is not closed under mod m multiplication.

The proof of the converse requires an important property of prime numbers, namely if p is a prime and a, b are two integers such that $p \mid ab$ then either $p \mid a$ or $p \mid b$. Since a more general result will be proved in the next chapter, for the moment we take this fact for granted. Now suppose m is a prime. Let $[a], [b]$ be non-zero elements of Z_m. We have to show $[ab]$ is also a non-zero residue class. If not then $ab \equiv 0$ (modulo m) which means m divides ab. But then, by the fact just mentioned, either $m \mid a$ or $m \mid b$. In the first case, $[a] = 0$ while, in the second, $[b] = [0]$. In either case we have a contradiction. Thus $[a \cdot b] \neq [0]$, or in other words, the set $Z_m - \{[0]\}$ is closed under multiplication modulo m. ∎

Because of this proposition, whenever m is a prime, there is a well-defined binary operation of multiplication on the set of non-zero residue classes modulo m. This operation is already known to be associative. Also the residue class [1] is the identity element. Thus $Z_m - \{[0]\}$ is a monoid. To prove that it is in fact a group we use the following result which is of independent interest.

1.4 Proposition: In any group both the cancellation laws hold. Conversely any finite semigroup in which both the cancellation laws hold is a group.

Proof: The first assertion follows from the properties of inverses proved in Proposition (3.4.7). The second assertion is just a rewording of Theorem (3.4.10). ▪

Now to apply this result to $Z_m - \{[0]\}$ when m is a prime. We simply have to show that the cancellation laws hold in it. Because of commutativity. only one cancellation law needs to be proved. So let $[x]$, $[y]$ and $[z] \in Z_m - \{[0]\}$ and suppose $[x][z] = [y][z]$. Then $xz - yz \equiv 0$ (mod m). Hence m divides the product $(x - y)z$. Since m is a prime either $m \mid (x - y)$ or $m \mid z$, (by the property of primes mentioned above). The latter possibility means $[z] = [0]$, contradicting the hypothesis. So m divides $x - y$ which means $[x] = [y]$. Thus cancellation holds in $Z_m - \{[0]\}$. So it is a group. Note that it has $m - 1$ elements, (all but one residue classes modulo m,). To record our work we introduce a terminology. Let $(G, \cdot)$ be a group. If the set G is finite then the group G is said to be of finite order and $|G|$ is called the **order** of G and is often denoted by $\circ(G)$. (If G is not finite, the group is said to be of infinite order.) With this terminology, our result is as follows:

1.5 Theorem: For every positive integer m, Z_m is an abelian group of order m under addition modulo m If m is a prime, then $Z_m - \{[0]\}$ is an abelian group of order $m - 1$ under multiplication modulo m. ▪

Further examples of groups can be obtained by taking cartesian products of groups, sets of functions from any set into a group and by taking subgroups. The last concept is especially interesting because various subgroups of the group of permutations of a set are important in various fields of study. Moreover, we shall show in Section 3 that every abstract group is isomorphic to a subgroup of a group of permutations of some set. Let us therefore study the concept of a subgroup carefully. Although it is only a special case of the general concept of a substructure of an algebraic structure defined in Definition (3.4.14), in view of its special importance we restate it. A characterisation of it will be given in Exercise (1.6).

1.6 Definition: Let $(G, \cdot)$ be a group. Then a non-empty subset H of G is called **subgroup** of G if H is closed under $\cdot$ and inversion, that is, for all $x, y \in H$ we have, (i) $x \cdot y \in H$ and (ii) $x^{-1} \in H$.

Trivially, if e is the identity element of G then $\{e\}$ is a subgroup of G. It is called the trivial subgroup*. G is also a subgroup of itself. It should be noted that the conditions (i) and (ii) above are independent of each other. For example, let Z be the group of all integers under the usual addition, $+$. Then the set of all positive integers satisfies (i) but not (ii) while the set of all odd integers satisfies (ii) but not (i). The set of all even integers satisfies both (i) and (ii) and hence is a subgroup of Z. More generally it is easy to show that for any positive integer m, the set of all multiples of m, that is, the set $\{mx : x \in Z\}$ is a subgroup of Z. It is denoted by mZ. It is interesting to note that these are the only non-trivial subgroups of Z, as we now show.

1.7   Proposition: The only subgroups of $\mathbb{Z}$ are $\{0\}$ and those of the form $m\mathbb{Z}$ for some positive integer m.

Proof: Let H be a subgroup of $\mathbb{Z}$. If $H \neq \{0\}$, then H contains some non-zero integer x. Then $-x$ is also in H. Since one of x and $-x$ is positive, H contains at least one positive integer. Let m be the smallest positive integer in H. (The existence of such m follows from the fact that the set of positive integers is well ordered, see Exercise 3.3.21). We claim $H = m\mathbb{Z}$. First, to show $m\mathbb{Z} \subset H$, we consider any multiple mx of m. If x is 0, then $mx = 0 \in H$. If x is a positive integer we can show by induction on x that $mx \in H$. (Clearly $m \cdot 1 \in H$ and if $m \cdot (x-1) \in H$ then $mx = m \cdot (x-1) + m \in H$.) If x is a negative integer then $-x$ is positive. We write $m \cdot x = -m \cdot (-x)$ which belongs to H since $-m$ is already in H by condition (ii) above. In all cases $m \cdot x \in H$, that is $m\mathbb{Z} \subset H$. For the converse, we need the euclidean algorithm mentioned in the solution to Problem 2.2.10. Let $h \in H$. Write $h = mp + q$ for some integers p and q with $0 \leqslant q \leqslant m - 1$. Then

$$q = h - mp = h + [m \cdot (-p)] \in H$$

since $h \in H$ and $m \cdot (-p) \in H$. Now if $q = 0$, then $h = mp$ and so $h \in m\mathbb{Z}$. If $q \neq 0$, then q is a positive integer in H and $q < m$. But this contradicts the very definition of m as the least positive integer in H. So $q = 0$ must hold and hence $h = mp \in m\mathbb{Z}$. Thus we have shown that $H \subset m\mathbb{Z}$. This proves that H, unless it is the trivial subgroup, $\{0\}$, is of the form $m\mathbb{Z}$ for some positive integer m. ∎

The last result is exceptional in that in general, it is far from easy to characterise all possible subgroups of a given group. (The reason we could do it for $\mathbb{Z}$ is that it is a cyclic group, a concept we shall define shortly). For example let $\mathbb{R}$ be additive group of real numbers. Then $\mathbb{R}$ has as subgroups, $\mathbb{Z}$, Q (the set of all rational numbers), the set $\mathbb{Z} + \sqrt{2}\,\mathbb{Z}$ defined as the set of all numbers of the form $m + \sqrt{2}n : m, n \in \mathbb{Z}$ and many other subgroups. All these are also subgroups of $\mathbb{C}$, the group of all complex

---

*A group with only one element in it is called a **trivial group**.

numbers under addition. Let $\mathbb{C}^*$ denote the group of non-zero complex numbers under multiplication. Let

$$S^1 = \{x + iy \in \mathbb{C} : x^2 + y^2 = 1\}.$$

Geometrically S^1 consists of all complex numbers which lie on a circle of radius 1 centered at the origin. It is easy to show, using elementary properties of absolute values of complex numbers, that S^1 is a subgroup of $\mathbb{C}^*$. When regarded as a group by itself S^1 is called the **circle group**. This group plays a crucial role in the study of certain other abelian groups. Note that it is abelian because every subgroup of an abelian group is abelian.

Let us now take the permutation group $S(X)$ of a set X. It often happens that X carries some additional structure. Then instead of taking all bijections of X into itself, we take only those which preserve this additional structure. In most cases, such bijections constitute a subsgroup of $S(X)$. This subgroup, when regarded as a group in its own right, is an important tool in the study of that additional structure because it depends on that structure. For example there may be some partial order $\leqslant$ on the set X. Then the set of those bijections $f : X \to X$ which are order preserving (that is, for all all $x, y \in X,\ x \leqslant y$ iff $f(x) \leqslant f(y)$) is a subgroup of $S(X)$.

An especially interesting example of a subgroup so constructed is obtained by taking X to be set of points in a plane or in a space. The additional structure is that given by the euclidean distance. We take only such permutations of X which preserve this distance, that is, bijections $f : X \to X$ which have the property that for all $x, y \in X$,

$$d(x, y) = d(f(x), f(y)),$$

where d denotes the euclidean distance between two points. In Chapter 3, Section 2 we called such a bijection as an isometry. Since the composite of two iosmetries is an isometry and the inverse of an isometry is an isometry, it follows that the set of all isometries of X is a subgroup of $S(X)$. It is called the **group of isometries** or the **group of symmetries** of X. To justify the second name we note that the larger this group, the greater is the degree of symmetry in the figure X. For example, suppose X consists of the three vertices, A, B and C of a triangle. If no two sides of the triangle ABC are equal, then there can be only one isometry from X into itself, namely, the identity function 1_X. Suppose, however, that the triangle ABC is isosceles with the sides AB, AC having equal length. Then the triangle is symmetric about the altitude through A. If we define $f : X \to X$ by

$$f(A) = A, f(B) = C \text{ and } f(C) = B$$

then f is an isometry. So here the group of symmetries of X has two elements, f and 1_X. If further, ABC is an equilateral triangle then the group of symmetries of X is the entire permutation group $S(X)$ of order 6. All the

six elements in this group have geometric interpretations. There are three
elements which represent rotations around the centre through angles of
0°, 120° and 240° (a rotation through angle 0° is of course the identity
function). The three other isometries correspond to the reflections through
the three altitudes of the triangle. All these six elements along with their
geometric representations are shown in Figure 5.1, where (a) represents the
identity function or the original state of the triangle. The function represented by any other triangle is obtained by sending every labelled vertex
to the vertex at the corresponding position in (a). For example the triangle
in (b) represents the function which takes *A* to *C*, *B* to *A* and *C* to *B*.

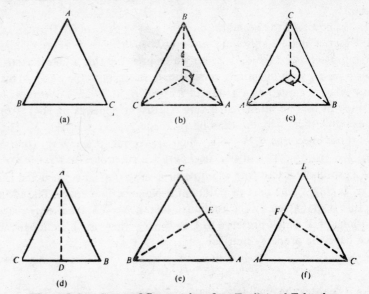

Figure 5.1: **Group of Symmetries of an Equilateral Triangle**

The groups of symmetries are very important in problems of placement
where we do not want to distinguish between symmetric arrangements, i.e
those which can be obtained from each other by applying some isometry
We shall compute many such groups when we shall study group action
(see the Epilogue). For the moment let us compute the group of symmetrie
of a regular pentagon. Let us name its vertices clockwise as v_1, v_2, v_3, v_4
and v_5. Let C be its centre. Let L_i be the line joining v_i to C for $i = 1,..., 5$
Let G be the group of all symmetries of the pentagon. Then G contain
rotations through angles which are multiples of 72° and also reflection
(or 'flips') through the lines $L_1 ..., L_5$. To see that these are the only
possible isometries. suppose $f : \{v_1, ..., v_5\} \rightarrow \{v_1, ..., v_5\}$ is an isometry. The
$f(v_1)$ can be any of the five vertices v_1 to v_5. However, the moment $f(v_1)$ i
fixed as v_i (say) then $f(v_2)$ must be one of the two 'neighbouring' vertice
of v_i. As soon as $f(v_1)$ and $f(v_2)$ are fixed, we have no choice in definin
$f(v_3), f(v_4), f(v_5)$ because of the requirement that all the distances b

preserved. Hence there can be only 10 distinct isometries at the most. Since we already have listed 10 isometries, we have exhausted the group *G*. Thus *G* is a group of order 10. To study its structure further, let r_i denote clockwise rotation through an angle of $72i$ degrees for $i = 1, 2, 3, 4, 5$; and let f_i denote the reflection through the line L_i. For example, as functions from $\{v_1 \ldots, v_5\}$ to itself, r_2 is given by

$$r_2(v_1) = v_3, \; r_2(v_2) = v_4, \; r_2(v_3) = v_5, \; r_2(v_4) = v_1 \text{ and } r_2(v_5) = v_2,$$

while f_4 is given by $f_4(v_1) = v_2, f_4(v_2) = v_1, f_4(v_3) = v_5, \; f_4(v_4) = v_4$ and $f_4(v_5) = v_3$. By actual computation, we see that the composite $r_2 \circ f_4$ equals f_5 while the composite $f_4 \circ r_2$ equals f_3. (The reader is also urged to verify these by actually doing physical experiments with a labelled pentagon.) Thus the group is non-abelian. More generally, it can be shown that for all $n \geqslant 3$, the group of symmetries of a regular *n*-gon is a non-abelian group of order $2n$. It is called the **dihedral group** of order $2n$ (or sometimes of order *n*) and is denoted by D_n.

For applications to organic chemistry, the group of isometries of a regular tetrahedron is very important because in the tetrahedral model of a carbon atom, a carbon atom is situated at the centre of a regular tetrahedron and is attached, by covalent bonds, to four radicals situated at the vertices. All the 24 permutations of the four vertices of a regular tetrahedron are isometries. But, for figures in space the question of orientation is important because we can distinguish between an object and its mirror image (that is, reflection through some plane), as the two have opposite orientations. This is shown in Figure 5.2 where the regular tetrahedron

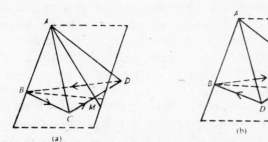

(a)   (b)

Figure 5.2: Reflection in a Plane Reverses Orientation

ABCD in (a), after undergoing reflection in the plane containing *A*, *B*, and *M* (the midpoint of *CD*) looks as in (b). The corresponding function $f : \{A, B, C, D\} \rightarrow \{A, B, C, D\}$ is given by $f(A) = A, f(B) = B, f(C) = D$, $f(D) = C$. Clearly *f* is an isometry. But it is not orientation preserving. To see why it is so, suppose we traverse the boundary of the face *BCD* in the sense $B - C - D - B$ so that the interior of the face is on our left. Then the fourth vertex *A* will be towards the head in (a). But if we traverse the face *BCD* in (b) in the same manner then *A* will be towards the opposite side. (The same idea can be expressed in terms of the right handed and the left handed screw, or, more analytically, in terms of the cross product of certain vectors.) Clearly the composite of two isometries is orientation preserving if both of them are orientation preserving or both are orientation reversing. When one of the two isometries preserves and the other reverses the orientation, their composite is orientation reversing. So the orientation preserving isometries form a subgroup of the group of all symmetries. Generally, for figures in space, by the group of their symmetries we mean the group of all orientation preserving symmetries. For plane figures the question of orientation does not arise because an orientation reversing transformation corresponds to a reflection in some line L in the plane. But we can think of it in space as a rotation through 180° around the axis L. Hence, as a transformation in space it is orientation preserving. Physically this amounts to realising a reflection through a flipping of the figure. However, where flipping is not allowed, the question of orientation does matter even for a plane figure. For example, in the case of a regular pentagon which is laid on a table and is not to be lifted from it, the group of isometries would comprise only of the five rotations.

Besides the groups discussed so far, there are two other groups which are important in applications to physics. One of them is called the **group of quarterions** and is denoted often by Q. It has eight elements denoted by $1, i, j, k, -1, -i, -j,$ and $-k$. The full table of multiplication is given in Figure 5.3. But there is a very easy way to remember it. As far as the six elements $\pm i, \pm j, \pm k$ are concerned, the multiplication is precisely the cross product for the three co-ordinate vectors **i, j, k** and their negatives (except for $i^2 = -1$ etc.). The element 1 is the identity. Multiplication by -1 reverses the sign. The verification of the associativity is fairly tedious. As for inverses, 1 and -1 are their own inverses while the inverse of every other element is its negative. This group is non-abelian, because $ij = k$ but $ji = -k$.

The other important group is variously known as the **Klein group,** the **axial group, the quadratic group** or the **four group** (because it has four elements). Let us denote it by K. It has four elements e, a, b, c and the group operation is defined so as to make e the identity element. As for the remaining elements, we let $a \cdot a = b \cdot b = c \cdot c = e$ and $a \cdot b = b \cdot a = c,$ $b \cdot c = c \cdot b = a, c \cdot a = a \cdot c = b.$ It is an abelian group of order 4. The reader will notice a certain relationship between Q, the quaternions group and K, the Klein group because of the cyclical manner in which the multiplication is defined for three elements. Indeed we may say that if we ignore the minus sign in the quaternion group then we get precisely the Klein group. Formally we identify 1 and -1 to a single element and call it e. Similarly we let a, b, c correspond to $\pm i, \pm j, \pm k$ respectively. Then we get K from

Q. This is a special case of what is known as the construction of a quotient group and will be studied in Section 2.

·	1	−1	i	j	k	−i	−j	−k
1	1	−1	i	j	k	−i	−j	−k
−1	−1	1	−i	−j	−k	i	j	k
i	i	−i	−1	k	−j	1	−k	j
j	j	−j	−k	−1	i	k	1	−i
k	k	−k	j	−i	−1	−j	i	1
−i	−i	i	1	−k	j	−1	k	−j
−j	−j	j	k	1	−i	−k	−1	−i
−k	−k	k	−j	i	1	j	−i	−1

Figure 5.3: Multiplication Table for Quaternions

It is clear that the intersection of two (and in fact, any number of) subgroups of a group G is again a subgroup of G. However, the union of two subgroups need not be a subgroup. For example in the quaternion group just defined, let $H_1 = \{\pm 1, \pm i\}$ and $H_2 = \{\pm 1, \pm j\}$ Then $H_1 \cap H_2 = \{\pm 1\}$ which is a subgroup of Q. But $H_1 \cup H_2 = \{\pm 1, \pm i, \pm j\}$ is not a subgroup because it is not closed under multiplication.

We often have a group G and a subset S of G. We then want to consider the smallest subgroup of G containing S. It is called the subgroup **generated** by S, and we denote it by $G(S)$. Of course, if S itself is a subgroup of G then $G(S) = S$. Clearly, if S is the empty set, then $G(S)$ is the trivial subgroup $\{e\}$ of G. More generally, if S a subset of G then we let $\mathcal{H}$ be the collection of all subgroups of G which contain S. In symbols, $\mathcal{H} = \{H; S \subset H$ and H is a subgroup of $G\}$. Then $\mathcal{H}$ is non-empty since the whole group G is a member of $\mathcal{H}$. If G is a finite group then obviously $\mathcal{H}$ will have only finitely many elements, say, $H_1, H_2, \ldots, H_n$ (one of which will be G). We now let $H = \bigcap_{i=1}^{n} H_i$. Then H is a subgroup of G. Also $S \subset H_i$ for all $i = 1, \ldots, n$

and hence $S \subset H$. So H is a subgroup of G containing S. To show that it is the smallest such subgroup, suppose K is any other subgroup of G containing S. Then $K = H_i$ for some i. But $H \subset H_i$ for all $i = 1,...,n$. So $H \subset K$. Therefore $H = G(S)$.

Thus the existence of the subgroup generated by a subset has been established at least for the case where the group is finite. The same argument actually applies even for infinite groups. The collection $\mathcal{H}$ considered above is possibly infinite. But we can take intersections even of infinite families of sets. However, we do not want to go into it, because anyway the construction given above for $G(S)$ is only of theoretical importance. If we want to apply it for a concrete set S, the difficulty starts right from deciding which subgroups of G contains S. Moreover, most of the work done in finding these subgroups is superfluous since we want only the smallest among them. We therefore look for some intrinsic construction for $G(S)$, that is, some construction in which elements of $G(S)$ are expressed concretely in terms of S. The situation is analogous to the problem of finding the equivalence relation generated by a given relation on a set (See Proposition 3.2.15 and Exercise 3.2.11). In the proposition below we give this 'internal' description of $G(S)$.

1.8 Proposition: Let $(G, \cdot)$ be a group and let S be a nonempty subset of G. Let $H = \{x_1^{n_1} x_2^{n_2} ... x_r^{n_r} : x_i \in S, n_i \in \mathbf{Z}, r \geqslant 0\}$. Then H is a subgroup of G and it is the subgroup generated by S.

Proof: H consists of all possible products of powers of elements of S. (Negative powers are also included). These products need not be all distinct. Clearly $S \subset H$ since every element $x \in S$ can be written as x^1. To show H is a subgroup, by very definition H is closed under multiplication. As for closure under inversion, we simply note that

$$(x_1^{n_1} x_2^{n_2} ... x_r^{r^r})^{-1} = x_r^{-n_r} x_{r-1}^{-n_{r-1}} ... x_2^{-n_2} x_1^{-n_1}$$

(by Proposition (3.4.7)) which is again an element of H. Thus H is a subgroup of G containing S. To show it is the smallest such subgroup, suppose K is a subgroup of G and $S \subset K$. Then we claim $H \subset K$. Take a typical element $x_1^{n_1} x_2^{n_2} ... x_r^{n_r}$ of H. Since $x_i \in S$, $x_i \in K$ for all $i = 1, ..., r$. But K, being a subgroup, is closed under multiplication and inversion. Hence $x_i^{n_i} \in K$ for $i = 1,..., r$ and finally $x_1^{n_1} x_2^{n_2} ... x_r^{n_r} \in K$. So $H \subset K$. Thus H is the smallest subgroup of G containing S. ∎

This proposition not only establishes the existence of the subgroup $G(S)$ generated by a subset S but also gives the nature of its elements in terms of the elements of S. It may of course happen that the same element of $G(S)$ can be expressed in more than one way, say, $x_1^{n_1} ... x_r^{n_r}$ and $y_1^{m_1} ... y_s^{m_s}$. Note also that we are not requiring the x_i's to be distinct, because the elements $x_1 x_2 x_1$ and $x_1^2 x_2$ need not be the same. However, in an abelian group, we may require the x_i's to be distinct. As examples, we see that in the quaternion group Q, the subgroup generated by the singleton set $\{i\}$ is $\{\pm 1, \pm i\}$. Note here that $i^2 = -1$, $i^3 = -i$, $i^4 = 1$ and the other powers of i do not give rise to new elements, for example $i^5 = i^4 \cdot i = i$. The subgroup generated by i, j is the entire group Q. In $\mathbb{Z}$, the singleton set $\{1\}$ generates the whole group. There is a special name for groups with this property.

1.9 Definition: A group G is called **cyclic** if there exists $x \in G$ such that G is generated by the singleton set $\{x\}$. Any such x is called **a generator** of G. (We often say 'x generates G' instead of '$\{x\}$ generates G').

Thus, the group $(\mathbb{Z}, +)$ is cyclic with 1 and -1 as generators. For every positive integer m, the group $\mathbb{Z}_m$ of residue classes under addition is also cyclic, with $[1]$ as a generator. Note that if $n \, (>1)$ divides m then $[n]$ cannot be a generator of $\mathbb{Z}_m$ because if $k = m/n$ then, the subgroup generated by $[n]$ will contain only k elements, namely, $[n]$, $[2n], \ldots, [nk-n]$ and $[nk] \, (=[m] = [0])$ For example, $[2]$ is not a generator for $\mathbb{Z}_4$ but $[1]$ and $[3]$ are. In the next chapter we shall determine all the generators of a cyclic group.

Since a cyclic group consists of all powers of an element and any two powers of the same element commute with each other (by the laws of indices, Proposition (3.4 8)), it follows that every cyclic group is abelian. The converse is false. As a simple counter-example, the Klein group, K, defined above is not cyclic. The subgroups generated by $\{e\}$, $\{a\}$, $\{b\}$ and $\{c\}$ are respectively, $\{e\}$, $\{e, a\}$, $\{e, b\}$ $\{e, c\}$. Note of these is the entire group K.

Although a cyclic group consists of all powers of its generator, these powers need not be distinct. Indeed if the group is finite then they are bound to repeat (by the pigeon-hole principle). It turns out that this repetition follows a regular pattern. The precise result is given below: The reader is asked to first verify its truth for the cyclic group Z_n.

1.10 Theorem: Let G be a cyclic group of order n with a generator x. Then n is the smallest positive integer such that $x^n = e$, the identity of the group. For any two integers i and j, $x^i = x^j$ if and only if $i \equiv j$ (modulo n). The group G consists of e, x, $x^2, \ldots, x^{n-1}$.

Proof: First we show that there exists a positive integer r such that $x^r = e$. Consider the successive powers, x, x^2, x^3, $x^4, \ldots$. These are all elements of G. But since G is finite, these elements cannot be all distinct. So there exist positive integers i and j with $i < j$ such that $x^i = x^j$. Let $r = j - i$. Then $x^r = x^{j-i} = x^j \cdot (x^i)^{-1}$ (by Proposition (3.4.8)). Hence $x^r = e$. Since there exists at least one positive integer r with $x^r = e$, there exists a least positive integer, say m, with this property. (Here again we are using the fact that the set of positive integers is well-ordered, see Exercise (3.3.21).) We have to show that $m = n$. First note that the elements

$e, x, x^2, ..., x^{m-1}$ are all distinct. (For if $x^p = x^q$ (say) with $0 \leqslant p < q < m$ then once again $x^{q-p} = e$ with $q - p < m$ contradicting the minimality of m.) But G has only n distinct elements. So $m \leqslant n$. Note further that e, $x, ..., x^{m-1}$ are the only distinct powers of x. To see this suppose x^k is any power of x. By the euclidean algorithm, we can write $k = sm + t$ where s, t are integers, with $0 \leqslant t < m$. Then $x^k = x^{sm+t} = x^{ms} \cdot x^t = (x^m)^s \cdot x^t = e^s \cdot x^t = e \cdot x^t = x^t$. Thus x^k equals x^t which is in the set $\{e, x, ..., x^{m-1}\}$. But we are assuming that x is a generator for G. Hence the powers of x exhaust the set G. So $G \subset \{e, x, ..., x^{m-1}\}$, giving $n \leqslant m$ (by comparison of cardinalities). Thus we have shown $m = n$ which proves the first assertion, For the second assertion, if $i \equiv j$ (modulo n) then we write $i = j + km$ for some integer k and show with laws of indices, that $x^i = x^j$. Conversely suppose $x^i = x^j$. Then $x^{i-j} = e$. Once again we use the euclidean algorithm and write $i - j = un + v$ where u, v are integers and $0 \leqslant v < n$. We then get $x^v = e$. If $v \neq 0$, then this is a contradiction to the fact that n is the smallest positive integer such that $x^n = e$. This shows $v = 0$ and hence $i \equiv j$ (modulo n). The last assertion has already been proved. ∎

This theorem completely describes the structure of finite cyclic groups. Even when a group G is not cyclic, the subgroup generated by an element, say x, of G is a cyclic group when regarded as a group by itself. This subgroup is often denoted by (x). If (x) is finite then we say the element x is of **finite order** (or sometimes a **torsion element**). Further the order of the group (x) is called the **order** of x. Obviously if the group G is finite then every element is of finite order. But even in an infinite group some elements may be of finite order. Trivially, the identity element is one such element. As a non-trivial (and interesting) example, consider the circle group S^1 defined above. This consists of all complex numbers of absolute value 1 and the group operation is that of multiplication of complex numbers. In polar form, every element of S^1 can be written as $e^{i\theta}$ (which equals $\cos \theta + i \sin \theta$) where θ is a real number. Note that $e^{i\theta_1} = e^{i\theta_2}$ iff $\theta_1 - \theta_2$ is an integral multiple of 2π. Also $e^{i\theta_1} \cdot e^{i\theta_2} = e^{i(\theta_1 + \theta_2)}$. Now let n be any positive integer. Let $z = e^{2\pi i/n}$. Then z is an element of order n. All powers of z are complex nth roots of 1. Hence the group (z) is called the **group of nth roots of unity** and any generator for it is called a **primitive** nth *root* of unity. (Unity is just another name for the number 1). Note that S^1 also contains elements of infinite order. We leave it to the reader to prove that $e^{i\theta}$ is of finite order iff θ is a rational multiple of π.

Note that if x is an element of order n then x^{n-1} is the inverse of x. In other words, the inverse of x can be expressed as a positive power of x. As a simple consequence of this fact we get the following result which, although not profound, makes it a little easier to check whether a subset of a finite group is a subgroup or not.

1.11 Proposition: Let G be a finite group and H a non-empty subset of

G. Then *H* is a subgroup of *G* if and only if *H* is closed under multiplication.

Proof: The necessity of the condition is obvious and does not require that *G* is finite. For sufficiency, suppose *G* is finite and *H* is closed under multiplication. To show *H* is a subgroup of *G* we merely have to verify that *H* is closed under inversion, that is if $x \in H$ then $x^{-1} \in H$. Now every element of *G* is of finite order since *G* is itself finite. Let $x \in H$ and *n* be the order of *x*. If $n = 1$, then $x = e$ and $x^{-1} = e \in H$. Suppose $n > 1$. Then x^{n-1} is the inverse of *x* Since *H* is closed under multiplication and $x \in H$, $x \cdot x \in H$. Then again $(x \cdot x) \cdot x \in H$. Continuing, $x^m \in H$ for all positive integers *m*. In particular $x^{n-1} \in H$. But this means $x^{-1} \in H$. This completes the proof. ◼

As an illustration, in the group of isometries of a regular pentagon considered above, the five rotations form a subgroup because the composite of two clockwise rotations through angles $72 i°$ and $72 j°$ is a clockwise rotation through $72 (i + j)°$.

So far we studied subgroups in general. Among all subgroups, certain subgroups called normal subgroups stand out because of the important properties they have. Although these properties will be studied in the next two sections, we introduce the concept of normal subgroups here and give some examples. Admittedly, the definition below may appear without motivation. But this would come later.

1.12 Definition : A subgroup *N* of a group *G* is called **normal** in *G* if for all $n \in N$ and all $g \in G$, $gng^{-1} \in N$.

Note that in this definition *n* ranges over *N* but *g* ranges over the whole group *G*. If $g \in N$ and $n \in N$ then we certainly have $gng^{-1} \in N$, whether *N* is normal or not (because *N* is closed under inversion and multiplication). Therefore, whenever we want to test whether a subgroup *N* of *G* is normal or not, it suffices to verify the condition $gng^{-1} \in N$ for all $n \in N$ and for all $g \in G - N$. Note also that normality is relative to the ambient group. We may have subgroups *N*, *H* with $N \subset H \subset G$ and *N* normal in *H* but not in *G*.

In case the group *G* is abelian, $gng^{-1} = gg^{-1}n = en = n$ and thus every subgroup of an abelian group is normal. So the concept of a normal subgroup is interesting only when the group is not abelian. Trivially, for any group *G*, the trivial subgroup $\{e\}$ and the whole group *G* are normal in *G*. As some non-trivial examples let *G* be the group of symmetries of a regular pentagon. We saw above that *G* consists of 5 rotations, r_1, r_2, r_3, r_4 and r_5 and 5 reflections (or flips) $f_1, f_2, ..., f_5$. Let $N = \{r_1, r_2, ..., r_5\}$. Then as noted above, *N* is a subgroup of *G*. We claim it is a normal subgroup of *G*. By the remark made above, it suffices to show that for all $i, j = 1,$ $2, ..., 5$ $f_i \circ r_j \circ f_i^{-1}$ equals some r_k. Note further that f_i^{-1} equals f_i. As a typical case consider $f_3 \circ r_2 \circ f_3$, Here f_3 as a function from $\{v_1, ..., v_5\}$ to itself is given by $f_3(v_1) = v_5, f_3(v_2) = v_4, f_3(v_3) = v_3, f_3(v_4) = v_2$ and $f_3(v_5) = v_1$. Also r_2 is given by $r_2(v_1) = v_3, r_2(v_2) = v_4, r_2(v_3) = v_5, r_2(v_4) = v_1$ and $r_2(v_5) = v_2$. Let g be the composite $f_3 \circ r_2 \circ f_3$. By direct computation we see that $g(v_1) = v_4, g(v_2) = v_5, g(v_3) = v_1, g(v_4) = v_2$ and $g(v_5) = v_3$. Thus g is precisely the rotation r_3. So $f_3 \, r_2 f_3^{-1} = r_3 \in N$. Similarly the other cases can be verified. So N is a normal subgroup of G. (In the next section we shall see a much quicker way of showing that N is normal in G.) Similar reasonings shows that for each i, the subgroup (f_i) is not normal in G.

A few other examples of normal subgroups will be given in the exercises. We conclude this section with two characterisations of normality. Although both of them are mere reformulations of the definition, we state them because of certain important concepts involved in them. Other characterisations of normal subgroups will be given in the next two sections.

The first characterisation involves the process of extending a binary operation on a set to a binary operation on its power set (cf. Exercise (3.4.15)). Let $*$ be a binary operation on a set X and let $P(X)$ be the power set of X. If $A, B \in P(X)$, we defined $A \circledast B$ to be the set $\{a * b : a \in A, b \in B\}$. This is a subset of X. It consists of all possible elements which can be expressed as products of two elements, one from A and the other from B. Such expressions need not be unique, For example, if $X = \mathbf{N}$, the set of positive integers, $*$ is the usual multiplication, $A = \{1, 3, 4, 6\}$ and $B = \{2, 3, 5\}$ then $A \circledcirc B$ is the set $\{2, 3, 5, 6, 8, 9, 12, 15, 18, 20, 30\}$. Thus $\circledast$ is a binary operation on $P(X)$, which is often denoted by $*$ as well. This operation inherits many of the properties such as commutativity, associativity and presence of identities of the original binary operation. However, very few elements of $P(X)$ have inverses. If A is a singleton set, say $\{x\}$, then $A * B$ is often denoted by $x * B$. This set is often called the left translate of the set B by the element x. Similariy if $B = y$ then $A * B$ is denoted by $A * y$ and is called the right translate of A by y, (cf. Exercise (3.4.2)). This terminology has a geometric origin. If $*$ denotes the usual addition for vectors in a euclidean space then a translate of a set is nothing but what is popularly called its 'parallel translate', because it is obtained by shifting every element of the original set by the same vector. In the next section, we shall have a lot to do with the various translates of a subgroup of a group. They will be called the cosets of the subgroup. For the time being, we characterise normality of a subgroup in terms of certain sets obtained from it by translating it on the left by an element of the group and on the right by the inverse of that element.

1.13 Proposition: Let G be a group and H a subgroup of G. Then for every $g \in G$, gHg^{-1} is a subgroup of G. H is normal in G iff for all $g \in G$, gHg^{-1} coincides with H.

Proof: By definition gHg^{-1} consists of all elements of the form gxg^{-1} as x varies over H. If G were abelian, then gHg^{-1} would be the same as H. In general, it could differ from H. To show it is a subgroup of G, we first show it is closed under multiplication. Two typical elements of gHg^{-1} are of the form gxg^{-1} and gyg^{-1} for some $x, y \in H$. Then

$$(gxg^{-1}) \cdot (gyg^{-1}) = gx\,(g^{-1}g)\,yg^{-1} = gxeyg^{-1} = g(xy)g^{-1}.$$

But $xy \in H$ since H is a subgroup. So $g(xy)g^{-1} \in gHg^{-1}$, that is, $(gxg^{-1})(gyg^{-1}) \in H$. As for inversion, if $x \in H$, then $(gxg^{-1})^{-1} = (g^{-1})^{-1} x^{-1} g^{-1} = gx^{-1}g^{-1}$ (by Proposition (3.4.8) again). But $x^{-1} \in H$. So $gx^{-1}g^{-1} \in gHg^{-1}$. Hence $(gxg^{-1})^{-1}$ is in H. Therefore gHg^{-1} is closed both under multiplication and inversion. So it is a subgroup of G. If H is normal in G, then for all $x \in H$ and $g \in G$, $gxg^{-1} \in H$. Hence $gHg^{-1} \subset H$. But actually, equality holds, because if $h \in H$ we can write $h = g(g^{-1}hg)g^{-1} = g(xhx^{-1})g^{-1}$ where $x = g^{-1}$. By normality of H, $xhx^{-1} \in H$ and hence $g(xhx^{-1})g^{-1} \in gHg^{-1}$. So $h \in gHg^{-1}$. Thus we have shown that if H is normal in G then gHg^{-1} equals H for all $g \in G$. The converse is clear because if gHg^{-1} coincides with H for all $g \in G$ then certainly for all $x \in H$ and $g \in G$, $gxg^{-1} \in H$. ∎

As an application of this result we have the following interesting corollary which is sometimes useful in proving that a subgroup is normal.

1.14 Corollary: Let G be a group. Suppose for some positive integer r, G has a unique subgroup of order r. Then the subgroup is normal in G. (G is not required to be finite.)

Proof: Let H be the unique snbgroup of order r and let $g \in G$. Then by the Proposition above, gHg^{-1} is a subgroup of G. We claim that gHg^{-1} is also of order r. For this, we define a bijection from H onto gHg^{-1}. Define $f : H \rightarrow gHg^{-1}$ by $f(x) = gxg^{-1}$. If $f(x) = f(y)$ for $x, y \in H$, then $gxg^{-1} = gyg^{-1}$ which implies $x = y$ by the cancellation laws (Proposition (1.4)). So f is one-to-one. And by the very definition of gHg^{-1}, f is onto. So f is a bijection and hence gHg^{-1} has cardinality r. But H is the only subgroup of order r. So gHg^{-1} must equal H. By the last proposition, this means that H is normal in G. ∎

For example, it can be shown that in the group G of all symmetries of a regular pentagon, there is only one subgroup of order 5, namely the subgroup N consisting of the five rotations. This gives an alternate proof that N is normal in G. Note however, that this method is not always applicable, because the converse of the corollary is false. For example in the quaternions group Q, the subgroups (i), (j), (k) are all of order 4 each and all of them are normal in Q.

The second characterisation of normality is based on the concept of conjugacy which we now define.

1.15 Definition: Let G be a group. If x, y are elements of G, we say y is conjugate to x (or a conjugate of x) in G if there exists $g \in G$ such that $y = gxg^{-1}$. Similarly if H, K are subgroups of G, we say K is conjugate to H in G if there exists $g \in G$ such that $K = gHg^{-1}$. (Sometimes we say that y is a **conjugate** of x **by** g and K is a **conjugate** of H **by** g.)

The following proposition is an immediate consequence of the definitions.

1.16 Proposition: Let G be a group and H a subgroup of G. Then the following statements are equivalent:

(i) H is normal in G.
(ii) All conjugates of elements in H are in H.
(iii) The only conjugate of H in the set of all subgroups of G is H itself.

Proof: The equivalence of (i) and (ii) is immediate from the definitions of normality and conjugacy. The equivalence of (i) and (iii) follows from Proposition (1.12) and the definition of conjugacy. ∎

Although the last proposition is far from profound, the concept of conjugacy is very important. After developing some machinery needed to apply it, we shall see its power in the chapter on group actions (see the Epilogue). For the moment we show that it induces an equivalence relation.

1.17 Proposition: Let G be a group. Define a binary relation R on G by $x \, Ry$ iff y is conjugate to x in G. Then R is an equivalence relation. Similarly conjugacy defines an equivalence relation on the set of all subgroups of G.

Proof: Reflexivity of R is clear since $x = e \cdot x \cdot e^{-1}$ for all $x \in G$. For symmetry suppose xRy. Then there exists $g \in G$ such that $y = gxg^{-1}$. But then $x = g^{-1}yg = g^{-1}y(g^{-1})^{-1}$, showing that x is a conjugate of y in G, that is, yRx. For transitivity, suppose x, y, $z \in G$ and xRy and yRz. Then there exists $g, h \in G$ such that $y = gxg^{-1}$ and $z = hyg^{-1}$. Then by substitution we get $z = h(gxg^{-1})h^{-1} = (hg) x (g^{-1}h^{-1}) = (hg) x (hg)^{-1}$ showing that xRz. So R is an equivalence relation on G. The proof of the second assertion is similar and left as an exercise. ∎

1.18 Definition: The equivalence classes of the set G under the relation of conjugacy are called **conjugacy** classes. The decomposition of G into conjugacy classes is called the **class decomposition** of G.

In an abelian group, no two distinct elements can be conjugates of each other. Consequently, its class decomposition is trivial. In a non-abelian group, it is in general not easy to tell whether two given elements x and y are in the same conjugacy class. A necessary condition is that they be of equal order. But this condition is far from sufficient. In the next section we shall find a formula for the size of the conjugacy class of an element.

Exercises

1.1 Prove that the set of all invertible elements of a monoid forms a group.

1.2 Give an example of a binary operation $*$ on a set X such that $*$ has a two sided identity and every element of X is invertible but $*$ is not associative. (Hint: Take X to be a finite set and prepare the table for $*$.)

1.3 Suppose $\cdot$ is an associative binary operation on a set G such that $\cdot$ has a right identity e and for every $x \in G$ there is some y such that $x \cdot y = e$. Prove that $(G, \cdot)$ is a group. (Hint: First show that the right cancellation law holds. Then show that e is a left identity and finally that $x \cdot y = e$ implies $y \cdot x = e$.)

1.4 Show by an example that the last result fails if we assume e to be a left identity (with no other change).

1.5 Suppose G is a group in which $(x \cdot y)^{-1} = x^{-1}y^{-1}$ for all $x, y \in G$. Prove that G is abelian. More generally, prove that if the equality $(xy)^n = x^n y^n$ holds for $x, y \in G$ for any three consecutive values of the integer n then G is abelian.

1.6 Prove that a non-empty subset H of a group G is a subgroup of G iff for all $x, y \in H$, $xy^{-1} \in H$. (In other words, the two conditions in Definition (1.6) can be replaced by a single condition. This often shortens the work needed to verify that a subset is a subgroup.)

1.7 Prove that the groups $Z_5 - \{[0]\}$, $Z_7 - \{[0]\}$ under respective modular multiplications are cyclic.

**1.8 Prove that for every prime p, the group $Z_p - \{[0]\}$ under modulo p multiplication is cyclic.

1.9 Prove that if G is an abelian group, and n is a positive integer then the set of all solutions of the equation $x^n = e$ in G, that is, the set $\{y \in G : y^n = e\}$ is a subgroup of G. Show by an example that this need not hold if G is non-abelian.

1.10 Let G be a cyclic group of order n. Suppose a positive integer m divides n. Prove that the equation $x^m = e$ has exactly m solutions in G and find them.

1.11 Prove that a subgroup of a cyclic group is cyclic.

1.12 Define $f : N \to N$ by $f(n) = n + 1$ if n is odd and $f(n) = n - 1$ if n is even. Define $g : N \to N$ by

$$g(n) = \begin{cases} 1 & \text{if } n = 1 \\ n+1 & \text{if } n \text{ is even} \\ n-1 & \text{if } n \text{ is odd and } n > 1 \end{cases}$$

Prove that both f, g are bijections. In the group $S(\mathbf{N})$ of all bijections of $\mathbf{N}$ onto itself, prove that both f, g are of finite order but neither $f \circ g$ nor $g \circ f$ is of finite order.

1.13   In an abelian group prove that the product of two elements of finite order is of finite order. Obtain an upper bound for its order in terms of the orders of the two elements and show by an example that it is sharp.

1.14   If x and y are two elements of any group G, prove that xy and yx have the same order. (Hint: For every positive integer n prove that $(xy)^{n+1} = x(yx)^n y$.)

1.15   Suppose G is a group of order n and $x_1, x_2, ..., x_n$ is a sequence of (not necessarily distinct) elements of G. Prove that there exist integers i, j with $1 \leqslant i \leqslant j \leqslant n$ such that $x_i x_{i+1} ... x_j = e$. (Hint : Use the pigeon hole principle.)

1.16   Prove that every finite group of even order contains at least one element of order 2. (Hint : Pair off every element with its inverse.)

1.17   A function $f : \mathbf{R} \to \mathbf{R}$ of the form $f(x) = ax + b$ where a, b are real constants with $a \neq 0$, is called a non-singular linear function. Prove that the set of all such functions is a subgroup of $S(\mathbf{R})$. Is it a normal subgroup?

1.18   Let $\mathbb{C}$ be the set of complex numbers, ∞ a symbol not in $\mathbb{C}$ and $\mathbf{C}^* = \mathbb{C} \cup \{\infty\}$. $\mathbf{C}^*$ is called the *extended complex number system*. A function $f : \mathbb{C}^* \to \mathbb{C}^*$ is called a *linear fractional transformation* (or L.F.T.) if it is of the form $f(z) = \dfrac{az + b}{cz + d}$ where a, b, c, d are complex numbers with $ad - bc \neq 0$ with the understanding that $f(\infty) = a/c$ and $f(z) = \infty$ if $cz + d = 0$. Prove that the set of all such linear fractional transformations is a subgroup of $S(\mathbb{C}^*)$.

1.19   Prove that the dihedral group D_n contains a normal, cyclic subgroup of order n. Prove that D_n can be generated by two elements.

1.20   Prove that the group of all orientation preserving isometries of a space figure is a normal subgroup of the group of all isometries of it.

1.21   Let S_3 be the group of all permutations of the set $X = \{1, 2, 3\}$. Let $f : X \to X$ be $f(1) = 1, f(2) = 3, f(3) = 2$ and let $g : X \to X$ be $g(1) = 3, g(2) = 2, g(3) = 1$. Let H, K be the subgroups of S_3 generated by f, g respectively. Prove that the sets HK and KH are not equal and neither is a subgroup of S_3.

1.22   Let H, K be subgroups of a group G. Prove that the set HK is a subgroup of G if and only if $HK = KH$. Prove further that when either one of H and K is normal in G, this condition is always satisfied and hence HK is a subgroup of G. If both H and K are normal in G, prove that HK is normal in G.

1.23 Prove that every subgroup of the quaternion group is normal.

1.24 Let H be the subgroup generated by a subset S of a group G. If $gxg^{-1} \in H$ for all $x \in S$ and $g \in G$, prove that H is normal in G.

1.25 Prove that two elements which are conjugate to each other have the same order. (A direct proof is not difficult. But the result can also be derived slickly from Exercise (1.14) above.)

1.26 The **centre** $Z(G)$ of a group G is defined as the set of all elements of G which commute with every element of G, that is, $Z(G) = \{z \in G : gz = zg\}$ for all $g \in G$. Prove that

   (i) $Z(G)$ is a normal subgroup of G. In fact every subgroup of G contained in Z is normal in G.

   (ii) G is abelian iff $Z(G) = G$. (Thus the centre of a group provides a measure of 'how abelian' the group is.)

   (iii) An element $z \in G$ is in the centre iff the conjugacy class of z in G consists only of z.

   (iv)   The centre of S_3 is $\{e\}$ and that of Q (the quaternion group) is $\{1, -1\}$.

1.27 Another measure of how abelian a group G is provided by the so-called commutator subgroup $C(G)$ of G defined as follows. For two elements x, y in G, their **commutator**, denoted by $[x, y]$, is defined as the element $xyx^{-1}y^{-1}$. The subgroup generated by all such commutators is called the **commutator subgroup** of G. Denote it by $C(G)$. Prove that:

   (i) x and y commute with each other iff $[x, y] = e$.

   (ii) G is abelian iff $C(G) = \{e\}$.

   (iii) $C(G)$ is a normal subgroup of G. (Hint : Use Exercise (1.24))

   (iv) $C(Q) = \{1, -1\}$.

   (v) $C(D_n)$ is a subgroup of order n.

1.28 Let G and H be groups. The cartesian product $G \times H$ is made a group under co-ordinatewise multiplication. The purpose of this exercise is to relate the properties of $G \times H$ to the corresponding properties of G and H. Prove that :

   (i) $G \times H$ is abelian iff both G, H are abelian.

   (ii)   If A, B are subgroups of G, H respectively then $A \times B$ is a subgroup of $G \times H$, which is normal iff A is normal in G and B is normal in H.

   (iii) If $G \times H$ is cyclic, so are G and H.

    (iv) Even if both G and H are cyclic, $G \times H$ need not be cyclic.

    (v)   If S, T are subsets of G, H respectively and they generate subgroups A, B then $S \times \{e\} \cup \{e\} \times T$ generates the subgroup $A \times B$.

1.29   Let G be a group, S any set and G^S the set of all functions from S to G. Under pointwise multiplication, G^S is a group. Study which properties of G pass over to G^S.

1.30   Let X be a set and $S(X)$ the permutation group of X. For a subset A of X, let $F_A = \{f \in S(X) : f(x) = x \text{ for all } x \in A\}$ and let $G_A = \{f \in S(X) : f(A) = A\}$. In other words, elements of F_A leave every element of A fixed individually while those of G_A leave the set A fixed as a whole. Prove that both F_A and G_A are subgroups of $S(X)$ and $F_A \subset G_A$. Prove that neither F_A nor G_A is normal in $S(X)$ unless either the set A is ϕ or $|X - A| \leqslant 1$.

1.31   Give an example of an infinite group in which every element is of finite order.

1.32   Prove that a group cannot be expressed as a union of two of its proper subgroups. Give an example of a group which can be expressed as the union of three proper subgroups.

Notes and Guide to Literature

The literature on groups is so vast and diverse that it is hardly possible to cite even typical references. Standards books on groups are Burnside [1], Zassenhaus [1], or Kurosh [1]. Groups are often studied with some other structure on them compatible with the group structure. The most important among these are the so-called topological groups. A classic reference on them is Pontryagin [1]. Groups, from the point of applications to physics, are studied in Hamermesh [1] or G.G. Hall [1]. For applications to chemistry see Bishop, D.M. [1].

The algebraic structures of a semigroup and a monoid are considerably weaker than that of a group. Still, recently, they have been studied because of their applicability, especially to finite state machines. See, for example, Dornhoff and Hohn [1].

Exercise 1.3 shows that there is some redundancy in the definition of a group as we have given. Some authors do adopt the definition given by Exercise 1.3.

For a generalisation of Exercise 1.5, see Problem No. E 2411 in the American Mathematical Monthly, Vol. 81, p. 410 (1974). For a group theoretic solution to the Rubik's cube puzzle, see Larsen [1].

2. Cosets of Subgroups

Every subgroup of a group induces a certain equivalence relation on the underlying set of that group. The equivalence classes are called cosets and they turn out to be nothing but the various translates of the subgroup. They all have the same cardinality. If the subgroup is normal, these cosets themselves form a group whose properties are obviously related to those of the original group and the normal subgroup. In this section we study this coset decomposition and derive some interesting consequences.

Before we give the general definition of the equivalence relation induced by a subgroup, let us take a couple of examples. Let us first consider $\mathbb{Z}$, the group of integers under the usual addition. For every positive integer m, $m\mathbb{Z}$ is a subgroup of $\mathbb{Z}$ as we saw in the last section $im\mathbb{Z}$ consists of all multiples of m, that is,

$$m\mathbb{Z} = \{..., -3m, -2m, -m, 0, m, 2m, ...\}$$

Consider the translates of $m\mathbb{Z}$ by various integers. For fixation of ideas take $m = 10$, Then $10\mathbb{Z} + 1$ is the set

$$\{..., -29, -19, -9, 1, 11, 21, ...\}.$$

Similarly, $10\mathbb{Z} + (-44)$ is the set

$$\{..., -54, -44, -34, -24, -14, -4, 6, 16, 26, ...\}$$

which also equals the translate $10\mathbb{Z} + 6$ or the translate $10\mathbb{Z} + 76$. In general we see that the translates $m\mathbb{Z} + x$ and $m\mathbb{Z} + y$ are equal iff $x - y$ is a multiple of $m \cdot$ It follows that there are exactly m distinct translates of $m\mathbb{Z}$, namely, $m\mathbb{Z}$ itself and

$$m\mathbb{Z} + 1, m\mathbb{Z} + 2, ..., m\mathbb{Z} + (m - 1).$$

Moreover, these are precisely the equivalence classes under the equivalence relation of congruency modulo m.

The second example is geometric. Let $\mathbb{R}^2$ be the cartesian plane

$$\{(x, y): x \in \mathbb{R}, y \in \mathbb{R}\}.$$

Under coordinatewise addition, $\mathbb{R}^2$ is an abelian group. Let L be a straight line passing through the origin, say $L = \{(x, y): y = 2x\}$. Then L is a subgroup of $\mathbb{R}^2$. (See Figure 5.4). L consists of points of the form $(t, 2t)$ for $t \in \mathbb{R}$. If (x_0, y_0) is any point of the plane then $L + (x_0, y_0)$ is the set

$$\{(t + x_0, 2t + y_0): t \in \mathbb{R}\}.$$

Eliminating t, this is precisely the set

$$\{(x, y) \in \mathbb{R}^2: y - y_0 = 2x - 2x_0\}.$$

Geometrically, this represents the straight line through (x_0, y_0) parallel to L. The whole plane is decomposed into the family of all lines parallel to L (including L itself). Through every point of $\mathbb{R}^2$ there passes one and only one such line. Two points say P and Q are on the same line if and only if

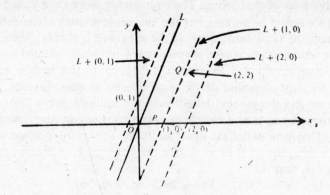

Figure 5.4:   Cosets of a Subgroup

the difference $Q - P$ (which geometrically represents the vector joining P to Q) is in the subgroup L.

We now want to extend the common feature of these two examples to the case of a subgroup H of an arbitrary group G. In the two examples above, the groups were abelian and so it did not matter whether we took the left translates or the right translates. However, in a non-abelian group we have to be careful. We develop the theory for left translates. An entirely analogous theory holds for right translates and is left as an exercise. First we give a name for the translates.

2.1 Definition: Let H be a subgroup of a group G. Then the left translate of H by an element x of G (that is, the set $xH = \{xh : h \in H\}$) is called the **left coset** of H by x.

We already saw two examples of this concept above. As another exmaple (which we shall visit frequently in this section), let G be the group of symmetries of a regular pentagon discussed in the last section. G consists of 5 rotations $r_1,..., r_5$ and 5 flips $f_1,..., f_5$. The rotation r_5 is the identity element $e \cdot$ For every $i = 1, .., 5$, $\{e, f_i\}$ is a subgroup of G since $f_i \circ f_i = e \cdot$ Let H be one of these subgroups, say, $H = \{e, f_1\}$. Let us determine the left coset $r_1 H$. By definition it is the set $\{r_1 \circ e, r_1 \circ f_1\} \cdot r_1 \circ e$ is of course r_1. A direct computation shows that $r_1 \circ f_1$ is f_4. So the left coset $r_1 \circ H$ is the set $\{r_1, f_4\}$. Note that this is also the left coset $f_4 H$ because $f_4 \circ e = f_4$ and $f_4 \circ f_1 = r_1$. Thus, two distinct elements of G can determine the same left coset of H. We leave it to the reader to show that there are four other left cosets (H itself being one of them since $eH = H$).

In the next proposition we characterise the left cosets of a subgroup as the equivalence classes of cetain equivalence relations on G.

2.2 Proposition: Let H be a subgroup of a group G. Define a binary relation R on G by xRy iff $x^{-1} y \in H$ for x, $y \in G$. Then R is an equivalence relation on G and its equivalence classes are precisely the left cosets of H in G.

Proof: Reflexivity of R is trivial since for every $x \in G$, $x^{-1}x = e$ and every subgroup contains the identity element. For symmetry we note that if $x^{-1}y \in H$ then $(x^{-1}y)^{-1} \in H$ (since every subgroup is closed under inversion). But $(x^{-1}y)^{-1} = y^{-1}(x^{-1})^{-1} = y^{-1}x$. So xRy implies yRx. Similarly transivity of R follows from the fact that H is closed under multiplication $(x^{-1}y \cdot y^{-1}z = x^{-1}z$ for all $x, y, z \in G)$. So R is an equivalence relation. It remains to show that the equivalence classes are precisely the left cosets of H in G. Let $x \in G$ and let S be the equivalence class under R containing x. Then, by definition, $S = \{y \in G : x^{-1}y \in H\}$ and we have to show that $xH = S$. A typical element of xH is of the form xh for some $h \in H$. Then $x^{-1}(xh) = (x^{-1}x)h = h \in H$ and so $xh \in S$. Hence $xH \subset S$. Conversely if $y \in S$ then $x^{-1}y = h$ for some $h \in H$. But then $y = x(x^{-1}y) = xh$ showing that $y \in xX$. Hence $xH = S$ as was to be shown. ∎

Because of this proposition, the results about equivalence relations become applicable to left cosets. We have, in particular,

2.3 Corollary: Two left cosets are either identical or mutually disjoint. The group is the disjoint union of all the distinct left cosets.

Proof: This follows by applying Proposition 3.2.6 to the equivalence relation above. ∎

As remarked earlier, an entirely analogous propositition holds for right cosets of a subgroup H in a subgroup G. Two right cosets Hx and Hy are equal if and only if $xy^{-1} \in H$. It should be noted that in general the left cosets themselves are different from the right cosets (although for an abelian group they are obviously equal). Two elements which are in the same left coset need not necessarily be in same right coset. For example, we considered above the group G of all symmetries of a regular pentagon and let H be the subgroup $\{e, f_1\}$. The left coset r_1H is the set $\{r_1, f_4\}$. But the right coset Hr_1 comes out to be the set $\{r_1, f_3\}$. Although r_1 and f_4 are in the same left cosets, they are in different right cosets. (The right coset Hf_4 is the set $\{r_4, f_4\}$.) The precise relationship between the left cosets and the right cosets is given by the following proposition.

2.4 Proposition: Let H be a subgroup of a group G. Then two elements $x, y \in G$ are in the same left coset of H in G iff their inverses x^{-1}, y^{-1} are in the same right coset of H. Consequently every right coset of H is obtained by taking the inverses of all elements of some left coset of H and vice versa.

Proof: Let $y, x \in G$. Then $xH = yH$ iff $x^{-1}y \in H$. The right cosets Hx^{-1}, Hy^{-1} are equal iff $x^{-1}(y^{-1})^{-1} \in H$ that is, iff $x^{-1}y \in H$. So $xH = yH$ iff $Hx^{-1} = Hy^{-1}$. Similarly $Hx = Hy$ iff $x^{-1}H = y^{-1}H$. Now let T be a right coset of H, say $T = Hx$. Let S be the left coset $x^{-1}H$. We claim T consists of precisely the inverses of elements of S. Let $y \in T$. Then $Hx = Hy$. So $y^{-1}H = x^{-1}H$ by what we proved earlier. But then $y^{-1} \in S$. So y is the inverse of some element (namely y^{-1}) of S. Conversely if $z \in S$ then $x^{-1}H = zH$, giving $Hx = Hz^{-1}$ and hence $z^{-1} \in T$. Thus the right coset T consists of precisely the inverses of the elements of the left coset S. Similarly every left coset is obtained by inverting all the elements of a right coset. ∎

As an illustration, in the example above $r_4 = r_1^{-1}$. The right coset Hr_4 therefore, should consist of the inverses of the elements of the left coset r_1H. This is indeed so, because $r_1H = \{r_1, f_4\}$, $Hr_4 = \{r_4, f_4\}$, $r_4 = r_1^{-1}$ and $f_4 = f_4^{-1}$.

As a consequence of the preceding proposition, we get,

2.5 Corollary: The number of distinct left cosets of a subgroup equals the number of its distinct right costs.

Proof: Let $\mathcal{L}$ be the set of all distinct left cosets of a subgroup H of a group G and let $\mathcal{R}$ be the set of all distinct right cosets of H in G. Define a function $g: \mathcal{L} \to \mathcal{R}$ by $g(xH) = Hx^{-1}$ for all $x \in G$. By the last proposition, g is well-defined because $xH = yH$ implies $Hx^{-1} = Hy^{-1}$. Conversely $Hx^{-1} = Hy^{-1}$ implies $xH = yH$. So g is one-to-one. Finally, given any right coset $Hy \in \mathcal{R}$, $Hy = g(y^{-1}H)$. So g is also onto. Therefore g is a bijection and hence $\mathcal{L}$ and $\mathcal{R}$ have the same cardinality. ∎

2.6 Definition: The number of distinct left (or right) cosets of a subgroup H in a group G is called the **index** of H in G and is generally denoted by $(G : H)$ (or by $[G : H]$).

For example, for every positive integer m, $(\mathbb{Z}, m\mathbb{Z}) = m$. If G is the group of symmetries of a regular pentagon and H is the subgroup considered above then $(G : H) = 5$. If $G = \mathbb{R}^2$ and H is the line L in Figure 5.4, then $(G : H)$ is infinite. In case G is a finite group, there is a simple formula for the index of a subgroup which we shall prove a little later.

We just remarked that the left coset xH is in general not equal to the right coset Hx. Of course this will be the case if the group is abelian. But this is too strong a condition. It will suffice if, instead of the whole group G being abelian, elements of H commute with all elements of G, that is, H is contained in the centre of G (see Exercise (1.26)). Even this requirement turns out to be stronger than is really needed. In order that the sets xH and Hx should coincide, all we need is that for every $h_1 \in H$ there exists $h_2 \in H$, such that $xh_1 = h_2x$ and vice versa. This condition will be trivially met if x and h_1 commute with each other (for then we can set $h_2 = h_1$). But even if they do not commute, it will still be satisfied if the element xh_1x^{-1} is in H (for then we can set $h_2 = xh_1x^{-1}$). This suggests that normality of H might do the trick. This is a good guess. As a matter of fact, this provides an important characterisation of normality which we now prove.

2.7 Theorem: Let H be a subgroup of a group G. Then the following conditions are equivalent:

(i) H is normal in G.
(ii) For every $x \in G$, $xH = Hx$.
(iii) Every left coset of H in G equals some right coset of H in G and and vice versa. (That is, the left and the right cosets of H in G coincide.)

Proof: (i) $\Rightarrow$ (ii). Suppose H is normal in G and $x \in G$. Then for any $h \in H$, $xhx^{-1} \in H$, and hence

$$xh = xh \cdot (x^{-1}x) = (xhx^{-1}) \cdot x \in Hx.$$

Thus

$$xH \subset Hx.$$

Similarly for any $h \in H$,

$$hx = x(x^{-1}hx) = x[x^{-1}h(x^{-1})^{-1}] \in xH$$

by normality of H, showing $Hx \subset xH$. Hence $xH = Hx$.

(ii) $\Rightarrow$ (iii). This hardly requires any proof because (ii) is more specific than (iii).

(iii) $\Rightarrow$ (ii). Let $x \in G$. Then the left coset xH equals some right cosets Hy. We are not given, nor is it necessarily true that $x = y$. Still we claim $Hy = Hx$. This follows from the fact that $x = x \cdot e \in xH = Hy$. So x is in the right coset Hy. But x is also in the right coset Hx since $x = e \cdot x$. So $Hx \cap Hy \neq \phi$ and hence $Hx = Hy$. Since $Hy = xH$ it follows that $xH = Hx$.

(ii) $\Rightarrow$ (i). Let $x \in G$, $h \in H$. Then $xh \in xH$. But $xH = Hx$. So $xh \in Hx$, i.e., there exists $h_1 \in H$ such that $xh = h_1x$. But then $xhx^{-1} = h_1 \in H$, showing that H is normal in G. ∎

As a simple application, we get the following result which is often useful to show that a subgroup is normal.

2.8 Theorem: A subgroup of index 2 is always normal.

Proof: Let H be a subgroup of a group G. Suppose $(G: H) = 2$. This means there are two distinct left cosets of H in G and also that there are two distinct right cosets of H in G. Now, in any group, the subgroup itself constitutes one of the left cosets and also one of the right cosets because

$H = eH = He$. Since the two left cosets are mutually disjoint, it follows that the other left coset must be precisely the complement of H in G. The same holds for the other right coset. Thus we conclude that the left and the right cosets of H in G coincide. Therefore H is normal in G. ∎

For example, if G is the group of symmetries of a regular pentagon and N is the subgroup consisting of all five rotations then $(G:N) = 2$. This can be proved directly by showing that N has only two left cosets, one consisting of all 5 rotations (that is, N itself) and the other consisting of all 5 flips. However, this will come much more easily from the formula for the index which we shall prove later. We then see that N is normal in G. Similarly let G be the group of isometries of a figure in space and let H be the subgroup of orientation preserving isometries. We claim that H is a normal subgroup of G. If there is no orientation reversing isometry in G, then $H = G$ and so H is normal in G. Suppose there is some orientation reversing isometry θ. Then we claim that H and θH are the only two left cosets of H in G. Note that all elements of θH are orientation reversing because the composite of an orientation reversing and an orientation preserving isometry is orientation reversing. Conversely if ψ is any orientation reversing isometry then we write $\psi = \psi \, (\psi^{-1} \circ \theta)$. Note that $\phi^{-1} \circ \theta$ is orientation preserving, being the composite of two orientation reversing isometries. So $\psi^{-1} \circ \theta \in H$. Hence $\psi \in \theta H$. Thus all orientation preserving isometries constitute the set H while all orientation reversing isometries constitute the set θH. Since there can be no other isometries, H and θH exhaust the group G. So H has index 2 in G. By the theorem above H is normal in G. (This could also be done directly as in Exercise (1.20). But we want to illustrate Theorem (2.9).) Since the right and left cosets of a normal subgroup are equal, from now onwards we shall simply call them as cosets.

It is instructive to reformulate Theorem (2.8). In the last section we saw how a binary operation $*$ on a set X induces a binary operation (which we continue to denote by $*$) on the power set $P(X)$. Let us do this construction for the group operation $\cdot$ on a group G. If H is a subgroup of G, and $x \in G$, then by definition, the left and right cosets xH and Hx are respectively the elements $\{x\} \cdot H$ and $H \cdot \{x\}$ of $P(G)$. If G is abelian then the operation $\cdot$ on $P(G)$ is commutative. Then Theorem (2.8) says that the subgroup H is normal in G iff the element H of $P(G)$ commutes with all singleton subsets of G. Thus normality is a weaker form of commutativity.

Let us now see whether $P(G)$ is a group under this operation $\cdot$. Associativity of $\cdot$ follows from that of the group operation. The singleton set $\{e\}$ is also easily seen to be an identity for $\cdot$ So $(P(G),.)$ is a monoid. However, we are in trouble when we look for inverses. If A is a subset of G with at least two elements then A, as an element of $P(G)$ can never be invertible under $\cdot$. If B is any subset of G, then $A \cdot B$ is empty (when $B = \phi$) or else has at least two distinct elements (if $a_1, a_1 \in A$, $a_1 \neq a_2$ and $b \in B$

then $a_1b \neq a_2b$ by cancellation law). In either case $A \cdot B$ cannot equal the identity element $\{e\}$. If follows that the only invertible elements of $P(G)$ are the singleton sets.

It turns out that if H is a normal subgroup of G then the set of all cosets of H in G is closed under the operation $\cdot$ on $P(G)$. Moreover, under the induced binary operation, it forms a group. This construction is important and we examine it in detail.

Let H be a normal subgroup of a group G. Let G/H be the set of all distinct cosets of H in G. Element of G/H are of the form xH (which is the same as Hx by normality) for $x \in G$. Then G/H is a subset of $P(G)$. We first show that G/H is closed under the binary operation $\cdot$ on $P(G)$.

2.9 Proposition: The product of any two cosets of a normal subgroup is again a coset of it.

Proof: Let H be a normal subgroup of a group G. Let xH, yH be two cosets of H in G. Then $(xH) \cdot (yH)$ equals $(\{x\} \cdot H) \cdot (\{y\} \cdot H)$ by definition which further equals $\{x\} \cdot (H \cdot \{y\}) \cdot H$ by associativity. But since H is normal, $H \cdot \{y\}$ equals $\{y\} \cdot H$. So $(xH) \cdot (yH)$ equals $(xy) \cdot (H \cdot H)$. Now $H \cdot H \subset H$ (since H is closed under the group operation). Also $H = H \cdot \{e\} \subset H \cdot H$. Therefore, $H \cdot H = H$. This shows $(xH) \cdot (yH)$ equals $(xy)H$ which is again a coset of H. (A more direct argument would be to show that every element of the form $xh_1 yh_2$ for $h_1, h_2 \in H$ can be expressed as xyh_3 for some $h_3 \in H$. This would show that $(xH) \cdot (yH) \subset (xy) H$. Conversely, for any $h_3 \in H$, $xyh_3 = (x \cdot e) \cdot (yh_3)$ showing $(xy)H \subset (xH) \cdot (yH)$.) ∎

Note that normality of H was crucially used in the proof above. If we once again take G to be the group of symmetries of a regular pentagon and H be the subgroup $\{e, f_1\}$. Then the left coset r_1H is $\{r_1, f_3\}$. The product $(r_1H) \circ (r_1H)$ consists of four elements $r_1 \circ r_1, r_1 \circ f_3, f_3 \circ r_1$ and $f_3 \circ f_3$. Computing these composites we see that $(r_1H) \circ (r_1H)$ is the set $\{r_1, f_1, f_5, r_5\}$ which is not equal to any left coset of H in G. As a matter of fact we leave it to the reader to prove that if a subgroup H has the property that for all $x, y \in G$, $(xH) \cdot (yH)$ equals $(xy)H$ then H is normal in G. (This gives yet another characterisation of normal subgroups.)

Because of this proposition, we get a well-defined binary operation on the set of all cosets of a normal subgroup. This operation is called, quite appropriately, the **coset multiplication**. It is generally denoted by the same symbol as the original group operation. We now proceed to show that under this operation, the set of cosets is a group.

2.10 Theorem: Let H be a normal subgroup of a group G. Then under coset multiplication, the set of cosets, G/H, is a group.

Proof: Recall once again that the coset multiplication is given by

$(xH) \cdot (yH) = (xy) H$ for $x, y \in G$. Associativity of this operation follows from the associativity of the group operation G. The coset H serves as a two sided identity because for $x \in G, (xH) \cdot H = (xH) \cdot (eH) = (xe) H = xH$ and similarly $H \cdot (xH) = xH$. Finally for inverses, we merely note that for any $x \in G$, the coset $x^{-1}H$ is the inverse of xH because,

$$(xH) \cdot (x^{-1} H) = (xx^{-1}) H = eH = H$$

and similarly $(x^{-1}H) \cdot (xH) = H$. Therefore G/H is a group under coset multiplication. ∎

2.11 Definition: The group G/H constructed in the last theorem is called the **group of cosets** of H in G or the **quotient group** of G by H.

The first name requires little elaboration. Justification for the second name will come a little later. Let us study two examples of quotient groups. Let $\mathbb{Z}$ be the group of integers under the usual addition. For every positive integer $m, m\mathbb{Z}$ is a subgroup of $\mathbb{Z}$ of index m and the cosets are the congruence classes modulo m as we saw above. In the last section, we saw that the set of these residue classes, $\mathbb{Z}_m$, is a group under residue addition. If $[x]$ and $[y]$ are two residue classes then $[x] + [y]$ is the residue class $[x + y]$. But since the residue classes $[x]$, $[y]$ and $[x + y]$ are nothing but the cosets $m\mathbb{Z} + x, m\mathbb{Z} + y$ and $m\mathbb{Z} + (x + y)$ we see that the residue addition coincides with the coset addition. Therefore, in this case, the quotient group $\mathbb{Z}/m\mathbb{Z}$ is the familiar group $\mathbb{Z}_m$.

As another example, take the quaternion group Q. It has eight elements, $\pm 1, \pm i, \pm j$ and $\pm k$. The elements 1 and -1 form a subgroup H of Q. This subgroup is in fact the centre of Q (see Exercise (1.26) (iv)). So H is normal in Q. The cosets of H in Q are the sets $\{1, -1\}, \{i, -i\}, \{j, -j\}$ and $\{k, -k\}$. For brevity let us denote these by e, a, b and c respectively. Then we see that e (which also equals H) is the identity for the coset multiplication. We also see $a \cdot b = \{i, -i\} \cdot \{j, -j\} = \{k, -k\} = c$ and similarly $b \cdot a = c$. Also $a \cdot a = \{i, -i\} \cdot \{i, -i\} = \{1, -1\} = e$. Completing these computations we see that the quotient group Q/H is nothing but the Klein group. This justifies our comment in the last section that the Klein group is obtained from the quaternion group by ignoring the minus sign.

We studied one more example of coset decomposition where the group G was the cartesian plane $\mathbb{R}^2$ and the subgroup H was a line L through the origin. The identification of the quotient group G/H will be greatly simplified by the concept of a group homomorphism to be studied in the next section. There is, in fact, an intimate relationship between homomorphisms and quotient groups. So we shall visit them again in the next section. For the moment let us see what properties of a group pass over to its quotient groups. It is clear that if G is abelian, so is the quotient group G/H. Moreover, if G is cyclic so is G/H. To see this, suppose x is a generator for G. We claim that the coset xH is a generator for G/H. Let yH be any element of G/H, where $y \in G$. Then y equals some power, say x^n of x. But then $yH = x^n H$ which equals $(xH) \cdot (xH) \cdot - \cdot (xH)$ (n times) if n is a positive integer and $(x^{-1}H) \cdot (x^{-1}H) \cdot - \cdot (x^{-1}H)(- n$ times) if n is a negative integer. In the first case, $yH = (xH)^n$ and in the second case $yH = (x^{-1}H)^{-n}$ which again equals $(xH)^n$ (since $x^{-1}H$ and xH are inverses of each other in G/H). If $n = 0$, then $y = e$ and $yH = H = (xH)^o$. In all cases we have expressed yH as a power of xH. So G/H is generated by xH and hence is cyclic.

It is tempting to think that the converse is also true, at least with the additional hypothesis that H is cyclic. Such an expectation is intuitively justified because knowing G/H is like knowing G modulo or upto H. This knolwedge, combined with the knowledge of H, ought to give the full information about G. Unfortunately, this does not come out to be quite true. For example let Q be the quaternion group once again and let H be $\{1, - 1\}$. Then H is abelian. Also the quotient group G/H is the Klein group which is also abelian. But Q is not abelian. In the same example let $N = \{1, - 1, i, - i\}$. Then N is a subgroup of Q and it is cyclic, because N is generated by the element i. N has only two cosets in Q namely N itself and $jN = \{j, -j, k, - k\}$. So N is normal in Q by Theorem 2.9. The quotient group Q/N has only two elements and is generated by the coset jN. Thus we see that both N and G/N are cyclic. But Q is not only not cyclic, it is not even abelian.

The discussion in this section has so far been independent of the cardinality of the groups. From the point of view of discrete mathematics, finite groups are especially important. So we now consider what additional results hold for finite groups. We begin with the promised formula for the index of a subgroup of a finite group. The result is due to **Lagrange** and is one of the basic theorems about finite groups.

2.12 Theorem : For finite groups, the order of a subgroup divides the order of the group and the ratio equals the index of that subgroup. In symbols, if G is a finite group and H a subgroup of G then $\circ (G)/\circ (H) = (G : H)$.

Proof : Note that $\circ(G)$ and $\circ(H)$ are simply the cardinalities of the sets G, H respectively. Now consider the decomposition of G into the left cosets of H in G, as given by Corollary (2.3). Let r be the number of distinct left cosets. Then $r = (G : H)$. We claim that all left cosets have the same cardinality, namely $\circ(H)$. Let xH be a left coset. Define $f : H \to xH$ by $f(h) = xh$ for $h \in H$. By cancellation laws in a group, f is one-to-one. Also, by very definition of the set xH, f is onto. So f is a bijection, proving that $|xH| = |H| = \circ(H)$. The set G is now decomposed into r mutually disjoint subsets, each having cardinality $\circ(H)$. By Corollary (2.2.4), $\circ(G) = |G| = r \circ(H) = (G : H) \circ (H)$. In particular we see that $\circ(H)$ divides $\circ(G)$ since r is an integer. ∎

For finite groups, the index of a subgroup is sometimes *defined* as the ratio of the order of the group to that of the subgroup. Because of Lagrange's theorem, the index is then an integer. However, we have not adopted this definition because it is not applicable for infinite groups. The definition we have given is applicable for all groups, whether finite or not. It can happen that an infinite group may have a subgroup of finite index. For example the index of $m\mathbb{Z}$ in $\mathbb{Z}$ is finite, but it cannot be expressed as the ratio of $\circ(\mathbb{Z})$ and $\circ(m\mathbb{Z})$ because both are infinite.

Lagrange's theorem has a host of applications. To begin with, let G be the group of all symmetries of a regular pentagon and let N be the subgroup of the 5 rotations. Then $\circ(G) = 10$ and $\circ(N) = 5$. So $(G : N) = 10/5 = 2$. Hence N is of index 2 in G and consequently is normal in G by Theorem 2.9. This gives a truly effortless way of proving that N is normal in G.

It is interesting to note that in many applications of Lagrange's theorem, the index of the subgroup is not very important. All that matters is that the order of a subgroup is a divisor of the order of the group. Note that there is no restriction on the subgroup (such as it be normal). Choosing the subgroup appropriately, we get various applications. We begin with the,

2.13 Proposition : The order of every element divides the order of a group.

Proof : Let G be a finite group and $x \in G$. Then the order of x is, by definition, the order of (x), the subgroup of G generated by x. The result now follows immediately from the last theorem. ▉

As a consequence we get the following result :

2.14 Corollary: If G is a finite group of order n then for every $x \in G$, $x^n = e$, the identity element of G.

Proof : Let m be the order of x. Then by the last Proposition m divides n. So $n = mr$ for some integer r. Now $x^m = e$ by Theorem 1.10. Hence $x^n = x^{mr} = (x^m)^r$ (by Proposition (3.4.8)) $= e^r = e$. ▉

Applying this corollary to a particular group, we get the following interesting theorem due to **Fermat**.

2.15 Theorem : If p is a prime number then for every integer x, $x^p \equiv x$ (modulo p).

Proof : We consider two cases, x is divisible by p and x is not divisible by p. If x is divisible by p then so is x^p. Therefore both x^p and x are congruent to 0 modulo p and the result holds trivially. Alternatively we can factorise $x^p - x$ as $x(x^{p-1} - 1)$ and see directly that it is divisble by p (since p divides the first factor, x).

It is the other case that is more interesting. If x is not divisible by p then the residue class, $[x]$, of x modulo p is non-zero. So $[x] \in \mathbb{Z}_p - \{[0]\}$. Since p is a prime, by Proposition (1.3), $\mathbb{Z}_p - \{[0]\}$ is a group under mod p multiplication. The order of this group is obviously $p - 1$. So by the corollary above, $[x]^{p-1}$ equals the identity element of this group, which is simply the residue class $[1]$. Since $[x]^{p-1}$ equals $[x^{p-1}]$ we have $x^{p-1} \equiv 1$ (mod p). This means $x^{p-1} - 1$ is divisible by p. But then so is $x(x^{p-1} - 1)$ which equals $x^p - x$. This completes the proof. ∎

For example, if $p = 5$ and $x = 3$ then $x^5 = 243$ and we indeed see that it is congruent to 3 modulo 5 because $243 - 3 = 240 = 5 \cdot 48$. We remark that Fermat's theorem can also be proved directly by induction on x. (In case x is negative, we work with $- x$.) But as with the solution to the Tournaments Problem, the inductive proof is more like a verification. The group theoretic proof given above, on the other hand, really 'explains' why $x^{p-1} - 1$ is divisible by p when x is not a multiple of p.

Lagrange's theorem puts a restriction on the order of a subgroup of a fininte group. For example, if G is a group of order 60 then a subgroup of G must have for its order 1, 2, 3, 4, 5, 6, 10, 12, 15, 20, 30 or 60, because these are the only positive divisors of 60. But it does not say that G will necessarily contain subgroups of all these orders. In Section 4 we shall show that the group of all orientation preserving isometries of a regular tetrahedron has order 12 but contains no subgroup of order 6, even though 6 is a divisor of 12. Thus in general the converse of Lagrange's theorem is false. It does hold for abelian groups. However, to prove it will require considerable machinery from the theory of group actions (see the Epilogue.). For the time being, we prove it for a very special type of abelian groups, namely, cyclic groups.

2.16 Theorem: Let G be a cyclic group of order n and let m be a positive integer dividing n. Then G contains a unique subgroup of order m.

Proof: By Theorem 1.10, G consists of $e, x, x, x^2, ..., x^{n-1}$ where x is any generator of G. Now let $d = n/m$. Then d is a positive integer. Let $H = (x^d)$, the subgroup generated by x^d. Clearly

$$H = \{e, x^d, x^{2d}, ..., x^{(m-1)d}\},$$

because the next power of x^d is x^{md} which equals x^n ($= e$). Hence $\circ(H) = m$. So G has a subgroup of order m. To show it is unique, suppose K is another subgroup of G of order m. Let r be the smallest positive integer such that $x^r \in K$. We claim x^r generates K. Let $y \in K$. Then $y = x^s$ for some $0 < s \leq n - 1$. By the euclidean algorithm, write s as $ru + v$ where u, v are integers with $0 \leq v < r$. Then

$$x^v = x^{s-ru} = x^s (x^r)^{-u} = y (x^r)^{-u} \in K$$

since $y \in K$ and $x^r \in K$. So v must be 0, for otherwise v will be a positive integer less than r such that $x^v \in K$, contradicting the definition of r. Therefore, $y = x^{rn} = (x^r)^u$, showing that $y \in (x^r)$. Hence $K = (x^r)$. Now we claim that $r = d$, which would of course prove that $H = K$ since $H = (x^d)$ and $K = (x^r)$. Since K is a cyclic group of order m generated by x^r, $(x^r)^m = e$, that is $x^{rm} = e$. By theorem 1.10, this means that rm is a multiple of n, set $n = rmp$. But $n = md$. So we get $d = rp$. In particular $d \geqslant r$. If $d > r$, then $n = md > mr$ and the $m + 1$ distinct elements $e, x^r, x^{2r}, \ldots, x^{mr}$ would all be in K, contradicting that the order of K is m. (Alternatively, we can simply interchange the roles of r and d and prove $r \geqslant d$.) Thus $r = d$. As noted before, this means $H = K$. So there is one and only one subgroup of order m in G. ∎

As the last application of Lagrange's theorem in this section, we apply it to a subgroup whose significance may not be obvious at first sight. However, later we shall relate it to the conjugacy classes defined in the last section and as a consequence will get some interesting results about groups whose powers are orders of a prime.

2.17 Definition: Let G be a group. For an element $x \in G$, the set $\{g \in G : gx = xg\}$ is called the **normaliser** of x in G and is denoted by N_x or by $N(x)$.

In other words, the normaliser of an element is the set of all elements which commute with it. Obviously $N_x = G$ if and only if x is in the centre of G. In an abelian group, $N_x = G$ for all elements x in G. In a non-abelian group, N_x is the measure of the extent to which x commutes with other elements of G. Note that N_x always contains the subgroup (x) generated by x, because all powers of x commute with x. In Q, we see that $N_i = \{1, -1, i, -i\}$ because none of the remaining elements commute with i, but $N_{-1} = Q$ because -1 is in the centre of Q.

Note that for any two elements x, y of a group G, $x \in N_y$ if and only if $y \in N_x$. This simple fact is often useful.

2.18 Proposition: Let G be a group and $x \in G$. Then N_x is a subgroup of G. Further $(x) \subset N_x$ and (x) is a normal subgroup of N_x (although not necessarily of G).

Proof: Let $g, h \in N_x$. Then $gx = xg$ and $hx = xh$. So $ghx = gxh = xgh$ showing that $gh \in N_x$. Hence N_x is closed under multiplication. Similarly if $g \in N_x$, then $g^{-1}x = g^{-1}xgg^{-1} = g^{-1}gxg^{-1} = xg^{-1}$ showing $g^{-1} \in N_x$. Thus N_x is also closed under inversion, proving that N_x is a subgroup of G. We already noted $(x) \subset N_x$. Hence (x) is a subgroup of N_x. To show that (x) is normal in N_x, let $y \in (x)$ and $g \in N_x$. Then $y = x^r$ for some integer r. Now $g \in N_x$ implies $x \in N_g$. But N_g is subgroup by what we proved just now. So $x^r \in N_g$ that is, $y \in N_g$. This means $gy = yg$ or $gyg^{-1} = y \in (x)$. Thus (x) is normal in N_x. ∎

The preceding proposition justifies the name 'normaliser'. Let us now see how big the normaliser of an element x is. The condition $gx = xg$ is equivalent to $gxg^{-1} = x$. This means that the conjugate of x by g equals x (see Definition 1.14). If this happens for a large number of elements g then N_x would be large and we would expect x to have very few conjugates. (In the extreme case where x commutes with every element of G, $N_x = G$ and x has only one conjugate, namely x itself.) Thus we are led to believe that the larger the normaliser is, the smaller would be the number of conjugates. This guess turns out to be quite correct as seen from the following theorem.

2.19 **Theorem:** Let x be an element of a finite group G. Then the number of (distinct) conjugates of x is the index of the normaliser N_x in G, and hence equals $\circ (G)/ \circ (N_x)$.

Proof: Let $\mathcal{L}$ be the set of all left cosets of N_x in G and let S be the set of all elements in G which are conjugate to x. (In order words, S is the conjugacy class of x.) We have to prove that the sets $\mathcal{L}$ and S have the same cardinality. We do so by establishing a bijection between them. Define $f: \mathcal{L} \to S$ by $f(gN_x) = gxg^{-1}$. We must first verify that f is well-defined because it has been defined in terms of a representative of a coset. Suppose the same coset gN_x is represented as hN_x where $h \in G$. Then by Proposition 2.2, $g^{-1}h \in N_x$, which means $g^{-1}hx = xg^{-1}h$. This gives $hx = gxg^{-1}h$ and finally, $gxg^{-1} = hxh^{-1}$. Thus $f(gN_x) = f(hN_x)$ and so the function f is well-defined. The same argument can be read backwards to show that if $f(gN_x) = f(hN_x)$ then $gN_x = hN_x$. This means f is one-to-one. Also f is obviously onto because every element of S is of the form gxg^{-1} for some $g \in G$ and hence equals $f(gN_x)$ for some $g \in G$. Therefore we have that f is a bijection and hence $|\mathcal{L}| = |S|$. But $|\mathcal{L}|$ is, by very definition, the index of N_x in G. By Lagrange's theorem $|\mathcal{L}|$ equals $\circ (G)/ \circ (N_x)$ and the proof is complete. ∎

Note that in the proof above, finiteness of the group G was used only in the last step. Thus, even for an infinite group, if an element has only finitely many conjugates then its normaliser will be of finite index.

As a corollary, we have

2.20 **Corollary:** Let G be a finite group. If two elements $x, y \in G$ are conjugates of each other then $\circ (N_x) = \circ (N_y)$.

Proof: Let S be the conjugacy class containing x and y. By the theorem above,

$$\circ (N_x) = \frac{\circ (G)}{|S|} = \circ (N_y). \quad ∎$$

What is really interesting is that the cardinality of each conjugacy class of a finite group G is a divisor of the order of G, even though these conjugacy classes are generally not subgroups of G. When $\circ(G)$ is such that its divisors are only of a certain type, this information is very valuable. To illustrate its power, we prove the following result.

2.21 Theorem: Every group whose order is the power of a prime has a non-trivial centre.

Proof: Let G be a finite group with $\circ(G) = p^m$ for some prime number p and some positive integer m. (If $m = 0$, the group is trivial and so is the result.) Let Z be the centre of G. Then Z is a subgroup of G (cf. Exercise (1.26)). We have to show $\circ(Z) > 1$. Let $r = \circ(Z)$.

Now consider the class decomposition of the set G, that is, the decomposition of G into mutually disjoint conjugacy classes, say, S_1, $S_2,...,S_k$. By what we said above, for every i, $|S_i|$ is a divisors of $\circ(G)$ which equals p^m. But the only possible divisors of p^m are powers of p (with $p^0 = 1$ included as a possible divisor). Note also that a conjugacy class S_i is a singleton set, say $\{x\}$ if and only if $x \in Z$ (see Exercise (1.26) again). Therefore, there are exactly r singleton conjugacy classes. Without loss of generality we may suppose that they are $S_1, S_2,..., S_r$ where $r \leqslant k$. Now, for $i > r$, $|S_i| = p^{m_i}$ for some $m_i > 0$. Hence p divides $|S_i|$. Now, by Proposition (2.2.3) we have

$$p^m = |G|$$

$$= \sum_{i=1}^{k} |S_i|$$

$$= \sum_{i=1}^{r} |S_i| + \sum_{>r} |S_i|$$

$$= r + \sum_{i>r} |S_i|$$

Hence $r = p^m - \sum_{i>r} |S_i|$. Now $p \mid p^m$ (since $m > 0$). Also $p \mid |S_i|$ for all $i > r$, as we say just now. So p divides r. Since r is positive, r is at least equal to p. This shows $r > 1$ as was to be proved. ∎

As an application of this result, we want to show that every group of order p^2 where p is a prime is abelian. First we need subsidiary results. The first is interesting by itself.

2.22 Proposition: Every group of prime order is cyclic.

Proof: Let G be a group of order p where p is a prime. Let x be any element of G other than the identity. Let $H = (x)$, the subgroup generated by x. Then $\circ(H)$ divides $\circ(G)$. But the only possible divisors of p are 1 and

p. Since $e, x \in H$, $\circ(H) > 1$. So $\circ(H) = p$. In other words $H = G$. Thus $G = (x)$, showing that G is cyclic. ∎

2.24 Proposition: If G is a group with centre Z and the quotient group G/Z is cyclic, then G is abelian.

Proof: The quotient group, as we recall, consists of the cosets of Z in G and the operation is that of coset multiplication. (By Exercise (1.26), Z is normal in G and so the quotient group G/Z is defined.) We are given that G/Z is cyclic. Let a coset xZ be a generator for G/Z, for some $x \in G$. Then for every $y \in G$, $yZ = (xZ)^r = x^r Z$ for some integer r. But this means that $y = x^r z$ for some $z \in Z$. Now let a, b be any two elements of G. By what we showed just now, there exist integers r, s and elements z, w in Z such that $a = x^r z$ and $b = x^s w$. Now, $ab = x^r z x^s w = x^r x^s z w$ because $z x^s = x^s z$ (since z is in the centre). So $ab = x^{r+s} z w$. But $ba = x^s w x^r z = x^s x^r w z = x^{r+s} z w$ (since w also commutes with all element of G). So $ab = ba$. That is, G is abelian. ∎

There is a sort of vacuousity about the way the last result is formulated. Its conclusion says that G is abelian which means $Z = G$ and the quotient group G/Z is trivial. Thus the proposition in effect says that if Z is the centre of a group and G/Z is cyclic then G/Z must be trivial. In other words, the conclusion shows that the hypothesis can hold only vacuously except in one case. It would be better to word the result as 'The quotient group G/Z cannot be cyclic except in the trivial case when $Z = G$.' But the formulation given above is fairly standard.

It should not, however, be supposed that the preceding proposition is useless. It so happens sometimes that we do not know beforehand that G is abelian. Still we may be in a position to prove that G/Z is cyclic. Then it follows from the last proposition that G is abelian. This is exactly what we do in the proof of the following theorem.

2.25 Theorem: Every group of order p^2 where p is a prime, is abelian.

Proof: Let G be a group of order p^2. Let Z be the centre of G. Then $\circ(Z) = 1, p$ or p^2, since these are the only divisors of p^2. The first (possibility is ruled out by Theorem (2.22). In the second case, $\circ(G/Z) = \circ(G)/\circ(Z) = p^2/p = p$. Hence by Proposition (2.23), G/Z is cyclic. So by the last proposition, G is abelian. If $\circ(Z) = p^2$, then of course $Z = G$ and so G is abelian. In any case the assertion holds. ∎

Thus we see that all groups of orders $4, 9, 25,\ldots$ are abelian. It is not true that every group of order p^3 is necessarily abelian. We already had two non-abelian groups of order $8(= 2^3)$, the quaternion group Q and the dihedral group D_4, that is, the group of isometries of a square. Of course, by Theorem (2.22) every group of order p^3 has a non-trivial centre and this fact helps in determining the structure of such a group. The converse of Lagrange's theorem can be proved for groups whose orders are prime powers. A proof can be based upon induction. A critical step is the following result.

2.26 Proposition: Every group of order p^m where p is a prime and m is a positive integer, contains a normal subgroup of order p.

Proof: Let G be a group with $\circ(G) = p^m$. Let Z be the centre of G. Then $\circ(Z) = p^r$ for some r with $1 \leqslant r \leqslant m$. Note that every subgroup of Z is normal in G by Exercise (1.26) (i). So it suffices to show that Z contains some subgroup of order p. Let x be any element of Z other than the identity. Let $H = (x)$. Then $\circ(H) = p^k$ for some k with $1 \leqslant k \leqslant r$. Now H is a cyclic group and p is a divisor of $\circ(H)$. So by Theorem (2.17) H contains a subgroup, say N, of order p. N is also a subgroup of Z and, as noted earlier, this completes the proof. ▮

Having proved that G has a normal subgroup N of order p, we now consider the quotient group G/N, whose order is $\circ(G)/\circ(N) = p^{m-1}$. So by induction, we can assume that G/N contains a subgroup of any order which is a divisor of p^{m-1}. If we could relate subgroups of G/N to those of G, we would get the existence of a desired subgroup of G. Such a relationship indeed exists and will be studied in the next section, where we shall prove the converse of Lagrange's theorem for groups whose orders are prime powers.

Before closing this section we remark that the concepts, 'conjugacy class' and 'normaliser' are particular cases of what are called 'orbits' and 'isotropy subgroup' respectively, of a group action. When we shall study group actions*, we shall again visit them, to motivate the general concepts, if for nothing else. These more general concepts will also enable us to prove deeper results about the structure of finite groups, one of which will be the converse of Lagrange's theorem for abelian groups.

Exercises

2.1   Let G be the group $\mathbb{R}^3$, under coordinatewise addition. Let H be the set $\{(x, y, z) \in \mathbb{R}^3 : 2x + 3y - 4z = 0\}$. Prove that H is a subgroup of G. Find the various cosets of H in G.

2.2   With G as above let $K = \{(x, y, z) \in \mathbb{R}^3 : 2x + 3y - 4z = 0, x - y + 2z = 0\}$. Prove that K is also a subgroup of G and find its cosets.

2.3   Prove that the cosets of the subgroup S^1 of the group of non-zero complex numbers are circles centred at the origin. Describe the

---

*See the Epilogue.

coset multiplication geometrically. (Hint : Use polar form of complex numbers.)

2.4 Let H, K be subgroups of a group G. Prove that the intersection of a left coset of H and left coset of K is either empty or a left coset of $H \cap K$.

2.5 Prove that the intersection of two subgroups of finite indices is a subgroup of finite index.

2.6 Let G be a group. The function $f : G \to G$ defined by $f(x) = x^{-1}$ is called inversion. Prove that f is a bijection. Give an alternate formulation of the proof of Proposition (2.4) by applying Exercise (3.2.21) to f.

2.7 Let G be a group and H, L subgroups of G with $L \subset H$. If L is of finite index in H and H is of finite index in G then prove that L is of finite index in G, and moreover, $(G : L) = (G : H)(H : L)$. Use this result to give an alternate solution to Exercise (2.5).

2.8 Show by an example that a subgroup of index 3 need not be normal.

2.9 Let $\mathcal{L}$ be the set of all left cosets of a subgroup H of a group G. Suppose we attempt to define a binary operation on $\mathcal{L}$ by $(xH) \cdot (yH) = (xy)H$ for $x, y \in G$. Prove that this is well-defined (that is, independent of the left coset representatives x and y) if and only if H is a normal subgroup of G.

2.10 Let $(G, \cdot)$ be a group with H a normal subgroup. When we defined the operation $\odot$ on the entire power set $P(G)$, we remarked that the singleton set $\{e\}$ is the identity for this operation. When we take the restriction of this operation to the set of all cosets of H in G (and call it coset multiplication), we find that H is the identity for it. Why does this not contradict the uniqueness of identities, established in Proposition (3.4.5)?

2.11 Prove Theorem (2.16) by induction on x.

2.12 If p is a prime and m is any positive integer, prove that for all integers x, $x^{p^m} \equiv x \pmod{p}$.

2.13 Let G be a group and $C(G)$ its commutator subgroup (Exercise (1.27)). Prove that the quotient group $G/C(G)$ is abelian. Prove further that if N is any normal subgroup of G such that G/N is abelian then $C(G) \subset N$. (Thus $C(G)$ is the smallest normal subgroup whose quotient group in abelian. The quotient group $G/C(G)$ is sometimes called the *abelianised* group G.)

2.14 Let H be a subgroup of index 2 in a group G. Suppose every element of $G - H$ is of order 2. Prove that H is abelian. (Hint : For $h \in H$ and $x \in G - H$, prove that $hxh = x$). Show by an example that G need not be abelian.

2.15 Suppose two elements x and y of a group G are conjugates of each other. Prove that their normalisers N_x and N_y are also conjugates of each other. (This gives an alternate proof of Corollary (2.21).)

2.16   Prove that the centre of a non-abelian group of order p^3 where p is a prime is of order p.

2.17   Let H, K be subgroups of a finite group G. Prove that the cardinality of the set HK equal $\circ(H) \circ (K)/\circ(H \cap K)$. (Hint: Consider the function $f: H \times K \to HK$ defined by $f(x, y) = xy$ for $x \in H$, $y \in K$. Prove that every element of HK has exactly $\circ(H \cap K)$ pre-images and apply Proposition (2.2.8).)

2.18   Suppose G is a group of order pq where p, q are primes with $p > q$ and that G contains a subgroup H of order p. Prove H is normal in G. (Hint : Apply the last exercise together with Corollary (1.13).).

2.19   Prove that the centre of a group of order pq where p, q are primes is either trivial or else the whole group.

2.20   Suppose H and K are normal subgroups of a finite group G. Suppose $\circ(H)$ and $\circ(K)$ are relatively prime, that is, have no common divisor except 1. Prove that every element of H commutes with every element of K. (Hint: Show that their commutator is in $H \cap K$ which must be only $\{e\}\cdot$).

2.21   Suppose G is a finite group and S is a subset of G with $|S| > \frac{3}{4} o(G)$. Prove that for any four elements x_1, x_2, x_3, $x_4 \in G$, there exists some $g \in G$ such that $x_i g \in S$ for $i = 1, 2, 3, 4$. (Hint: Apply the principle of inclusion and exclusion to the four left translates of S by x_1^{-1}, x_2^{-1}, x_3^{-1}, x_4^{-1}).

2.22   Suppose G is a finite group in which the number of solutions to the equation $x^2 = e$ exceeds $(3/4) \circ (G)$. Prove that every element of G is a solution of this equation and hence that G is abelian. (Hint: Apply the last exercise to the set of all solutions of this equation.)

2.23   Show by an example that the last result need not hold if the number of solutions to $x^2 = e$ equals $(3/4) \circ (G)$ (Hint: Consider the dihedral group of a suitable order).

2.24   If G is a group of order p^m, where p is a prime and m a positive integer and N is a normal subgroup of G with $\circ(N) > 1$, prove that $\circ(N \cap Z) > 1$. (Hint: Argue as in the proof of Theorem (2.22), after noting that all conjugates of elements of N are also in N and thereby obtaining a decomposition of N into conjugacy classes.)

2.25   Suppose G is a finite abelian group with n elements x_1, x_2, ..., x_n. Suppose G has exactly one element, say y, of order 2. Prove that the product $x_1 x_2 ... x_n$ equals y. (Hint : cf. the hint to Exercise (1.16).).

*2.26   Using the last exercise prove that for every prime p, $(p - 1)! + 1$ is divisible by p. (This is called **Wilson's theorem**.)

2·27   Suppose G is a group which has no nontrivial, proper subgroup (i.e., a subgroup other than $\{e\}$ and G). Prove that G is a finite, cyclic group of prime order.

2.28 Suppose G is a group of order $2n$ with trivial centre and an element x of order n. Prove that x cannot commute with any element of G except its own powers. (Hint: Consider $\circ(N_x)$.)

2.29 Analogous to the normaliser of an element, we can define the normaliser of a subgroup which is sometimes helpful in checking normality. Let G be a group and H is a subgroup. Let
$$N(H) = \{g \in G : gHg^{-1} = H\}.$$
$N(H)$ is called the **normaliser** of H in G. Prove that:

(i) $N(H)$ is a subgroup of G and contains H.
(ii) $N(H)$ is the largest subgroup of G containing H as a normal subgroup.
(iii) H is normal in G if and only if $N(H) = G$.
(iv) the number of distinct conjugates of H in G equals the index of $N(H)$ in G. (Hint: The argument resembles the proof of Theorem (2.20).)

Notes and Guide to Literature

The material in this section is basic for any study of structure of groups, especially finite groups. The theorems of Fermat and Wilson are illustrations of applications of group theory to number theory. Apparently, Wilson's theorem was already proved by Leibnitz.

Fermat (1601–1667), although a lawyer by profession, made significant contributions to number theory. But he is most famous for something which he claimed he could do but never actually published. The equation $x^2 + y^2 = z^2$ has many solutions for positive integers x, y, z (for example, $x = 3, y = 4, z = 5$). Fermat claimed that for no integer $n > 2$, the equation $x^n + y^n = z^n$ has solutions for positive integers x, y, z but did not write the proof down. Although this has been proved for many values of n, whether it holds for all $n > 2$ is still not known. Equations of this type are called **diophantine equations.** Fermat's conjecture is popularly called **Fermat's last theorem**. Although its solution is of no particular significance from the point of view of applications, no other conjecture in mathematics has engaged so many mathematicians for so long.

It is customary to define a quotient group in the manner of Exercise (2.9). We have preferred to define it through a binary operation defined for all subsets of the group, because this binary operation figures elsewhere too (for example, Exercise (2.17).).

3. Group Homomorphisms

Our discussion of groups has so far, involved only one group at a time except when we constructed new groups from old ones. In this section we study a particular relationship which exists between certain pairs of groups.

Whenever it does, the properties of one of the groups often throw some light on those of the other.

This relationship is known as a homomorphism. We already defined it (see Definition (3.4.15)) for general algebraic structures. But it is only when the algebraic structure is sufficiently strong that non-trivial results can be proved about homomorphisms. A group structure is such a structure. So we recapitulate the definition of a group homomorphism. In the next chapter we shall study an algebraic structure which is even stronger than a group structure. So results proved here will have their analogues in the next chapter.

3.1 Definition: Let G, H be groups. A function $f: G \to H$ is called a **group homomorphism** (or simply a homomorphism) if for all $x, y \in G$, we have $f(x \cdot y) = f(x) \cdot f(y)$ (where the same symbol is used to denote the group operation in both G and H). A homomorphism which is injective, surjective or bijective is respectively called a **monomorphism,** an **epimorphism,** and an **isomorphism.** In the last case, we say G is **isomorphic to** H and denote this symbolically by $G \cong H$. An isomorphism of a group onto itself is called an **automorphism.**

A comment is in order about the terminology. The common part 'morphism' comes from 'morphos' which means structure. All the four concepts defined above deal with the structures of the two groups. The prefixes 'homo' and 'iso' both mean 'similar' but while 'homo' indicates likeness 'iso' stresses an equality. The prefixes 'mono', 'epi' and 'auto' mean respectively 'one', 'on' and 'self'. Two groups (or more generally any two algebraic structures) which are isomorphic to each other are indistinguishable from each other. They are like replicas of the same object. A homomorphism is not such a strong concept as an isomorphism. It simply means that the function $f: G \to H$ commutes with or is compatible with the group multiplication. Given two elements x, y in G whether we first multiply them (in G) and apply f to the product xy or whether we first apply f to the elements x, y separately and multiply their images $f(x), f(y)$ in H, we get the same result. This condition can be graphically represented by requiring that the following diagram is commutative

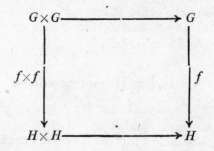

where the horizontal arrows represent the binary operations on G and H respectively and $f \times f : G \times G \to H \times H$ is the function which sends $(x, y) \in G \times G$ to $(f(x), f(y))$ in $H \times H$.

Of course, the concept of a homomorphism makes sense even if G, H are algebraic structures of a weaker type than a group, say semigroups or monoids. But, for a semigroup or a monoid homomorphism, there is not much we can infer from the definition. For example, let $\mathbb{R}$ be the monoid of real numbers with usual multiplication. If we define $f : \mathbb{R} \to \mathbb{R}$ by $f(x) = 0$ for all $x \in \mathbb{R}$ then f is a monoid homomorphism. But f does not take the identity element of $\mathbb{R}$ (namely 1) to the identity element of $\mathbb{R}$. With group homomorphisms things are much better as we now show.

3.2 Proposition: Let $f : G \to H$ be a group homomorphism. Then f maps the identity element of G to the identity element of H. Moreover, if $x \in G$ then $f(x^{-1})$ is the inverse of $f(x)$ in H. In other words, f is compatible with or preserves inverses.

Proof: Let e, e' denote, respectively, the identity elements of G, H. We have to show $f(e) = e'$. Denote $f(e)$ by a. Then $a \cdot e' = a = f(e) = f(e \cdot e) = f(e) \cdot f(e) = a \cdot a$. So $a \cdot e' = a \cdot a$. Since cancellation laws hold in a group, it follows that $e' = a$ as was to be shown. As for inverses, let $x \in G$. Denote $f(x)$, $f(x^{-1})$ by y, z respectively. We have to show $z = y^{-1}$. Now

$$yz = f(x) \cdot f(x^{-1}) = f(x \cdot x^{-1}) = f(e) = e'$$

(as shown just now). So

$$z = e'z = y^{-1} yz = y^{-1} e' = y^{-1}. \quad \blacksquare$$

The proof of the following proposition is left as an exercise. From now onwards 'homomorphism' means 'group homomorphism' in this chapter.

3.3 Proposition: The composite of two homomorphisms is a homomorphism. The composite of two isomorphisms as well as the inverse of an isomorphism are isomorphisms. If $f : G \to H$ is a homomorphism, $f(G)$ is a subgroup of H. If f is monomorphism then $f(G) \cong G$. (For this reason, a monomorphism is also called an **embedding** or an **imbedding**.) ▨

We now give some examples of group homomorphisms.

(1) If G is any group, then the identity function $1_G : G \to G$ is an isomorphism. Although this, by itself, is hardly profound, this fact combined with the last proposition means that in any collection of groups 'being isomorphic to' is an equivalence relation. A property of groups is said to **invariant under isomorphisms** if whenever a group has it so does every group isomorphic to it. For example, being an abelian group is such a property. Suppose $f : G \to H$ is an isomorphism and G is abelian. Given any $x, y \in H$, let

$a = f^{-1}(x)$, $b = f^{-1}(y)$. Then $xy = f(a) f(b) = f(ab) = f(ba) = f(b) f(a) = yx$. So H is abelian. Similarly being a cyclic group is a property which is invariant under isomorphisms. The property of being equal to a given group is not invariant under isomorphisms.

(2) Let G, H be any groups. The constant function from G to H which sends every element of G to the identity element of H is a homomorphism. It is called the **trivial homomorphism**. If H is an abelian group denoted additively, this homomorphism is also called the **zero homomorphism**. Note that the composite of two homomorphisms is trivial if either one of them is trivial. The converse is false in general.

(3) Let G be an abelian group and n a fixed integer. Define $f: G \to G$ by $f(x) = x^n$. Then f is a homomorphism. Even if G is not abelian f may be a homomorphism for some values of n. If f is a homomorphism for $n = 2$ or $n = 1$ then G must be abelian (cf. Exercise (1.5)).

(4) Let G be the additive group of real numbers. For a fixed $\lambda \in \mathbb{R}$ the function $f: \mathbb{R} \to \mathbb{R}$ defined by $f(x) = \lambda x$ for $x \in \mathbb{R}$ is a homomorphism. If $\lambda = 0$, this is the trivial homomorphism. If $\lambda \neq 0$, it is an isomorphism. Similar homomorphisms can be defined for $\mathbb{Q}$ or $\mathbb{C}$ instead of $\mathbb{R}$. We may also replace $\mathbb{R}$ by a euclidean vector space $\mathbb{R}^n$ and for a fixed $\lambda \in \mathbb{R}$ define $f: \mathbb{R}^n \to \mathbb{R}^n$ by $f(x) = \lambda x$ (that is, the product of the scalar λ with the vector x). Note that in all these examples, the fact that f is a homomorphism follows from distributivity of multiplication (by λ) over the addition.

(5) Let $\mathbb{R}$ be the additive group of real numbers and $\mathbb{R}^*$ the multiplicative group of non-zero real numbers. Define $f: \mathbb{R} \to \mathbb{R}^*$, by $f(x) = e^x$. Because of properties of the exponential function,

$$f(x + y) = e^{x+y} = e^x \cdot e^y = f(x) \cdot f(y).$$

Thus f is a homomorphism. Note that f is a monomorphism but not an epimorphism. Its range is the set $\mathbb{R}_+^*$ of all positive real numbers. This is a subgroup of $\mathbb{R}^*$ and if we view f as a homomorphism from $\mathbb{R}$ to $\mathbb{R}_+^*$, it is an isomorphism. In other words, the additive group of real numbers is isomorphic to the multiplicative group of positive real numbers.

(6) As the complex version of the last example, let $\mathbb{C}$, $\mathbb{C}^*$ be respectively the additive group of complex numbers and the multiplicative group of all non-zero complex numbers. Define $f: \mathbb{C} \to \mathbb{C}^*$ by $f(z) = \exp(2\pi i z)$. Then f is homomorphism. Note that f is not a monomorphism because of the periodicity of the complex exponential function. However, it is an epimorphism. The reason for putting the coefficient $2\pi i$ in the exponent is that if we restrict f to $\mathbb{R}$, which is a subgroup of $\mathbb{C}$, we get a homomorphism of $\mathbb{R}$ onto the circle group S^1 which sends a real number x to the complex number $\exp(2\pi i x) = \cos 2\pi x + i \sin 2\pi x$. Note that the entire subgroup $\mathbb{Z}$ of $\mathbb{R}$ is mapped to the identity element 1 of S^1.

(7) Let m be a positive integer and $\mathbb{Z}_m$ the group of all residue classes modulo m. These residue classes are the equivalence classes of $\mathbb{Z}$ under

the relation of congruency modulo m. We let $p : \mathbb{Z} \to \mathbb{Z}_m$ be the quotient (or the projection) function, (see Definition (3.2.12)). Then p is a homomorphism, by the very definition of addition in $\mathbb{Z}_m$. More generally, let G be any group and H any normal subgroup of G. In the last section we defined G/H as the group of all cosets of H in G, under coset multiplication. Define $p : G \to G/H$ by $p(x) = xH$ for $x \in G$. The function p assigns to each element of G, the coset to which it belongs. Then p is a group homomorphism, called the **quotient homomorphism**. It is an epimorphism. This example is very important from a theoretical point of view, because as we shall see below, essentially every epimorphism is of this type.

(8)   Let G be a group. Fix any element $g \in G$. Define $\theta_g : G \to G$ by $\theta_g(x) = gxg^{-1}$. In other words, the function θ_g is the conjugation by the element g. Since

$$\theta_g(xy) = g(xy)g^{-1} = (gxg^{-1})(gyg^{-1}) = \theta_g(x)\theta_g(y)$$

we see that θ_g is a homomorphism. It is easy to show that θ_g is a bijection (cf. the proof of Corollary (1.13)). So θ_g is an automorphism of G. An automorphism of this type called an **inner automorphism**.

(9)   Certain homomorphisms associated with products of two (or more) groups are very important. Let $G_1, G_2, ..., G_n$ be any groups. Let G be the set $G_1 \times G_2 \times ... \times G_n$. Then G becomes a group under co-ordinatewise multiplication. For each i, we define the function

$$\pi_i : G \to G_i, \pi_i(x_1, x_2, ..., x_n) = x_i$$

called the **projection** onto the ith factor. It is easily seen that π_i is an epimorphism. Let us denote the identity elements of all G_i's by e. Then define $\lambda_i : G \to G$ by $\lambda_i(x) = (e, e, ..., e, x, e, ..., e)$ for $x \in G_i$ where the x in the n-tuple is in the ith place. Then λ_i is a monomorphism. The range of λ_i is the subgroup $\{e\} \times ... \times \{e\} \times G_i \times \{e\} \times ... \times \{e\}$ of G. Let us call this subgroup H_i. Then by Proposition (3.3), G_i is isomorphic to H_i. Thus we see that every factor group in a product of groups is isomorphic to a subgroup of the product.

(10)   It is very easy to characterise which functions into a product of groups are homomorphisms. Let the groups G_i and G be as in (9). Let H be any group and $f : H \to G$ a function. Then for each $i = 1, 2, ..., n$, the composite $\pi_i \circ f$ is a function from H to G_i. Clearly if f is a homomorphism, so is $\pi_i \circ f$ for all $i = 1, ..., n$ by Proposition (3.3). Interestingly, the converse is also true. For if $\pi_i \circ f$ is a homomorphism for all i, then given any $x, y \in H$ we let $f(x)$ and $f(y)$ be respectively $(x_1, ..., x_n)$ and $(y_1, ..., y_n)$. Then $\pi_i(f(x)) = x_i$ and $\pi_i(f(y)) = y_i$ for all $i = 1, ..., n$. Let $f(xy) = (z_1, z_2 ..., z_n)$. Then $z_i = \pi_i(f(xy)) = \pi_i(f(x)) \cdot \pi_i(f(y)) = x_i y_i$ for all $i = 1, ..., n$. It follows that $f(xy) = f(x) \cdot f(y)$ since $f(x) \cdot f(y)$ is obtained by co-ordinatewise multiplication.

(11)   Let G, H be any two groups and suppose $f, g : G \to H$ are both homomorphisms. As usual we define the pointwise multiplication of f

and g and get a new function $h = fg : G \to H$ defined by $h(x) = f(x)g(x$
for $x \in G$. In general, h is not a homomorphism because if $x, y \in G$ then
$h(xy) = f(xy)g(xy)$ which equals $f(x)f(y)g(x)g(y)$ because f and g are homo
morphisms but $h(x)h(y)$ equals $f(x)g(x)f(y)g(y)$. However, if H is abelian
then we see that h is a homomorphism In that case the set of all homomor
phisms from G to H is itself a group under the operation just defined. This
group is denoted by Hom (G, H),

(12) Let G be the group $\mathbb{R}^2$ under co-ordinatewise addition. Define
$f : G \to \mathbb{R}$ by $f(x, y) = y - 2x$. Then f is an epimorphism. Note that $f^{-1}(\{0\}$
is the set $\{(x, y) \in \mathbb{R}^2 : y - 2x = 0\}$. We already considered this set in
Figure 5.4 and saw that it is a subgroup of $\mathbb{R}^2$.

We could continue the list further. But let us now turn to some basic
theorems about group homomorphisms. The key concept is the following

3.4 Definition: Let $f : G \to H$ be a group homomorphism. Let e be the
identity element of H. Then the set $f^{-1}(\{e\})$ is called the **kernel** of f.

For example, the kernel of the trivial homomorphism is the whole group
G. The kernel of a monomorphism consists only of the identity element
Actually, this property characterises monomorphisms as we shall see shortly
The kernel of the homomorphism in Example (6) above is the group of
integers. In Example (7), the kernel is precisely the subgroup $m\mathbb{Z}$ of $\mathbb{Z}$
We saw in the last section, that the quotient group $\mathbb{Z}/m\mathbb{Z}$ is precisely the
group $\mathbb{Z}_m$, which is the range of the homomorphism. We now show that
upto an isomorphism the range of every homomorphism is the quotient of
its domain by its kernel. The following result is called the **fundamental
theorem about group homomorphisms.**

3.5 Theorem: Let K be the kernel of a group homomorphism $f : G \to H$
Let R be the range of f. Then K is a normal subgroup of G. Moreover
there exists a unique isomorphism $\theta : G/K \to R$ such that $f = \theta \circ p$ where
$p : G \to G/H$ is the quotient homomorphism, in other words, in the following diagram

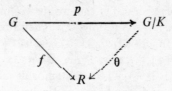

there is a unique way to fill the dotted arrow by a homomorphism, so as
to make it commutative.

Proof: First we show that K is a subgroup of G. If $x, y \in K$ then
$f(xy) = f(x)f(y) = ee = e$ showing that $xy \in K$. Similarly $f(x^{-1})$ equals
$[f(x)]^{-1}$ by Proposition (3.2). So if $x \in K$, then $f(x^{-1}) = [e]^{-1} = e$, showing

that $x^{-1} \in K$. Thus, we see that K is a subgroup of G. As for its normality, if $g \in G$ and $x \in K$ then $f(gxg^{-1}) = f(g)f(x)f(g^{-1}) = f(g)e[f(g)]^{-1}$ $= f(g)[f(g)]^{-1} = e$, showing that $xgx^{-1} \in K$. So K is a normal subgroup of G, and G/K, the group of cosets is well-defined.

Now we come to the more interesting part of showing that as a group, G/K is isomorphic to R, the range of f. (By Proposition (3.3), R is a subgroup of H and hence itself a group). We define $\theta : G/K \to R$ by $\theta(xK) = f(x)$. We must of course verify that this is well-defined. So suppose xK and yK are the same cosets. Then $xy^{-1} \in K$. So $f(xy^{-1}) = e$. But this means that $f(x)[f(y)]^{-1} = e$ and so $f(x) = f(y)$. Therefore $\theta(xK) = \theta(yK)$. This argument can be read backwards and proves that θ is one-to-one. Also since R is the range of f, every element of R is of the form $f(x)$ for some $x \in G$ and so θ is onto. It only remains to show that θ is a group homomorphism. Recall that the group structure on G/K was defined by coset multiplication, $(xK)(yK)=(xy)K$, for $x, y \in G$. So, $\theta[(xK)(yK)]=\theta[(xy)K]=$ $f(xy) = f(x)f(y) = \theta(xK)\theta(yK)$. This proves that θ is a group homomorphism and hence an isomorphism. The construction of θ is the same as that of the function g in Proposition (3.2.14). So it follows that $\theta \circ p = f$ and also that θ is the only isomorphism (in fact the only function) from G/K to R with this property. ∎

This theorem is of profound significance. Given a homomorphism $f: G \to H$, obviously it is only the range of f that is directly related to G and f. The entire group H plays no role. The theorem says that, upto an isomorphism, this range R may be treated as a quotient group of G and when so treated, the function f coincides with the quotient function from G onto this quotient group. This justifies our comment in Example (7) above, that essentially every epimorphism is a quotient homomorphism. The isomorphism θ is often called the **canonical** isomorphism. ('Canonical' is a formal synonym for 'standard' or 'simple'.)

Let us apply this theorem to a few examples. In Example (6), we considered the homomorphism $f: \mathbb{R} \to \mathbb{C}$ defined by $f(x) = e^{2\pi ix}$ for $x \in \mathbb{R}$. The range of f is S^1, the circle group. The kernel is $\mathbb{Z}$. Thus we see that the quotient group $\mathbb{R}/\mathbb{Z}$ is isomorphic to the circle group S^1. There is a vivid, geometric way to visualise f and this isomorphism. Represent $\mathbb{R}$ by a straight line. Then under f, a point at a distance α(say) from the origin goes to a point on S^1 whose argument is $2\pi\alpha$. (Here α can also take negative values.) Intuitively f wraps the real line around the circle an infinite number of times, with every interval of length 1 getting wrapped once around the full circle S^1. The coset $\alpha + \mathbb{Z}$ consists of all real numbers whose distance from α is an integer. All these points are taken to the same point of S^1. The construction of S^1 from $\mathbb{R}$ can be visualised as follows. Cut the line at all points marked by integers as in (a). Glue the end points of each segment of unit length. This results in an infinite family of circles as in (b). Now 'fuse together' all these circles into a single circle

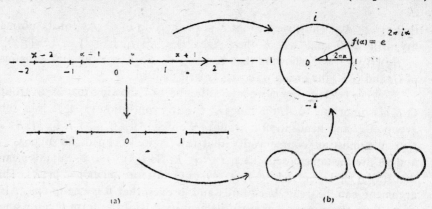

(a)   (b)

Figure 5.5: $\mathbb{R}/\mathbb{Z}$ Isomorphic to S^1

so that points which differ by an integer originally in $\mathbb{R}$ are fused together. Then every coset of $\mathbb{Z}$ in $\mathbb{R}$ corresponds to a point of S^1 and vice versa.

The second example is also geometric. In Example (12) above we defined $f: \mathbb{R}^2 \to \mathbb{R}$ by $f(x, y) = y - 2x$. Then f is an epimorphism with kernel $L = \{(x, y) \in \mathbb{R}^2 : y - 2x = 0\}$. So by Theorem (3.5) the quotient group $\mathbb{R}^2/L$ is isomorphic to $\mathbb{R}$. To see this isomorphism visually, draw a line M perpendicular to L. Then the various cosets of L in $\mathbb{R}^2$ are lines perpendicular to M. Smash the plane to the line M by perpedicular projection. Then every line perpendicular to M gets smashed to a point of M, namely the point of its intersection with M, the line L itself is smashed to the origin, while any line parallel to L at a perpendicular distance α is smashed to the point on M at a distance α from the origin where the distance α may be positive or negative, (Figure 5.6). Thus if we regard each coset of L as a single point we get M as $\mathbb{R}^2/L$. M is, of course, isomorphic to $\mathbb{R}$ as a group with a point at a distance α from 0 corresponding to

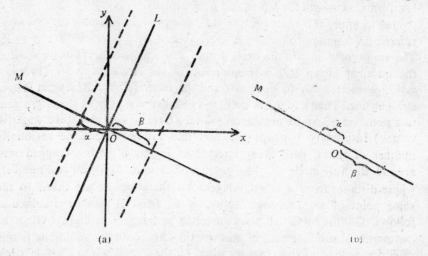

(a)   (b)

Figure 5.6: $\mathbb{R}^2/L$ Isomorphic to $\mathbb{R}$.

the real number a. (Instead of M we could take any line through O other than L, but the 'smashing' will still have to be done parallel to L.)

As the third example, we justify the term 'quotient group'. As used in algebra, 'quotient' is the opposite of 'product'. If 12 is the product of 3 and 4 then 3 is the quotient of 12 by 4 and 4 is the quotient of 12 by 3. Let G, H be two groups and $G \times H$ their product group. Denote the projection on G by π (see Example (9) above). Then $\pi(x, y) = x$ for all $x \in G$, $y \in H$. π is a homomorphism, in fact an epimorphism. The kernel K (say) of π is the subgroup $\{e\} \times H$. Clearly K is isomorphic to H. (We simply define $f : H \to K$ by $f(y) = (e, y)$ for $y \in H$.) So upto an isomorphism, we identify K with H. By Theorem (3.5), the quotient group $(G \times H)/H$ is isomorphic to G. Similarly $(G \times H)/G$ is isomorphic to H. This justifies the term 'quotient'. (Another possible justification is simply that for finite groups, the order of the quotient group equals the quotient of the order of the group by that of the subgroup.)

Although every factor group of a product appears, upto an isomorphism, as its quotient, the converse is not true. If G is a group and K is a normal subgroup then we can form the quotient group G/K. But G need not always be isomorphic to the product group $K \times G/K$. As a simple counterexample, let G be the quaternion group Q and let K be its centre. Then K is abelian. Also G/K is the Klein group which is also abelian. Hence the product $K \times G/K$ is abelian and so cannot be isomorphic to Q, which is non-abelian.

In the course of the proof of Theorem (3.5), we have proved the following result which is worth isolating.

3.6 Proposition: Let $f : G \to H$ be a group homomorphism with kernel K and range R. Let $y_0 \in H$. Then the equation $f(x) = y_0$ has a solution for x (in G) if and only if $y_0 \in R$. If x_0 is any solution, then every solution is of the form $x_0 k$ for some $k \in K$. Also f is one-to-one if and only if $K = \{e\}$.

Proof: The first assertion is obvious. For the second, we have to show that if $f(x_0) = y_0$ then $f^{-1}(\{y_0\})$ is the coset $x_0 K$. This can be proved directly. However, using the notation of Theorem (3.5), we have $p(x_0) = x_0 K \in G/K$ and $p^{-1}(\{x_0 K\}) = x_0 K \subset G$. Now $f = \theta \circ p$ gives, $f^{-1}(\{y_0\}) = p^{-1}(\theta^{-1}(\{y_0\})) = p^{-1}(\{x_0 K\})$ since $\theta(x_0 K) = f(x_0) = y_0$ and θ is one-to-one. So we get $f^{-1}(\{y_0\}) = x_0 K$ as was to be shown. The last statement is now obvious, because if $K = \{e\}$ then for every $y_0 \in H$, the equation $f(x) = y_0$ has at most one solution. The converse was already proved. ∎

Two special cases of this result must be already familiar to the reader. Consider a system of m linear equations in n real variables, say,

$$a_{11}x_1 + a_{12}x_2 + \dots + a_{1n}x_n = b_1$$
$$a_{12}x_1 + a_{22}x_2 + \dots + a_{2n}x_n = b_2$$
$$. \quad . \quad . \quad . \quad . \quad . \quad . \quad . \quad . \quad .$$
$$a_{m1}x_1 + a_{m2}x_2 + \dots + a_{mn}x_n = b_m$$

$$(1)$$

In the next chapter we shall see that the general solution of this system is obtained by finding any one particular solution and adding it to the general solution of the corresponding homogeneous system (which is obtained by replacing all the b_i's by 0). We can interpret this in terms of the last proposition as follows:

Consider the groups $\mathbb{R}^n$ and $\mathbb{R}^m$, each under coordinatewise addition. Denote their points by vectors. Thus $\bar{x} = (x_1, \dots, x_n) \in \mathbb{R}^n$ and $\bar{y} = (y_1, \dots, y_m) \in \mathbb{R}^m$. Define $\bar{f}: \mathbb{R}^n - \mathbb{R}^m$ by $\bar{f}(x_1, \dots, x_n) = \bar{y} = (y_1, \dots, y_m)$ where for each $i = 1, 2, \dots, m$, $y_i = \sum_{j=1}^{n} a_{ij}x_j$. Then $\bar{f}$ is easily seen to be a group homomorphism, (cf. Example (10) above). Let

$$\bar{b} = (b_1, \dots, b_m) \in \mathbb{R}^m.$$

Then solving (1) amounts to solving $\bar{f}(\bar{x}) = \bar{b}$. The corresponding homogeneous system is given by $\bar{f}(\bar{x}) = \bar{0}$. The kernel of $\bar{f}$ is precisely the set of all solutions of the homogeneous system. By adding these solutions to any one particular solution of (1), we get a coset of the kernel of $\bar{f}$ and this coset consists precisely of all solutions of (1). For $n = 2$, $m = 1$, the solutions of (1) form a line in the plane, parallel to the line which represents the kernel of the homomorphism. The result then geometrically means that in order to know a line completely, it suffices to know any one point on it and any one line parallel to it.

Similarly it can be shown that the general solution of a linear differential equation is obtained by finding the general solution of the corresponding homogeneous equation and adding to it any one particular solution of the original equation. This is a special case of Proposition (3.6).

Let us now see what a homomorphism between two groups does to their subgroups. Let $f: G \to H$ be a homomorphism with kernel K and range R. Let A, B be subgroups of G, H respectively. If $A \subset K$, then $f(A) = \{e\}$, the trivial subgroup of H. More generally, if A is any subgroup of G then the part of it which is common with K (that is, the intersection $A \cap K$) will be taken to $\{e\}$ by f. Intuitively whatever happens inside K will be masked by f. Similarly for any subgroup B of H, only the part it has in common with R (that is, $R \cap B$) will have any bearing with f. (If f is an epimorphism then $H = R$ and so the entire subgroup B will be governed by f.) It follows that in order to have a non-trivial relationship between subgroups of G and those of H, we must require the former to ontain K and the latter to be contained in R. With this restriction, we do have an important relationship proved below.

3.7 Theorem: Let $f: G \to H$ be a group homomorphism with kernel K and range R. Let $\mathcal{G}$ be the collection of subgroups of G containing K and $\mathcal{H}$ be the collection of subgroups of H contained in R. Then there is a one-to-one correspondence between $\mathcal{G}$ and $\mathcal{H}$. This correspondence preserves subgroups, normality and quotient groups. (The meaning of this statement will be clear in the course of the proof.)

Proof: We leave it as an exercise to prove that if A is any subgroup of G (not necessarily containing K) then $f(A)$ is a subgroup of H. Also obviously $f(A) \subset f(G) = R$. So $f(A) \in \mathcal{H}$. Similarly if B is any subgroup of H, then $f^{-1}(B) \in \mathcal{G}$. Note that in general, $f^{-1}(f(A))$ may be larger than A and $f(f^{-1}(B))$ may be smaller than B. We claim that if $A \in \mathcal{G}$ and $B \in \mathcal{H}$ then this cannot happen.

Let $A \in \mathcal{G}$. That is, $K \subset A$ and A is a subgroup of G. Certainly, $f^{-1}(f(A))$ contains A. To show that equality holds, suppose $x \in f^{-1}(f(A))$. Then $f(x) \in f(A)$, which means $f(x) = f(y)$ for some $Y \in A$. But then $xy^{-1} \in K$ as we saw in the proof of Theorem (3.5). Since $K \subset A$, $xy^{-1} \in A$. But $y \in A$. Hence $x = (xy^{-1}) \cdot y \in A$. Thus $f^{-1}(f(A)) \subset A$ as was to be shown. Similarly let $B \in \mathcal{H}$. Then $B \subset R$ and the equality $f(f^{-1}(B)) = B$ follows by a purely set theoretic argument.

So, if we define $\theta: \mathcal{G} \to \mathcal{H}$ by $\theta(A) = f(A)$ and $\psi: \mathcal{H} \to \mathcal{G}$ by $\psi(B) = f^{-1}(B)$, then these functions are inverses to each other. Hence each is a one-to-one correspondence, that is, a bijection. This proves the first assertion of the theorem. To say that the correspondence between $\mathcal{G}$ and $\mathcal{H}$ preserves subgroups means that whenever $A_1, A_2 \in \mathcal{G}$ with $A_1 \subset A_2$ then $f(A_1) \subset f(A_2)$ and similarly whenever $B_1, B_2 \in \mathcal{H}$ with $B_1 \subset B_3$, then $f^{-1}(B_1) \subset f^{-1}(B_2)$. This is a purely set-theoretic result. Suppose further A_1 is normal in A_2. Then we claim $f(A_1)$ is normal in $f(A_2)$. Let $x \in A_1$, $y \in A_2$. We have to show $f(y) f(x) [f(y)]^{-1} \in f(A_1)$. This follows since $yxy^{-1} \in A_1$ and $f(y) f(x) [f(y)]^{-1} = f(yxy^{-1})$. Similarly if B_1 is normal in B_2 then $f^{-1}(B_1)$ is normal in $f^{-1}(B_2)$. This is what is meant by preservation of normality. Finally, we show that if A_1 is normal in A_2 for $A_1, A_2 \in \mathcal{G}$ then the quotient group A_2/A_1 is isomorphic to the quotient group $f(A_2)/f(A_1)$. We could do this directly. But, an application of Theorem (3.5) would save a lot of work. Simply define $g: A_2 \to f(A_2)/f(A_1)$ by $g(x) = f(x) f(A_1)$, that is, $g(x)$ is that coset of $f(A_1)$ in $f(A_2)$ which contains $f(x)$. That g is a homomorphism follows from the fact that f is a homomorphism (and the definition of coset multiplication). g is also obviously onto. The kernel of g is the set $\{x \in A_2: g(x) = \text{the identity element of } f(A_2)/f(A_1)\}$. But the identity element of $f(A_2)/f(A_1)$ is the coset $f(A_1)$. Now, $f(x) f(A_1) = f(A_1)$ if and only if $f(x) \in f(A_1)$. This is equivalent to saying that $x \in f^{-1}(f(A_1))$. But

$f^{-1}(f(A_1)) = A_1$, as we saw above (since A_1 contains K). Thus the kernel of g is precisely A_1. So by Theorem (3.5), the quotient group A_2/A_1 is isomorphic to the range of g, which equals $f(A_2)/f(A_1)$. Conversely if B_1 is normal in B_2, then $f^{-1}(B_1)$ is normal in $f^{-1}(B_2)$. So, by what we proved just now, $f^{-1}(B_2)/f^{-1}(B_1)$ is isomorphic to $f(f^{-1})(B_2)/f(f^{-1}(B_1))$ which is nothing but B_2/B_1. We have now completely established the theorem. ▉

The following corollary is often useful.

3.8 Corollary: Let K be a normal subgroup of a group G. Then there is a one-to-one correspondence between subgroups of the quotient group G/K and subgroups of G containing K. This correspondence preserves subgroups, normality and quotient groups.

Proof: Let $p: G \to G/K$ be the quotient homomorphism. Then p has kernel K and is onto. So the result follows directly from the last theorem. ▉

As a concrete example, let $G = Q$, the group of quaternions and let K be its centre, $\{1, -1\}$. Then the quotient group G/K is the Klein group $H = \{e, a, b, c\}$ where $a = \{\pm i\}$, $b = \{\pm j\}$, $c = \{\pm k\}$ and $e = \{\pm 1\} \cdot H$ has five subgroups, $\{e\}$, $\{e, a\}$, $\{e, b\}$, $\{e, c\}$ and H itself. Under the correspondence in the corollary the corresponding subgroups of Q are K, (i), (j), (k) and Q respectively.

As an application of this theorem, we prove the converse of Lagrange's theorem for groups whose orders are powers of a prime, as promised in the last section. ▉

3.9 Theorem: Let G be a group with $\circ(G) = p^m$ where p is a prime number. Then for r with $0 \leqslant r \leqslant m$, G contains a subgroup of order p^r.

Proof: We argue by induction on m. If $m = 0$, G is the trivial group and the assertion holds. Let $m > 0$. The case $r = 0$ is also trivial. So assume $r > 0$. By Proposition (2.26), G contains a normal subgroup K of order p. Then the quotient group G/K has order p^{m-1}. By induction hypothesis, G/K has a subgroup B of order p^{r-1}. Let $f: G \to G/K$ be the quotient homomorphism. By Theorem (3.7), $f^{-1}(B)$ is a subgroup of G containing K and $f^{-1}(B)/K$ is isomorphic to B. Since $\circ(B) = \circ(f^{-1}(B))/\circ(K)$ and $\circ(K) = p$, it follows that $\circ(f^{-1}(B)) = p^r$. So G contains a subgroup of order p^r, for all $0 \leqslant r \leqslant m$. This completes the inductive step and the proof. ▉

As remarked earlier, two groups which are isomorphic to each other may be regarded as identical from the point of view of group theory. One of the central problems in group theory is to show that any two groups with certain common properties are isomorphic to each other, or that every group with certain properties must be isomorphic to some standard familiar group. This is known as the problem of *group classification*. The most trivial result of this type is that any two trivial groups (that is, groups having only one element each) are isomorphic to each other. This means that upto isomorphism there is only one trivial group. That is why we speak of '*the* trivial group' (or '*the* zero group' when we are dealing with abelian groups).

As a slightly less trivial example, we show that when we form the permutation group of a set, it is only the cardinality of that set that matters.

3.10 Proposition: Suppose $f : X \to Y$ is a bijection of sets. Then the permutation groups, $S(X)$ and $S(Y)$ of X, Y respectively are isomorphic to each other. In other words, two sets of the same cardinality have isomorphic permutation groups.

Proof: Recall that $S(X)$ consists of all bijections from X to itself. Given any such bijection $\sigma : X \to X, f \circ \sigma \circ f^{-1}$ is a bijection of Y onto itself (because f and f^{-1} are bijections). So we have function $\lambda : S(X) \to S(Y)$ defined by $\lambda(\sigma) = f \circ \sigma \circ f^{-1}$ for $\sigma \in S(X) \cdot$ (λ is like conjugation by f, but note that f is not an element of $S(X)$ in general.) Then λ is itself a bijection because the function $\mu : S(Y) \to S(X)$ defined by $\mu(\tau) = f^{-1} \circ \tau \circ f$ is clearly the inverse function of λ. Also if $\theta, \psi \in S(X)$ then $\lambda(\theta\psi) = f \circ (\theta \circ \psi)f^{-1} = (f \circ \theta) \circ (\psi \circ f^{-1}) = (f \circ \theta \circ f^{-1}) \circ (f \circ \psi \circ f^{-1}) = \lambda(\theta) \lambda(\psi)$ showing λ is a group homomorphism. Hence λ is a group isomorphism. That is, $S(X)$ and $S(Y)$ are isomorphic to each other. ▮

Because of this proposition, when we consider the permutations group of a set with n elements, it does not matter which particular elements we choose. The most standard choice is to take them as the positive integers from 1 to n. The group of permutations of the set $\{1, 2, \ldots, n\}$ is called the **symmetric group** of degree n and is denoted by S_n. We already know that $\circ(S_n) = n!$. We shall consider these groups in the next section.

The problem of classification of groups is fairly involved, even when restricted to finite groups. In the last chapter we saw that any two finite Boolean algebras with the same number of elements are isomorphic to each other. This is far from the case for groups. The groups S_3 and $\mathbb{Z}_6$ both have order 6. But $\mathbb{Z}_6$ is abelian while S_3 is not (see section 1). So S_3 cannot be isomorphic to $\mathbb{Z}_6$. Thus equality of orders is necessary but far from sufficient for two groups to be isomorphic. Later (see the Epilogue), we shall obtain a criterion for two finite abelian groups of the same order to be isomorphic. As a forerunner, we have the following simple result which completely settles the case of cyclic groups.

3.11 Proposition: Any two finite cyclic groups of the same order are isomorphic to each other. Also any two infinite cyclic groups are isomorphic to each other.

Proof: Let G be a cyclic group generated by an element x of it. Let $\mathbb{Z}$ be the additive group of integers. Define $f: \mathbb{Z} \to G$ by $f(n) = x^n$ for $n \in \mathbb{Z}$. Because of the laws of indices (Proposition (3.4.8)), f is a homomorphism. Also f is onto because every element of G is some power of x. So by Theorem (3.5) again, G is isomorphic to $\mathbb{Z}/K$ where K is the kernel of f. The proof now reduces to the computation of K.

If G is finite of order m (say) then by Theorem (1.10), $x^n = e$ if and only if n is a multiple of m. So in this case K is simply the subgroup $m\mathbb{Z}$ of $\mathbb{Z}$. But we have already seen that $\mathbb{Z}/m\mathbb{Z}$ is the group $\mathbb{Z}_m$ of residues modulo m. Thus every cyclic group of order m is isomorphic to $\mathbb{Z}_m$. Hence any two such groups are isomorphic to each other.

Suppose now that G is infinite. Then the kernel K contains only 0. Otherwise, by Proposition (1.7), K would equal $m\mathbb{Z}$ for some positive integer m. But then $\mathbb{Z}/K$, would be finite and isomorphic to G, contradicting that G is infinite. So $K = \{0\}$, that is, f is a monomorphism. Since f is also an epimorphism, it is in fact an isomorphism. Thus every infinite cyclic group is isomorphic to $\mathbb{Z}$ and so any two such groups are mutually isomorphic. ∎

In the course of the proof we have done something more than the proposition asserts. Let $\mathscr{C}$ be the class of all cyclic groups. We have shown that if G is any member of $\mathscr{C}$ then G is isomorphic either to $\mathbb{Z}$ or to $\mathbb{Z}_m$ for some unique positive integer m. In other words the collection of the groups $\{\mathbb{Z}, \mathbb{Z}_1, \mathbb{Z}_2, ...\}$ is a complete set of representatives (upto isomorphism) for the class $\mathscr{C}$. More generally, a collection $\mathscr{G}$ of groups is called a *complete set of representatives* for a class $\mathscr{D}$ of groups if every member of $\mathscr{D}$ is isomorphic to one and only one member of $\mathscr{G}$. It follows, in particular, that no two distinct members of $\mathscr{G}$ are isomorphic to each other, which means, intuitively, that $\mathscr{G}$ is free of redundancy.

The problem of finding a complete set of representatives for a given class of groups is an important one and varies in difficulty depending upon the class. We have just solved it for the class of all cyclic groups. The classes that are interesting include those which consist of all groups of a given order. For a positive integer n, let $\mathscr{C}_n$ be the class of all groups of order n. If n is a prime then a complete set of representatives for $\mathscr{C}_n$ is given by the following proposition.

3.12 Proposition: Every group of a prime order p is isomorphic to $\mathbb{Z}_p$. That is $\{\mathbb{Z}_p\}$ is a complete set of representatives for the class $\mathscr{C}_p$ of all groups of order p.

Proof: Let G be a group of order p. Let x be any element of G other than e. Let $H = (x)$. Then $o(H) > 1$. But by Lagrange's theorem $o(H)$ is a divisor of p. So $o(H) = p$. That is, $G = H = (x)$, proving that G is a cyclic group of order p. By the last proposition, G is isomorphic to $\mathbb{Z}_p$. ∎

When n is not a prime but a product of two primes, a complete set of representatives for n will be obtained later. In general, however, there is no way that will work for all n. Considerable ingenuity is needed even for particular values of n. By way of illustration, we settle the cases $n = 4$ and $n = 6$ (the first two cases not covered by the last proposition).

3.13 Proposition: For $\mathscr{C}_4$, the Klein group and the group $\mathbb{Z}_4$ form a complete set of representatives. For $\mathscr{C}_6$, the groups S_3 and $\mathbb{Z}_4$ form a complete set.

Proof: Let G be a group of order 4. If G contains an element of order 4 then G is cyclic, and isomorphic to $\mathbb{Z}_4$. Suppose G contains no element of order 4. Let the elements of G be e, x, y, z with e as the identity. Now the order of x is either 1, 2 or 4 by Proposition (2.14) If $o(x) = 1$, $x = e$. Also $o(x) \neq 4$ by assumption. So $o(x) = 2$. Similarly y, z are of order 2 each. Now $xy \neq e$, x or y as this would respectively imply $y = x$, $y = e$, or $x = e$ none of which is true. So $xy = z$. Similarly $yx = z$. (Or we could use Theorem (2.25) by which G is abelian). Similar reasoning gives $yz = zy = x$ and $xz = xz = x$. If we define $f: G \to H$ (where H is the Klein group) by $f(e) = e, f(x) = a, f(y) = b$ and $f(z) = c$ then f is an isomorphism. Thus we have shown that every member of $\mathscr{C}_4$ is isomorphic either or to $\mathbb{Z}_4$ or the Klein group. Since these two groups are not themselves isomorphic to each other, they constitute a complete set of representatives for $\mathscr{C}_4$.

Next, suppose G is a group of order 6. We consider the cases G abelian and G non-abelian separately. First suppose G is abelian. We show that G has an element of order 6 which would imply G is isomorphic to $\mathbb{Z}_6$. By Exercise (1.16), G has at least one element say x of order 2. We claim G has no other element of order 2. Because if y were such element then $\{e, x, y, xy\}$ would be a subgroup of G. This contradicts Lagrange's theorem since 4 does not divide 6. Now let z be any element of G other than e and x. By what we have shown, the order of z is 6 or 3. In the first case, we are done. In the second, we leave it to the reader to show that zx has order 6. In any case G has an element of order 6 as was to be shown.

Finally, suppose G is a non-abelian group of order 6. Then G has no elements of order 6. So every element other than the identity has order 2 or 3. We claim that there is at least one element of order 3. If not, then the square of every element of G will be e and hence G would be abelian (cf. Exercise (1.5)). So G has at least one element say y of order 3. Let $H = (y)$. Then H has index 2 in G and hence is normal in G, by Theorem (2.9). Now let x be an element in G of order 2 (which exists by Exercise (1.16)). Then certainly $x \notin H$. Now $x^{-1} = x$ and since H is normal in G, $xyx^{-1} = xyx \in H$. So $xyx = e$ or $xyx = y$ or $xyx = y^2$. The first equality implies $xy = x$ which is impossible. The second equality, $xyx = y$ implies

$xy = yx$ (since $x^2 = e$). But this means y commutes with x. Then xy would have order 6, as can be easily shown, implying that G is cyclic. So the second possibility is also ruled out, leaving $xyx = y^2$, or $xy = y^2x$. The six distinct elements of G are now seen to be $e, y, y^2, x, xy \, (= y^2x)$ and $yx \, (= xy^2)$. With this knowledge about the structure of G, it is now easy to establish an isomorphism $f: G \to S_3$. We think of S_3 as the group of isometries of an equilateral triangle (Figure 5.1). We let $f(y)$ be the clock-wise rotation through 120° and $f(x)$ be the reflection in any one of the altitudes. It is easy to extend f to the remaining elements of G so as to get an isomorphism of G onto S_3.

Summing up, every group of order 6 is isomorphic to $\mathbb{Z}_6$ or to S_3 depending upon whether it is abelian or not. So $\mathbb{Z}_6$ and S_3 constitute a complete set of representatives of $\mathscr{C}_6$. ∎

We remark that with the machinery to be developed later in this book, the proposition can be proved much more easily. The direct proof above is meant to illustrate the type of reasoning needed. It also gives a good opportunity to review most of the basic concepts introduced in the last two sections.

The problem of group classification seeks to find, for a given group G, some known group H which is isomorphic to G. A related problem is that of group representation. Here we are interested in finding a homomorphism f from a given group G to a familiar group H. This homomorphism must of course be non-trivial so as to cast some light on the properties of G in terms of those of H. If f can be chosen to be a monomorphism then G is isomorphic to a subgroup of H, namely $f(G)$. When this is the case, we know G upto an isomorphism, as soon as we know H and all its subgroups.

Thus the problem of group representation is to express an 'abstract' group as isomorphic to a subgroup of some 'concrete' group. This problem is surprisingly simple as shown by the following theorem due to **Cayley**.

3.14 Theorem: Every group is isomorphic to a subgroup of a group of permutations.

Proof: Let G be a given group. We consider the permutations group of the set G itself. We denote it as usual by $S(G)$. It consists of all bijections from G to itself (they need not be isomorphisms) and the group operation is the composition of functions. For each $g \in G$ define a function $T_g: G \to G$ by $T_g(x) = gx$ for $x \in G$. In other words, T_g is nothing but the left transla-tion by the element g (cf. Exercise (3.4.2)). Because of the cancellation laws in groups, T_g is one-to-one. Also if $y \in G$ then $T_g(g^{-1}y) = gg^{-1}y = y$ showing that T_g is also onto. So T_g is a bijection of the set G onto itself and hence is an element of $S(G)$. (Note that we are not claiming, nor is it necessarily true that T_g is a group homomorphism. Actually, T_g will be a group homomorphism iff $g = e$.) Now define $f: G \to S(G)$ by $f(g) = T_g$. We assert that f is a monomorphism. First, let $g, h \in G$. We claim

$T_{gh} = T \circ T_h$. Since both sides are functions from G into itself, it snffices to show that for every $x \in G$, $T_{gh}(x) = (T_g \circ T_h)(x)$. But the left hand side is, by definition, $(gh)x$, which by associativity equals $g(hx)$, which is precisely the right hand side. So $T_{gh} = T_g \circ T_h$, that is $f(gh) = f(g)f(h)$ for all $g, h \in G$. This shows that f is a group homomorphism. As for its kernel, suppose $g \in G$ and $f(g)$ is the identity element of $S(G)$. This means, T_g is the identity function on G, that is, $T_g(x) = x$ for all $x \in G$. But then $gx = x = ex$ for all x, giving $g = e$ by the cancellation law. So the kernel of f consists only of the identity element. Therefore f is a monomorphism, showing that G is isomorphic to $f(G)$ which is a subgroup of $S(G)$. ∎

Cayley's theorem is of the same spirit as the Stone representation theorem for Boolean algebras mentioned in the last chapter. Both assert that certain abstract algebraic structures are isomorphic to some very concrete structures of the same category. There is, however, considerable difference in the degree of their depth. While Cayley's theorem is easy to prove, the Stone representation theorem, even in the finite case required some work and in the infinite case (which we did not prove) requires the use of the axiom of choice.

For finite sets, Cayley's theorem can be given a still more concrete form.

3.15 Theorem: Every finite group of order n is isomorphic to a subgroup of S_n.

Proof: Let G be a group of order n. By the last theorem G is isomorphic to a subgroup of $S(G)$, the group of permutations of the set G. But by Proposition (3.10), the permutation group of any set with n elements is isomorphic to $S(G)$. We take this set to be $\{1, 2,..., n\}$. Then $S(G)$ is isomorphic to S_n. Consequently G is isomorphic to a subgroup of S_n. ∎

3.16 Corollary: For a positive integer n, there are only finitely many distinct groups of orders n upto isomorphism.

Proof: The group S_n is finite (having order $n!$) and so has only finitely many subgroups. Since every group of order n is isomorphic to at least one of these, it follows, that upto isomorphism, there can be only finitely many distinct groups of order n. ∎

Worded differently the corollary says that the class $\mathscr{C}_n$ of all groups of order n has a finite, complete set of representatives. It should be noted that the corollary does not, by itself, give any particular set of representatives for $\mathscr{C}_n$. To find such a set, we would have to find all subgroups of S_n having order n, and then decide which of them are isomorphic to each other. By picking exactly one from each type, we would get a complete set of representatives for the class $\mathscr{C}_n$. This procedure is theoretically possible but highly inefficient. First, even for relatively small values of n, $n!$ is very large.

So it is impracticable to list down all *n*-subsets of S_n and find out which of them are subgroups. Even if we manage to do that, we still have to decide which of these subgroups are isomorphic to each other. This is another difficult problem. Given two groups say *G* and *H* of order *n* each, there is no easy way known to tell whether *G* and *H* are isomorphic to each other. Of course, we could consider all possible bijections from *G* onto *H* and examine them one by one to see if at least one of them is an isomorphism. This is theoretically a finitistic process. But once again, because there are *n*! bijections, this is not a practical way. True, we could weed out quite a few bijections right away, for example those which do not take the identity element of *G* to that of *H*. But even then we are still left with a large number of them. It is often easier to show that *G* and *H* are *not* isomorphic to each other. We simply take a suitable property (invariant under isomorphisms) which one of the groups has and the other does not. For example if *G* is abelian and *H* is not, we see at once that *G* cannot be isomorphic to *H*, without examining any bijection between them. Similarly if *G* has an element of a particular order and *H* has no such element, it follows that *G* is not isomorphic to *H*. Such a property is said to *distinguish* the two groups. Of course a property which serves to distinguish one pair of groups may not work for some other pair. This is, in fact, the trouble. Given two groups we may go on comparing their properties. If we come across a property which distinguishes them we are done. But if not, we cannot still say that the groups are isomorphic. We may of course be convinced that they are so. But an actual isomorphism will have to be constructed, which is again not so easy. Interestingly, for finite abelian groups, there is indeed a way to tell whether two groups are isomorphic by simply examining a few of their properties (or *invariants* as they are technically called). We shall study them later (see the Epilogue).

We conclude this section with another representation of groups which is sort of dual to that given by Cayley's theorem. In Cayley's theorem, we express a given group, upto isomorphism, as a subgroup of a group of a specific type, namely, a permutation group. In the representation to be studied now, we shall express it as a quotient group of a group of a specific type, called free groups. We begin by defining what they are.

In Section 1, we studied the concept of a subgroup generated by a subset *S* of a group *G*. Suppose this subgroup happens to be *G* itself. By Proposition (1.8), every element of *G* can be expressed in the form $x_1^{n_1} x_2^{n_2} \ldots x_r^{n_r}$ where $x_i \in S$, $n_i \in \mathbb{Z}$ and *r* is a non-negative integer. This expression need not be unique. For example, the quaternion group *Q* is generated by the set $S = \{i, j\}$. We can write $k \in Q$ as ij or as $i^3 j^7$ and also as ji^{-1} or as $j^3 i$, and in many other ways. Such multiple expressions result because of certain equalities which hold in a particular group. (For example in the quaternions group, $i^4 = e$ and so $i^5 = i$ etc.). It is possible to conceive of a group which is 'free' of such equalities. The only equalities that hold in such a group are those that are implied by the laws of indices. By making a convention that no two adjacent x_i's in the expression $x_1^{n_1} \dots x_r^{n_r}$ are equal and also by requiring that no index n_i is zero, we then get a unique expression* for every element of G, as a product of powers of elements of S. The following definition gives the name for such a group.

3.17 Definition: A group G is said to be **free** if there exists a subset $S \subset G$ such that every element of G can be uniquely expressed as

$$x_1^{n_1} x_2^{n_2} \dots x_r^{n_r}$$

where $r \geqslant 0$, $x_i \in S$ and n_i is a non-zero integer for $i = 1, \dots, r$ and no two adjacent x_i's are equal ($r = 0$ corresponds to the identity element of G). We also say that the set S **freely generates** the group G. By convention, we take the trivial group to be a free group generated by the empty subset.

The simplest non-trivial example of a free group is the group $\mathbb{Z}$ of integers under addition. Here the set S may be taken as $\{1\}$ or as $\{-1\}$. None of other groups we have considered so far is free. Indeed it is easy to show, by the pigeon hole principle, that no finite group (except the trivial group) can be free. Moreover, if the set S in the definition above happens to contain at least two elements, say a and b then G cannot be abelian, or else ab and ba would give two different expressions for the same element of G. Thus no abelian group, other than $\mathbb{Z}$ and the trivial group can be free. However, free groups do exist. In fact given any set, we can construct a group which is freely generated by it as we show now.

3.18 Theorem: Let S be any set. Then there exists group G having S as a subset such that S freely generates G.

Proof: We have little choice in defining G. We let G be the set of all monomials of the form $x_1^{n_1} x_2^{n_2} \dots x_r^{n_r}$ where $r \geqslant 0$ ($r = 0$ gives the null monomial which is also an element of G, to be denoted by e), $x_i \in S$, n_i are non-zero integers and no two adjacent x_i's are equal. Here $x_i^{n_i}$ is only a formal symbol. It is not to be interpreted as the n_ith power of x_i. Such an interpretation is meaningless at the moment because as yet no binary operation has been defined. To avoid confusion, we might denote the element $x_1^{n_1} \dots x_r^{n_r}$ as a sequence of ordered pairs $((x_1, n_1), (x_2, n_2), \dots, (x_r, n_r))$. In other words, elements of G are all finite sequences taking values in the set $S \times (\mathbb{Z} - \{0\})$, including the sequence of length 0, that is, the null

---

*Another equivalent way to ensure uniqueness of expression is to allow the adjacent x_i's to be equal, but to restrict each exponent n_i to be either 1 or -1 and require that in case $x_i = x_{i+1}$ then $n_i = n_{i+1}$. This avoids x and x^{-1} appearing adjacently and thereby ensures uniqueness. With this convention $a^3 b^{-2} a^2$ will have to be written as $aaab^{-1}b^{-1}aa$. An expression of the former type is called the *normal form* while the latter is called the *reduced word form* of an element.

sequence. We identify each $x \in S$ with the element x^1, (or equivalently, the sequence $(x, 1)$) in the set G. This way S is a subset of G.

We now turn to defining a binary operation on G which will make G into a group freely generated by G. Let $x = x_1^{n_1}, \ldots, x_r^{n_r}$ and $y = y_1^{m_1}, \ldots y_k^{m_k}$ be two elements of G. We shall define xy by induction on their 'lengths', that is on the integers r and k. When $r = 0$, $x = e$, the null sequence and we simply let $xy = y$ for any y. Similarly if $k = 0$, $xy = x$ for all x. Suppose $r > 0$ and $k > 0$. If $x_r \neq y_1$ we simply juxtapose x and y and define xy as $x_1^{'1}x_2^{n_2} \ldots x_r^{n_r}y_1^{m_1}y_2^{m_2} \ldots y_k^{mk}$ which is indeed an element of G. However, if $x_r = y_1$, this definition will not be valid because of the restriction on the form of the elements of G. In this case, we proceed as follows: Suppose $n_r + m_1 \neq 0$. Then we simply let

$$xy = x_1^{n_1} \ldots x_{r-1}^{n_{r-1}}x_r^{n_r+m_1}y_2^{m_2} \ldots y_k^{mk}.$$

This is a sequence of length $r + k - 1$ and is in G because $x_r \neq y_2$ (since $x_r = y_1$, and $y_1 \neq y_2$). The only remaining case is when $x_r = y_1$ and $n_r + m_1 = 0$. (Intuitively in this case, x_r^n and $y_1^{m_1}$ should cancel each other.) Formally, in this case we let xy be the product of $x_1^{n_1} \ldots x_{r-1}^{n_{r-1}}$ and $y_2^{m_2} \ldots y_k^{mk}$ which is a product of sequences of lengths $r-1$ and $k-1$ and hence is already supposed to be defined by induction hypothesis. Thus we have completely defined a binary operation on G. (The definition is not as clumsy as it appears. The manner in which the product is defined is exactly the manner in which the product of two such monomials in a group would be simplified until no further reduction is possible.)

Now we prove that with this binary operation G is indeed a group. By very definition, the null sequence e is an identity element. The verification of associativity is a little tedious and has to be done inductively. Let $x = x_1^{n_1} \ldots x_r^{n_r}$, $y = y_1^{m_1} \ldots y_k^{mk}$ and $z = z_1^{p_1} \ldots z_t^{p_t}$ be any three elements of G. If $x_r \neq y_1$ and $y_k \neq z_1$, both $(xy)z$ and $x(yz)$ come out to be equal to $x_1^{n_1} \ldots x_r^{n_r}y_1^{m_1} \ldots y_k^{mk}z_1^{p_1} \ldots z_t^{p_t}$. When either $x_r = y_1$ or $y_k = z_1$ we have to make several cases. Suppose for example, that $x_r = y_1$ and $y_k \neq z_1$. If $n_r + m_1 \neq 0$ then both $(xy)z$ and $x(yz)$ equal $x_1^{n_1} \ldots x_r^{n_r+m_1}y_2^{m_2} \ldots y_k^{mk}z_1^{p_1} \ldots z_t^{p_t}$. On the other hand if $n_r + m_1 = 0$, then xy is defined inductively as $x'y'$ where $x' = x_1^{n_1} \ldots x_{r-1}^{n_{r-1}}$; $y' = y_2^{m_2} \ldots y_k^{mk}$. In this case $(xy)z = (x'y')z$ and $x(yz) = x'(y'z)$. Since x', y' are shorter than x, y by induction hypothesis we have $(x'y')z = x'(y'z)$. Hence $(xy)z = x(yz)$. Other cases are handled similarly, resorting to induction when needed. This establishes the associativity of the multiplication. Inverses are very easy to obtain. The inverse of $x_1^{n_1} \ldots x_r^{n_r}$ is simply $x_r^{-n_r}x_{r-1}^{-n_{r-1}} \ldots x_2^{-n_2}x_1^{-n_1}$. Thus we see that G is a group. That it is freely generated by S follows by the very construction of G. ∎

3.19 Definition: The group G constructed above is called the **free group** on the set S and is denoted by $F(S)$.

Although Theorem (3 18) establishes the existence of free groups on any set, unless the set S is empty or a singleton they are not very familiar as noted above. Still they have some very interesting theoretical properties. Just as the subgroups of permutation groups provide a complete picture of all groups (upto isomorphisms), the quotient groups of free groups give a complete picture of all groups (upto isomorphisms). In other words, every group is isomorphic to a quotient group of a free group. A special case of this result is already familiar to us. Every cyclic group is isomorphic to a quotient group of $\mathbb{Z}$, which is the free group on a singleton set. To prove the general result, we first need a property of free groups which is important in its own rights.

3.20 Theorem: Let S be a set and $F(S)$ the free group on S. Let H be any group and $f: S \to H$ any function. Then f can be uniquely extended to a group homomorphism from $F(S)$ to H. That is, there exists a unique homomorphism $\theta: F(S) \to H$ such that $\theta(x) = f(x)$ for all $x \in S$.

Proof: Every element of $F(S)$ is of the form $x_1^{n_1} \ldots x_r^{n_r}$, where $x_i \in S$, $n_i \in \mathbb{Z} - \{0\}$ with the restrictions noted above. Now $\theta(x_i)$ has to be $f(x_i)$ which is already defined for $i = 1, \ldots, r$. So if θ is to be a group homomorphism, we have to choice but to define $\theta(x_1^{n_1} \ldots x_r^{n_r})$ as $[f(x_1)]^{n_1} \ldots \ldots [f(x_r)]^{n_1} \in H$. $\theta(e)$ is defined as the identity element of the group H. To show that θ is a homomorphism, let $x = x_1^{n_1} \ldots x_r^{n_r}$ and $y = y_1^{m_1} \ldots \ldots y_k^{m_k} \in G$. To prove that $\theta(xy) = \theta(x)\theta(y)$, we would again have to proceed by induction on the lengths of x and y, keeping in mind the definition of xy, just as we did while proving associativity. (This, by the way, is a common feature in mathematics. Whenever something is defined inductively, many of its properties, at least those where the definition is directly involved, have to be proved by an inductive argument.) So we get a well-defined homomorphism $\theta: F(S) \to H$. By its very construction, $\theta(x) = f(x)$ for all $x \in S$. Also, because we had no choice in defining θ, it follows that θ is the only homomorphism extending f. ∎

The result we are after is now an immediate consequence.

3.21 Theorem: Every group is isomorphic to a quotient group of a free group.

Proof: Let G be a group. Let S be any subset of G which generates G (not necessarily freely). We may even take S to be G itself. Let $F(S)$ be the free group on the set S. Elements of $F(S)$ should not be confused with those of G, even though both can be represented as products of powers of elements of S. For example G may be abelian. In that case, for $x_1, x_2 \in S$, $x_1 x_2$ and $x_2 x_1$ are the same elements in G but as elements of $F(S)$, they are distinct. (In case the reader finds it necessary to avoid confusion, elements of S may be denoted by putting bars over them while representing elements of G. Thus $\bar{x}_1\bar{x}_2$ may equal $\bar{x}_2\bar{x}_1$ but $x_1 x_2 \neq x_2 x_1$. But with a little care, this would not be necessary.) Now define $f : S \to G$ by $f(x) = x$ for $x \in S$ (or as $f(x) = \bar{x}$ if we put bars to denote elements of G). In case $S = G$, f is merely the identity function on the set G. By the theorem above there exists a unique group homomorphism $\theta : F(S) \to G$ which extends f. In fact θ is given by $\theta\,(x_1^{n_1} \ldots x_r^{n_r}) = x_1^{n_1} \ldots\ldots x_r^{n_r}$ where on the left hand side $x_1^{n_1} \ldots x_r^{n_r}$ represents an element of $F(S)$ and on the right hand side it represents an element of G. Since S generates the group G, it follows that θ is onto. Let K be the kernel of θ. Then by Theorem (3.5), G is isomorphic to the quotient group $F(S)/K$. ∎

Although this theorem is of great theoretical significance, it is rarely used in the form it is stated. Given a known group G, it is not particularly helpful to express it as the quotient group of $F(S)$ (S being some generating set for G) by some normal subgroup K. In practice, we often use the theorem the other way. We specify a free group and some normal subgroup of it and use this information to *define* the group G. The theorem above says that this method does define every group upto isomorphism As a concrete example, let us see how the group G of isometries of a regular pentagon (also called the dihedral group D_5, see Section 1) can be defined by this procedure. We already know the structure of G completely. We know that G can be generated by two elements (for example, by a rotation and a flip). So we consider a set S with two elements, say, $S = \{a, b\}$ and take the free group $F(S)$ on S. We define $f : S \to G$ by $f(a) = r_1$, $f(b) = f_1$ (see Section 1). Then by Theorem (3.20), there is a unique group homomorphism $\theta : F(S) \to G$ which extends f. Let K be the kernel of θ. If we could compute the kernel K in some independent manner, then we could define G, the dihedral group D_5, as the quotient group $F(S)/K$. Let us therefore see what K looks like. In G we have the relations $r_1^5 = f_1^2 =$ identity and so a^5, $b^2 \in K$. Moreover, in G we also have $r_1 \circ f_1 = f_4$ which is an element of order 2. So $(ab)^2 \in K$. K contains many other elements. But we claim that the three elements, a^5, b^2 and $(ab)^2$ of $F(S)$ generate K as a normal subgroup, that is, K is the smallest normal subgroup of G which contains these three elements. To show this, suppose N is any normal subgroup of G which contains a^5, b^2 and $abab$. We show $K \subset N$. First note that $baba \in N$, because we can write $baba$ as $b\,(abab)\,b^{-1}$ which must be in N by normality. Next we claim that $(a^2 b)^2 \in N$. We write $(a^2b)^2$ as $[a(abab)a^{-1}]\,ab^{-1}ab$. Since $abab \in N$, $a(abab)a^{-1} \in N$. Also $ab^{-1}ab = ab^{-2}a^{-1} abab$ which is again in N since $b^{-2} \in N$ gives $ab^2a^{-1} \in N$ and $(ab)^2 \in N$. Putting it together, we get that $(a^2b)^2 \in N$. Similarly $(ba^2)^2 \in N$. By induction on n, it can be shown that for every positive integer n, $(a^nb)^2$ and $(ba^n)^2$ are in N. As for the negative powers, we already saw above that $ab^{-1}ab \in N$. Since $b^{-2} \in N$, we get $(ab^{-1})^2 = ab^{-1}ab\ b^{-2} \in N$. Similarly $(b^{-1}a)^2 \in N$. It is then seen by induction that $(a^nb^{-1})^2$ and $(b^{-1}a^n)^2$ are in N for all positive integers n. By taking inverses, $(a^nb)^2$ and $(ba^n)^2$ are in N for all integral values of n.

With this spadework, we are now in a position to show that K, the kernel of $\theta : F(S) \to G$ is contained in N. Let $x = x_1^{n_1} x_2^{n_2} \ldots x_r^{n_r}$ be a typical element of K. We are given that $\theta(x) = e$ and we have to show that $x \in N$. The proof will be by induction on r, the length of x. Each x_i is a or b and since no two adjacent x_i's are equal, they are alternatively a's and b's. Suppose $r = 1$. Then x is either a^{n_1} or b^{n_1} and accordingly $\theta(x)$ is either $r_1^{n_1}$ or $f_1^{n_1}$. We are given that $\theta(x) = e$. Recalling that the orders of the elements r_1 and f_1 in the group G are 5 and 2 respectively, it follows that in the first case n_1 is a multiple of 5 while in the second case n_1 is a multiple of 2. In either case $x_1^{n_1} \in N$ (since $a^5 \in N$ and $b^2 \in N$) Now suppose $r > 1$ and the result holds for all elements of length less than r. First consider the case r is even. Then either $x_1 = b$ or $x_r = b$. In the first case, $x = b^{n_1} a^{n_2} x_3^{n_3} \ldots x_r^{n_r}$, i.e., $x = b^{n_1} y$ where $y = a^{n_2} x_3^{n_3} \ldots x_r^{n_r}$. We now make two subcases. First let n_1 be even. Then $\theta(x) = e$ gives $[\theta(b)]^{n_1} \theta(y) = e$ which implies $\theta(y) = e$ since $(f_1)^{n_1} = e$ (n_1 being even). So $y \in K$. Since y has length less than r, $y \in N$ by induction hypothesis. But $b^{n_1} \in N$ since n_1 is even and $b^2 \in N$. Thus $x \in N$ in case n_1 is even. Now suppose n_1 is odd, say, $n_1 = 2p + 1$. Then we write $x = b^{2p} z$ where $z = ba^{n_2} x_3^{n_3} \ldots x_r^{n_r}$. We again get $\theta(z) = e$. Now $z = (ba^{n_2})^2 w$ where $w = a^{-n_2} b^{-1} x_3^{n_3} \ldots x_r^{n_r} = a^{-n_2} x_3^{n_3 - 1} x_4^{n_4} \ldots x_r^{n_r}$ (since $x_3 = b$). Once again we show that $\theta[(ba^{n_2})^2] = e$ and hence $\theta(w) = e$, that is $w \in K$. The length of w is $r - 1$ (or even less in case $n_3 = 1$). So by induction hypothesis, $w \in N$. Since $(ba^{n_2})^2 \in N$ (as proved above) and $b^{2p} \in N$ (since $b^2 \in N$), it follows that $y \in N$ and finally $x \in N$. We are still in the case r is even. We have completely disposed off the subcase $x_1 = b$. If $x_1 = a$ then $x_r = b$ (since r is even) and a similar argument holds. Alternatively, we take x^{-1} which equals $x_r^{-n_r} \ldots x_1^{-n_1}$ and apply the argument above to show that $x^{-1} \in N$, whence $x \in N$. We still have to consider the case r is odd. This is much easier. We simply write $x = x_1^{n_1} x_2^{n_2} \ldots x_r^{n_r}$ as $x_1^{n_1}(u) x_1^{-n_1}$ where u is $x_2^{n_2} \ldots x_r^{n_r} x_1^{n_1}$ which equals $x_2^{n_2} \ldots x_r^{n_r + n_1}$ (since $x_r = x_1$, r being odd). $\theta(x) = e$ gives $[\theta(x_1)]^{n_1} \theta(u)[\theta(x_1)]^{-n_1} = e$ which implies $\theta(u) = e$, showing that $u \in K$. Now the length of u is $r - 1$ (or even less in case $n_r = -n_1$). So by induction hypothesis, $u \in N$ But N is normal. So $x = x_1^{n_1} u x_1^{-n_1} \in N$. Thus in all cases, we have shown that $x \in N$. This completes the proof that $K \subset N$ and shows that the kernel K of θ is the smallest normal subgroup of $F(S)$ containing the three elements a^5, b^2 and $(ab)^2$.

Let us pause to recapitulate what we have achieved through this rather sticky argument. We started with a known group G, the group of isometries of a regular pentagon. We then expressed G as a homomorphic image of the free group $F(S)$ where S is a set with two elements a and b. We let K be the kernel of this homomorphism. We then showed that K is the normal subgroup of $F(S)$ generated by a subset R of $F(S)$ namely, $R = \{a^5, b^2, abab\}$. Now suppose we do not start with the group G. We can still consider the set $S = \{a, b\}$, the free group $F(S)$ and the subset $R = \{a^5, b^2, abab\}$. We then let K be the normal subgroup of $F(S)$ generated by R, that is, the smallest normal subgroup of $F(S)$ containing R. We could still define the quotient group $F(S)/K$. Our work shows that this quotient group is isomorphic to the group of isometries of a regular pentagon. Strictly speaking elements of $F(S)/K$ are cosets of K by elements of $F(S)$. However, it is customary to denote them by the same symbols as elements of $F(S)$. In particular, we denote the cosets aK, bK (which are elements of $F(S)/K$) by a and b respectively. Then $F(S)/K$ is generated (although not freely) by these elements a and b. The fact that a^5, b^2 and $abab$ are elements of K means that in $F(S)/K$, the following identities hold, (i) $a^5 = e$ (ii) $b^2 = e$ and (iii) $abab = e$. In presence of (i) and (ii), (iii) is equivalent to $ba = a^4b$. These identities are called *relations*. It is customary to describe $F(S)/K$ as the group generated by two elements a and b satisfying (or subject to) the relations (i), (ii) and (iii). Using these relations it is easy to see that $F(S)/K$ consists of elements of the form a^ib^j where i, j are integers and where a^ib^j is to be regarded as the same element as a^mb^n if $i \equiv m$ (modulo 5) and $j \equiv n$ (modulo 2). To multiply two such elements, say, a^ib^j and a^rb^s we reduce $a^ib^ja^rb^s$ by repeated applications of the relation $ba = a^4b$. For example, $abab = a(ba)b = aa^4b^2 = a^5b^2 = e$. It is clear that $F(S)/K$ is a non-abelian group of order 10, generated by two elements.

This procedure can be generalised. We let S be any set whatsoever and form the free group on S, $F(S)$. We also let R be any subset of $F(S)$ and define K to be the smallest normal subgroup of $F(S)$ containing R. Then the quotient group $F(S)/K$ is said to have elements of S as **generators**. The elements of R are called **relations** (often the corresponding equalities which hold in $F(S)/K$ are called relations). This method of specifying a group is known as presenting a group through generators and relations. It is a very compact way of specifying a group. For example, instead of defining the quaternions group Q in terms of the elaborate table of multiplication shown in Figure 5.3 (and then carrying out the horrendous verification of associativity) we may simply *define* the quaternion group as the group generated by two elements a and b subject to the relations $a^4 = b^4 = e$, $ba = a^3b = ab^3$. We leave it to the reader to verify that the group so defined is indeed isomorphic to Q. As another example, the relations $a^7 = b^3 = e$, $ba = a^2b$ give rise to a non-abelian group of order 21.

If the set S is finite, the group $F(S)/K$ is said to be **finitely generated**. If the set R is also finite, it is called **finitely presented**. Obviously every finite group is finitely generated. It is not necessarily true that a subgroup of a finitely generated group is also finitely generated. It is true but non-trivial to prove that a subgroup of a free group is free. A proof can be given using graph theory (see the Epilogue).

Theorem (3.21) is applicable to all groups, abelian as well as non-abelian. Even if G is an abelian group, however, the free group whose quotient group is isomorphic to G need not be abelian. This is sometimes undesirable. Given an abelian group G, we often want to express G as a quotient group of some abelian group whose properties are analogous to those of free groups. Such groups are called **free abelian** groups. They turn out to be the abelianised free groups (see Exercise (2.13)). The term 'free abelian' should not be interpreted as 'free and abelian'. Indeed the only free groups which are abelian are $\mathbb{Z}$ and the trivial group. The correct meaning of 'free abelian' is the counterpart of free groups for the world of abelian groups. Properties of free abelian groups, as well as the analogue of Theorem (3.21) will be developed through exercises.

While closing the section, we remark that the topic of group representations is intimately related to that of group actions (see the Epilogue).

Exercises

3.1 For any three groups G_1, G_2, G_3, prove that $G_1 \times G_2$ is isomorphic to $G_2 \times G_1$ and $(G_1 \times G_2) \times G_3$ and $G_1 \times (G_2 \times G_3)$ are both isomorphic to $G_1 \times G_2 \times G_3$.

3.2 Suppose $f: G \to H$ is a group homomorphism and $x \in G$ has order n. Prove that the order of $f(x)$ divides n. If f is onto and K is a subgroup of index m in H, prove that $f^{-1}(K)$ has index m in G.

3.3 Suppose G, H are finite groups and $f: G \to H$ a homomorphism with range R. Prove that $\circ(R)$ is a common divisor of $\circ(G)$ and $\circ(H)$. Deduce that if $\circ(G)$ and $\circ(H)$ are relatively prime then the only homomorphism from G to H is the trivial one.

3.4 Let G be any group. Let $A(G)$ be the set of all automorphisms of G onto itself. Prove that $A(G)$ is a subgroup of $S(G)$. (The group $A(G)$ is called the **automorphism group** of G.)

3.5 Let G, H be two groups. If G is isomorphic to H, prove that $A(G)$ is isomorphic to $A(H)$. Show by an example that the converse is false.

3.6 Prove that $A(\mathbb{Z}_p)$ is isomorphic to the group $\mathbb{Z}_p - \{([0]\}$ under modulo p multiplication for any prime p. Also find $A(\mathbb{Z})$.

3.7 Let $I(G)$ be the set of all inner automorphisms (see Example (8) in the text) of a group G. Prove that $I(G)$ is a subgroup of G and as a group $I(G)$ isomorphic to the quotient group G/Z where Z is the centre of G.

3.8 Prove that the inversion function for a group (Exercise (2.6)) is an automorphism if and only if the group is abelian.

**3.9 Prove that every group with more than two elements has a non-trivial automorophism group, that is, has an automorphism other than the identity function.

3.10 Suppose G is a finite group and $f: G \to G$ is an automorphism which sends more than three fourths of the elements of G to their inverses. Prove that $f(x) = x^{-1}$ for all $x \in G$ and G is abelian (cf. Exercise (2.22)).

3.11 Let G be a group and K a normal subgroup of G. For any subgroup H of G, HK is a subgroup of G by Exercise (1.22). Prove that $H \cap K$ is normal in H and that the quotient group $H/H \cap K$ is isomorphic to the quotient group HK/K. (This is called the **Noether isomorphism theorem.**) Illustrate this with $G = \mathbb{R}^2$ under co-ordinatewise addition and H, K two lines through the origin.

3.12 Let T be a subset of a set S. Let G, H be the permutation groups of T, S respectively. Define $f: G \to H$ as follows. Let $\theta \in G$. Define $f(\theta)$ to be the function $\tau: S \to S$ such that $\tau(x) = \theta(x)$ if $x \in T$ and $\tau(x) = x$ if $x \in S - T$. Prove that f is a monomorphism. (See also Exercise (1.30).).

3.13 Let $\mathbb{C}^*$ be the multiplicative group of non zero complex numbers and $\mathbb{R}^*$ the multiplicative group of non-zero real numbers. Define $f: \mathbb{C}^* \to \mathbb{R}^*$ by $f(z) = |z|$ for $z \in \mathbb{C}^*$. (Here, if $z = x + iy$ with x, y real then $|z|$ means $\sqrt{x^2 + y^2}$.) Prove that f is a homomorphism with kernel S^1. Verify the truth of Theorem (3.5) for this homomorphism in the light of Exercise (2.3).

3.14 Prove that the Klein group is isomorphic to $\mathbb{Z}_2 \times \mathbb{Z}_2$.

3.15 Let G be an abelian group, H a subgroup of G and $p: G \to G/H$ the quotient homomorphism. Suppose there exists a homomorphism $j: G/H \to G$ such that for all $xH \in G/H$, $p(j(xH)) = xH$. (In other words p has a right inverse which is a group homomorphism). Prove that G is isomorphic to the product group $H \times G/H$. (Hint: For $x \in G$, let $h(x) = x - j(p(x))$. Prove that h is a homomorphism of G into itself with range H.)

3.16 Suppose G, H are groups with K, L as normal subgroups respectively. Suppose $f: G \to H$ is a homomorphism which takes K into L (that is, $f(K) \subset L$). Prove that there exists a homomorphism $\psi: G/K \to H/L$ such that the following diagram (in which the horizontal arrows represent the respective quotient homomorphisms) is commutative:

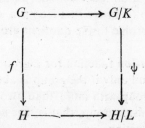

3.17 Suppose $f: G \to H$ is a homomorphism where H is abelian. Prove that the kernel of f must contain $C(G)$, the commutator subgroup of G. Using this give an alternate solution to Exercise (2.13).

3.18 Let f and g be homomorphisms from a group G to a group H. Suppose S is a subset of G and K is the subgroup of G generated by S. If $f(x) = g(x)$ for all $x \in S$, prove that $f(x) = g(x)$ for all $x \in K$. (Consequently a homomorphism is completely determined by its values on any set of generators for the domain group.)

3.19 Prove that every group of order 10 is isomorphic either to $\mathbb{Z}_{10}$ or to the group of isometries of a regular pentagon.

3.20 Prove that every group of order p^2 where p is a prime is isomorphic either to $\mathbb{Z}_{p^2}$ or to $\mathbb{Z}_p \times \mathbb{Z}_p$.

*3.21 Prove that every non-abelian group of order eight is isomorphic either to Q, the group of quaternions or to D_4, the group of isometries of a square. (Hint: start with Exercise (2.16). Note that the quotient group by the centre cannot contain an element of order 4.)

3.22 Prove that every abelian group of order 8 is isomorphic to either $\mathbb{Z}_8$, $\mathbb{Z}_2 \times \mathbb{Z}_4$ or $\mathbb{Z}_2 \times \mathbb{Z}_2 \times \mathbb{Z}_2$.

3.23 In the proof of Cayley's theorem, what would happen if we associated an element g to the right translation by g? Prove that an alternate proof of Cayley's theorem is possible if we associate an element g to the right translation by g^{-1}.

3.24 Give an alternate proof of Corollary (3.16) (that is one not based on Theorem (3.15)).

3.25 Let G, H, K be abelian groups. Let $f: G \to H$ be a homomorphism. Define $f^{\#}: \text{Hom}(H, K) \to \text{Hom}(G, K)$ by $f^{\#}(\lambda) = \lambda \circ f$ for $\lambda \in \text{Hom}(H, K)$. Similarly define $f_{\#}: \text{Hom}(K, G) \to \text{Hom}(K, H)$ by $f_{\#}(\mu) = f \circ \mu$ for $\mu \in \text{Hom}(K, G)$. Prove that $f^{\#}$ and $f_{\#}$ are homomorphisms.

3.26 Let S and T be sets of the same cardinality. Prove that the free groups $F(S)$ and $F(T)$ are isomorphic to each other. (The converse is true but not so easy to prove. It will be postponed to Exercise (6.3.31).)

3.27 Prove that a group with two generators a, b with relations $a^5 = b^2 = e$ and $ab = ba$ is isomorphic to $\mathbb{Z}_5 \times \mathbb{Z}_2$ and hence to $\mathbb{Z}_{10}$.

3.28 Suppose G is a group with two generators a and b with relations $a^m = b^n = e$ and $ba = a^k b$, where m, n, k are positive integers. Prove that in G, $(a^i b^j)(a^r b^s)$ equals $a^{i+rk^j} b^{j+s}$ for all integers i, j, r, s.

3.29 In the last exercise suppose $m = 7$ and $n = 3$. Prove that G is a non-abelian group of order 21 if $k = 2$ or 4, is the cyclic group of order 21 if $k = 1$ and is the group $\mathbb{Z}_3$ if $k = 3$. (Hint: In each case consider a homomorphism from $F(\{a, b\})$ onto the appropriate group and show that its kernel is generated by the three relations.)

3.30   Prove that the group generated by two elements a and b with relations $a^4 = b^4 = e$, $ab^3 = a^3b = ba$ is isomorphic to the group of quaternions.

3.31   Let S be any set and $F(S)$ the free group on S. Let $C(S)$ be the commutator subgroup of $F(S)$. The quotient group $F(S)/C(S)$ is called the **free abelian group** on the set S. Denote this group by $A(S)$. We think of S as a subset of $A(S)$ by identifying $x \in S$ with the coset $xC(S)$ in $A(S)$. If G is an abelian group and $f : S \to G$ any function prove that there exists a unique homomorphism $\psi : A(S) \to G$ which extends f. Hence prove that every abelian group is isomorphic to a quotient group of a free abelian group.

*3.32   Let S be any set. Let $\mathbb{Z}$ be the additive group of integers. Let $\mathbb{Z}^S$ be the set of all functions from S to $\mathbb{Z}$. Then $\mathbb{Z}^S$ is an abelian group under pointwise addition of functions. Let $B(S)$ be the subset $\{f \in \mathbb{Z}^S : f$ vanishes at all except finitely many points of $S\}$. Prove that $B(S)$ is a subgroup of $\mathbb{Z}^S$ and that it is isomorphic to $A(S)$, the free abelian group on S. (Hint : Define $f : S \to B(S)$ by $f(s)$ to be the function $f_s : S \to \mathbb{Z}$ which has value 1 at s and 0 everywhere else. Extend f to a homomorphism from $F(S)$ to $B(S)$ and show its kernel to be precisely $C(S)$).

3.33   Identify an element $s \in S$ with the function f_s defined in the hint to the last exercise. With this identification S is a subset of $B(S)$. Prove that every element of $B(S)$ can be expressed as a sum

$$n_1 s_1 + n_2 s_2 + \ldots + n_r s_r$$

where

$$r \geqslant 0, \ s_i \in S, \ n_i \in \mathbb{Z} - \{0\}$$

and no two s_i's are the same and such an expression is unique except for the order of the summands. (This gives a direct construction for the free abelian group on a set S.)

3.34   Prove that the product of two free abelian groups is a free abelian group. (In particular $\mathbb{Z} \times \mathbb{Z}$, $\mathbb{Z} \times \mathbb{Z} \times \mathbb{Z} \ldots$ are free abelian groups. So the free abelian groups are not as 'abstract' as the free group).

3.35   Let G be the free group with two generators, a and b. Let H be the set of those elements of G in which the sum of the exponents of a equals the sum of exponents of b. (Thus, ab, ba, $a^{-1}b^{-1}$, ab^2a, $abab$, $ba^4b^{-7}a^{-2}b^8$ are in H but aba, $abab^{-1}$, $a^4b^{-3}a^2b^{-1}$ are not in H.) Prove that:

(i) H is a normal subgroup of G,

(ii) H is a free group generated freely by the set

$$\{a^n b^n : n \in \mathbb{N}\} \cup \{b^n a^n : n \in \mathbb{N}\};$$

another set of generators being $\{a^r b^n : n = \pm 1, \pm 2, \pm 3.$

(iii) Let $\mathbb{Z} \times \mathbb{Z}$ be the group of all ordered pairs of integers under coordinatewise addition. Consider the unique homomorphism $f : G \to \mathbb{Z} \times \mathbb{Z}$ which takes a to $(1, 0)$ and b to $(0, 1)$

(which exists by Theorem (3.20)). Then H is precisely $f^{-1}(Z)$ where $Z = \{(x, x) : x \in \mathbb{Z}\}$.

(iv) H is not finitely generated. (Hint: Otherwise a finite subset from (ii) would generate H.)

Notes and Guide to Literature

All the concepts in this section are of fundamental nature. Their analogues for other algebraic structures will come in the next chapter. Some of the examples of homomorphisms we have given are actually linear transformations and will be visited again when we study vector spaces.

Free groups are important in many branches of mathematics, such as 'knot theory'. For more on them, see for example Robinson [1]. For knot theory, see Crowell and Fox [1].

Although the terminology used in this section is by now standard, the reader should be wary in consulting old literature. For example, Carmichael [1] calls 'simple isomorphism' what we call as isomorphism.

4. Permutation Groups

As proved in the last section, the permutation groups of finite sets, along with their subgroups, give a complete picture of all finite groups. For this reason alone they deserve to be studied in detail. But there are other reasons too. In applications we often consider the various transformations of a set into itself. These form groups which are subgroups of the permutation groups (not just isomorphic to them). In the chapter on group actions*, we shall see many instances of this. Secondly, some of the algebraic concepts we study in various branches of mathematics can be expressed very simply in terms of the permutation groups. We shall illustrate this in the next chapter by showing how the definition of a determinant of a square matrix can be given in terms of permutation groups. Finally, some of the properties of the permutation groups proved here are needed when we want to prove the well-known impossibility of finding a formula for the roots of a polynomial of degree 5 or more.** Apart from these theoretical applications, permutation groups also provide interesting counter-examples.

As before, by S_n we denote the group, under composition, of all bijections of an n-element set, generally taken to be the set $\{1, 2, ..., n\}$ where n is a positive integer. Elements of S_n are functions from $\{1, 2, ..., n\}$ to itself and will generally be denoted by σ, τ, θ etc. It is rather clumsy to

---

*See the Epilogue.
**See the Epilogue.

denote these functions in the usual functional notation, in the form $f(1) =$ ..., ..., ..., $f(n) = ...$. It is convenient to have a compact notation. One method is to agree to list $f(1), ..., f(n)$ always in this order and simply write this arrangement. For example 23541 is the (unique) permutation say σ, of 1, 2, 3, 4, 5 which takes 1 to 2, 2 to 3, 3 to 5, 4 to 4 and 5 to 1. To stress that we are thinking of 23541 as a function rather than as an arrangement of symbols (although the two are equivalent), it is helpful to sacrifice a little space and write σ as $\begin{pmatrix} 1 & 2 & 3 & 4 & 5 \\ 2 & 3 & 5 & 4 & 1 \end{pmatrix}$. An advantage of this notation is that it is not necessary to have the top row always in the increasing order (although it is generally so) provided the entries in the bottom row are also shifted correspondingly. Thus σ can as well be expressed as $\begin{pmatrix} 1 & 3 & 5 & 2 & 4 \\ 2 & 5 & 1 & 3 & 4 \end{pmatrix}$ or as $\begin{pmatrix} 1 & 2 & 3 & 5 & 4 \\ 2 & 3 & 5 & 1 & 4 \end{pmatrix}$ or in many other similar ways. This helps in writing down the inverse of a permutation as well as the composite of two permutation. For inversion we simply interchange the top row with the bottom row and then reshuffle it, if necessary, to bring it to the standard form. Thus, in the example above, σ^{-1} is the permutation $\begin{pmatrix} 2 & 3 & 5 & 4 & 1 \\ 1 & 2 & 3 & 4 & 5 \end{pmatrix}$, which in the standard from equals $\begin{pmatrix} 1 & 2 & 3 & 4 & 5 \\ 5 & 1 & 2 & 4 & 3 \end{pmatrix}$ or 51243. Similarly let τ be some other element of S_5, say $\tau = \begin{pmatrix} 1 & 2 & 3 & 4 & 5 \\ 5 & 2 & 1 & 3 & 4 \end{pmatrix}$. To find the composite $\sigma \circ \tau$ we rewrite σ so that its top row coincides with the bottom row of τ, that is, $\sigma = \begin{pmatrix} 5 & 2 & 1 & 3 & 4 \\ 1 & 3 & 2 & 5 & 4 \end{pmatrix}$. Then $\sigma \circ \tau$ is simply $\begin{pmatrix} 1 & 2 & 3 & 4 & 5 \\ 1 & 3 & 2 & 5 & 4 \end{pmatrix}$ which is obtained by taking the bottom row of σ (in the new form). The procedure given here amounts to taking each element of $\{1, 2, 3, 4, 5\}$ and tracing it under the action of τ and then that of σ. For example 3 goes to 1 under τ and $1(= \tau(3))$ goes to 2 under σ. So 3 goes to 2 under the composite $\sigma \circ \tau$.

We now turn to another method of expressing a permutation, which is important not only notationally but conceptually as well. We begin by defining a special type of permutations.

4.1 Definition: Let n, r be integers with $1 \leqslant r \leqslant n$. A permutation $\sigma \in S_n$ is said to be a cycle of length r or an r-cycle if there exists distinct elements $i_1, i_2,, ..., i_r \in \{1, 2, ..., n\}$ such that

(i) $\sigma(i_1) = i_2,\ \sigma(i_2) = i_3,\ldots\ \sigma(i_{r-1}) = i_r,\ \sigma(i_r) = i_1$

and

(ii) $\sigma(i) = i$ for all $i \in \{1, 2,\ldots, n\} - \{i_1,\ldots.\ i_r\}$.

In other words, a cycle of length r permutes the elements of some r-subset of $1, 2,\ldots, n$ in a cyclic manner and leaves all other elements unaffected. A cycle of length 1 is of course the identity permutation. A cycle of length 2 is a permutation which interchanges two elements of $\{1, 2,\ldots, n\}$ and leaves other elements fixed. These are the simplest non-trivial cycles and are called **transpositions**. Their importance in the structure of S_n will be brought out later. As examples, for $n = 4$, the permutations $\begin{pmatrix} 1 & 2 & 3 & 4 \\ 1 & 3 & 2 & 4 \end{pmatrix}$, $\begin{pmatrix} 1 & 2 & 3 & 4 \\ 4 & 2 & 1 & 3 \end{pmatrix}$ and $\begin{pmatrix} 1 & 2 & 3 & 4 \\ 4 & 3 & 1 & 2 \end{pmatrix}$ are cycles of length 2, 3 and 4 respectively (the corresponding permuted subsets being $\{2, 3\}$, $(1, 4, 3\}$ and $\{1, 4, 2, 3\}$). However, the permutation $\begin{pmatrix} 1 & 2 & 3 & 4 \\ 4 & 3 & 2 & 1 \end{pmatrix}$ is not a cycle by itself. It is, however, the product of the two 2-cycles $\begin{pmatrix} 1 & 2 & 3 & 4 \\ 4 & 2 & 3 & 1 \end{pmatrix}$ and $\begin{pmatrix} 1 & 2 & 3 & 4 \\ 1 & 3 & 2 & 4 \end{pmatrix}$. It is obvious that the order of an r-cycle as an element of S_n is r.

There is a very compact way to denote permutations which are cycles. Suppose σ is a cycle of length r and $i_1, i_2,\ldots, i_r$ are the elements which are cyclically permuted by σ (in the order of their listing). Then σ is denoted by $(i_1, i_2,\ldots, i_r)$ or simply by $(i_1 i_2 \ldots i_r)$. We could start with any i_k and the same r-cycle would be denoted by $(i_k i_{k+1}\ldots i_r i_1 \ldots i_{k-1})$. It follows that the same r-cycle can be denoted in r ways depending on the starting point. For example the 4-cycle $\begin{pmatrix} 1 & 2 & 3 & 4 \\ 4 & 3 & 1 & 2 \end{pmatrix}$ can be denoted by $(1\ 4\ 2\ 3)$, $(4\ 2\ 3\ 1)$, $(2\ 3\ 1\ 4)$ or by $(3\ 1\ 4\ 2)$. From this, it is easy to compute the number of distinct cycles of length r. A transposition (that is a 2-cycle) which interchanges, say, i and j will be denoted by (ij) or (ji).

What happens when we compose two cycles ? The answer depends on the two cycles. Suppose σ, $\tau \in S_n$ are cycles of length r, s respectively, say, $\sigma = (i_1 i_2\ldots i_r)$ and $\tau = (j_1 j_2 \ldots j_s)$. If these two cycles are disjoint, (that is the sets $\{i_1,\ldots, i_r\}$ and $\{j_1,\ldots, j_s\}$ are disjoint) then it is easy to see that they commute with each other, that is $\sigma \circ \tau = \tau \circ \sigma$, because under both of them, i_k goes to i_{k+1} (with i_r going i_1), j_p goes to j_{p+1} (with j_s going to j_1) and all other elements of $\{1, 2,\ldots, n\}$ remain unchanged. Note, however, that this permutation is not a cycle (unless $r = 1$ or $s = 1$). Two non-disjoint cycles do not in general commute with other. For example, $(12) \circ (13)$ equals $(1\ 3\ 2)$ while $(13) \circ (12)$ equals $(1\ 2\ 3)$.

Cycles are the building blocks for elements of S_n and behave, to some extent, like the atoms in a Boolean algebra. The following theorem is the analogue of Proposition (4.1.10), Part (ii).

4.2 Theorem: Every permutation of n symbols, other than the identity, is the product of mutually disjoint cycles of length greater than 1. This expression is unique except for the order of the cycles (and except for the different ways of expressing the same cycle).

Proof: We proceed by induction on n. For $n = 1$, there is no permutation other than the identity permutation and so the result holds vacuously. Suppose $n > 1$ and the result holds for all $m < n$. Let σ be a permutation of n symbols, say, $1, 2, ..., n$. Since $\sigma \neq$ identity, there exists some x such that $\sigma(x) \neq x$. Set $i_1 = x$. Define $i_2 = \sigma(i_1) = \sigma(x)$, $i_3 = \sigma(i_2) = \sigma^2(i_1) = \sigma^2(x)$,... and in general $i_{k+1} = \sigma(i_k) = \sigma^k(i_1)$ for every positive integer k. This gives an infinite sequence $(i_1, i_2, i_3, ..., i_k, ...)$. However, this sequence takes values in the finite set $\{1, 2, ..., n\}$. So its terms cannot be all distinct. Thus there exist integers p, q with $p < q$ such that $i_p = i_q$. Let $q - p = s$. Then $i_q = i_{p+s} = \sigma^{p-1+s}(i_1) = \sigma^{p-1}(\sigma^s(i_1)) = \sigma^{p-1}(i_{s+1}))$ while $i_p = \sigma^{p-1}(i_1)$. So $i_p = i_q$ implies $i_{s+1} = i_1$, since σ^{p-1}, as a function from $\{1, 2, ..., n\}$ into itself is injective. Thus we have shown that there exists a positive integer s such that $i_{s+1} = i_1$. Let r be the least positive integer with this property. Then $i_1, i_2, ..., i_r$ are all distinct (by an argument analogous to the proof of Theorem (1.10)). It is then clear that $(i_1\ i_2...i_r)$ is an r-cycle. Call this r-cycle as σ_1. Define $\tau: \{1, 2, ..., n\} \to \{1, 2, ..., n\}$ by $\tau(i_k) = i_k$ for all k and $\tau(x) = \sigma(x)$ if $x \neq i_1, ..., i_r$. In other words, τ behaves exactly as σ except on the set $\{i_1, i_2, ..., i_r\}$ which is left invariant by τ. Clearly τ and σ_1 commute with each other and the product $\sigma_1 \tau$ (or $\tau \sigma_1$) equals σ. Now we may as well think of τ as a permutation of the set T, where $T = \{1, 2, ..., n\} - \{i_1, i_2, ..., i_r\}$. Now $| T | = n - r < n$. If τ is the identity then $\sigma = \sigma_1$ which is a cycle. If $\tau \neq$ identity, then by induction hypothesis τ can be expressed as a product of disjoint cycles of length greater than 1, say, $\tau = \sigma_2 \sigma_3 ... \sigma_k$. Evidently $\sigma_2, ..., \sigma_k$ are disjoint from σ_1. Every permutation of T can also be regarded as a permutation of $\{1, 2, ..., n\}$ (cf. Exercise (3.12)). So $\sigma = \sigma_1 \sigma_2 ... \sigma_k$, a product of cycles. As for uniqueness suppose σ is also expressed as $\theta_1 \theta_2 ... \theta_u$ where each θ_i is a cycle of length greater than 1 and every two θ's are mutually disjoint. We started with x such that $\sigma(x) \neq x$. Clearly x must be in one of these cycles, as otherwise $\sigma(x) = x$. Because disjoint cycles commute with each other we may suppose that x appears in θ_1 (otherwise we reshuffle and reindex the cycles). We may also suppose that the cycle θ_1 starts with x. But then θ_1 must be $(x \sigma(x)\ \sigma^2(x)...\sigma^{r-1}(x))$ which is precisely σ_1. So $\theta_1 = \sigma_1$, giving $\sigma_1 \sigma_2 ... \sigma_k = \theta_1 \theta_2 ... \theta_u$. By cancellation law, $\sigma_2 \sigma_3 ... \sigma_k =$

$\theta_2...\theta_u$. Both sides are permutations of the set T above. Since $|T| < n$, we apply induction hypothesis, because of which the factorisation is unique. So $k = u$ and with a re-indexing and reshuffling of the θ's,

$$\sigma_2 = \theta_2,..., \sigma_k = \theta_k.$$

So the factorisation of σ as $\sigma_1\sigma_2...\sigma_k$ is unique modulo the order of the cycles. ▓

In the theorem above, we ignored the 1-cycles (and therefore had to disallow the identity permutation) because they are redundant elements in a factorisation. However, for certain purposes which we do not mention here, it is important to ensure that every element of $\{1, 2, ..., n\}$ belongs to at least one cycle even if it is a cycle of length 1. If we do so, we get the following reformulation of Theorem (4.2).

4.3 Theorem: Every permutation (including the identity) of n symbols can be expressed as a product of mutually disjoint cycles the sum of whose lengths is n. This factorisation is unique modulo the same restrictions as above.

Proof: There is basically no change in the proof. But since we are allowing cycles of length 1 also, the argument can be given a little more systematically by an algorithm. We illustrate this with the permutation

$$\sigma = \begin{pmatrix} 1 & 2 & 3 & 4 & 5 & 6 & 7 \\ 4 & 2 & 1 & 3 & 7 & 6 & 5 \end{pmatrix} \text{ in } S_7.$$

We start with 1 and consider $\sigma(1)$, $\sigma^2(1)$, $\sigma^3(1)$, ... till we get 1. In this case we get $(1\ 4\ 3)$ as the cycle containing 1. Now we take the first element not covered so far. This element is 2. Since $\sigma(2) = 2$, (2) is the 1-cycle containing 2. The next element not yet covered is 5 and $(5\ 7)$ is the 2-cycle containing it. Finally, the only element left is 6 which is in the 1-cycle (6). So σ is expressed as the product of 4 cycles, (143) (2) (57) (6). Since the process is carried till all symbols are exhausted it is obvious that the sum of the lengths of all cycles is n. For uniqueness, we apply induction again. ▓

4.4 Definition: The factorisation of a permutation into mutually disjoint cycles is called its **cyclic decomposition.**

Whether to include cycles of length 1 in a cyclic decomposition or not is really a matter of convention. As remarked above, sometimes it is important to include them. For the moment, however, we consider only cycles of length greater than one in the factorisation. The cyclic decomposition provides the most compact way of representing a permutation in S_n. For example, let G be the group of isometries of a regular pentagon, considered in Section 1. If we number its vertices as 1, 2, 3, 4, 5 clockwise then G is a subgroup of S_5. A clockwise rotation through 72 degrees is a cycle of length

5, namely (1 2 3 4 5). Similarly, clockwise rotations through 144, 216 and 288 degrees are given by the cycles (1 3 5 2 4), (1 4 2 5 3) and (1 5 4 3 2) respectively. Rotation through 360 degrees is of course the identity permutation. As for reflections, or 'flips', let f_1 be the reflection in the line passing through the centre of the pentagon and the vertex 1. Then the cyclic decomposition of f_1 is (25) (34). Similarly the flips f_2, f_3, f_4, f_5 are given by (13) (45), (24) (15), (12) (35) and (14) (23) respectively as can be verified with a diagram (cf. Figure 5.1).

When a permutation is expressed in its cyclic decomposition form, its inverse is also obtained immediately in its cyclic decomposition form. If $(i_1, i_2, ..., i_r)$ is an r-cycle, its inverse permutation is the r-cycle read backwards, that is, the r-cycle $(i_r, i_{r-1}, ..., i_2, i_1)$. (The proof that these two are inverses of each other is trivial.) Since mutually disjoint cycles commute with each other, the product of their inverses (in the same order) gives the inverse of their product. As an example, in the proof of Theorem (4.3), we considered $\sigma = \begin{pmatrix} 1 & 2 & 3 & 4 & 5 & 6 & 7 \\ 4 & 2 & 1 & 3 & 7 & 6 & 5 \end{pmatrix}$ whose cycle decomposition is (143) (57). Hence $\sigma^{-1} = $ (341) (75). There is, however, no easy way in general to get the cyclic decomposition of the product of two permutations in terms of their cyclic decompositions. We simply have to compute the product and find its cyclic decomposition by the method above.

Because of Theorem (4.2), (or its companion, Theorem (4.3)), when we want to prove something about an arbitrary permutation in S_n, the problem can often be reduced to proving it for a cycle. As an illustration, we have the following result which underscores the importance of transpositions (that is, cycles of length 2).

4.5 Theorem: Every permutation of n symbols can be expressed as a product of transpositions. Consequently, the subgroup of S_n generated by all transpositions is the entire group S_n.

Proof: The identity permutation is the product of zero 2-cycles. (It is the empty product. We already treated empty sums as 0 and empty products as 1 in Chapter 4. However, if this sounds absurd, we write the identity permutation as (12) (21) assuming $n > 1$.) By Theorem (4.2) every permutation σ, other than the identity permutation is expressed as the product of cycles of lengths ≥ 2. So it suffices to show that every cycle of length 2 or more can be expressed as a product of transpositions. This is very easy because a direct calculation shows that an r-cycle $(i_1 i_2 ... i_r)$ equals the product $(i_1 i_r) (i_1 i_{r-1}) ... (i_1 i_2)$. (Given $x \in \{1, 2, ..., n\}$, we consider two cases, $x = i_k$ for some $1 \leq k \leq r$ and $x \neq i_1, i_2, ..., i_r$ and verify that both the sides take x to the same element in both the cases.) Thus it follows that every element of S_n can be expressed as a product of transpositions. The last assertion is now clear because the subgroup generated by a subset must contain all possible products of elements of that subset. ▨

Note that it is not claimed, nor is it true, that the factorisation of a permutation as a product of transpositions is unique. For example, the permutation (2 3 1) of S_3 can be written as (21) (23) or as (32) (31) or as (13) (12). Even the number of factors need not be the same as we see from the equation (23) = (12) (13) (12).

As a simple application of the last theorem, we make good a promise given in the last section about symmetric Boolean functions. If $f(x_1, ..., x_n)$ is a Boolean function of n Boolean variables, the definition of symmetry required only that f be invariant for all possible interchanges of two variables at a time. However, in applications we needed that f is invariant under any permutation of the variables. This can be proved as follows. Let H be the set of those permutations of the variables $x_1, ..., x_n$ under which f is invariant. In other words

$$H = \{\sigma \in S_n : f(x_1, ..., x_n) = f(x_{\sigma(1)}, x_{\sigma(2)}, ..., x_{\sigma(n)})\}.$$

We have to show that H is the entire S_n. In any case H is a subgroup of S_n, because if f is invariant under σ and τ it would be so under $\sigma \circ \tau$ and under σ^{-1}. Now, by the definition of symmetry, every transposition (i, j) belongs to H. So the subgroup generated by the set of all transpositions in S_n is contained in H. But by Theorem (4.5) this subgroup is the entire S_n. So $H = S_n$, which means f is invariant under any permutation of the variables.

We remark that in order to generate S_n, we do not need all transpositions. For example, the $n - 1$ transpositions (12), (13), ..., (1n) also generate S_n, because every other transpositions, say, (i, j) can be written as $(1, i)(1, j)(1, i)$.

Although the number of transpositions into which a permutation is factorised is not unique, as we saw above, interestingly, the parity of this number is the same. This means that if we factorise a permutation σ as a product of transpositions as $\sigma = \tau_1 .... \tau_r$ and also as $\tau = \theta_1 ... \theta_s$ then r and s need not be equal but either both are even or both are odd. This fact is not immediately obvious. We prove it by first constructing a suitable homomorphism defined on S_n. Let $P_2(n)$ denote the set of all 2-subsets of $\{1, 2, ..., n\}$. Note that $|P_2(n)| = \binom{n}{2} = \dfrac{n(n-1)}{2}$.

4.6 Proposition: Let $\sigma \in S_n$. Define $f(\sigma)$ be the product of all real numbers of the form $\dfrac{\sigma(i) - \sigma(j)}{i - j}$ as $\{i, j\}$ ranges over all 2-subsets of $\{1, 2, ..., n\}$, that is, $f(\sigma) = \prod\limits_{\{i, j\} \in P_2(n)} \dfrac{\sigma(i) - \sigma(j)}{i - j}$. Then $f(\sigma)$ is either 1 or -1. The function $f : S_n \rightarrow \{1, -1\}$ is a homomorphism of the group S_n into the group $\{1, -1\}$ under multiplication.

Proof: $f(\sigma)$ is a product of $\binom{n}{2}$ factors. Each factor is non-zero since $i \neq j$ implies $\sigma(i) \neq \sigma(j)$, σ being one-to-one. Each 2-subset of $\{1, 2, ..., n\}$ gives rise to one factor. Note that we are considering 2-subsets $\{i, j\}$ (with $i \neq j$) and not ordered pairs (i, j). Whether we take $\dfrac{\sigma(i) - \sigma(j)}{i - j}$ or $\dfrac{\sigma(j) - \sigma(i)}{j - i}$ we get the same factor. Now as $\{i, j\}$ runs over all possible 2-subsets of $\{1, 2, ..., n\}$, so does $\{\sigma(i), \sigma(j)\}$. Therefore, in the expression for $f(\sigma)$, the product of the numerators numerically equals the product of the denominators. Hence $f(\sigma) = 1$ or -1. (If we recall the concept of an inversion of a permutation from Definition (3.3.12), we see that every inversion gives a negative factor of $f(\sigma)$. So $f(\sigma)$ is 1 or -1 according as the total number of inversions of σ is even or odd.)

Now to show that f is a homomorphism suppose σ, $\tau \in S_n$ and $\theta = \sigma \circ \tau$.

Then
$$f(\theta) = \prod_{\{i, j\} \in P_2(n)} \frac{\theta(i) - \theta(j)}{i - j} = \prod_{\{i, j\} \in P_2(n)} \frac{\sigma(\tau(i)) - \sigma(\tau(j))}{i - j}$$

$$= \prod_{\{i, j\} \in P_2(n)} \frac{\sigma(\tau(i) - \sigma(\tau(j))}{\tau(i) - \tau(j)} \; \frac{\tau(i) - \tau(j)}{i - j}$$

$$= \left(\prod_{\{i, j\} \in P_2(n)} \frac{\sigma(\tau(i)) - \sigma(\tau(j))}{\tau(i) - \tau(j)} \right) \left(\prod_{\{i, j\} \in P_2(n)} \frac{\tau(i) - \tau(j)}{i - j} \right)$$

Now the second factor equals $f(\tau)$ by definition. In the first factor, i, j are only dummy variables. As $\{i, j\}$ runs through every element of $P_2(n)$, so does $\{\tau(i), \tau(j)\}$, except possibly in a different order. But the multiplication for real numbers is commutative as well as associative. So the first factor is the same as $\displaystyle\prod_{\{\tau(i), \tau(j)\} \in P_2(n)} \frac{\sigma(\tau(i)) - \sigma(\tau(j))}{\tau(i) - \tau(j)}$ which is nothing but $f(\sigma)$. So we see that $f(\sigma \circ \tau) = f(\sigma) f(\tau)$. This proves that f is a homomorphism. ∎

4.7 Theorem: When a permutation is expressed as a product of transpositions, the number of factors has the same parity in all such expressions, namely, the parity of the number of inversions in that permutation.

Proof: Let $f: S_n \to \{1, -1\}$ be the homomorphism defined in the last proposition. If τ is any transposition, say, (r, s) then we claim $f(\tau) = -1$. To see this we count the number of inversions in τ. Without loss of generality suppose $r < s$. Then τ is given by

$$\begin{pmatrix} 1 & 2...r-1 & r & r+1...s-1 & s & s+1...n \\ 1 & 2 \quad r-1 & s & r+1...s-1 & r & s+1...n \end{pmatrix}$$

Suppose (i, j) is an inversion pair for τ, that is $i < j$ but $\tau(i) > \tau(j)$. The only way this can happen is, (i) $i = r$ and $r + 1 \leqslant j \leqslant s$ and (ii) $r \leqslant i \leqslant s - 1$ and $j = s$. Both (i) and (ii) can occur $s - r$ times each. But the case $i = r$ and $j = s$ is common to both (i) and (ii). It follows that the number of inversions of τ is $2(s - r) - 1$, which is odd. So in the expression of $f(\tau)$ as

$$\prod_{\{i, j\} \in P_2(n)} \frac{\tau(i) - \tau(j)}{i - j}$$ an odd number of factors is negative. Hence $f(\tau)$ is

negative. But $f(\tau) = 1$ or -1. So $f(\tau) = -1$ as we wanted to show. Thus f takes every transposition to -1.

It is now easy to complete the proof. Suppose a permutation σ is expressed as a product of transpositions in two ways, say, $\sigma = \tau_1\tau_2...\tau_p$ and $\sigma = \theta_1\theta_2...\theta_q$. Since f is a homomorphism and $f(\tau_i) = f(\theta_j) = -1$ for all i, j, we have $f(\sigma) = (-1)^p = (-1)^q$. Also from the proof of the last Proposition, $f(\sigma) = (-1)^r$ where r is the number of inversions of σ. So p, q, r are either all even or all odd. ■

Because of this theorem, we can classify all the permutations in S_n into two well-defined categories as follows:

4.8 Definition: A permutation is called **even** or **odd** according as the number of factors in its expression as a product of transposition is even or odd.

Thus all transpositions are odd permutations. The product of any two transpositions is an even permutation. If $(i_1, i_2,..., i_r)$ is an r-cycle then, as we saw above, $(i_1, i_2,..., i_r)$ equals the product $(i_1 i_r)(i_1, i_{r-1})...(i_1 i_2)$. So an r-cycle is an even permutation if r is odd and an odd permutation if r is even.

We have the following simple proposition which expresses the parity of permutations obtained from other permutations.

4.9 Proposition: The product of two even or two odd permutations is even while the product of an even permutation and an odd permutation is odd. The inverse of a permutation has the same parity as that permutation.

Proof: If σ is expressed as a product of r transpositions and τ as a product of s transpositions (say), then $\sigma \circ \tau$ is the product of $r + s$ transpositions, where r, s are positive integers. Now $r + s$ is even if r, s have the same parity and odd if they have oppositive parity. This proves the first assertion. For the second, we note first that each transposition is its own inverse. So if $\sigma = \tau_1\tau_2...\tau_r$ where each τ_i is a transposition then $\sigma^{-1} = \tau_r\tau_{r-1} ...\tau_2\tau_1$. Hence σ and σ^{-1} are either both odd or both even. ■

As a consequence, we have the following simple but important result.

4.10 Proposition: For every integer $n > 1$ the set of all even permutations of n symbols is a subgroup of S_n. It has index 2 in S_n and is a normal subgroup of S_n.

Proof: Let H be the set of all even permutations of n symbols. By the last proposition, H is closed under composition and under inversion. So H is a subgroup of S_n. To find its index in S_n fix any transposition $\tau \in S_n$. (Since $n > 1$, such a transposition exists.) Then $\tau \notin H$. We claim that H and τH are the only left cosets of H in S_n. This amounts to showing that τH equals the set of all odd permutations in S_n. For this, if $\sigma \in H$ then $\tau\sigma$ is odd by the last proposition. Conversely, if θ is an odd permutation then $\tau\theta$ is even, again by the last proposition. But $\theta = \tau(\tau\theta)$. So $\theta \in \tau H$. Then H has index 2 in S_n. From Theorem (2.9), H is normal in S_n. (A direct proof of normality is also easy, using the last proposition.)   ▊

4.11 Definition: The group of all even permutations of n symbols called the **alternating group** of degree n and denoted by A_n.

For $n = 1$, $A_n = S_n$. For $n > 1$, $\circ(A_n) = \frac{1}{2} \circ(S_n) = \frac{n\,!}{2}$ by the proposi-

tion. Thus A_2 is the trivial group. A_3 is a group of order 3. It is the cyclic group generated by the 3-cycle (123). The group A_4 has order 12 and because of its interesting structure deserves to be studied in detail. Its elements can be classified as follows: (i) the identity element, (ii) eight 3-cycles of the form (123) and (iii) three elements, each of which is a product of two disjoint transpositions, namely, $(12)(34)$, $(13)(24)$ and $(14)(23)$. It is easy to see that if we take elements of type (i) and (iii) then these four elements form a subgroup K of A_4. Further K is isomorphic to the Klein group. (To see this, note that $(12)(34)(13)(24)$ equals $(14)(23)$ and similarly the product of any two elements of type (iii) is the third element.) We claim that K is a normal subgroup of A_4. Curiously it is easier to show that K is normal in the larger group S_4. For this we consider $N(K)$, the normaliser of K in S_4 (cf. Exercise (2.29)). We have to show that $N(K) = S_4$. In view of Theorem (4.5), it suffices to show that every transposition is in $N(K)$. Take a typical transposition $\tau = (12)$. Then $\tau e \tau^{-1} = e \in K$ where e is the identity permutation. Also by direct calculation we see that $\tau(12)(34)\tau^{-1} = (12)(12)(34)(12) = (34)(12) = (12)(34) \in K$ and similarly $\tau(13)(24)\tau^{-1} = (12)(13)(24)(12) = (14)(23) \in K$ and $\tau(14)(23)\tau^{-1} = (13)(24) \in K$. So $\tau K \tau^{-1} = K$, that is $\tau \in N(K)$. Similarly every transposition is in $N(K)$. So $N(K) = S_4$, which means K is normal in S_4 and *a fortiori* in A_4.

The alternating group A_4 provides certain interesting counterexamples. First we show it contains no subgroup of order 6, which would prove that the converse of Lagrange's theorem is not true in general. Suppose, if possible, L is a subgroup of A_4 with $\circ(L) = 6$. Then L has index 2 in A_4 and so L is a normal subgroup of A_4 by Theorem (2.9). Now by

Exercise (1.16), L must contain at least one element of order 2. But the only elements of order 2 in A_4 are the three elements of type (iii) listed above. So L must contain at least one of them. Without loss of generality we may suppose that (12) (34) $\in L$. Let $\tau = (123)$. Then $\tau \in A_4$ and $\tau^{-1} = (132)$. By direct calculation we see that $\tau(12)$ $(34)\tau^{-1} = (123)$ (12) (34) (132) = (14) (23). But L is normal in A_4. So (14) (23) $\in L$. Hence (13) (24) which is the product of (12) (34) and (14) (23) is also in L. In other words L contains the subgroup K. But this contradicts Lagrange's theorem because $\circ(K) = 4$, $\circ(L) = 6$ and 4 does not divide 6. This contradiction shows that A_4 cannot contain a subgroup of order 6, even though 6 is a divisor of $\circ(A_4)$.

Another interesting counter-example given by A_4 is regarding the normality of a subgroup. As noted above K is a normal subgroup of A_4. Let $M = \{e, (12) (14)\}$. Then M is a subgroup of K and is normal in K since K is abelian (alternatively, M has index 2 in K and hence is normal in K). So we have K as a normal subgroup of A_4 and M as a normal subgroup of K. Still M is not a normal subgroup of A_4! Indeed, as we saw above, the element (14) (23) is a conjugate of (12) (34) by an element of A_4 (namely the 3-cycle (123)). If M were normal in A_4 then it would have to contain all the conjugates of its elements. This example can be interpreted as saying that 'being a normal subgroup' is not a transitive relation on the set of all subgroups of a group.

The group A_4 also has an interesting geometric interpretation. In Section 1 we considered the group of isometries of a regular tetrahedron. If we label the four vertices of it as 1, 2, 3, 4 then every permutation of $\{1, 2, 3, 4\}$ is an isometry. So the group of isometries is the entire group S_4. However, some of these isometries do not preserve orientation. In terms of even and odd permutation, it is very easy to tell which isometries are orientation preserving and which are orientation reversing. Consider a transposition, say, (34) This represents the reflection in the plane passing through the vertices 1 and 2 and the midpoint of the edge joining 3 and 4, (see Figure 5.2). Such a reflection reverses orientation. Similarly every other transposition is orientation reversing. The composite of two orientation preserving or two orientation reversing isometries is orientation preserving while the composite of an orientation preserving and an orientation reversing isometry is orientation reversing. These facts combined with Proposition (4.2) imply that the group of orientation preserving isometries of a regular tetrahedron is precisely A_4. It is easy to see the geometric transformations represented by various elements of A_4. The identity element of course represents the identity transformation. A typical element of type (ii), say, the 3-cycle (234) represents a rotation around the altitude of the tetrahedron passing through the vertex 1. A typical element, say, (12) (34) of type (iii) represents a 180° rotation around the axis passing through the midpoints of the edges 12 and 34.

The alternating groups of higher degree (that is A_n for $n > 4$) can also be interpreted as groups of orientation preserving isometries of higher dimensional figures. Such figures are called **simplexes**. A two dimensional simplex is a triangle; a three dimensional simplex is a tetrahedron. However, we shall not pursue this line further. Instead, we turn to a property for which these groups are most famous.

4.12 Definition: A group G is called **simple** if it has no non-trivial, proper normal subgroups, that is no normal subgroup other than $\{e\}$ and G.

A group of prime order is simple because it not only has no proper normal subgroups, it has no proper, non-trivial subgroups at all, by Lagrange's theorem. However, these are trivial examples. When we look for non-trivial examples of simple groups, abelian groups are no help. Let G be an abelian group. Let x be any element of x other than the identity and let $H = (x)$, the subgroup generated by x. If $H \neq G$, then H is a proper normal subgroup of G (since G is abelian). On the other hand if $H = G$, then G is cyclic and so has a proper subgroup except when its order is a prime. (See Theorem (2.17). If G is an infinite cyclic group generated by x then the subgroup (x^2) is a proper subgroup.)

So, to get non-trivial examples of simple groups, we have to turn to non-abelian groups. Using a case-by-case argument, it can be shown that no group of order 59 or less is simple, except when its order is a prime. The group A_5, of order 60 is simple as we now show. This group is, therefore, the smallest non-trivial, simple group. Incidentally, the adjective 'simple' in the definition above does not mean that the structure of a simple group is very easy to determine. On the contrary, characterising and classifying simple groups has been one of the most formidable problems in group theory. The term 'simple' is used by an analogy with other algebraic structures (such as simple rings, simple algebras which we shall not study). If a group G has a proper normal subgroup N then the structure of N and of the quotient group G/N give some clue about the structure of G (although, as we saw earlier, they do not completely determine it). In case of a simple group no such simplification is possible. So 'simple' here simply means something which cannot be further simplified and not something which is very easy to handle.

4.13 Theorem: A_5, the alternating group of degree 5 is simple.

Proof: Let N be a normal subgroup of A_5 other than the identity subgroup. We have to show that N is the entire group A_5. The proof will proceed in three steps. First we shall show that N must contain at least one 3-cycle. Then we shall show that N contains all 3-cycles. And finally we shall prove $N = A_5$.

We begin by listing the elements of A_5 other than the identity element.

Keeping in mind that A_5 consists of all even permutations of the set $\{1, 2, 3, 4, 5\}$ and Theorem (4.2), elements of A_5 (other than the identity) fall into 3 categories:

(i) 3-cycles, that is, elements of the form (123). There are 20 such elements. (Ostensibly there are 60 3-cycles in a set with 5 elements, but every 3-cycle can be represented in 3 ways).

(ii) Products of two mutually disjoint transpositions, that is, elements of the form (12) (34). In all there are 15 such elements. (Prove!).

(iii) 5-cycles, that is, elements of the form (12345). There are 24 such elements.

The first step of our proof is to show that N contains at least one elemement of type (i), Since N is non-trivial, it suffices to show that whenever N contains an element of type (ii) or type (iii) then it must also contain an element of type (i). We do these two separately. First suppose, without loss of generality that, (12) (34) $\in N$. Let $\theta = $ (12) (45). Then θ is an even permutation. So $\theta \in A_5$. Also $\theta^{-1} = \theta = $ (45) (12) Since (12) (34) $\in N$ and N is normal in A_5, $\theta(12)(34)\theta^{-1}$ is in N. By a direct calculation, this element comes out as (12)(35). So (12)(35) $\in N$. But N is also closed under multiplication. So (12) (34)(12) (35) $\in N$. This element is precisely the 3-cycle (354). So N contains a 3-cycle. In the second case, suppose N contains an element of type (iii) which, without loss of generality may be taken as the 5-cycle (1 2 3 4 5). Let $\sigma = $ (23) (45). Then $\sigma \in A_5$ and $\sigma(12345)\sigma^{-1}$ comes out as (13254) which is in N by normality. But then (12345) (13254) which equals (142) is also in N. Thus we have shown that in all cases N must contain at least one 3-cycle, completing the first step.

Now in the second step we have to show that N contains every 3-cycle, that is, all elements of type (i). For this, it suffices to show that any two 3-cycles are conjugate to each other in A_5, for then, by normality whenever N contains any one 3-cycle it will contain all 3-cycles. Let σ, τ be two 3-cycles. For proving that they are conjugate in A_5, the fact that conjugacy is an equivalence relation (see Proposition (1.16)) will be very useful. If $\sigma = \tau$ then they are conjugate to each other by reflexity. So we suppose $\sigma \neq \tau$. First consider the case where σ and τ consist of the same three symbols. Then they can be written down in the form typified by $\sigma = $ (123) and $\tau = $ (132). Take $\theta = $ (23) (45). Then σ equals $\theta\tau\theta^{-1}$, proving that σ is conjugate to τ in A_5. Next, suppose that σ and τ have two symbols in common. Then we can suppose without of loss of generality that $\sigma = $ (123), $\tau = $ (124) or $\sigma = $ (123), $\tau = $ (214). In the first case let $\lambda = $ (345). Then $\tau = \lambda\sigma\lambda^{-1}$. In the second case $\tau = \mu\sigma\mu^{-1}$ where $\mu = $ (12) (34). In either case σ, τ are conjugate to each other in A_5. So two 3-cycles which differ in one symbol (and possibly in the order of the common symbols) are conjugate to each other. This fact, used repeatedly along with transitivity of the conjugacy relation, implies that any two 3-cycles are conjugate to each other in A_5. As noted before, this completes the second step of our proof, namely, that the subgroup N contains all 3-cycles.

For the third step, we must now show that the set of all 3-cycles generates the entire group A_5. For this, it suffices to show that all elements of type (ii) and (iii) above can be expressed as products of 3-cycles. This is very easy. A typical element of type (ii), say, (12) (34) equals (132) (134) while a typical element of type (iii), say, (12345) equals (123) (345).

We have thus shown that the normal subgroup N must be entire A_5 unless $N = (e)$. So A_5 is simple. ■

For $n > 5$ also, A_n is a simple group. (The proof is similar to the one for A_5 but a little more elaborate.) However, the result above will be good enough for our purpose. The groups A_1, A_2 and A_3 are also simple (although trivially so). However, A_4 is not simple because we saw that it has a normal subgroup K of order 4. In this sense, the group A_4 behaves exceptionally among all alternating groups.

A few other standard results about permutation groups will be given as exercises.

Exercises

4.1 In S_n let $H = \{\sigma \in S_n : \sigma(n) = n\}$. Prove that H is isomorphic to S_{n-1}. This way we regard S_{n-1} as a subgroup of S_n. Prove that when so done, $S_{n-1} \cap A_n$ equals A_{n-1}. Is S_{n-1} a normal subgroup of S_n?

4.2 In S_n prove that the number of distinct r-cycles is

$$\frac{n(n-1)\dots(n-r+1)}{r}.$$

4.3 Suppose $(i_1 i_2 \dots i_r)$ is an r-cycle in S_n. Let σ be any permutation in S_n. Suppose $j_k = \sigma(i_k)$ for $k = 1, \dots, r$. Prove that $\sigma(i_1 i_2 \dots i_r)\sigma^{-1}$ is the r-cycle $(j_1 \dots j_r)$.

4 4 Suppose two elements, say, τ and θ are conjugate to each other in S_n. Prove that for every $r = 1, \dots, n$ the number of r-cycles in the cyclic decompositions of τ and θ are the same.

*4.5 Prove that the converse of the last result is also true. (Hint: Write down the complete cyclic decompositions of τ and θ, in descending order of lengths. Using Exercise (4.3), find a permutation σ such that $\tau = \sigma\theta\sigma^{-1}$.)

4.6 Prove that the number of distinct conjugacy classes in S_n equals $p(n)$, the number of partitions of the integer n (see comments after Proposition (2.3.15).)

4.7 Let $\theta = (1, 2, \dots, n) \in S_n$. Prove that θ commutes only with its own powers. (Hint: Consider the number of elements conjugate to θ and apply Theorem (2.20).)

4.8   Prove that for $n > 2$, the group S_n has a trivial centre.

4.9   Prove that for all $n \geqslant 3$, A_n is generated by all 3-cycles. (This is in fact, the third step in the proof of Theorem (4.13) for $n = 5$. Instead of listing all elements of A_n, a simpler proof can be based upon the fact that every element of S_n can be written as a product of transpositions of the form $(1, i), i = 2,..., n$.)

4.10   In fact, show that, A_n for $n \geqslant 3$, can be generated by the $(n-2)$ 3-cycles of the form $(1, 2, i)$, $i = 3,..., n$. (Hint: Express $(1, r, s)$ in terms of such cycles.)

4.11   Prove that S_n can be generated with just two elements, say by $(1, 2)$ and $(1, 2, 3,..., n)$.

4.12   Prove that K is the only normal subgroup of A_4 other than $\{e\}$ and A_4.

4.13   Prove that the group of orientation preserving isometries of a cube is a group of order 24. Describe its elements geometrically as permutations of the set of its 8 vertices.

4.14   Prove that the group in the last exercise is in fact isomorphic to S_4. (Hint: Note that every isometry of the cube must permute its diagonals among themselves.)

4.15   If N is a normal subgroup of S_5, prove that $N = \{e\}$, A_5 or S_5. (Hint: Consider $N \cap A_5$.)

4.16   A group G is a said to be **solvable** if there exists a finite sequence of subgroups $G = N_0 \supset N_1 \supset N_2 \supset ... \supset N_{r-1} \supset N_r = \{e\}$ such that, for each $i = 1,..., r$, N_i is a normal subgroup of N_{i-1} and the quotient group N_{i-1}/N_i is abelian. Prove that all abelian groups, all dihedral groups D_n, the quaternion group Q and the groups S_1, S_2, S_3, S_4 are solvable but the groups A_5 and S_5 are not solvable.

4.17   Prove that a subgroup of a solvable group is solvable. Prove also that every quotient group of a solvable group is solvable.

4·18   Prove that for $n \geqslant 5$, A_n and S_n are not solvable.

4.19   Let H be a normal subgroup of a group G. Prove that if H and G/H are solvable, so is G. (Hint: Use Corollary (3.8).)

4.20   Let G be a group of order p^m where p is a prime. Prove that G contains subgroups $G_0 \subset G_1 \subset ... \subset G_m = G$ with $o(G_i) = p^i$, $i = 0,..., m$ and G_{i-1} normal in G_i. In particular G is solvable. [Hint: Apply induction on m and Theorem (2.22). Consider G/H where $\circ(H) = p$ and H is a subgroup of the centre of G.]

*4.21   In S_9, let θ and σ be respectively the permutations (147) (258) (369) and (456) (798). Let G be the subgroup of S_9 generated by θ and σ. Prove that G is a non-abelian group of order 27 in which every element except the identity has order 3. (Hint: Interpret θ and σ in terms of $\mathbb{Z}_3$.)

4.22   Prove that the group $\mathbb{Z}_3 \times \mathbb{Z}_3 \times \mathbb{Z}_3$ also has 1 element of order 1 and 26 elements of order 3 but that it is not isomorphic to the group G of the last exercise.

*4.23   Prove that A_n is simple for all $n > 5$. (Hint: Apply induction on n.)

4.24   Prove that the order of a permutation (as an element of the permutation group) equals the least common multiple of the lengths of its cycles.

Notes and Guide to Literature

The permutation groups have been studied for a long time, long before 'abstract' groups were defined. The cyclic decomposition of a permutation is needed in what is called 'Polya's theory of counting'*.

The term 'solvable' defined in Exercise (4.16) may appear strange. Its justification will come in later* where it will be related to the solvability of a polynomial equation. The fact that S_n is not solvable for $n \geqslant 5$, will be used to give a proof, due to Galois, that it is impossible to give a formula for the roots of a general polynomial of degree $\geqslant 5$.

Exercise (4.21) is adapted from Carmichael [1], where the reader will also find more about permutation group. The significance of Exercise (4.21) and (4.22) is note-worthy. If two groups G and H are isomorphic then, trivially, for every positive integer n, the number of elements of order n in G is the same as the number of elements of order n in H. The two exercises show that the converse is false. In other words, in order to check whether two groups are isomorphic it will not suffice to compare the numbers of elements of each order. This illustrates the remark made in the last section, that in general there is no efficient criterion to test whether two given groups are isomorphic or not. However, for finite abelian groups, it turns out that the converse does hold. The theory of finite abelian groups is therefore considerably simpler than that of all finite groups.

The alternating group A_5 is the 'smallest' simple group, in the sense of order. There are groups, other than A_n ($n \geqslant 5$), which are simple. The search for and the classification of simple groups has occupied group theorists for several decades. See Gorenstein [1] for a recent comprehensive work on the subject.

* See the Epilogue.

Six

Rings, Fields and Vector Spaces

In the last chapter we studied, rather thoroughly, groups, which are algebraic structures that arise from a single binary operation on a set. In this chapter we study algebraic structures which are richer than groups, because they arise by putting certain additional structures on a group (actually an abelian group). The first such structure is a ring, in which we have one more binary operation on an abelian group. These will be defined and studied in the first section. The foremost example of a ring is the familiar ring of integers and in the second section we shall study a class of rings which share some of the pleasant properties of the ring of integers. Fields are a very important type of rings. Field theory has several interesting applications (see the Epilogue) and the present chapter will develop the machinery needed.

The other algebraic structure that arises from an abelian group is that of a vector space, which is obtained by specifying a rule whereby elements of the group are multiplied by elements of some field, often called scalars. They will be studied in Section 3. The theory of vector spaces is a generalisation of certain aspects of the euclidean spaces. Homomorphisms of vector spaces are called linear transformations. They can be represented most conveniently by matrices. The theory of matrices (along with determinants) will be briefly developed in Section 4.

1. Basic Concepts and Examples

Just as the definition of an abstract group was modelled after the permutation groups, some familiar rings, notably the ring of integers, serve as the model for the definition of a ring. On $\mathbb{Z}$, the set of integers, we have the binary operations $+$ and $\cdot$. Of course $\mathbb{Z}$ has a lot more other structures, for example, that imposed by the usual order relation $\leqslant$. But from the point of view of algebra, the binary operations are most important. Under addition, $\mathbb{Z}$ is an abelian group, in fact a cyclic group and we already abstracted this structure in the last chapter. If we now include the other binary operation, namely, multiplication, on $\mathbb{Z}$, then we get an example of what is known as a ring. In order to define an abstract ring, we take an abelian group $(R, +)$ and suppose that there is another binary operation $\cdot$ on R. The properties that we want to assume about $\cdot$ must come from those which are true for the usual multiplication on $\mathbb{Z}$. And once again we have to strike a balance between depth and generality. As with group structures, associativity of $\cdot$ turns out to be vital, while its commtuativity is, although desirable, not indispensible. Moreover, we must assume some inter-relationship between the two operations $+$ and $\cdot$. For, without it, the study of rings would merely amount to two separate studies, one of the group $(R, +)$ and one of the semigroup $(R, \cdot)$.

These considerations motivate the following definitions:

1.1 Definition: A **ring** is a triple $(R, +, \cdot)$, where $(R, +)$ is an abelian group, $(R, \cdot)$ is a semigroup and $\cdot$ is distributive over $+$. If further, $\cdot$ is commutative, the ring is called **commutative**. If $\cdot$ has an identity element (almost always denoted by 1), then the ring is said to have an **identity**.

As a foremost example, $\mathbb{Z}$, the set of integers with the usual addition and multiplication is a ring. In fact, it is a commutative ring and also has an identity element. The sets of real numbers, rational numbers, complex numbers each with the usual operations are also commu ative rings with identity elements. Many other examples will be given soon. But first we comment upon the definition. The definitions of a ring and of a commutative ring are standard. As for the identity element, some authors require that it be different from the zero element. As with Boolean algebra, it can be shown that if $0=1$ then the ring consists of only one element. Such a ring is called a **trivial ring**. For every positive integer m, $\mathbb{Z}_m$, the set of residue classes modulo m, is a commutative ring with identity under modulo m addition and multiplication. All the ring axioms follow from the corresponding properties of $\mathbb{Z}$ (see Chapter 3 Section 4).

Before giving other examples of rings, let us prove a few simple properties about rings. In any ring, the underlying abelian group will always be denoted additively and its identity element by 0. The additive inverse of an element x will be denoted by $-x$ As usual, when multiplication operation is denoted by $\cdot$, $a \cdot b$ will be denoted by ab. As with Boolean algebras, addition will be a deeper operation than multiplication. This means, an expression like $ab + c$ will be interpreted as $(a \cdot b) + c$ and not as $a \cdot (b + c)$.

The first result establishes the expected behaviour (which we generally take for granted because of over-familiarity) of the additive identity and inverses w.r.t. multiplication.

1.2 Proposition: Let R be a ring. Then

    (i) $x \cdot 0 = 0$ $x = 0$ for all $x \in R$
    (ii) $(-x)y = x(-y) = -(xy)$ and $(-x)(-y) = xy$ for all $x, y \in R$
    (iii) if R has an identity then $-x = (-1)x = x(-1)$ for all $x \in R$.
    (iv) if R has an identity 1 and $1 = 0$, then R has only one element.

Proof:

    (i)   Let $x \cdot 0$ be a. Now, $x(0 + 0)$ equals, by left distributivity, $x0 + x0$ or $a + a$. But $0 + 0 = 0$. So $x(0 + 0) = x0 = a$. Thus $a + a = a + 0$. Since $(R, +)$ is a group, this implies $a = 0$. That is, $x0 = a$. Similarly using the other distributive law, $0x = 0$ for all $x \in R$.

    (ii) $-(xy)$ is, by definition, the additive inverse of xy. To show that it equals $(-x)y$ amounts to showing that $(-x)y + xy = 0$. But $(-x)y + xy = (-x + x)y = 0y = 0$ by (i). Similarly $x(-y) + xy = x(-y + y) = x0 = 0$ by (i), showing that $x(-y) = -(xy)$. Since the equation $x(-y) = -(xy)$ holds for all x, y it holds if we replace x by $-x$, giving $(-x)(-y) = -[(-x)y]$. Call $-x$ as z. Then $-[(-x)y]$ is $-(zy)$ which equals $(-z)y$. But $-z$ is simply x. So $(-x)(-y) = xy$.

    (iii) We have $1 + (-1) = 0$. So $x[1 + (-1)] = x0 = 0$ by (i). By distributivity, the left hand side equals $x \cdot 1 + x(-1)$, that is, $x + x(-1)$. Thus $x + x(-1) = 0$, which means $x(-1)$ is the additive inverse of x. So $x(-1) = -x$. Similarly $(-1)x = -x$. For (iv), if $1 = 0$, then $x = x \cdot 1 = x \cdot 0 = 0$ for all $x \in R$. So $R = \{0\}$. ∎

Note the vital use of the distributive law made throughout the proof. In its absence, we can expect nothing about the multiplicative properties of the additive identities and inverses. We proved a result similar to (i) for Boolean algebras (see Theorem (4.1.3)). But there are two important differences. First, in a ring, we do not have the symmetry between $+$ and $\cdot$ which holds in a Boolean algebra. The dual result $x + 1 = x$ is in fact false except in the trivial ring. Also the proof of (i) differs from that of the corresponding result for Boolean algebras. In the case of a ring, the operation $+$ is fairly strong algebraically. To some extent, complements in a Boolean algebra play the role of additive inverses in a ring. This similarity should not, of course, be stretched too far. For example, $(a + b)'$ does not equal $a' + b'$ but equals $a'b'$ in a Boolean algebra.

There is, nevertheless, a definite relationship between Boolean algebras and certain types of rings defined below:

1.3 Definition: A ring R with identity is called a **Boolean ring** if $x^2 = x$ for all $x \in R$.

As the name suggests, there must be some intimate relationship between Boolean algebras and Boolean rings. There is indeed one and to establish it we first prove a result which is interesting in its own right.

1.4 Proposition: A ring R in which $x^2 = x$ for all $x \in R$, is commutative. Moreover, in such a ring $x + x = 0$ for all $x \in R$. (In other words $x = -x$ and consequently the plus and minus signs can be used interchangeably in such rings.)

Proof: Let $x, y \in R$. Then $(x + y)^2 = x + y$. On the other hand, we expand $(x + y)^2$ using the distributive law as $x^2 + xy + yx + y^2$. (Note that we do not know the ring to be commutative yet, so we cannot combine xy and yx to write $2xy$.) Since $x^2 = x$ and $y^2 = y$ we get $x + y = x + xy + yx + y$, which further means $xy + yx = 0$ for all $x, y \in R$. In particular putting $y = x$, $x^2 + x^2 = 0$, that is, $x + x = 0$ for all $x \in R$, proving the second part. Now to get the first assertion let $x, y \in R$. We already know $xy + yx = 0$ and so $xy = -yx$. But because of what we just proved, $-yx$ is the same as yx. So $xy = yx$, that is, the ring is commutative. ∎

We remark that a ring in which $x^3 = x$ for all elements is also commutative. (The proof is left as an interesting exercise.) Actually a much more general result is true, but it is not relevant to our discussion.

We are now ready to prove the relationship between Boolean algebras and Boolean rings. To avoid confusion, for the purpose of this theorem only, the addition in a ring will be denoted by $\oplus$ as $+$ will be needed to denote the addition in a Boolean algebra.

1.5 Theorem: Let $(B, +, \cdot, ')$ be a Boolean algebra. Define a binary operation $\oplus$ on B by $a \oplus b = ab' + ba'$ for $a, b \in B$. Then $(B, \oplus, \cdot)$ is a Boolean ring. Conversely, let $(R, \oplus, \cdot)$ be a Boolean ring. Define a binary operation $+$ on R by $a + b = a \oplus b \oplus ab$ for $a, b \in R$ and a unary operation $': R \to R$ by $x' = 1 \oplus x$ for $x \in R$. Then $(R, +, \cdot, ')$ is a Boolean algebra.

Proof: The first assertion is precisely the content of Exercise (4.1.19) and will not be proved. We shall prove only the converse implication. Suppose $(R, \oplus, \cdot)$ is a Boolean ring and the operations $+, '$ are defined as indicated. Then $\cdot$ is commutative by the last proposition. Since $\oplus$ is also commutative it follows that $+$ is commutative. This verifies the first axiom of Boolean algebras. As for the presence of identities, $\cdot$ is already given to have 1 as its identity element. For $+$, 0 is the identity because for all $a \in R$, $a + 0 = a \oplus 0 \oplus a0$ which, in view of Proposition (1.2) equals $a \oplus 0 \oplus 0$, that is a, since 0 is given to be the identity for $\oplus$. Thus another axiom of

Boolean algebras holds. We now prove that both · and + are distributive over each other. Both the proofs are by direct computation. Let $x, y, z \in R$. Then we have

$$x \cdot (y + z) = x \cdot (y \oplus z \oplus yz)$$

$$= xy \oplus xz \oplus xyz \text{ (since } \cdot \text{ is distributive over } \oplus)$$

$$= xy \oplus xz \oplus x^2yz \text{ (since } x^2 = x)$$

$$= xy \oplus xz \oplus (xy)(xz)$$

(since R is commutative by Proposition (1.4))

$$= xy + xz$$

which proves distributivity of · over +. The other distributivity follows from

$$(x + y)(x + z) = (x \oplus y \oplus xy) \ (x \oplus z \oplus xz)$$

$$= x \oplus xy \oplus xy \oplus xz \oplus yz \oplus xyz \oplus xz \oplus xyz \oplus xyz$$

(since $x^2 = x$ and · is commutative)

$$= x \oplus yz \oplus xyz \text{ (since } xy \oplus xy = 0, \ xz \oplus xz = 0 \text{ and}$$

$$xyz \oplus xyz = 0)$$

$$= x + yz.$$

It only remains to verify the fourth axiom of a Boolean algebra, regarding complementation. Let $x \in R$. We have to show $x + x' = 1$ and $xx' = 0$ where $x' = 1 \oplus x$. Now, by definition, $xx' = x(1 \oplus x) = x \oplus x^2 = x \oplus x = 0$ by Proposition (1.4). Also $x + x' = x \oplus x' \oplus xx' = x \oplus x' \oplus 0$ by what we just proved. So $x + x' = x \oplus x' = x \oplus 1 \oplus x = x \oplus x \oplus 1 = 0 + 1 = 1$, again by Proposition (1.4). Thus we have verified that $(R, +, \cdot, ')$ is a Boolean algebra.

This theorem is significant for several reasons. It provides us with some examples of rings, because every Boolean algebra gives rise to a commutative ring with identity. Secondly, because of the converse statement in the theorem, we get an equivalent formulation of the concept of a Boolean algebra. (Another equivalent formulation, in terms of lattices, was given in Chapter 4). We could as well work with Boolean rings instead of Boolean algebras. Many algebraists actually do this. From their point of view, then, Boolean algebras occupy only a corner in the world of all rings. However, from the point of view of applications, this corner is an important one and that is why we have devoted one whole chapter to it.

After this digression to Boolean algebras, let us return to the study of rings in general. Because of property (i) in Proposition 1.2, the operation of multiplication can never satisfy the cancellation laws except in the trivial ring, because cancellation by 0 is impossible. In many rings, however, cancellation by non-zero elements is possible. Such rings are important and we give them a special name. To put the definition in the form in which it is generally given, we introduce another term.

1.6 Definition: An element x in a ring R is called a **zero-divisor** if there exists $y \in R$ such that $y \neq 0$ and either $xy = 0$ or $yx = 0$. (Depending upon which of these equality holds we may call x as a left zero-divisor or a right zero-divisor. However, since for most part, we shall study this concept only for commutative rings we may avoid the hair-splitting.)

The element 0 is of course a zero-divisor in any ring. (Other than the trivial ring. From now onwards, we may not explicitly mention the trivial exceptions that arise out of trivial rings. That is why, some authors require, as in the case of a Boolean algebra, that a ring should have at least two elements.) In the ring of integers, there are no zero divisors except 0. In a ring obtained from a Boolean algebra, on the other hand, every element except the identity is a zero-divisor.

The concept of a zero-divisor is intimately related to cancellation law as we see in the following proposition.

1.7 Proposition: Let R be a ring and $x \in R$. Then for all $y, z \in R$, either of the equations $xy = xz$ or $yx = zx$ implies $y = z$ if and only if x is not a zero divisor. In other words, cancellation by an element is possible iff it is not a zero-divisor.

Proof: If x is a zero divisor then there exists $y \neq 0$ such that $xy = 0$ (or $yx = 0$). Take $z = 0$. Then $xz = xy$ (or $yx = zx$) but $y \neq z$, so cancellation by x does not hold. Conversely suppose x is not a zero-divisor. Let $y, z \in R$ and suppose $xy = xz$ (the other case, $yx = zx$ is similarly treated). Then $xy - xz = 0$. But by Proposition (1.2), $-xz = x(-z)$. So by distributivity once again, $x(y-z) = 0$. Since x is not a zero-divisor, this forces $y - z$ to vanish, i.e., $y = z$. ∎

We are now ready to define the kind of rings where cancellation by non-zero elements is possible.

1.8 Definition: A commutative ring with no non-zero zero-divisors is called an **integral domain** or an **entire ring**.

Before giving examples of integral domains, we remark that there is a little lack of unanimity about this definition. Some authors do not require an integral domain to be commutative while some authors require it not only to be commutative but also to have an identity element. The essence of all the three definitions is the absence of zero-divisors. We shall stick to the definition given above. As the foremost example of an integral domain we have the ring of integers (which in fact justifies the word 'integral' in the definition). Since we do not require an integral domain to have an identity element, for every positive integer m, the set $m\mathbb{Z}$ is also an integral domain. Many other examples will be given later on. In fact, the next section will deal with certain special types of integral domains. For the moment we define a concept which is stronger than an integral domain.

We saw in Chapter 3, Section 4, that for associative operations, the presence of an inverse is a stronger condition than cancellability. If we assume this stronger condition, we get the following definition.

1.9 Definition: Let R be a commutative ring with identity. If every non-zero element of R has a multiplicative inverse then R is called a **field**.

The most standard examples of fields are those of rational numbers, real numbers and complex numbers. However, there are many other 'abstract' examples of fields. Indeed, the construction of fields with suitable properties will occupy us for quite some time. Every field is an integral domain, because if an element x has a multiplicative inverse x^{-1}, then x cannot be a zero divisor, since $xy = 0$ would imply $x^{-1}xy = 0$, i.e., $1 \cdot y = 0$ or $y = 0$. That the converse is false is proved by $\mathbb{Z}$, the ring of integers. However, for finite rings the converse does hold. To see this, we paraphrase the two definitions slightly. Let R^* be the set of all non-zero elements in a ring R. To say that R has no non-zero zero-divisors is equivalent to saying that R^* is closed under multiplication. It is therefore a semigroup. If R is a field then $(R^*, \cdot)$ is an abelian group while if R is an integral domain then $(R^*, \cdot)$ is merely a commutative semigroup which obeys cancellation laws. We proved in Proposition (5.1.4) that every finite semigroup satisfying cancellation laws is a group. Using this proposition, we get the following result.

1.10 Theorem: Every finite integral domain is a field. ∎

As an example, let m be a positive integer and let $\mathbb{Z}_m$ be the ring of residue classes modulo m. If m is not a prime then m can be expressed as xy where $0 < x, y < m$. Then $[x]$ and $[y]$ are non-zero elements of $\mathbb{Z}_m$ whose product is $[xy]$ which is the zero element of $\mathbb{Z}_m$. Thus $\mathbb{Z}_m$ is not an integral domain if m is a composite number. However, if m equals some prime p, then, whenever xy is divisible by p, either x or y is divisible by p, as will be proved later. So $\mathbb{Z}_p$ is an integral domain and hence a field by the theorem above. (Alternatively, we could use Theorem (5.1.5) in which it was proved that the non-zero residue classes modulo a prime form a group. Both the arguments are, however, essentially the same because the crux of both of them is Proposition (5.1.4).)

Note that unike the fields of rational, real and complex numbers, the field $\mathbb{Z}_p$, where p is a prime is finite. It has p elements. Finite fields are quite important in applications (see the Epilogue). For the time being we only introduce an important concept about fields, or more generally, about integral domains. We begin with a preliminary result.

1.11 Proposition: Let $(R, +, \cdot)$ be an integral domain. Suppose for some non-zero $x \in R$ and some positive integer n, $nx = 0$. Then $ny = 0$ for all $y \in R$.

Proof: From the last chapter, recall that nx means the sum $x + ... + x$
(n times). For any $y \in R$, $n(xy) = (nx)y$ by distributivity. But by commutativity $n(xy) = x(ny)$. So if $nx = 0$, we get $x(ny) = 0$ for all $y \in R$.
Since R has no non-zero zero-divisors, this means $ny = 0$ as was to be
proved. ∎

1.12 Corollary: In any integral domain $(R, +, \cdot)$ the order of every non-
zero element in the group $(R, +)$ is the same. Moreover, this order, if
finite, is a prime.

Proof: By the last proposition, either all non-zero elements are of infinite
order or else all of them are of finite order. In the first case the assertion
holds. In the second case, let x, y be any two non-zero elements of R.
Suppose their orders in the group $(R, +)$ are m, n respectively where m, n
are positive integers. Then by the last proposition, $nx = 0$ and so m divides
n (see Theorem (5.1.10)). Similarly $my = 0$ and so n divides m. But then
$m = n$. So all non-zero elements are of the same order say n. If n is
composite, say, $n = ij$ where i, j are proper divisors of n, then for any
non-zero x in R we have ix as a non-zero element of order $j < n$, a contradiction. So n is a prime. ∎

1.13 Definition: An integral domain in which every non-zero element is
of infinite order is said to be of **characteristic** 0. If every non-zero element
has order p (which must be a prime) then the integral domain is said to
be of **characteristic** p.

Thus the fields of real, complex and rational numbers, as well as the
ring of integers have characteristic 0, while the field $\mathbb{Z}_p$ is of characteristic
p. Obviously, every field of characteristic 0 is infinite. Put differently,
every finite field is of prime characteristic. There do exist, however,
infinite fields of prime characteristic. The theory of the two types of fields
differs considerably. In the case of characteristic 0, in an equation $nx = ny$,
where n is a positive integer, we can cancel n and get $x = y$. If the
characteristic is a prime p, this cancellation will be valid only if n is not a
multiple of p.

We remarked that in a ring R with no non-zero, zero-divisors, the set
R^* of all non-zero elements is a semigroup obeying both the cancellation
laws. In case of a field it is an abelian group. If we drop the commutativity requirement, we get the following definition.

1.14 Definition: A ring with identity in which every non-zero element
has a multiplicative inverse is called a **division ring**.

The name is justified by the fact that in a division ring, division by
non-zero elements is possible, that is equations of the form $ax = b$ (as
well as those of the form $xa = b$) can be solved uniquely for x whenever

$a \neq 0$. A division ring is just like a field except for the commutativity requirement and is therefore also called a **skew field** or a **non-commutative field**. (In the old literature, a division ring was called a field and what we call as a field was described as a commutative field). Every field is a division ring. A classic example of a division ring which is not a field is the **ring of quaternions**. It is defined analogously to the group of quaternions and in fact we shall denote it by the same symbol Q. As a set it consists of all formal expressions of the form $a_0 + a_1 i + a_2 j + a_3 k$ where a_0, a_1, a_2, a_3 are real numbers and i, j, k are some unknown symbols. Let $a = a_0 + a_1 i + a_2 j + a_3 k$ and $b = b_0 + b_1 i + b_2 j + b_3 k$ be two quaternions, that is, elements of Q. Then $a + b$ is defined to be the quaternion $(a_0 + b_0) + (a_1 + b_1) i + (a_2 + b_2) j + (a_3 + b_3) k$. It is easily verified that $(Q, +)$ is an abelian group. In fact $(Q, +)$, as a group, is isomorphic to the group $\mathbb{R}^4$ under coordinatewise addition. To define multiplication of quaternions, we keep in mind the multplication table for quaternions group and collect the coefficients of like terms. Thus for a, b as above, $a \cdot b$ is defined as $c_0 + c_1 i + c_2 j + c_3 k$ where $c_0 = a_0 b_0 - a_1 b_1 - a_2 b_2 - a_3 b_3$, $c_1 = a_0 b_1 + a_1 b_0 + a_2 b_3 - a_3 b_2$, $c_2 = a_0 b_2 + a_2 b_0 + a_3 b_1 - a_1 b_3$ and $c_3 = a_0 b_3 + a_3 b_0 + a_1 b_2 - a_2 b_1$. That $(Q, +, \cdot)$ is a ring is a little tedious, but routine verification. The element $1 + 0i + 0j + 0k$ (denoted by 1) is the identity. The definition of quaternionic multiplication must have reminded the reader of the cross product of two vectors in $\mathbb{R}^3$. There is actually a connection between the two. In the third section (Exercise (3.26)) we shall give a definition of quaternionic multiplication in terms of the dot and the cross product of vectors. This definition will also make it easier to verify the ring axioms for Q. If $a = a_0 + a_1 i + a_2 j + a_3 k \in Q$, then its **norm** $N(a)$ is defined as the real number $\sqrt{a_0^2 + a_1^2 + a_2^2 + a_3^2}$. If $a \neq 0$, it is clear that $N(a) > 0$. For a non-zero quaternion a, its inverse is given by $b_0 + b_1 i + b_2 j + b_3 k$ where $b_0 = a_0/N(a)$, $b_1 = -a_1/N(a)$, $b_2 = -a_2/N(a)$ and $b_3 = -a_3/N(a)$ as can be verified by direct multiplication. However, the quaternionic multiplication is not commutative (for example, $ij \neq ji$ where $i = 0 + 1i + 0j + 0k$ etc.). Thus $(Q, +, \cdot)$ is division ring which is not a field. This ring is important for applications in number theory and physics.

As with other algebraic structures we have the concept of a ring isomorphism. Two rings R, S are said to be **isomorphic** if there exists a bijection $f: R \to S$ such that for all $x, y \in R$, $f(x + y) = f(x) + f(y)$ and $f(xy) = f(x) f(y)$. Isomorphic rings may be regarded as replicas of each other and have identical ring theoretic properties. As isomorphism between two rings is also an isomorphism of their additive groups. But the converse is false, as will be seen by many examples below. The more general concept of a ring homomorphism will be studied through exercises.

We now study some of the most standard ways of constructing new rings from old ones. This study will be instructive for two reasons. First, along the way it will introduce us to many basic concepts. Secondly, with suitable combinations of these methods, we get many interesting examples of rings (for example, infinite fields of prime characteristics).

(1) *Product of Rings*: Let R_1 and R_2 be two rings. Let us denote the binary operations in both by the same symbols. (This practice will generally be followed whenever we consider more than one ring at a time. The natural exceptions are, of course, those situations where this would lead to confusion as in Theorem 1.5 for example.) Let $R = R_1 \times R_2$. As usual, we define two binary operations on R coordinatewise. Then R becomes a ring. Clearly R is commutative if and only if R_1 and R_2 are commutative. Also if R_1, R_2 have identity elements (both of which to be denoted by 1), then $(1, 1)$ is the identity for R. Note, however, that in general R always has zero-divisors. Let x_1, x_2 be non-zero elements of R_1, R_2 respectively. Then neither $(x_1, 0)$ nor $(0, x_2)$ is the zero element of R (which is $(0, 0)$). But $(x_1, 0) \cdot (0, x_2) = (x_1 \cdot 0, 0 \cdot x_2) = (0, 0)$ by Proposition (1.2). It follows that even if R_1, R_2 are fields, $R_1 \times R_2$ is not even an integral domain. This does not mean that $R_1 \times R_2$ cannot be made into a field with some other binary operation instead of coordinatewise multiplication. For example, let R_1, R_2 each be $\mathbb{R}$, the field of real numbers. Then $\mathbb{R} \times \mathbb{R}$ may be identified with $\mathbb{C}$, the set of complex numbers, by thinking of an element (x, y) of $\mathbb{R} \times \mathbb{R}$ as the complex number $x + iy$. Under this identification, the coordinatewise addition of elements of $\mathbb{R} \times \mathbb{R}$ does correspond to the usual addition of complex numbers. But the multiplicative structures are quite different. In $\mathbb{R} \times \mathbb{R}$ the product of (x_1, y_1) and (x_2, y_2) is $(x_1 y_1, x_2 y_2)$. But the product of the corresponding complex numbers $x_1 + iy_1$ and $x_2 + iy_2$ is $(x_1 x_2 - y_1 y_2) + i(x_1 y_2 + x_2 y_1)$. As we saw above $\mathbb{R} \times \mathbb{R}$ is not a field. But $\mathbb{C}$ is a field. So as rings, they are not isomorphic. But they were isomorphic as groups.

The construction of product rings can be obviously generalized to the case of the product of any finite sequence of rings.

(2) *Subrings*: As with other algebraic structures, we can consider substructures of rings. They are called subrings. Formally, if $(R, +, \cdot)$ is a ring and $S \subset R$, then S is said to be a **subring** of R if first of all S is a subgroup of the group $(R, +)$ and secondly S is closed under multiplication. These conditions can be succinctly put by saying that for all $x, y \in S$, both $x - y$ and xy are in S (cf. Exercise 5.1.6). Clearly with the restrictions of the binary operations, S itself is a ring. If R is commutative, so is S. However, even if R has an identity, S need not. For example, for every positive integer $m > 1$, $m\mathbb{Z}$ is a subring of $\mathbb{Z}$ but has no identity element. Sometimes the subring may have its own identity element but it may be different from that of the ring R (if any). For example let $R = \mathbb{R} \times \mathbb{R}$ under coordinatewise operations. Let $S = R \times \{0\}$. Then S is a subring of R. S has $(1, 0)$ as its identity element while the identity of R is $(1, 1)$. If we take the product of $\mathbb{R}$ with some ring not having an identity, for example $\mathbb{R} \times 2\mathbb{Z}$, then $\mathbb{R} \times \{0\}$ is a subring with an identity element $(1, 0)$ but the ambient ring $\mathbb{R} \times 2\mathbb{Z}$ has no identity. (We remark that some authors do require a subring to contain the identity element, if any, of the original ring.)

It is easily seen that a subring of an integral domain is an integral domain. In particular if R is a field and S is a subring of R then S is an integral domain. However, S need not be a field as we see from the fact that $\mathbb{Z}$ is a subring of $\mathbb{R}$. In case the subring is also a field, it is called a **subfield**. If F is a subfield of a field K, we also say K is an **extension field** of F. Extension fields will be extensively studied later on (see the Epilogue).

(3) *Rings of Functions*: This is yet another instance of generating new algebraic structures. Let $(R, +, \cdot)$ be a ring. X any set and F be the set R^X, consisting of all functions from X to R. Then under pointwise addition and multiplication of functions F is a ring. If R is commutative, so is F. If R has an identity element 1, then the constant function which assumes the value 1 at every point of X is an identity element for F. Note, however, that when R is an integral domain F is not so, unless the set X is a singleton (in which case we might as well identify F with R). For let Y be a proper non-empty subset of X. Let $f : X \to R$ be a function which vanishes on Y and assumes some non-zero value on the complement, $X - Y$. Let $g : X \to R$ be a function which vanishes on $X - Y$ but not anywhere on Y. Then f, g are both non-zero, but $fg = 0$.

We can identify R as a subring of F in various ways. For an element $r \in R$, let c_r be the constant function which takes the value r at all points of X. Let C be the set of all such constant functions. Then C is a subring of R. The function $\theta : R \to C$ defined by $\theta(r) = c_r$ for $r \in R$ is clearly a ring isomorphism. So upto an isomorphism, R is a subring of F. Another way to embed the ring R into F is to fix some $x_0 \in X$. For each $r \in R$, let $d_r : X \to R$ be the function which takes the value r at x_0 and the value 0 at all $x \neq x_0$ in X. Let D be the set of all such functions d_r as r varies over R. Then D is a subring of F which is isomorphic to R.

If X is a finite set we can think of F as the product of copies of R. Specifically, let $X = \{1, 2,..., n\}$. Then a function $f : X \to R$ determines a unique n-tuple of elements of R, namely, $(f(1), f(2),..., f(n))$. This gives a function ψ from F to the cartesian product $R \times R \times ... \times R$ (n-times). This function is a ring isomorphism.

It often happens that the sets X and R have some other structures in addition to the ring structure on R. We then consider only such functions from X to R, which are interesting from the point of view of that structure. Such functions often form subrings of F. For example let X be the unit interval $[0, 1]$ $(= \{x \in \mathbb{R} : 0 \leqslant x \leqslant 1\})$ and let $R = \mathbb{R}$, the field of real numbers. Let $C([0, 1])$ be the set of all continuous functions from $[0, 1]$ to $\mathbb{R}$. Since the sum and the product of two continuous functions are continuous, it follows that $C([0, 1])$ is a subring of the ring of all functions from $[0, 1]$ to $\mathbb{R}$. Similarly we have the ring of all differentiable functions from $[0, 1]$ to $\mathbb{R}$. Such rings are important in the study of that particular structure on the set X.

(4) *Rings of Matrices*: In the last construction, both the addition and multiplication of functions were defined pointwise. Depending upon the nature of the set X, it is often possible to define multiplication of functions in some other manner (without changing the definition of addition of functions) and this gives new rings. We already had an instance of this. As we just saw, we may identify the product $R \times R$ with the set of all functions from the set $\{1, 2\}$ to the set R. The ring structure on $R \times R$ corresponds to that obtained by pointwise addition and multiplication of functions from $\{1, 2\}$ to R. But suppose we define a new multiplication of two functions f, g by $(f * g)(1) = f(1) g(1) - f(2) g(2)$ and $(f * g)(2) = f(1) g(2) + f(2) g(1)$, then $(R \times R, +, *)$ becomes a ring which is isomorphic to the field of complex numbers, because $*$ corresponds to the multiplication of complex numbers if we identify the function f with the complex number $f(1) + i f(2)$. Similarly we ask the reader to view the ring of quaternions as obtained from defining a suitable multiplication on the set of all functions from the set $\{0, 1, 2, 3\}$ to R.

Now suppose X is the set $\{1, 2, ..., m\} \times \{1, 2, ..., n\}$ where m, n are some fixed positive integers. As remarked in Chapter 2, Section 1, a function from X into R is nothing but a matrix of order $m \times n$ (that is, with m rows and n columns) with entries coming from the set R. A typical such matrix, say A, is denoted as shown below* or, in a more compact form by $(a_{ij})_{\substack{1 \leqslant i \leqslant m \\ 1 \leqslant j \leqslant n}}$ or simply by (a_{ij}). Here a_{ij} is the value of the function A at the point (i, j) in the domain set $\{1, ..., m\} \times \{1, ..., n\}$, for $1 \leqslant i \leqslant m$, $1 \leqslant j \leqslant n$. It is called the (i, j)th entry of the matrix. The matrix A is said to be a matrix of order $m \times n$ **over** the ring R. If $m = n$, the matrix is called a **square matrix** of order n.

$$
\begin{array}{c}
\\
\\
\\
ith\ row \rightarrow \\
\\
\\
\end{array}
\left[
\begin{array}{cccccc}
a_{11} & a_{12} & \cdots & a_{1j} & \cdots & a_{1n} \\
a_{21} & a_{22} & \cdots & a_{2j} & \cdots & a_{2n} \\
\cdots & \cdots & \cdots & \cdots & \cdots & \cdots \\
a_{i1} & a_{i2} & \cdots & a_{1j} & \cdots & a_{in} \\
\cdots & \cdots & \cdots & \cdots & \cdots & \cdots \\
a_{m1} & a_{m2} & \cdots & a_{mj} & \cdots & a_{mn}
\end{array}
\right]
$$

$$\uparrow$$
$$j\text{th column}$$

* It is equally common (although typographically a little clumsy) to enclose a matrix by parentheses instead of brackets.

Given two matrices, say, $A = (a_{ij})$ and $B = (b_{ij})$ of the same order (that is with the same number of rows and the same number of columns) over the same ring R we can add them entrywise to get another matrix of the same order. Symbolically, $(a_{ij}) + (b_{ij}) = (a_{ij} + b_{ij})$. This amounts to pointwise addition of the corresponding functions into the ring R. We can also define the multiplication of two matrices of the same order entrywise, that is, $(a_{ij})(b_{ij}) = (a_{ij} b_{ij})$. But this would correspond to the pointwise multiplication of the corresponding functions and the resulting ring would be the same as the ring of functions considered above. To get a different ring structure we define a new multiplication called the matrix product. In order that the matrix product of two matrices, say A and B, (in this order) is defined it is necessary that the number of columns of A should equal the number of rows of A. So suppose $A = (a_{ij})$ is an $m \times n$ matrix and $B = (b_{jk})$ is an $n \times p$ matrix where, m, n, p are some positive integers. Then we define their **matrix product** (denoted by $A \cdot B$ or AB) to be the $m \times p$ matrix whose $(i, k)^{th}$ entry is the sum $\sum_{j=1}^{n} a_{ij} b_{jk}$ for $1 \leqslant i \leqslant m$, $1 \leqslant k \leqslant p$. For example

$$\begin{bmatrix} 1 & 2 & 3 & -1 \\ 0 & 1 & 2 & -2 \\ 1 & -1 & 4 & 0 \end{bmatrix} \begin{bmatrix} 2 & \sqrt{2} \\ 1 & -3 \\ 0 & \pi \\ -3 & -2 \end{bmatrix} = \begin{bmatrix} 7 & \sqrt{2} - 4 + 3\pi \\ 7 & 1 + 2\pi \\ 1 & \sqrt{2} + 3 + 4\pi \end{bmatrix}$$

$$A \qquad\qquad B \qquad\qquad A \cdot B$$

The easiest way to visualise the matrix product AB is to note that the entry in its ith row and kth column is obtained by taking the ith row of A and the kth column of B (each of which has n entries), multiplying their corresponding entries and summing the products. For example, in the illustration above, the third row of A has entries 1, -1, 4, 0 and the corresponding entries in the second column of B are $\sqrt{2}$, -3, π and -2. Multiplying them in that order and adding gives $\sqrt{2} + 3 + 4\pi$ which is the entry in the third row and the second column of the product matrix AB.

Matrix multiplication is not, strictly speaking a binary operation on the set of all matrices over a ring R, because not every pair of matrices can be multiplied (we may even have matrices A, B such that AB is defined but BA is not defined). Nevertheless matrix multiplication has properties similar to certain attributes of binary operations. We list three such properties:

(i) *Associativity*: Let A, B, C be matrices of orders $m \times n$, $n \times p$ and

$p \times q$ (say). Then all the products AB, BC, $(AB)C$ and $A(BC)$ are defined and the last two are equal. This is so because the (i, r)th entries in them are

$$\sum_{k=1}^{p} \left(\sum_{j=1}^{n} a_{ij} b_{jk} \right) c_{kr} \text{ and } \sum_{j=1}^{n} a_{ij} \left(\sum_{k=1}^{p} b_{jk} c_{kr} \right)$$

respectively and they are equal because of the associativity and distributivity of multiplication in the ring R.

(ii) *Distributivity*: Let A, B, C, D be matrices of orders $m \times n$, $n \times p$, $n \times p$ and $p \times q$ respectively. Then $A(B + C)$ equals $AB + AC$ and $(B + C)D$ equals $BD + CD$. This is also proved by straightforward verification that the corresponding entries are equal.

(iii) *Identity matrices*: Suppose the ring R has an identity element 1. For every positive integer n, let I_n be the $n \times n$ matrix in which all diagonal entries are 1 and all other entries are 0. For example, I_5 is the matrix

$$\begin{bmatrix} 1 & 0 & 0 & 0 & 0 \\ 0 & 1 & 0 & 0 & 0 \\ 0 & 0 & 1 & 0 & 0 \\ 0 & 0 & 0 & 1 & 0 \\ 0 & 0 & 0 & 0 & 1 \end{bmatrix}$$

Then for any $m \times n$ matrix A, $AI_n = A$ and for any $n \times p$ matrix B, $I_n B = B$.

For every positive integer n, let $M_n(R)$ be the set of all $n \times n$ matrices over R. Then matrix multiplication is a well-defined binary operation on $M_n(R)$. This operation is associative, distributive over matrix addition and, in case R has an identity, has I_n has the identity element. Thus we see that $M_n(R)$ is a ring, and has an identity in case R does. It is called the **ring of** $n \times n$ **matrices** over R. For $n = 1$, it is isomorphic to R. For $n > 1$, it is in general non-commutative even when R is commutative.

We shall return to matrices later in this chapter. Here we are interested in them primarily as means of constructing new rings. Various subrings of the ring of matrices often provide a 'concrete' representation of some abstract rings. For example, the classic view of complex numbers is that they are obtained by 'adjoining' to the real number system, the 'imaginary' square root of -1. Instead of this we can as well define a complex number to be a 2×2 matrix over $\mathbf{R}$, of the form $\begin{bmatrix} x & -y \\ y & x \end{bmatrix}$ (i.e., a 2×2 real matrix whose $(1, 1)$th and $(2, 2)$th entries are identical and $(1, 2)$th and $(2, 1)$th entries are the negatives of each other). Let S be the set of all matrices of this form as x, y vary over all real numbers. Then it is easily seen that S is a subring of

$M_2(\mathbb{R})$. Further, under the correspondence which takes $\begin{bmatrix} x & -y \\ y & x \end{bmatrix}$ to $x + iy$, S is isomorphic to the field of complex numbers. The number $i(= 0 + i1)$ corresponds to the matrix $\begin{bmatrix} 0 & -1 \\ 1 & 0 \end{bmatrix}$, while a real number $x(= x + i0)$ corresponds to $\begin{bmatrix} x & 0 \\ 0 & x \end{bmatrix}$. So we could as well define a complex number as a 2×2 matrix of a certain form. This representation is very important because we shall consider many extension fields which are obtained by adjoining certain new elements to some 'ground' or 'base' field. In all such cases, the elements of the new fields can be 'concretely' realised as certain square matrices of appropriate order over the ground field, and the extension field turns out to be isomorphic to a subring of the ring of matrices.

(5) *Rings of Polynomials*: This is another important class of rings that arise from defining multiplication of functions in a manner other than pointwise. Let $\mathbb{N}_0$ be the set $\{0, 1, 2, ...\}$ and R be a ring with identity. Let f, g be two functions from $\mathbb{N}_0$ to R. We define $f + g$ pointwise. But we define fg to be the function whose value at an element $n \in \mathbb{N}_0$ is the sum $\sum_{i=0}^{n} f(i) g(n - i)$. We leave it to the reader to verify that this operation makes the set F of all functions from $\mathbb{N}_0$ to R into a ring. This ring has an identity element, namely, the function which takes the value 1 at 0 and the value 0 at all other points of $\mathbb{N}_0$.

Lest this multiplication sound too bizarre, let us view it a little differently. Note that every function f from $\mathbb{N}_0$ to R is a sequence $\{a_n\}_{n=0}^{\infty}$ where $a_i = f(i)$, for $i = 0, 1, 2, ...$. Let x be some symbol not in R. Consider the power series $\sum_{i=0}^{\infty} a_i x^i$. This power series is purely formal, that is, notational and does not signify convergence or limiting process of any kind. Two such power series are added term by term, that is,

$$\left(\sum_{i=0}^{\infty} a_i x^i \right) + \left(\sum_{i=0}^{\infty} b_i x^i \right) = \sum_{i=0}^{\infty} (a_i + b_i) x^i.$$

But to multiply them we use the law of indices $x^{i+j} = x^i \cdot x^j$. So in the product $\left(\sum_{i=0}^{\infty} a_i x^i \right) \left(\sum_{i=0}^{\infty} b_i x^i \right)$ the coefficient of x^n will be $\sum_{i=0}^{n} a_i b_{n-i}$. This is precisely $\sum_{i=0}^{n} f(i) g(n - i)$ where f, g are functions from $\mathbb{N}_0$ to R defined by $f(i) = a_i$ and $g(i) = b_i$ for $i \in \mathbb{N}_0$. Thus we see that the definition of multiplication of two functions is simply the multiplication for power series.* So far we

---

* Another interpretation is that it is simply the convolution as defined in Exercise (3.4.25) if the poset is $\mathbb{N}_0$ with the usual order, where we think of $f(m, n)$ as $f(n - m)$, for $m \leqslant n$.

treated x as some unknown symbol (or an '**indeterminate**' as it is techni-cally called). But we can identify it with the power series $\sum_{i=0}^{\infty} a_i x^i$ in which $a_1 = 1$ and $a_i = 0$ for all $i \neq 1$. The ring defined above is called the **ring of power series** in x over the ring R. We shall denote it by $R\{x\}$. The symbol x may of course be replaced by any other symbol, say y. It is clear that $R\{x\}$ is commutative iff R is commutative.

Far more interesting is the subring consisting of the so-called finite power series, or more popularly, polynomials. Formally, a **polynomial** in x with coefficients in R is an expression of the form $a_0 + a_1 x \ldots + a_i x^i + \ldots$ in which all a_i's after some i are 0. The largest n for which $a_n \neq 0$ is called the **degree** of the polynomial and this element a_n is called the **leading coefficient** of the polynomial. (If $a_i = 0$ for all i, then the polynomial is the zero polynomial and we assign no degree to it.) The set of polynomials is easily seen to be a subring of the power series ring over R and is denoted by $R[x]$. Note that if R is an integral domain, so is $R[x]$ because if $f(x)$, $g(x)$ are non-zero polynomials of degrees m and n then $f(x) \, g(x)$ is a polynomial with degree $m + n$, whose leading coefficient is the product of the leading coefficients of $f(x)$ and $g(x)$. However, even if R is a field, $R[x]$ is not a field, it is only an integral domain.

The ring R itself may be identified with a subring of $R[x]$, consisting of the constant polynomials. Thus $R[x]$ is an extension ring of R and will be very important in the construction of extension fields.

We can also consider the ring of polynomials in two or more indeter-minates, say, $x_1, x_2, \ldots, x_k$. It is denoted by $R[x_1, x_2, \ldots x_k]$. It consists of all finite sums of the form $\sum x_1^{n_1} x_2^{n_2} \ldots x_k^{n_k}$ where n_i are non-negative integers. Their multiplication is defined using the laws of indices for powers of each indeterminate. Alternatively, $R[x_1, \ldots, x_k]$ may be defined inductively as the polynomial ring in x_k over $R[x_1, \ldots, x_{k-1}]$ for $k > 1$.

(6) *Quotient Rings*: Just as we form quotient groups of groups we can form quotient rings of rings. Recall that the quotient group consists of cosets of a subgroup. In order to have a well-defined coset multiplication, we have to have some restriction on the subgroup, namely that it be nor-mal. Similarly, in the case of a ring, we consider the cosets of a subring but to define a ring structure on the set of these cosets, the subring must satisfy certain additional property. We define what it is.

1.15   Definition: Let $(R, +, \cdot)$ be a ring. A subset I of R is called a **left ideal** (or a **right ideal**) if I is a subgroup of $(R, +)$ and for all $x \in I$, $r \in R$, $rx \in I$ (respectively $xr \in I$). If I is both a left ideal and a right ideal then it is called a two-sided ideal or simply an **ideal**.

Evidently every left ideal and every right ideal is a subring because the second condition implies that it is closed under multiplication. The con-verse is not true in general. For example, $\mathbb{Z}$ is a subring of $\mathbb{R}$ but it is not an ideal. Indeed we see that if a ring R has an identity element 1 then no proper left or right ideal can contain 1, for if it does then since $r \cdot 1 = r = 1 \cdot r$ for all $r \in R$, it would contain every $r \in R$. If a is any element of a ring R then we let $Ra = \{ra : r \in R\}$. It is easily seen that Ra is a left ideal. It is called the **principal left ideal** generated by a. Similarly $aR = \{ar : r \in R\}$ is a right ideal, called the principal right ideal generated by a. If R is commutative then of course Ra equals aR and is an ideal of R called the principal ideal generated by a. If R has an identity then aR contains a, but otherwise it need not. For example if $R = 2\mathbb{Z}$ and $a = 2$ then $2R = 4\mathbb{Z}$ which does not contain 2. So the term ideal 'generated' by a is somewhat misleading. Generally, it is used only for rings having identity elements.

To see how quotient rings are obtained from ideals, let I be a two-sided ideal in a ring R. Then I is also a subgroup of the additive group of R. We consider its cosets in R and denote them additively, that is, if $x \in R$ then $x + I$ denotes the set $\{x + y : y \in I\}$ which is the coset of the subgroup I containing x. Since R is an abelian group, I is normal in R and so there is no difficulty in defining the coset addition. In fact for this purpose it would suffice if I were merely a subring of R. But when we try to define coset multiplication by $(x + I)(y + I) = xy + I$, in order to show it is well-defined, the fact that I is an ideal is needed. For suppose $x + I = z + I$ and $y + I = w + I$. Then $x - z \in I$ and $y - w \in I$. We have to show that $xy + I = zw + I$, that is, $xy - zw \in I$. For this we write $xy - zw$ as $x(y - w) + (x - z)w$. Since $x \in R$ and $y - w \in I$, $x(y - w) \in I$. Similarly, since $(x - z) \in I$ and $w \in R$, $(x - z)w \in I$. I is also closed under addition. So $x(y - w) + (x - z)w \in I$, that is, $xy - zw \in I$ as was to be shown. Having obtained well-defined operations of addition and multiplication, it is now a routine matter to verify that R/I is a ring. It is commutative if R is. If R has an identity 1 then the coset $1 + I$ is the identity element of R/I. As a simple example of quotient rings let $R = \mathbb{Z}$ and $I = m\mathbb{Z}$ where m is some positive integer. Then R/I is precisely $\mathbb{Z}_m$, the ring of residue classes modulo m, under modulo m addition and multiplication. When m is not a prime, $\mathbb{Z}_m$ is not an integral domain as we saw earlier. Thus we see that the quotient ring of an integral domain need not be an integral domain. Actually, when R/I will be an integral domain (or a field) depends more on the ideal I than on the ring R. Since we shall construct many fields as quotient rings of certain rings by suitable ideals, we characterise those ideals whose quotient rings are fields.

1.16 Definition: An ideal I of a ring R is called **maximal** if it is a proper ideal (that is $I \underset{\neq}{\subset} R$) and is not properly contained in any proper ideal of R, that is whenever J is an ideal of R such that $I \subset J \subset R$ then either $J = I$ or $J = R$.

If we consider the collection of all proper ideals of R and partially order it by inclusion, then a maximal element of it is a maximal ideal and conversely. This justifies the name. If $R = \mathbb{Z}$, and m is a positive integer, then $m\mathbb{Z}$ is maximal if and only if m is a prime. (This will be proved in the next section). The importance of maximal ideals comes from the following result.

1.17   Proposition: Let R be a commutative ring with identity and I an ideal of R. Then the quotient ring R/I is a field if and only if I is a maximal ideal of R.

Proof: Let I be a maximal ideal. R/I is already a commutative ring with identity $1 + I$. To prove it is a field we have to show that every non-zero element of it has a multiplicative inverse. A non-zero element of R/I is a coset of the form $x + I$, where $x \notin I$. Now let J be the set $I + Rx$, that is the set of all elements of the form $y + rx$ where $y \in I$ and $r \in R$. Taking $r = 0$, we see that $I \subset J$. Also taking $y = 0$ and $r = 1$, $x \in J$. Since $x \notin I$, I is a proper subset of J. We claim J is an ideal of R. For let $y_1 + r_1 x$ and $y_2 + r_2 x \in J$. Then $(y_1 + r_1 x) - (y_2 + r_2 x) = (y_1 - y_2) + (r_1 - r_2)x \in J$ since $y_1 - y_2 \in I$ and $r_1 - r_2 \in R$. This proves that J is a subgroup of $(R, +)$ (cf. Exercise (5.1.6)). For proving that J is an ideal, suppose $z \in R$ and $y + rx \in J$ where $y \in I$, $r \in R$. Then $z(y + rx) = zy + zrx \in J$ since $zy \in I$ (I being an ideal). So J is an ideal of R which properly contains I. By maximality of I, J must equal R. In particular, $1 = y + rx$ for some $y \in I$, $r \in R$. Then $1 + I = (y + I) + (r + I)(x + I)$. $y + I$ is the zero element of R/I. Hence $1 + I = (r + I)(x + I)$. This means that the element $x + I$ has $r + I$ as its multiplicative inverse. As noted before, this proves that R/I is a field.

Conversely, suppose R/I is a field. We have to show that the ideal I is maximal. Let J be an ideal of R which properly contains I. We have to show $J = R$. As noted above, it suffices to show that $1 \in J$. Now since $I \underset{\neq}{\subset} J$, there exists $x \in J$ such that $x \notin I$. This means the coset $x + I$ is a non-zero element of R/I. Since R/I is a field, $x + I$ has an inverse, say, $y + I$ for some $y \in R$. This means $(x + I)(y + I)$ equals $1 + I$. But then $xy - 1 \in I$. Let $z = xy - 1$. Then $z \in J$ (since $I \subset J$). Also $xy \in J$ since $x \in J$ and J is an ideal. So $1 = xy - z \in J$ as was to be shown ∎

The last proposition will be applied to the polynomial ring $F[x]$ over a field F, to get an extension field which will contain F as a subfield (upto an isomorphism). Of course we must have some way of telling which ideals in $F[x]$ are maximal. This will be found in the next section.

Just as normal subgroups are intimately related to group homomorphisms, ideals are related to ring homomorphisms We shall define this concept and give its properties (which are straightforward analogues of the corresponding properties for groups) through the exercises.

(7)   *Field of Quotients:*   The last construction of forming new rings by taking quotients of some rings by ideals is applicable for all types of rings

and results in fields for certain types of ideals. The construction to be given now is applicable only for integral domains and in spite of the word 'quotient' in it has nothing to do with quotient rings. It is motivated by the construction of rational numbers as ratios or quotients of integers. (The word 'rational' in fact comes from 'ratio'.) Suppose R is an integral domain and $x, y \in R$, with $x \neq 0$. We would like to define the ratio or the quotient of y by x to be an element z such that $xz = y$. Because of Proposition (1.7), such an element z, if at all it exists, is unique. But it need not always exists. If it does, we say x **divides** y and write $x \mid y$. For example, in $\mathbb{Z}$, 6 divides 30 but 8 does not divide 30. Of course we can think of 30/8 as a rational number. We can say that the field of rational numbers is obtained by adding to $\mathbb{Z}$ all such 'missing quotients' of pairs of integers. We associate the pair (30, 8) with the rational number 30/8. Note that 30/8 is the same rational number as 15/4. In other words, the pair (30, 8) corresponds to the same rational number as the pair (15, 4). More generally, if $a, b, c, d \in \mathbb{Z}$ with $b, d \neq 0$ then the pairs (a, b) and (c, d) correspond to the same rational number iff $ad = bc$. The addition and multiplication of rational numbers can also be described in terms of the corresponding ordered pairs. Thus $(a, b) + (c, d) = (ad + bc, bd)$ and $(a, b) \cdot (c, d) = (ac, bd)$ (These are nothing but the rules we learn in early school, if we think of the first member of the pair as numerator and the second member as denominator.)

We now show how this construction can be generalized to the case of an 'abstract' integral domain R. First we consider the set $S = R \times (R - \{0\})$, consisting of all ordered pairs (a, b) of elements of R with $b \neq 0$. On S we define a relation $\sim$ by $(a, b) \sim (c, d)$ iff $ad = bc$. It is easy to show that $\sim$ is an equivalence relation. The equivalence class containing an element (a, b) of S will be denoted by $[a, b]$. Let Q be the set of all equivalence classes of S under the equivalence relation $\sim$. We are going to define a field structure on Q. First we define two binary operations, still denoted by $+$ and $\cdot$ **by** $[a, b] + [c, d] = [ad + bc, bd]$ and $[a, b] \cdot [c, d] = [ac, bd]$. We must, of course, verify that these are well-defined. We leave this as an exercise and turn to proving that with these operations, Q is indeed a field (where, too, we shall be very sketchy in the proof).

1.18 Theorem: With the operations defined above, Q is a field which contains a subring T which is isomorphic to R. Every element of Q can be expressed as a quotient of two elements of T.

Proof: The properties which are necessary to make Q a commutative ring follow one-by-one from the corresponding properties of R. We assert that even if R has no identity element, Q always does. We assume R is not trivial. (If R has only one element, the set S is empty, so Q has only one element and the theorem holds trivially.) Let x be any non-zero element of

R. We claim that $[x, x]$ is the identity for multiplication on Q. For, let $[a, b] \in Q$. Then $[a, b] \cdot [x, x] = [ax, bx]$. But $[ax, bx] = [a, b]$ because $abx = axb$. (Note that if we took some other non-zero element y in R, then we would get the same identity element, because $[x, x] = [y, y]$.)

For proving that Q is a field, suppose $[a, b]$ is a non-zero element of Q. Then $a \neq 0$ (because, ordered pairs of the form $(0, b)$, $b \in R - \{0\}$ constitute an equivalence class under $\sim$ which represents the zero element of Q). So $(b, a) \in S$. Now $[a, b] \cdot [b, a] = [ab, ab]$ which is the identity element of Q, showing that $[b, a]$ is the multiplicative inverse of $[a, b]$. Thus Q is a field. Fix some non-zero $x \in R$. Let $T = \{[ax, x] : a \in R\}$. It is easy to show that T is a subring of Q. Moreover, the function $f \colon R \to T$ defined by $f(a) = [ax, x]$ for $a \in R$ is a ring isomorphism. For the last assertion, let $[a, b] \in Q$. Note that $[b, bx]$ is the multiplicative inverse of $[bx, b]$. So the quotient $[ax, x]/[bx, b]$ equals $[ax, x] \cdot [x, bx]$ which is the same as $[a, b]$. ∎

1.19 Definition: The field Q constructed above is called the **field of quotients** of the integral domain R. The element $[a, b]$ is generally denoted by a/b. Also R is identified with T and hence treated as a subring of Q.

The field of quotients is sometimes called the quotient field. It should not be confused with the concept of a quotient ring. It is not obtained from any ideal. Instead it is constructed by adjoining to the integral domain R, the missing quotients. If in R, some element b does divide some other element a with ratio c (say), then the pair (a, b) is equivalent to (cx, x) for any non-zero x, and to $(c, 1)$ in particular if R has an identity. In other words, in such a case $[a, b]$ is in T which is just a replica of R. In particular if R itself is a field to start with then Q would coincide with T and so we do not get anything new. The construction of Q from R typifies a common theme underlying many apparently diverse constructions in mathematics. Whenever you would like something to exist (in this case the ratios or the quotients) and it does not exist in the domain you have, enlarge the domain by adjoining certain hypothetical objects (in this case, equivalence classes of pairs). Figuratively, these extra objects serve to fill the 'holes' which arise because of non-existence of certain elements in the original domain. Another instance of this philosophy is the construction of complex numbers. In the real number system, the number -1 has no square root. So we adjoin the hypothetical (or 'imaginary') number i such that $i^2 = -1$. In order to have a field which contains all real numbers, plus this new number i, we must also have numbers of the form $a + ib$ where a, b are real and this leads to the field of complex numbers.

Suppose now that the integral domain R is itself the ring $F[x]$, that is, the ring of polynomials over some field F (see Construction (5) above). Then the field of quotients of $F[x]$ is denoted by $F(x)$. Elements of this field are ratios of the form $p(x)/q(x)$ where $p(x)$, $q(x)$ are polynomials with coefficients in F and $q(x) \neq 0$. Such ratios are often called **rational functions** of x. Consequently, the field $F(x)$ is called the field of rational functions in x. Note that even when F is a finite field, the polynomial ring $F[x]$ is infinite. So $F(x)$ is also infinite, since it contains a replica of $F[x]$. Taking $F = \mathbb{Z}_p$ where p is a prime, we get $\mathbb{Z}_p(x)$ as an example of an infinite field of a prime characteristic.

Exercises

1.1 Prove that in a ring with identity, the commutative law for addition is redundant, that is, can be derived from the remaining axioms of a ring. (Hint: cf. Proposition 3.4.12.)

1.2 An element x of a ring is called **idempotent** if $x^2 = x$. (Actually, the definition makes sense for any binary operation on a set; see Exercise 3.4.23.) Prove that an integral domain has at least one and at most two idempotents. Prove that a Boolean ring cannot be an integral domain unless it is trivial or isomorphic to $\mathbb{Z}_2$.

1.3 The **centre** Z of a ring R is defined as the set $\{z \in R : zr = rz$ for all $r \in R\}$. Prove that the centre is a subring. Find the centres of the ring of quaternions and the ring of 2×2 matrices over $\mathbb{R}$.

1.4 Prove that a ring in which $x^3 = x$ for all x is commutative. (Hint: First show that the square of every element is in the centre. Then rewrite $(xy)^3$ as $x(yx)^2y$.)

1.5 For any two quaternions a, b, prove that
$$N(a + b) \leqslant N(a) + N(b) \text{ and } N(ab) = N(a)N(b).$$

1.6 Prove that the ring of quaternions contains subrings isomorphic to $\mathbb{R}$ and $\mathbb{C}$.

1.7 Obtain a representation of the quaternion ring as a subring of $M_4(\mathbb{R})$.

*1.8 Obtain a repreresentation of the quaternion ring as a subring of $M_2(\mathbb{C})$.

1.9 Prove that a field has characteristic 0 iff it contains a subfield isomorphic to $\mathbb{Q}$ (the field of rationals) and that it has characteristic p iff it contains a subfield isomorphic to $\mathbb{Z}_p$. (Hint: In both cases, consider the subfield generated by the element 1.)

1.10 Prove that the additive group in a field is not cyclic unless the field is isomorphic to $\mathbb{Z}_p$ for some prime p.

1.11 Let $R = R_1 \times R_2$ where R_1, R_2 are rings. If S_1, S_2 are subrings of R_1, R_2 respectively, prove that $S_1 \times S_2$ is a subring of R. Is every subring of R of this type? Prove that the subring $R_1 \times \{0\}$ is an ideal of R, which is isomorphic to R_1 and the quotient ring $R/(R_1 \times \{0\})$ is isomorphic to R_2.

1.12 Let R be a ring and R_1, $R_2 \subset R$. Prove that if R_1, R_2 are subrings, or left ideals or right ideals of R, so is $R_1 \cap R_2$. Show by an example that the intersection of a left ideal and a right ideal need not be either a left or a right ideal. (It is however a subring).

1.13   An $n \times n$ matrix A is called a **diagonal** matrix if all the entries off its principal diagonal are zero, that is $a_{ij} = 0$ for $i \neq j$. Let R be a ring. Prove that the set of all diagonal $n \times n$ matrices is a subring of $M_n(R)$. Is it an ideal?

1.14   Let R be the ring of all continuous, real valued functions on the unit interval $[0, 1]$. Fix some subset $A \subset [0, 1]$. Let $M_A = \{f \in R : f(x) = 0$ for all $x \in A\}$. Prove that M_A is an ideal of R and that it is maximal iff the set A is a singleton. (Hint: Note that if $B \subset A$, then $M_A \subset M_B$.)

**1.15   In the last exercise, prove that every maximal ideal of R is of the form $M_{\{x_0\}}$ for some $x_0 \in [0, 1]$. (The proof requires the fact that $[0, 1]$ is compact.)

*1.16   Let R be the set of all analytic functions from $\mathbb{C}$ to $\mathbb{C}$. Prove that R is a subring of the ring of all functions from $\mathbb{C}$ to $\mathbb{C}$. Prove further that R is an integral domain. (The proof requires that the set of zeros of a non-zero analytic function is discrete, that is, has no limit points.)

1.17   Prove that the sets $\{a + \sqrt{2}b : a, b \in \mathbb{Z}\}$ and $\{a + i\sqrt{2}b : a, b \in \mathbb{Z}\}$ are subrings of $\mathbb{R}$ and $\mathbb{C}$ respectively. Give other examples of this type.

1.18   (a) For any ring R with identity verify that the multiplication of power series with coefficients in R is indeed associative and hence that $R\{x\}$ is a ring.

(b) Let R be a ring with identity, n a positive integer and x an indeterminate. Prove that the ring $M_n(R[x])$ of $n \times n$ matrices over the polynomial ring $R[x]$ may be identified with the ring $(M_n(R))[x]$ of polynomials over the ring $M_n(R)$ of $n \times n$ matrices over R.

1.19   Let F be a field and $F\{x\}$ the ring of power series over F. Prove that the power series $1 - x$ is invertible and has for its inverse the power series $1 + x + x^2 + x^3 + \ldots + x^4 + \ldots$. (Caution: This does not mean that the geometric series $1 + x + x^2 + \ldots$ 'converges' to $1/1 - x$. Actually, we have not defined the concept of convergence in an abstract ring. The result to be proved simply means that the product of the two power series $1 - x$ (i.e., $1 - x + 0x^2 + 0x^3 + 0x^4 + \ldots$) and $1 + x + x^2 + x^3 + \ldots$ is the power series 1 (that is, $1 + 0x + 0x^2 + 0x^3 + \ldots$). More generally, prove that a power series $\sum_{n=0}^{\infty} a_n x^n$ is invertible iff $a_0 \neq 0$.)

1.20   Just as the last exercise deals with formal inversion of power series, we can study formal derivatives of power series, that is, derivatives which arise purely from a particular form and not from any limiting process as in calculus. If $f(x) = \sum\limits_{n=0}^{\infty} a_n x^n$ is a power series in x, we define its **derivative** (denoted by $f'(x)$) to be the power series $\sum\limits_{n=0}^{\infty} (n+1)a_{n+1}x^n$. If $f(x)$, $g(x)$ are any two power series, prove that

$$(f+g)'(x) = f'(x) + g'(x) \text{ and } (fg)'(x) = f(x)g'(x) + f'(x)g(x).$$

1.21 Prove that a commutative ring with identity is a field if and only if it has no ideals other than the zero ideal and the whole ring.

1.22 Let R, S be rings. A function $f: R \to S$ is called a **ring homomorphism** if for all x, $y \in R$, $f(x+y) = f(x) + f(y)$ and $f(xy) = f(x)f(y)$. The kernel of f is defined as the set $\{x \in R : f(x) = 0\}$. Prove that the kernel of a ring homomorphism is an ideal and conversely every ideal is the kernel of some ring homomorphism. If $f: R \to S$ is a ring homomorphism with kernel K and range T, prove that T is a subring of S and is isomorphic, as a ring, to the quotient ring R/K. Also state and prove the analogue of Theorem 5.3.7.

1.23 Using the last two exercises, give an alternate proof of Proposition 1.16.

1.24 Let $f: R \to S$ be a ring homomorphism where R is field. Prove that f is either identically zero or else an isomorphism of R onto the range of f.

1.25 Let Q be the field of quotients of an integral domain R. Let F be any field. Given any one-to-one homomorphism $f: R \to F$, prove that there exists a unique ring homomorphism $g: Q \to F$ which extends f (that is, $g(x) = f(x)$ for all $x \in R$).

1.26 An **ordered field** is a field, which is totally ordered in a manner compatible with the field structure. Formally it is a quadruple $(F, +, \cdot, <)$ such that $(F, +, \cdot)$ is a field and $<$ is a strict linear order on the set F satisfying

(i) if $a < b$ then $a + b < a + c$ for all $a, b, c, \in F$ and
(ii) if $a < b$ and $0 < c$ then $ac < bc$ for all $a, b, c, \in F$.

(The definition would make sense for any ring, not just for a field. But it is only for integral domains and fields that the concept of ordering has important consequences.)

In an ordered field an element x is called **positive** if $x > 0$ and **negative** if $x < 0$. Prove that the square of every non-zero element is positive. Hence show that an ordered field must be of characteristic 0.

1.27 Show by an example, that in a ring, $(x + y)(x - y)$ need not equal

$x^2 - y^2$. Prove, however, that this holds in a commutative ring. Prove also, that in a commutative ring, the **binomial theorem** holds, that is, for any two elements x and y and any positive integer n, $(x + y)^n$ equals

$$x^n + \binom{n}{1} x^{n-1}y + \binom{n}{2} x^{n-2}y^2 + \cdots +$$

$$+ \binom{n}{r} x^{n-r}y^r + \cdots + \binom{n}{n-1} xy^{n-1} + y^n.$$

Notes and Guide to Literature

Ring theory may very well be considered to be the centre of modern algebra. Jacobson's three volumes [1], provide a thorough treatment. Special references for commutative and non-commutative rings are Zariski and Samuel [1] and Herstein [2] respectively.

The origin of the terms 'ring' and ideal' is somewhat obscure. The former is attributed to the geometric representation of the ring $\mathbb{Z}_m$ by the complex mth roots of unity. These points lie on a circle or a 'ring'. 'Ideal' probably comes from the so called 'ideal' points in projective geometry.

The quaternions were invented by Hamilton. For more on them see Herstein [1], where the reader will also find a proof of a remarkable theorem due to Wedderburn that every finite division ring is a field. That is why, although we can form quaternion rings over $\mathbb{Z}_p$ exactly as over $\mathbb{R}$, they are not division rings.

Exercises (1.15) and (1.16) highlight how the structure of certain rings of functions is affected by the structure on the domain set. A good reference on the rings of continuous functions is Gillman and Jerison [1].

The binomial theorem (Exercise 1.27) explains why the binomial coefficients are called that way. It is one of the most well-known theorems of algebra. For historical references to it and to the binomial coefficients see Knuth [1], Vol. 1.

2. Special Types of Integral Domains

It was remarked in the last section that the ring of integers is a foremost example of a ring. Still, the definition of a ring is far too general to capture any of the interesting properties of $\mathbb{Z}$. If we confine ourselves to integral domains, some of the phenomena that take place in $\mathbb{Z}$ can be imitated, such as the construction of the field of quotients. But the class of all integral domains is still too large to permit generalisation of the deeper properties of integers. So we want to restrict it further by putting some more conditions on integral domains. Of course, these additional restrictions must not be so strong that, upto isomorphism, $\mathbb{Z}$ is the only ring satisfying them! For, in that case there is little point in the generalisation. In other words, we are once again trying to gain some depth in our abstraction, without having to sacrifice generality altogether.

In this section we consider three progressively stronger conditions on an integral domain which make it resemble $\mathbb{Z}$ more and more. Yet, the class of rings satisfying even the strongest of these three conditions, will be large enough to include in it, besides $\mathbb{Z}$, the ring of polynomials over a field and many other rings. The properties to be proved for rings satisfying this condition will hold in particular for the ring of integers. These properties will then be used to fill some gaps which were left in the earlier chapters (for example, finding the number of generators for a cyclic group).

We begin with the strongest of the three conditions. It is intended to imitate the euclidean algorithm for integers (which we already used without proof a few times, for example in Problem (2.2.10) and Theorem (5.1.7)). Recall that, according to this algorithm, given integers a, b with $b > 0$, we can find integers q, r such that $a = qb + r$ with $0 \leqslant r < b$. The integers q and r are easily seen to be unique and are respectively called the **quotient** and the **remainder** (or **residue**) obtained when a is divided by b. This algorithm was originally used by Euclid to give a procedure for finding the greatest common divisor of the two integers a and b. The next step is to replace a by b and b by r and carry out the division once again to get a new remainder, r_1 (say). We then continue, everytime letting the old divisor be the new dividend and the old remainder be the new divisor. This is a familiar procedure which we all learn in schools. But what guarantee is there that it will terminate with a remainder 0 (and give the divisor at this stage as the greatest common divisor)? The crucial point is that the remainder, if non-zero, is always less than the divisor. This, combined with the fact that the set of positive integers is well-ordered (see Exercise (3.3.21)) ensures that the process of successive divisions, with the remainders getting smaller every time will not go on indefinitely.

Now, how do we carry this out in an 'abstract' integral domain R? A direct imitation would require that on R we should assume an order structure with properties similar to the usual ordering on $\mathbb{Z}$. Then, first of all, the order relation has to be compatible with the binary operations on R (cf. Exercise (1.26)). Such ordered integral domains can indeed be defined and studied But if we further require that the set of positive elements in R be well-ordered (which is true in case of $\mathbb{Z}$), then this requirement would be too strong and there would not be any 'naturally occurring' examples of rings besides $\mathbb{Z}$ and its subrings that will satisfy it.

So we look for another generalisation. Instead of ordering the elements of R directly we consider a suitable function d from R into the set of non-negative integers (which is well-ordered) and work with the values assumed by this function. We then make the following definition.

2.1 Definition: An integral domain R is called a **euclidean ring** (or a euclidean domain) if there exists a function $d: R - \{0\} \to \mathbb{N}_0$ (the set of non-negative integers), such that the following condition (∗) holds.

(∗) for all $a, b \in R$, with $b \neq 0$, there exist $q, r \in R$ satisfying, $a = qb + r$ where either $r = 0$ or $d(r) < d(b)$.

The function d is often called the **degree** function and $d(a)$ is read as 'degree of a'. This terminology comes from the ring of polynomials over a field, which we shall show to be a euclidean ring a little later. A trivial example of a euclidean ring is any field. If R is a field and we let $d(a) = 0$ for $a \in F, a \neq 0$, then (∗) holds because, since b^{-1} exists we can take $q = ab^{-1}$ and $r = 0$.

As the first non-trivial example of a euclidean ring, we have,

2.2 Theorem: The ring of integers is a euclidean ring.

Proof: Define $d: \mathbb{Z} - \{0\} \to \mathbb{N}_0$ by $d(x) = |x|$, that is, the absolute value of x. We have to show that (∗) is satisfied. If $a = 0$, then we set $q = r = 0$. Suppose a, b are both non-zero. We make four cases.

(I) $a > 0, b > 0$. In this case we argue by induction on a. For $a = 1$, we set $q = 1, r = 0$ if $b = 1$ and $q = 0, r = 1$ if $b > 1$. Now suppose $a = n > 1$ and the result holds for all positive values of a less than n. If $n < b$, we set $q = 0$ and $r = n$. Suppose $n \geqslant b$. Then $n - b$ is either 0 or a positive integer less than n. So, in either case we can find q_0, r_0 such that $n - b = q_0 b + r_0$ with $r_0 = 0$ or $d(r_0) < b$. We then have $n = (q_0 + 1) b + r_0$ and so can set $q = q_0 + 1$ and $r = r_0$.

(II) $a > 0, b < 0$. Let $c = -b$. Then $a > 0, c > 0$ and so by Case I; there exist $q_1, r_1 \in \mathbb{Z}$ such that $a = q_1 c + r_1$ with $r_1 = 0$ or $|r_1| < |c|$. Let $q = -q_1$ and $r = r_1$. Then $a = qb + r$ and either $r = 0$ or $|r| = |r_1| < |c| = |b|$, as desired.

(III) $a < 0, b > 0$. Replace a by $-a$ and apply Case I, to get $-a = q_2 b + r_2$ (say). Then set $q = -q_2$ and $r = -r_2$.

(IV) $a < 0, b < 0$. Here too, replace a, b by $-a, -b$ and convert to Case I. ∎

Note that what we have called euclidean algorithm is slightly different from the condition (∗) in Definition (2.1) applied to the ring $\mathbb{Z}$. We assume that $b > 0$ and require the remainder to be non-negative, whereas condition (∗) only gives $|r| < |b|$ (unless $r = 0$). This is only a minor difference and can be easily straightened out. Let a, b be integers with $b > 0$. By the last theorem, we can find q, r such that $a = qb + r$ with $|r| < |b|$. If $r \geqslant 0$, then of course we are done. If $r < 0$, then, r must be in the set $\{-b + 1, ..., -1\}$

(because $|r| < b$). But then $b + r$ is in the set $\{1,..., b - 1\}$ and $a = (q - 1)b + (b + r)$ gives the desired representation of a with the remainder, (that is, $b + r$) non negative. Thus we have completely proved the euclidean algorithm and thereby sanctified its uses in the previous chapter. (We could have as well given the proof the first time the algorithm was used. But we deferred it till now because the definition of a euclidean ring involves it crucially.) Note that the essence of the inductive step in Case I of the proof above is the same as finding the largest multiple of b which can be subtracted from a. The trouble is how to find it. The inductive argument really shows that the largest multiple of b which can be subtracted from a is the next larger multiple than the largest multiple which can be subtracted from $a - b$.

Because of the last theorem, whatever properties are true of all euclidean rings would hold, in particular, for $\mathbb{Z}$. But before proving such properties, let us show that the class of euclidean rings is large enough to make it worthwhile to prove theorems about them. In this vein we prove:

2.3 Theorem: For every field F, the polynomial ring $F[x]$ is a euclidean ring.

Proof: We define $d : F[x] - \{0\} \to \mathbb{N}_0$ to be function which assigns to each polynomial its degree as defined in the last section. Thus if $f(x) = a_0 + a_1 x + ... + a_n x^n$ with $a_n \neq 0$, then $d(f(x)) = n$. We assign no degree to the zero-polynomial. For the condition (*) let $f(x)$, $g(x)$ be polynomials of degree m, n respectively, say, $f(x) = a_0 + a_1 x + ... + a_m x^m$ and $g(x) = b_0 + b_1 x + ... + b_n x^n$ with $a_m \neq 0 \neq b_n$. If $m < n$ we write $f(x) = 0 \cdot g(x) + f(x)$ and this proves (*). If $m \geqslant n$, then we subtract a suitable multiple of $g(x)$ from $f(x)$. Specifically, let $h(x) = f(x) - (a_m/b_n) g(x) x^{m-n}$. Note that here for the first time we are using that F is a field (whence a_m/b_n that is, $a_m(b_n)^{-1}$ exists). If $h(x) = 0$ then $g(x)$ divides $f(x)$ and we are done. Otherwise $h(x)$ is a polynomial of degree less than m. So we can proceed by induction on m and write $h(x)$ as $q_0(x) g(x) + r_0(x)$ where $r_0(x)$ is either 0 or a polynomial of degree less than n. Then we simply write

$$f(x) = \left[q_0(x) + \frac{a_m}{b_n} x^{m-n} \right] g(x) + r_0(x)$$

as desired. Thus condition (*) in the definition is satisfied. So $F[x]$ is a euclidean ring. ∎

We mention yet another example of a euclidean ring, called the ring of **Gaussian integers**. It consists of complex numbers whose real and imaginary parts are integers, that is, $\{x + iy : x, y \in \mathbb{Z}\}$. It is easily seen to be a sub-ring of the field of complex numbers (cf. Exercise (1.17)) and hence is an integral domain. Let us denote this ring by G. If we define $d : G - \{0\} \to \mathbb{N}_0$ by $d(x + iy) = x^2 + y^2$, then it can be shown that with this function, G becomes a euclidean ring. This ring has interesting applications to number theory a few of which will be given as exercises.

We now turn to proving properties of euclidean rings. The foremost is the following:

2.4   Theorem: Every ideal of a euclidean ring is a principal ideal (that is, consists precisely of all multiples of some element).

Proof: Let R be a euclidean ring and I an ideal of R. We have to show that I is of the form Rx for some $x \in R$. Certainly if $I = \{0\}$, then we may take $x = 0$. So we suppose I contains some non-zero elements. Let $d : R - \{0\} \to \mathbb{N}_0$ be the degree function for R. Since the set $\mathbb{N}_0$ is well ordered, among the non-zero elements of I, there exists some element, say x, such that $d(x)$ is the minimum value of d on $I - \{0\}$, that is for all $y \in I$, $y \neq 0$, $d(x) \leqslant d(y)$. We claim that the principal ideal Rx coincides with I. Certainly $Rx \subset I$ because $x \in I$ and I is an ideal. For the converse, suppose $y \in I$. Since $x \neq 0$, by condition (∗) in **Definition** (2.1), there exist q, $r \in R$ such that $y = qx + r$ where $r = 0$ or $d(r) < d(x)$. We claim that only the first possibility holds. For, since $y \in I$ and $qx \in I$, we have that $r \in I$. If $r \neq 0$, then $d(r) < d(x)$ would contradict the choice of x as an element of I of the smallest possible degree. So $r = 0$. Hence $y = qx \in Rx$. That is, $I \subset Rx$. Therefore I equals the principal ideal Rx.   ▇

The theorem as well as the proof must have reminded the perceptive reader of Proposition (5.1.7), in which all subgroups of $\mathbb{Z}$ were obtained. Actually, the two proofs are not substantially different, because both of them are based on the fact that $\mathbb{N}$ (or $\mathbb{N}_0$) is well-ordered and the euclidean algorithm.

Note also that the proof above not only shows that every ideal of a euclidean ring is a principal ideal, but also tells how to find a generator (that is, the element x) of it. Any element of the smallest possible degree among the non zero elements of an ideal will be a generator for that ideal. The importance of this theorem will be clear when we study some of its consequences. There is a name for rings which have the property asserted by its conclusion.

2.5   Definition: An integral domain in which every ideal is principal is called a **principal ideal domain** (abbreviated p.i.d.) or a principal ideal ring.

This is the second of the three conditions we are studying. The preceding theorem may be reworded as 'Every euclidean ring is a p.i.d.'. The converse if false. Motzkin has shown that the ring $\{a + i \sqrt{19}\, b : a, b \in \mathbb{Z}\}$ is a p.i.d. but not a euclidean-ring. The proof is beyond our scope.

Not every integral domain is a principal ideal ring. In fact, as we are about to prove, a p.i.d. must contain a unit element. Since we are not requiring all integral domains to have identity elements, we can get many integral domains such as $2\mathbb{Z}$, $3\mathbb{Z}$, ... which are not p.i.d.'s. As a less trivial example, consider $\mathbb{Z}[x]$, the ring of polynomials over $\mathbb{Z}$. Let I be the set of those polynomials in $\mathbb{Z}[x]$, the sum of whose coefficients is even, that is, $I = \{a_0 + a_1 x + ... + a_n x^n : \sum_{i=0}^{n} a_i \text{ is even}\}$ It is easy to show that I is an ideal of $\mathbb{Z}[x]$. However, it is not a principal ideal. For, if it were, then there would exist a polynomial $f(x)$ in I, such that every element of I is a multiple of $f(x)$. What can be the degree of $f(x)$? Since the constant polynomial (of degree 0) 2 is in I, there exists a polynomial $g(x)$ in $\mathbb{Z}[x]$ such that $2 = f(x) g(x)$. But since degree of $f(x) g(x) =$ degree of $f(x)$ + degree of $g(x)$, this forces $f(x)$ (and also $g(x)$) to be constants. Further $f(x)$ must be either 2 or -2 (since these are the only even integers which divide 2). But then $1 + x$, which is in I, cannot be expressed as a multiple of $f(x)$.

Let us now study properties of principal ideal domains.

2.6 Proposition: Every principal ideal domain has an identity.

Proof: Let R be a p.i.d. . Then R itself is an ideal of R. So $R = Rx$ for some $x \in R$. Then every element of R is a multiple of x. In particular there exists $e \in R$ such that $x = ex$. We claim that e is the identity for R. For let $y \in R$ Then $yex = yx$. Obviously $x \neq 0$ (as otherwise $R = \{0\}$, the trivial ring). So cancelling by x, $ye = y$. By commutativity, ey also equals y. So e is the identity element of R. ◼

In particular it follows that all euclidean rings have an identity element. A minor but pleasant consequence of this proposition is that whenever an ideal I of a p.i.d is generated by an element x then x is in I, because $x = x \cdot 1 \in xR = I$.

Before proving other properties of principal ideal domains, we prove a few simple but useful results which show how certain concepts associated with divisibility of one element by another can be characterised in terms of the principal ideals generated by the two elements. First we characterise divisibility itself. For the remainder of this section, unless otherwise stated, R will denote an integral domain, having an identity element. If $a \in R$, (a) will denote the ideal Ra.

2.7 Proposition: Let $a, b \in R$. The a divides* b if and only if $(b) \subset (a)$.

Proof: Suppose a divides b. Then there exists $c \in R$ such that $b = ac$. Let $x \in (b)$. Then $x = rb$ for some $r \in R$. But then $x = r(ca) = (rc)a$ showing that $x \in (a)$. Thus $(b) \subset (a)$. Conversely suppose $(b) \subset (a)$. Since R has an identity, $b \in (b)$. So $b \in (a)$, that is $b = ca$ for some $c \in R$. In other words, a divides b. ▨

*Note that the relation 'a divides b' depends crucially on the ring R. In $\mathbb{Z}$, 3 does not divide 5. But viewed as elements of Q (which is an extension ring of e), 3 does divide 5. To stress the role of the ring R, we should perhaps write 'a divides b in R' instead of just 'a divides b'. But, by context we always understand the ring R. This also applies to all concepts involving divisibility such as g.c.d., units, primes, associates to be studied later.

From this proposition we see at once that the relation | defined by '*a* | *b*' iff *a* divides *b* is reflexive and transitive. (Of course, a direct proof is equally simple.)

2.8 Definition: An element *u* of *R* is called a **unit** if it divides the identity element of *R*.

For example, in $\mathbb{Z}$, 1 and −1 are the only two units. In the ring of Gaussian integers, 1, −1, *i* and −*i* are the units. In the ring of polynomials over a field, all non-zero constant polynomials are units. It is easily seen that all units of *R* form a multiplicative group. A unit should not be confused with the unit element (which is just another name for the identity). The name 'unit' probably comes from one of the characterisations listed below. In mensuration problems, we try to choose the unit of measurement in such a way that it divides all the quantities to be measured.

2.9 Proposition: Let $u \in R$. Then the following statement are equivalent:

   (i)   *u* is a unit.
  (ii)   *u* is invertible (i.e., has a multiplicative inverse in *R*).
 (iii)   *u* divides every element of *R*.
 (iv) $(u) = R$.

Proof: The equivalence of (i) and (ii) is immediate from the definition. If *u* is a unit then there exists some *v* such that $uv = 1$. Then for any $r \in R$, $r = r \cdot 1 = (rv)u$. So *u* divides *r*. This proves (i) ⇒ (iii). The implication (iii) ⇒ (i) needs no proof. Finally, the equivalence of (iii) and (iv) is an immediate consequence of the definition of (u). ▊

Because a unit divides every element of a ring, it is to be considered as a trivial factor. A factorisation in which one of the factors is a unit is, therefore, a trivial factorisation. There is a name for those elements of *R* which have no nontrivial factorisations.

2.10 Definition: An element *x* is called a **prime** element, if whenever $x = ab$ then either *a* or *b* is a unit.

A few comments are in order about the definition. First, according to this definition every unit is a prime element, because every factor of a unit is itself a unit. In particular 1 is a prime element. Some authors specifically exclude units from primes. This is really a matter of convention. Secondly, in $\mathbb{Z}$, the ring of integers, all the prime numbers, namely, 2, 3, 5, 7, ... are prime elements. But the converse is false! Because −2, −3, −5, ... are also prime elements. In general, if *x* is a prime element and *u* is any unit then *xu* is also a prime element. Our goal is to prove a unique factorisation theorem for principal ideal domains. For this, we would obviously not like to treat *x* and *xu* (where *u* is a unit) as different factors. Such pairs of elements are given a name.

2.11 Definition: An element a is called an **associate** of b if there exists a unit u such that $a = bu$.

The following proposition characterises this concept.

2.12 Proposition: Let $a, b \in R$. Then the following statements are equivalent:

(i) a is an associate of b

(ii) a and b divide each other

(iii) $(a) = (b)$.

Proof: (i) $\Rightarrow$ (ii). Suppose a is an associate of b. Then there exists a unit u such that $a = bu$. So certainly b divides a. But since u is a unit, there exists $v \in R$ such that $uv = 1$. Then $av = buv = b$ which shows that a also divides b. So (ii) holds.

(ii) $\Rightarrow$ (i). Suppose a, b divide each other. Then there exist $u, v \in R$ such that $a = bu$ and $b = av$. Then $a = auv$. Assuming $a \neq 0$, this gives $uv = 1$ and so u is a unit. But then a is an associate of b. If $a = 0$, then b is also 0. Then $a = 0 = b \cdot 1$. Since 1 is a unit, a is an associate of b. So (i) holds.

The equivalence of (ii) with (iii) is immediate from Proposition (2.7). ▮

2.13 Proposition: Associateship is an equivalence relation. Divisibility defines a partial order on the set of the equivalence classes.

Proof: The first part follows from the characterisation (iii) above. For $a, b \in R$, define $a \leqslant b$ to mean a divides b. Then $\leqslant$ is reflexive and transitive. However, it need to not be antisymmetric. In view of (ii) above, $a \leqslant b$ and $b \leqslant a$ holds iff a is an associate of b. From Proposition (3 3.4) it follows that on the set of equivalence classes we have a well-defined partial order. ▮

Let us see if the poset just obtained is a lattice.

2.14 Theorem: Every two elements in a principal ideal domain have a greatest common divisor (abbreviated g.c.d.). It is unique upto associateship and is expressible as a linear combination of the two elements. Every pair of non-zero elements has a least common multiple which is also unique upto associateship.

Proof: Let a, b be two elements of a p.i.d. R. Let I be the set of all linear combinations of a and b with coefficients from R, that is, $I = \{xa + yb : x, y \in R\}$. It is easy to show that I is an ideal of R. I is evidently closed under subtraction. Also, if $xa + yb \in I$ and $r \in R$, then $r(xa + yb) = (rx)a + (ry)b$ which is in I. So I is an ideal and since R is a p.i.d., I is of the form Rc for some $c \in I$. We claim that c is a g.c.d. of a and b. First,

$a \in I$ since $a = 1.a + 0.b$. So $a = \lambda c$ for some $\lambda \in R$. This means $c \mid a$. Similarly $c \mid b$. Thus c is a common divisor of a and b. To show it is the greatest common divisor, let d be a common divisor of a and b, say $a = md$ and $b = nd$ for $m, n \in R$. Now c, being an element of I is a linear combination of a and b, say $c = x_0 a + y_0 b$ for $x_0, y_0 \in R$. Then $c = (x_0 m + y_0 n)d$, showing that d divides c. So c is the greatest common divisor of a and b. (Here 'greatest' is in the sense of the partial order defined above.) As for its uniqueness, suppose c' is any other g.c.d of a and b. Then c must divide c' and c' must also divide c. So c and c' are associates of each other. Thus the g.c.d. is unique upto associateship. That it is expressible as a linear combination of a and b is already proved.

As for the least common multiple (l.c.m.), let a, b be non-zero elements of R. Let c be their g.c.d.. Then $c \neq 0$. Let $a = \lambda c$ and $b = \mu c$, where $\lambda, \mu \in R$. Let $d = \lambda \mu c$. We claim that d is the least common multiple of a and b. Now, $d = \mu a = \lambda b$ and so d is a common multiple of a and b. To show that it is the least common multiple, note first, that c is a linear combination of a and b, say $c = \alpha a + \beta b$ with $\alpha, \beta \in R$. Then $c = (\alpha \lambda + \beta \mu)c$ which gives $\alpha \lambda + \beta \mu = 1$. Now suppose e is a common multiple of a and b, say $e = xa = yb$ with $x, y \in R$. Then $e = \lambda xc = \mu yc$, giving $\lambda x = \mu y$, since $c \neq 0$. From $\alpha \lambda + \beta \mu = 1$ we get $x = \alpha \lambda x + \beta \mu x = \mu(\alpha y + \beta x)$ and hence $e = ax = a\mu(\alpha y + \beta x) = d(xy + \beta x)$ (since $d = a\mu = b\lambda$). This shows that d divides e. Thus d is the least common multiple of a and b. Uniqueness of d (upto associateship) follows by the same argument as that for uniqueness of g.c.d. ∎

This theorem shows that if R is a p.i.d., then the set of equivalence classes (under associateship) of non-zero elements of R is a lattice.

The crux of the last theorem is not just the existence of a greatest common divisor for every two elements of a p.i.d., but also the fact that it can be expressed as a linear combination of the two elements. However, the proof does not say how to express the g.c.d as a linear combination of the two elements. This is to be expected since the very definition of a principal ideal domain is non-constructive in the sense that it merely says that every ideal is generated by a single element, without saying anything as to how to find this generator. However, in the case of a euclidean ring, the things are much better. The algorithm for finding the g.c.d. of two positive integers, given at the beginning of this section can be obviously extended to any euclidean ring. The crucial step in proving that the algorithm actually works is the following.

2.15 Proposition: Let R be a euclidean ring and $a, b \in R$. If $a = qb + r$ then the g.c.d. of a and b is the g.c.d. of b and r. If $r = 0$, then b is the g.c.d. of a and b. (Here equality is upto associateship.)

Proof: Let c, d be, respectively, the g.c.d.'s of a and b and of b and r. Then d divides b and r so it divides $qb + r$, i.e. d divides a. So d is a common divisor of a and b. Hence $d \mid c$. Similarly, c, being a common divisor of a and b is a divisor of r (since $r = a - qb$). So c divides both b and r, and hence $c \mid d$. Thus c and d are associates of each other. For the second assertion, if $r = 0$, then b itself is a common divisor of a and b and obviously is their greatest common divisor. ∎

It is now very easy to give an algorithm for finding the g.c.d. of two elements a, b of a euclidean ring R. If $b = 0$ then a is their g.c.d. Otherwise we set $a_0 = a$, $b_0 = b$ and write $a_0 = q_0 b_0 + r_0$ where $r_0 = 0$ or $d(r_0) < d(b_0)$. If $r_0 = 0$, then b_0 is the g.c.d. of a and b and it is a linear combination of a and b since $b_0 = b = 1 \cdot b + 0 \cdot a$. If $r_0 \neq 0$ we set $a_1 = b_0$, $b_1 = r_0$ and repeat the process, that is, write $a_1 = q_1 b_1 + r_1$ where $r_1 = 0$ or $d(r_1) < d(b_1) = d(r_0)$. In the first case, b_1 is the g.c.d. of a_1 and b_1, and hence also of a and b by the last proposition. Also $b_1 = r_0 = a_0 - q_0 b_0 = a - q_0 b$, which is a linear combination of a and b. If $r_1 \neq 0$, we continue the procedure setting $a_2 = b_1$ and $b_2 = r_1$. In general, we continue inductively. If $r_i \neq 0$, we set $a_{i+1} = b_i$ and $b_{i+1} = r_i$ and write $a_{i+1} = q_{i+1} b_{i+1} + r_{i+1}$ which $r_{i+1} = 0$ or $d(r_{i+1}) < d(r_i)$. By induction, it is seen that each r_i is a linear combination of a and b. Since the degree function takes only non-negative integers as values, there must exist some n such that $r_n = 0$. Then r_{n-1} is the g.c.d. of a and b. As a numerical example we compute the g.c.d. of 1092 and 195 as follows :

$$1092 = 5 \times 195 + 117$$
$$195 = 1 \times 117 + 78$$
$$117 = 1 \times 78 + 39$$
$$78 = 2 \times 39 + 0$$

So 39 is the g.c.d. of 1092 and 195. To express it as a linear combination, we write $39 = 117 - 1 \times 78 = 117 - (195 - 117) = 2 \times 117 - 195 = 2(1092 - 5 \times 195) - 195 = 2 \times 1092 - 11 \times 195$.

As noted above, for principal ideal domains which are not euclidean rings, there is no algorithm for expressing the g.c.d. of two elements as a linear combination of them. Still, the mere existence of such an expression has some useful consequences. We prove one of them here. First we need a definition.

2.16 Definition: Two elements a, b of R are said to be **relatively prime** (or **coprime**) to each other if they have 1 (and hence any unit) as their g.c.d.

For example, in $\mathbb{Z}$, 6 and 25 are relatively prime to each other. It is immediate from the definition that two elements are relatively prime iff they have no common factors except the units.

2.17 Proposition: If $a, b, c \in R$, $a \mid bc$ and a, b are relatively prime then $a \mid c$, provided R is a principal ideal domain.

Proof: Let $bc = ax$ for some $x \in R$. Since a, b are relatively prime, their g.c.d. is 1 and since R is a p.i.d., we can write 1 as $\lambda a + \mu b$ for some $\lambda, \mu \in R$. Multiplying by c we get $c = \lambda ac + \mu bc = \lambda ac + \mu ax = a(\lambda c + \mu x)$, showing that a divides c as desired. ∎

The following corollary, although a special case of the last result, is worth singling out.

2.18 Corollary: Suppose R is a p.i.d. and p is a prime element of R. Then, whenever p divides bc for $b, c \in R$, p must divide either b or c.

Proof: Suppose p does not divide b. The only factors of p other than units are p and its associates. None of these divides b. So the only common factors of p and b are units. Hence p and b are relatively prime. By the last proposition, $p \mid c$. Hence if p divides bc then either $p \mid b$ or $p \mid c$. ∎

We had already used this corollary in the proof of Theorem (5.1.5) in which it was shown that if p is a prime then the set of non-zero residue classes modulo p is a group of order $p - 1$ under multiplication modulo p. The same fact was also used (without proof) in showing that $\mathbb{Z}_p$ is field, in the last section. Now that we have proved it, we have filled the gap. Actually, now that we have at our disposal Proposition (2.17), which is more general than Corollary (2.18), we can prove a more general result.

2.19 Proposition: Let m be a positive integer. Then the set of residue classes modulo m, which are relatively prime to m, that is, the set $\{[x]: x$ is relatively prime to $m\}$ is a group under modulo m multiplication.

Proof: Let $\mathbb{Z}_m$ be the ring of residue classes modulo m. Let S be the set of residue classes relatively prime to m. (Note that if $x \equiv y \pmod{m}$, then m and x are relatively prime iff m and y are relatively prime. So the concept of a residue class being relatively prime to m, is well-defined.) We assert that S is closed under modulo m multiplication. Let $[x], [y] \in S$. Then $x, y \in \mathbb{Z}$ and both are relatively prime to m. So there exist $a, b, c, d \in \mathbb{Z}$ such that $am + bx = 1$ and $cm + dy = 1$. Then we see that $(acm + bcx + ady)m + bdxy = 1$. So m and xy are relatively prime (any common factor of them will be a factor of 1 since 1 is a linear combination of xy and m). So $[xy] \in S$, that is $[x][y] \in S$. So S is closed under modulo m multiplication. Therefore, mod m multiplication defines a commutative and associative binary operation on S. We claim that the cancellation law holds for this operation. For suppose $[x], [y], [z]$ are in S and $[x][z] = [y][z]$. Then $(x - y)z$ is divisible by m. But m is relatively prime to z. So by Proposition (2.17), m divides $x - y$. Thus $x \equiv y \pmod{m}$ or $[x] = [y]$. We have shown that S is a semigroup in which the cancellation laws hold. Obviously S is finite, since

$S \subset \mathbb{Z}_m$ and $\mathbb{Z}_m$ is finite. So by Proposition (5.1.4), S is a group. This completes the proof. But it is instructive to go a little further. Every element of S is a unit in the ring $\mathbb{Z}_m$, because it has a multiplicative inverse. Conversely, if $[x]$ is a unit in $\mathbb{Z}_m$, then there exists $[y] \in \mathbb{Z}_m$ such that $xy \equiv 1$ (modulo m). But then x must be relatively prime to m. (Prove.) So $[x] \in S$. In other words, the elements of S are precisely the units of the ring $\mathbb{Z}_m$. ∎

Let us denote this group by R_m. Elements of R_m can be represented by integers between 1 and $m - 1$ which are relatively prime to m. The number of such integers is $\phi(m)$, where ϕ is the Euler ϕ-function (see Exercise (2.4.15)).

Integers which are relatively prime to a given integer figure many times in various contexts. We cite one example.

2.20 Proposition: Let G be a cyclic group of order n. Then G has $\phi(n)$ distinct generators.

Proof: Let x be a generator for G. Then G consists of $x, x^2, ..., x^{n-1}$ and $x^n (= e)$. We claim that for $1 \leqslant m < n$, x^m is a generator of G iff m is relatively prime to n. Suppose x^m is a generator. Then every element of G is expressible as some power of x. In particular $(x^m)^r = x$ for some $r \in \mathbb{Z}$. But then $x^{mr} = x$ which shows $mr \equiv 1$ (modulo n), by Theorem (5.1.10). So $[m][r] = [1]$ in the ring $\mathbb{Z}_n$. This means $[m]$ is a unit in $\mathbb{Z}_n$ and so m is relatively prime to n. Conversely if m is relatively prime to n, we trace this argument backwards and get x as $(x^m)^r$ for some r. But then every x^t can be expressed as $(x^m)^{ri}$ and hence is in the subgroup generated by x^m. This establishes our claim and shows that the number of generators of G equals the number of positive integers which are less than n and are relatively prime to n. By definition, this number equals $\phi(n)$. ∎

We apply this result to the group of all complex nth roots of unity. Recall from Chapter 5, Section 1 that a generator for this group is called a primitive nth root of unity. So we see that for every positive integer n, the number of primitive nth roots of unity is $\phi(n)$.

A few other results where the g.c.d. of two integers is needed will be given as exercises. Let us go back to principal ideal domains in general. From Proposition (2.19) it follows that if p is a prime then $\mathbb{Z}_p$ is a field. (We had already mentioned this in the last chapter. But the proof was incomplete since it involved the use of Corollary (2.18).) Note that $\mathbb{Z}_p$ is the quotient ring of $\mathbb{Z}$ by the ideal $p\mathbb{Z}$, which is the principal ideal generated by p which is a prime element. It would be natural to inquire if this can be generalised to all principal ideal domains. The affirmative answer is provided by the following result.

2.21 Theorem: Let R be a principal ideal domain and $p \in R$. Then p is a prime element if and only if the quotient ring $R/(p)$ is a field.

Proof: If p is a unit then $(p) = R$ and the quotient ring is trivial. It is, therefore, the trivial field. Conversely, if $R/(p)$ is the trivial ring then $R=(p)$ and so p is a unit. By our convention, units are included among prime elements. So the result holds trivially when p is a unit.

Suppose p is not a unit. Then (p) is a proper ideal of R. In view of Proposition (1.16), we are reduced to proving that (p) is a maximal ideal iff p is a prime element. Suppose (p) is a maximal ideal and let $p = ab$ be a factorisation of p. Then $p \in aR = (a)$, So $(p) \subset (a)$. By maximality of (p), $(a) = R$ or else $(a) = (p)$. In the first case a is a unit. In the second case, p is an associate of a by proposition (2.12). So $a = pu$ for some unit u. But then $p = ab = pub$ whence $ub = 1$ and so b is a unit. Thus in any factorisation of p at least one factor must be a unit. So p is a prime element.

Conversely suppose p is a prime element. Let I be an ideal such that $(p) \subsetneq I \subset R$. We have to show that $I = R$. Since R is a p.i.d., I is of the form (q) for some $q \in R$. Since $(p) \subset (q)$, by proposition (2.7), q divides p. So $p = aq$ for some $a \in R$. Now a cannot be a unit, for otherwise, q will be an associate of p and hence $(p) = (q)$, contradicting that (p) is properly contained in I. Since p is a prime element, $p = aq$ and a is not a unit, the other factor, q, must be a unit. But then $(q) = R$, showing $I = R$, as was to be proved.   ▨

This theorem is of great importance in the construction of field extensions. Let F be a field. Then $F[x]$, the ring of polynomials over F is a euclidean ring and hence a p.i.d. The prime elements of this ring are called **irreducible polynomials** over F. Suppose $f(x)$ is an irreducible polynomial over F. Let I be the ideal generated by $f(x)$. I consists of all polynomials of the form $f(x) g(x)$ where $g(x) \in F[x]$. Let K be the quotient ring $F[x]/I$. Then by the theorem above K is a field. If $f(x)$ is a constant (i.e., a degree zero) polynomial then of course $I = F[x]$ and K is trivial. But suppose $f(x)$ is of positive degree. Then the function $\theta : F \to K$ defined by $\theta(a) = a + I$ for $a \in F$ is a ring homomorphism with kernel 0, because the ideal I contains no constant polynomial except 0. So the range of θ is a subring of K and is isomorphic to F. As usual we identify F as a subfield of K. Thus K is a field extension F. This extension has a very interesting property. To study it, we introduce a concept which itself is of great importance.

2.22   Definition: Let $f(x) = a_0 + a_1 x + \ldots + a_n x^n$ be a polynomial with coefficients in some ring R (not necessarily an integral domain). Then an element α of R is called a **root** (or a **zero**) of $f(x)$ if

$$a_0 + a_1 \alpha + a_2 \alpha^2 + \ldots + a_n \alpha^n = 0.$$

There is a handy way to look at this concept. In the polynomial $f(x)$, x is just a symbol, an indeterminate. If we substitute for x, some element, say, a of the ring R, then we get an element of R which we denote by $f(a)$. This way we get a function from R to itself. We may denote this function by f. A root of the polynomial $f(x)$ is then simply an element at which the function f takes the value 0. It should be noted, however, that a polynomial is conceptually different from the function it induces. Two different polynomials may induce the same function. For example, let $R = \mathbb{Z}_2$, the ring of mod 2 residue classes. Let $f(x) = x^2 + x + 1$ and $g(x) = 1$. Then $f(x)$ and $g(x)$ are different polynomials but they determine the same function from $\mathbb{Z}_2$ to itself because, $f(0) = 0 + 0 + 1 = 1 = g(0)$ and $f(1) = 1 + 1 + 1 = 1 = g(1)$.

2.23 Proposition: Suppose R is a commutative ring with identity. Let $f(x)$, $g(x)$ be polynomials in $R[x]$. Let $h(x) = f(x) + g(x)$ and $p(x) = f(x)g(x)$, Let f, g, h, p be the functions from R to R defined by the polynomials $f(x)$, $g(x)$, $h(x)$ and $p(x)$ respectively. Then h is the pointwise sum of f and g and h is the pointwise product of f and g.

Proof: We have to show that for every $a \in R$, $h(a) = f(a) + g(a)$ and $p(a) = f(a)g(a)$. This may seem obvious and is indeed straightforward to prove, keeping in mind the way polynomials are added and multiplied. Note, however, that for proving $p(a) = f(a)g(a)$, commutativity of R is vitally needed. If R is the ring of quaternions, $f(x) = x$ and $g(x) = ix$ then $p(x) = ix^2$. But $f(j)g(j) = jk = i$ while $p(j) = ij^2 = -i$. Thus $f(j)g(j) \neq p(j)$ even though as polynomials, $f(x)g(x) = p(x)$. The crux of the matter is that when we think of R as a subring of $R[x]$, the indeterminate x commutes with all elements of R but when x is replaced by some element a (say) of R, this element need not commute with other elements of R. If R is commutative this difficulty does not arise and the result holds. ∎

Using the last proposition and the euclidean algorithm we get the following characterisation of roots, sometimes called the 'remainder theorem'.

2.24 Theorem: Let F be a field and $f(x) \in F[x]$. Then an element α of F is a root of $f(x)$ if and only if $(x - \alpha)$ is a factor of $f(x)$ in $F[x]$.

Proof: Suppose first that $x - \alpha$ is a factor of $f(x)$. Then $f(x) = (x - \alpha)g(x)$, for some $g(x) \in F[x]$. By the last proposition, $f(\alpha) = (\alpha - \alpha)g(\alpha) = 0$. So α is a root of $f(x)$. Conversely suppose α is a root of $f(x)$. Then $f(\alpha) = 0$. By the euclidean algorithm for $F[x]$, there exist polynomials $q(x)$, $r(x)$ in $F[x]$ such that $f(x) = q(x)(x - \alpha) + r(x)$ where $r(x) = 0$ or $\deg(r(x)) < \deg(x - \alpha) = 1$. In any case it follows that $r(x)$ is a constant, say β. Now, by the last proposition again, $f(\alpha) = q(\alpha)(\alpha - \alpha) + r(\alpha)$. But $f(\alpha) = 0$, since α is a root of $f(x)$. So $r(\alpha) = \beta = 0$. Hence $r(x)$ is the zero polynomial. Thus $f(x) = q(x)(x - \alpha)$, showing that $x - \alpha$ is a factor of $f(x)$. ∎

A consequence of this result is noteworthy.

2.25 Proposition: A polynomial of degree n over a field can have at most n distinct roots in that field.

Proof: Let F be a field and suppose $f(x) \in F[x]$ has degree n. We prove the result by induction on n. For $n = 0$, the polynomial is a non-zero constant polynomial and so has no roots. Suppose $n > 0$ and the result has been proved for all polynomials of degree less than n. If $f(x)$ has no root the result certainly holds. So suppose $f(x)$ has at least one root say α. Then by the last theorem, we can write $f(x) = (x - \alpha)\, g(x)$ where $g(x)$ is a polynomial over F. Clearly $g(x)$ has degree $n - 1$. If α_1 is any root of $f(x)$ other than α, then $f(\alpha_1) = 0 = (\alpha_1 - \alpha)\, g(\alpha_1)$ which shows $g(\alpha_1) = 0$ since $\alpha_1 - \alpha \neq 0$. Thus all roots of $f(x)$ except α are roots of $g(x)$. By induction hypothesis, $g(x)$ has at most $n - 1$ distinct roots in F. These roots, along with α, are the only roots of $f(x)$. So $f(x)$ has at most n distinct roots.   ▮

This proposition can be strengthened in two ways. First, it remains true even if each root is counted as many times as its multiplicity. This version will be given as an exercise. Secondly, it holds even for integral domains. If R is an integral domain we let F be its field of quotients. Then $R \subset F$ and so $R[x] \subset F[x]$. If $f(x) \in R[x]$ then a root of $f(x)$ in R is also a root of it in F. So the number of distinct roots of $f(x)$ in R cannot exceed the number of distinct roots of $f(x)$ in F, which, by the proposition, is at most the degree of f. Thus the hypothesis that the ground ring be a field can be relaxed a little. However, commutativity cannot be dispensed with. In the ring of quaternions, the polynomial $x^2 + 1$ of degree 2 has at least 6 distinct roots, namely $\pm i$, $\pm j$ and $\pm k$. Similarly, absence of zero divisors is necessary. In a Boolean ring, every element is a root of the polynomial $x^2 - x$ (which also equals $x^2 + x$, since $-1 = 1$).

After this disgression about roots, let us go back to the field extension obtained through an irreducible polynomial. Suppose $f(x)$ is an irreducible polynomial of degree $n\,(> 0)$ over a field F. Then $f(x)$ can have no roots in F, because if α were a root, then $(x-\alpha)$ would be a nontrivial factor of $f(x)$, contradicting that it is a prime element in $F[x]$. (The converse is not true. The polynomial $(x^2 + 1)^2$ has no root in $\mathbb{R}$, but is not irreducible in in $\mathbb{R}[x]$, as it is the product of $x^2 + 1$ with itself.) We let I be the ideal of $F[x]$, generated by $f(x)$, and K be the quotient ring $F[x]/I$. We already saw that K is a field. We identify F with a subfield of K, by letting $a \in F$ correspond to the coset $a + I$. Since $F \subset K$, the polynomial $f(x)$ might as well be considered as a polynomial over K. But, as a polynomial in $K[x]$, $f(x)$ is not irreducible. As a matter of fact it has a root and this root is none other than the element $x + I$. To see this, suppose $f(x) = a_0 + a_1 x + \cdots + a_n x^n$ where $a_0, a_1, \ldots, a_n \in F$. Let α be the coset $x + I$ in $F[x]/I$. By definition of coset multiplication, $\alpha^2 = (x + I)(x + I) = x^2 + I$. More generally, for all $k \geqslant 1$, $\alpha^k = x^k + I$. Also for any $a \in F$, $a\alpha$ really means $(a + I)$ $(x + I)$ which equals $ax + I$. Thus we see that $f(\alpha) = a_0 + a_1\alpha + \cdots + a_n\alpha^n$

is simply $(a_0 + a_1x + ... + a_nx^n) + I$. But this is the zero coset, I, because $a_0 + a_1x + ... + a_nx^n = f(x) \in I$. So $f(\alpha) = 0$. That is, $x + I$ is a root of $f(x)$ in $K[x]$. We have thus proved the first half of the following important result.

2.26 Theorem: Every irreducible polynomial of positive degree over a field has a root in some extension field of it. Moreover, no such root can satisfy a non-zero polynomial of lower degree with coefficients in the given field.

Proof: The first part was already proved above. For the second assertion, suppose F is a field and $f(x)$ is an irreducible polynomial of degree n in $F[x]$. Let α be a root of $f(x)$ in some extension field, say K, of F. Let $J = \{g(x) \in F[x] : g(x) = 0\}$. Clearly $f(x) \in F[x]$ and the assertion amounts to showing that J cannot contain any non-zero polynomial of degree less than n. It is easily seen that J is an ideal of $F[x]$. Since $F[x]$ is a euclidean ring, by Theorem 2.4, J is a principal ideal, i.e., there is some polynomial $h(x) \in F[x]$, such that J consists precisely of multiples of $h(x)$. In particular $f(x)$ is a multiple of $h(x)$. But since $f(x)$ is irreducible, this can happen only if $h(x)$ is an associate of $f(x)$. In that case, J is also generated by $f(x)$. Consequently every non-zero element of J is a multiple of $f(x)$ and hence has degree at least n. ∎

Lest the construction appear too abstract, we illustrate it with a particular example. Let $F = \mathbb{R}$, the field of real numbers and let $f(x) = x^2 + 1$. Then $f(x)$ is an irreducible polynomial over $\mathbb{R}$, because if $f(x) = p(x)q(x)$ where neither $p(x)$ nor $q(x)$ is a unit then both $p(x)$ and $q(x)$ would be polynomials of degree 1. Since polynomials of degree 1 always have roots, and any root of $p(x)$ (or of $q(x)$) is also a root of $f(x)$, this would imply that $f(x)$ has a root in $\mathbb{R}$. But we know this not to be the case, since, in $\mathbb{R}$, the square of every element is non-negative (cf. Exercise 1.26). So $f(x)$ is irreducible over $\mathbb{R}$. Let $K = \mathbb{R}[x]/I$ as constructed above. We claim $\mathbb{R}[x]/I$ is isomorphic to $\mathbb{C}$, the field of complex numbers. Define $\theta : \mathbb{R}[x] \to \mathbb{C}$ by $\theta(g(x)) = g(i)$, for $g(x) \in \mathbb{R}[x]$. In other words, if $g(x) = b_0 + b_1x + ... + b_mx^m$ where $b_0, ..., b_m \in \mathbb{R}$, then $\theta(g(x))$ is the complex number $b_0 + b_1i + ... + b_mi^m$. Clearly θ is a ring homomorphism and θ is onto since every complex number, say, $a + ib$ can be expressed as $\theta(g(x))$ where $g(x) = a + bx$. We claim that the kernel of θ is precisely the ideal I, generated by $x^2 + 1$. If $g(x) \in I$ then $g(x) = h(x)(x^2 + 1)$ for some $h(x) \in \mathbb{R}[x]$. But then $g(i) = h(i)(i^2 + 1) = h(i)0 = 0$ showing that $g(x) \in \text{Ker}(\theta)$. Conversely, suppose $g(x) \in \text{Ker}(\theta)$. Then $g(i) = 0$. By the euclidean algorithm write $g(x) = q(x)(x^2 + 1) + r(x)$. Then $r(i) = 0$. Also $r(x)$ is either 0 or a polynomial of degree 0 or 1. But the complex number i satisfies no polynomial equation of degree 0 or 1 with real coefficients. (If it did, i would be a real number.) So $r(i) = 0$ forces $r(x)$ to be identically zero. Hence $g(x) = q(x)$

$(x^2 + 1)$, showing $g(x) \in I$. Thus we have shown that the kernel of θ is precisely the ideal I. By the fundamental theorem about ring homomorphisms (cf. Exercise 1.22), it follows that $\mathbb{R}[x]/I$ is isomorphic to $\mathbb{C}$. Thus the extension field we constructed is isomorphic to the field of complex numbers. In this isomorphism the root $x + I$ goes to the number i. We could have as well defined $\psi \colon \mathbb{R}[x] \to \mathbb{C}$ by $\psi(g(x)) = g(-i)$. In that case $x + I$ would correspond to $-i$.

The argument above shows that we might as well have *defined* complex numbers as elements of the extension field $\mathbb{R}[x]/I$ constructed above. As mentioned in the last section, an alternate approach is to regard a complex number as a 2×2 real matrix of a certain form. With this representation the complex number i corresponds to the matrix $\begin{bmatrix} 0 & -1 \\ 1 & 0 \end{bmatrix}$. The smallest subring of $M_2(\mathbb{R})$ containing all matrices of the form $\begin{pmatrix} b & 0 \\ 0 & b \end{pmatrix}$ for $b \in \mathbb{R}$ and the matrix $\begin{pmatrix} 0 & -1 \\ 1 & 0 \end{pmatrix}$ is isomorphic to $\mathbb{C}$, and hence to $\mathbb{R}[x]/I$. A direct isomorphism between $\mathbb{C}$ and this subring can be constructed by first defining a homomorphism $\phi \colon \mathbb{R}[x] \to M_2(\mathbb{R})$ which takes a polynomial

$$b_0 + b_1 x + \cdots + b_m x^m$$

to the matrix

$$\begin{pmatrix} b_0 & 0 \\ 0 & b_0 \end{pmatrix} + \begin{pmatrix} b_1 & 0 \\ 0 & b_1 \end{pmatrix} \begin{pmatrix} 0 & -1 \\ 1 & 0 \end{pmatrix} + \begin{pmatrix} b_2 & 0 \\ 0 & b_2 \end{pmatrix} \begin{pmatrix} 0 & -1 \\ 1 & 0 \end{pmatrix}^2 + \cdots +$$

$$+ \begin{pmatrix} b_m & 0 \\ 0 & b_m \end{pmatrix} \begin{pmatrix} 0 & -1 \\ 1 & 0 \end{pmatrix}^m.$$

It can be shown that the kernel of ϕ is precisely the ideal I generated by the polynomial $x^2 + 1$. The matrix $\begin{pmatrix} 0 & -1 \\ 1 & 0 \end{pmatrix}$ is called the companion matrix of the polynomial $x^2 + 1$. More generally, for every irreducible polynomial $f(x)$ of degree n over F, we can define its companion matrix as a certain $n \times n$ matrix. Then the extension field formed from $f(x)$ comes out to be isomorphic to the subring of $M_n(F)$, generated by the companion matrix. This provides a 'concrete' representation of the somewhat 'abstract' extension field constructed above. We could do this now, but we defer it to the fourth section where we would be in a better position to do it.

It is natural to inquire whether Theorem (2.26) can be extended to any non-constant polynomial. Suppose $f(x)$ is a polynomial of degree $n > 0$, over a field F. Can we always find some extension field of K in which $f(x)$, (regarded as a polynomial over K now) has a root? It $f(x)$ is irreducible, the answer is 'yes' by Theorem (2.26). If $f(x)$ is not irreducible but has some non-trivial factor $g(x) \in F[x]$ which is irreducible, then also the answer is 'yes', because, any root of $g(x)$ (in whatever extension field of F) is evidently a root of $f(x)$ also. So the problem reduces to proving that the polynomial $f(x)$ has at least one irreducible factor. This can be done by repeated factorisation, or what amounts to the same, by induction on the degree of $f(x)$. If $f(x)$ is not irreducible, then it has a proper factor (i.e. a factor which is neither a unit nor an associate of $f(x)$, say $f_1(x)$. Note that the degree of $f_1(x)$ will be strictly less than $f(x)$. So by induction hypothesis, $f_1(x)$ has an irreducible polynomial as a factor. This polynomial will also be a factor of $f(x)$. Alternatively, we repeat the argument for $f_1(x)$. If $f_1(x)$ is not irreducible, then it has a proper factor $f_2(x)$. Again deg $(f_2(x)) < \deg(f_1(x))$. We now apply the same argument to $f_2(x)$. This process has to stop because the degree function cannot go on strictly descending infinitely many times. Thus eventually we shall find some irreducible factor of $f(x)$.

Although this settles the question with which we started (namely, whether Theorem (2.26) can be generalised to any non-constant polynomial), it is instructive to see whether the argument given above for the existence of irreducible factors can be generalised from the ring $F[x]$ to some other types of rings. For a euclidean ring R, there is little difficulty once we show that the degree of a proper factor is less than that of the original element. However, if we try to generalise it to a still wider class of integral domains, namely, principal ideal domains, then we have to reformulate the fact that a sequence of elements with strictly descending degrees must be finite (i.e. cannot go on *ad infinitum*). As we have done many times before, if we replace the elements with ideals then we get the appropriate reformulation.

2.27 Proposition: In a principal ideal domain, every properly ascending chain of ideals must be finite.

Proof: Let R be a p.i.d. We have to prove that there cannot exist an infinite sequence of ideals of R, $I_1, I_2, I_3, \ldots, I_n, \ldots$, in which $I_n \underset{\neq}{\subset} I_{n+1}$ for all $n = 1, 2, \ldots$. Suppose such a sequence exists. Then we shall get a contradiction as follows: Let $I = \bigcup_{n=1}^{\infty} I_n$. The set I is the union of the infinite family of sets $\{I_1, I_2, I_3, \ldots, I_n, \ldots\}$. It is defined exactly like the union of finitely many sets. That is, I consists of those elements of R which belong to at least one I_n or $I = \{x \in R : x \in I_n$ for some n, which may depend on $x\}$. Note that each I_n is contained in I. Now, in general, the union of ideals need not be an ideal. However, in this case we show that I is an ideal. First, let $x, y \in I$. We have to show $x - y \in I$. By definition there exist positive integers m, n such that $x \in I_m$ and $y \in I_n$. Now either $m \leqslant n$ or $n \leqslant m$. In the first case, $x \in I_n$ (since $I_m \subset I_n$). Also $y \in I_n$ So I_n being an ideal, $x - y \in I_n$. But then $x - y \in I$. In the second case x, y are both in I_m, whence $x - y \in I_m$ and hence $x - y \in I$. In either case $x - y \in I$. For the second condition of an ideal, let $x \in I$ and $r \in R$. Then $x \in I_n$ for some n. But then $rx \in I_n$, since I_n is an ideal. So $rx \in I$. Thus we have shown that I is an ideal of R. Now we use the fact that R is a p.i.d. So there is some $x \in I$ such that $I = Rx$. By definition, $x \in I_n$ for some n. But then $Rx \subset I_n$, I_n being an ideal. So $I \subset I_n$. We already have that $I_{n+1} \subset I$. So $I_{n+1} \subset I_n$. This contradicts our assumption that I_n is a proper subset of I_{n+1}. This contradiction establishes our assertion, namely that the given sequence of ideals must terminate. ▨

This proposition is the key step in proving the existence of a factorisation of an element of a p.i.d. as a product of powers of distinct prime elements. By distinct primes here, we mean primes which are not associates of each other. It also turns out that the factorisation in unique, where uniqueness is again, upto associateship. We state and prove both these results in the following theorem which may be called the **unique factorisation theorem**.

2.28 Theorem: Let R be a principal ideal domain and $x \in R$. Suppose x is not zero and not a unit. Then there exists a unit u and distinct prime elements $p_1, \ldots, p_r$ in R (which are not units) and positive integers $n_1, n_2 \ldots, n_r$ such that $x = up_1^{n_1} p_2^{n_2} \ldots p_r^{n_r}$. If x also equals $vq_1^{m_1} q_2^{m_2} \ldots q_s^{m_s}$ where v is a unit and the q's are distinct prime elements other than units and m's are positive integers then $r = s$ and with a re-indexing of the q_i's, we have that for each $i = 1, \ldots, r$, $m_i = n_i$ and p_i is an associate of q_i.

Proof. Let S be the set of those elements of R which are not zero, not units and for which a factorisation into a unit times powers of distinct prime elements is possible. We have to show that $x \in S$. Note first that S is closed under multiplication. For if a, $b \in S$ then certainly $ab \neq 0$ and ab cannot be a unit (the factors of a unit are themselves units). Moreover, from factorisations of a and b as products of units times prime powers, we easily get a similar factorisation for ab. (If some prime appearing in a is an associate of a prime appearing in b, then we replace one of them with the other, add the exponents and adjust the unital coefficient suitably.) Note also that S contains all prime elements of R.

Now suppose $x \notin S$. Call x as x_1 and let $I_1 = (x_1)$, the ideal generated by x_1. Then x_1 is not a prime element and so has a proper factorisation say, bc where neither b nor c is an associate of x_1. Since $x_1 \notin S$ and S is closed under multiplication it follows that either $b \notin S$ or $c \notin S$. Let x_2 be the factor which is not in S. (If both $b \notin S$ and $c \notin S$, make an arbitrary choice.) Let $I_2 = (x_2)$. Since x_2 divides x_1, $I_1 \subset I_2$. Also $I_1 \neq I_2$ as otherwise

x_2 will be an associate of x_1. Now $x_2 \notin S$. So we repeat the argument with x_2 replacing x_1. This gives an element $x_3 \notin S$ such that $I_2 \underset{\neq}{\subset} I_3$ where $I_3 = (x_3)$

Continuing in this manner we get a strictly ascending infinite sequence of ideals. This is an outright contradiction to the last proposition. So $x \in S$, proving the first part.

For uniqueness of factorisation suppose $x = up_1^{n_1} p_2^{n_2} \dots p_r^{n_r} = vq_1^{m_1} \dots q_s^{m_s}$ where u, v are units, p_i's are distinct primes and so are q_i's, and n's, m's are positive integers. Then the prime p_1 divides the product of the elements $v, q_1, \dots, q_1$ (m_1 times),$\dots$, $q_s, \dots, q_s$ (m_s times). Applying Corollary (2.18) repeatedly, we see that p_1 must divide one of these elements. Since p_1 is not a unit, it cannot divide v, which is a unit. So p_1 must divide one of the q's. With suitable re-indexing, if necessary, we suppose p_1 divides q_1, say $q_1 = w_1 p_1$. Since q_1 is a prime, w_1 must be a unit. Cancelling p_1, we get $up_1^{n_1-1} p_2^{n_2} \dots p_r^{n_r} = vw_1 q_1^{m_1-1} q_2^{n_2} \dots q_s^{m_s}$. We now repeat the argument (if $n_1 = 1$, we repeat it with p_2 instead of p_1). It follows eventually that m_1 must equal n_1 and hence $up_2^{n_2} \dots p_r^{n_r} = vw_1^{n_1} q_2^{m_2} \dots q_s^{ms}$. We again repeat the argument with p_2 and write one of the q_i's as an associate of p_2, say, $q_2 = w_2 p_2$ with w_2 a unit. Continuing in this manner till all p_i's are exhausted, we see $r \leqslant s$. By symmetry $s \leqslant r$. So $r = s$. Also in the course of the argument we already showed that if p_i and q_i are associates of each other then $n_i = m_i$. This gives the desired result. ∎

In particular, the last theorem is true for euclidean rings, of which $\mathbb{Z}$ is an example. This gives the unique factorisation theorem for positive integers. It also shows that every non-zero polynomial over a field can be factorised as a product of irreducible polynomials.

As happens with many important theorems, we give a name for the property appearing in the conclusion of the last theorem.

2.29 Definition: An integral domain R with identity is called a **unique factorisation domain** (or u.f.d. for short) if every non-zero, non-unit element of R can be expressed as a product of a unit and powers of distinct prime elements and any two such factorisations are equal upto associateships.

The last theorem can be reworded as saying that every p.i.d. is a u.f.d. This is the last of the three conditions on an integral domain which seek to capture some of the properties of $\mathbb{Z}$, the ring of integers. The three conditions, namely being a euclidean ring, being a p.i.d. and being a u.f.d. are progressively weaker, and so the classes of rings satisfying them are progressively larger. The relationship between various classes is shown diagrammatically in Figure 6 1.

To show that the class of p.i.d.'s is a proper sub-class of u.f.d.'s, consider $\mathbb{Z}[x]$, the ring of polynomials over $\mathbb{Z}$. We saw earlier that $\mathbb{Z}[x]$ is not a p.i.d , However, it can be shown that $\mathbb{Z}[x]$ is a u.f.d.. Given a polynomial $f(x)$ in $\mathbb{Z}[x]$, we consider it as an element of $\mathbb{Q}[x]$ (since $\mathbb{Z}[x] \subset \mathbb{Q}[x]$). Now $\mathbb{Q}[x]$ is a u.f.d., in fact a euclidean ring. So we can factorise $f(x)$ into

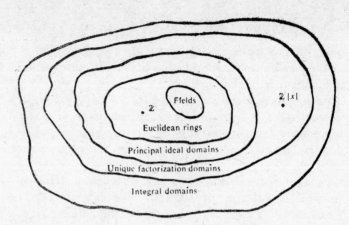

Figure 6.1: Various Classes of Rings

irreducible polynomials with coefficients as rational numbers. Taking out
the common denominators of these coefficients, it can be shown that $f(x)$
has a factorisation into polynomials with integer coefficients. The details
will be given through exercises.

Some of the properties of p.i.d.'s such as the existence of g.c.d. and
l.c.m. hold for u.f.d.'s. (Indeed, as we recall from school, one of the
methods for finding the g.c.d. is to factorise the two elements into powers
of primes.) These results will also be given as exercises, where the reader
will also find examples of integral domains which are not u.f.d.'s.

Exercises

2.1   For $\mathbb{Z}$, prove that a stronger version of the euclidean algorithm is
      true. Given $a, b \in \mathbb{Z}$ with $b \neq 0$, there exist integers q, r such
      that $a = qb + r$ where $|r| \leqslant \frac{1}{2}|b|$.

2.2   Let G be the ring of Gaussian integers, that is $G = \{x + iy : x,$
      $y \in \mathbb{Z}\}$. Prove that if $z_1, z_2 \in G$, then $d(z_1 z_2) = d(z_1) d(z_2)$. Suppose
      $x + iy \in G$ and n is a positive Integer. Prove that there exists
      $p + iq \in G$ and $r + is \in G$ such that

$$x + iy = n(p + iq) + (r + is) \text{ and } d(r + is) < n^2.$$

      (Hint: Apply the last exercise separately to the real and imaginary
      parts of $x + iy$.)

2.3   Prove that G is a euclidean ring. (Hint: In order to divide $x + iy$
      by $a + ib$, divide $(x + iy)(a - ib)$ by $a^2 + b^2$ and use the last
      exercise.)

2.4   Prove that 2 is not a prime element of G. More generally, prove
      that any integer which is expressible as a sum of two squares of
      integers is not a prime element of G.

2.5   Suppose p is a prime integer and x, y are relatively prime integers such that $x^2 + y^2 = cp$ for some integer c. Prove that p is not a prime element of G. [Hint: Factor $x^2 + y^2$ as $(x + iy)(x - iy)$ and apply Corollary 2.18.]

2.6   In the last exercise, prove further that p can be expressed as a sum of two squares of positive integers. [Hint: If $a + ib$ is a prime divisor of p, so is $a - ib$.]

2.7   If p is a prime of the form $4n + 1$, prove that there exists an integer x, $1 \leqslant x \leqslant p - 1$, such that $x^2 + 1$ is divisible by p. (Hint: Let $y = 1.2.\ldots.q$ where $q = \dfrac{p-1}{2}$. Using Wilson's theorem (Exercise 5.2.26), prove that $y^2 \equiv -1$ modulo p. Now reduce y modulo p to a desired x) This result is sometimes expressed by saying that -1 is a **quadratic residue** modulo p.

2.8   Prove that every prime of the form $4n + 1$ can be expressed as a sum of two squares of integers (e.g. $5 = 2^2 + 1^2$; $29 = 5^2 + 2^2$ etc.). Prove further that this expression is essentially unique, that is if $p = x^2 + y^2 = a^2 + b^2$ where x, y, a, b are positive integers then $\{x, y\} = \{a, b\}$. [Hint: Use the last three exercises. Note that if $p = x^2 + y^2$ then $x + iy$ must be a prime element of G.]

2.9   Suppose p is a prime of the form $4n + 3$. Prove that:

   (i)   there does not exist an integer x such that $x^2 = -1 \pmod p$, or in other words, -1 is not a quadratic residue moduldo p. (Hint: If it is, then show that $[x]$ is an element of order 4 in the group $\mathbb{Z}_p - \{[0]\}$ under modulo p multiplication.)

   (ii) p cannot be expressed as a sum of two squares of integers.

   (iii) p is a prime element of G.

   (Primes greater than 2 fall into two categories. Those of the form $4n + 1$ and those of the form $4n + 3$. These exercises show that the two types of primes behave differently and this difference is brought out fully by the ring of Gaussian integers.)

2.10 Suppose a positive integer n is expressed as $2^k p_1^{m_1} \ldots p_r^{m_r} q_1^{n_1} \ldots q_s^{n_s}$ where $k, r, s \geqslant 0$, $p_1, \ldots, p_r$ are distinct primes of the form $4n + 1$; $q_1, \ldots, q_s$ are distinct primes of the form $4n + 3$ and $m_1, \ldots, m_r, n_1, \ldots, n_s$ are positive integers. Prove that n can be expressed a sum of two squares of integers if and only if all the n_i's are even.

2.11 Let R be a principal ideal domain and let a, $b \in R$. Find generators, in terms of a and b, for (i) the ideal $(a) \cap (b)$ (ii) the ideal generated by $(a) \cup (b)$.

2.12 Let R be a commutative ring with identity. Two ideals A and B of R are said to be **relatively prime** if there exist $a \in A$, $b \in B$ such that $a + b = 1$. Suppose R is a p.i.d. and x, $y \in R$. Prove that x and y are relatively prime iff the ideals (x) and (y) are relatively prime.

2.13   Let $m_1,..., m_k$ be positive integers which are pairwise relatively prime, that is for every $i \neq j$, m_i and m_j are relatively prime. Let $m = m_1 m_2 ... m_k$. Prove that the ring $\mathbb{Z}_m$ is isomorphic to the product ring $\mathbb{Z}_{m_1} \times \mathbb{Z}_{m_2} \times ... \times \mathbb{Z}_{m_k}$. (Hint: Define $\theta : \mathbb{Z}_m \to \mathbb{Z}_{m_1} \times ... \times \mathbb{Z}_{m_k}$ by $\theta([x]) = ([x], [x], ..., [x])$ where the same symbol $[x]$ is used to denote the residue classes of x modulo different integers. There is little difficulty in showing that θ is a well-defined ring homomorphism and θ is one-to-one. Then use the pigeon hole principle to show that θ is onto. This result is called the **Chinese Remainder Theorem**).

2.14   Using the Chinese Remainder Theorem, find 5832×6639 modulo 840. (Hint : $840 = 8.3.5.7$. Perform the multiplication modulo 8, 3, 5 and 7 separately to get answers x_1, x_2, x_3, x_4 (say), respectively. By the Chinese Remainder Theorem there exists an integer x, $0 \leqslant x < 840$ such that $x \equiv x_1 \pmod 8$, $x \equiv x_2 \pmod 3$, $x \equiv x_3 \pmod 5$ and $x \equiv x_4 \pmod 7$. The trouble is that there is no easy algorithm for finding such x, short of searching. If we have to do a very large number of arithmetical operations modulo 840, it is advantageous to prepare tables which give such x for each quadruple $(x_1, x_2, x_3, x_4) \in \mathbb{Z}_8 \times \mathbb{Z}_3 \times \mathbb{Z}_5 \times \mathbb{Z}_7$.)

*2.15   Generalise the Chinese Remainder Theorem to any commutative ring, R, with identity as follows. Let $A_1, ..., A_k$ be ideals of R which are pairwise relatively prime. Then prove that R/A is isomorphic to $R/A_1 \times R/A_2 \times ... \times R/A_k$. (Hint: proceed as in Exercise (2.13). The only difficulty is in showing that θ is onto. For this, prove by induction on k, that given $x_1, ..., x_k \in R$, there exists $x \in R$ such that $x - x_i \in A_i$ for all $i = 1, 2, ..., k$.)

2.16   Characterise the units of $R_1 \times R_2 \times ... \times R_k$ where $R_1, ..., R_k$ are rings, with identities. Combining this with the Chinese Remainder Theorem, give an alternate proof of the fact that the Euler ϕ-function is multiplicative, (cf. Exercise (2.4.15)).

2.17   Prove that the group of automorphisms of a cyclic group of order m (cf. Exercise (5.3.6)) is isomorphic to the group of modulo m residue classes relatively prime to m.

2.18   Suppose x is an element of a group G and m, n are relatively prime integers. If $x^m = x^n = e$ prove that $x = e$.

2.19   Suppose x, y are elements of a field F and m, n are relatively prime positive integers such that $x^m = y^m$ and $x^n = y^n$. Prove that $x = y$. Show that this still holds if F is only an integral domain but not necessarily if it is any arbitrary ring.

2.20   Generalise Proposition (2.23) as follows. Let S be a ring with identity. Let Z be the centre of S. Suppose R is a subring of S such that $1 \in R$ and $R \subset Z$ (Such a subring is called a **central subring**.) Then for any $a \in S$ and any two polynomials $f(x)$, $g(x)$ in $R[x]$, show that $h(a) = f(a) + g(a)$ and $p(a) = f(a)g(a)$, where $h(x) = f(x) + g(x)$ and $p(x) = f(x) g(x)$.

2.21 Prove that a polynomial of degree 3 or less over a field is irreducible if and only if it has no roots in that field. (Hint: In any nontrivial factorisation, there will be at least one linear factor.)

2.22 The **multiplicity** of a root α of a polynomial $f(x)$ over a field F is defined as the largest integer m such that $(x - \alpha)^m$ divides $f(x)$ in $F[x]$. A root is called **simple** or **multiple** according as its multiplicity equals or exceeds 1. Prove that if $f(x) = a_0 + a_1 x + \ldots\ldots + a_n x^n$ $(a_n \neq 0)$ and has distinct roots $\alpha_1, \ldots, \alpha_k$ in F, with multiplicities $m_1, \ldots, m_k$ respectively then $\sum_{i=1}^{k} m_i \leqslant n$ and equality holds iff $f(x)$ can be expressed as a product of n polynomials of degree 1 each. (Such a factoring is called a **complete splitting** of $f(x)$). In the case of equality prove further that

$$\sum_{i=1}^{k} m_i \alpha_i = -\frac{a_{n-1}}{a_n} \text{ and } \prod_{i=1}^{k} \alpha_i^{m_i} = \frac{(-1)^n a_0}{a_n}.$$

2.23 Let $f(x)$ be a polynomial of degree $n \ (> 0)$ over a field F and let $f'(x)$ be its formal derivative (cf. Exercise (1.20)). Prove that α is a multiple root of $f(x)$ in F if and only if it is a common root of $f(x)$ and $f'(x)$.

2.24 In a unique factorisation domain, obtain a criterion for divisibility of one element by another in terms of the factorisations of the two elements into prime powers.

2.25 Prove that in a u.f.d., every two elements have a g.c.d. and every two non-zero elements have an l.c.m.. Prove also that Proposition (2.17) holds for a u.f.d..

2.26 In $\mathbb{Z}[x]$, prove that the g.c.d. of 2 and x cannot be expressed as a linear combination of 2 and x. (Thus the behaviour of the g.c.d. of two elements of an integeral domain improves in the hierarchy of the three conditions. In a u.f.d. it simply exists, in a p.i.d. it equals some linear combination of the two elements and in a euclidean ring there is an algorithm for expressing it this way.)

2.27 Let R be the ring, under pointwise operations, of all complex-valued functions which are analytic in the entire complex plane $\mathbb{C}$ (see Exercise (1.16)). Prove that the units of R are those functions which have no zero in $\mathbb{C}$, while the prime elements of R are those functions which have at most one zero (of multiplicity 1) in $\mathbb{C}$. Prove that functions with infinitely many zeros (such as $\sin z$, $\cos z$) cannot be factorised in R as products of finitely many primes. (The solutions require only one basic property of analytic functions, namely, if α is a zero of multiplicity m of an analytic function $f(z)$ then there exists an analytic function $g(z)$ such that $f(z) = (z - \alpha)^m g(z)$ and $g(\alpha) \neq 0$.)

2.28   Prove that in the ring $\mathbb{Z} + i\sqrt{3}\mathbb{Z}$ (that is, the subring of $\mathbb{C}$, consisting of elements of the form $a + i\sqrt{3}b$ where a, b are integers), 4 can be expressed as a product of primes in two different ways. (Thus we have an example of an integral domain with identity which fails to be a u.f.d. because of non-uniqueness of factorisation. The ring R in the last exercise, on the other hand, fails to be a u.f.d. because of absence of factorisation).

2.29   A polynomial $a_0 + a_1x + \ldots + a_nx^n$ in $\mathbb{Z}[x]$ is said to be **primitive** if the g.c.d. of its coefficients $a_0, \ldots, a_n$ is 2, (equivalently, there is no prime number which divides all of them). Prove that the product of two primitive polynomials is a primitive polynomial. [Hint: Given a prime p, consider the terms of the smallest degree in each polynomial, whose coefficients are not divisible by p.]

2.30   Suppose $f(x)$ is a primitive polynomial in $\mathbb{Z}[x]$ and $f(x) = g(x)h(x)$ in $\mathbb{Q}[x]$. Then prove that $f(x)$ can be factorised in $\mathbb{Z}[x]$ (Hint: Write $g(x)$ as $\dfrac{p}{q} g_1(x)$ where p, q are integers and $g_1(x)$ is a primitive polynomial in $\mathbb{Z}(x)$. Do the same for $h(x)$ and apply the last exercise. This result is called **Gauss' Lemma**.)

2.31   Using the last exercise prove that $\mathbb{Z}[x]$ is a unique factorisation domain. (Hint: Note that $\mathbb{Z} \subset \mathbb{Z}[x] \subset \mathbb{Q}[x]$. Intuitively there are two types of primality in $\mathbb{Z}[x]$, one that comes from the fact that the coefficients are integers and the other that comes from the fact that the elements are polynomials. Because of Gauss' Lemma these two types of primalities can be separated from each other.)

2.32   Let F be a field and $f(x) = x^2 + x + 1$. Prove that if $F = \mathbb{R}$ or $\mathbb{Z}_2$, then $f(x)$ is irreducible, but if $F = \mathbb{Z}_3$ then it is reducible over F. In the first two cases let I be the ideal $(f(x))$. If $F = \mathbb{R}$, prove that $F[x]/I$ is isomorphic to the field of complex numbers, while if $F = \mathbb{Z}_2$ prove that $F[x]/I$ is a field with 4 elements.

*2.33   Using the field with 4 elements constructed above, solve the Religious Conference Problem for $n = 4$. (These two exercises give an idea of the construction and applicability of finite fields. It is possible to study both in greater details, see the Epilogue.)

2.34   Let $f(x) = a_0 + a_1x + a_2x^2 + \ldots + a_nx^n \in \mathbb{Z}[x]$ and suppose p/q is a root of $f(x)$ where p, q are relatively prime integers. Prove that p divides a_0 and q divides a_n. (This result, called the **rational root test** gives the possible rational roots of a polynomial with integer coefficients. Each one of them is then checked individually to see if it is in fact a root.)

2.35   In Proposition (2.20), show more generally that if d is a positive integer which divides n then G has exactly $\phi(d)$ elements of order d. [Hint: Apply the possition to a subgroup of G of order d, which exists and is unique by Theorem (5.2.17).]

2.36 Using the last exercise give an alternate proof of Exercise (3.4.24), (viii).

2.37 Let H be a group of order n. Prove that H is cyclic iff it has the property that for every positive integer d dividing n, H contains at most d elements satisfying the equation $x^d = e$. [Hint: For the converse, for every positive divisor d of n, let

$$H_d = \{x \in H: \circ(x) = d\}.$$

From the condition H satisfies, show that if x is an element of order d then the only elements of order d in H are x^m where m is relatively prime to d. Hence $|H_d| \leqslant \phi(d)$ for all $d \mid n$. So

$$n = \circ(H) = \sum_{d \mid n} |H_d| \leqslant \sum_{d \mid n} \phi(d) = n.$$

So for every d, we must have $|H_d| = \phi(d)$. In particular $H_n \neq \phi$.]

2.38 Using the last exercise and Proposition (2.25), show that the multiplicative group of a finite field is cyclic. More generally show that every finite subgroup of the multiplicative group of any field is cyclic.

2.39 Let p be an odd prime. Let $f(x) = x^{p-1} - 1$. Considering $f(x)$ as a polynomial over the field $\mathbb{Z}_p$ and applying Theorem (5.2.16) and the result of Exercise (2.22), obtain another proof of Wilson's theorem, $(p - 1)! \equiv -1 \pmod{p}$. [Hint: Every non-zero element of $\mathbb{Z}_p$ is a root of $f(x)$.]

Notes and Guide to Literature

The treatment in this section follows largely the pattern of Herstein [1], Chapter 3, except for one major difference. Herstein adds one more condition to the definition of a euclidean ring, to the effect that the degree of every element is greater than that of its proper divisor. This makes it easier to apply induction on the degree and thereby give direct proofs of the results. However, this extra condition seems superfluous, because even without it, every euclidean ring is a principal ideal domain. Another point of departure is that unlike Herstein, who cofines to euclidean rings most of the time, we have proved the results for p.i.d.'s. It is perhaps debatable whether we have achieved any real generality in doing so, especially because we have not given any examples of rings which are p.i.d.'s but not euclidean rings. We have only quoted an example which is due to Motzkin [1]. The reason we have preferred to work with p.i.d.'s rather than euclidean rings is an important difference of technique. In a euclidean ring we directly imitate what goes in $\mathbb{Z}$. In a principal ideal domain on the other hand, we first translate the concepts in terms of ideals. When so translated, they are often meaningful even for rings which are not p.i.d.'s, for example, the concept of relatively prime ideals in Exercise (2.12). In mathematics it often happens that a concept cannot be generalised as it is to a wider setting, but when appropriately translated, the generalisation becomes self-evident. Art lies in finding such 'appropriate' translations.

The Chinese Remainder Theorem is said to have been known to the ancient Chinese. For more on its applications to computations, see Knuth [1], Volume 2 or Tremblay and Manohar [1].

The result of Exercise (2.8) is due to Fermat. For an elementary proof, see for example, Chandrasekhar [1]. See Ahlfors [1] for properties of complex analytic functions.

3. Vector Spaces

Compared to the last two sections, the material in this section is relatively light. Accordingly our presentation will be somewhat sketchy. Another reason for this is that we expect the reader to be familiar at least with the algebra of the two and the three dimensional vectors as used in physics. Indeed we count on this familiarity to motivate 'abstract' vector spaces to be studied in this section.

In physics we take a vector as a quantity which has both a magnitude and a direction and which obeys the parallelogram law for addition. Common examples of vectors are velocity, force, acceleration, electrostatic field vector and so on. Scalars on the other hand, have only magnitude and no direction. They are added as real numbers. Speed, work, pressure etc. are scalars. The vector spaces to be studied in this section are actually abstraction of the vectors in physics. But unlike the definition of a ring (which is modelled after the ring of integers), the linkage between abstract vectors and their forerunners, namely, the vectors in physics is a rather remote one. So instead of giving right away the definition of an abstract vector space, we proceed to show how it evolves from the vectors in physics.

As a first step we have a convenient mathematical representation of physical vectors by directed line segments in the plane or in space. (Sometimes a vector is *defined* as a directed line segment). Here two directed line segments which are of equal length and have the same direction are to be regarded as representing the same vector. So, formally, a vector is an equivalence class of directed line segments, under the equivalence relation in which two line segments with the same length and direction are to be regarded as equivalent. An alternate approach is to consider only those directed line segments which start at the origin O. If P is any point then the directed line segment OP joining O to P is called the **position vector** of P. Position vectors are added according to the parallelogram law (Figure 6.2 (a)). Similarly if P is a point and λ is a real number (a 'scalar' as it is called) them λOP if the position vector of a point Q which lies on the same ray through O as P or on the opposite ray according as $\lambda \geqslant 0$ or $\lambda \leqslant 0$, and whose distance from O is $|\lambda|$ times that of P (Figure 6.2. (b)).

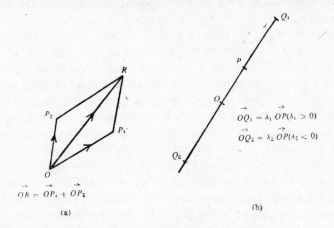

Figure 6.2 : Elementary Operations on Vectors.

Besides the two operations of vector addition and scalar multiplication, we have two types of products, the scalar (or the dot) product of two vectors and the vector (or the cross) product of two vectors. A number of identities about these products hold true. Most of the basic concepts of geometry such as distance, angle, lines, planes, areas, volumes etc. can be expressed in terms of vectors. This study is known as vector algebra. In vector calculus, on the other hand, we study functions whose domains and/or codomains consist of vectors. We study differentiability of such functions, define things like the gradient, the divergence and the curl and prove and apply such classic results as the theorems of Gauss and Stoke. This study constitutes one of the richest applications of mathematics to physical problems. Indeed many concepts of vector calculus owe their origin to physical problems. (This vestige can be seen even from the nomenclature. For example, a vector valued function is called a 'vector field' because the electric field due to a charge is an important example of such a function. Of course, 'field' as used here has little to do with the way we are using the term 'field', namely to denote a certain kind of rings.)

In terms of depth, the theory of vector spaces which we are going to develop will not come anywhere near the theory outlined above. First, we shall not deal with any limiting process and this takes away the entire vector calculus from our purview. Consequently, the abstraction we are after will only be of the algebraic aspect of vectors. But even here, we shall be sacrificing a large part, namely, the structure that arises because of the dot and the cross product. It is indeed possible to abstract and generalise the concept of the dot product. The structure so formed is called an **inner product space** and is important in a branch of mathematics called functional analysis. But we shall not pursue it. As for the cross product, it is a feature peculiar to vectors in space, that is, three dimensional vectors. It cannot be generalised even to four dimensional vectors, let alone for abstract vectors.

So, summing it up, the definition of an abstract vector space is designed to generalise only those aspects which depend on the addition of vectors and on the multiplication of a vector by a scalar. As usual, we list a few properties of these two operations and assume some of them as axioms. The set of vectors forms an abelian group under addition. We already defined abelian groups in Chapter 5. Just as a ring is obtained by putting an extra structure on an abelian group (and inter-relating it to the group structure by means of the distributive law), an abstract vector space is obtained from an abelian group by endowing it with an additional structure corresponding to scalar multiplication. To see what this structure should be, let us list some of the properties of the familiar scalar multiplication for the vectors considered above. Let V be the set of such vectors. Then the scalar multiplication is really a function from $\mathbf{R} \times V$ into V which assigns to an ordered pair (λ, u) where $\lambda \in \mathbf{R}$ and $u \in V$, the vector λu in V. This function is *not* a binary operation on V (because it is not a function from $V \times V$ into V). However, it is compatible with the binary operation of addition in V in the sense that for all $\lambda \in \mathbf{R}$ and u, v in V, $\lambda(u + v)$ equals $(\lambda u) + (\lambda v)$. The scalar multiplication is also compatible with the ring operations in $\mathbf{R}$ in the sense that for all $\lambda, \mu \in \mathbf{R}$ and for all $u \in V$, we have $(\lambda + \mu) u = (\lambda u) + (\mu u)$ and $(\lambda \mu)u = \lambda(\mu u)$. Moreover, the identity element of $\mathbf{R}$, namely 1, also behaves like an identity element for scalar multiplication in the sense that for all $u \in V$, $1u = u$.

To define an abstract vector space, we start with an abelian group V, a field F and a function from $F \times V$ into V which plays the role of scalar multiplication. The formal definition is as follows:

3.1 Definition: A vector space over a field F is a triple $(V, +, \cdot)$ where $(V, +)$ is an abelian group and $\cdot : F \times V \to V$ is a function satisfying the following properties for all $\lambda, \mu \in F$ and all $u, v \in V$.

(i) $\lambda \cdot (u + v) = (\lambda \cdot u) + (\lambda \cdot v)$

(ii) $(\lambda + \mu) \cdot u = (\lambda \cdot u) + (\mu \cdot u)$

(iii) $(\lambda \mu) \cdot u = \lambda \cdot (\mu \cdot u)$

(iv) $1 \cdot u = u$, 1 being the identity element of F.

Note that in (ii), $+$ denotes the addition in F on the left hand side and the addition in V on the right hand side. Also the same symbol $\cdot$ is used for the multiplication in F and for the scalar multiplication. This double usage causes no confusion. We emphasize, however, that the scalar multiplication is not a binary operation either on F or on V. It is a sort of a hybrid. In technical terms, the scalar multiplication represents the action of the field F on the group V. We shall study other examples of such actions of algebraic structures later (see the Epilogue).

The reader may wonder where we needed that F is a field. Will the definition not make sense if instead of the field F we merely have a ring R with identity? The answer is 'yes'. The algebraic structure that arises this way is called a **module** (actually a **left module**) over the ring R. Thus vector spaces are special cases of modules. The theory of modules is also an interesting one. But we shall not study it here. (A few indicative results will be given as exercises.) Because a field is a very special type of a ring, it is but natural that there would be some results which hold good for vector spaces but not for all modules. These results would involve the special properties of a field namely, commutativity and, more importantly, the presence of multiplicative inverses for non-zero elements. As a simple illustration, we have the following proposition, where the first three parts hold good for modules as well but the last one is true for vector spaces only. (As with other places where a dot is used to denote a binary operation, we shall suppress it while denoting the scalar multiplication. Thus if $\lambda \in F$ and $u \in V$ we write λu instead of $\lambda \cdot u$. Another minor point about notation is that in physics the vectors are almost invariably denoted by bold face letters so as to distinguish them from scalars. We shall not do so. Generally, field elements will be denoted by lower case Greek letters and vectors by small case English letters. However, this is not a uniform convention since we shall have occasions to consider vectors which are themselves elements of some field. Ultimately, it is only through the context that a particular symbol will represent a scalar or a vector. The symbol 0 will simultaneously represent the zero element of the field F and also the zero vector, that is, the identity for addition of vectors. By the way, the zero vector, also called the null vector, should not be thought of as 'a vector with zero length' or as 'a vector having no particular direction' as is done with the ordinary vectors. The concepts of length and direction are meaningless for abstract vectors.)

3.2 Proposition: Let V be a vector space over a field F. Then we have, for all $\lambda \in F$ and $u \in V$,

  (i) $\lambda 0 = 0$

  (ii) $0u = 0$

  (iii) $(-\lambda)u = -(\lambda u) = \lambda(-u)$

  (iv) if $\lambda u = 0$ then either $\lambda = 0$ or $u = 0$.

Proof: The proof of the first three parts resembles that of Proposition (1.2) and hence is left as an exercise. For the last part, suppose $\lambda u = 0$. If $\lambda \neq 0$, then λ^{-1} exists in F. By (i), $\lambda^{-1}(\lambda u) = \lambda^{-1} 0 = 0$. But on the other hand, $(\lambda^{-1})\lambda u = (\lambda^{-1}\lambda)u = 1u = u$, using the axioms of a vector space. So $u = 0$. ∎
  The last part says, in a way, that there are no zero divisors for the

scalar multiplication in a vector space. As a corollary we have that if $\lambda \neq 0$ then the equation $\lambda u = \lambda v$ forces $u = v$. Similarly, if $u \neq 0$, then $\lambda u = \mu u$ implies $\lambda = \mu$.

The concepts of a subspace of a vector space, the quotient space of a vector space and homomorphism of vector spaces are defined in the expected manner. Specifically, let V be a vector space over a field F. Then a **subspace** of V is a subset W of V such that (i) W is a subgroup of the abelian group V and (ii) W is closed under scalar multiplication, that is, for all $\lambda \in F$ and $u \in W$, $\lambda u \in W$. It is clear that a subspace of a vector space is itself a vector space over the same field. Also if W is a subspace of V then the set V/W, consisting of the cosets of W in V, is already an abelian group under coset addition. If we define $\lambda (u + W)$ to be $(\lambda u) + W$ for $\lambda \in F$, $u \in V$ then we get a well-defined scalar multiplication on V/W and it makes V/W a vector space called the **quotient space** of V by W. If V, W are any two vector spaces over the same field F, then a **vector space homomorphism** from V to W is a function $T: V \to W$ such that for all $u, v \in V$ and $\lambda \in F$, $T(u + v) = T(u) + T(v)$ and $T(\lambda u) = \lambda T(u)$. Vector space homomorphisms are more popularly called **linear transformations**. The reason for this term will be explained in the exercises. A linear transformation which is also a bijection is called a **vector space isomorphism**. It is easy to show that the kernel and the range of a linear transformation are subspaces of its domain and co-domain respectively. The analogues of the theorems about group homomorphisms hold for linear transformations. We leave it to the reader to state and prove them.

We now give a few examples of vector spaces. Embodied in every vector space, there is an abelian group. So, not surprisingly, some of the examples here have already appeared as examples of groups. Still we state them again, emphasising this time, the scalar multiplication.

1) The vectors consisting of directed line segments originating at some fixed point O form a vector space over $\mathbf{R}$, the field of real numbers. We considered above only the two and the three dimensional vectors. But more generally, we can take the position vectors of points in any higher dimensional euclidean space. In fact we may as well identify a point in a euclidean space with its position vector and say that a euclidean space is itself a vector space over the field of real numbers. Any line or plane passing through the origin is a subspace of this vector space (cf. Figure 5.4, where one such subspace was pictured only as a subgroup).

2) Trivially, every field F is a vector space over itself. The scalar multiplication $\cdot : F \times F \to F$ is simply the field multiplication. More generally, suppose R is a (not necessarily commutative) ring with identity 1 and F is a subfield of R such that $1 \in F$. Then R is a vector space over F, with the scalar multiplication $\cdot : F \times R \to \mathbf{R}$ being merely the restriction of the binary operation $\cdot : R \times R \to R$. It is of course not necessary that R should actually contain the field F. It suffices if R contains an isomorphic

replica of F. For example, let R to be the ring $M_n(F)$ of all $n \times n$ matrices over a field F. We identify an element $\lambda \in F$ with the $n \times n$ matrix whose diagonal entries are λ each and whose all other entries are 0. If $A = (a_{ij})$ is any $n \times n$ matrix over F then for any $\lambda \in F$, we have

$$
\begin{pmatrix} \lambda & 0 & 0...0 \\ 0 & \lambda & & 0 \\ 0 & & \ddots & \vdots \\ \vdots & & & \\ 0 & 0 & ... & \lambda \end{pmatrix}
\begin{pmatrix} a_{11} & a_{12}...a_{1n} \\ a_{21} & a_{21}...a_{2n} \\ .... \\ a_{n1} & a_{n2}...a_{nn} \end{pmatrix}
=
\begin{pmatrix} \lambda a_{11} & \lambda a_{12}...\lambda a_{1n} \\ \lambda a_{21} & \lambda a_{22}...\lambda a_{2n} \\ .... \\ \lambda a_{n1} & \lambda a_{n2}...\lambda a_{nn} \end{pmatrix}
$$

Therefore, when $M_n(F)$ is regarded as a vector space over F, the scalar multiplication of an $n \times n$ matrix with an element of F simply amounts to multiplying each entry of the matrix by λ. In view of this, we can even consider matrices which are not square matrices. Let m, n be fixed positive integers and let $M_{m,n}(F)$ be the set of all $m \times n$ matrices over F. Then $M_{m,n}(F)$ is a vector space over F with scalar multiplication defined entry-wise. If we take only such matrices which have 0 in a particular place (say the second row and the third column) then we get a subspace of $M_{m,n}(F)$. Similarly the set of all matrices with the property that the sum of the entries in a particular row (or column) is 0 is a subspace of $M_{m,n}(F)$.

3) As another example of an extension ring forming a vector space, let $F\{x\}$ be the ring of all formal power series in an indeterminate x over a field F. Then $F\{x\}$ is a vector space over F. The scalar multiplication by $\lambda \in F$ amounts to multiplying the coefficient of every power of x by λ. An important subspace of this is the set of all polynomials in x, $F[x]$. If we take the set of all power series in which the coefficient of a particular power of x (say of x^{50}) is 0 we get a subspace of $F\{x\}$. Another important class of subspaces is the sets of those polynomials whose degrees are bounded by some fixed positive integers (with the zero polynomial included). For example all polynomials of degree 5 or less (with the zero polynomial included) form a subspace of $F[x]$ and hence of $F\{x\}$. Note that this subspace is generated by the set $\{1, x, x^2, x^3, x^4, x^5\}$ because it is clearly the smallest subspace of $F\{x\}$ which contains this set. $F(x)$, the field of rational functions in x over F is also a vector space over F.

4) As with other algebraic structures, the product of two (or more) vector spaces over the same field is a vector space over that field. The scalar multiplication is defined coordinatewise. Similarly, if V is a vector space over a field F, S is any set and W is the set of all functions from S into V then W is a vector space over F. The scalar multiplication is defined pointwise. That is, if $f: S \to V$ is an element of W and $\lambda \in F$ then λf is the function from S to V which assigns to a point $s \in S$, the vector $\lambda f(s)$. A particular vector space, which can be looked at as an instance of either of these two constructions is noteworthy. Let n be any positive integer. Let F^n be the cartesian product $F \times F \times ... \times F$ (n times). Elements of F^n are ordered

n-tuples of elements of F. They can also be considered as functions from the set $S = \{1, 2,..., n\}$ into F. F^n is a vector space over F. By convention we let F^o be the trivial vector space. The importance of vector spaces of this type is that every finitely generated vector space over F is isomorphic to F^n for some n. We shall prove this later. But as a special case, we can see that the vector space of all directed line segments in space is isomorphic to $\mathbf{R}^3$, that is, to $\mathbf{R} \times \mathbf{R} \times \mathbf{R}$. An explicit isomorphism is obtained by setting any fixed rectangular frame of coordinates say x, y, z. If a point P has co-ordinates x, y and z, we associate the triple (x, y, z) to the position vector of P. It may be recalled from geometry that the real numbers x, y, z represent the components of the vector $\overrightarrow{OP}$ along the three axes. (It is customary to let $\mathbf{i}$, $\mathbf{j}$, $\mathbf{k}$ be unit vectors along these axes. Then $\overrightarrow{OP} = x\mathbf{i} + y\mathbf{j} + z\mathbf{k}$). Something very similar will be used in the proof of the general isomorphism theorem just mentioned.

5)   Let V, W be vector spaces over a field F, $L(V, W)$ be the set of all linear transformations of from V to W. $L(V, W)$ is a subset of the set of all functions from V to W, which is a vector space under pointwise operations. We claim that $L(V, W)$ is a subspace and hence that it is a vector space over F. For this we have to show that whenever $T_1, T_2 \in L(V, W)$ and $\lambda \in F$, then $T_1 + T_2 \in L(V, W)$ and $\lambda T_1 \in L(V, W)$, that is, we have to show $T_1 + T_2$ and λT_1 are linear transformations from V to W. Recall that $T_1 + T_2$ is defined by pointwise addition of values of T_1 and T_2. So, if $u, v, \in V$ and $\lambda \in F$ then $(T_1+T_2)(u + v) = T_1(u + v) + T_2(u+v) = T_1(u) + T_1(v) + T_2(u) + T_2(v)$ (since T_1, T_2 are linear) $= [T_1(u) + T_2(u)] + [T_1(v) + T_2(v)]$ (since V is an abelian group under $+$) $= (T_1 + T_2)(u) + (T_1 + T_2)(v)$, showing, that $T_1 + T_2$ preserves addition. Also $(T_1+T_2)(\lambda u) = T_1(\lambda u) + T_2(\lambda u) = \lambda T_1(u) + \lambda T_2(u) = \lambda[T_1(u) + T_2(u)] = \lambda[(T_1 + T_2)(u)]$. This proves that $T_1 + T_2 \in L(V, W)$. The proof that $\lambda T_1 \in L(V, W)$ is even simpler and left to the reader. Thus $L(V, W)$ is a vector space over F. It is also sometimes denoted by $\text{Hom}_F(V, W)$ so as to distinguish it from $\text{Hom}(V, W)$ which simply consists of all group homomorphisms from the abelian group V to the abelian group W (see Example (11) in Chapter 5, Section 3).

A particularly interesting case of $\text{Hom}_F(V, W)$ arises when the vector space W is just the field F. Elements of $\text{Hom}_F(V, F)$ are functions $T: V \to F$ such that for all $u, v \in V$ and $\lambda \in F$, $T(u + v) = T(u) + T(v)$ and $T(\lambda u) = \lambda T(u)$. Such functions are called **linear functionals** on V and $\text{Hom}_F(V, F)$ is called the **dual space** of V. The dual of V is often denoted of $\hat{V}$ or V^*.

6)   Let $(B, +, \cdot, ')$ be a Boolean algebra. By Theorem (1.5), $(B, \oplus, \cdot)$ is a Boolean ring. We can make $(B, \oplus)$ a vector space over $\mathbb{Z}_2$ (which is a field as proved in Section 1). Since $\mathbb{Z}_2$ has only two elements, namely 0 and 1, we have little choice in defining the scalar multiplication. We define

$0 \cdot x = 0$ and $1 \cdot x = x$ for all $x \in B$. The axioms of a vector space are easily verified. Thus we see that a Boolean algebra (or rather, the corresponding Boolean ring) is a vector space over $\mathbb{Z}_2$. (Actually, the ring structure is not very vital here. All that is needed is that $x \oplus x = 0$ for every $x \in B$. This will be pointed out in an exercise.)

We now turn to the determination of the structure of abstract vector spaces. In Chapter 5, we remarked that the problem of classifying groups according to isomorphism is an extremely difficult one. The same is true for rings. As compared to this, vector spaces over a field behave remarkably simply. Upto isomorphism, there are very few distinct vector spaces over a field F. Moreover, their form can be easily written down. The situation is comparable to the structure theorem for Boolean algebras (Theorem 4.1.11) where we classified all finite Boolean algebras. In a somewhat similar manner we classify all finitely generated vector spaces over a field F. We begin with a few basic definitions. Throughout, V will denote a vector space over a field F. An expression of the form $\lambda_1 v_1 + \lambda_2 v_2 + \ldots + \lambda_n v_n$ where n is a non-negative integer, $v_1, \ldots, v_n \in V$ and $\lambda_1, \ldots, \lambda_n \in F$ is called a **linear combination** of the v_i's with the coefficients λ_i's. The term 'linear' comes from the fact that the equation of a straight line through the origin in cartesian coordinates is obtained by equating to zero some linear combination of the coordinates with constant coefficients. We consider only finite linear combinations. Consideration of infinite linear combinations would require some type of a limiting process. Thus even if S is an infinite subset of V, by a linear combination in S, we shall mean a linear combination of a finite number of vectors from S. In terms of linear combinations we can easily describe the subspace generated by a subset.

3.3 Proposition: Let $S \subset V$ and let $L(S)$ be the set of all linear combinations of elements of S. ($L(S)$ is often called the **linear span** of S.) Then $L(S)$ is a subspace of V. Moreover, it is the smallest subspace of V containing S. (If $S = \phi$, $L(S) = \{0\}$. This conforms to the convention that a sum with no terms equals 0.)

Proof: First we show that $L(S)$ is a subspace of V. If $u, v \in L(S)$, then each u, v is a linear combination of vectors in S, say $u = \lambda_1 v_1 + \ldots + \lambda_n v_n$ and $v = \mu_1 w_1 + \ldots + \mu_m w_m$ where v_i's, w_j's are in S and λ_i's and μ_j's are in F. Now we simply write $u + v$ as $\lambda_1 v_1 + \ldots + \lambda_n v_n + \mu_1 w_1 + \ldots + \mu_m w_m$ and see that it is in $L(S)$. (Some of the v's may overlap with some of the w's. But we are not requiring that the vectors appearing in a linear combination be distinct. If we do, then we can sort them out and add together the coefficients of equal vectors.) Similarly for any $\lambda \in F$, $\lambda u = \lambda \lambda_1 u_1 + \ldots + \lambda \lambda_n v_n$ and so $\lambda u \in L(S)$. Thus $L(S)$ is a subspace of V. If $u \in S$, then $u = 1 \cdot u$ is a linear combination in S and so $u \in L(S)$. Thus $S \subset L(S)$. Finally, suppose W is any subspace of V containing S. By induction on n, it is easy to show that for $v_1, \ldots, v_n \in S$ and $\lambda_1, \ldots, \lambda_n \in F$, $\lambda_1 v_1 + \lambda_2 v_2 + \ldots + \lambda_n v_n \in W$. So $L(S) \subset W$. Summing it up, $L(S)$ is the smallest subspace of V containing S. ∎

Because of this proposition, the subspace $L(S)$ is said to be **spanned** (rather than generated) by S. By way of an example, let $V = \mathbb{R}^3$ and let $S = \{(1, 0, -1), (-2, 1, 3)\}$. Suppose $(x_1, x_2, x_3) \in L(S)$, Then there exist $\lambda_1, \lambda_2 \in \mathbb{R}$ such that $(x_1, x_2, x_3) = \lambda_1(1, 0, -1) + \lambda_2(-2, 1, 3)$. This is equivalent to three simultaneous equations $x_1 = \lambda_1 - 2\lambda_2$, $x_2 = \lambda_2$ and $x_3 = -\lambda_1 + 3\lambda_2$. Eliminating λ_1 and λ_2 we get $x_1 - x_2 + x_3 = 0$. Conversely, if x_1, x_2, x_3 satisfy $x_1 - x_2 + x_3 = 0$ then we can solve for λ_1, λ_2 and express (x_1, x_2, x_3) as a linear combination of $(1, 0, -1)$ and $(-2, 1, 3)$. Thus $L(S)$ is the subspace $\{(x_1, x_2, x_3) \in \mathbb{R}^3 : x_1 - x_2 + x_3 = 0\}$. Suppose we add the vector $(0, 2, 1)$ to S. Then it is very easy to show that the span of the set $\{(1, 0, -1), (-2, 1, 3), (0, 2, 1)\}$ is the entire space $\mathbb{R}^3$. However, if we add the vector $(1, 4, 3)$ to S then the linear span of the set $\{(1, 0, -1), (-2, 1, 3), (1, 4, 3)\}$ is the same as that of the original set S. The reason is that the new vector added is already a linear combination of elements in S, because $(1, 4, 3) = 9(1, 0, -1) + 4(-2, 1, 3)$. So $(1, 4, 3)$ is redundant in the sense that it does not contribute anything new to the linear span of S.

Given a vector space V over a field F, our intention is to find a subset S of V which spans V. Obviously we would like this spanning subset S to be as small as possible (or else we might simply take S to be V itself). To ensure this, we must avoid redundancies of the kind illustrated above. There is a very handy way to tell whether a set is free of such redundancies. But first we need a definition.

3.4 Definition: A subset $S \subset V$ is said to be **linearly dependent** (over F) if there exist distinct vectors $v_1, v_2, \ldots, v_n \in S$ and $\lambda_1, \ldots, \lambda_n \in F$, not all 0, such that $\lambda_1 v_1 + \lambda_2 v_2 + \ldots + \lambda_n v_n = 0$. A set which is not linearly dependent is called **linearly independent**.

In this definition, S is a set of vectors. According to our convention about sets, when the elements of a set are listed, neither their order nor their repetition matters. As was seen in Chapter 3, Section 1, if we want to take into account the repetitions, the appropriate concept is a multiset. In the study of vector spaces, we often encounter multisets of vectors. It is convenient to extend the preceding definition of linear dependence when S is a multiset. In that case we do not require that the elements $v_1, \ldots, v_n$ be distinct. We simply require that none of them appears more frequently than it does in the multiset S. Note that if some element, say, v appears at least twice in S, then S is linearly dependent because we can write $1v + (-1)v = 0$. The linear dependence of a multiset is conceptually different from that of a set. If v is a non-zero vector, then the multiset $\{v, v\}$ is linearly dependent but the set $\{v, v\}$, which is the same as the set

$\{v\}$, is linearly independent! Still, to avoid pedantic fussiness, we shall speak only of linear dependence of a set, even though at times we mean the linear dependence of a multiset. The context will always make it clear which way we mean. By a further abuse of language, when S is a finite set or a multiset, say $S = \{w_1, ..., w_k\}$ then instead of saying that S is linearly dependent (or independent) we may say that the vectors $w_1, ..., w_k$ are linearly dependent (or independent).

Worded differently, a subset S is linearly independent iff no non-trivial linear combination of elements of S vanishes. In the example given above, the set $\{(1, 0, -1), (-2, 1, 3), (0, 2, 1)\}$ is linearly independent while the set $\{(1, 0, -1), (-2, 1, 3), (1, 4, 3)\}$ is not because we have

$$9(1, 0, -1) + 4(-2, 1, 3) - 1(1, 4, 3) = (0, 0, 0) = 0.$$

In the vector space of all polynomials over a field, the set $\{1, x, x^2, x^3, ...\}$ is linearly independent. But if we add any one more polynomial to it, it becomes linearly dependent. As with any other concepts about vector spaces, the role of the ground field should not be ignored. Sometimes the same abelian group may be made into a vector space over two fields. It may happen that a subset S is linearly dependent over one of these fields but not over the other. For example, consider $\mathbb{C}$, the set of complex numbers as a vector space over itself. Then the set $\{1, i\}$ is linearly dependent since $i \cdot 1 + (-1)i = 0$. Since $\mathbb{C}$ is an extension field of $\mathbb{R}$, we may think of $\mathbb{C}$ as a vector space over $\mathbb{R}$. But then $\{1, i\}$ is linearly independent over $\mathbb{R}$ because if $\lambda, \mu \in \mathbb{R}$ and $\lambda \cdot 1 + \mu i = 0$ then λ, μ must be both 0. Similarly the set $\{1, \sqrt{2}, \sqrt{3}, \sqrt{5}\}$ is linearly dependent over $\mathbb{R}$. But it is linearly independent over $\mathbb{Q}$, the field of rational numbers. (The proof of this fact is not so immediate. We shall indicate a proof in the exercises.)

The relationship between linear dependence and the redundancy discussed above is brought out in the following proposition.

3.5 Proposition: For a subset S of V, the following statements are equivalent:

(1) S is linearly dependent over F.
(2) There exists $v \in S$ such that v is in the linear span of $S - \{v\}$.
(3) There exists a proper subset T of S such that $L(S) = L(T)$.

Proof: (1) $\Rightarrow$ (2). Suppose S is linearly dependent over F. Then, by definition there exist distinct vectors, say, $v_1, v_2, ..., v_n \in S$ and scalars $\lambda_1, ..., \lambda_n \in F$, not all 0, such that $\lambda_1 v_1 + \lambda_2 v_2 + ... + \lambda_n v_n = 0$. Since not all λ_i's are 0, we may suppose, without loss of generality, that $\lambda_1 \neq 0$. Then $v_1 = \mu_2 v_2 + ... + \mu_n v_n$ where $\mu_i = -\dfrac{\lambda_i}{\lambda_1}$ for $i = 2, ..., n$. Let $v = v_1$. Then $v_2, ..., v_n \in S - \{v\}$ and we see that v is in the linear span of $S - \{v\}$. So (2) holds.

(2) ⇒ (3). Suppose $v \in S$ can be expressed as $v = \lambda_1 v_1 + \ldots + \lambda_n v_n$ where $v_1, \ldots, v_n \in S - \{v\}$ and $\lambda_1, \ldots, \lambda_n \in F$. Let $T = S - \{v\}$. Then T is a proper subset of S. We claim that $L(T) = L(S)$. Certainly, since $T \subset S$, we have $L(T) \subset L(S)$. For the other way inclusion let $u \in L(S)$. Then u can be expressed as $\mu_1 w_1 + \ldots + \mu_m w_m$ for some $w_1, \ldots, w_m \in S$ and $\mu_1, \ldots, \mu_m \in F$. Now, if none of the w_i's equals v, then u is a linear combination of elements of T and hence is in $L(T)$. If $w_i = v$ for some values of i, we substitute each such w_i by $\lambda_1 v_1 + \ldots + \lambda_n v_n$, in the expression $u = \mu_1 w_1 + \ldots + \mu_m w_m$ and thus get u as a linear combination of elements of S other than v. But then $u \in L(T)$. So $L(S) \subset L(T)$ and hence $L(S) = L(T)$.

(3) ⇒ (1). Suppose T is a proper subset of S and $L(T) = L(S)$. There exists some $v \in S$ such that $v \notin T$. Since $v \in S$, $v \in L(S)$ and hence $v \in L(T)$. But this means there exist $v_1, \ldots, v_n \in T$ and $\lambda_1, \ldots, \lambda_n \in F$ such that $v = \lambda_1 v_1 + \ldots + \lambda_n \lambda_n$. We may assume that the v_i's are distinct, for otherwise we can add the coefficients of equal v_i's. Since $v \notin T$, it follows that $v, v_1, \ldots, v_n$ are distinct vectors in S. Also $\lambda_1 v_1 + \lambda_2 v_2 + \ldots + \lambda_n v_n + (-1)v = 0$. Since the coefficient -1 is non-zero, it follows that S is linearly dependent. ▊

We are now in a position to prove the existence of a spanning set which is free of redundancies. The idea is to go on 'shrinking' a given spanning set, while retaining its linear span till we reach a stage where no more shrinking is possible. By the proposition above, the spanning set at this stage will be linearly independent. Now, how do we ensure that the process of shrinking the spanning set will stop eventually? For finite sets there is little difficulty as we see in the proof of the following proposition.

3.6 Proposition: Suppose S is a finite subset of V. Then there exists a linearly independent subset T of S such at $L(T) = L(S)$.

Proof: Let $W = L(S)$. If S itself is linearly independent, we set $T = S$. If not, then by the last proposition, there exists a subset $S_1 \subsetneq S$ such that $L(S_1) = W$. Again if S_1 is linearly independent we set $T = S_1$. If not there exists $S_2 \subsetneq S_1$ such that $L(S_2) = W$. Continuing in this manner we get a strictly descending sequence of sets $S \supsetneq S_1 \supsetneq S_2 \supsetneq S_3 \supsetneq \ldots$ each of which spans W. But S is a finite set. So this sequence must terminate at some stage. Suppose it terminates at S_k. Then S_k is linearly independent and spans W. We set $T = S_k$. ▊

This proposition is actually true without the hypothesis that S be finite. But the argument given above breaks down. In an infinite set we may have an infinite chain of strictly descending sets, say

$$S \supsetneq S_1 \supsetneq S_2 \supsetneq, \ldots, \supsetneq S_n \supsetneq S_n + 1 \ldots.$$

It is tempting to let T be the intersection of all these S_i's, that is,

$$T = \bigcap_{n=1}^{\infty} S_n.$$

(Something similar was done in the proof of Proposition (2.27), where we took the union of a strictly ascending sequence of ideals.) But the trouble is that this T may not be linearly independent. So we may have to trim it still further. The existence of the desired subset of S has to be established by the use of axiom of choice. The manner in which it is used is analogous to the manner in which it is used for proving that every partial order can be extended to a linear order (see Chapter 3, Section 3). But we do not pursue this line further. So we leave the result as it is. Its applicability is thereby restricted, but adequate for our purpose. The type of spaces for which it is applicable is defined below.

3.7 Definition: A vector space V over a field F is said to be **finite-dimensional** (over F) if there exists a finite subset S of V such that $L(S) = V$.

For example, the space F^n consisting of all ordered n-tuples of elements of F is finite-dimensional. A finite spanning set is given by $e_1, e_2, ..., e_n$ where for each i, e_i is the n-tuple whose ith entry is 1 and all other entries are 0. The space of all polynomials in x over F is not finite dimensional. However, the subspace of it consisting of all polynomials of degree 5 or less, say, is finite-dimensional with a spanning set $\{1, x, x^2, x^3, x^4, x^5\}$, We could have as well taken $\{1, 1 + x, 1 + x + x^2, 1 + x + x^2 + x^3, 1 + x + x^2 + x^3 + x^4, 1 + x + x^2 + x^3 + x^4 + x^5\}$ as a spanning set. Many other choices are also possible. Note that although we have defined a finite dimensional vector space, we have not yet defined the dimension of a vector space. We shall do this a little later. For the time being, therefore, 'finite dimensional' simply means 'finitely generated',

Proposition (3.6) provides the key to obtain the structure of finite dimensional vector spaces. First let us paraphrase it in terms of an extremely important concept, which we define.

3.8 Definition: A basis for a vector space V over a field F is a subset of V which spans V and which is linearly independent.

For example, for F^n, the set $\{e_1, e_2, ..., e_n\}$ is a basis. It is called the **standard basis** for F^n. Many other bases are possible. For example, in $\mathbb{R}^3$, the set $\{(1, 0, -1), (-2, 1, 3), (0, 2, 1)\}$ is a basis. But the set $\{(1, 0, -1), (-2, 1, 3)\}$ is not a basis because although it is linearly independent, its span, as calculated above, is a proper subspace of $\mathbb{R}^3$. On the other hand, the set $\{(1, 0, -1), (-2, 1, 3), (0, 2, 1), (1, 1, 1)\}$ is not a basis, because although it spans $\mathbb{R}^3$ it is not linearly independent. Thus the two requirements of a basis are mutually independent.

Proposition (3.6) can be paraphrased to say that every finite dimensional vector space has a basis. As remarked after its proof, this holds for all vector spaces. Although we shall not prove it, a few examples are noteworthy. For the space $F[x]$ of all polynomials over F, it is easy to see that the set $\{1, x, x^2, ..., x^n, ...\}$ is a basis. Since $\mathbb{R}$ is an extension field of $\mathbb{Q}$, it is a vector space over $\mathbb{Q}$. Any basis of $\mathbb{R}$ over $\mathbb{Q}$ is called a **Hamel basis**. It is interesting that no particular Hamel basis has yet been constructed. But its mere existence is useful in many contexts. This may sound puzzling to a beginner. But it is fairly typical of situations where the axiom of choice is applied. We simply get the existence of something without any explicit description of it.

We now prove what role a basis plays in the structure of a vector space. In doing so we prove various simple characterisations of a basis. Many authors take one of these characterisations as the definition of a basis.

3.9   Theorem: Let B be a subset of a vector space V over a field F. Then the following conditions are equivalent.

(1) B is a basis for V.

(2) B spans V and no proper subset of B spans V. (In other words B is a minimal spanning set for V.)

(3) B is linearly independent and no proper superset of B is linearly independent. (In other words, B is a maximal linearly independent set.)

(4)   Every element of V can be uniquely expressed as a linear combination of elements of B. (The meaning of uniqueness will be clarified in the proof).

Proof: (1) $\Rightarrow$ (2). Suppose B is a basis for V. Then by definition, B spans V. Also B is linearly independent. So by Proposition (3.5), no proper subset of B can have the same span as B. So (2) holds,

(2) $\Rightarrow$ (3). Suppose B is a minimal spanning set for V. Then no proper subset of B has the same span as B and so again by proposition (3.5), B is linearly independent. To show it is a maximal such set suppose C is a proper superset of B. Then $L(B) \subset L(C)$. But $L(B) = V$. So $L(C) = V$. This means that C has the same linear span as that of its proper subset B. So by Proposition (3.5), C is linearly dependent. This proves (3).

(3) $\Rightarrow$ (4). Suppose B is maximally linearly independent. We assert that $L(B) = V$. For let $v \in V$. If $v \in B$ then certainly $v \in L(B)$. If $v \notin B$, then the set $B \cup \{v\}$ is a proper superset of B and hence by assumption, is linearly dependent. So there exist distinct vectors $v_1, v_2, ..., v_n \in B$ and $\lambda_1, ..., \lambda_n \in F$ not all 0 such that $\lambda_1 v_1 + ... + \lambda_n v_n = 0$. One of these v_i's must be v, for otherwise all v_i's would be in B proving that B is linearly dependent, a

contradiction. Without loss of generality suppose $v = v_1$. Then $\lambda_1 \neq 0$, for otherwise $\lambda_2 v_2 + \ldots + \lambda_n v_n = 0$, again contradicting that B is linearly independent. Since $\lambda_1 \neq 0$, we can write $v = v_1 = \mu_2 v_2 + \ldots + \mu_n v_n$ where $\mu_i = -\dfrac{\lambda_i}{\lambda_1}$ for $i = 2, \ldots, n$. This shows $v \in L(B)$. Thus every vector in V is in $L(B)$. So $L(B) = V$, or in other words every element of V can be expressed as a linear combination of elements of B. We have further to show that this expression is unique. Here uniqueness means the following. Suppose the same vector v is expressed both as $\lambda_1 v_1 + \ldots + \lambda_n v_n$ and also as $\mu_1 v_1 + \ldots + \mu_n v_n$ where the v_i's are distinct elements of B and all λ_i's and μ_j's are in F. Then, $\lambda_i = \mu_i$ for all $i = 1, 2, \ldots, n$. To prove this suppose v admits the two expressions $\lambda_1 v_1 + \ldots + \lambda_n v_n$ and $\mu_1 v_1 + \ldots + \mu_n v_n$. Let $\alpha_i = \lambda_i - \mu_i$ for $i = 1, \ldots, n$. Then $\alpha_1 v_1 + \ldots + \alpha_n v_n = 0$. Since B is linearly independent, this means each α_i is 0, or $\lambda_i = \mu_i$ as was to be proved.

$(4) \Rightarrow (1)$ Suppose every element of V has a unique expression as a linear combination of elements of $\bar{B}$. Then certainly B spans V. To show B is linearly independent, suppose $v_1, \ldots, v_n$ are distinct elements of B and $\lambda_1 v_1 + \ldots + \lambda_n v_n = 0$ for some $\lambda_1, \ldots, \lambda_n \in F$. Now the zero vector 0 also equals $0v_1 + 0v_2 + \ldots + 0v_n$. So by uniqueness of expression, $\lambda_i = 0$ for all $i = 1, 2, \ldots, n$. In other words, no non-trivial linear combination of distinct vectors from B can vanish. So B is linearly independent. Thus we have shown that B spans V and also that B is linearly independent. By definition, B is a basis for V. ∎

As a corollary, we get the following structure theorem for finite dimensional vector spaces.

3.10 Corollary: Every finite dimensional vector space over a field F is isomorphic to F^n for some integer n.

Proof: Let V be a finite-dimensional vector space over F. Then there exists a finite set S such that $L(S) = V$. By Proposition (3.6), S contains a linearly independent set B which also spans V. By definition, B is a basis for V. Let $n = |B|$. If $n = 0$, then $B = \phi$ and V consists of only the zero vector. In this case F^0 is also the trivial vector space by convention and so the result holds. So suppose $n > 0$. Let $v_1, v_2, \ldots, v_n$ be the distinct elements of B. Define $h : F^n \to V$ by $h(\lambda_1, \lambda_2, \ldots, \lambda_n) = \sum\limits_{i=1}^{n} \lambda_i v_i$. Then h is a linear transformation. For, suppose $(\lambda_1, \ldots, \lambda_n)$, $(\mu_1, \ldots, \mu_n)$ are in F^n. Then $h[(\lambda_1, \ldots, \lambda_n) + (\mu_1, \ldots, \mu_n)] = h(\lambda_1 + \mu_1, \ldots, \lambda_n + \mu_n) = \sum\limits_{i=1}^{n} (\lambda_i + \mu_i) v_i = \sum\limits_{i=1}^{n} \lambda_i v_i + \sum\limits_{i=1}^{n} \mu_i v_i = h(\lambda_1, \ldots, \lambda_n) + h(\mu_1, \mu_2, \ldots, \mu_n)$. Similarly, for $\alpha \in F$, $h(\alpha(\lambda_1, \ldots, \lambda_n)) = \sum\limits_{i=1}^{n} \alpha \lambda_i v_i = \alpha h(\lambda_1, \ldots, \lambda_n)$. So h is a vector space homomorphism. To prove that it is an isomorphism we have to show that it is a bijection. But this is exactly what statement (4) in the last theorem says. ■

In case the field F is finite, the cardinality of the set F^n is $|F|^n$ by Proposition (2.2.13). This gives an alternate proof of Corollary (4.1.12), about the cardinality of a finite Boolean algebra. We already saw how a Boolean algebra may be regarded as a vector space over $\mathbb{Z}_2$, a field with elements. If the Boolean algebra is finite, then as a vector space it is certainly finite dimensional. So its cardinality is a power of 2. In the same vein, we get the following result about cardinalities of finite fields.

3.11   Corollary: The number of elements in a finite field is a power of some prime.

Proof: Let K be a finite field. In Section 1 we saw that the characteristic of K must be a prime, say, p. Using the result of Exercise (1.9), K contains a subfield F which is isomorphic to $\mathbb{Z}_p$. Since K is an extension field of F, it is also a vector space over F. Also it is finite dimensional, because we may take K itself as a finite set which spans K. By the last Corollary, K is isomorphic to F^n for some n. But then $|K| = |F^n| = p^n$. ■

This corollary puts a restriction on the cardinality of finite fields. For example, we see that there can be no field with 6 or with 10 elements. However, the corollary does not say that given any prime power p^n, there exists a field with p^n elements. The existence of such fields can indeed be proved, using the method of construction of fields starting from irreducible polynomials, given in the last section (cf. Exercise (2.32)). Moreover, two fields with the same number of elements turn out to be isomorphic to each other. So for every prime power p^n, upto isomorphism there is one and only one field with p^n elements. The proof of this fact, as well as other properties of finite fields will be given when we study applications of fields (see the Epilogue).

Interesting as these corollaries are, there is still something missing. We have not yet defined the dimension of a finite dimensional vector space. According to Corollary (3.10), if V is a finite dimensional vector space over a field F then V is isomorphic to F^n for some non-negative integer n. It is tempting to define the dimension of V as this integer n. But there is a catch here. Corollary (3.10) does not assert that the integer n is unique. If we go into its proof we see that the integer n was the cardinality of the basis B for V. There was nothing canonical about this basis B. Suppose we had some other basis, say, C for V with cardinality m. Then by the same reasoning, V would come out as isomorphic to F^m. So the dimension of V would be m as well as n. This would defeat our expectation that the dimension of a vector space be uniquely defined.

Fortunately, this does not happen. The same vector space may have many different bases, but they all have the same cardinality as we show. First we prove a preliminary result.

3.12 Proposition: In any vector space, any set of $n+1$ (or more) vectors in the linear span of n vectors is linearly dependent, where n is any positive integer.

Proof: Let V be a vector space over a field F. Let $v_1, v_2, ..., v_n \in V$ and let $S = \{v_1, v_2, ..., v_n\}$. The statement of the proposition means that if we take any $n+1$ (or more) vectors, say, $w_1, \cdots, w_{n+1}$ in $L(S)$ then the set $\{w_1, w_2, ..., w_{n+1}\}$ is linearly dependent. Since every superset of a linearly dependent set is linearly dependent, once we prove the assertion for $n+1$ vectors, it automatically holds for more than $n+1$ vectors. Our task is therefore to show that the set $\{w_1, ..., w_{n+1}\}$ where each $w_i \in L(\{v_1, ..., v_n\})$ is linearly dependent. For this, we apply induction on n.

Suppose $n = 1$. Then $w_1 = \lambda_1 v_1$ and $w_2 = \lambda_2 v_1$ for some $\lambda_1, \lambda_2 \in F$. If $\lambda_2 = 0$ then $w_2 = 0$ and $0 \cdot w_1 + 1 \cdot w_2 = 0$ gives a non-trivial, vanishing linear combination of w_2 and w_2. If $\lambda_2 \neq 0$, then $\lambda_2 w_1 + (-\lambda_1) w_2 = 0$. In either case we see that $\{w_1, w_2\}$ is linearly dependent.

Suppose now that the result holds for all values of n less than some positive integer k. We prove it for $n = k$. So let $w_1, ..., w_{k+1} \in L(\{v_1, ... v_k\})$. Then we can find elements $\lambda_{ij} \in F$ such that,

$$w_1 = \lambda_{11}v_1 + \lambda_{12}v_2 + \cdots\cdots + \lambda_{1k}v_k$$

$$w_2 = \lambda_{21}v_1 + \lambda_{22}v_2 + \cdots\cdots + \lambda_{2k}v_k$$

$$\cdots\cdots\cdots\cdots\cdots\cdots\cdots\cdots$$

$$w_i = \lambda_{i1}v_1 + \lambda_{i2}v_2 + \cdots\cdots + \lambda_{ik}v_k$$

$$\cdots\cdots\cdots\cdots\cdots\cdots\cdots\cdots$$

$$w_k = \lambda_{k1}v_1 + \lambda_{k2}v_2 + \cdots\cdots + \lambda_{kk}v_k$$

and
$$w_{k+1} = \lambda_{k+1,1}v_1 + \lambda_{k+1,2}v_2 + \cdots\cdots + \lambda_{k+1,k}v_k.$$

We may suppose that the coefficients in the last equation are not all 0. If they are, then $w_{k+1} = 0$ and we can write $0w_1 + 0w_2 + ... + 0w_k + 1w_{k+1} = 0$, proving the linear dependence of $\{w_1, ..., w_k, w_{k+1}\}$. So we assume that $\lambda_{k+1, j} = 0$ for some j. We may further suppose that $j = k$, that is, $\lambda_{k+1, k} \neq 0$; as otherwise we merely re-index the v's. Now, for each $i = 1, 2, ..., k$ we let $u_i = w_i - \mu_i w_{k+1}$, where $\mu_i = \lambda_{ik}/\lambda_{k+1, k}$. Then we see that u_i is a linear combination of $v_1, ..., v_{k-1}$. So $u_1, u_2 ..., u_k$ are k vectors in the linear span of the $k-1$ vectors $v_1, ..., v_{k-1}$. By induction hypothesis, the set $\{u_1, ... u_k\}$ is linearly dependent. Hence, there exist $\alpha_1, ..., \alpha_k \in F$ not all 0 such that $\alpha_1 u_1 + ... + \alpha_k u_k = 0$. Substituting $u_i = w_i - \mu_i w_{k-1}$, we get $\beta_1 w_1 + \alpha_2 w_2 + ... + \beta_k w_k + \beta_{k+1} w_{k+1} = 0$ where $\beta_i = \alpha_i$ for $i = 1, ..., k$ and $\beta_{k+1} = -(\alpha_1 \mu_1 + \alpha_2 \mu_2 + ... + \alpha_k \mu_k)$. Since not all α's are 0, not all

β's are 0. So $\{w_1, \ldots, w_k, w_{k+1}\}$ is linearly dependent. This completes the inductive step and proves the proposition. ▮

Worded differently, this proposition says that if $B \subset L(S)$ where S is a finite set, and B is linearly independent then $|B|$ cannot exceed $|S|$. Using this result, we are now in a position to settle the possible ambiguity about dimension.

3.13 Theorem: Any two bases of a finite dimensional vector space V over a field F have the same cardinality. This cardinality is also the unique integer n such that V is isomorphic to F^n.

Proof: Suppose B and C are two bases for V. Since V is finite dimensional there exists a finite set $S \subset V$ such that $L(S) = V$. Then elements of B are in the linear span of S. But B is also linearly independent. So by the last proposition, $|B| \leqslant |S|$. In particular B is finite. Similarly C is finite. To show that B and C must have the same cardinality we note that C is linearly independent and $C \subset L(B)$. So again by the last proposition, $|C| \leqslant |B|$. Interchanging the roles of B and C, we get $|B| \leqslant |C|$. Hence $|B| = |C|$. Thus we have shown that any two bases of V have the same number of elements. Let this number be n. We already saw in Corollary (3.10) that V is isomorphic to F^n. To complete the proof we must show that V is not isomorphic to F^m for $m \neq n$. Since 'being isomorphic to' is an equivalence relation, this ultimately reduces to proving that for $m \neq n$, F^n is not isomorphic to F^m. Let, if possible, $h : F^n \to F^m$ be an isomorphism. Let $B = \{e_1, e_2, \ldots, e_n\}$ be the standard basis for F^n discussed earlier. Let $C = h(B)$. We leave it to the reader to prove, from the fact that h is an isomorphism, that C is a basis for F^m. Since h is a bijection, $|C| = |h(B)| = |B| = n$. So F^m has a basis, C, with n elements. But the standard basis for F^m has m elements. So by the first assertion, $m = n$ a contradiction. Thus F^m and F^n are not isomorphic and hence n is the unique integer such that V is isomorphic to F^n. ▮

With this theorem at hand, we can now make the following definition:

3.14 Definition: The **dimension** of a finite dimensional vector space V is the cardinality of any basis for it. It is denoted by $\dim_F (V)$ or simply by $\dim (V)$.

It is only now that we can say that a finite dimensional vector space has finite dimension. Because of the standard basis for F^n, we see that its dimension is n. Again the role of the ground field should not be forgotten. For example, $\mathbb{C}$, as a vector space over itself has dimension 1. But as a vector space over $\mathbb{R}$, its dimension is 2, because $\{1, i\}$ is a basis for $\mathbb{C}$ over $\mathbb{R}$. If $\mathbb{C}$ is regarded as a vector space over $\mathbb{Q}$, the field of rational numbers, then its dimension is infinite. (A proof of this is based on the fact that the set $\mathbb{Q}$ is countable while $\mathbb{C}$ is not.) Intuitively, when the elements of a vector space V over a field F can be expressed in terms of the elements of F, then the dimension of V equals the number of free choices that we can make in specifying elements of V. As the simplest example, in F^n the elements are ordered n-tuples of elements of F. Each of the n entires in such n-tuples can be chosen freely, that is, without any restriction. So there are n free choices, equal to the dimension of F^n over F. By a similar reasoning, in an $m \times n$ matrix (a_{ij}) over F, each a_{ij} can be chosen independently of the others. So the dimension of $M_{m,n}(F)$ over F is mn. However, consider the subspace V of $M_{m,n}(F)$ consisting of those matrices in which the sum of the entries in the first row is 0. In such a matrix the entries in all rows except the first one can be chosen freely. However, in the first row, any $n - 1$ entries may be chosen arbitrarily. Once these are chosen, the remaining entry is determined completely. So the number of free choices in all is $(m-1)n + n-1$ that is $mn-1$. This is the dimension of the subspace V. As yet another example, the dimension of $\mathbf{R}^5$ is 5. Let V be the subspace of $\mathbf{R}^5$ defined by $V = \{(x_1, x_2, x_3, x_4, x_5) : 2x_1 - x_4 + \pi x_5 = 0\}$. For elements of V we can choose x_1, x_2, x_3 and x_4 arbitrarily. But once they are chosen x_5 must be defined by $(x_4 - 2x_1)/\pi$. So dim $(V) = 4$. (We could also have chosen x_1, x_2, x_3 and x_5 arbitrarily and defined x_4 in terms of them.) If we add one more restriction and define W as $\{(x_1, x_2, x_3, x_4, x_5) : 2x_1 - x_4 + \pi x_5 = 0$ and $3x_2 - x_3 + x_4 + x_5 = 0\}$ then dimension would go down by one. In W, only x_1, x_2 and x_3 can be chosen freely and so dim $(W) = 3$. As a general rule, each additional restriction reduces the dimension by 1. Of course, this additional restriction must be 'really new', that is, something not implied by the other restrictions. For example, if we add one more restriction to W, namely, $2x_1 + 6x_2 - 2x_3 + x_4 + (\pi + 2)x_5 = 0$, then the dimension of W does not change, because this new restriction can be expressed in terms of the earlier restrictions as, $(2x_1 - x_4 + \pi x_5) + 2(3x_2 - x_3 + x_4 + x_5) = 0$. This topic will be further taken up in the next section where we shall solve systems of linear equations.

For dimensions of subspaces we have the following result. It conforms to our intuition. But it is not trivial, because a similar result does not hold for all algebraic structures. For example, a subgroup of a finitely generated group need not be finitely generated. (A counter-example was given in Exercise (5.3.35)).

3.15 Proposition: Let W be a subspace of a finite dimensional vector space V. Then W is also finite dimensional and dim $(W) \leqslant$ dim V. Moreover, any basis of W can be extended to a basis for V.

Proof: Suppose W is not finite dimensional. Then no finite subset can span W. Let w_1 be any non-zero vector in W. Let $S_1 = \{w_1\}$. Then $L(S_1) \subsetneq W$. So there exists $w_2 \in W$ such that $w_2 \notin L(S_1)$. Let $S_2 = S_1 \cup \{w_2\}$. Then again $L(S_2) \subsetneq W$. So find $w_3 \in W - L(S_2)$. Let $S_3 = S_2 \cup \{w_3\}$. Continue this process indefinitely. It is easily seen, by induction, that each S_n is linearly independent. Now let B be a basis for V. Then $L(B) = V$. $S_n \subset L(\bar{B})$ for all n. But by Proposition (3.12), S_n is linearly dependent for all $n > |B|$. This is a contradiction. So W must be finite dimensional. Let A be a basis for W. Then A is linearly independent and $A \subset L(B)$. So again by Proposition (3.12), $|A| \leqslant |B|$. So dim $(W) \leqslant$ dim (V). For the last assertion, suppose $C = \{u_1, ..., u_m\}$ is a basis for W. We go on adding vectors to C one by one till we get a basis for V. Let $T_m = C$. If $L(T_m) = V$ then T_m is a basis for V. Otherwise let u_{m+1} be any vector in V which is not in $L(T_m)$. Let $T_{m+1} = T_m \cup \{u_{m+1}\}$. It is easily seen that T_{m+1} is linearly independent. If it spans V, it is a basis for V. If not, enlarge it to a linearly independent set T_{m+2}. This process must stop at T_n where $n = $ dim (V). For otherwise T_{n+1} would be a linearly independent set of cardinality $n + 1$ in the n-dimensional space V, contradicting Proposition (3.12) once again. Thus we have extended the given basis, C, for W to a basis T_n for V. ∎

The way the basis for V was constructed in this proof is exactly opposite to that based on Proposition (3.6). There we start with a spanning set S for V and go on shrinking it till it cannot be shrunk any further. This minimal spanning set is then a basis for V. On the other hand, in the construction here, we start with a linearly independent subset of V and go on enlarging it till it is impossible to enlarge it any more without making it linearly dependent. This maximal linearly independent subset of V is then a basis for V. As we just saw, this construction is convenient where subspaces are involved. Not surprisingly then, the earlier construction is more convenient for quotient spaces. Using it, it can be shown that dim $(V/W) \leqslant$ dim V. The exact relationship between the dimensions of V, W and V/W is that dim $(V) = $ dim $(W) + $ dim (V/W). This fact and a few other results about dimensions will be given as exercises. We conclude the present section with an application of the results proved in this section to field extensions. Throughout the remainder of this section let F denote a field, R will be a ring with identity containing F as a central subring. (That is, $F \subset R$, elements of F commute with those of R and the element 1 in F is also the identity for R. R itself need not be commutative.) If $\alpha \in R$ then $F[\alpha]$ will denote the smallest subring of R containing F and α. It is easily seen that $F[\alpha]$ consists of all elements of the form $f(\alpha)$ where $f(x) \in F[x]$. (cf. Exercise (2.20)). More generally, if $\alpha_1, \alpha_2, ..., \alpha_n \in R$ then $F[\alpha_1, ..., \alpha_n]$ will denote the smallest subring of R containing F and $\alpha_1, ..., \alpha_n$. We shall mostly deal with the case where R is a field. But some of the concepts, such as the following, make sense for any R.

3.16 Definition: An element $\alpha \in R$ is called **algebraic** over F if there exists a non-zero polynomial $p(x)$ over F such that $p(\alpha) = 0$. Otherwise it is called a **transcendental** element.

For example, let $F = \mathbf{Q}$, the field of rationals and let $R = \mathbf{R}$, the field of real numbers. Then $\sqrt{2}$, $\sqrt[3]{5}$ are algebraic over Q because $\sqrt{2}$ is a root of the polynomial $x^2 - 2$ which has rational coefficients and similarly $\sqrt[3]{5}$ is a root of $x^3 - 5$. Trivially, every element of Q is algebraic over itself. Given a real number, it is extremely difficult in general to decide whether it is algebraic or transcendental over Q. It is known that e and π are transcendental over Q. But about $e + \pi$ it is not known whether it is algebraic or not. In fact it is not even known whether $e + \pi$ is rational or irrational. On this background, any result which asserts that a certain number is algebraic is valuable. We shall prove that if R is a field extension of F and α, $\beta \in R$ are algebraic over F then $\alpha + \beta$ and $\alpha\beta$ are algebraic over F. There seems no obvious way of doing this. Even if we are given polynomials $p(x)$, $q(x)$ in $F[x]$ having α, β as roots, it is far from clear how to construct from these a polynomial having $\alpha + \beta$ as a root.

However, with a suitable characterisation, the result can be proved easily. The characterisation we need is the following:

3.17 Theorem: Let K be a field extension of F. Then an element $\alpha \in K$ is algebraic over F if and only if the vector space $F[\alpha]$ is finite dimensional over F.

Proof: Suppose α is algebraic over F. Then there exists a polynomial $p(x)$ of degree n (say) over F such that $p(\alpha) = 0$. Let $S = \{1, \alpha, \alpha^2, ..., \alpha^{n-1}\}$. We claim that the vector space $F[\alpha]$ is spanned by S. As noted above, a typical element of $F[\alpha]$ is of the form $f(\alpha)$ for some $f(x) \in F[x]$. Now $F[x]$ is a euclidean ring by Theorem (2.3). So there exist polynomials $q(x), r(x) \in F[x]$ such that $f(x) = p(x) \, q(x) + r(x)$ where $r(x) = 0$ or degree of $r(x)$ is less than n. This gives $f(\alpha) = p(\alpha)q(\alpha) + r(\alpha) = r(\alpha)$, since $p(\alpha) = 0$, (cf. Proposition (2.23).) If $r(x) = 0$ then $r(\alpha) = 0$. But then $f(\alpha) = 0$ and 0 is always in the linear span of S. So suppose $r(x)$ is not identically 0. Then $r(x)$ is a polynomial of degree $k \leqslant n - 1$. Let $r(x) = a_0 + a_1 x + ... + a_k x^k$ where $a_0, ..., a_k \in F$. Then $f(\alpha) = r(\alpha) = a_0 1 + a_1 \alpha + ... + a_k \alpha^k$. Since $k \leqslant n - 1$, we see that $f(\alpha)$ is in the linear span of S. So $F[\alpha]$ is finite dimensional over F.

Conversely suppose $F[\alpha]$ is finite dimensional over F. Let n be the dimension of $F[\alpha]$ over F. Then $F[\alpha]$ is spanned by a set S of cardinality n. By Proposition (3.12), the $n + 1$ elements $1, \alpha, \alpha^2, ..., \alpha^n$ in the span of S are linearly dependent. So there exist $a_0, a_1, ..., a_n \in F$, not all 0, such that $a_0 1 + a_1 \alpha + a_2 \alpha^2 + ... + a_n \alpha^n = 0$. But this means $p(\alpha) = 0$ where $p(x)$ is the polynomial $a_0 + a_1 x + ... + a_n x^n$ in $F[x]$. So α is algebraic over F. ∎

In order to use this characterisation effectively, we need a definition.

3.18 Definition: An extension field K of a field F is said to be a **finite extension** if K, as a vector space over F, is finite dimensional. The dimension of K over F is called the **degree** of K over F and is denoted by $[K: F]$.

It should be noted that here the adjective 'finite' refers to the dimension of K over F and not to the field K itself. The field K may very well be infinite. There is a reason why this dimension is called the 'degree' of K over F. In the proof of Theorem (2.26), we saw that if $f(x)$ is an irreducible polynomial of degree n in $F[x]$ then there exists an extension field K of F such that in K, $f(x)$ has a root α. Using essentially the same argument as in the last theorem, it is easy to show that the field K constructed there has dimension n over F, the same as the degree of the polynomial $f(x)$ from which it was constructed. Hence the name 'degree'.

As a consequence of the last theorem we have the following result:

3.19   Corollary: If K is a finite extension field of F then every element of K is algebraic over F.

Proof: Let $\alpha \in K$. Consider the subring $F[\alpha]$ of K generated by F and α. Then $F[\alpha]$ is a subspace of K, regarded as a vector space over F. By assumption, K is finite dimensional over F. So by Proposition (3.15), $F[\alpha]$ is finite dimensional over F. Hence by theorem (3.17), α is algebraic over F. ■

The leverage which Corollary (3.19) has over Theorem (3.17) in proving that α is algebraic over F is that it may not be so easy to show directly that $F[\alpha]$ is finite dimensional over F. But by Corollary (3.15) it is sufficient to find some finite extension, say, K, of F which contains α. As a further aid, the following theorem shows that in order to prove that a field extension is finite, we may use a chain of intermediate fields.

3.20   Theorem: Suppose F, L, M are fields with $F \subset L \subset M$. Then if L is a finite extension of F and M is a finite extension of L then M is a finite extension of F. Moreover, $[M : F] = [M : L] \cdot [L : F]$.

Proof: It is only the first assertion that will be of immediate use to us. The second assertion is much stronger. Curiously, however, it is easier to prove because it is also more specific and hence provides a clue for its proof. (See the comments after the solution of Problem $(2.2.11)\cdot$) So we shall proceed to prove the second assertion. Let $\{x_1, ..., x_m\}$ be a basis of L over F where $m = [L : F]$ and $\{y_1, ..., y_n\}$ be a basis of M over L where $n = [M : L]$. We want to find a basis of M over F having mn elements. The most natural candidate would be the set $S = \{x_i y_j, 1 \leqslant i \leqslant m, i \leqslant j \leqslant n\}$. We show, in fact, that S is a basis for M over F by showing that it spans M and secondly that it is linearly independent over F (which would also show that S has mn distinct elements). First, let $\alpha \in M$. Then there exist $b_1, ..., b_n \in L$ such that $\alpha = \sum_{j=1}^{n} b_j y_j$. Also for each $j = 1, ..., n$, there exist $a_{1j}, a_{2j}, ..., a_{mj} \in F$ such that $b_j = \sum_{i=1}^{m} a_{ij} x_i$. Substitution gives $\alpha = \sum\limits_{i=1}^{m} \sum\limits_{j=1}^{\;} a_{ij}x_iy_j$. Thus we have expressed α as a linear combination of the x_iy_j's with coefficients from F. This shows that the linear span of S (over F) is M. As for linear independence of S, suppose $\sum\limits_{i=1}^{m} \sum\limits_{j=1}^{n} b_{ij}x_iy_j = 0$ where $b_{ij} \in F$ for $i \leqslant i \leqslant m, 1 \leqslant j \leqslant n$. We now work backwards, that is, we group together the terms. For each $j = 1, \ldots, n$, let $c_j = \sum\limits_{i=1}^{m} a_{ij}x_i$. Then $c_j \in L$, also $\sum\limits_{j=1}^{n} c_jy_j = 0$. But $\{y_1, \ldots, y_n\}$ is linearly independent over L. So $c_j = 0$ for all $j = 1, \ldots, n$. This means $\sum\limits_{i=1}^{n} a_{ij}x_i = 0$. which, by linear independence of x_i's over F, forces each a_{ij} to be 0. So S is linearly independent over F. As noted before, this completes the proof. ∎

None of the results we have proved so far about field extensions is very profound. But, as it happens many times, simple results, when ingeneously combined, yield non-trivial results. We are now ready to prove that the sum and the product of two algebraic elements are algebraic, something which is not easy to prove directly.

3.21 **Theorem:** Let K be an extension field of a field F. Suppose $\alpha, \beta \in K$ are algebraic over F. Then $\alpha + \beta$ and $\alpha\beta$ are algebraic over F.

Proof: Let L, M be respectively $F[\alpha]$ and $F[\alpha, \beta]$. By definition, these are, respectively, the subrings of K generated by $F \cup \{\alpha\}$ and $F \cup \{\alpha, \beta\}$. But we claim they are in fact subfields of K. First consider $F[\alpha]$. Since α is algebraic over F, it satisfies some non-zero polynomial equation. Let I be the set of all polynomials $f(x)$ in $F[x]$ such that $f(\alpha) = 0$. It is easily seen that I is an ideal of $F[x]$. Also it is a non-zero ideal because it contains at least one non-zero polynomial. Now $F[x]$ is a euclidean ring by Theorem (2.3). Let $p(x)$ be a polynomial of minimum possible degree in I. Then by Theorem (2.4) (or rather its proof), I is generated by $p(x)$. That is, $I = (p(x))$.

The next step is to show that $p(x)$ is irreducible over F. Let if possible, $p(x) = f(x)g(x)$ be a proper factorization of $p(x)$ in $F[x]$. Then $f(x)$ and $g(x)$ are of lower degrees than $p(x)$. Now, $f(\alpha)g(\alpha) = p(\alpha) = 0$. But K is a field and so has no zero divisors. (This is the first time that we are crucially using that K is a field and not just a ring extension of F.) So $f(\alpha) = 0$ or $g(\alpha) = 0$. In either case we get a polynomial in $F[x]$ of a lower degree than $p(x)$ having α as a root, contradicting the choice of $p(x)$. So $p(x)$ is irreducible over $F[x]$. (Caution: $p(x)$ is not irreducible over $K[x]$. In fact since $p(\alpha) = 0$, $p(x)$ has a factor $x - \alpha$ by Theorem (2.24).).

The remainder is basically a duplication of the argument in Theorem (2.26). Define $\theta : F[x] \to K$ by $\theta(f(x)) = f(\alpha)$. Then θ is a ring homomorphism (see again Exercise (2.20)). By definition, I is the kernel of θ. Also the range of θ is $F[\alpha]$. So by the fundamental theorem about ring homomorphisms (Exercise (1.22)), $F[\alpha]$, as a ring, is isomorphic to the quotient ring $F[x]/I$. But I is generated by $p(x)$, which is a prime element in $F[x]$. So by Theorem (2.21), $F[x]/I$ is a field. Hence $F[\alpha]$, which is isomorphic to it is also a field.

Therefore, L (that is, $F[\alpha]$) is not only a subring of K but a subfield of K. By theorem (3.17), L is a finite extension of F. Next, consider $M = F[\alpha, \beta]$. Clearly $M = [F[\alpha]] [\beta] = L [\beta]$. Now β is algebraic over F and hence all the more so over L (since $F \subset L$). So by exactly the same argument as was used in showing that $F[\alpha]$ is a field, it follows that $L[\beta]$, that is, M is a subfield of K. Also it is a finite extension of L. So by Theorem (3.20), M is a finite field extension of F. Hence by Corollary (3.19), (applied to M and not to K), every element of M is algebraic over F. In particular, $\alpha + \beta$, $\alpha\beta$ are elements of M (since $\alpha, \beta \in M$). So they are algebraic over F. ∎

Exercises

3.1 In the definition of a vector space, prove that it is enough to assume that $(V, +)$ is a group (not necessarily an abelian group). In other words show that in presence of all other axioms, the commutativity of $+$ can be proved and need not be assumed as an axiom (Hint: See Exercise (1.1)).

3.2 Which of the following subsets are subspaces of $\mathbb{R}^3$?

    (i) $\{(x_1, x_2, x_3) \in \mathbb{R}^3 : 2x_1 + 3x_2 + 4x_3 = 0\}$
    (ii) $\{(x_1, x_2, x_3) \in \mathbb{R}^3 : 2x_1 + 3x_2 + 4x_3 = 1\}$
    (iii) $\{(x_1, x_2, x_3) \in \mathbb{R}^3 : x_1 > 0, \, x_2 > 0, \, x_3 < 0\}$
    (iv) $\{(x_1, x_2, x_3) \in \mathbb{R}^3 : x_1, x_2, x_3 \text{ are rational}\}$
    (v) $\{(x_1, x_2, x_3) \in \mathbb{R}^3 : x_1^2 + x_2^2 + x_3^2 = 1\}$.

3.3 Prove that a subset W of a vector space V is a subspace of V iff W is closed under linear combinations, that is, all linear combinations of elements of W are in W.

3.4 Let V be a vector space over a field F. Prove :
    (a) the intersection of any family of subspaces of V is again a subspace of V.
    (b) if F is infinite, then the union of finitely many subspaces of V is a subspace of V only if one of these subspaces contains all others.

3.5 Let $T: V \to W$ be a function, where V, W are vector spaces over the same field F. Prove that T is a linear transformation iff T preserves linear combinations, that is, for all $v_1, ..., v_n \in V$ and

$\lambda_1, ..., \lambda_n \in F$, $T(\lambda_1 v_1 + ... + \lambda_n v_n) = \lambda_1 T(v_1) + ... + \lambda_n T(v_n)$. Hence the name 'linear' transformation).

3.6 Let W_1, W_2 be vector spaces over a field F. Let $W = W_1 \times W_2$. Let π_1, π_2 be projections of W onto W_1, W_2 respectively, that is, $\pi_1(x_1, x_2) = x_1$ and $\pi_2(x_1, x_2) = x_2$ for $(x_1, x_2) \in W$. Prove that π_1, π_2 are linear transformations. What are their kernels and ranges ?

3.7 If V, W_1, W_2 are vector spaces over a field F, prove that $\text{Hom}_F(V, W_1 \times W_2)$ is isomorphic to $\text{Hom}_F(V, W_1) \times \text{Hom}_F(V, W_2)$.

3.8 Prove that if $T : V \to W$ is a linear transformation with kernel K and range R then R is isomorphic to the quotient space V/K. (The kernel of a linear transformation is also often called its **null space**.) Let $v_0 \in V$ and $w_0 = T(v_0)$. Prove that the set of solutions to the equation $T(v) = w_0$ is precisely $K + \{v_0\}$. (cf. Proposition (5.3.6).)

3.9 Give an alternate proof of Proposition (3.6) as follows. Let $S = \{v_1, v_2, ..., v_k\}$. Go on picking these vectors one-by-one. After v_i is picked, see if v_i is in the linear span of $\{v_1, ..., v_{i-1}\}$. If it is not, include it in T. Prove that the set T that is obtained after all elements of S are picked is linearly independent and has the same span as S. (This proof is in the line of the construction in the proof of Proposition (3.15). From the point of view of mechanical implementation, this proof is better than the proof in the text.)

3.10 Let $V = \{(x_1, x_2, x_3, x_4, x_5) \in \mathbb{R}^5 : 2x_1 - x_2 + \pi x_3 = 0, x_2 + x_3 - x_4 - x_5 = 0\}$. Obtain a basis for V.

3.11 Let V be a vector space of dimension n over a field F. Let S be a finite subset of V. Prove that any two of the following three statements, taken together, imply the third

   (i) $|S| = n$.
   (ii) S spans V.
   (iii) S is linearly independent.

3.12 Using the fact $\sqrt{2}$, $\sqrt{3}$, $\sqrt{6}$ and $\sqrt{5}$ are all irrational show that:

   (i) $\{1, \sqrt{2}\}$ is linearly independent over $\mathbb{Q}$,
   (ii) $\{1, \sqrt{2}, \sqrt{3}\}$ is also linearly independent over $\mathbb{Q}$,
   (iii) $\{1, \sqrt{2}, \sqrt{3}, \sqrt{6}\}$ is linearly independent over $\mathbb{Q}$ and
   (iv) $\{1, \sqrt{2}, \sqrt{3}, \sqrt{5}\}$ is linearly independent over $\mathbb{Q}$.

3.13 Let X, Y be subspaces of a finite dimensional vector space V. Let $X + Y = \{x + y : x \in X, y \in Y\}$. Prove that $X + Y$ is a subspace of V and that it is generated by $X \cup Y$. Prove further that $\dim(X+Y) = \dim(X) + \dim(Y) - \dim(X \cap Y)$. (Hint: Start with a basis for $X \cap Y$. Extend it to a basis for X and also to a basis for Y. Show that the union of these two bases is a basis for $X + Y$.)

3.14　Prove that $\dim (V \times W) = \dim V + \dim W$.

3.15　Let W be a subspace of a finite-dimensional vector space V. Prove that if $\{v_1 + W, v_2 + W, ..., v_k + W\}$ is a basis for V/W then $\{v_1, ..., v_k\}$ is linearly independent. Hence show that $\dim (V) = \dim (W) + \dim (V/W)$.

3.16　Let $T: V \to W$ be a linear transformation of finite dimensional vector spaces, with kernel K and range R. The dimensions of K and R are called respectively the **nullity** and the **rank** of T and are often denoted by $n(T)$ and $r(T)$. Prove that $n(T) + r(T) = \dim (V)$. If $T_1: V \to W$ and $T_2: W \to X$ are linear transformations, prove that $r(T_2 \circ T_1) \leqslant r(T_2)$ and also $r(T_2 \circ T_1) \leqslant r(T_1)$.

3.17　Prove that a linear transformation $T: V \to W$ is one-to-one iff it takes linearly independent sets in V to linearly independent sets in W.

3.18　Let V, W be vector spaces of equal dimension. Prove that a linear transformation $T: V \to W$ is one-to-one iff it is onto. (Thus, to some extent, finite dimensional vector spaces behave like finite sets.)

3.19　Let X be a subspace of a finite dimensional vector space V. Prove that there exists a subspace Y of V such that (i) $X \cap Y = \{0\}$ and (ii) $\dim (Y) = \dim (V) - \dim (X)$. For any such space Y prove that $X + Y = V$ and $X \times Y$ is isomorphic to V. (In a sense, X and Y are complementary subspaces. However, for a given X, Y need not be unique in general.)

3.20　Let G be an abelian group. Prove that G can be regarded as a module over $\mathbb{Z}$, the ring of integers.

3.21　Let R be a ring and suppose M is a left ideal of R. Prove that M can be regarded as a left module over R. Show by an example that it may happen that $rm = 0$ but neither r nor m is 0.

3.22　Let K be a field extension of a field F and $[K:F] = n$. Let $\alpha \in K$ be a root of an irreducible polynomial $p(x)$ in $F[x]$. Prove that the degree of $p(x)$ is a divisor of n. If n is prime, show that $F[\alpha]$ is either F or K.

3.23　Prove that the set of those real numbers which are algebraic over $\mathbb{Q}$ is countable. (Hint: Prove that there are only countably many polynomials in $\mathbb{Q}[x]$. Each of them can have only finitely many roots.)

3.24　Prove that transcendental real numbers exist and in fact their set is uncountable. (See Exercise (2.2.24) and the comments on it.)

3.25　Let $V_1, V_2, ..., V_k$ and W be vector spaces. A function $f: V_1 \times V_2 \times ... \times V_k \to W$ is called **multilinear** if for every $i = 1, ..., k, f(v_1, ..., v_{i-1}, \alpha u_i + \beta v_i, v_{i+1}, ..., v_k) = \alpha f(v_1, ..., u_i, ..., v_k) + \beta f(v_1, ..., v_i, ..., v_k)$ for all $v_1 \in V_1, ..., u_i, v_i \in V_i, ..., v_k \in V_k$. In other words, a multilinear function is linear in each variable when all other variables are held constant. Compute $f(u_1 + v_1, ..., u_k + v_k)$ for a multilinear function f. For $k = 2$, a multilinear function is called **bilinear**. If F is a field, prove that the function $f : F^n \times F^n \to F$ defined by $f((x_1, ..., x_n), (y_1, ..., y_n)) = x_1 y_1 + x_2 y_2 + ... + x_n y_n$ is bilinear.

3.26 Although we are not studying inner product spaces, we assume the reader is familiar with the definitions and properties of the usual dot and cross product of vectors in $\mathbb{R}^3$. For the purpose of this exercise we shall denote vectors in $\mathbb{R}^3$ by bold faced letters. **i, j, k** will denote three unit vectors forming a right handed, orthogonal system as usual. If $\mathbf{u} = u_1\mathbf{i} + u_2\mathbf{j} + u_3\mathbf{k}$ and $v = v_1\mathbf{i} + v_2\mathbf{j} + v_3\mathbf{k} \in \mathbb{R}^3$ then $\mathbf{u} \cdot \mathbf{v}$ denotes their dot product $u_1 v_1 + u_2 v_2 + u_3 v_3$ and $\mathbf{u} \times \mathbf{v}$ denotes their cross product, $(u_2 v_3 - u_3 v_2)\,\mathbf{i} + (u_3 v_1 - u_1 v_3)\,\mathbf{j} + (u_1 v_2 - u_2 v_1)\,\mathbf{k}$. Given a quaternion $a = a_0 + a_1 i + a_2 j + a_3 k$, we associate to it the ordered pair $(a_0, \mathbf{a})$ where $\mathbf{a}$ is the vector $a_1\mathbf{i} + a_2\mathbf{j} + a_3\mathbf{k}$. This sets up a bijection between $\mathbb{Q}$ and $\mathbb{R} \times \mathbb{R}^3$. In terms of this bijection, prove that the quaternionic multiplication takes the form $(a_0, \mathbf{a}) \cdot (b_0, \mathbf{b}) = (a_0 b_0 - \mathbf{a} \cdot \mathbf{b}, \; a_0 \mathbf{b} + b_0 \mathbf{a} + \mathbf{a} \times \mathbf{b})$. This is highly analogous to complex multiplication. (The quaternionic addition shows no remarkable change of form, $(a_0, \mathbf{a}) + (b_0, \mathbf{b})$ is simply $(a_0 + b_0, \mathbf{a} + \mathbf{b})$.) In this formulation the norm of a, $N(a)$ is simply $\sqrt{|a_0|^2 + |\mathbf{a}|^2}$, note again the close formal resemblance with the absolute value of a complex number. Using this representation of quaternions and the various identities about the dot and the cross product (which may be found in any elementary textbook on vector algebra), verify that $(Q, +, \cdot)$ is a division ring. Also give an easier proof of the multiplicative property of the norm, that is, $N(ab) = N(a)\,N(b)$ for all $a, b \in Q$. This exercise also shows why the quaternions are denoted in a peculiar manner using the symbols i, j, k and not merely as ordered 4-tuples (a_0, a_1, a_2, a_3) which seems more logical.

3.27 Let G be an abelian group in which $2x = 0$ for all $x \in G$. Prove that G can be considered as a vector space over the field $\mathbb{Z}_2$ in a unique way. (This generalises Example (6) of vector spaces.)

3.28 Let G be any abelian group. For any positive integer n, let $nG = \{ng : g \in G\}$. Prove that nG is a subgroup of G and that in the quotient group G/nG, every element has an order which is a divisor of n.

3.29 Let S be a finite set and G be the free abelian group on S (see Exercise (5.3.32)). Prove that $G/2G$ is a vector space of dimension $|S|$ over $\mathbb{Z}_2$.

3.30 Let S, T be finite sets and let G, H be free abelian groups on S, T respectively. Prove that G is isomorphic to H if and only if $|S| = |T|$. (In view of this exercise the **rank** of a finitely generated free abelian group G is well-defined. It is the cardinality of any finite set S such that G is isomorphic to the free abelian group on the set S.)

3.31   Prove that the last exercise also holds for free groups. [Hint: Combine the last exercise with Exercise (5.3.31). Thus the **rank** of a finitely generated free group is also well-defined; it equals the number of elements in any set which freely generates it]

Notes and Guide to Literature

The results in this section are elementary because we are confining ourselves to finite dimensional vector spaces. For existence of a basis for an infinite dimensional space, see Lang [1], where the reader will also find more about modules. The theory of modules over a ring depends crucially on the properties of that ring. (Vector spaces correspond to the case where the ring is a field.) See, for example, Herstein [2].

For a proof of transcendence of e see Herstein [1]. It is much more difficult to prove that π is transcendental. This was proved by Lindemann in 1882. See Niven [1]. While proving that a particular number is transcendental can be quite difficult, interestingly enough, the mere existence of such numbers (without showing a particular number to be transcendental) can be established by a simple cardinality argument, as in Exercise (3.24). This is another instance of proving existence through abundance.

The inner product spaces, meant to abstract the geometric structure of euclidean spaces, form the starting point of a large area of a branch of mathematics called functional analysis. In this, we study generalisations of such topics as Fourier series and Bessel functions. There are numerous treatises on functional analysis, for example Rudin [2] or Limaye [1].

The aspects of euclidean geometry which do not depend on the concepts of distance and angles can be generalised to vector spaces over fields other than the field of real numbers. Strange as it may seem, even after these two basic concepts are removed, a surprisingly large amount of geometry can be still salvaged and leads to the theory of what are called projective geometries. We shall briefly touch them while studying applications of finite fields (see the Epilogue).

Multilinear functions are more popularly called **multilinear forms.** Bilinear forms are especially important. See Lang [1] or Hoffman and Kunze [1].

4.   Matrices and Determinants

In Section 1, we introduced matrices to give examples of rings. In this section we study one of the standard applications of matrices, namely, to provide a handy representation for linear transformations of finite dimensional vector spaces. In the study of properties of a square matrix, the determinant is an invaluable tool. We assume that the reader is familiar with the properties of 3×3 determinants. We shall show how they can be extended to

$n \times n$ determinants, using the permutation groups, studied in the last chapter. We shall also study the concept of the rank of a matrix, which is also intimately related to the solution of a system of linear equations. Throughout this section we shall consider only finite dimensional vector spaces over a field F. If n is a positive integer, then the vector space F^n, as defined in the last section, consists of all ordered n-tuples $(x_1, ..., x_n)$ of elements of F^n. Each such element can be thought of as a matrix with one row (and n columns) and hence is also called a **row vector**. Occasionally, we shall find it convenient to represent elements of F^n by **column vectors**, that is by $n \times 1$

matrices $\begin{pmatrix} x_1 \\ x_2 \\ \vdots \\ x_n \end{pmatrix}$. We can do so because the vector space of all row vectors

(of length n) is clearly isomorphic to the vector space of all column vectors of length n. An $m \times n$ matrix may be thought of either as an ordered collection of its row vectors or as an ordered collection of its column vectors.

We begin with a simple proposition whose proof is a matter of straightforward verification and hence left to the reader.

4.1 Proposition: Let $A = (a_{ij})$ be an $m \times n$ matrix over a field F. Let F^m, F^n be respectively the vector spaces of column vectors of lengths m and

n over F. Define $T: F^n \to F^m$ by $T(x) = Ax$ for $x \in F^n$. (If $x = \begin{pmatrix} x_1 \\ \vdots \\ x_n \end{pmatrix}$ then

$$Ax = \begin{pmatrix} y_1 \\ \vdots \\ y_m \end{pmatrix}$$

where $\begin{pmatrix} y_1 \\ \vdots \\ \vdots \\ y_m \end{pmatrix} = \begin{pmatrix} a_{11} & a_{12}...a_{1n} \\ a_{21} & a_{22}...a_{2n} \\ \cdots\cdots\cdots \\ a_{m1} & a_{m2}...a_{mn} \end{pmatrix} \begin{pmatrix} x_1 \\ \vdots \\ x_n \end{pmatrix}$.)

Then T is a linear a transformation from F^n to F^m. ■

Thus we see that every matrix gives rise to a linear transformation. In the last section we saw that F^n has $\{e_1, e_2, ..., e_n\}$ as a standard basis, where e_i is a column vector of length n which has 1 in the i-th place and 0 everywhere else. Note that, under the linear transformation T, the image of e_i is precisely the i-th column vector of the matrix A. Let us denote this column vector by v_i. Now a typical element, say u, of F^n is of the form $\lambda_1 e_1 + ... + \lambda_n e_n$ for $\lambda_1, ..., \lambda_n \in F$. Then $T(u) = \lambda_1 T(e_1) + ... + \lambda_n T(e_n) = \lambda_1 v_1 + ... + \lambda_n v_n$. Therefore, if R denotes the range of T, then R is the subspace of F^m spanned by the set $\{v_1, ..., v_n\}$. This set need not be linearly independent. However, by Proposition (3.6), we can get a linearly independent subset, say,

$\{v_{i_1}, ..., v_{i_r}\}$ of it which will be a basis for R. Then dimension of R is r. The dimension of the range of a linear transformation is called its **rank**. (Exercise (3.16)). We therefore have the following result for the linear transformation constructed in the last proposition.

4.2 Proposition: The rank of the linear transformation T obtained from the matrix A as above equals the maximum number of linearly independent columns of A. ∎

This result is often expressed by saying that the rank of T equals the column rank of A, where the **column rank** of a matrix is defined as the maximum number of its linearly independent columns. Similarly the **row rank** of a matrix A is defined as the maximum number of linearly independent rows in it. The row rank can also be interpreted as the rank of a suitable linear transformation as follows:

4.3 Proposition: Let A be an $m \times n$ matrix over a field F. Let F^m, F^n be the vector spaces of row vectors of lengths m, n respectively over F. Define $T: F^m \to F^n$ by $T(x) = xA$ for $x = (x_1, ..., x_m) \in F^m$. Then T is a linear transformation. Moreover, the rank of T is the row rank of A. ∎

The proof is completely dual to that of the last two propositions. This duality, which results by an interchange of rows and columns is called transposition. Formally, if $A = (a_{ij})$ is an $m \times n$ matrix over any ring R then its **transpose**, denoted by A' (or sometimes by A^T or A^t) is defined as the $n \times m$ matrix whose entry in the ith row and jth column is a_{ji}. In symbols, $A' = (a'_{ij})$ where $a'_{ij} = a_{ji}$ for $i = 1, ..., n$; $j = 1, ..., m$. For example, the transpose of the matrix

$$\begin{pmatrix} 1 & 2 & 0 & -1 & 4 \\ \pi & e & -2 & \sqrt{3} & 9 \\ 0 & 1 & 2 & 13 & 1/2 \end{pmatrix} \text{ is the } 5 \times 3 \text{ matrix } \begin{pmatrix} 1 & \pi & 0 \\ 2 & e & 1 \\ 0 & -2 & 2 \\ -1 & \sqrt{3} & 13 \\ 4 & 9 & 1/2 \end{pmatrix}$$

Similarly the transpose of a row vector is a column vector and vice versa. A few simple properties of transposes are listed below. The simple proof is again omitted.

4.4 Proposition: Let A, B be two $m \times n$ matrices over a commutative ring R. Then for any α, $\beta \in R$, $(\alpha A + \beta B)' = \alpha A' + \beta B'$. If C is any $n \times p$ matrix over R then the matrix products AC and $C'A'$ are both defined and moreover, $(AC)' = C'A'$. Finally, for every matrix A, $(A')' = A$. ∎

Transposition is a convenient device by which concepts about rows of a matrix can be translated to concepts about columns of the transpose, and vice versa. For example, the row rank of a matrix A equals the column rank of A'. What is not so obvious is that the row rank of A equals the column rank of A itself. This common value is then called its rank. A direct proof of this fact is somewhat awkward. After studying determinants we would be in a position to prove it.

Propositions (4.1) and (4.3) show how matrices give rise to linear transformations. We now want to proceed the other way. That is, given a linear transformation $T: V \to W$ where V, W are vector spaces of dimensions n, m respectively, we want to show that T can be represented by an $m \times n$ matrix over F. This representation makes it possible to study linear transformations through matrices (which are easier to operate on machines). But besides its utility, the very existence of such a representation is remarkable, especially because there is no similar representation for homomorphisms of other algebraic structures such as groups and rings. What is so nice about vector spaces? The answer is the presence of bases. Every element of a vector space can be uniquely expressed in terms of the basis elements. In this respect vector spaces behave like free groups, where every element can be uniquely expressed in terms of the generators (see Chapter 5, Section 3, see also Exercise (5.3.33) where a similar expression is obtained for elements of a free abelian group). Because of this property of a basis, we have the following simple but useful result, which is the analogue of Theorem (5.3.20):

4.5 Theorem: Let B be a basis for a finite dimensional vector space V over a field F. Let W be any vector space over F. Then any function $f: B \to W$ can be uniquely extended to a linear transformation from V to W. (Note, in particular, that this says that if two linear transformations agree on a basis, then they agree everywhere.)

Proof: Let $v_1, \ldots, v_n$ be the distinct element of B. Then $f(v_i)$ is given as an element of W for each $i = 1, \ldots, n$ and we want a linear transformation $T: V \to W$ such that $T(v_i) = f(v_i)$ for $i = 1, \ldots, n$. Let $v \in V$. By Theorem (3.9), there exist unique $\lambda_1, \ldots, \lambda_n \in F$ such that

$$v = \sum_{i=1}^{n} \lambda_i v_i.$$

We define

$$T(v) = \sum_{i=1}^{n} \lambda_i f(v_i).$$

Since the λ_i's are unique there is no problem about T being well-defined. To prove T is a linear transformation, suppose, $v, w \in V$ with $v = \sum_{i=1}^{n} \lambda_i v_i$ and

$$w = \sum_{i=1}^{n} \mu_i v_i, \ \lambda_i, \mu_i \in F, \ i = 1, \ldots, n.$$

Then $v + w = \sum\limits_{i=1}^{n} (\lambda_i + \mu_i)v_i$. So $T(v + w) = \sum\limits_{i=1}^{n} (\lambda_i + \mu_i)f(v_i) = \sum\limits_{i=1}^{n} \lambda_i f(v_i)$

$+ \sum\limits_{i=1}^{n} \mu_i f(v_i) = T(u) + T(v)$. Similarly for any $\alpha \in F$, $T(\alpha v) = \sum\limits_{i=1}^{n} \alpha \lambda_i f(v_i)$

$= \alpha \sum\limits_{i=1}^{n} \lambda_i f(v_i) = \alpha\, T(v)$. Thus T is a linear transformation. Moreover, for for every i, $v_i = 0v_1 + \ldots + 0v_{i-1} + 1v_i + 0v_{i+1} + \ldots + 0v_n$. So $T(v_i) = 0f(v_1) + \ldots + 0f(v_{i-1}) + 1f(v_i) + 0f(v_{i+1}) + \ldots + 0f(v_n) = f(v_i)$. Thus T is a desired extension of f. As for uniqueness of T, suppose $S: V \to W$ is another linear extension of f. Then for $v = \sum\limits_{i=1}^{n} \lambda_i v_i$, by linearity of S (see Exercise (3.5)),

$S(v) = \sum\limits_{i=1}^{n} \lambda_i S(v_i) = \sum\limits_{i=1}^{n} \lambda_i f(v_i)$, since $S(v_i) = f(v_i)$ for all $i = 1, \ldots, n$. This shows $S(v) = T(v)$ for all $v \in V$. ∎

Verbally, a linear transformation is uniquely determined by its action on a basis. Thus, to define a linear transformation from V to W it suffices to take any basis B for V and any function $f: B \to W$. We are free to choose this function anyway we like. A judicious choice leads to interesting consequences, one of which is given below:

4.6   Theorem: Let V, W be vector spaces of dimensions n, m respectively. Then $L(V, W)$, that is, the vector space of all linear transformations from V into W has dimension nm.

Proof: Let $B = \{v_1, \ldots, v_n\}$ and $C = \{w_1, \ldots, w_m\}$ bases for V, W respectively. For each $i = 1, \ldots, n$ and $j = 1, \ldots, m$ let $f_{ij}: B \to W$ be the function which takes v_i to w_j and v_k to 0 for $k \neq i$. (f_{ij} is somewhat like the characteristic function of the singleton set $\{v_i\}$, except that the value at v_i is not 1 but w_j.) By the last theorem, each f_{ij} can be uniquely extended to a linear transformation T_{ij} from V to W. We claim that the set $\{T_{ij}: 1 \leqslant i \leqslant n, 1 \leqslant j \leqslant m\}$ of nm linear transformations is a basis for $L(V, W)$. This would of course imply that $L(V, W)$ has dimension nm.

First, let $T \in L(V, W)$. Then T is a linear transformation from V to W. For each $i = 1, \ldots, n$, $T(v_i)$ is an element of W. Since $\{w_1, \ldots, w_m\}$ is a basis for W, there exist unique scalars $a_{1i}, a_{2i}, \ldots, a_{mi}$ such that

$$T(v_i) = \sum\limits_{j=1}^{m} a_{ji} w_j.$$

Now let S be the linear transformation $\sum\limits_{j=1}^{m} \sum\limits_{i=1}^{n} a_{ji} T_{ij}$. We claim that $S = T$. For this, in view of the last theorem again, it suffices to show that for every $k = 1, \ldots, n$, $S(v_k) = T(v_k)$. Now, $S(v_k) = \sum\limits_{j=1}^{m} \sum\limits_{i=1}^{n} a_{ji} T_{ij}(v_k) = \sum\limits_{j=1}^{m} (a_{jk} w_j)$ because $T_{ij}(v_k) = 0$ for $i \neq k$ and for $i = k$, $T_{ij}(v_k) = w_j$. So $S(v_k) = T(v_k)$.

Hence $S = T$. That is, T is a linear combination of the T_{ij}'s, showing that $L(V, W)$ is spanned by them.

We only have to show now that the T_{ij}'s form a linearly independent set. Suppose a_{ij} are scalars with $\sum\limits_{j=1}^{m} \sum\limits_{i=1}^{n} a_{ji}T_{ij} = 0$. Then for every k,

the value of $\sum\limits_{j=1}^{m} \sum\limits_{i=1}^{n} a_{ji}T_{ij}$ on the element v_k is 0. By the same computation

as before, this value is $\sum\limits_{j=1}^{m} a_{jk}w_j$. But $\{w_1, ..., w_m\}$ is linearly independent.

So $a_{jk} = 0$ for all $j = 1, ..., m$. Since this holds for every $k = 1, ..., n$ it follows that $a_{ji} = 0$ for all i, j. Thus the set $\{T_{ij} : i = 1, ..., n; j = 1, ..., m\}$ is linearly independent. ∎

In particular we have the following corollary:

4.7 Corollary: The dimension of a space is the same as that of its dual.

Proof: We simply recall that the dual of a vector space V over a field F is $L(V, F)$ where F is regarded as a vector space over itself. Since the dimension of F (over itself) is 1, the result follows by taking $m = 1$, in the last theorem. ∎

The nm numbers a_{ij} constructed in the proof of Theorem (4.6) determine the transformation T completely in terms of the bases B and C. We can get an $m \times n$ matrix by arranging these numbers into m rows and n columns. The matrix so obtained will give a complete representation of the transformation T. However, this matrix depends not only on the transformation but also on the bases B and C. If we change either one of them, we could get a different matrix for the same transformation. Actually, the order of the elements in B and C also matters. If we reshuffle the elements of B then the columns of A would be permuted and if we reshuffle the elements of C then the rows of A will be permuted.

To assign a unique matrix to a linear transformation, we introduce the concept of an ordered basis. This is the same as a basis, except that the order of the elements matters. Formally, an **ordered basis** for a vector space V over a field may be defined as a finite sequence $(v_1, v_2,...,v_n)$ of distinct elements of V such that the set $\{v_1,...v_n\}$ is a basis for V over F. Thus, $(v_1, v_2,..., v_n)$ is not the same ordered basis as $(v_2, v_1,..., v_n)$ even though, as sets, $\{v_1, v_2,..., v_n\}$ is the same as $\{v_2, v_1,..., v_n\}$. With this concept, we are now ready to give the definition of the matrix associated with a linear transformation.

4.8 Definition: Let $(v_1,..., v_n)$ and $(w_1,..., w_m)$ be ordered bases for vector spaces V, W respectively over a field F. Let $T: V \to W$ be a linear transformation. For each $k = 1,..., n$ let $T(v_k) = \sum\limits_{j=1}^{m} a_{jk} w_j$. Then the $m \times n$ matrix

$A = (a_{ij})$ is called the **matrix** of T w.r.t. the ordered bases $(v_1,..., v_n)$ and $(w_1,..., w_m)$.

The easiest way to remember this matrix is to note that its kth column is obtained by writing down the coefficients of $T(v_k)$ when it is expressed as a linear combination of $w_1,..., w_m$. If we start with an $m \times n$ matrix A and construct the linear function $T: F^n \rightarrow F^m$ given by Proposition (4.1), then the matrix of T w.r.t. the standard ordered bases for F^n and F^m comes out to be A itself. But with a different choice of bases, it will be different. To illustrate this, we work out a numerical example.

4.9 Problem: Define $T: \mathbb{R}^2 \rightarrow \mathbb{R}^3$ by $T\begin{pmatrix} x_1 \\ x_2 \end{pmatrix} = \begin{pmatrix} y_1 \\ y_2 \\ y_3 \end{pmatrix}$ where $y_1 = 2x_1 - x_2$, $y_2 = -x_1 + x_2$ and $y_3 = x_1 + x_2$. Find the matrix of T w.r.t. the ordered basis $\left(\begin{pmatrix} 1 \\ 2 \end{pmatrix}, \begin{pmatrix} -1 \\ 1 \end{pmatrix} \right)$ for $\mathbb{R}^2$ and the ordered basis $\left(\begin{pmatrix} 1 \\ 0 \\ 1 \end{pmatrix}, \begin{pmatrix} 2 \\ -1 \\ 2 \end{pmatrix}, \begin{pmatrix} 3 \\ 1 \\ 0 \end{pmatrix} \right)$ for $\mathbb{R}^3$.

Solution: The matrix of T w.r.t. the standard ordered bases would be $\begin{pmatrix} 2 & -1 \\ -1 & 1 \\ 1 & 1 \end{pmatrix}$. But we have to find the matrix w.r.t. the given bases. First $T\begin{pmatrix} 1 \\ 2 \end{pmatrix} = \begin{pmatrix} 2 & -1 \\ -1 & 1 \\ 1 & 1 \end{pmatrix}\begin{pmatrix} 1 \\ 2 \end{pmatrix} = \begin{pmatrix} 0 \\ 1 \\ 3 \end{pmatrix}$. We want to express this as a linear combination of the given vector in $\mathbb{R}^3$. Suppose the coefficients are α, β, γ. Then we get three equations:

$$\alpha + 2\beta + 3\gamma = 0$$

$$-\beta + \gamma = 1$$

$$\alpha + 2\beta = 3$$

Systematic methods for solving such equations will be discussed later on. For the moment we solve them heuristically to get $\alpha = 7$, $\beta = -2$,

$\gamma = -1$. This gives $\begin{pmatrix} 7 \\ -2 \\ -1 \end{pmatrix}$ as the first column of the desired matrix. A

similar procedure (which involves solving the system $\alpha + 2\beta + 3\gamma = -3$,

$-\beta + \gamma = 2$ and $\alpha + 2\beta = 0$) gives the second column as $\begin{pmatrix} 2 \\ -1 \\ -1 \end{pmatrix}$. So

the matrix of T w.r.t. the given ordered bases is $\begin{pmatrix} 7 & 2 \\ -2 & -1 \\ -1 & -1 \end{pmatrix}$. ▮

For a fixed choice of ordered bases for V and W the correspondence from $L(V, W)$ to $M_{m, n}(F)$ which associates to $T \in L(V, W)$ its matrix A, is a bijection. The inverse of this bijection is determined by essentially the same construction as used in Proposition (4.1). Once we know the matrix, A, of a linear transformation $T: V \to W$ w.r.t. ordered bases $(v_1, v_2, \ldots, v_n)$ and $(w_1, \ldots, w_m)$, it is easy to evaluate $T(v)$ for any $v \in V$. First write v as $\sum\limits_{i=1}^{n} \lambda_i v_i$ where $\lambda_1, \ldots, \lambda_n \in F$. Then $T(v) = \sum\limits_{j=1}^{m} \mu_j w_j$ where the coefficients $\mu_1 \ldots, \mu_m$ are given most conveniently by the matrix equation,

$$\begin{pmatrix} \mu_1 \\ \vdots \\ \mu_m \end{pmatrix} = A \begin{pmatrix} \lambda_1 \\ \vdots \\ \lambda_n \end{pmatrix}.$$

The proof of this fact is simple and hence omitted.

The importance of the assignment of matrices to linear transformations comes from the fact it is compatible with addition and scalar multiplication. Moreover, matrix multiplication corresponds to composition of linear transformations. We state these results precisely in the following proposition.

4.10 Proposition: Let V, W, X be vector spaces with ordered bases $(v_1, \ldots, v_n)$, $(w_1, \ldots, w_m)$ and $(x_1, \ldots, x_p)$ respectively. Let $T: V \to W$, $S: V \to W$ and $U: W \to X$ be linear transformations with matrices A, B, C respectively w.r.t. these ordered bases. Then

(i) $A + B$ is the matrix of the linear transformation $T + S: V \to W$
(ii) For every scalar λ, λA is the matrix of λT and
(iii) CA is the matrix of $U \circ T : V \to X$.

(All matrices are w.r.t. the given ordered bases.)

Proof: The first two statements follow immediately from the definitions. We prove the third statement. Let $D = (d_{ij})$ be the matrix of the composite linear transformation $U \circ T$ from V to X. We have to show that $D = CA$. Now for any $j = 1, 2, ..., n$, we have, by definition, $(U \circ T)(v_j) = \sum_{l=1}^{p} d_{lj} x_l$. But $(U \circ T)(v_j)$ also equals $U(T(v_j))$. From $T(v_j) = \sum_{k=1}^{m} a_{kj} w_k$ and the linearity of U, $U(T(v_j))$ equals $\sum_{k=1}^{m} a_{kj} U(w_k)$. But, once again $U(w_k) = \sum_{i=1}^{p} c_{ik} x_l$. So we get $\sum_{l=1}^{p} d_{lj} x_i = \sum_{k=1}^{m} a_{kj} (\sum_{i=1}^{p} c_{ik} x_l)$. Since $\{x_1, ..., x_p\}$ is linearly independent, the coefficients of x_i ($i = 1, ..., p$) on both the sides must be equal. So $d_{ij} = \sum_{k=1}^{p} c_{ik} a_{kj}$. But the right hand side is, by very definition, the (i, j)th entry in the product matrix CA. So $D = CA$ as was to be shown. ■

The preceding proposition may be used both ways. Sometimes a result is easier to prove for linear transformations than for matrices or vice versa. In such cases, we prove it where it is easier and transfer it to the other context. For example, associativity of composition of linear transformations is a simple general property of functions. Using it, we can prove the associativity of matrix multiplication (for matrices with entries in a field F) as follows. Let A, B, C be $m \times n$, $n \times p$ and $p \times q$ matrices. Let S, T, U be the linear transformations obtained from them by Proposition (4.1). Then $(S \circ T) \circ U$ is a linear transformation from F^q to F^m and its matrix w.r.t. the standard ordered bases is $(AB)C$. Similarly $S \circ (T \circ U)$ has $A(BC)$ as its matrix w.r.t. the same ordered bases. Since $(S \circ T) \circ U = S \circ (T \circ U)$ and the bases are the same, it follows that $(AB)C = A(BC)$. This proves associativity of matrix multiplication a little more elegantly than the direct proof in Section 1. (Note however, that this proof is applicable only for matrices over a field, while the earlier proof was valid for matrices over any ring.)

As another application of the interplay between matrices and linear transformations, we make good a promise given in Section 2. If F is a field, $p(x)$ is an irreducible polynomial in $F[x]$ and I is the ideal generated by $p(x)$, then in the proof of Theorem (2.26) we saw that the quotient ring $F[x]/I$ is an extension field of F and that in this extension field the polynomial $p(x)$ has a root (namely, the coset $x + I$). There we promised that using matrices we can get a 'concrete' representation of this extension field. We now show how this representation is obtained. First we need a couple of definitions.

4.11   Definition: A polynomial whose leading coefficient is 1 is called a **monic** polynomial. ('Mono' means one.).

Thus a monic polynomial of degree n over a field F is of the form

$$a_0 + a_1 x + \ldots + a_{n-1} x^{n-1} + x^n, \text{ where } a_0, \ldots, a_{n-1} \in F.$$

If $f(x)$ is any non-zero polynomial in $F[x]$, we can write it uniquely as $c\, g(x)$ where $c \in F$ and $g(x)$ is a monic polynomial. (We simply let c be the leading coefficient of $f(x)$ and let $g(x) = \dfrac{1}{c} f(x)$.) It is obvious that $f(x)$ and $g(x)$ have the same degree. Any root of $f(x)$ is also a root of $g(x)$ and vice versa. Also since non-zero elements of F are the units in $F[x]$, $f(x)$ and $g(x)$ are associates of each other. So $f(x)$ is irreducible iff $g(x)$ is so. Also the ideals of $F[x]$ generated by $f(x)$ and $g(x)$ are the same (see Proposition (2.12)). So, passing from an arbitrary polynomial to a monic polynomial over a field, does not materially change its properties. It is merely a standardisation device. The number of constants needed to determine a polynomial of degree n is $n + 1$ in general. But if it is a monic polynomial it is only n, because one of the constants (namely the leading coefficient) is already known to be 1. Using this simple fact, we associate an $n \times n$ matrix to a monic polynomial as follows.

4.12 Definition: Let $f(x) = a_0 + a_1 x + \ldots + a_{n-1} x^{n-1} + x^n$ be a monic polynomial. Then its **companion matrix, $C(f(x))$,** is defined to be the $n \times n$ matrix

$$\begin{pmatrix} 0 & 0 & 0 & \cdots\cdots & 0 & -a_0 \\ 1 & 0 & 0 & \cdots\cdots & 0 & -a_1 \\ 0 & 1 & 0 & \cdots\cdots & 0 & -a_2 \\ \cdots & \cdots & \cdots & \cdots\cdots & \cdots & \cdots \\ 0 & 0 & 0 & & 1 & -a_{n-1} \end{pmatrix}$$

In the comments following Theorem (2.26), we remarked that the matrix $\begin{pmatrix} 0 & -1 \\ 1 & 0 \end{pmatrix}$ corresponds to the complex number i, which is a root of the polynomial $1 + x^2$ in $\mathbf{R}[x]$. Note that $\begin{pmatrix} 0 & -1 \\ 1 & 0 \end{pmatrix}$ is precisely the companion matrix of the monic polynomial $1 + x^2$. If we identify an element $\lambda \in \mathbf{R}$ with the matrix $\begin{pmatrix} \lambda & 0 \\ 0 & \lambda \end{pmatrix}$ then we see that the matrix $\begin{pmatrix} 0 & -1 \\ 1 & 0 \end{pmatrix}$ satisfies the equation $1 + \begin{pmatrix} 0 & -1 \\ 1 & 0 \end{pmatrix}^2 = 0$, because by direct computation,

$$\begin{pmatrix} 0 & -1 \\ 1 & 0 \end{pmatrix}^2 = \begin{pmatrix} -1 & 0 \\ 0 & -1 \end{pmatrix}.$$

Thus the companion matrix of $1 + x^2$ is a root of the polynomial $1 + x^2$, regarded now as a polynomial over the ring $M_2(\mathbf{R})$ of all 2×2 matrices over F. We now generalise this and show that the companion matrix of every monic polynomial over a fields is a root of that polynomial, regarded as a polynomial over the ring of matrices over that field.

4.13 Proposition: Let $f(x) = a_0 + a_1 x + \ldots + a_{n-1} x^{n-1} + x^n$ be a monic polynomial over a field F and let $A = C(f(x))$ be its companion matrix. Then A is a root of $f(x)$ in the ring $M_n(F)$, that is,

$$a_0 + a_1 A + a_2 A^2 + \ldots + a_{n-1} A^{n-1} + A^n = 0.$$

Moreover A cannot be a root of any polynomial in $F[x]$ of degree less than n.

Proof: We could, of course, prove this by directly computing the various powers of A and forming their linear combination. But there is a better way out. Let $T : F^n \to F^n$ be the linear transformation obtained from A as in Proposition (4.1). Then $a_0 + a_1 A + a_2 A^2 + \ldots + a_{n-1} A^{n-1} + A^n$ is the matrix of the transformation $a_0 + a_1 T + A_2 T^2 + \ldots + a_{n-1} T^{n-1} + T^n$, by Proposition (4.10). (Here a_0 really means a_0 times the identity transformation.). Let us call this transformation S. We have to show that S is identically 0. Since S is a linear transformation, it suffices to show that S vanishes on some basis for F^n (by Theorem (4.5)). We take the standard basis $\{e_1, e_2, \ldots, e_n\}$ for F^n. First consider $T(e_1)$. This is obtained by looking at the first column of A and taking the linear combination of the e_i's with coefficients coming from this column. But the first column of A is $\begin{pmatrix} 0 \\ 1 \\ 0 \\ . \\ 0 \end{pmatrix}$. So $T(e_1) = e_2$. Similarly

$T(e_2) = e_3$, $T(e_3) = e_4$, $\ldots$, $T(e_{n-1}) = e_n$. However $T(e_n)$ equals $- a_0 e_1 - a_1 e_2 - \ldots - a_{n-1} e_n$. Now $T^2(e_1) = T(T(e_1)) = T(e_2) = e_3$. $T^3(e_1) = T(T^2(e_1)) = T(e_3) = e_4$. In general $T^i(e_1) = e_{i+1}$ for $i = 1, \ldots, n-1$. However, $T^n(e_1) = T(T^{n-1}(e_1)) = T(e_n) = -a_0 e_1 - a_1 e_2 \ldots - a_{n-1} e_n$. It now follows that $S(e_1) = (a_0 + a_1 T + \ldots + a_{n-1} T^{n-1} + T^n)(e_1) = a_0 e_1 + a_1 T(e_1) + a_2 T^2(e_1) + \ldots + a_{n-1} T^{n-1}(e_1) + T^n(e_1) = 0$. Similarly we can compute $S(e_2)$ and show that it is 0. But there is a much better way. Note that $S(e_2) = S(T(e_1))$. But S and T commute with each other, since S is a linear combination of powers of T and T commutes with its own powers. So $S(T(e_1))$ is the same as $T(S(e_1))$. But $S(e_1) = 0$ as was shown just now. So, $T(S(e_1)) = 0$ and hence $S(e_2) = 0$. Similarly $S(e_3) = 0$ because $S(e_3) = S(T(e_2)) = T(S(e_2)) = T(0) = 0$. Continuing like this, we get $S(e_4) = 0$, $\ldots$, $S(e_n) = 0$. Thus S is identically 0. This proves the first assertion, namely that T (and hence A) is a root of $f(x)$.

For the second assertion, suppose $g(x) \in F[x]$ is a polynomial of degree m, with $m < n$. Then we have to show that $g(A) \neq 0$ or equivalently $g(T) \neq 0$. Let $g(x) = b_0 + b_1 x + \ldots + b_m x^m$, with $b_m \neq 0$ where $b_0, b_1 \ldots, b_m \in F$. Let

$U = b_0 + b_1 T + ... + b_m T^m$. To show that U is not identically 0, it suffices to show that $U(e_1) \neq 0$. Using the computation done above, $U(e_1) = b_0 e_1 + b_1 e_2 + ... + b_m e_{m+1}$ (we are using here that $m < n$). But the set $\{e_1, ..., e_{m+1}\}$ is linearly independent. So $U(e_1) = 0$ would force $b_i = 0$ for all $i = 0, ..., m$ contradicting that $b_m \neq 0$. So T and hence A cannot be a root of $g(x)$. ▮

We are now ready to give the 'concrete' representation of the extension field obtained from an irreducible polynomial.

4.14 Theorem: Let $p(x)$ be a monic, irreducible polynomial of degree n over a field F. Let A be its companion matrix. In the ring $M_n(F)$, let $F(A)$ be the subring generated by F and A (i.e., the smallest subring of $M_n(F)$ containing $F \cup \{A\}$). Then $F(A)$ is a field, containing F as a sub-field. Moreover, in this field, A is a root of $p(x)$. Every element of $F(A)$ is a unique linear combination of $A^0, A^1, A^2, ..., A^{n-1}$, where $A^0 = I_n = $ the $n \times n$ identity matrix.

Proof: The argument is essentially a duplication of the first part of the proof of Theorem (3.21). Let I be the set of all polynomials in $F[x]$ which have A as a root in $M_n(F)$. Then I is an ideal of $F[x]$. (We need Exercise (2.20) here, because of which if $k(x) = f(x) g(x)$ in $F[x]$, then $k(A) = f(A) g(A)$.) Now by the last proposition, $p(x)$ is a polynomial of the least possible degree among all non-zero polynomials in I. So, by the proof of Theorem (2.4), the ideal I is generated by $p(x)$. Here we are given that $p(x)$ is irreducible. In Theorem (3.21), we had to prove this. This is the only point of difference. The rest of the proof is identical. Define $\theta : F[x] \to M_n(F)$ by $\theta(f(x)) = f(A)$. Then θ is a ring homomorphism with kernel I and range $F(A)$. So $F(A)$ is isomorphic to the quotient ring $F[x]/I$ which is a field. Under this isomorphism, the element $x + I$ of $F[x]/I$ goes to A. From Theorem (2.26), we already know that $F[x]/I$ is an extension field of F and $x + I$ is a root of $p(x)$. So the subring $F(A)$ is a field, containing F, and A is a root of $p(x)$. Finally, for the last assertion, a typical element of $F(A)$ is of the form $f(A)$ for some $f(x) \in F[x]$. (This $f(x)$ need not be unique.) By division algorithm in $F[x]$, write $f(x)$ as $p(x) q(x) + r(x)$ where $r(x) = 0$ or $\deg(r(x)) < n$. In any case $r(x)$ is a linear combination of $1, x, ..., x^{n-1}$. So $r(A)$ is some linear combination of $A^0, A^1, ..., A^{n-1}$. But, $f(A) = p(A) q(A) + r(A) = 0 + r(A) = r(A)$. So $f(A)$ is a linear combination of

$$A^0, A^1, A^2, ..., A^{n-1}.$$

As for uniqueness suppose $b_0 A^0 + b_1 A^1 + ... + b_{n-1} A^{n-1} = c_0 A^0 + c_1 A^1 + ... + c_{n-1} A^{n-1}$. Then A is a root of $(b_0 - c_0) + (b_1 - c_1) x + ... + (b_{n-1} - c_{n-1}) x^{n-1}$. By the last proposition, this polynomial must be the zero polynomial. So $b_i = c_i$ for all $i = 0, 1, ..., n-1$. ▮

This theorem provides a handy method for constructing field extensions. We take any irreducible polynomial of degree n (say) over a field F, convert it to a monic polynomial, take its companion matrix A and take all possible linear combinations of the powers of A, namely A^0, A^1. $A^2,...,A^{n-1}$ with coefficients from F. Unfortunately, there is no easy way, in general, to tell whether a polynomial is irreducible or not. One sufficient condition will be given in the exercises.

When we defined the matrix of a linear transformation $T: V \to W$ we emphasised that it depends on what ordered bases we take for V and W and that it would change if either of the two ordered bases is changed. Let us now see the manner in which it changes. First we need a definition.

4.15 Definition: Let $(v_1, v_2,....v_n)$ and (x_1, x_2, x_n) be two ordered bases for a vector space V. Then the matrix of the identity transformation $1_V: V \to V$, where the domain V has the basis $(v_1,...,v_n)$ and the codomain V has the basis $(x_1,...,x_n)$, is called the **matrix of change of basis** from $(v_1,...,v_n)$ to $(x_1,...,x_n)$.

This is actually a special case of Definition (4.8) and therefore will not be much elaborated. It is a square matrix whose jth column consists of the coefficients of v_j when it is expressed as a linear combination of $x_1, x_2,...,x_n$. The name 'change of basis matrix' may be justified as follows. Call this matrix A. Suppose an element of v is expressed as $\sum_{i=1}^{n} \lambda_i v_i$ and also as $\sum_{i=1}^{n} \mu_i x_i$ where λ's and μ's are in F. Then the relationship between the λ's and μ's is given by

$$\begin{pmatrix} \mu_1 \\ \mu_2 \\ \vdots \\ \mu_n \end{pmatrix} = A \begin{pmatrix} \lambda_1 \\ \lambda_2 \\ \vdots \\ \lambda_n \end{pmatrix}$$

Thus if we know the 'components' of a vector w.r.t. the 'old' basis then we can easily determine its components w.r.t. the 'new' basis.

If we interchange the two ordered bases we have the following expected result.

4.16 Proposition: Given two ordered bases $(v_1,...,v_n)$ and $(x_1,...,x_n)$ for V, the matrix of change of basis from $(v_1,...,v_n)$ to $(x_1,...,x_n)$ is the inverse of the matrix of change of basis from $(x_1,...,x_n)$ to $(v_1,...,v_n)$.

Proof: Let A and B denote these two matrices respectively. Then by Proposition (4.10). BA is the matrix of the identity transformation $1_V: V \to V$ w.r.t. the ordered basis $(v_1,...,v_n)$ for *both* the domain and the codomain. So BA must be the identity matrix I_n. Similarly, AB, being the matrix of $1_V: V \to V$ w.r.t. the ordred basis $(x_1,...,x_n)$ for the domain as well as codomain, equals I_n. So A and B are inverses of each other. ∎

We are now ready to study the effect of change of bases in general.

4.17 Proposition: Let $T: V \to W$ be a linear transformation. Let A be the matrix of T w.r.t. the ordered bases $(v_1,...,v_n)$ for V and $(w_1,...,w_m)$ for W. Let B be the matrix of T w.r.t. the ordered bases $(x_1,...,x_n)$ for V and $(y_1,...,y_m)$ for W. Then $B = DAC^{-1}$ where C is the matrix of change of basis from $(v_1,..., v_n)$ to $(x_1,..., x_n)$ and D is the matrix of change of basis from $(w_1,...,w_m)$ to $(y_1,...,y_m)$.

Proof: The proof is short despite the prolix statement of the result. Consider the following commutative diagram where near each vector space we write an ordered basis for it and for each arrow, on one side we write the linear transformation it represents and on the other side the matrix of this linear transformation w.r.t. the given bases.

$$
\begin{array}{ccc}
 & T & \\
(v_1...,v_n)\ V & \!\!\!\!-\!\!\!\!-\!\!\!\!-\!\!\!\!-\!\!\!\!-\!\!\!\!- & W\ (w_1,...,w_m) \\
 & A & \\
1_V \ \Big| \ C & \quad 1_W \ \Big| \ D & \\
 & T & \\
(x_1,...,x_n)\ V & \!\!\!\!-\!\!\!\!-\!\!\!\!-\!\!\!\!-\!\!\!\!-\!\!\!\!- & W\ (y_1,...,y_m) \\
 & B &
\end{array}
$$

Commutativity of this diagram, along with Proposition (4.10), implies $BC = DA$. But by the last proposition, C is invertible. So $B = DAC^{-1}$. ∎

When we represent a linear transformation from a vector space into itself by a matrix, almost invariably the same ordered basis is chosen both for the domain and the codomain (except for some theoretical purposes such as above). In this case the preceding proposition reduces to the following.

4.18 Proposition: Let $T: V \to V$ be a linear transformation. Let A be the matrix of T w.r.t. an ordered basis $(v_1,...,v_n)$ (which is understood to be both for the domain and the codomain). Let B be the matrix of T w.r.t. the ordered basis $(x_1,..,x_n)$. Then $B = CAC^{-1}$, where C is the matrix of change of basis from $(v_1,...,v_n)$ to $(x_1,...,x_n)$.

Proof: We merely take $W = V$, $m = n$, $(w_1,...,w_m) = (v_1,...,v_n)$ and $(y_1,...,y_m) = (x_1,...,x_n)$ in the last proposition. Then $D = C$ and so $B = CAC^{-1}$. ∎

For many purposes, the matrix C in the equation $B = CAC^{-1}$ is not very important. What matters is that the matrices A and B are related to each other like this by some invertible matrix C. We have encountered this concept earlier in group theory, where two elements, say x and y are said to be conjugate to each other if there exists some g such that $y = gxg^{-1}$ (see Definition (5.1.14)). There is no reason why the same term cannot be used to describe the analogous relationship for matrices. But a different name is already too standard to be changed. We give it below:

4.19 Definition: Given $n \times n$ matrices A, B over a field F, we say B is **similar** to A if there exists an invertible $n \times n$ matrix C such that

$$B = CAC^{-1}.$$

Using the same reasoning as in Proposition (5.1.16) it follows that similarity is an equivalence relation on $M_n(F)$, the set of all $n \times n$ matrices over F. Proposition (4.18) shows that although a change of the ordered basis for V changes the matrix of a linear transformation from V to itself, this variation is confined to the similarity class of the original matrix. One of the major problems in linear algebra is to find, for a given linear transformation $T : V \to V$, a suitable ordered basis for V w.r.t. which the matrix of T will be of some simple, or canonical, form. In the language of matrices, the problem amounts to finding, for a given square matrix A over a field F, a matrix B which is similar to A and which is in a canonical form. The search for such **canonical forms** leads to many interesting developments involving the properties not only of the matrix A but also of the field F. But we shall not pursue them.

An attribute of square matrices which is invariant under similarity is called a **similarity invariant**. For example, invertibility is a similarity invariant. For suppose A, B are similar matrices and A is invertible. Then $B = CAC^{-1}$ for some matrix C. By the general properties of inverses w.r.t. an associative binary operation (Proposition (3.4.7)), it follows that B is invertible with $CA^{-1}C^{-1}$ as its inverse. When a property of matrices is invariant under similarity, it can be defined unambiguously for a linear transformation of a finite dimensional vector space into itself. If $T : V \to V$ is such a linear transformation, we choose any ordered basis for V and let A be the matrix of T w.r.t. this ordered basis. If we had chosen a different basis the matrix would possibly be different, say B. Still, A and B are similar by Proposition (4.18). So, since the property in question is a similarity invariant, either both A and B have it or else neither does. Accordingly we say that T has or lacks that property.

Among the most important similarity invariants of a square matrix are its eigenvalues, determinant, characteristic polynomial and trace. Of these, we shall discuss only the determinant. In the exercises we shall briefly mention eigenvalues and trace. (Actually, an eigenvalue is the basic concept. The others can be defined in terms of eigenvalues. But we shall not go into this.)

The reader has undoubtedly studied determinants of real numbers of order 2 and 3. (A determinant of order 1, $|a|$, is trivially the number a itself.) A 2×2 determinant $\begin{vmatrix} a & b \\ c & d \end{vmatrix}$ is defined as $ad - bc$. A typical 3×3 determinant has the form

$$D = \begin{vmatrix} a_{11} & a_{12} & a_{13} \\ a_{21} & a_{22} & a_{23} \\ a_{31} & a_{32} & a_{33} \end{vmatrix}$$

and its value equals

$$a_{11} \begin{vmatrix} a_{22} & a_{23} \\ a_{32} & a_{33} \end{vmatrix} - a_{12} \begin{vmatrix} a_{21} & a_{23} \\ a_{31} & a_{33} \end{vmatrix} + a_{13} \begin{vmatrix} a_{21} & a_{22} \\ a_{31} & a_{32} \end{vmatrix},$$

which comes out to be the sum of six terms

$$a_{11}a_{22}a_{33} - a_{11}a_{23}a_{32} - a_{12}a_{21}a_{33} + a_{12}a_{23}a_{31} + a_{13}a_{21}a_{32} - a_{13}a_{22}a_{31}.$$

The theory of $n \times n$ determinants which we shall develop will be a straightforward extension of this. But there is a minor conceptual difference. A determinant, as defined above, is a number which is expressed in a peculiar way namely as a square array of several numbers. For our purpose, on the other hand, the determinant will be a function defined on the set of square matrices, abbreviated as det. For example, the determinant D above will be written as det (A) where A is the 3×3 matrix

$$\begin{pmatrix} a_{11} & a_{12} & a_{13} \\ a_{21} & a_{22} & a_{23} \\ a_{31} & a_{32} & a_{33} \end{pmatrix}$$

Thus, the determinant of a square matrix of real numbers (say) is a real number, but the determinant will be a function.

As a first measure of generalisation, we replace the field of real numbers by a commutative ring R with identity. Many of the results about determinants hold for this case too. A few results, however, require that R be a field. Till we encounter such results, we shall assume we are dealing with matrices over a commutative ring with identity.

Now comes the crucial question. How to define det (A) when A is an $n \times n$ matrix if $n > 3$? For possible answers we look at the definition of a 3×3 determinant above. In the first expression it was written as a linear combination of three 2×2 determinants, with coefficients coming from the first row, with alternating $+$ and $-$ signs. A similar approach is possible for any n. Thus, given an $n \times n$ matrix $A = (a_{ij})$, we consider its $(n-1) \times (n-1)$ submatrices obtained by removing the first row and one of the columns. Let A_j be the submatrix obtained by removing the first row and the jth column of A, $j = 1, ..., n$. Then define det (A) as $\sum_{j=1}^{n} (-1)^{1+j} a_{1j}$ det (A_j).

This is known as the **inductive definition** of the determinant. Its main advantage is its simplicity. Also certain properties of determinants, where an inductive argument is needed, are more amenable with this definition. Its disadvantage is that the first row is given a special role, when there is in fact nothing special about any row in particular. Also it is not clear that the determinants of a matrix and its transpose are equal. We shall therefore adopt another definition, which is based on the permutation group S_n of n symbols, which was studied in detail in Section 4 of the last chapter. Later we shall show that our definition is equivalent to the inductive definition. Of course, our definition too has its disadvantages, one of which is that it requires the knowledge of S_n. So ultimately it is a matter of taste as to which definition one prefers.

Let us take a closer look at the determinant D of the 3×3 matrix $A = (a_{ij})$ given above. It consists of 6 terms three of which are with a $+$ sign and 3 with a $-$ sign. Moreover, each term is a product of 3 factors each of which is an entry of the matrix A. These entries are such that in each term there is precisely one entry from each row and each column. So every term is of the form $a_{1x}a_{2y}a_{3z}$ where $x\ y\ z$ is a permutation of $\{1, 2, 3\}$. If we denote this permutation by σ, we can write the term as $a_{1\sigma(1)}\ a_{2\sigma(2)}\ a_{3\sigma(3)}$. We see further that the term appears with a $+$ or $-$ sign according as the permutation σ is even or odd. For example, if $\sigma = \begin{pmatrix} 1 & 2 & 3 \\ 2 & 3 & 1 \end{pmatrix}$ then σ is even and the term $a_{12}a_{23}a_{31}$ indeed appears with a $+$ sign, while if $\tau = \begin{pmatrix} 1 & 2 & 3 \\ 3 & 2 & 1 \end{pmatrix}$ then τ is odd and term $a_{13}a_{22}a_{31}$ appears with a negative sign. Thus, we see that for the 3×3 matrix $A = (a_{ij})$, det (A) equals $\sum\limits_{\sigma \in S_3} (-1)^\sigma\ a_{1\sigma(1)}\ a_{2\sigma(2)}\ a_{3\sigma(3)}$ where $(-1)^\sigma = 1$ or -1 according as σ is even or odd. The case of the 2×2 determinant is similar and in fact simpler.

In S_2 there are only two permutations, one of which is even and the other is odd. So $\begin{vmatrix} a_{11} & a_{12} \\ a_{21} & a_{22} \end{vmatrix}$ also equals $\sum\limits_{\sigma \in S_2} (-1)^\sigma a_{1\sigma(1)}\ a_{2\sigma(2)}$.

It is now clear how to define the determinant of any $n \times n$ matrix over R.

4.20 Definition: Let A be an $n \times n$ matrix over a commutative ring R with identity. Then the determinant of A (denoted by det (A), $d(A)$ or by $|A|$), is defined as the element, $\sum\limits_{\sigma \in S_n} (-1)^\sigma a_{1\sigma(1)} a_{2\sigma(2)}...a_{n\sigma(n)}$ of R, where $(-1)^\sigma = 1$ or -1 according as the permutation σ is even or odd.

It is obvious that in proving properties of determinants, the group S_n will figure many times. As a typical example, we prove that A and its transpose have the same determinant.

4.21 Proposition: Let B be the transpose of an $n \times n$ matrix A over R. Then $\det(B) = \det (A)$.

Proof: Let $A = (a_{ij})$. Then $B = (b_{ij})$ where $b_{ij} = a_{ji}$ for $i = 1, \ldots, n$; $j = 1, \ldots, n$. By definition, $\det (B) = \sum\limits_{\sigma \in S_n} (-1)^\sigma b_{1\sigma(1)} \ldots b_{n\sigma(n)} = \sum\limits_{\sigma \in S_n} (-1)^\sigma a_{\sigma(1)1}$ $a_{\sigma(2)2} \ldots a_{\sigma(n)n}$. Now σ is a permutation of $\{1, \ldots, n\}$. So as i varies from 1 to n, so does $\sigma(i)$, except possibly in a different order. Also the ring R is commutative. So we can write $a_{\sigma(1)1} a_{\sigma(2)2} \ldots a_{\sigma(n)n}$ as $a_{1\tau(1)} a_{2\tau(2)} \ldots a_{n\tau(n)}$ where $\tau = \sigma^{-1}$. Note further that as σ ranges over S_n so does τ. Moreover σ and τ have the same parity. Therefore $\det (B)$ equals $\sum\limits_{\tau \in S_n} (-1)^\tau a_{1\tau(1)} a_{2\tau(2)} \ldots$ $a_{n\tau(n)}$, which, by definition, is $\det (A)$. ∎

Because of this proposition, whenever we prove a result regarding the rows of a determinant, the corresponding result for columns also holds. The technique in the proof above is noteworthy. The variable σ is a dummy ranging over S_n. If we replace it by any variable τ which also ranges over S_n, the sum will be unchanged. The choice of τ is made as a function of σ; in the last proof τ was σ^{-1}. Of course the function taking σ to τ must be one-to-one and onto, that is, it must be a permutation of S_n. It is this technique of change of variable of summation (along with properties of the permutation groups) that makes our definition of determinant easy to manipulate despite its gigantic size. (Note that there are $n!$ terms each term being a product of n elements.) As another illustration of this technique, we prove the following result which tells what happens when two rows (or columns) are interchanged.

4.22 Proposition: Let B be a matrix obtained from an $n \times n$ matrix A by interchanging two rows (or two columns) of A. Then $\det (B) = -\det (A)$.

Proof: In view of the last proposition we prove the result only for rows. Let B be obtained from A by interchanging the rth row with the sth row where $r < s$. Then $b_{ij} = a_{ij}$ for all j and for all $i \neq r, s$ and $b_{rj} = a_{sj}$, $b_{sj} = a_{rj}$ for all j. Now $\det (B) = \sum\limits_{\sigma \in S_n} (-1)^\sigma b_{1\sigma(1)} \ldots b_{r\sigma(r)} \ldots b_{s\sigma(s)} \ldots b_{n\sigma(n)}$ $= \sum\limits_{\tau \in S_n} (-1)^\sigma a_{1\sigma(1)} \ldots a_{r\sigma(s)} \ldots a_{r\sigma(r)} \ldots a_{n\sigma(n)}$. Let θ be the transposition $(r\ s)$, and let $\tau = \sigma \circ \theta$ for $\sigma \in S_n$. Then as σ ranges over S_n so does τ. However, σ and τ are always of opposite parity. So $(-1)^\sigma = -(-1)^\tau$. Now, in the expression for $\det (B)$ obtained above, $a_{1\sigma(1)} \ldots a_{r\sigma(s)} \ldots a_{s\sigma(r)} \ldots a_{n\sigma(n)}$ is nothing but $a_{1\tau(1)} \ldots a_{r\tau(r)} \ldots a_{s\tau(s)} \ldots a_{n\tau(n)}$. So $\det (B)$ equals $- \sum\limits_{\sigma \in S_n} (-1)^\tau a_{1\tau(1)} \ldots a_{n\tau(n)}$. Thus $\det (B) = -\det (A)$. ∎

4.23 Corollary: Suppose B is obtained from A by a permutation of rows. Let θ be this permutation. Then $\det (B) = \det A$ if θ is even and $\det (B) = -\det (A)$ if θ is odd. (Similarly for permutation of columns.)

Proof: By Theorem (5.4.5), express θ as $\theta_1 \theta_2 \ldots \theta_k$ (say) where each θ_i is a transposition. Now the permutation θ of rows of A can be effected by applying, in succession, the interchanges of rows given by $\theta_k, \theta_{k-1}, \ldots, \theta_2, \theta_1$ to A. Every time the sign of the determinant changes. So det $(B) = (-1)^k$ det (A). But k is even or odd according as θ is even or odd. Hence the result. ∎

Let us now prove the well-known result that if two rows (or columns) of a matrix A are identical then det $(A) = 0$. Suppose the rth and sth rows of A are identical. The familiar proof is by interchanging these two rows. Then the new matrix is also A. So by Proposition (4.22), det $(A) = -$ det (A). But we have to be wary in concluding from this that det $(A) = 0$. This is valid for matrices over $\mathbf{R}$, and more generally over any commutative ring R in which $2x = 0$ implies $x = 0$. But not all rings are of this type. In fact, in a Boolean ring, and also in any integral domain of characteristic 2, $x = -x$ for all x. So we give a different argument which is applicable for all cases.

4.24 Proposition: If two rows (or columns) of a square matrix are identical then its determinant is 0.

Proof: Let A be $n \times n$ matrix whose rth and sth rows are identical with $r < s$. Then $a_{rj} = a_{sj}$ for all $j = 1, 2, \ldots, n$. Then,

$$\det(A) = \sum_{\sigma \in S_n} (-1)^\sigma a_{1\sigma(1)} \ldots a_{r\sigma(r)} \ldots a_{s\sigma(s)} \ldots a_{n\sigma(n)}$$

As before, let θ be the transposition $(r\,s)$, and for $\sigma \in S_n$, let $\tau = \sigma \circ \theta$. Then $a_{1\tau(1)} \ldots a_{r\tau(r)} \ldots a_{s\tau(s)} \ldots a_{n\tau(n)}$ is the same $a_{1\sigma(1)} \ldots a_{r\sigma(r)} \ldots a_{s\sigma(s)} \ldots a_{n\sigma(n)}$ because $a_{r\tau(r)} = a_{s\tau(r)} = a_{s\sigma(s)}$ and similarly $a_{s\tau(s)} = a_{r\tau(s)} = a_{r\sigma(r)}$, while for $i \neq r, s, \sigma(i) = \tau(i)$. As noted above, σ and τ have opposite parity. So $(-1)^\sigma = -(-1)^\tau$. Thus we see that in the summation above every term appears twice, but with opposite signs. (It should be noted that if we further convert τ to $\tau \circ \theta$ then we get back σ. So these two terms cancel each other.) Hence det $(A) = 0$. ∎

This proposition is useful in evaluation of determinants. Its utility is enhanced by combining it with another property of determinants, called its linearity in each row (or column). To state what it means, let $u_1, u_2, \ldots, u_n$ be the rows of an $n \times n$ matrix A over R. Each u_i is an ordered n-tuple, $(a_{i1}, a_{i2}, \ldots, a_{in})$ and hence is an element of R^n. Since we are not yet assuming that R is a field, R^n is not quite a vector space. Still the concept of a linear combination is meaningful. So, suppose, $u_i = \lambda v_i + \mu w_i$ for some $\lambda, \mu \in R$, where $v_i, w_i \in R^n$, say $v_i = (b_{i1}, b_{i2}, \ldots, b_{in})$ and $w_i = (c_{i1}, c_{i2}, \ldots, c_{in})$. This means, for each $j = 1, \ldots, n$, $a_{ij} = \lambda b_{ij} + \mu c_{ij}$. Now let us keep all other rows of A unchanged and form two matrices B and C, where the rows of B are $u_1, \ldots, u_{i-1}, v_i, u_{i+1}, \ldots, u_n$ and those of C are

$$u_1, \ldots, u_{i-1}, w_i, u_{i+1}, \ldots, u_n.$$

The following proposition relates the determinants of A, B and C.

4.25 Proposition: With the notation above,

$$\det(A) = \lambda \det(B) + \mu \det(C).$$

Proof: Once the statement is understood, its proof is very simple. By definition,

$$\det(A) = \sum_{\sigma \in S_n} (-1)^{\sigma} a_{1\sigma(1)} \dots a_{i-1,\, \sigma(i-1)} a_{i\sigma(i)} a_{i+1,\, \sigma(i+1)} \dots a_{n\sigma(n)}$$

$$= \sum_{\sigma \in S_n} (-1)^{\sigma} a_{1\sigma(1)} \dots a_{i-1,\, \sigma(i-1)} (\lambda b_{i\sigma(i)} + \mu\, c_{i\sigma(i)}) a_{i+1,\, \sigma(i+1)} \dots a_{n\sigma(n)}$$

$$= \lambda \sum_{\sigma \in S_n} (-1)^{\sigma} a_{1\sigma(1)} \dots b_{i\sigma(i)} \dots a_{n\sigma(n)}$$

$$+ \mu \sum_{\sigma \in S_n} (-1)^{n} a_{1\sigma(1)} \dots c_{i\sigma(i)} \dots a_{n\sigma(n)}$$

$$= \lambda \det(B) + u \det(C). \quad\blacksquare$$

The last two propositions are the basis of one of the most frequently used methods of simplifying a determinant, namely adding a multiple of some row (or column) to another row (or column), (Exercise (4.14)). We may very well represent an $n \times n$ matrix A as an ordered n-tuple of its rows, $(u_1, u_2, \dots, u_n)$ where u_i is the ith row of A. (Of course, u_i itself is an ordered n-tuple of elements of R.) We may therefore regard the determinant of an $n \times n$ matrix as a function of n variables $u_1, \dots, u_n$ each ranging over R^n. For $\det(A)$, we write $\det(u_1, u_2, \dots, u_n)$. The preceding result then says that the det function is separately linear in each variable, or in the language of Exercise (3.25), it is a multilinear function. When more than one row is expressed as a linear combination of row vectors, we can reduce it, applying Proposition (4.25) to only one row at a time. For example, suppose $u_1 = \lambda_1 v_1 + \mu_1 w_1$ and $u_2 = \lambda_2 v_2 + \mu_2 w_2$. Then $\det(u_1, u_2, u_3, \dots, u_n)$

$$= \det(\lambda_1 v_1 + \mu_1 w_1, u_2, u_3, \dots, u_n)$$

$$= \lambda_1 \det(v_1, u_2, u_3, \dots, u_n) + \mu_1 \det(w_1, u_2, u_3, \dots, u_n)$$

which further reduces to,

$$\lambda_1 \lambda_2 \det(v_1, v_2, u_3, \dots, u_n) + \lambda_1 \mu_2 \det(v_1, w_2, u_2, \dots, u_n)$$
$$+ \lambda_2 \mu_1 \det(w_1, v_2, u_3, \dots, u_n) + \mu_1 \mu_2 \det(w_1, w_2, u_3, \dots, u_n).$$

Repeated applications of multilinearity, coupled with earlier results about determinants, yield the following result about determinant of the product of two matrices.

4.26 Theorem: If A, B, are square matrices of order n, then

$$\det(AB) = \det(A) \det(B).$$

Proof: Let $A = (a_{ij})$, $B = (b_{ij})$. Denote AB by $C = (c_{ij})$. Let $v_1, ..., v_n$ be the rows of B and $w_1, ..., w_n$ be rows of C. From the definition of matrix multiplication, it is easily seen that each w_i is a linear combination of the v's with coefficients coming from the ith row of A. Specifically,

$$w_i = \sum_{j=1}^{n} a_{ij} v_j. \text{ So } \det(AB) = \det(C) = \det\left(\sum_{j=1}^{n} a_{1j} v_j, \sum_{j=1}^{n} a_{2j} v_j, ..., \sum_{j=1}^{n} a_{nj} v_j\right).$$

Since each variable is a linear combination of n rows and there are n such variables, if we apply the last proposition, $\det(C)$ will be a linear combination of n^n determinants in all, formed from the various v's. This looks like too large a number to manipulate. But most of these n^n determinants vanish. A typical term in the expression for $\det(C)$ will be

$$a_{1j_1} a_{2j_2} ... a_{rj_r} ... a_{nj_n} \det(v_{j_1}, v_{j_2} ..., v_{j_n}),$$

where each $j_1, ..., j_n$ varies from 1 to n, independently of the others. Whenever $j_r = j_s$ for some $r \neq s$, this determinant is 0 by Proposition (4.24). So we have to consider only those terms where $j_1, ..., j_n$ are all distinct. This means $j_1, j_2 ..., j_n$ is a permutation of $\{1, ..., n\}$. Thus there will be only $n!$ terms which appear in the expression for $\det(C)$ as a linear combination. In a slightly different notation,

$$\det(C) = \sum_{\sigma \in S_n} a_{1\sigma(1)} a_{2\sigma(2)} ... a_{n\sigma(n)} \det(v_{\sigma(1)}, v_{\sigma(2)}, ..., v_{\sigma(n)})$$

But by Proposition (4.23),

$$\det(v_{\sigma(1)}, v_{\sigma(2)}, ..., v_{\sigma(n)}) = (-1)^\sigma \det(v_1, ..., v_n) = (-1)^\sigma \det(B).$$

So $\qquad \det(C) = \det(B) \sum_{\sigma \in S_n} (-1)^\sigma a_{1\sigma(1)} a_{2\sigma(2)} ... a_{n\sigma(n)}$

which equals $\det(A) \det(B)$ as desired. ∎

An important generalisation of this theorem will be given as an exercise. It is proved by an analogous argument.

This result can also be expressed by saying that the determinant function is a monoid homomorphism from the monoid $M_n(R)$ to the monoid R (both under multiplication). Note, however, that determinant is not a ring homomorphism, $\det(A + B)$ is generally different from $\det(A) + \det(B)$. Each row of $A + B$ is the sum of the corresponding rows of A and B. So if we apply Proposition (4.25), $\det(A + B)$ will be a sum of 2^n determinants, of which $\det(A)$ and $\det(B)$ will be only two terms. Still, the fact that determinant preserves multiplication has important consequences as the following two corollaries show.

4.27 Corollary: If a square matrix A is invertible so is its determinant (as an element of the ring R). Also $\det(A^{-1}) = [\det(A)]^{-1}$.

Proof: Let A be an $n \times n$ matrix over R. Suppose $B \in M_n(R)$ is an inverse of A. Then $AB = I_n$ where I_n is the identity matrix of order n. Then by the last theorem, det (A) det $(B) =$ det (I_n). But det (I_n) is evidently 1. So det (B) is the inverse of det (A) in R. (Note that R is commutative. However, $M_n(R)$ need not be commutative. The argument here applies even for one-sided inverse.) ▨

4.28 Corollary: Similar matrices have the same determinant. In other words, the determinant is a similarity invariant.

Proof: Suppose A, B, C are $n \times n$ matrices with $B = CAC^{-1}$. Then by Theorem (4.26) and the last corollary, det $(B) = [\det(C)]^{-1} \det(A) \det(C) = = \det(A)$ since R is commutative. ▨

The converse of this corollary is false. For example the 2×2 matrices $\begin{pmatrix} 1 & 0 \\ 0 & 1 \end{pmatrix}$ and $\begin{pmatrix} 1 & 1 \\ 0 & 1 \end{pmatrix}$ have the same determinant but are not similar.

(since $\begin{pmatrix} 1 & 0 \\ 0 & 1 \end{pmatrix}$ commutes with every 2×2 matrix, it cannot be similar to any other matrix.) However, the converse of corollary (4.27) is true. That is, if det (A) has an inverse in R, then A is invertible. To prove this, we shall first prove a result, which will also show that our definition of a determinant of an $n \times n$ matrix is equivalent to the inductive definition. First we define the gadgets needed in 'expanding a determinant w.r.t. a particular row'.

4.29 Definition: Let $A = (a_{ij})$ be an $n \times n$ matrix. Then the determinant of the $(n-1) \times (n-1)$ submatrix of A obtained by removing the ith row and jth column of A is called the (i, j) the **minor** of A and is denoted by M_{ij}. The element $(-1)^{i+j} M_{ij}$ is called the (i, j)th **cofactor** of A and will be denoted by C_{ij}.

In particular if $i = 1$, then C_{ij} is the coefficient of a_{ij} in the inductive definition of the determinant. So the equivalence of the two definitions will follow as a consequence of the following theorem.

4.30 Theorem: For any $r = 1, \ldots, n$, det $(A) = \sum_{j=1}^{n} a_{rj} C_{rj}$. Similarly,

$$\det(A) = \sum_{j=1}^{n} a_{jr} C_{jr}.$$

Proof: Let for $j = 1, \ldots, n$, $T_j = \{\sigma \in S_n : \sigma(r) = j\}$. Then each T_j has $(n-1)!$ elements, $T_j \cap T_k = \phi$ for $j \neq k$ and $S_n = \bigcup_{j=1}^{n} T_j$. Hence

$$\det(A) = \sum_{\sigma \in S_n} (-1)^\sigma a_{1\sigma(1)} \ldots a_{n\sigma(n)} = \sum_{j=1}^{n} \sum_{\sigma \in T_j} (-1)^\sigma a_{1\sigma(1)} \cdots a_{r\sigma(r)} \cdots a_{n\sigma(n)})$$

$$= \sum_{j=1}^{n} a_{rj} \sum_{\sigma \in T_j} (-1)^\sigma a_{1\sigma(1)} \ldots a_{r-1, \, \sigma(r-1)} a_{r+1, \, \sigma(r+1)} \ldots a_{n\sigma(n)}$$

So the proof of the first assertion will be complete if we show that for each j,

$$C_{rj} = \sum_{\sigma \in T_j} (-1)^\sigma a_{1\sigma(1)} \ldots a_{r-1, \, \sigma(r-1)} a_{r+1, \sigma(r+1)} \ldots a_{n\sigma(n)} \tag{1}$$

Let B be the $(n-1) \times (n-1)$ submatrix of A obtained by removing the rth row and jth column. Note that for $i > r$, the ith row of A becomes the $(i-1)$th row B (with deletion of one element namely a_{ij}) and similarly for $k > j$, the kth column of A becomes the $(k-1)$th column of B. In other words, for $i = 1, \ldots, n-1$ and $k = 1, 2, \ldots, n-1$, the (i, k)th element, b_{ik} of B is given by

$$b_{ik} = \begin{cases} a_{ik} & \text{if} \quad i < r \quad \text{and} \quad k < j \\ a_{i+1, k} & \text{if} \quad i \geqslant r \quad \text{and} \quad k < j \\ a_{i, k+1} & \text{if} \quad i < r \quad \text{and} \quad k \geqslant j \\ a_{i+1, k+1} & \text{if} \quad i \geqslant r \quad \text{and} \quad k \geqslant j \end{cases} \tag{2}$$

Now, by definition, $C_{rj} = (-1)^{r+j} \det (B)$

$$= (-1)^{r+j} \sum_{\tau \in S_{n-1}} (-1)^\tau b_{1\tau(1)} b_{2\tau(2)} \ldots b_{n-1, \, \tau(n-1)}.$$

So, in (1) both sides are summations of $(n-1)!$ terms each. The result will be established if we show that the terms on the two sides match one by one. For this define $f \colon T_j \to S_{n-1}$ by $f(\sigma) = \tau$ for $\sigma \in T_j$, where

$$\tau \colon \{1, 2, \ldots, n-1\} \to \{1, 2, \ldots, n-1\} \text{ is defined by}$$

$$\tau(i) = \begin{cases} \sigma(i) & \text{if} \quad i < r \quad \text{and} \quad\quad \sigma(i) < j \\ \sigma(i+1) & \text{if} \quad i \geqslant r \quad \text{and} \quad \sigma(i+1) < j \\ \sigma(i) - 1 & \text{if} \quad i < r \quad \text{and} \quad\quad \sigma(i) > j \\ \sigma(i+1) - 1 & \text{if} \quad i \geqslant r \quad \text{and} \quad \sigma(i+1) > j \end{cases} \tag{3}$$

τ is essentially the same permutation as σ, except that indices greater than r in the domain and those greater than j in codomain are lowered by 1. For example, if $n = 9$, $r = 7$, $j = 3$ and $\sigma = \begin{pmatrix} 1 & 2 & 3 & 4 & 5 & 6 & 7 & 8 & 9 \\ 2 & 9 & 1 & 4 & 5 & 8 & 3 & 7 & 6 \end{pmatrix}$

then $\tau = \begin{pmatrix} 1 & 2 & 3 & 4 & 5 & 6 & 7 & 8 \\ 2 & 8 & 1 & 3 & 4 & 7 & 6 & 5 \end{pmatrix}$. Clearly the function f taking σ to τ is a bijection. Also from (2) and (3) it follows that for every $\sigma \in T_j$, $a_{1\sigma(1)} \ldots a_{r-1, \sigma(r-1)} a_{r+1, \sigma(r+1)} \ldots a_{n\sigma(n)}$ equals $b_{1\tau(1)} b_{2\tau(2)} \ldots b_{n-1, \tau(n-1)}$. Thus the terms on the two sides of (1) are the same except possibly for sign. To complete the proof we now merely have to show that $(-1)^\tau = (-1)^{r+j}$ $(-1)^\sigma$ for all $\sigma \in T_k$. For this we compare the inversion pairs in σ and in τ. If (p, q) is an inversion pair of τ (i.e., if $p < q$ and $\tau(p) > \tau(q)$) then there is a corresponding inversion pair in σ, which may be (p, q) or $(p + 1, q)$ or $(p, q + 1)$ or $(p + 1, q + 1)$ depending upon how p is related to r and how q is related to j. Conversely, to every inversion pair in σ, there is a corresponding inversion pair in τ, except for such inversion pairs (p, q) in σ for which p or q equals r. Let x be the number of inversion pairs in σ for which $p = r$ and y the number of inversion pairs in σ for which $q = r$. Then, the number of inversion pairs in $\sigma = x + y +$ number of inversion pairs in τ. From Theorem (5.4.7), a permutation is even or odd according as the number of inversion pairs in it is even or odd. We see that $(-1)^\sigma = (-1)^{x+y}(-1)^\tau$. We are given $\sigma(r) = j$. Let us write σ in the form

$$\sigma = \begin{pmatrix} 1 & 2...r-1 & r & r+1...n \\ \sigma(1) & \sigma(2)...\sigma(r-1) & j & \sigma(r+1)...\sigma(n) \end{pmatrix}$$

Then x is the number of entries in the second row which lie to the left of j and are greater than j. Similarly y is the number of entries in the second row which lie to the right of j and which are less than j. So the number of entries in the second row which are greater than j is $x + (n - r - y)$. But this number obviously is $n - j$. Thus we get $x - y = r - j$. (In the numerical example given above, $r = 7, j = 3, x = 4, y = 0$.) We have not succeeded in computing $x + y$. But it does not matter, because we are interested only in its parity. Now $x + y$ has the same parity as $x - y$. Also $r - j$ has the same parity as $r + j$. So $(-1)^{x+y} = (-1)^{r+j}$. Thus for every $\sigma \in T_k$, the term $(-1)^\sigma a_{1\sigma(1)} \cdots a_{r-1, \sigma(r-1)} a_{r+1, \sigma(r+1)} \cdots a_{n\sigma(n)}$ equals $(-1)^{r+j}(-1)^\tau b_{1\tau(1)} \cdots$ $\cdots b_{n-1, \tau(n-1)}$. As noted before, this establishes (1) and completes the proof of the first assertion. The second assertion is just the 'transposed' version of the first assertion. ∎

We are now ready to prove the converse of Corollary (4.27).

4.31 Theorem: A square matrix over a commutative ring with identity is invertible if and only if its determinant is an invertible element of that ring.

Proof: The direct implication was proved in Corollary 4.27. For the converse, suppose $A = (a_{ij})$ is an $n \times n$ matrix over a commutative ring R with identity. Let D be the transpose of the matrix of cofactors of A, that is D is $n \times n$ matrix in which $d_{ij} = C_{ji}$. Let us compute the product matrix AD.

A typical diagonal entry of the product is of the form $\sum_{j=1}^{n} a_{rj}d_{jr}$. But this equals $\sum_{j=1}^{n} a_{rj}C_{rj}$ which is det (A) by the last theorem. What about non-diagonal entries? A typical such entry is $\sum_{j=1}^{n} a_{rj}d_{js}$, where $r \neq s$. We claim this is 0. $\sum_{j=1}^{n} a_{rj}d_{js} = \sum_{j=1}^{n} a_{rj}C_{sj}$. Let B be the matrix (b_{ij}) where $b_{ij} = a_{ij}$ for all $i \neq s$ and all j and $b_{sj} = a_{rj}$ for $j = 1, ..., n$. In other words B is obtained from A by replacing its sth row with the rth row leaving all other rows unaffected. It follows that the cofactors C_{sj} are the same for A as well as for B. So $\sum_{j=1}^{n} a_{rj}C_{sj} = \sum_{j=1}^{n} b_{sj}C_{sj}$ which equals det (B) by the last theorem. But det $(B) = 0$ since two rows of B (namely, the rth row and the sth row) are identical. Thus we have shown that all non-diagonal entries of D are 0. Putting it all together AD equals (det A) I_n. Similarly DA equals (det A) I_n. We are assuming that det (A) is an invertible element of R. It follows that $\dfrac{1}{\text{det } (A)} D$ is the inverse of A. ∎

Note that in the proof of the converse, the fact that det A is invertible in R was used only at the end. Even without it, we still have $AD = DA = $ (det A) I_n for any $n \times n$ matrix A over a commutative ring R. The matrix D is called the **adjoint matrix** of A and denoted by adj (A).

Using this theorem we can get a method for solving a system of n linear equations in n unknowns, known as Cramer's rule, which will be given as an exercise. But it is a very inefficient method because it requires the computation of so many $n \times n$ determinants, which is time-consuming. Later we shall study a method, due to Gauss, which is efficient and also more general in that it applies to a system of m linear equations in n unknowns, where m may be different from n.

Nevertheless, the last theorem has interesting theoretical consequences and we proceed to study them. Our treatment of the determinant so far was applicable to square matrices over any commutative ring with identity. From now onwards we assume that we are dealing with matrices over a field F. As before, if A is an $n \times n$ matrix over F we shall think of its rows and columns as elements of F^n. Note, however, that now F^n is a vector space over F and so the results of the last section become applicable. The following proposition gives a handy characterisation of the linear dependence or otherwise of the rows (or columns) of a square matrix.

4.32   Proposition: Let A be an $n \times n$ matrix over a field F. Then the following statements are equivalent:

(1)   The rows of A are linearly dependent over F (as elements of F^n).
(2)   The columns of A are linearly dependent over F (also as elements of F^n).

(3) $\det (A) = 0$.

Proof: We shall prove first that (2) and (3) are equivalent. We shall then apply this equivalence to A', the transpose of A. Since $\det (A) = \det (A')$ by Proposition 4.21, it follows that (1) and (3) are equivalent, because the columns of A' are precisely the rows of A.

(2) $\Rightarrow$ (3). Let the columns of A be $c_1, c_2, ..., c_n$. If they are linearly dependent, then one of them can be expressed as a linear combination of the remaining columns (see Proposition 3.5). Suppose that the jth column c_j equals $\lambda_1 c_1 + ... + \lambda_{j-1} c_{j-1} + \lambda_{j+1} c_{j+1} + ... + \lambda_n c_n$ for some $\lambda_1, ..., \lambda_{j-1}$, $\lambda_{j+1}, ... \lambda_n \in F$. We form a matrix B from A by adding $-\lambda_i c_i$ to c_j for $i = 1, ..., j - 1, j + 1, ... n$. By Proposition 4.25 (or rather, the comment following it), $\det (B) = \det (A)$. But in B, the jth column is 0. So $\det (B) = 0$. Hence $\det (A) = 0$.

(3) $\Rightarrow$ (2). Suppose $\det (A) = 0$. Let $T : F^n \to F^n$ be the linear transformation obtained from A by Proposition 4.1. Let R be the range of T. As noted while proving Proposition 4.2, R is spanned by the columns of A. If the columns of A are linearly independent, then dim (R) would be n, which would mean that $R = F^n$ that is, T is onto. But then T would also be one-to-one (see Exercise 3.18). So T would have an inverse $T^{-1}: F^n \to F^n$ and the matrix of T^{-1} w.r.t. the standard basis would be the inverse of the matrix of T. But the matrix of T is A. So A would be invertible. This would contradict Theorem 4.31, according to which the determinant of an invertible matrix cannot be 0. So the columns of A are linearly dependent.

This proves the equivalence of (2) and (3) and, as noted before, completes the proof. ■

It is convenient to paraphrase this proposition. A square matrix A is called **non-singular** (or **regular**) if it is invertible and **singular** otherwise. (This is another instance of how a general concept is given a special name in a particular context. Another example was the use of 'similar' to convey the same meaning as 'conjugate'). The following proposition characterises non-singular matrices.

4.33 Proposition: Let A be an $n \times n$ matrix over a field F. Then the following statements are equivalent.

(1) A is non-singular.

(2) A has a right inverse, that is, there exists an $n \times n$ matrix B such that $AB = I_n$.

(3) A has a left inverse.

(4) $\det (A) \neq 0$.

(5) The row rank of A is n.

(6) The column rank of A is n.

(7) $Ax = 0 \Rightarrow x = 0$ for every column vector x,

(8) $xA = 0 = 0 \Rightarrow x = 0$ for every row vector x.

Proof: Clearly (1) $\Rightarrow$ (2). If (2) holds, then det (A) det $(B) = \det (I_n) = 1$ showing that det $(A) \neq 0$. So (2) $\Rightarrow$ (4). From Theorem (4.31), (4) $\Rightarrow$ (1), because, in a field every non-zero element is invertible. Thus we see that (1), (2) and (4) are equivalent. By a similar argument, (1), (3) and (4) are equivalent. The equivalence of (4), (5) and (6) follows from the last proposition. It only remains to show that the last two statements are equivalent to the rest.

Clearly (1) $\Rightarrow$ (7), because $Ax = 0 \Rightarrow A^{-1} Ax = 0 \Rightarrow I_n x = 0 \Rightarrow x = 0$ for every column vector x (of length n). Similarly (1) $\Rightarrow$ (8).

To show that (7) $\Rightarrow$ (1), we consider once again the linear transformation $T: F^n \to F^n$ defined by $T(x) = Ax$. (7) is equivalent to saying that T is one-to-one. But then by Exercise (3.18), T is also onto and hence has an inverse, T^{-1}. As before the matrix of T^{-1} is an inverse of A. So (1) holds.

To complete the proof, we only have to show that (8) $\Rightarrow$ (1). For this, we define $T : F^n \to F^n$ by $T(x) = xA$ and argue as above. Alternatively, let B be the transpose of A. Then (8) is equivalent to saying that $By = 0 \Rightarrow y = 0$ for all column vectors y. So from the equivalence of (1) and (7) applied to B, B is invertible. But then the transpose of the inverse of B would be the inverse of A. So (1) holds. ▉

We are now ready to prove the equality of the row rank and the column rank of a matrix. A special case of this, where the matrix is a square one and the row rank is the highest possible was proved in the last proposition. We first extend this result to any rectangular (that is, not necessarily square) matrix.

4.34 Proposition: Suppose B is an $r \times n$ matrix of row rank r, over a field F. Then the column rank of B is also r.

Proof: Let $u_1, \ldots, u_r$ be the rows of B, regarded as vectors in F^n. Let $c_1, c_2, \ldots, c_n$ be the columns of B, regarded as vector in F^r. The hypothesis means that $u_1, \ldots, u_r$ are linearly independent. In view of Proposition (3.12), this implies $n \geqslant r$. Now let k be the column rank of B. Then some k columns of B are linearly independent and every other column is a linear combination of these k columns. For notational simplicity, suppose these columns are $c_1, \ldots, c_k$. We have to show that $k = r$. Certainly, since $\{c_1, .., c_k\}$ is a linearly independent set of vectors in F^r, it follows, again from Proposition (3.12) that k cannot exceed r. To complete the proof, we show k cannot be less than r.

Suppose $k < r$. Let C be the $r \times r$ submatrix of B formed by the columns $c_1, \ldots, c_r$. Let $v_1, \ldots, v_r$ be the rows of C. Note that each v_i is a truncation of u_i. Now $\{c_1, \ldots, c_r\}$ is linearly dependent since c_r can be expressed as a linear combination of $c_1, \ldots, c_k$ (and $k < r$). So by Proposition (4.32), the rows of C, that is, $v_1 ..., v_r$ are also linearly dependent. This by itself does not mean that $u_1,..., u_r$ are linearly dependent, because v_i's are merely truncations of u_i's. However, we show that $\{u_1,..., u_r\}$ is linearly dependent. This will contradict our hypothesis and thereby establish our result.

Since $v_1 .., v_r$ is linearly dependent, there exist $\lambda_1,..., \lambda_r \in F$, not all 0, such that $\lambda_1 v_1 +...+ \lambda_r v_r = 0$. This means that for all $j = 1,..., r$, $\lambda_1 b_{1j} + \lambda_2 b_{2j} +...+ \lambda_r b_{rj} = 0$. We now show that this holds for every j, that is even for $r < j \leqslant n$. Fix such j. Then c_j is a linear combination of $c_1 ..., c_k$, say $c_j = \mu_1 c_1 + \mu_2 c_2 +...+ \mu_k c_k$. This means that for every $i = 1,..., r$, $b_{ij} = \mu_1 b_{i_1} + \mu_2 b_{i_2} +,..+ \mu_k b_{ik}$. So,

$$\lambda_1 b_{1j} + \lambda_2 b_{2j} +...+ \lambda_r b_{rj}$$

$$= \lambda_1 \left(\sum_{p=1}^{k} \mu_p\, b_{1p} \right) + \lambda_2 \left(\sum_{p=1}^{k} \mu_p\, b_{2p} \right) + ... + \lambda_r \left(\sum_{p=1}^{k} \mu_p\, b_{rp} \right)$$

$$= \mu_1 \left(\sum_{q=1}^{r} \lambda_q\, b_{q1} \right) + \mu_2 \left(\sum_{q=1}^{r} \lambda_q\, b_{q2} \right) +...+ \mu_k \left(\sum_{q=1}^{r} \lambda_q\, b_{qk} \right)$$

$$\text{(by regrouping of terms)}$$

$$= \mu_1 0 + \mu_2 0 +...+ \mu_k 0 \text{ (since } \sum_{q=1}^{r} \lambda_q\, b_{qp} = 0 \text{ for all } p = 1,2,...,k) = 0.$$

This shows that $\lambda_1 b_{1j} + \lambda_2 b_{1j} +...+ \lambda_r b_{rj} = 0$ for *all* $j = 1,2,..., n$. This indeed means that the rows of B, that is $u_1,..., u_r$ (and not merely their truncations $v_1,..., v_r$) are linearly dependent. ∎

Now we are ready to prove the equality of the two ranks for all matrices. The proof is by linking both of them to a third, common number.

4.35 Theorem: The row rank and the column rank of an $m \times n$ matrix A are equal. This common number also equals the order of the largest non-singular square submatrix of A.

Proof: Let r be the row rank of A. We claim that A contains a non-singular $r \times r$ submatrix and also that it does not contain a non-singular square submatrix of a higher order. Let $\{u_{i_1} u_{i_2}, ..., u_{i_r}\}$ be a linearly independent set of rows of A and let B be the $r \times n$ submatrix of A formed by these rows. Then the row rank of B is r and so by the last proposition its column rank is also r. So there exist some columns $c_{j_1}, c_{j_2},....c_{j_r}$ of B which are linearly independent. Let C be the matrix formed by these r columns (of B). Then C is an $r \times r$ submatrix of A. Also C is non-singular by Proposition (4.33). Thus A contains a non-singular square submatrix of order r. Now suppose $s > r$ and D is a square submatrix of A of order s. We show D is singular. Suppose D is obtained by rows $u_{p_1}, u_{p_2},..., u_{p_s}$ and columns $c_{q_1}, c_{q_2},..., c_{q_s}$ of A. Then the rows of D are truncations of $u_{p_1},..., u_{p_s}$. Now, since $s > $ row rank of A, $u_{p_1},..., u_{p_s}$ are linearly dependent and *a fortiori* so are their truncations. Thus the rows of D are linearly dependent. By Proposition (4.33), D is singular.

Thus we have shown that the row rank of every matrix equals the order of the largest non-singular square submatrix. Let us apply this to A', the transpose of A. Then the row rank of A' equals the order of the largest non-singular square submatrix of A'. Now, a square submatrix of A' is the transpose of a square submatrix of the same order of A. Also a square matrix in non-singular iff its transpose is non-singular (since the two have the same determinants). It follows that the row rank of A' equals the order of the largest non-singular square submatrix of A. So row rank of $A' =$ row rank of A. But row rank of $A' =$ column rank of A. This completes the proof. ▨

The second assertion has an interesting theoretical consequence. In our treatment so far, we have emphasised the role of the ground field. However, as far as the rank of a matrix is concerned, it turns out to be independent of the ground field, as long as the ground field is large enough to include all the entries of the matrix. Let A be an $m \times n$ matrix over a field F. Suppose K is an extension field of F. Then A can also be thought of as a matrix over K. Conceivably, the rank of A as a matrix over F could differ from its rank as a matrix over K. But the theorem obove shows that this cannot happen. For, suppose B is a square submatrix of A. Then det (B), being a sum of the products of the entries of B is independent of whether A is regarded as a matrix over F or over K. Consequently, B is non-singular as a matrix over F iff it is non-singular as a matrix over K. Hence the order of the largest non-singular submatrix of A is independent of whether A is treated as a matrix over F or over K.

Determination of the rank of a matrix is important in various connections. For example, if we are given vectors u_1, u_2...., u_m in R^n and are asked to find the dimension of the subspace of R^n spanned by these vectors, this is equivalent to finding the rank of the $m \times n$ matrix whose rows are $u_1,...,u_m$. The rank also plays a crucial role in solving systems of linear equations as we shall see shortly. So it is desirable to have a method for finding the rank of a given matrix. It is not a practicable proposition to evaluate the determinants of all square submatrices and then apply the last theorem. The method we give below is due to Gauss and called the **row reduction.** In this method, given an $m \times n$ matrix A, we apply a series of operations to its rows so as to get another $m \times n$ matrix B which has the same rank as A, but which is in a standard form called the **echelon form** and consequently whose rank can be determined by inspection. Let us first define this form. (The word 'echelon' means a steplike formation and the name will be justified from the form).

4.36 Definition: An $m \times n$ matrix $B = (b_{ij})$ over a field F is said to be in the **echelon form,** if there exists an integer r, $0 \leqslant r \leqslant m$ and a strictly monotonically increasing sequence of positive integers $1 \leqslant j_1 < j_2 < ... < j_r \leqslant n$ such that

(i) for every $i = 1,...,r$, $b_{ik} = 0$ for $k < j_i$ and $b_{ik} = 1$ for $k = j_i$

(ii) for every i with $r < i \leqslant m$, and for all $k = 1, 2,...,n$, $b_{ik} = 0$.

A matrix in the echelon form looks like

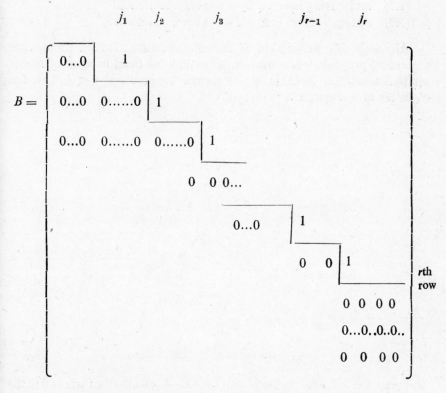

Here all entries below the 'staircase' are 0. (Hence the name.) We now compute the rank of this matrix,

4.37 Proposition: The rank of the matrix B above in the echelon form is r.

Proof: Let $v_1, ..., v_r, ..., v_m$ be the rows of B. Then for $i > r$, v_i is identically 0. We claim that $v_1, ..., v_r$ are linearly independent. Let $\lambda_1, ..., \lambda_r \in F$ and suppose $\lambda_1 v_1 + \lambda_2 v_2 + ... + \lambda_r v_r = 0$. Then for every k, $\lambda_1 b_{1k} + \lambda_2 b_{2k} + ... + \lambda_r b_{rk} = 0$. Put $k = j_1$. Then $b_{2k} = b_{3k} = ... = b_{rk} = 0$ and $b_{1k} = 1$. This gives $\lambda_1 = 0$. Hence $\lambda_2 b_{2k} + \lambda_3 b_{3k} + ... + \lambda_r b_{rk} = 0$ for all k. Now put $k = j_2$ and apply the same reasoning to get $\lambda_2 = 0$. Continuing in this manner, we get $\lambda_1 = \lambda_2 = ... = \lambda_r = 0$. Thus the first r rows of B are linearly independent. But all the remaining rows are identically 0. So r is the maximum number of linearly independent rows of B and therefore equals its rank. ∎

Now suppose A is an $m \times n$ matrix. We claim that by performing certain row operations on A, it can be reduced to a matrix in the echelon form, without affecting its rank. These row operations are of three types:

(R1) interchange of any two rows,
(R2) multiplying one row by a non-zero scalar and
(R3) adding a scalar multiple of one row to another.

Although this description of the row operations is clear enough, for theoretical pruposes it is convenient to paraphrase them in terms of matrix multiplication. Let A be an $m \times n$ matrix. Suppose $1 \leqslant i < j \leqslant m$. Let P_1 be the $m \times n$ matrix

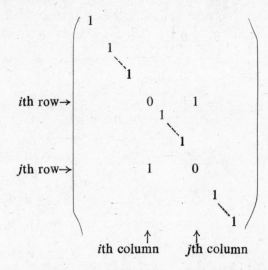

This matrix is the same as the identity matrix I_m except that there is 0 in the (i, i)th and (j, j)th place and there is 1 in the (i, j)th and in the (j, i)th place. It is clear that $P_1 A$ is the same matrix as A with the ith and jth row interchanged. Thus the row operation R1 is equivalent to multiplication by P_1 on the left. Note that P_1 is its own inverse. Similarly, R2, that is multiplying one of the rows, say, the ith row, by λ is equivalent to multiplication by P_2 where P_2 is like I_m except that the (i, i)th place is λ. If $\lambda \neq 0$, then P_2 is invertible (with the inverse having $1/\lambda$ in the (i, i)th place). For the third operation R3, suppose λ times the ith row of A is added to the jth row. Then it is easily seen that this is equivalent to forming $P_3 A$ where P_3 is the $m \times m$ matrix which is the same as I_m except that the (j, i)th entry is λ. P_3 is also invertible (with the inverse having $-\lambda$ in the (j, i)th place).

We leave it to the reader to verify (using this interpretation or directly) that the rank of a matrix is unaffected under any of these operations and hence under any composition of them in any order. We show that by performing these operations in a suitable order, every matrix can be reduced to an echelon form.

4.38 Proposition: Let $A = (a_{ij})$ be an $m \times n$ matrix. Then by applying the row operations above in a suitable order, A can be reduced to a matrix (of the same rank as A) which is in the echelon form.

Proof: If all columns of A are identically 0 then A is the zero matrix which is already in the echelon form (with $r = 0$), Otherwise, let j_1 be the smallest integer such that the j_1th column of A is non-zero. Then $a_{ij_1} \neq 0$ for some i. Pick any such i and interchange the ith row with the first row (i.e. apply $R1$). This gives a matrix C (say) in which all columns upto (j_1-1)th column are 0 and $c_{1j_1} \neq 0$. Divide first row of C by c_{1j_1} (i.e. apply R2). This gives a matrix D, in which the first j_1-1 columns are 0 and $d_{1j_1} = 1$. Now apply R3 to D, $(m-1)$ times. That is, for each $i = 2, m$, subtract d_{ij_1} times the first row from the ith row. This gives a new matrix E in which the j_1th column has 1 at the top and 0's everywhere else. Let A_1 be the $(m-1) \times (n-j_1)$ submatrix of E obtained by removing the first row and the first j_1 columns of E. If $A_1 = 0$, we stop and let $B = E$ because the matrix E is in the echelon form. If A_1 is not 0, we subject it to the same procedure as A. A is then reduced to a matrix say G of the form

$$
\begin{array}{c}
\quad j_1\text{th column} \quad j_2\text{th coiumn} \\
\qquad\quad \downarrow \qquad\qquad \downarrow \\
G = \begin{pmatrix}
0 & 0...0 & 1.................... \\
0 & 0...0 & 0.........0\ 1... \\
0 & 0 & 0 \\
\vdots & \vdots & \quad\vdots \quad A_2 \\
0 & 0...0 & 0
\end{pmatrix}
\end{array}
$$

for some $j_2 > j_1$. Let A_2 be the submatrix of G obtained by removing the first two rows and the first j_2 columns of G. Once again if $A_2 = 0$, we stop and let $B = G$ which is in the echelon form. Otherwise subject A_2 to the same procedure. Note that A_i has $m-i$ rows for $i = 1, 2,....$ So this process cannot go on beyond m steps. So it must stop at some stage with an $m \times n$ matrix B in the echelon form. ∎

The construction is reader simple and will be illustrated in Problem (4.41). The reader is also urged to illustrate it on his own with one or two examples. Since the rank of a matrix in the echelon form is known by Proposition (4.37), we now have a systematic procedure for computing the rank of a matrix. The procedure is in fact, so algorithmic in nature that a computer program can be written to implement it.

Finally, we study how the row reduction of a matrix plays a crucial role in the solution of a system of linear equations. Let us consider a system of m equations in n unknowns, $x_1, ..., x_n$ given by

$$a_{11}x_1 + \dots + a_{1n}x_n = b_1$$
$$a_{21}x_1 + \dots + a_{2n}x_n = b_2$$
$$\dots\dots\dots\dots\dots\dots\dots\dots\dots \qquad (*)$$
$$a_{m1}x_1 + \dots + a_{mn}x_n = b_m$$

We represent this as $Ax = b$ where x, b are column vectors of length n and m respectively and A is the $m \times n$ matrix (a_{ij}). If $b = 0$, then (*) becomes

$$Ax = 0 \qquad (**)$$

This is called a **homogeneous system**. From Proposition (4.2), the solutions of (**) from a vector subspace of $\mathbf{R}^n$ (namely the kernel of the linear transformation T associated with A). This subspace is called the **solution space** of (**). If rank of A is r, then by Exercise (3.10), the solution space has dimension $n - r$. Also, from the discussion following Proposition (5.3.6), the solution space of (*) is either empty or it is a coset of the solution space of (**). So, to find the general solution of (*), we first find the general solution of (**) and then any one particular solution of (*). We do these two parts separately.

The solution of (**) is simplified by the following Proposition.

4.39 Proposition: Suppose the matrix A is reduced to a matrix B in the echelon form by row reduction. Then the solution space of (**) is the same as that of $Bx = 0$.

Proof: We noted earlier that each row operation on a matrix corresponds to multiplying it on the left by a non-singular matrix. So $B = PA$ where P is an $m \times m$ matrix which is the product of a number of matrices of the form P_1, P_2, P_3 (corresponding to the row operations $R1$, $R2$ and $R3$ respectively). All these matrices are invertible. So P is invertible, that is non-singular. Now any solution of $Ax = 0$ is also a solution of $PAx = 0$, i.e. $Bx = 0$. Here we do not need that P is non-singular. But for the converse, suppose x is such that $Bx = 0$, i.e. $PAx = 0$. Now we apply (7) in Proposition (4.33) to the non-singular matrix P and to the column vector Ax. Then $PAx = 0$ gives $Ax = 0$. So x is a solution of (**). Thus solutions of (**) are the same as that of $Bx = 0$. ∎

When B is in the echelon form it is very easy to write down the solutions of $Bx = 0$. Let us adopt the notation of Definition (4.36). Then for

$$j \neq j_1, \dots, j_r,$$

x_j can be given any value. Once these values are fixed, x_j for $j = j_1, \dots, j_r$ is uniquely determined as follows. Consider v_r, the rth row of B. (This is the last non-zero row of B). Since $Bx = 0$, we have in particular, that $v_r x = 0$. But $v_r x$ is nothing but $x_{j_r} + b_{r, j_r+1} x_{j_r+1} + \dots + b_{rn}x_n$. Since $x_{j_r+1}, \dots, x_n$ are already determined, we must set $x_{j_r} = - (b_{r, j_r+1} x_{j_r+1} + \ldots + b_{rn}x_n)$. Now consider the $(r - 1)$th row, v_{r-1} of B. Then $v_{r-1} x = 0$. This gives

$$x_{j_{r-1}} + b_{r-1, j_{r-1}+1} x_{j_{r-1}+1} + \ldots + b_{r-1, n} x_{r-1, n} = 0.$$

In this equation, x_i is determined already for all $i > j_{r-1}$. So $x_{j_{r-1}}$ is uniquely determined by this equation. Next consider the $(r - 2)$th row of B and determine $x_{j_{r-2}}$. Continuing in this manner, x_{j_2} and finally x_{j_1} is determined.

The original matrix A has the same rank as B, namely, r. So we see that in the general solution of (**), there are $n - r$ arbitrary constants because $(n - r)$ variables out of $x_1, \ldots, x_n$ can be assigned values arbitrarily and the remaining x_j's are determined in terms of them. Since the dimension of the solution space of (**) is $n - r$, this is consistent with the intuitive meaning of dimension (given in the last section) as the number of free choices possible.

Having obtained the general solution of the homogeneous system (**), let us now solve the original system (*). that is $Ax = b$. Here b is a column vector of length m. We think of b as an $m \times 1$ matrix and perform exactly the same row operations on it as on A to reduce (*) to

$$Bx = c \qquad\qquad (*')$$

where B is in the echelon form and c is a column vector with entries $c_1, c_2, \ldots, c_m$ (say). More specifically, suppose $B = PA$ where P is a product of a number of matrices of the form P_1, P_2, P_3 corresponding to the row operations $R1, R2, R3$ respectively. Then $c = Pb$, and (*') is the same as $PAx = Pb$, or equivalently, $P(Ax - b) = 0$. Since P is non-singular, it follows that solutions of (*) and (*') are identical. So the problem reduces to solving (*'). The corresponding homogeneous system, $Bx = 0$ has already been solved. So we simply have to find any one solution for (*') or else prove that no solution exists. The following proposition answers this question completely.

4.40 Proposition: Suppose the matrix B in (*') has the form in Definition (4.36). Then (*') has a solution iff $c_i = 0$ for all $i > r$.

Proof: If $u_i = (b_{i1}, b_{i2}, \ldots, b_{in})$ is the ith row of B and x is a solution of (*') then we have $c_i = b_{i1} x_1 + b_{i2} x_2 + \ldots + b_{in}x_n$. But for $i > r$, u_i is identically 0. So if a solution exists for (*') then $c_i = 0$ for all $i > r$. Conversely, suppose $c_i = 0$ for all $i > r$. Then a solution can be obtained in the manner analogous to that for the homogeneous system. Set $x_j = 0$ for all $j \neq j_1, \ldots, j_r$. For $j = j_1, \ldots, j_r$ determine x_j one-by-one starting with $j = j_r$. We have,

$$x_{j_r} + b_{r, j_r+1} x_{j_r+1} + \ldots + b_{rn}x_n = c_r.$$

This gives $x_{j_r} = c_r$. Next, we have,

$$x_{j_{r-1}} + b_{r-1, j_{r-1}+1} x_{j_{r-1}+1} + \ldots + b_{r-1, j_r} x_{j_r} + \ldots + b_{r-1, n} x_n = c_{r-1}$$

This gives $x_{j_{r-1}} + b_{r-1,\,j_r}\,c_r = c_{r-1}$ and determines. $x_{j_{r-1}}$. Similarly we determine $x_{j_{r-2}}, \ldots, x_{j_2}$ and finally x_{j_1}. This way we get a solution for (*'). ∎
We illustrate this with a numerical example.

4.41 Problem: Find the general solutions of the following systems in the real unknowns $x_1,\ x_2,\ x_3,\ x_4,\ x_5$.

(i)
$$2x_3 - 2x_4 + x_5 = 2$$
$$2x_2 - 8x_3 + 14x_4 - 5x_5 = 2$$
$$x_2 + 3x_3 + x_5 = 8$$

(ii)
$$2x_3 - 2x_4 + x_5 = 2$$
$$2x_2 - 8x_3 + 14x_4 - 5x_5 = 0$$
$$x_2 + 3x_3 + x_5 = 8$$

Solution: (i) We write the system in the form $Ax = \begin{pmatrix} 2 \\ 2 \\ 8 \end{pmatrix}$ where A is the

matrix $\begin{pmatrix} 0 & 0 & 2 & -2 & 1 \\ 0 & 2 & -8 & 14 & -5 \\ 0 & 1 & 3 & 0 & 1 \end{pmatrix}$. We reduce A to a matrix in the echelon

form through the following sequence of matrices. We also indicate the row operation by which each matrix is obtained from the preceding one.

$$\begin{pmatrix} 0 & 0 & 2 & -2 & 1 \\ 0 & 2 & -8 & 14 & -5 \\ 0 & 1 & 3 & 0 & 1 \end{pmatrix} \text{(given matrix } A)$$

$$\begin{pmatrix} 0 & 2 & -8 & 14 & -5 \\ 0 & 0 & 2 & -2 & 1 \\ 0 & 1 & 3 & 0 & 1 \end{pmatrix} \text{(interchange of first two rows)}$$

$$\begin{pmatrix} 0 & 1 & -4 & 7 & -5/2 \\ 0 & 0 & 2 & -2 & 1 \\ 0 & 1 & 3 & 0 & 1 \end{pmatrix} \text{(multiplying first row by 1/2)}$$

$$\begin{pmatrix} 0 & 1 & -4 & 7 & -5/2 \\ 0 & 0 & 2 & -2 & 1 \\ 0 & 0 & 7 & -7 & 7/2 \end{pmatrix} \begin{array}{l}\text{(subtracting 1 times the first row}\\ \text{from the third)}\end{array}$$

$$\begin{pmatrix} 0 & 1 & -4 & 7 & -5/2 \\ 0 & 0 & 1 & -1 & 1/2 \\ 0 & 0 & 7 & -7 & 7/2 \end{pmatrix} \text{(multiplying the second row by 1/2)}$$

$$\begin{pmatrix} 0 & 1 & -4 & 7 & -5/2 \\ 0 & 0 & 1 & -1 & 1/2 \\ 0 & 0 & 0 & 0 & 0 \end{pmatrix} \begin{array}{l} \text{(subtracting 7 times the second row} \\ \qquad \text{from the third row)} \end{array}$$

This is in the echelon form. So this is the matrix B. Here $r = 2, j_1 = 2$ and $j_2 = 3$. So, the general solution of the homogeneous system is obtained by setting $x_1 = \lambda_1, x_4 = \lambda_4$ and $x_5 = \lambda_5$ arbitrarily and solving for x_3 and then for x_2. We have $x_3 - x_4 + \frac{1}{2}x_5 = 0$ from the second row. This gives $x_3 = \lambda_4 - \frac{1}{2}\lambda_5$. Also from the first row, $x_2 - 4x_3 + 7x_4 - \frac{5}{2}x_5 = 0$, giving $x_2 = 4x_3 - 7x_4 + \frac{5}{2}x_5 = 4(\lambda_4 - \frac{1}{2}\lambda_5) - 7\lambda_4 + \frac{5}{2}\lambda_5 = -3\lambda_4 + \frac{1}{2}\lambda_5$. Hence the general solution of the homogeneous system is $x_1 = \lambda_1, x_2 = -3\lambda_4 + \frac{1}{2}\lambda_5$, $x_3 = \lambda_4 - \frac{1}{2}\lambda_5, x_4 = \lambda_4$ and $x_5 = \lambda_5$ where $\lambda_1, \lambda_4, \lambda_5$ are arbitrary constants.

Now to solve (i), we apply the same row reduction to $\begin{pmatrix} 2 \\ 2 \\ 8 \end{pmatrix}$ and get suc-

cessively, $\begin{pmatrix} 2 \\ 2 \\ 8 \end{pmatrix}, \begin{pmatrix} 1 \\ 2 \\ 8 \end{pmatrix}, \begin{pmatrix} 1 \\ 2 \\ 7 \end{pmatrix}, \begin{pmatrix} 1 \\ 1 \\ 7 \end{pmatrix}$ and finally $\begin{pmatrix} 1 \\ 1 \\ 0 \end{pmatrix}$. In this column

vector all the entries after the 2nd row (note, $r = 2$) are 0. So a solution exists. To obtain a particular solution, we set $x_1 = x_4 = x_5 = 0$ and solve for x_3 and then for x_2. Once again, the second row of B gives $x_3 - x_4 + \frac{1}{2}x_5 = 1$. Hence $x_3 = 1$. Next, the first row of B gives $x_2 - 4x_3 + 7x_4 - \frac{5}{2}x_5 = 1$ which implies $x_2 = 5$. This gives a particular solution of (i). By adding it to the general solution of the homogeneous system, we get the general solution of (i) as

$$\begin{pmatrix} x_1 \\ x_2 \\ x_3 \\ x_4 \\ x_5 \end{pmatrix} = \begin{pmatrix} \lambda_1 \\ -3\lambda_4 + \frac{1}{2}\lambda_5 + 5 \\ \lambda_4 \quad - \frac{1}{2}\lambda_5 + 1 \\ \lambda_4 \\ \lambda_5 \end{pmatrix}$$

where λ_1, λ_4 and λ_5 are arbitrary constants, ranging over **R**, the real field.

(ii) Here the coefficient matrix is the same as in (i). So it reduces to the same matrix B in the echelon form. When we apply the same row operations to $\begin{pmatrix} 2 \\ 0 \\ 8 \end{pmatrix}$, it reduces, successively, to $\begin{pmatrix} 0 \\ 2 \\ 8 \end{pmatrix}, \begin{pmatrix} 0 \\ 2 \\ 8 \end{pmatrix}, \begin{pmatrix} 0 \\ 2 \\ 8 \end{pmatrix}, \begin{pmatrix} 0 \\ 1 \\ 8 \end{pmatrix}$

and finally to $\begin{pmatrix} 0 \\ 1 \\ 1 \end{pmatrix}$. Here there is a non-zero entry in the third row. Since $3 > 2 = $ rank of A, by Proposition (4.40), the system has no solution. Such a system is therefore called **inconsistent.** ▨

To conclude we show how the method of row reduction can be used to find the inverse (if any) of a square $n \times n$ matrix. Let X be the (unknown) inverse of A. Here X is an $n \times n$ matrix. Finding X is equivalent to solving the equation $AX = I_n$. For each j, let x_j be the jth column of X and let e_j be the column vector of length n in which only the jth entry is 1 and the other entries are all 0. Then solving for X really amounts to solving n systems of linear equations, namely, $Ax_j = e_j$ for $j = 1, ..., n$. These systems are independent of each other. But since the coefficient matrix is the same, namely A, a lot of work is common. So we apply row reduction to both the sides of $AX = I_n$ and get a non-singular matrix P such that $PAX = PI_n = P$, with $PA = B$ in the echelon form. If $r = $ rank of $B < n$, then there will be at least one j for which the equation $Ax_j = e_j$ cannot be solved. If $r = n$, then B will be of the form

$$\begin{pmatrix} 1 & \cdots\cdots\cdots\cdots \\ 0 & 1 & \cdots\cdots\cdots \\ 0 & 0 & 1 & \cdots\cdots \\ 0 & & 0 & \\ & & & 1\cdots \\ 0 & 0 & 0 & & 1 \end{pmatrix}$$

where all entries below the diagonal are 0. Now for each j, we determine, successively, $x_{nj}, x_{(n-1)j}, ..., x_{2j}, x_{1j}$ from the system of equations

$$B \begin{pmatrix} x_{1j} \\ x_{2j} \\ \vdots \\ x_{nj} \end{pmatrix} = Pe_j.$$

The technique of determining $x_{j_r}, x_{j_{r-1}}, ... x_{j_2}$ and finally x_{j_1} from (*') is called **back substitution.** A slightly different approach will be indicated in Exercise (4.40).

Exercises

4.1 Prove proposition (4.4).

4.2 A matrix $A = (a_{ij})$ is called **symmetric** if $a_{ij} = a_{ji}$ for all i, j. It is called **skew-symmetric**, if $a_{ij} = -a_{ji}$ for all i, j. Obviously such matrices must be square matrices. Prove that a matrix A is symmetric iff $A = A'$, the transpose of A and that it is skew-symmetric iff $A = -A'$. For any square matrix A, prove that $A + A'$ is symmetric and $A - A'$ is skew-symmetric.

4.3 Prove that every square matrix over a field of characteristic different from 2 can be expressed as a sum of two matrices one of which is symmetric and the other skew-symmetric.

4.4 Prove that the property of the basis in Theorem (4.5) actually characterises a basis. In other words, suppose B is a subset of a finite-dimensional vector space with the property that for every vector space W and for every function $f: V \to W$, there exists a unique linear transformation $T: V \to W$ which extends f. Prove that B is a basis for V.

4.5 Prove that two vector spaces over a field are isomorphic iff they have the same dimension. In particular prove that every vector space is isomorphic to its dual.

*4.6 Let V be a vector space over F, V^* its dual and V^{**} its double dual (i.e., V^{**} is the dual of V^*). For each $v \in V$, define $e_v: V^* \to F$ by $e_v(T) = T(v)$. This function e_v is called, quite appropriately, the **evaluation** at v. Prove that e_v is a linear transformation and hence $e_v \in V^{**}$. Define $\theta: V \to V^{**}$ by $\theta(v) = e_v$ for $v \in V$. Prove that θ is a vector space isomorphism.

4.7 Define $T: \mathbb{R}^3 \to \mathbb{R}^3$ by $T\begin{pmatrix} x_1 \\ x_2 \\ x_3 \end{pmatrix} = \begin{pmatrix} 2x_1 - x_2 \\ x_1 + x_2 + x_3 \\ x_2 \quad -x_3 \end{pmatrix}$. Find the matrix of T w.r.t. the ordered basis $\left(\begin{pmatrix} 1 \\ 0 \\ -1 \end{pmatrix}, \begin{pmatrix} 2 \\ 1 \\ 3 \end{pmatrix}, \begin{pmatrix} 1 \\ -1 \\ 1 \end{pmatrix} \right)$, Find the rank of T.

4.8 Let V be the set of all polynomials over a field F, of degree $\leqslant 4$, including the zero polynomial. Define $D: V \to V$ by $D(f(x)) = f'(x)$, the formal derivative of $f(x)$, (see Exercise (1.20)). Prove that D is a linear transformation. (This is often expressed by saying that differentiation is a linear operator.) Find the matrix of D w.r.t. the ordered basis $(1, x, x^2, x^3, x^4)$ and also w.r.t. some other ordered basis.

4.9 Let $\alpha, \beta, \gamma, \delta$ be elements of a field F. Define $T \colon M_2(F) \to M_2(F)$ by $T(A) = \begin{pmatrix} \alpha & \beta \\ \gamma & \delta \end{pmatrix} A$ for $A \in M_2(F)$. Prove that T is a linear transformation and find its matrix w.r.t. the ordered basis

$$\left(\begin{pmatrix} 1 & 0 \\ 0 & 0 \end{pmatrix}, \begin{pmatrix} 0 & 1 \\ 0 & 0 \end{pmatrix}, \begin{pmatrix} 0 & 0 \\ 1 & 0 \end{pmatrix}, \begin{pmatrix} 0 & 0 \\ 0 & 1 \end{pmatrix} \right).$$

4.10 Let $\sigma \in S_n$, i.e. σ is a permutation of the set $\{1, 2, ..., n\}$. Let $(v_1, ..., v_n)$ be an ordered basis for a vector space V. Let T_σ be the unique linear transformation which takes v_i to $v_{\sigma(i)}$ for $i = 1, ..., n$. Let P_σ be the matrix of T_σ w.r.t. the ordered basis $(v_1, ..., v_n)$. P_σ is called the **permutation matrix** of σ. Prove that in every row and column of P_σ there is exactly one 1 and all remaining entries are 0. If $\sigma, \tau \in S_n$ prove that $P_{\sigma \circ \tau} = P_\sigma P_\tau$ and $P_{\sigma^{-1}} = (P_\sigma)^{-1}$. Show that every $n \times n$ matrix with exactly one 1 in each row and column and zeros elsewhere equals P_σ for a unique $\sigma \in S_n$. Finally, prove that $\det (P_\sigma) = (-1)^\sigma$.

4.11 Let F be a field and n a positive integer. Let $GL(n, F)$ and $SL(n, F)$ be respectively the sets $\{A \in M_n(F) : A \text{ is non-singular}\}$ and $\{A \in M_n(F) : \det (A) = 1\}$. Prove that:

(i) Both $GL(n, F)$ and $SL(n, F)$ are groups under matrix multiplication

(ii) $SL(n, F)$ is a normal subgroup of $GL(n, F)$ and that the quotient group is isomorphic to the multiplicative group F^*, i.e., the group of non-zero elements of F under multiplication. (Hint: Note that the determinant function gives a group homomorphism from $GL(n, F)$ to F^*.)

(iii) If $P \colon S_n \to GL(n, F)$ is defined by $P(\sigma) = P_\sigma$, then P is a monomorphism.

4.12 Given a finite group G and a field F, prove that for some integer n, G is isomorphic to a subgroup of $GL(n, F)$. (Theorem (4.14) shows how matrices provide a concrete representation of a field extension. This exercise shows how every finite group can be represented as a group of matrices.)

4.13 A square matrix is called **upper triangular** or (simply triangular) if all the entries below its diagonal are 0, i.e. $a_{ij} = 0$ for $j < t$. A lower triangular matrix is defined dually (or using the transpose). Prove that if $A = (a_{ij})$ is an $n \times n$ triangular matrix, then

$$\det (A) = a_{11} a_{22} ... a_{nn}.$$

4.14 Prove that when a scalar multiple of a row (or column) of a square matrix is added to some other row (or column), its determinant is unchanged.

4.15 Generalise Theorem (4.26) as follows. Let A be an $m \times n$ matrix with $m \leqslant n$. Let B be an $n \times m$ matrix. Then AB is an $m \times m$ matrix. If $j_1,\ldots, j_m$ are integers with $1 \leqslant j_1 < j_2 < \ldots < j_m \leqslant n$, let $A_{j_1,\ldots, j_m}$ denote the $m \times n$ submatrix of A formed by its j_1th, j_2th,..., and j_mth columns and let $B_{j_1,\ldots, j_m}$, denote the $m \times m$ submatrix of B formed by its j_1th,..., j_mth rows. Prove that

$$\det (AB) = \sum_{1 \leqslant j_1 < \ldots < j_m \leqslant n} \det (A_{j_1,\ldots, j_m}) \det (B_{j_1,\ldots, j_m}).$$

4.16 Prove **Cramer's rule.** Suppose A is a non-singular $n \times n$ matrix. Then show that for every column vector b of length n, the equation $Ax = b$ has a unique solution for the column vector x, and further that this solution is given by $x_i = \dfrac{\det (A_i)}{\det (A)}$ where A_i is the matrix obtained from A by replacing its ith column by the column vector b.

4.17 The **trace** of an $n \times n$ matrix A is defined as the sum of its diagonal elements, that is, as $\sum_{i=1}^{n} a_{ii}$. If A, B are two $n \times n$ matrices prove that (i) trace $(A) =$ trace (A'), (ii) trace $(\lambda A + \mu B) = \lambda$ trace $(A) + \mu$ trace (B) for any scalars λ, μ and (iii) trace $(AB) =$ trace (BA).

4.18 Prove that the trace is a similarity invariant. (Hint: In (iii) in the last exercise, let $A = CD$ and $B = C^{-1}$.)

4.19 Let $T : V \to V$ be a linear transformation where V is a vector space over a field F. An element $\lambda \in F$, is called an **eigenvalue** of T if there exists a non-zero vector $v \in V$ such that $Tv = \lambda v$. Any such vector is called an **eigenvector** of T, corresponding to λ. Prove that all the eigenvectors of T corresponding to λ, along with the zero vector, form a subspace V_λ of V, called the **eigenspace** corresponding to λ. If V has a basis consisting of eigenvectors of T (not necessarily belonging to the same eigenvalue), prove that the matrix of T w.r.t. this basis is diagonal.

4.20 Define $T : \mathbf{R}^2 \to \mathbf{R}^2$ by $T \begin{pmatrix} x_1 \\ x_2 \end{pmatrix} = \begin{pmatrix} 2 & 1 \\ 3 & 0 \end{pmatrix} \begin{pmatrix} x_1 \\ x_2 \end{pmatrix}$. Prove that T has two eigenvalues -1 and 3. Find corresponding eigenspaces.

4.21 Let V be set of all functions $f : \mathbf{R} \to \mathbf{R}$ which have derivatives of all orders. Prove that V is a vector space over $\mathbf{R}$, under pointwise addition and scalar multiplication. Prove that the function $D : V \to V$ defined by $D(f) = f'$, the derivative function of f, is a linear transformation. Prove that every real number is an eigenvalue of D. If $\lambda \in \mathbf{R}$, find the eigenvectors corresponding to λ. (Such eigenvectors are often called **eigenfunctions.** This term appears frequently in the solution of linear differential equations).

4.22   Let $T:V \to V$ be a linear transformation where V is an n-dimensional vector space over a field F. Let A be the matrix of T w.r.t. some fixed ordered basis for V. Prove that an element $\lambda \in F$ is an eigenvalue of T if and only if the matrix $A - \lambda$ (i.e. the matrix $A - \lambda I_n$) is singular. (Quite often, λ is called an eigenvalue of A.)

4.23   Let A be an $n \times n$ matrix over a field F. Then $x I_n - A$ is an $n \times n$ matrix over the polynomial ring $F[x]$. Prove that the determinant of this matrix is a monic polynomial of degree n in $F[x]$. This polynomial is called the **characteristic polynomial of A**. Prove that an element $\lambda \in F$ is an eigenvalue of A iff it is a root of the characteristic polynomial of A. Verify this, for Exercise (4.20) by showing that the characteristic polynomial of the matrix $\begin{pmatrix} 2 & 1 \\ 3 & 0 \end{pmatrix}$ is $x^2 - 2x - 3$.

4.24   Prove that a linear transformation of an n-dimensional vector space into itself cannot have more than n distinct eigenvalues.

*4.25   Let $f(x)$ be a monic polynomial of degree n over a field F and let A be its companion matrix. Prove that the characteristic polynomial of A is $f(x)$.

**4.26   Prove that every matrix is a root of its characteristic polynomial.

4.27   Prove that the characteristic polynomial is a similarity invariant. (Hence if T is a linear transformation of a finite dimensional vector space into itself, we can unambiguously define its characteristic polynomial).

*4.28   Suppose a linear transformation T of an n-dimensional vector space V into itself has n distinct eigenvalues $\lambda_1, ..., \lambda_n$ with $v_1, v_2, ..., v_n$ as corresponding eigenvectors. Show that $\{v_1, ..., v_n\}$ is a basis for V, using induction on n. Then deduce that:

(i)   the characteristic polynomial of T is
   $(x - \lambda_1)(x - \lambda_2)...(x - \lambda_n)$.

(ii)   the trace of T is $\lambda_1 + \lambda_2 + ... + \lambda_n$.

(iii)   the determinant of T is $\lambda_1 \lambda_2 ... \lambda_n$.

4.29   If A, B are $m \times n$ and $n \times p$ matrices, prove that the rank of AB can exceed neither the rank of A nor that of B. (cf. Exercise (3.16).) Prove however that if $m = n$ and A is non-singular then rank (AB) = rank (B). [Hint: Apply Exercise (3.16), to the product $A^{-1}(AB)$, to get rank $(B) \leqslant$ rank (AB).] Similarly if $p = n$ and B is non-singular then rank (AB) = rank (A). Deduce that the rank of a linear transformation equals the rank of its matrix w.r.t. any ordered bases for its domain and codomain.

4.30 For an $n \times n$ matrix A prove that the following statements are equivalent:

    (1) A is non-singular

    (2) For every $n \times p$ matrix B, rank $(AB) =$ rank (B)

    (3) For every $m \times n$ matrix B, rank $(BA) =$ rank (B).

(This gives two more characterisations of non-singular matrices.)

4.31 Using the last exercise, prove that the row operations $R1$, $R2$ and $R3$ do not affect the rank of a matrix.

4.32 Consider the system (*) (that is, $Ax = b$) of m linear equations in n unknowns. Let $[A : b]$ be the $m \times (n+1)$ matrix obtained by adding the column vector b as one more column to A. Prove that (*) has a solution iff rank of A is the same as that of the matrix $[A : b]$. (This is essentially a paraphrase of Proposition (4.40). The matrix $[A : b]$ is often called the **augmented matrix** of (*).)

4.33 If $a_1, \ldots, a_n$ are elements of a commutative ring R with identity then the determinant of the matrix.

$$\begin{pmatrix} 1 & 1 & \cdots & 1 \\ a_1 & a_2 & \cdots & a_n \\ a_1^2 & a_2^2 & \cdots & a_n^2 \\ \cdots & \cdots & \cdots & \cdots \\ a_1^{n-1} & a_2^{n-1} & \cdots & a_n^{n-1} \end{pmatrix}$$

is called a **Vandermonde's determinant**. Prove that this determinant equals the product $\prod\limits_{1 \leqslant i < j \leqslant n} (a_j - a_i)$ and that it is non-zero when $a_1, \ldots, a_n$ are distinct elements of a field. (Hint: From each row, subtract a_1 times the preceding row, starting from the last row.)

4.34 A matrix of the form $\begin{pmatrix} 1 & 1/2 & 1/3 & \cdots & 1/n \\ 1/2 & 1/3 & 1/4 & \cdots & 1/(n+1) \\ \cdots & \cdots & \cdots & \cdots & \cdots \\ 1/n & 1/(n+1) & 1/(n+2) & \cdots & 1/(2n+1) \end{pmatrix}$

is called a **Hilbert matrix**. Using the method of row reduction find its inverse for $n = 3$ and 4. Can you guess a formula for its inverse for the general case?

4.35 Let V be a vector space of dimension n over a field F. Suppose $f : V \times V \to F$ is a bilinear function (also called a bilinear form, see Exercise (3.25) for definition.) Let $(v_1, \ldots, v_n)$ be an ordered basis for V. Let A be the $n \times n$ matrix (a_{ij}) where $a_{ij} = f(v_i, v_j)$. Prove that

f is uniquely determined by A. A is called the **matrix of the bilinear form** f w.r.t. the ordered basis $(v_1, ..., v_n)$. If $f: F^n \times F^n \to F$ is defined by $f((x_1, ..., x_n), (y_1, ..., y_n)) = \sum_{i=1}^{n} x_i y_i$, prove that the matrix of f w.r.t. the standard basis is I_n. How are the matrices of the same bilinear form w.r.t. two different bases related to each other?

4.36 Suppose $f(x) = a_0 + a_1 x + ... + a_n x^n$ is a primitive polynomial with integer coefficients. Suppose there is a prime number p such that p divides $a_0, a_1, ..., a_{n-1}$ but p does not divide a_n and moreover p^2 does not divide a_0. Prove that $f(x)$ is irreducible in $\mathbb{Z}[x]$. (Hint: Assume a factorisation and check divisibility by p of the coefficients of the factors, using Corollary (2.18) to get a contradiction.)

4.37 In the exercise above prove that $f(x)$ is, in fact, irreducible over $\mathbb{Q}[x]$. (Hint: Apply Exercise (2.30).)

4.38 If p is a prime number prove that the polynomial $x^{p-1} + x^{p-2} + ... + x + 1$ is irreducible in $\mathbb{Z}[x]$ and hence in $\mathbb{Q}[x]$. (Hint: Substitute $x = y + 1$ and show that the resulting polynomial in y is irreducible. Then deduce that the original polynomial is also irreducible.)

4.39 An $m \times n$ matrix $A = (a_{ij})$ over a field F is a said to be **partitioned** if there exist positive integers $p < q$ with $p < m$, $q < n$ such that $a_{ij} = 0$ for (i) all $1 \leqslant i \leqslant p$, $q < j \leqslant n$ and (ii) all $p < i \leqslant m$, $1 \leqslant j \leqslant q$. Equivalently, A has the form $\begin{pmatrix} B & 0 \\ \hline 0 & C \end{pmatrix}$ where B, C are $p \times q$ and $(m-p) \times (n-q)$ matrices respectively and all other entries are 0. Prove that

(i) rank (A) = rank (B) + rank (C)
(ii) if $m = n$ and $p = q$ then det (A) = det (B) det (C).

4.40 In reducing the system of equations (*) to (*'), at various stages multiples of the ith row ($i = 1, ..., r$) were subtracted only from the subsequent rows. Suppose we subtract suitable multiples of the ith row from the earlier rows as well so that the j_ith column has a 1 in the ith row and 0 everywhere else. Show that from the resulting system, the general solution can be written down by mere inspection. Do problem (4.41) this way.

Notes and Guide to Literature

The enormous length of this section should not come as a surprise. Apart from the monstrous notations needed for matrices, the topic of linear algebra is vast enough to occupy not only a whole chapter but a whole book. Our treatment constitutes only a glimpse at some important aspects of it. For more on canonical forms, eigenvalues etc. see Herstein [1].

We have discussed vector spaces over a general field F. When F is the field of real (or complex) numbers, an additional structure can be imposed through the concept of an inner product, as noted in the last section. Among all linear transformations from one inner product space to another, if we take those which behave 'nicely' w.r.t. the inner product, we can prove deeper results about them. Among such transformations are the so-called unitary and Hermitian transformations. See again Herstein [1].

The construction of the isomorphism θ in Exercise (4.6) may appear rather contrived. But it has a profound significance. The mere existence of an isomorphism between V and V^{**} follows from Exercise (3.5), since both have the same dimension. But such an isomorphism would depend on the choice of particular bases for V and V^{**}. The isomorphism θ, on the other hand, does not depend on the choice of any particular basis for V. For this reason it is called the **canonical isomorphism** from V to V^{**}. Elements of V^* are linear functions on V, while those of V^{**} are linear functionals on V^*. Identifying V with V^{**} (through θ) amounts to interchanging the roles of functions and their arguments. Intuitively, we are recovering V from V^* by taking linear functionals on V^{**}. This technique is used in many branches of mathematics. To cite one instance, the Stone representation theorem for Boolean algebras, is proved using this technique.

The result of Exercise (4.15) is called **Cauchy-Binet** theorem. It is useful in enumeration of certain graphs. (See the Epilogue)

Exercise (4.26) is the famous **Cayley-Hamilton theorem.** A truly instructive proof of it requires a good dip into canonical forms. But a slick, elementary proof using only properties of determinants over the ring of polynomials is also possible. For such a proof and a generalisation see Greenberg [1].

Determinants are among the most classical algebraic forms and have been heavily studied. It is not our intention to give a drill in the manipulation of determinants. By way of sample we have mentioned only the Vandermonde determinant and the Hilbert matrix. For more exercises on determinants see Knuth [1], Vol. I.

Exercise (4.36) is called **Eisenstein's criterion.** It is one of the few criteria available for proving primality of a polynomial.

We recall that in this section we have confined ourselves to finite dimensional vector spaces. Nearly all concepts here can be generalised to infinite dimensional spaces. But the generalisations are far from trivial.

For more extensive coverage of linear algebra and its applications, see Noble [1].

The method given in Exercise (4.40) is called the **Gauss-Jordan** method.

Seven

Advanced Counting Techniques

One of the most fascinating features of mathematics is the interplay between two apparently unrelated branches of it. In Chapter 1 we stressed the essential difference between discrete and continuous mathematics, namely that in the former there is no limiting process. That is why, in the problems of discrete mathematics the sets involved are generally finite. Handling infinite sets usually require some kind of limiting process. We remarked, however, that even in discrete mathematics, an element of the infinite comes from the fact that the set of positive integers is infinite. Most of the counting problems involve an integer variable, often several such variables. For example, in the Shares Problem, b_n is the number of shares during the nth year. Here the integer n can take any value $1, 2, 3, \ldots$. In any particular instance, we may be interested only in one value of n, say $n = 50$. In that case, b_{50} can be computed, in theory, by sheer arithmetical computations, starting from $b_0 = 0$, $b_1 = 1$ and using the recurrence relation $b_{n+2} = b_n + b_{n+1}$ with $n = 0, 1, \ldots, 48$. But this is hardly satisfying. Very often we have to know b_n for several values of n simultaneously. We would also like to know how rapidly b_n grows with n. To answer questions like this, we have to study the sequence $\{b_n\}_{n=0}^{\infty}$ as a whole. We would also like to 'solve' the recurrence relation $b_n = b_{n-1} + b_{n-2}$, i.e., get the value of b_n directly from n, without having to know b_{n-1}, b_{n-2}.

Generating functions provide a means of studying sequences of real (or complex) numbers and for solving recurrence relations. Given a sequence, say $\{a_n\}_{n=0}^{\infty}$, we define its ordinary generating function as $\sum_{n=0}^{\infty} a_n x^n = a_0 + a_1 x + a_2 x^2 + \ldots + a_n x^n + \ldots$. This definition is meaningless unless x is specified. We could let x be simply an indeterminate. In that case, the generating function is just a formal power series in x, discussed in Example 5 of Chapter 6, Section 1, where we defined the ring of formal power series. This approach has limited utility because the properties of the formal series have to be derived first and doing this may entail the same amount of work as proving results directly about the given sequence b_n in the first place. There is nevertheless some notational simplification and sometimes seeing things in a familiar notation can inspire ideas leading to solutions.

A far more fruitful approach is to let x be a real variable (or sometimes a complex variable, in which case it is often denoted by z). When this is done, the power series $\sum_{n=0}^{\infty} a_n x^n$ represents a function of the variable x which can range continuously over the entire interval of convergence of the power series, (or the disc of convergence in the case of a complex variable). This function, which is analytic, can be manipulated, differentiated, integrated and the terms of the original sequence $\{a_n\}$ can be recovered from it if desired. In other words, the rich machinery of continuous mathematics becomes available. Of course. before applying it, we must ensure that it is in fact applicable, that is, the hypotheses of its theorems are satisfied. Usually, this requires establishing the uniform convergence of the series in question. This care is often not exercised in elementary treatises on the subject. In other words, an attempt is made to tap the fruits of the theory of analytic functions without paying the price, namely, adhering to its rigour. This is probably keeping in line with the historical development of the subject. It is a fact that as great a mathematician as Euler, who proved numerous identities using generating functions, rarely bothered about the kind of logical discipline that would be expected of a mathematician today. What is truly remarkable is that despite the consequent vulnerability to going wrong, every single result of Euler is true. In other words, even though the method used was formal and lacked in rigour, the results themselves were correct and today they can all be substantiated with rigourous justifications.

The approach we shall take will be a combination of the two, although more inclined towards the naive approach. We shall take x to be a real (sometimes a complex) variable and use facts from the theory of power series. Although we shall omit the theoretical justifications, we shall point out where they are needed and where they can be found. In the first section we define generating functions and initiate their applications. More elaborate applications to combinatorial problems (including a solution to the Postage Problem) will be taken up in the second section. The third section deals with the methods of solving recurrence relations, one of which is based on generating functions. In the last section we present a number of problems which are solved by recurrence relations.

1.   Generating Functions of Sequences

How do you remember a sequence of letters? If the sequence constitutes a 'meaningful' word like 'MAGIC' or 'INDUCTION', there is little difficulty in remembering and reproducing it. But what if it is a crazy sequence like '$A - S - T - C$' or '$A - H - G - H - B - F - G - F - C$'*? In such cases, a most frequently used mnemonic device is to form a phrase or sentence the first letters of whose words are the members of the given sequence. For example, in the examples just given, two popular 'long forms' are 'All Silver Tea Cups' and 'A Handsome Guy Having Brave Features Goes For Cocacola': On the face of it, we are making the problem more complicated because instead of having to remember a sequence of single letters, we now have to remember whole words. Still, these memory tricks work because the phrase or the sentence has some meaning and hence is easier to remember. It is necessary, of course, that this meaning should appeal to us in some way or else we shall have a situation where the cure is worse than the disease. Further, in order that we can recover the original sequence of letters uniquely, the choice of the words should be such that the same meaning cannot be conveyed in more than one way. For example, if we agree that 'John is poor' means the same as 'John is not rich' then there would be an ambiguity as to whether the original sequence of letters was '$J - I - P$' or '$J - I - N - R$'. If we make a stipulation that the word 'Not' shall not be used, the second possibility is ruled out.

The underlying idea behind generating functions of sequences of real numbers is similar, although, of course, they go far deeper than cheap memory aids. We select a family of functions $f_0(x), f_1(x), f_2(x), \ldots$ of a real variable x. These functions are called **indicator functions**. Given a sequence $\{a_n\}_{n=0}^{\infty}$ of real numbers, we form the sum $\sum\limits_{n=0}^{\infty} a_n f_n(x)$. In general this sum is infinite and so the question of its convergence is important. Frequently, the indicator functions are real-valued and the convergence is the usual convergence for an infinite series of real numbers, namely, as the limit of the sequence of its partial sums. In other words

$$\sum_{n=0}^{\infty} a_n f_n(x) = \lim_{m \to \infty} S_m(x)$$

where

$$S_m(x) = \sum_{n=0}^{m} a_n f_n(x)$$

---

* Readers familiar with elementary trigonometry and coordinate geometry would recognise, of course, that neither sequence is all that crazy. The first is obtained from the first letters of 'All', 'Sine', 'Tangent' and 'Cosine' which describe which of the three trigonometric functions are positive in the various quadrants of the plane. The second sequence gives the entries in a 3×3 determinant which is crucial in analysing the general quadratic equation $Ax^2 + By^2 + 2Hxy + 2Gx + 2Fy + C = 0$.

for $m = 0, 1, 2, \dots$. This infinite sum represents a function of the variable x. Its domain is the set of all values of x for which the series $\sum\limits_{n=0}^{\infty} a_n f_n(x)$ is convergent. This function is called the generating function of the sequence $\{a_n\}$. Let us denote it by $A(x)$.

Sometimes, it is convenient to let the functions f_n have complex numbers as their domains and codomains. In that case, we often denote the generating function by $A(z)$. It is defined as $\sum\limits_{n=0}^{\infty} a_n f_n(z)$, where the convergence involved is that of a series of complex numbers. More generally, we could let the indicator functions take values in any vector space, say V, over $\mathbf{R}$ (or $\mathbf{C}$). If we have some sort of a convergence in V, then

$$\sum_{n=0}^{\infty} a_n f_n(x)$$

is a vector in V for every x for which the series is convergent. It is an infinite linear combination of the vectors $f_0(x), f_1(x), \dots$. But most of the time we shall confine ourselves to the case where the functions $f_n(x)$ are real-valued. To spell out the analogy with the mnemonic devices, the original sequence $\{a_n\}_{n=0}^{\infty}$ corresponds to the sequence of letters, the terms of the series $\sum\limits_{n=0}^{\infty} a_n f_n(x)$ are like the words whose first letters are from the sequence, while the sum function $A(x)$ plays the role of the phrase or the sentence formed by these words. Note that the generating function depends as much on the given sequence $\{a_n\}$ as it does on the choice of the indicator functions. We want to choose these generating functions once and for all. When we do so, we can represent a given sequence $\{a_n\}$ by its generating function.

Now, what would be a good choice of indicator functions? The answer is dictated by two requirements. First, the generating function $A(x)$ should be something easier to analyse, something whose properties are known to us. (In the analogy given above, this corresponds to the phrase or the sentence formed having a meaning which appeals to us.) Secondly, we should be able to reconstruct the sequence $\{a_n\}$ if we know its generating function $A(x)$. In other words, we should be able to 'resolve' $A(x)$, along the indicator functions. This condition is analogous to, but stronger than requiring that the indicator functions be linearly independent, as elements of the vector space of all real valued functions of x. (Linear independence deals only with finite linear combinations, whereas $A(x)$ is an infinite linear combination of the indicator functions). For example, we cannot take $f_0(x) = x^2, f_1(x) = 3x$ and $f_2(x) = 2x^2 + 6x$. Because then, no matter what the other f_n's are, the sequences $\{1, 1, 0, 0, 0, \dots\}$ and $\{0, 0, \frac{1}{2}, 0, 0, 0, 0, \dots\}$ would have the same generating function and so we cannot uniquely get back to a sequence from its generating function.

The two requirements just laid down severely restrict the choice of indicator functions. By far the most standard choice is to take $f_n(x) = x^n$ for $n = 0, 1, 2,...$ (with the understanding that $f_0(x) = 1$ for all x), or some variation thereof. This way the generating function $A(x)$ of a sequence $\{a_n\}$ is simply the power series $\sum\limits_{n=0}^{\infty} a_n x^n$. As such, all the nice theorems about power series (a few of which will be listed shortly) become available. Although we shall stick to this choice of the indicator functions in most of what is to come, we remark that taking $f_n(x) = \cos nx$ or $\sin nx$ is also a fruitful choice because the generating function so obtained is what is known as a Fourier series and next to power series, these are probably the most thoroughly studied series of functions.

With this rather lengthy introduction we now come to the most basic definition of this chapter.

1.1 Definition: Let $\{a_n\}_{n=0}^{\infty}$ be a sequence of real numbers. Then its **ordinary generating function** (O.G.F) is defined as $\sum\limits_{n=0}^{\infty} a_n x^n$ and its **exponential generating function** (E.G.F.) is defined as $\sum\limits_{n=0}^{\infty} \frac{a_n}{n!} x^n$.

For example, the O.G.F. of the sequence 1, 1, 1, 1,... is

$$1 + x + x^2 + ... + x^n + ... = \sum_{n=0}^{\infty} x^n$$

while its E.G.F. is

$$\sum_{n=0}^{\infty} \frac{x^n}{n!} = 1 + x + \frac{x^2}{2!} + \frac{x^3}{3!} + ... + \frac{x^n}{n!} + ...$$

Most of the time x will be a real variable. However, occasionally we shall need to assign it complex values, in which case we shall sometimes denote it by z. As a rule, if the terms of a sequence are denoted by putting suffixes on a small case letter, its O.G.F. will be denoted by the corresponding captial letter. Thus the O.G.F. of $\{a_n\}_{n=0}^{\infty}$ is $\sum\limits_{n=0}^{\infty} a_n x^n = A(x)$, that of $\{b_n\}$ is $B(x)$ and so on. There is no standard notation for E.G.F. of a sequence. Actually it is not an independent concept, because the E.F.G. of a sequence $\{a_n\}$ is precisely the O.G.F. of the sequence $\left\{\dfrac{a_n}{n!}\right\}$ and vice versa. The reason for the word 'exponential' is that the E.F.G. of the sequence 1, 1, 1,...1,... is $\sum\limits_{n=0}^{\infty} \dfrac{x^n}{n!}$, which is precisely the power series expansion of the exponential function e^x. Since we shall refer to the exponential generating function far less often than the ordinary generating function, by 'generating function' we shall mean O.G.F. unless otherwise stated.

We now summarise the basic properties of genergting functions.

1.2 Theorem: Let $A(x)$, $B(x)$ and $C(x)$ be respectively the O.G.F.'s of the sequences $\{a_n\}$, $\{b_n\}$, and $\{c_n\}$. Then

(i) *(Uniqueness)* $A(x) = B(x)$ if and only if $a_n = b_n$ for all n.

(ii) *(Linear Combinations)* Suppose λ, μ are constants such that $c_n = \lambda\, a_n + \mu\, b_n$ for all n. Then $C(x) = \lambda\, A(x) + \mu\, B(x)$.

(iii) *(Products)* Suppose for every n, $c_n = a_0 b_n + a_1 b_{n-1} + a_2 b_{n-2} + \ldots +$
$$a_{n-1} b_1 + a_n b_0 = \sum_{i=0}^{n} a_i b_{n-i}. \text{ Then } C(x) = A(x)\, B(x).$$

(iv) *(Differentiation)* $A(x)$ is an infinitely differentiable function of x and its derivatives can be obtained by term-by-term differentiation. That is,

$$A'(x) = a_1 + 2a_2 x + 3a_3 x^2 + \ldots + na_n x^{n-1} + \ldots + = \sum_{n=1}^{\infty} na_n x^{n-1}$$

$$A''(x) = 2a_2 + 6a_3 x + \ldots + n(n-1)a_n x^{n-3} + \ldots = \sum_{n=2}^{\infty} n(n-1)a_n x^{n-2}$$

$$\vdots$$

$$A^{(k)}(x) = \sum_{n=k}^{\infty} n(n-1)\ldots(n-k+1)a_n x^{n-k}.$$

Moreover, for each k, $a_k = \dfrac{A^{(k)}(0)}{k!}$ (with the understanding that $A^{(0)}(x) = A(x)$).

Proof: The meanings of and the arguments needed to prove these assertions depend on which approach is taken. If we let x be a formal variable (or an 'indeterminate'), then $A(x)$, $B(x)$ etc. are nothing more than formal power series. In this approach, the various concepts about power series are defined in such a way that the assertions (i) to (iv) would come out true. Thus two series $\sum_{n=0}^{\infty} a_n x^n$ and $\sum_{n=0}^{\infty} b_n x^n$ are *defined* to be equal if $a_n = b_n$ for all n. Similarly, their combinations and products are so defined that (ii) and (iii) hold (cf. Example 5 in Chapter 6, Section 1). Note that the rule for multiplication is the same as for multiplying polynomials and is based upon $x^i x^j = x^{i+j}$. Finally, for (iv), the formula for $A'(x)$ is the very definition of the formal derivative of $A(x)$ (cf. Exercise (6.1.20)). This derivative, unlike the derivatives in calculus, is not obtained through any limiting process. Even though it shares some of the properties of the derivatives in calculus, they have to be established independently. Theorems of calculus are useless. Note also that the very last statement is meaningless in the formal approach because once we let x be just a formal variable, we cannot assign it any value.

In the approach we have taken, however, x is a real variable and $A(x)$, $B(x)$ etc. are functions of x, defined by the respective power series in x. Their domains are the respective 'intervals of uniform convergence' (specifically, the interval of uniform convergence of $\sum_{n=0}^{\infty} a_n x^n$ is $(-R, R)$ where $R = 1/\overline{\lim} \mid a_n \mid^{1/n}$, (with the understanding that $1/0 = \infty$), R is called the radius of convergence of the power series.) and the statements to be proved here must be qualified by the phrase 'for all x in the intersection of the domains'. The proofs are based on properties of power series. (ii) is true for any series (not just power series). (iii) is also a consequence of a theorem about products of series. For (iv) to make sense, it is necessary to assume that the radius of convergence of $A(x)$ is positive. The proof of (iv) requires uniform convergence of the derived series. For notational uniformity, it is convenient to rewrite these formulae as

$$A'(x) = \sum_{n=0}^{\infty} (n+1) a_{n+1} x^n, \; A''(x) = \sum_{n=0}^{\infty} (n+1)(n+2) a_{n+2} x^n, \ldots,$$

$$A^{(k)}(x) = \sum_{n=0}^{\infty} (n+1)(n+2)\ldots(n+k) a_{n+k} x^n.$$

These expansions are valid in the interval of convergence of $A(x)$. Since this interval always contains 0, setting $x = 0$ we get $A^k(0) = a_k k!$ for all $k = 0$, $1, 2, \ldots$. This shows how we can recover the original sequence from its O.G.F. In particular it proves (i). We omit the details of these basic theorems because they belong to continuous mathematics and are available in any standard treatise of the subject. ▯

In this theorem, we started with sequences of real numbers and studied properties of their generating functions. For applications, it is convenient to paraphrase certain parts of this theorem starting from the other end. That is, we take a function $A(x)$ of x. Assuming that $A(x)$ can be expanded by a power series in x, say $A(x) = \sum_{n=0}^{\infty} a_n x^n$, it follows that $A(x)$ is the generating function of a sequence of real numbers, namely the sequence $\{a_n\}_{n=0}^{\infty}$. Uniqueness guarantees that this sequence is uniquely determined by the function $A(x)$. This means that the coefficient of each power of x is uniquely determined, So it makes sense to speak of 'the coefficient of x^n in $A(x)$' for every $n = 0, 1, 2, \ldots$. We shall often use this terminology. (For example, the coefficient of x^5 in $\dfrac{1}{1 - 2x}$ is 2^5.) Part (ii) of Theorem (1.2) can be paraphrased by saying that if $A(x)$, $B(x)$ admit power series expansions then for every $n = 0, 1, 2, \ldots$, the coefficient of x^n in $\lambda A(x) + \mu B(x)$ is the corresponding linear combination of the coefficients of x^n in $A(x)$ and $B(x)$. Similarly (iii) gives a formula for the coefficient of x^n in the product function $A(x)B(x)$. Part (iv) gives a formula for the coefficient of x^n in $A(x)$ in terms of the nth derivative of $A(x)$.

A crucial question now is how to tell if a given function of x can be expressed as a power series in x. Part (iv) gives a necessary condition for this, namely that the function be infinitely differentiable. For real functions, this condition is not sufficient. A classic counter-example is the function $A : R \rightarrow R$ defined by $A(x) = \exp(-1/x^2) = e^{-1/x^2}$ for $x \neq 0$ and $A(0) = 0$. Using properties of growth of exponential functions it is not hard to show that A is infinitely differentiable and that $A^{(k)}(0) = 0$ for all $k = 0, 1, 2, \ldots$. So $\sum_{k=0}^{\infty} A^{(k)}(0)x^k = 0 \neq A(x)$. Thus $A(x)$ cannot be expanded as a power series in $A(x)$. Surprisingly, the situation is much simpler for complex functions. A well-known theorem of complex analysis asserts that if a complex-valued function $f(z)$ of a complex variable is differentiable in a neighbourhood of 0, then it can be expanded as a power series in z and that this expansion is valid in any disc centered at 0 in which f is differentiable. This is why the theory of complex power series expansions is easier to handle than that of real power series expansions. The explanation for this apparently paradoxial situation is that complex differentiability is a much stronger condition than real differentiability. To stress this fact, complex differentiability is often called analyticity.

These difficulties need not worry us, however. The real functions that we shall consider will be expressible by power series since they will generally be restrictions of complex analytic functions.

A special case of (iii) is worth mentioning separately. Let r be a non-negative integer. Then the coefficient of x^n in $A(x)$ is the same as the coefficient of x^{n+r} in $x^r A(x)$. This follows by taking $B(x) = x^r$. In the formal approach, this follows more easily by multiplying the power series for $A(x)$ term by term by x^r.

Because of this theorem, we can start from the O.G.F.'s of a few standard sequences and build up those of many others. The next theorem lists the generating functions of some of the most standard sequences. Once again we omit the details of the proof because these results will be found in continuous mathematics except for a slight change of terminology. Instead of saying that $A(x)$ is the O.G.F. of a sequence $\{a_n\}$, a continuous mathematician is more apt to say that the power series expansion of $A(x)$ is $\sum_{n=0}^{\infty} a_n x^n$.

1.3 Theorem: (i) $1 + x + \dfrac{x^2}{2!} + \dfrac{x^3}{3!} + \cdots + \dfrac{x^n}{n!} + \ldots = \sum_{n=0}^{\infty} \dfrac{x^n}{n!} = e^x$ is valid for all x (ii) If n is a positive integer then $(1 + x)^n = 1 + \binom{n}{1}x + \binom{n}{2}x^2 + \cdots + \binom{n}{r}x^r + \cdots + \binom{n}{n-1}x^{n-1} + x^n = \sum_{r=0}^{n} \binom{n}{r}x^r$ is valid for all x. (iii) For any real n (not necessarily a positive integer) and any positive integer r, define $\binom{n}{r}$ to be $\dfrac{n(n-1)(n-2) \ldots (n-r+1)}{r!}$. Then $(1+x)^n$

$= \sum\limits_{r=0}^{\infty} \binom{n}{r} x^r$, is valid for all x with $|x| < 1$. (iv) For all $|x| < 1$, $\dfrac{1}{1-x}$
$= 1 + x + x^2 + \ldots + x^n + \ldots$. ∎

Comments: (ii) is a special case of the binomial theorem of algebra (see Exercise (6.1.28)). Note that the sum involved here is finite. By analogy (iii) is called the binomial theorem of calculus. Note, however, that here the numbers $\binom{n}{r}$ have no combinatorial significance. When n is in fact a positive integer, $\binom{n}{r}$ in (iii) does coincide with its usual meaning as the number of ways to select r objects out of n. Unlike (ii), (iii) is valid only for $|x| < 1$. (iv) is really a special case of (iii) obtained by taking $n = -1$ and replacing x by $-x$. It can also be derived directly from the formula for the sum of a geometric progression. More generally $\dfrac{1}{1-\alpha x} = 1 + \alpha x$
$+ \alpha^2 x^2 + \ldots + \alpha^n x^n + \ldots$ for all $|x| < \dfrac{1}{|\alpha|}$ where α is any non-zero real number. We shall occasionally need the complex version of this result.

The case of (*i*) is especially interesting. Sometimes the exponential function e^x is *defined* by the power series $\sum\limits_{n=0}^{\infty} \dfrac{x^n}{n!}$. In that case there is little to prove in (i) except to show that this series converges for all x. The price one has to pay is that all the properties of the exponential function (including the multiplicative property, namely, $e^{x+y} = e^x e^y$) have to be established from this series. Another approach is to define the exponential function as the inverse of the natural logarithm. Its properties then follow from those of the natural logarithm $\left(\text{which is usually defined as } \int\limits_{1}^{x} \dfrac{dt}{t}\right)$ and we shall use some of these properties. When this approach is taken, (i) is precisely the infinite Taylor series expansion of e^x at 0.

We now illustrate the applications of generating functions. The applications to problems of enumeration and to solving recurrence relations will be relegated to the next two sections. In this section we show how generating functions can be used to prove certain combinatorial identities. Generating functions have real variables for their arguments. Inasmuch as a real variable can assume infinitely many values, it would appear that we could generate a host of identities by merely assigning various values to the variable x. For example, by putting $x = \frac{1}{2}$ in (iv) of Theorem (1.3), we get $2 = 1 + \frac{1}{2} + \frac{1}{4} + \frac{1}{8} + \dots + \frac{1}{2^n} + \dots$. Care has to be taken to see that x is given only permissible values, i.e., values within the interval of convergence of the particular power series, or else some very absurd identities would result. For example, if we set $x = 2$ in (iv) of Theorem (1.3), we get the ridiculous result that $1 + 2 + 4 + 8 + \dots = -1$. Actually, it was absurdities like this that convinced mathematicians of the need to exercise caution in handling infinite series.

If we assign x a permissible value then the resulting identity would of course be mathematically valid. But that does not mean it will be worth stating separately. In order for such identities to have appeals of their own (other than as special cases of more general identities), it is necessary that at least one of the two sides should have some special significance or some elegance of form. Take for example, the binomial expansion in (ii) in Theorem (1.3), namely, $(1 + x)^n = 1 + \binom{n}{1} x + \binom{n}{2} x^2 + \dots + \binom{n}{n-1} x^{n-1} + x^n$, which is valid for all x. Putting $x = 10$ we get the identity $11^n = 1 + \binom{n}{1} 10 + \binom{n}{2} 100 + \dots + (10)^n$ which has no particular significance. But if we put $x = 1$, we get $2^n = 1 + \binom{n}{1} + \binom{n}{2} + \dots + \binom{n}{n-1} + 1$. Here the right hand side has a special significance. Since $\binom{n}{r}$ is the number of r-subsets of a set with n elements, the right hand side is the total number of subsets of a set with n elements. Thus we get an alternate proof of Theorem (2.2.15). (Cf. Exercise (2.2.16) (i) where this identity was to be established combinatorially.) Similarly setting $x = -1$, we get $0 = \sum_{r=0}^{n} (-1)^n \binom{n}{r}$ from which it follows that $\sum_{r \text{ even}} \binom{n}{r} = \sum_{r \text{ odd}} \binom{n}{r}$. Since the sum of these two sums equals 2^n (as we just saw), each sum equals 2^{n-1}, which means that the number of subsets of even cardinalities of a set with n elements is 2^{n-1}. This is equivalent to Proposition (2.3.3).

Interesting as this is, a far more frequent use of generating functions in proving combinatorial identities is through the uniqueness of power series expansions (Part (i) of Theorem (1.2)). The way this technique works is as follows. We take the given expression to be simplified (or one of the sides of the identity to be proved) and by some algebraic manipulations if necessary, identify it as the coefficient of some power of x, say of x^t in some function $A(x)$ of x. We then expand $A(x)$ by power series using a combination of some of the standard power series expansions. The coefficient of x^t in this expansion must equal the given expression.

As a typical illustration, we prove the following identity.

1.4   Proposition: For every positive integer n, $\displaystyle\sum_{r=0}^{n} \binom{n}{r}^2 = \binom{2n}{n}$.

Proof: We immediately recognise the right hand side as the coefficient of x^n in the expansion of $(1 + x)^{2n}$. If we could show that the left hand side equals the same, we would be through. The 'sum of the products' form of the left hand side suggests that Part (iii) of Theorem (1.2) may be useful. However, $\binom{n}{r}$ is the coefficient of x^r in $(1 + x)^n$ and if we want the result to come out as the coefficient of x^n, then the coefficient of x^r needs to be multiplied by that of x^{n-r}. This difficulty can be remedied by noting that $\binom{n}{r}$ is the same as $\binom{n}{n-r}$. Consequently we have,

$$\sum_{r=0}^{n} \binom{n}{r}^2 = \sum_{r=0}^{n} \binom{n}{r} \binom{n}{n-r}$$

$$= \sum_{r=0}^{n} [\text{coefficient of } x^r \text{ in } (1 + x)^n] \times [\text{coeff. of } x^{n-r} \text{ in } (1 + x)^n]$$

$$= \text{coefficient of } x^n \text{ in } (1 + x)^n (1 + x)^n \text{ by Theorem (1.2), (iii).}$$

Since $(1 + x)^n (1 + x)^n = (1 + x)^{2n}$, the result follows. ∎

As another example we do the following problem.

1.5   Problem: Let n, k, r be positive integers. Evaluate the sum

$$\binom{n}{k} + \binom{n+1}{k} + \ldots + \binom{n+r}{k}.$$

Solution: The given sum has for its summands the coefficients of x^k in $(1 + x)^n, (1 + x)^{n+1}, \ldots, (1 + x)^{n+r}$. By (ii) in Theorem (1.2), the given sum equals the coefficient of x^k in $(1 + x)^n + (1 + x)^{n+1} + \ldots + (1 + x)^{n+r}$. This observation would be of little help unless we have some way of handling this latter sum. Fortunately, we can do this by noting that it is the sum of the first $r + 1$ terms of a geometric progression with common ratio $(1 + x)$ and first term $(1 + x)^n$. Using the well-known summation formula for geometric progressions, we see that

$$(1 + x)^n + \ldots + (1 + x)^{n+r} = \frac{(1 + x)^{r+1} - 1}{1 + x - 1} (1 + x)^n$$

$$= \frac{1}{x} [(1 + x)^{n+r+1} - (1 + x)^n].$$

So the given sum equals the coefficient of x^k in $(1/x) [(1+x)^{n+r-1} - (1+x)^n]$, which is the same as the coefficients of x^{k+1} in $[(1 + x)^{n+r+1} - (1 + x)^n]$. By the binomial theorem the coefficients of x^{k+1} in $(1 + x)^{n+r+1}$ and $(1 + x)^n$ are, respectively, $\binom{n+r+1}{k+1}$ and $\binom{n}{k+1}$ and so by (ii) Theorem (1.2) again, their difference equals the coefficient of x^{k+1} in $[(1+x)^{n+r+1} - (1+x)^n]$. So the given sum equals $\binom{n+r+1}{k+1} - \binom{n}{k+1}$. ■

In this problem, all the summands were coefficients of the same power of x in various functions of x and hence Theorem (1.2) (ii) could be applied directly. Sometimes when the summands represent coefficients of different powers of x in various functions, we have to multiply these functions by suitable powers of x before applying (ii) of Theorem (1.2). We illustrate this technique in the next problem.

1.6 Problem: Let n, k, r be integers with $0 < r \leqslant k \leqslant n$. Evaluate the sum $\binom{n}{k} - \binom{n}{k-1} + \binom{n}{k-2} + - \binom{n}{k-3} + \ldots + (-1)^r \binom{n}{k-r}$.

Solution: For $i = 0, \ldots, r$, $\binom{n}{k-i}$ is the coefficient of x^{k-i} in $(1 + x)^n$, which is the same as the coefficient of x^k in $x^i (1 + x)^n$. So the given sum equals the coefficient of x^k in $\sum_{i=0}^{r} (-1)^i x^i (1 + x)^n$. This is again a geometric progression with common ratio $-x$. So

$$\sum_{i=0}^{r} (-x)^i (1 + x)^n = (1 + x)^n \frac{1 - (-1)^{r+1} x^{r+1}}{1 + x}$$

$$= [(1 + x)^{n-1} + (-1)^r x^{r+1} (1 + x)^{n-1}].$$

The coefficient of x^k in this function is $\binom{n-1}{k} + (-1)^r \binom{n-1}{k-r-1}$. This, therefore, is the given sum. ■

Note that the special case $r = 0$, gives Proposition (2.2.19), while if we put $n = r = k$, and use that $\binom{n}{i}$ equals $\binom{n}{n-i}$, we get

$$\binom{n}{0} - \binom{n}{1} + \binom{n}{2} - \binom{n}{3} + \ldots + (-1)^n \binom{n}{n} = 0,$$

which we obtained above by a different method, i.e. by setting $x = -1$ in the binomial expansion of $(1 + x)^n$.

So far we used the binomial expansion of $(1 + x)^n$. It should be noted that here x is a dummy variable. As such, we may replace it by any expression. In particular we may replace x throughout by $2x$, $x - 3$, x^2 or by other function of x. The same can be done for any other power series in x. Care has to be taken to see that this function of x takes only those values which lie in the interval of convergence of the original power series. For example, replacing x by $2x$ in (iii) in Theorem (1.3), we get $(1 + 2x)^n$

$$= \sum_{r=0}^{\infty} \binom{n}{r} (2x)^r = \sum_{r=0}^{\infty} 2^r \binom{n}{r} x^r.$$ This is valid whenever $|2x| < 1$,

i.e. whenever $|x| < \frac{1}{2}$. If the original expansion is valid for all x (as is the case with (ii) in Theorem (1.3)) then of course so is the new one.

We illustrate this technique of changing the variable in the following:

1.7 Problem: Let m, n be positive integers with $m \leqslant n$. Sum the series
$$\sum_{k=0}^{m} (-1)^k \binom{n}{k} \binom{n}{m-k}.$$

Solution: In the binomial expansion $(1 + x)^n = \sum_{k=0}^{n} \binom{n}{k} x^k$ if we replace

x by $-x$ we get $(1 - x)^n = \sum_{k=0}^{m} (-1)^k \binom{n}{k} x^k$. Using this and (iii) in

Theorem (1.2), we recognise the given sum as the coefficient of x^m in the product $(1 - x)^n (1 + x)^n$, i.e., in $(1 - x^2)^n$. Replacing x by $-x^2$ in the

binomial theorem, $(1 - x^2)^n = \sum_{k=0}^{n} (-1)^k \binom{n}{k} x^{2k}$. So we see that if m is

even, the given sum equals $(-1)^{m/2} \binom{n}{m/2}$ while if m is odd, it is 0. ∎

There is yet another powerful technique by which the validity of an identity can be extended considerably. It is based on Proposition (6.2.25) according to which the number of roots of a non-zero polynomial over a field cannot exceed its degree. As a consequence it follows that if $f(x)$ and $g(x)$ are polynomials of degree $\leqslant r$ (say) over a field F and if the polynomial equation $f(x) = g(x)$ holds for at least $r + 1$ distinct values of x in F, then it must hold identically over the field F. (Consider roots of the difference polynomial $h(x) = f(x) - g(x)$.) The way this technique works for proving identities is as follows. First we prove an identity for a variable n which takes only positive integral values. Usually, this is done by induction, a combinatorial argument or using generating functions as outlined above. If the two sides of the identity are polynomials in n (of not necessarily equal degree), then by what we just said, the identity must hold if we replace

n by a real variable x (or a complex variable z). We can then assign x any real value, often a fraction or a negative integer. Converting the two expressions we get a 'new' identity whose direct proof may not by so obvious.

We illustrate this technique in the following sequence of identities. The first is proved by a combinatorial argument.

1.8 Proposition: Let m, n be positive integers. Then

$$\sum_{k=0}^{m} \binom{m}{k}\binom{n+k}{m} = \sum_{k=0}^{m} \binom{m}{k}\binom{n}{k} 2^k$$

Proof: If $\binom{n+k}{m}$ were equal to $\binom{n}{k} 2^k$ then the identity would be trivial to prove. But of course in general $\binom{n+k}{m}$ need not equal $\binom{n}{k} 2^k$. The presence of the term 2^k suggests that a combinatorial argument might work since 2^k is the number of all subsets of a set with k elements. Let M, N be two mutually disjoint sets with m and n and n elements respectively. Let $\mathscr{S} = \{(X, Y): X \subset M, Y \subset N \cup X \text{ and } |Y| = m\}$. We apply a double counting argument. Letting $|X| = k$, where $k = 0, 1, ..., m$ and summing over k, we see that the left hand side of the identity is precisely $|\mathscr{S}|$. On the other hand, for $(X, Y) \in \mathscr{S}$, let $Y_1 = Y \cap M$ and $Y_2 = Y \cap N$. (See Figure 7.1). Note that $|Y_1| + |Y_2| = m$ and $Y_1 \subset X \subset M$. Conversely given any subsets Y_1, Y_2 of M, N respectively, with $|Y_1| + |Y_2| = m$ and a subset X with $Y_1 \subset X \subset M$, if we let $Y = Y_1 \cup Y_2$, we get an element $(X, Y) \in \mathscr{S}$. At this correspondence is obviously one-to-one, we can find

Figure 7.1: Proof of Proposition (1.8)

$|\mathscr{S}|$ by counting ordered triples (Y_1, Y_2, X) with $Y_2 \subset N$, $Y_1 \subset X \subset M$ and $|Y_1| + |Y_2| = m$. Let $j = |Y_2|$. Then $0 \leqslant j \leqslant m$, $|Y_1| = m - j$ and $|M - Y_1| = m - (m - j) = j$. For a fixed j, (Y_1, Y_2) can be chosen in $\binom{m}{m-j}\binom{n}{j}$ ways. For each such choice, the set X can be obtained by adding to Y_1 any subset of $M - Y_1$. Since there are, in all, 2^j subsets of

$M - Y_1$, it follows that $| \mathcal{S} | = \sum\limits_{j=0}^{m} \binom{m}{m-j} \binom{n}{j} 2^j$. Noting that

$\binom{m}{m-j} = \binom{m}{j}$ and changing the index of summation from j to k we

get the right hand side of the identity, proving the result. ∎

We now derive another identity from this, in the manner indicated above.

1.9 Proposition: With m, n as before,

$$\sum_{k=0}^{m} \binom{m}{k} \binom{n+k}{m} = (-1)^m \sum_{k=0}^{m} \binom{m}{k} \binom{n+k}{k} (-2)^k.$$

Proof: Although the left hand side of this side is identical to that in the last proposition, the nature of the right hand side makes it unlikely that a combinatorial argument would work. The two sides of the identity in the last proposition are polynomials in n. Since the identity holds for all positive integral values of n, it must hold if n is given any real value. In particular it holds if we replace n by $-n-1$, yielding

$$\sum_{k=0}^{m} \binom{m}{k} \binom{-n-1+k}{m} = \sum_{k=0}^{m} \binom{m}{k} \binom{-n-1}{k} 2^k \qquad (*)$$

Here the binomial coefficient $\binom{-n-1+k}{m}$ and $\binom{-n-1}{k}$ have no

combinatorial significance. They are formally defined by the formula

$$\binom{x}{r} = \frac{x(x-1)\ldots(x-r+1)}{r!}.$$

We leave it as an exercise to check that

$$\binom{-x}{r} = (-1)^r \binom{x+r-1}{r} \text{ for all } x.$$

So the left hand side becomes

$$(-1)^m \sum_{k=0}^{m} \binom{m}{k} \binom{n+m-k}{m}. \text{ Let } j = m - k. \text{ Then}$$

$$\binom{m}{k} = \binom{m}{j}.$$

(Here we do need that m, k are integers.) So

$$\sum_{k=0}^{m} \binom{m}{k} \binom{n+m-k}{m} = \sum_{j=0}^{m} \binom{m}{j} \binom{n+j}{m}.$$

which upon changing from j to k becomes $\sum\limits_{k=0}^{m} \binom{m}{k}\binom{n+k}{m}$. Thus the

left hand side of (*) reduces to $(-1)^m \sum\limits_{k=0}^{m} \binom{m}{k}\binom{n+k}{m}$. By a similar

reasoning, the right hand side of (*) becomes $\sum\limits_{k=0}^{m} \binom{m}{k}\binom{n+k}{k}(-2)^k$.

Multiplying both sums by $(-1)^m$ proves the result. ∎

In the applications so far, we did only algebraic manipulations to generating functions. We now study a couple of applications where term-by-term differentiation (Part (iv) of Theorem (1.2)) is used. Thus facts from continuous mathematics will figure for the first time. We begin by obtaining a formula for the sum $1 + 2 + \ldots + n$ where n is a positive integer. Although this formula is well-known and can be obtained by very elementary methods, we derive it here to illustrate the method used. First we prove a result which may be of independent interest.

1.10 Proposition: Let $A(x)$ be the O.G.F. of a sequence $\{a_n\}_{n=0}^{\infty}$. For each $n \geqslant 0$, let $c_n = a_0 + a_1 + \ldots + a_n$. Then the O.G.F. of the sequence $\{c_n\}$ is $\dfrac{A(x)}{1-x}$.

Proof: By (iv) in Theorem (1.3), $\dfrac{1}{1-x} = 1 + x + x^2 + \ldots + x^n + \ldots$. The result now follows from (iii) in Theorem (1.2), if we take $b_n = 1$ for all $n = 0$, 1, 2,... ∎

1.11 Theorem: For every positive integer n, $1 + 2 + \ldots + n = \dfrac{n(n+1)}{2}$.

Proof; In view of the last proposition, $1 + 2 + \ldots + n = $ coefficient of x^n in $\dfrac{A(x)}{1-x}$ where $A(x) = 0 + x + 2x^2 + 3x^2 + \ldots + nx^n + \ldots$. So we first need to find a closed form for $A(x)$. There is no function in our list so far which has $\sum\limits_{n=1}^{\infty} nx^n$ for its power series expansion. However, we start with

$$\frac{1}{1-x} = 1 + x + x^2 + \ldots + x^n + \ldots,$$

Differentiating both the sides w.r.t. x and using (iv) in Theorem (1.2), we get

$$\frac{d}{dx}\left(\frac{1}{1-x}\right) = \frac{1}{(1-x)^2} = \sum_{n=1}^{\infty} nx^{n-1}.$$

Multiplying both sides by x, we get $\dfrac{x}{(1-x)^2} = \sum\limits_{n=1}^{\infty} nx^n = A(x)$. So $1 + 2 + \ldots + n$ would be the coefficient of x^n in $\dfrac{x}{(1-x)^3}$, which is the same as the coefficient of x^{n-1} in $\dfrac{1}{(1-x)^3}$. To find it, we write $\dfrac{1}{(1-x)^3}$ as $(1-x)^{-3}$ and expand it by the binomial theorem (Part (iii) of Theorem (1.3)) as

$$\sum_{r=0}^{\infty} (-1)^r \binom{-3}{r} x^r \text{ where } \binom{-3}{r} = \frac{(-3)(-4)\ldots(-3-r+1)}{r!}$$

$$= (-1)^r \frac{(r+2)\ldots 4.3}{r!} = (-1)^r \binom{r+2}{2}. \text{ So, } (1-x)^{-3}$$

$$= \sum_{r=0}^{\infty} (-1)^{2r} \binom{r+2}{2} x^r = \sum_{r=0}^{\infty} \binom{r+2}{2} x^r. \text{ For } r = n-1,$$

$$\binom{r+2}{2} = \binom{n+1}{2} = \frac{n(n+1)}{2} \text{ and as noted before, this equals}$$

$1 + 2 + \ldots + n$. ∎

The crucial part in this proof was to get a closed form for a function whose power series expansion was known to us. This is not always possible to obtain and even when it is, considerable ingenuity may be needed to find it. In the next problem we encounter a situation where a differential equation has to be solved, hardly to be expected in discrete mathematics.

1.12   Problem: Evaluate the sum $\displaystyle\sum_{k=0}^{m} \binom{2k}{k} \binom{2m-2k}{m-k}$.

Solution: The nature of the sum indicates that it is the coefficient of x^m in $[A(x)]^2$ where $A(x) = \displaystyle\sum_{n=0}^{\infty} \binom{2n}{n} x^n$. We have to find a closed expression for $A(x)$. Differentiating we get,

$$A'(x) = \sum_{n=1}^{\infty} \binom{2n}{n} n x^{n-1} = \sum_{n=0}^{\infty} \binom{2n+2}{n+1} (n+1) x^n.$$

Now,

$$\binom{2n+2}{n+1} (n+1) = 2(2n+1) \binom{2n}{n} = 4n \binom{2n}{n} + 2 \binom{2n}{n}.$$

Multiplying by x^n are summing over n, we get

$$A'(x) = 4 \sum_{n=0}^{\infty} \binom{2n}{n} n x^n + 2 \sum_{n=0}^{\infty} \binom{2n}{n} x^n$$

$$= 4x \sum_{n=1}^{\infty} \binom{2n}{n} n x^{n-1} + 2 \sum_{n=0}^{\infty} \binom{2n}{n} x^n = 4x A'(x) + 2A(x).$$

Hence

$$A'(x) = \frac{2A(x)}{1-4x}.$$

If we let $y = A(x)$, this gives a differential equation $\dfrac{dy}{dx} = \dfrac{2y}{1 - 4x}$ whose general solution is $y^2 = \lambda(1 - 4x)^{-1}$, where λ is a constant. When $x = 0$,

$$y = \binom{0}{0} = 1, \text{ giving } \lambda = 1. \text{ So } y = (\sqrt{(1 - 4x)})^{-1} = (1 - 4x)^{-1/2}. \text{ (If we}$$

expand $(1 - 4x)^{-1/2}$ by the binomial theorem we indeed see it comes out as

$\displaystyle\sum_{r=0}^{\infty} \binom{2r}{r} x^r$. But we had to *arrive at* the answer rather than verify it.)

Since $A(x) = \dfrac{1}{\sqrt{1 - 4x}}$, $[A(x)]^2 = \dfrac{1}{1 - 4x} = \displaystyle\sum_{n=0}^{\infty} (4x)^n$ by (iv) in Theorem

(1.3). So the given sum equals 4^m. ∎

So far the variable x in the generating functions was assumed to be a real variable. As remarked after the proof Theorem (1.2), there are some theoretical advantages in considering generating functions of a complex variable. In the next problem we show that there are practical advantages too. Because of the fundamental theorem of algebra, a real polynomial can be split completely as a product of complex linear polynomials. This factorization is helpful while resolving into partial fractions, where the real factorisation may not be adequate. The problem also illustrates how sometimes it pays to treat finite sums as infinite sums.

1.13 Problem: Evaluate the sum $\displaystyle\sum_{m=\lceil n/2\rceil}^{n} (-1)^m \binom{m}{n-m}$ where n is a positive integer.

Solution: The binomial coefficient $\binom{m}{n-m}$ is 0 when $n - m < 0$ and also when $n - m > m$, i.e. when $m < \lceil n/2\rceil$. (Note that $\lceil n/2\rceil$ is the smallest integer $\geqslant n/2$.) So the given sum might as well be taken as the infinite sum

$\displaystyle\sum_{m=0}^{\infty} (-1)^m \binom{m}{n-m}$. Since $(-1)^m \binom{m}{n-m} = $ coefficient of x^{n-m} in $(-1)^m$

$(1 + x)^m = $ coefficient of x^n in $(-1)^m x^m (1+x)^m$, the given sum equals the

coefficient of $x^n \displaystyle\sum_{m=0}^{\infty} (-1)^m x^m (1 + x)^m$. By (iv) in Theorem (1.3), this equals,

$\dfrac{1}{1 + x + x^2}$. We resolve $\dfrac{1}{1 + x + x^2}$ into partial fractions. As a real polynomial, $1 + x + x^2$ is irreducible. But as a complex polynomial, it factors as $(1 - \omega x)(1 - \omega^2 x)$ where $\omega = \cos\dfrac{2\pi}{3} + i \sin\dfrac{2\pi}{3}$ is a primitive cube root of 1. Writing $\dfrac{1}{1 + x + x^2} = \dfrac{A}{1 - \omega x} + \dfrac{B}{1 - \omega^2 x}$, we get two equations for

A and B, namely $A + B = 1$ and $A\omega^2 + B\omega = 0$, which give $A = \dfrac{1}{1-w}$ and $B = \dfrac{1}{1-\omega^2}$. So $\dfrac{1}{1+x+x^2} = \dfrac{1}{1-\omega}\dfrac{1}{1-\omega x} + \dfrac{1}{1-\omega^2}\dfrac{1}{1-\omega^2 x}$. Expanding $\dfrac{1}{1-\omega x}$ and $\dfrac{1}{1-\omega^2 x}$ each by (iv) in Theorem (1.3) again, we see that the coefficient of x^n in $\dfrac{1}{1+x+x^2}$ is $\dfrac{\omega^n}{1-\omega} + \dfrac{\omega^{2n}}{1-\omega^2}$. This is, therefore, the value of the given sum. However, since the sum is real, it is desirable to write the answer in the real form. Since $\omega^3 = 1$, the answer will depend on the congruence class of n modulo 3. If $n = 0$ (mod 3), then $\omega^n = \omega^{2n} = 1$ and $\dfrac{\omega^n}{1-\omega} + \dfrac{\omega^{2n}}{1-\omega^2} = \dfrac{1}{1-\omega} + \dfrac{1}{1-\omega^2} = \dfrac{2-\omega-\omega^2}{1-\omega-\omega^2+\omega^3} = \dfrac{3}{3} = 1$. So the given sum equals 1 when $n \equiv 0$ (mod 3). We leave it to the reader to check that it comes out as -1 when $n \equiv 1$ (mod 3) and as 0 when $n \equiv 2$ (mod 3). (In this problem, we could have dispensed with complex numbers by writing $\dfrac{1}{1+x+x^2}$ as $\dfrac{1}{1-x^3} - \dfrac{x}{1-x^3}$. But we wanted to illustrate their use). ▪

Infinite series are by far the most common means to express a function as the limit of some more manageable functions such as polynomials. There are, however, other methods of doing the same, such as infinite products and continued fractions. We shall not treat continued fractions. But we include a brief discussion of infinite products because sometimes, in combination with generating functions, they yield certain identities.

Just as infinite series are defined as limits of their sequences of partial sums, an infinite product is defined as the limit of its sequence of partial products. Specifically, $\prod\limits_{n=1}^{\infty} a_n$ is defined as $\lim\limits_{n\to\infty} p_n$ where, for each n, $p_n = a_1 a_2 \ldots a_n = \prod\limits_{k=1}^{n} a_k$. If this limit exists and is non-zero, then it is easy to show that $a_n \to 1$ as $n \to \infty$. (This is analogous to the fact that the nth term of a convergent series must tend to 0 as $n \to \infty$.) It is customary to write a_n as $1 + b_n$. Then $b_n \to 0$ as $n \to \infty$.

The usual rules of algebra, such as distributivity, carry over from finite products to infinite products, if appropriate care is exercised regarding convergence. As with series, in the past numerous identities were derived through such manipulations, paying little heed to rigour. We prove one such identity here. Its combinatorial significance will be apparent in the next section.

1.14 Proposition: If $|x| < 1$ then $(1 + x)(1 + x^2)(1 + x^4)\ldots(1 + x^{2^n})\ldots$

$\ldots = \prod\limits_{n=0}^{\infty} (1 + x^{2^n}) = 1 + x + x^2 + \ldots + x^n + \ldots$.

Proof: By (iv) in Theorem (1.3), the right hand side equals $\dfrac{1}{1-x}$. We can rewrite this as $\dfrac{1+x}{1-x^2}$ and then as $\dfrac{(1+x)(1+x^2)}{1-x^4}$. Continuing, we have for every n,

$$\frac{1}{1-x} = \frac{(1+x)(1+x^2)\dots(1+x^{2^n})}{1-x^{2^{n+1}}}$$

or,

$$(1+x)\dots(1+x^{2^n}) = \frac{1-x^{2^{n+1}}}{1-x}.$$

Since $|x| < 1$, $x^{2^{n+1}} \to 0$ as $n \to \infty$. So the infinite product $\displaystyle\prod_{n=1}^{\infty}(1+x^{2^n})$ converges to $\dfrac{1}{1-x}$. ∎

So far we have illustrated some standard methods by which generating functions are applied to prove combinatorial identies. In a given problem it is often a combination of several techniques that is needed and ultimately it is only through practice that one learns to pull the right trick. It should be noted, however, that despite the multitude of formulas and tricks, it is not always possible to sum a given series, even when it consists of the first few terms of a sequence whose generating function has a nice closed form. For example, let $H_n = 1 + \dfrac{1}{2} + \dfrac{1}{3} + \dots + \dfrac{1}{n}$. It is not hard to find a closed form expression for $\displaystyle\sum_{n=1}^{\infty} \frac{x^n}{n}$ (cf. Exercise (1.3)). It would then appear that in view of proposition $(1 \cdot 10)$, we can get a handy formula for H_n. But this is not so. The trouble is that finding the coefficient of x^n in a function of x may lead to a summation. The numbers H_n are called **harmonic numbers**, because of the fact that $\dfrac{n}{H_n}$ is the harmonic mean of the numbers $1, 2, \dots, n$. It is well known that $H_n \to \infty$ as $n \to \infty$. Because of their frequent appearances in applications, the harmonic numbers have been extensively studied. There are some highly accurate estimates available for H_n. (It is known, for example, that $H_n \approx \ln n$ for large values of n; see the Epilogue.) Still, there is no closed form for H_n as there is for $1 + 2 + \dots + n$.

We conclude this section with a brief discussion of generating functions with two variables. A sequence is a function of an integer variable. In combinatorics, we frequently come across expressions having several integer variables (or 'parameters' as they are sometimes called). For example, the binomial coefficient $\dbinom{n}{r}$, which gives the number of ways to choose r objects from n objects, depends on n and r. Similarly in $_nP_r$, the number of r-permutations of n objects, there are two integer variables, n and r. (In both these examples, r is restriced by $0 \leqslant r \leqslant n$, but we may remove this restriction by setting $\begin{pmatrix} n \\ r \end{pmatrix} = {}_nP_r = 0$ for $r > n$.) The number of ways to distribute n identical objects into r distinct boxes so that no box contains more than, say, k objects is a function of three variables, namely, n, r and k.

Just as we studied sequences (i.e. functions of a single integer variable) by forming their generating functions w.r.t some indicator functions, we can handle functions of several integer variables by associating with them functions of the same number of real (or complex) variables. As with one variable, the most common choice of indicator functions is polynomials of several variables. We give below the definition for the case of two variables. The extension for the case of more variables is similar.

1.15 Definition: Let S be the set of all non-negative integers and let $f: S \times S \to \mathbf{R}$ be a function. For $(m, n) \in S \times S$, write a_{mn} for $f(m, n)$. Then the **ordinary generating function** of f, (or the O.G.F. of $\{a_{m,n}\}_{m,n=0}^{\infty}$) is defined as $\sum_{m=0}^{\infty} \sum_{n=0}^{\infty} a_{mn} x^m y^n$ and denoted by $A(x, y)$.

Again we may either let x and y be mere indeterminates in which case $A(x, y)$ is a formal power series, or we may let x and y be real (or complex) variables. In the latter case the question of convergence arises and in fact becomes more intriguing because although the order of summation is unimportant for finite double summations, it need not be so for infinite double summations. However, these questions can be answered satisfactorily so as to give a rigourous treatment of generating functions of several variables.

Let us obtain a closed form for the generating function $\sum_{m,n=0}^{\infty} a_{mn} x^m y^n$ where $a_{mn} = \begin{pmatrix} m \\ n \end{pmatrix}$. Since $\begin{pmatrix} m \\ n \end{pmatrix} = 0$ for $n > m$, for each fixed m, the sum $\sum_{n=0}^{\infty} \begin{pmatrix} m \\ n \end{pmatrix} x^m y^n$ is really finite and equals $x^m (1+y)^m$ by the binomial theorem.

So $A(x, y) = \sum_{m=0}^{\infty} x^m (1 + y)^m = \dfrac{1}{1 - x(1 + y)}$ by Theorem (1.3), part (iv).

As with generating functions of one variable, we can recover $\begin{pmatrix} m \\ n \end{pmatrix}$ from $A(x, y)$ by expanding it and taking the coefficient of $x^m y^n$.

An alternate way to handle functions of two integer variables is to fix one of them. For each fixed value of one of the two variables, we get a sequence. If $a_{mn} = f(m, n)$ then for each fixed m, $\{a_{mn}\}_{n=0}^{\infty}$ is a sequence. Denote its generating function by $F_m(x)$. That is, $F_m(x) = \sum_{n=0}^{\infty} a_{mn} x^n$. This way we get a sequence $\{F_m(x)\}_{m=0}^{\infty}$ of generating functions. The original $\{a_{mn}\}$ can be recovered as the coefficient of x^n in $F_m(x)$. For example, if $a_{mn} = \binom{m}{n}$, then for each $m = 0, 1, 2, ..., F_m(x) = (1 + x)^m$.

There is also a graphic way to represent functions of two integer variables which has little to do with generating functions. If $a_n = f(n)$, $n = 0, 1, 2, ...$ is a sequence, we can indicate it by marking the points $0, 1, 2, ...$ on a number line and writing the value a_n at the point marked n. Indeed this is what we do implicity when we list the sequence as $a_0, a_1, a_2, ...$. For a function of two integer variables, say $a_{mn} = f(m,n)$, we first identify the domain set as the set of all points in the cartesian plane both of whose coordinates are non-negative integers. (Such points are often called the lattice points. They form a subset of the ring of Gaussian integers discussed in Chapter 6, Section 2). We then simply write the value a_{mn} at the point (m,n) in the cartesian plane. It may happen that a_{mn} is defined only for some of these points. For example, the combinational definition of $\binom{m}{n}$ makes sense only for $n \leqslant m$ (although, of course, the algebraic definition, namely, $\dfrac{m(m-1)...(m-n+1)}{n!}$ makes sense for any m). In such cases it is customary to describe the subsets like $\{(m, n): m = n\}$ as 'boundaries' in analogy with functions of continuous variables, although, in the strict topological sense, every subset of the lattice points is discrete and hence has no boundary points.

In Figure 7.2(a), we give the graphic representation of $a_{mn} = \binom{m}{n}$ for $0 \leqslant n \leqslant m$. Note the 'boundary conditions', $a_{mn} = 1$ for $n = 0$ and $n = m$. In this particular example, another, although essentially equivalent, graphic representation of the binomial coefficients is more common. It is popularly

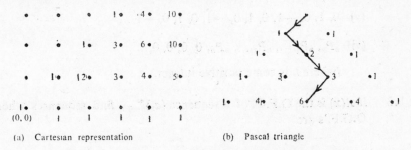

(a) Cartesian representation   (b) Pascal triangle

Figure 7.2: Graphic Representation of Binomial Coefficients.

known as the **Pascal triangle** and is shown in Fig. 7.2(b). Many properties of the binomial coefficients can be paraphrased vividly in terms of the geometric properties of the Pascal triangle. For example, the identity $\binom{m}{n} = \binom{m}{m-n}$ is reflected in the symmetry of the triangle about the vertical axis. The fact that $\binom{m}{0} + \binom{m}{1} + \dots + \binom{m}{m} = 2^m$ is equivalent to saying that the sum of the entries in the mth row is 2^m.

Exercises

1.1   Find a closed form for the ordinary generating functions of the following sequences:

    (i) $1, -1, 1, -1, 1, -1,\dots$

    (ii) $1, 0, 1, 0, 1, 0, 1, 0,\dots$

    (iii) $1, 0, 0, 1, 0, 0, 1, 0, 0, 1, 0, 0, 1,\dots$

    (iv) $1, -1, -1, 1, -1, -1, 1, -1, -1, 1,\dots$

    (v) $1, 2, 3, 4, 5,\dots$

1.2   Find a closed form for the exponential generating functions of the following sequences:

    (i) $1, 2, 4, 8, 16, 32,\dots$

    (ii) $1, -1, 1, -1, 1, -1, 1, -1,\dots$

    (iii) $1, 0, 1, 0, 1, 0, 1, 0,\dots$

    (iv) $1, 0, -1, 0, 1, 0, -1, 0, 1, 0, -1, 0,\dots$

    (v) $0, 1, 0, -1, 0, 1, 0, -1, 0, 1, 0, -1,\dots$

    (vi) $_nP_0, \; _nP_1,\dots, \; _nP_r,\dots, \; _nP_n, \; 0, 0, 0, 0,\dots$

    (where n is some positive integer.)

1.3   If $A(x)$ is the O.F.G. of a sequence $\{a_n\}_{n=0}^{\infty}$, find sequences whose O.G.F.'s are

    (i) $\dfrac{A(x) + A(-x)}{2}$ (ii) $\dfrac{A(x) - A(-x)}{2}$ (iii) $\displaystyle\int_0^x A(t)\, dt$.

Hence find the O.G.F. of the sequence $1, \frac{1}{2}, \frac{1}{3}, \frac{1}{4},\dots$.

1.4   Let ξ be a primitive complex m-th root of unity where m is a positive integer. Let $A(x) = \sum_{n=0}^{\infty} a_n x^n$. Prove that

$$a_0 + a_m x^m + a_{2m} x^{2m} + \ldots + a_{nm} x^{nm} + \ldots = \frac{1}{m} \sum_{j=1}^{m} A(\xi^j x).$$

1.5   Prove the following identities for binomial coefficients where n, p, q are positive integers.

(i) $\displaystyle\sum_{k=1}^{n} k \binom{n}{k} = n2^{n-1}$

(ii) $\displaystyle\sum_{k=1}^{n} k^2 \binom{n}{2} = n(n+1)2^{n-2}$

(iii) $\displaystyle\sum_{k=1}^{n} \frac{(-1)^{k-1}}{k} \binom{n}{k} = 1 + \frac{1}{2} + \ldots + \frac{1}{n}$

(iv) $\displaystyle\left[\binom{n}{0} - \binom{n}{2} + \binom{n}{4} - \binom{n}{6} + \ldots \right]^2$

$$+ \left[\binom{n}{1} - \binom{n}{3} + \binom{n}{5} - \binom{n}{7} + \ldots \right]^2 = 2^n$$

[Hint: Consider the absolute value of the complex number $(1 + i)^n$.]

(v) $\displaystyle\sum_{j=0}^{n} \binom{p}{j} \binom{q}{n-j} = \binom{p+q}{n}$

(vi) $\displaystyle\sum_{j=0}^{\infty} \binom{p}{j} \binom{q}{n+j} = \binom{p+q}{p+n}.$

1.6   Evaluate the following sums where n is a positive integer:

(i) $\displaystyle\sum_{j=0}^{n} 5^j \binom{n}{j}$ (ii) $\displaystyle\sum_{k=0}^{n} k \binom{n}{k}^2$

(iii) $\displaystyle\sum_{j=1}^{\infty} \frac{j}{2^j}$

1.7 If n is a positive integer, prove that the coefficients of x^n in $\dfrac{(1+x)^n}{1-x}$ and $\dfrac{(1+x)^{2n+1}}{1-x}$ are respectively, 2^n and 4^n. Hence evaluate

$$\binom{2n}{n} + 2\binom{2n-1}{n} + 2^2\binom{2n-2}{n} + \ldots + 2^k\binom{2n-k}{n}$$
$$+ \ldots + 2^n\binom{n}{n}.$$

1.8 Show that $\displaystyle\sum_{k=0}^{n} \binom{n+k}{n} \dfrac{1}{2^k} = 2^n$. (A simpler proof will be given in Section 4.)

1.9 Prove that $\dfrac{1}{(1-x)^n} = (1+x+x^2+\ldots)^n = \displaystyle\sum_{r=0}^{\infty} \binom{n+r-1}{r} x^r$, n a positive integer.

1.10 Let $f: X \to \mathbf{R}$ be a function, whose range is finite. For each $\lambda \in \mathbf{R}$, let p_λ be the probability that f assumes the value λ ($p_\lambda = 0$ if λ is not in the range of f.) Then the **expected value** of f is defined as $\displaystyle\sum_{\lambda \in \mathbf{R}} p_\lambda \lambda$ (which is a finite sum since the range of f is finite). Prove that in case X is finite, the expected value of f equals $\dfrac{1}{|X|} (\displaystyle\sum_{x \in X} f(x))$. (In other words, the expected value coincides with the average value of f. However, it is customary to use the former term when, instead of specifying the values of f at various points of X directly, we are more interested in the probability that f attains a particular value. Note that two different functions may have the same expected value.)

1.11 Three balls are drawn at random from an urn containing 4 red, 5 white and 6 green balls. What is the expected number of red balls in the draw?

1.12 Suppose n is a positive integer and p is a real number with $0 \leqslant p \leqslant 1$. Prove that

(a) $\displaystyle\sum_{r=0}^{n} r \binom{n}{r} p^r(1-p)^{n-r} = np$ and

(b) $\displaystyle\sum_{r \text{ even}} \binom{n}{r} p^r(1-p)^{n-r} = \frac{1}{2}[1+(1-2p)^n]$.

$\left[\text{Hint: Set } x = \dfrac{p}{1-p} \text{ in } \dfrac{d}{dx}(1+x)^n \text{ and in } \dfrac{(1+x)^n+(1-x)^n}{2} \text{ respectively.}\right]$

1.13 Suppose p is the probability that a head will show when a coin is tossed. (For an unbiased coin, $p = \frac{1}{2}$.) If this coin is tossed n times find (a) the expected number of occurrences of head (b) the probability that a head occurs an even number of times.

1.14 Prove Theorem (1.11), using the result of Problem (1.5). Using
$$n^2 = \left[2 \binom{n}{2} + \binom{n}{1} \right], \text{ similarly evaluate } 1^2 + 2^2 + \ldots + n^2,$$
where n is a positive integer. Verify your answer by induction on n. Also derive the same result from Proposition (1.10).

1.15 In the solution of Problem (1.12), verify that the power series expansion of $(1 - 4x)^{-1/2}$ is indeed $\sum\limits_{r=0}^{\infty} \binom{2r}{r} x^r$. Show that $(1 - 4x)^{-1/2}$ is also the E.G.F. of the sequence $\{_{2r}P_r\}_{r=0}^{\infty}$.

1.16 Find the O.G.F. of the sequence $\{a_n\}_{n=0}^{\infty}$, where $a_0 = 0$ and for $n \geqslant 1, a_n = \dfrac{1}{n} \binom{2n-2}{n-1}$. (The terms of this sequence are called **Catalan numbers**. We already encountered them in the solution to the Vendor Problem and also in Exercise (3.4.22).) Hence show that for $n > 1$, $a_n = \sum\limits_{k=1}^{n-1} a_k a_{n-k}$.

1.17 Verify the last assertion in the solution to Problem (1.13).

1.18 For a positive integer n, evaluate $\sum\limits_{m=\lceil n/2 \rceil}^{n} \binom{m}{n-m}$.

1.19 Suppose F is a field and $f(x) = a_0 + a_1 x + \ldots + a_n x^n \in F[x]$ has non-zero roots $\alpha_1, \alpha_2, \ldots, \alpha_n$ in F. Prove that
$$\frac{1}{\alpha_1} + \frac{1}{\alpha_2} + \ldots + \frac{1}{\alpha_n} = -\frac{a_1}{a_0}.$$

[Hint: Note that $f(x) = a_n(x - \alpha_1)(x - \alpha_2) \ldots (x - \alpha_n)$. Consider $f(1/x)$.]

1.20 Let $f(x) = 1 - \dfrac{x}{3!} + \dfrac{x^2}{5!} - \dfrac{x^3}{7!} + \dfrac{x^4}{9!} + \ldots + \dfrac{x^n}{(-1)^n(2n+1)!} + \ldots$. Show that the zeros of f are of the form $(n\pi)^2$ where n is a positive integer. [Hint: Clearly f has no negative zeros. For $x > 0$, show that $f(x) = \dfrac{\sin y}{y}$ where $y^2 = x$.]

1.21 Assuming that the result of Exercise (1.19) is valid for power series (i.e. for 'polynomials of infinite degree'), and applying it to the function $f(x)$ in the last exercise, prove that $\sum\limits_{n=1}^{\infty} \dfrac{1}{n^2} = \dfrac{\pi^2}{6}$.

1.22 If p, q, n are positive integers, prove that

$$\sum_{k=0}^{p}\binom{p}{k}\binom{q}{k}\binom{n+k}{p+q}=\binom{n}{p}\binom{n}{q}.$$

[Hint: Write $\binom{n+k}{p+q}$ as $\sum_{j=0}^{\infty}\binom{k}{j}\binom{n}{p+q-j}$. Interchange the order of summation and make several uses of Exercises (2.2.16) (ii) and (1.5) (v).]

1.23 Combining the result of the last exercise with the technique of the proof of Proposition (1.9), prove that

$$\sum_{k=0}^{p}\binom{p}{k}\binom{q}{k}\binom{n+p+q-k}{p+q}=\binom{n+p}{p}\binom{n+q}{q}$$

(The special case $p = q$ is called the **Li-Jen-Shu Formula**.)

1.24 Let a_{mn} be the number of ways to select n integers from $\{1, 2, ..., m\}$ so that no two consecutive integers are selected (cf. Exercise (2.3.5)). Verify that the O.G.F. of $\{a_{mn}\}_{m, n=0}^{\infty}$ is $\dfrac{1+xy}{1-x-yx^2}$.

What is the 'boundary' in this case?

1.25 Suppose you start at the top apex of the Pascal triangle and in each unit of time go to either of the two neighbouring points in the row immediately below. Prove that the number of such paths from the apex to any point in the Pascal triangle is precisely the entry at the point. (One such path is shown in Fig. 7.2 (b).) Using this, give a 'geometric' proof of the identity $\binom{m}{0}+\binom{m}{1}+...+\binom{m}{m}=2^m$.

1.26 Interpret and prove the following binomial identities in terms of the Pascal triangle

(i) $\displaystyle \binom{m}{n}=\binom{m-1}{n}+\binom{m-1}{n-1}$

(ii) $\displaystyle \sum_{k=0}^{n}\binom{p}{k}\binom{q}{n-k}=\binom{p+q}{n}$

(iii) $\displaystyle \sum_{k=0}^{\infty}\binom{m}{2k}=\sum_{k=0}^{\infty}\binom{m}{2k+1}=2^{m-1}.$

1.27 For each positive integer, m, let $A_m(x) = 1 + x^m + x^{m^2} + x^{m^3} + \cdots$.

Prove that $\dfrac{1}{1-x} = \prod\limits_{p} A_p(x)$ where in the right hand side, p ranges over all primes.

Notes and Guide to Literature

For the theorems about power series which we have referred to, see Kreyzig [1] or, for a more rigorous approach, Rudin [1]. The need for a rigorous approach was stressed by Weierstrass. The formula $\sum\limits_{n=1}^{\infty} \dfrac{1}{n^2} = \dfrac{\pi^2}{6}$ and its 'proof' in Exercise (1.21) are due to Euler. It is a masterpiece of an unscrupulous reasoning yielding a beautiful result. The result can, however, be established by rigorous methods, using, for example Fourier series; see again, Rudin [1].

There are numerous combinatorial identities, many of them being rediscovered from time to time all over the world. The binomial coefficients have a long history. The Pascal triangle was actually known before Pascal, but he used it for his work on probability theory. MacMahon [1] and Riordan [1] give a full discussion of the use of generating functions. Propositions (1.8) and (1.9) are from Lovasz [1].

2. Application to Enumeration Problems

In algebraic computations, we frequently expand an expression and then simplify it by grouping the 'like' terms together. The coefficient of each such term is the number of times it appears in the expansion and finding it is essentially a combinatorial process. Consider for example, the expansion of $(x + y)^n$, where x, y are elements of a commutative ring with identity (cf. Exercise (6.1.28)). Since this is a product of n factors, each having two terms, the expansion consists of 2^n terms in all, of the form $a_1 a_2 \ldots a_n$ where for each $i = 1, \ldots, n$, a_i is either x or y. Since x and y commute with each other, for every k, $0 \leqslant k \leqslant n$, all terms in which exactly k of the a_i's equal x (and the remaining $(n - k)$ a_i's are equal to y) are equal to $x^k y^{n-k}$. Thus for every choice of k symbols out of the n symbols $a_1, a_2, \ldots, a_n$, we get a term which equals $x^k y^{n-k}$ and vice versa. It follows that the coefficient of $x^k y^{n-k}$ is $\dbinom{n}{k}$ and hence that $(x + y)^n = \sum\limits_{k=0}^{n} \dbinom{n}{k} x^k y^{n-k}$. This is indeed the most direct way to prove the binomial theorem.

As another example of the use of combinatorial reasoning, let us do a simple problem.

2.1 Problem: Find the coefficient of x^7 in

$$(1 + x + x^3)(x + x^2 + x^3)(x^4 + x^6).$$

Solution: Although this problem has no special significance, we present several ways of doing it, because it will be used to motivate further discussion. First, there is the brute-force method which consists of fully multiplying out the three factors. This would give 18 terms, and we take those that equal x^7. But this is labourious and most of the labour is wasted since we want only the terms that equal x^7. So we look for a better solution. Note that, in the expansion, a term x^7 will arise as the product $x^{n_1} x^{n_2} x^{r_3}$ where x^{n_1}, x^{n_2} and x^{n_3} are one of the summands in the first, second and the third factor respectively. The nature of the factors indicates that n_1 has to be 0 (corresponding to the summand 1 which equals x^0), 1 or 3. Similarly n_2 has to be 1, 2 or 3 and n_3 has to be 4 or 6. Moreover,

$$x^7 = x^{n_1} x^{n_2} x^{n_3} = x^{n_1 + n_2 + n_3}$$

gives $n_1 + n_2 + n_3 = 7$. Thus the problem is equivalent to finding the number of ordered triples (n_1, n_2, n_3) of integers for which $n_1 + n_2 + n_3 = 7$, $n_1 = 0, 1$ or 3, $n_2 = 1, 2$ and 3 and $n_3 = 4$ or 6. For $n_3 = 4$, $n_1 + n_2$ must be 3 and this gives only two possibilities: $n_1 = 0$, $n_2 = 3$ and $n_1 = 1$, $n_2 = 2$. If $n_3 = 6$, then $n_1 + n_2 = 1$ which can happen only when $n_1 = 0$, $n_2 = 1$. So in all there are 3 triples satisfying the given conditions. Thus the coefficient of x^7 in the expansion is 3.

   This solution can be improved upon slightly. We note that the given expression equals

$$(1 + x + x^3)\, x\, (1 + x + x^2)\, x^4(1 + x^3) = x^5(1 + x + x^3)(1 + x + x^3)(1 + x^2).$$

So the coefficient of x^7 in it is the same as the coefficient of x^2 in

$$(1 + x + x^3)(1 + x + x^2)(1 + x^2).$$

As above, this equals the number of triples (m_1, m_2, m_3) such that

$$m_1 = 0, 1 \text{ or } 3;\ m_2 = 0, 1 \text{ or } 2;\ m_3 = 0 \text{ or } 2$$

and $m_1 + m_2 + m_3 = 2$. It is a little easier to count the number of such triples. The count again comes to 3, which is the desired coefficient. ▨

   The solution given above clearly indicates that the preceding algebraic problem is equivalent to the combinatorial problem of finding the number of ways to distribute 7 identical balls into three (distinct) boxes so that the first box contains 0, 1 or 3 balls, the second one contains 1, 2 or 3 balls and the third one contains 4 or 6 balls.

   We now turn the tables around. Instead of translating an algebraic problem into a combinatorial one, it is often more rewarding to go the other way. This is so because algebra is a more developed branch of mathematics.

Numerous tricks are available to manipulate algebraic expressions. Not all of these tricks may have combinatorial analogues. For example, if we encounter a factor $1/1-x$ we can replace it by the infinite series $1+x+x^2+ \ldots + x^n + \ldots$ and vice versa. It is difficult to see the combinatorial analogue of such a replacement. Even for those algebraic manipulations which do have combinatorial interpretations, it is often the case that the algebraic manipulations are easier to think of than their combinatorial equivalents. For example, in the 'improvement' to the solution to the last problem, the combinatorial interpretation is to start by placing one ball in the second box and 4 balls in the third box and then find the number of ways to distribute the remaining 2 balls. Although this trick is simple enough, a person is not likely to think of it as readily as its corresponding algebraic manipulation. So heavily drilled in algebra are we, that the moment we see the expression $(1 + x + x^3)(x + x^2 + x^3)(x^4 + x^6)$, the very first thing that comes to our mind is to take out x and x^4 as common factors from the second and the third brackets respectively.

It is worthwhile at this point to consider the relationship between algebra and geometry. Suppose we want to solve simultaneously, the system of equations: (i) $x - 7y + 25 = 0$ and (2) $x^2 + y^2 = 25$. We could translate this problem geometrically. If x and y denote the cartesian coordinates of a point in a plane w.r.t a fixed rectangular frame of reference, then, the solutions to (1) constitute a straight line L while (2) represents a circle C of radius 5 with centre at the origin. The common solutions to (1) and (2), therefore, correspond to the points of intersections of L and C. If we sketch C and L we find that they intersect at $(3,4)$ and $(-4, 3)$. So the solutions to the system (1) and (2) are: $x = 3$, $y = 4$ and $x = -4$, $y = 3$. However, this method is rarely used. First of all, it is impossible to sketch curves or to read the coordinates of points in the plane with absolute precision and even a slight inaccuracy would lead to erroneous solutions. secondly, as in the present case, it is often easier to solve a system of equations algebraically. That is why, instead of translating algebraic problems in terms of geometry, we often go the other way. As the reader might have experienced, many propositions in geometry which require considerable ingenuity to prove by the methods of Euclid, follow rather routinely from algebra with a suitable choice of coordinates. (This is not to suggest that geometry is useless. While geometry may not give the exact values of a solution, it provides deep insights into the nature of the solution).

The relationship between algebra and combinatorics is not quite so lop-sided. Because of the immense variety of combinatorial problems, there is no single algebraic model that can handle them all. There are many combinatorial results which are still best proved by combinatorics. There are even algebraic results where combinatorial arguments are needed, the binomial theorem being one such result. For the time being, however, we shall concentrate on the use of algebraic manipulations in combinatorics. Specifically, we shall study how generating functions can be used to formulate problems of enumeration, that is, problems where we have to count the number of ways to do a certain thing (such as distributing balls into boxes, selecting subsets of a given set, permuting objects, traversing paths in a given diagram.) subject to certain restrictions.

The way this technique works is as follows. In enumeration problems, there is usually an integer parameter, say n. (Sometimes there are more than one integer parameters.) In a given problem we may be interested only in one particular value of n. For example, in the problem discussed above we wanted to distribute 7 balls among 3 boxes subject to certain restrictions. But there is nothing special about 7 and we might as well consider the problem of enumerating the number of ways to distribute n balls into those boxes, where n is any non-negative integer. (We could also change the number of boxes, 3, to a variable, say m, but for the moment, let us not get into this kind of generalisation.) So let a_n be the number of ways to put n identical balls into three boxes so that the first box contains 0, 1 or 3 balls, the second contains 1, 2 or 3 balls and the third contains 4 or 6 balls. We now look for an expression $A(x)$ of a formal variable x, which will have the property that each way of doing the particular thing (in the present case, each distribution of the n balls) corresponds to one occurrence of the term x^n in $A(x)$. Consequently, a_n will be the coefficient of x^n in $A(x)$, and hence, $A(x)$ will be nothing but the generating function of the sequence $\{a_n\}_{n=0}^{\infty}$. Having found $A(x)$, we can expand it as a power series and recover a_n as the coefficient of x^n. We often call $A(x)$ the **enumerator** for the combinatorial problem concerned. More generally, the term enumerator is applied to any function (possibly of more than one variable) which stores, in some way, (usually as the coefficients of some terms) the number of ways to do something. As with generating functions, there are many kinds of enumerators. But we shall be concerned with only two of them, the ordinary enumerator (whlch we shall simply call as enumerator) and the exponential enumerator. As a rule of thumb, rhe ordinary enumerator is useful for problems involving combinations, while, for problems of permutation, the exponential enumerator is more convenient. In the literature, the term 'enumerator' is frequently used without a formal definition. The reason probably is that, as a concept, an enumerator is not substantially different from a generating function. Perhaps the best way to describe an enumerator is as the generating function of a sequence associated with an enumeration problem. In a nutshell, an enumerator is a generating function put into action!

In essence, what we are doing here is to obtain an algebraic 'coding' of each way of executing a combinatorial processs. Thus, in the problem at hand, putting n_1 balls in the first box, n_2 in the second and n_3 in the third is 'coded' by the expression $x^{n_1} x^{n_2} x^{n_3}$ which will equal x^n if and only if $n_1 + n_2 + n_3 = n$. The restrictions on the numbers of balls in the boxes translate into restrictions on the values n_1, n_2, n_3 can take.

Of course, merely paraphrasing a problem is no guarantee of its solution. In our case, the success of the technique of introducing generating functions in problems of enumeration depends on the answers to the following questions: (i) How easy is it to identify the enumerator say, $A(x)$? and (ii) How easy is it to expand $A(x)$? Let us answer these questions for the problem we are studying, namely, finding a_n, the number of ways to put n balls into 3 boxes, subject to the restrictions given above. $A(x)$ is the O.G.F. of the sequence $\{a_n\}_{n=0}^{\infty}$. Is there any way to get $A(x)$, other than by first computing a_n? The answer is 'yes'. Let us first consider enumerators for each box. Let $B(x)$ be the ordinary enumerator for the first box. This means $B(x)$ is the O.G.F. of the sequence $\{b_n\}_{n=0}^{\infty}$ where b_n is the number of ways to put n balls in the first box. Because of the restriction imposed, namely that the first box must contain 0, 1 or 3 balls we have $b_0 = b_1 = b_3 = 1$ and $b_n = 0$ for all other n. So $B(x) = 1 + x + x^3$. Similarly, if we let $C(x)$ and $D(x)$ be the enumerators for the second and the third box respectively, then $C(x) = x + x^2 + x^3$ and $D(x) = x^4 + x^6$. The trick is now to observe that $A(x) = B(x) C(x) D(x)$. This is so because every occurrence of x^n in $A(x)$, i.e., every distribution of n balls into the three boxes, arises from a product of the form $x^{n_1} x^{n_2} x^{n_3}$ where $n_1+n_2+n_3=n$ and x^{n_1}, x^{n_2}, x^{n_3} are terms appearing in $B(x)$, $C(x)$ and $D(x)$ respectively. Thus we see that question (i) above has an affirmative answer. Question (ii) reduces to doing a problem like (2.1) (which deals with the case $n = 7$). We can simplify $A(x)$ as $x^5(1 + x + x^3) (1 + x + x^2) (1 + x^2)$. From this we see that $a_n = 0$ for $n < 5$ and for $n > 12$. For $5 \leqslant n \leqslant 12$, a_n is the coefficient of x^{n-5} in $(1 + x + x^3) (1 + x + x^2) (1 + x^2)$. This expression, however, has to be expanded 'by hand'. It comes to $1 + 2x + 3x^2 + 4x^3 + 3x^4 + 3x^5 + x^6 + x^7$. So ultimately, $A(x) = x^5 + 2x^6 + 3x^7 + 4x^8 + 3x^9 3x^{10} + x^{11} + x^{12}$. We can now tell instantaneously, for every n, how many ways there are to put n balls into the 3 boxes subject to the given restrictions.

Summing up, in this problem we had a partial success. It was easy to identify the enumerator $A(x)$. But there was no slick way to expand it. In some problems, even the second step is easy to carry out and we have complete success. We shall see examples of such problems shortly. However, even when we have only a partial success, the method is worthwhile because once we obtain $A(x)$ in some form, expanding it is often a mechanical process and can be carried out on machines. Secondly, in some problems it is more important to get a closed form for the enumerator than to expand it, because the closed form can provide us with the information we need, as in the following problem.

2.2 Problem: At an international conference, 25 countries send teams of 4 delegates each. A committee is to be formed from these 100 delegates subject to the following rules: (i) the number of members in the committee shall be odd (ii) the committee shall include at least one and at most three delegates from each country. How many such committees can be formed?

Solution: Let us ignore the first rule for the moment. For $n = 0, 1, ...$, let a_n be the number of committees satisfying the second restriction. Let

$$A(x) = \sum_{n=0}^{\infty} a_n x^n.$$

Then $A(x)$ is the enumerator for the formation of committee, ignoring the first restriction. Every such committee is equivalent to 25 subcommittees, one from each country, put together. Because of (ii), each such subcommittee contains 1, 2 or 3 members from that country. From 4 delegates, we can have 4 subcommittees consisting of 1 member, 6 subcommittees consisting of 2 members and 4 subcommittees consisting of 3 members. It follows that $A_i(x)$, the enumerator for formation of subcommittee for the ith country, is $4x + 6x^2 + 4x^2$, for $i = 1, ..., 25$. Once again, we claim that the product $A_1(x)A_2(x) \ldots A_{25}(x)$ equals $A(x)$. This is so because for every $n = 0, 1, 2, ...$ each occurrence of x^n in this product is equivalent to an ordered 25-tuple, $(n_1, n_2, ..., n_{25})$ where x^{n_i} occurs in $A_i(x)$ and $n_1 + n_2 + ... + n_{25} = n$. So $A(x) = (4x + 6x^2 + 4x^3)^{25}$. It would be horrendous to expand $A(x)$. But we can get by without it. Note that $A(x)$ is a polynomial (of degree 75) and hence the expansion is valid for every value of x. In particular, setting $x = 1$, we would get $A(1) = (4 + 6 + 4)^{25} = 14^{25} = a_0 + a_1 + a_2 + ... + a_n + ...$. This will be the total number of committees that can be formed so as to satisfy (ii). (This number could also be obtained without generating functions, since for every country there are 14 possible ways its subcommittee can be formed and we could combine the subcommittees any way we like.)

However, in view of the restriction (i), namely, that the number of committee members be odd, the answer to the problem is not $a_0 + a_1 + a_2 + ...$ but $a_1 + a_3 + a_5 + ...$. There is a tricky way to evaluate this sum. Note that $A(-x) = a_0 - a_1 x + a_2 x^2 - a_3 x^3 + ...$. So $a_1 x + a_3 x^3 + a_5 x_5 + ...$ $= \dfrac{A(x) - A(-x)}{2}$. Setting $x = 1$, we get $a_1 + a_3 + a_5 + ... = \dfrac{A(1) - A(-1)}{2}$

$\dfrac{14^{25} - (-2)^{25}}{2} = \dfrac{1}{2}(14^{25} + 2^{25})$. This is the desired answer and there is no easy way to get to it without generating functions. ∎

In this problem, there is no easy formula for a_n because $A(x)$ cannot be expanded easily. Still, by factoring $A(x)$ and applying the binomial theorem, we get the number of committees with n members

$$= a_n = \text{coefficient of } x^n \text{ in } (4x^{25})\left(1 + \frac{3}{2}x + x^2\right)^{25}$$

$$= \text{coefficient of } x^{n-25} \text{ in } (4)^{25}\left[1 + x\left(x + \frac{3}{2}\right)\right]^{25}$$

$$= \text{coefficient of } x^{n-25} \text{ in } 4^{25} \sum_{r=0}^{25} \binom{25}{r} x^r \left(x + \frac{3}{2} \right)^r$$

$$= 4^{25} \sum_{r=0}^{25} \binom{25}{r} \left[\text{coefficient of } x^{n-25-r} \text{ in } \left(x + \frac{3}{2} \right)^r \right]$$

$$= 4^{25} \sum_{r=0}^{25} \binom{25}{r} \binom{r}{n-r-25} \left(\frac{3}{2} \right)^{2r+25-n}$$

$$= 6^{25} \sum_{r=0}^{25} \binom{25}{r} \binom{r}{n-r-25} \left(\frac{3}{2} \right)^{2r-n}$$

This has to be left in the summation form. Thus, although the theory of enumerators does not give a handy formula for a_n, it does express it as a sum which can be evaluated mechanically for a given value of n.

As an example where it is easy to identify the enumerator as well as to expand it, we give an alternate proof of Theorem (2.3.12) where it was shown that the number of r-selections of n types of objects, with unlimited repetitions allowed, is equal to $\binom{n+r-1}{r}$. The proof there was a rather tricky one. With enumerators, we can do the problem routinely. Let a_r be the desired number of selections (for a fixed n). Then $A(x) = \sum_{r=0}^{\infty} a_r x^r$ is the enumerator for the selection of objects of the given n types. Let $A_i(x)$ be the enumerator for the selection of objects of type i, $i = 1, \ldots, n$. Since there is no restriction on the number of objects of type i that can be chosen and since all objects of the same type are to be regarded as identical, it follows that for every $m = 0, 1\ 2, \ldots$ there is only one way to choose m objects of type i. So $A_i(x) = 1 + x + x^2 + \ldots + x^m + \ldots = 1/1-x$. Once again, $A(x) = A_1(x) A_2(x) \ldots A_n(x)$. So $A(x) = (1 + x + x^2 + \ldots)^n = (1 - x)^{-n}$. = Expanding this by the binomial theorem (cf. Exercise (1.9)) we get

$$a_r = \binom{n+r-1}{r} \text{ as desired.}$$

At the heart of the arguments in the preceding examples is the simple formula $x^i x^j = x^{i+j}$. Repeated applications of this formula allow us to express the desired enumerator $A(x)$ as the product of the enumerators $A_1(x)$, $A_2(x), \ldots$ for individual 'containers' (which were either the boxes, the countries or the piles of objects) because the nature of the problem in each case is such that the selections for each container can be made independently of the others and consequently every term in each $A_i(x)$ can be multiplied by every term in $A_j(x)$ for $i \neq j$. This will no longer be the case if the conditions of the problem implied some mutual dependence among the selections for some of the containers. In such cases, such inter-related containers have to be handled separately as we illustrate in the following variation of Problem (2.2).

2.3 Problem: Suppose in Problem (2.2), two of the countries are superpowers. How many committees can be formed if there is an additional restriction that these superpowers have an equal representation in the committee?

Solution: Denote the enumerator by $B(x)$ this time. Suppose the first two countries are the superpowers. Then for $i = 3, 4 ..., 25$, $A_i(x)$ remains the same, namely, $4x + 6x^2 + 4x^3$. However, we have to consider the two superpowers together. If one of them has m members in the committee so does the other. This means a term x^m for one of them can be multiplied only with a term x^m for the other and not with any other power of x. Since there are $\binom{4}{m}$ ways to choose m members from 4 delegates, it follows that the enumerator for the two superpowers together is no longer the entire $(4x + 6x^3 + 4x^3)^2$ but only the part $16x^2 + 36x^4 + 16x^6$ of it. Hence

$$B(x) = (16x^2 + 36x^4 + 16x^6)(4x + 6x^2 + 4x^3)^{23}.$$

As in Problem (2.2), the desired answer is $\frac{1}{2}[B(1) - B(-1)] = \frac{1}{2}[68. 14^{23} - 68(-2)^{23}] = 34(14^{23} + 2^{23})$. ∎

In the examples so far, the desired enumerator came out to be the product of finitely many expressions. As remarked in the last section, with a little care, it is possible to consider infinite products and in Proposition (1.14) we proved that the infinite product

$$(1 + x)(1 + x^2)(1 + x^4)(1 + x^8) ... (1 + x^{2^n}) ... \text{ equals } 1 + x + x^2 + x^3 + ...$$

By interpretting both sides of this identity as enumerators, we can prove a result from number theory. Although, the result itself can be proved by other methods and is not very deep, it illustrates how the theory of enumerators can be applied to prove results from other branches of mathematics.

2.4 Theorem: Every positive integer has a unique binary expansion, that is, an expansion with base 2.

Proof: In a binary expansion, the only possible digits are 0 and 1. So the assertion is equivalent to saying that every positive integer can be uniquely expressed as a sum of powers of 2. (Here $2^0 = 1$ is to be included as a power, but not the fractional powers $\frac{1}{2}, \frac{1}{4}, \frac{1}{8}, ...$) It is easy to interpret this assertion combinatorially. Suppose we have an infinite number of boxes, $B_0, B_1, B_2, ...$ Assume that the box B_i can contain either no ball or else 2^i balls for $i = 0, 1, 2, ...$ and consider the problem of distributing n identical balls into these boxes. Obviously for every n all except finitely many of these boxes will be empty. The enumerator $A_i(x)$ for B_i is $1 + x^{2^i}$, $i = 0, 1, 2,...n$. Let $A(x)$ be the enumerator for distributing balls into the boxes. Then $A(x) = \overset{\infty}{\underset{i=0}{\pi}} \, A_i(x)$ because the reasoning given earlier for finitely many boxes applies for this case as well. (Note that when $(1 + x)(1 + x^2)$ $(1 + x^4)(1 + x^8)...(1 + x^{2^i})...$ is expanded, each term will be an infinite product, all except finitely many factors of which are 1. For example, $x \cdot 1 \cdot x^4 \cdot x^8 \cdot 1 \cdot 1 \cdot x^{64} \cdot 1 \cdot 1 \cdot 1 \cdot 1...$ equals x^{77}. The terms of the form $x^{n_1} x^{n_2} x^{n_3} ...$ where infinitely many of the n_i's are greater than 0, are 0 if $\mid x \mid < 1$, as we assumed in (Proposition (1.14)).) By Proposition (1.14), $A(x) = 1 + x + x^2 + ...$ $+ x^n + ...$ But this means that for every n, there is only one way of putting n balls into the boxes $B_0, B_1, B_2,...$, which is equivalent to saying that there is only one way to express n as a sum of powers of 2. ∎

The argument given above is partition theoretic. It amounts to saying that every positive integer has only one partition in which the part sizes are distinct powers of 2. Analogous arguments yield interesting results about partitions. Recall (Definition (2.3.14)) that a partition of an integer n into m parts is a sequence $(n_1, n_2,..., n_m)$ such that $n_1 \geqslant n_2 \geqslant ... \geqslant n_m \geqslant 1$ and $n_1 + n_2 + ... + n_m = n$. The integers $n_1, n_2,..., n_m$ are called the part sizes. $P_{n, m}$ denotes the number of partitions of n into m parts and $p(n)$ denotes the total number of partitions of n (into whatever the number of parts). Clearly $p(n) = \overset{n}{\underset{m=1}{\Sigma}} P_{n, m}$. As mentioned in Chapter 2, Section 3, there is no easy formula either for $P_{n, m}$ or for $p(n)$. However, using enumerators, it is easy to get formulas for their generating functions. We begin by finding a closed form for $\overset{\infty}{\underset{n=0}{\Sigma}} P_{n,m} x^n$, for a fixed m.

2.5 Proposition: For a positive integer m,

$$\sum_{n=0}^{\infty} P_{n,m} x^n = \frac{x^m}{(1 - x)(1 - x^2)...(1 - x^n)}.$$

Proof: By Proposition (2.3.16), $P_{n, m}$ equals the number of partitions of n in which the largest part size is m. Consider m boxes, $B_1, B_2,..., B_m$. For $i = 1,..., m$ suppose B_i contains infinitely many bags each containing i balls. If we are not allowed to split the bags, then the number of balls that can be picked from B_i has to be a multiple of i, i.e. $0, i, 2i, 3i, 4i,....$ Any partition of an integer with part sizes not exceeding m is equivalent to a way to pick n balls by picking bags from the boxes $B_1,..., B_m$. All we have to do is to group together parts of size i and think of each of them as a bag from the box B_i. For example the partition $(5, 5, 4, 3, 3, 3, 3, 1, 1)$ of 28 corresponds to picking 2 bags from B_1, none from B_2, 4 bags from B_3, 1 from B_4 and 2 from B_5. The enumerator for the box B_i is

$$1 + x^i + x^{2i} + x^{3i} + \ldots = \frac{1}{1-x^i}.$$

It follows that the enumerator for picking balls from the m boxes is

$$\frac{1}{1-x} \cdot \frac{1}{1-x^2} \cdot \ldots \cdot \frac{1}{1-x^i} \cdot \ldots \cdot \frac{1}{1-x^m}.$$

This would be the generating function for the number of partitions in which all parts are of size $\leqslant m$. But we want the largest part size to be m. So there has to be at least one part of size m, which amounts to saying that at least one bag from the box B_m must be picked. Hence the enumerator for B_m will not be $1 + x^m + x^{2m} + x^{3m} + \ldots$ but $x^m + x^{2m} + x^{3m} + \ldots$ which equals $x^m (1 + x^m + x^{2m} + \ldots) = \dfrac{x^m}{1-x^m}$. The result is now clear. ▨

2.6 Corollary: For every positive integer n, $p(n)$, the number of partitions of n, equals the coefficient of x^n in the expression

$$\frac{x}{1-x} + \frac{x^2}{(1-x)(1-x^2)} + \frac{x^3}{(1-x)(1-x^2)(1-x^3)} + \ldots$$

$$\ldots + \frac{x^n}{(1-x)(1-x^2)\ldots(1-x^n)}.$$

Proof: This is an immediate consequence of the last proposition and the fact that $p(n) = \sum\limits_{m=1}^{n} P_{n,\,m}$. ▨

It should be noted that this corollary does *not* give the generating function for the sequence $\{p(n)\}_{n=1}^{\infty}$, because the expression appearing in it depends on n. However, using essentially the same argument as in the last proposition, we get the generating function for $(p(0),\ p(1),\ (P(2),\ldots)$ where we set $p(0) = 1$.

2.7 Theorem: For every non-negative integer n, $p(n)$ is the coefficient of x^n in the infinite product

$$\frac{1}{(1-x)} \frac{1}{(1-x^2)} \ldots \frac{1}{(1-x^m)} \ldots .$$

Proof: We duplicate the argument in the proof of Proposition (2.5), except that instead of m boxes we have an infinite collection of boxes $B_1,\ B_2,\ldots$ with the ith box containing infinitely many bags each containing i balls. ▨

As a corollary, for a fixed n we get an expression for $p(n)$, which is a little simpler.

2.8 Corollary: For every positive integer n, $p(n)$ is the coefficient of x^n in

$$\frac{1}{(1-x)(1-x^2)\ldots(1-x^n)}.$$

Proof: We split the infinite product $\prod\limits_{m=1}^{\infty} \dfrac{1}{1-x^m}$ as $A(x)\ B(x)$ where $A(x) =$

$= \prod\limits_{m=1}^{n} \dfrac{1}{1-x^m}$ and $B(x) = \prod\limits_{m=n+1}^{\infty} \dfrac{1}{1-x^m}$. When $B(x)$ is expanded it contains no terms of degree $\leqslant n$, other than 1. So the coefficient of x^n in $\prod\limits_{m=1}^{\infty} \dfrac{1}{1-x^m}$ is the same as that in $A(x)$. The result now follows from the last theorem. ∎

When it comes to actually finding $p(n)$ for a given n, the results above are of little help because expanding the expressions as power series in x so as to find the coefficient of x^n takes about the same amount of work. However, the arguments are useful when we are dealing with partitions with some restrictions on the part sizes. Theorem (2.7) easily generalizes to show that the number of partitions of n in which the part sizes have to be from among m_1, m_2, m_3,...(say) is the coefficient of x^n in the product $\prod\limits_{k} \dfrac{1}{1-x^{m_k}}$. Sometimes we may be able to expand this product by other means and thereby get a closed formula for the number of such partitions. As an illustration, we do the Postage Problem. Recall that in this problem we have to find the number of ways in which a postage of two rupees can be affixed on an envelope with stamps of denominations 10 paise, 20 paise and 30 paise.

We assume that the stamps of each denomination are indistinguishable from each other and that they are available in infinite supply. (In the present problem, it suffices to assume that at least 20, 10 and 6 stamps of the three denominations are available respectively.) As remarked earlier, taking 10 paise as a unit, the problem amounts to finding the number of partitions of 20 into parts with sizes 1,2, and 3. The enumerator for such partitions is $\dfrac{1}{(1-x)(1-x^2)(1-x^3)} = \dfrac{1}{(1-x)^3(1+x)(1+x+x^2)}$. We

can resolve this into partial fractions as follows. $(1+x+x^2) = (1-\omega x)$ $(1-\omega^2 x)$ where $\omega = -\dfrac{1}{2} + \dfrac{\sqrt{3}}{2} i$ (cf. Problem (1.13).) Now, let

$$(*) \quad \frac{1}{(1-x)(1-x^2)(1-x^3)} = \frac{A}{(1-x)^3} + \frac{B}{(1-x)^2} + \frac{C}{1-x} + \frac{D}{1+x}$$
$$+ \frac{E}{1-\omega x} + \frac{F}{1-\omega^2 x}$$

By expanding the right hand side of $(*)$ and comparing coefficients of numerators we would get a system of linear equations in the 6 unknowns A, B, C, D, E, F. Solving it would be a horrendous job. So we apply certain well-known tricks. Multiplying both sides of $(*)$ by $1 + x$ and then setting $x = -1$, we get $D = 1/8$. Similarly, we get $E = 1/9$ and $F = 1/9$. If we multiply $(*)$ by $(1-x)^3$ and set $x = 1$, we get $A = 1/6$. However, finding B and C is a little tricky. Multiplying $(*)$ by $(1-x)^3$, we get $f(x) = A +$

$+ B(1 - x) + C(1 - x)^2 + (1 - x)^3 g(x)$ where $f(x) = \dfrac{1}{(1+x)(1+x+x^2)} =$

$= \dfrac{1}{x^3 + 2x^2 + 2x + 1}$ and $g(x) = \dfrac{D}{1 + x} + \dfrac{E}{1 - \omega x} + \dfrac{F}{1 - \omega^2 x}$. Differentiat-

ing, $f'(x) = \dfrac{- 3x^2 - 4x - 2}{(1 + x)^2 (1 + x + x^2)^2} = - B + 2C(1 - x) - 3(1 - x)^2 g(x) +$

$+ (1 - x)^3 g'(x)$. We need not compute $g'(x)$. We set $x = 1$ and get $f'(1) =$
$= - B$, which gives $B = -f'(1) = 9/36 = 1/4$. Differentiating again,
we could get C from $f''(1)$. Alternatively, in (∗) we set $x = 0$, giving
$A + B + C + D + E + F = 1$. Since all other constants are known, C
comes out as $17/72$. Thus $\dfrac{1}{(1 - x)(1 - x^2)(1 - x^3)} = \dfrac{1}{6}\Big(1 - x\Big)^{-3} + \dfrac{1}{4}$

$\Big(1 - x\Big)^{-2} + \dfrac{17}{72}\dfrac{1}{1 - x} + \dfrac{1}{8}\dfrac{1}{1 + x} + \dfrac{1}{9}\dfrac{1}{1 - \omega x} + \dfrac{1}{9}\dfrac{1}{1 - \omega^2 x}$. We expand

each term on the right by Theorem (1.3), (iii) and (iv). The coefficient of

x^n in $(1 - x)^{-3}$ is $(- 1)^n \dfrac{(- 3)(- 4)...(- 3 - n + 1)}{n!} = \dfrac{3.4......(n + 2)}{n!} =$

$= \dbinom{n + 2}{2}$. Similarly the coefficient of x^n in $(1 - x)^{-2}$ is $\dbinom{n + 1}{1}$. The

other four terms are expanded by the formula $\dfrac{1}{1 + \alpha x} = \overset{\infty}{\underset{n=0}{\Sigma}} \alpha^n x^n$. Summing

it up, we get $\dfrac{1}{(1 - x)(1 - x^2)(1 - x^3)} = \overset{\infty}{\underset{n=0}{\Sigma}} a_n x^n$, where

$$a_n = \frac{1}{6}\binom{n + 2}{2} + \frac{1}{4}\Big(n + 1\Big) + \frac{17}{72} + \frac{1}{8}\Big(-1\Big)^n +$$

$\dfrac{1}{9}\omega^n + \dfrac{1}{9}\omega^{2n} = \dfrac{6(n+2)(n+1)+18(n+1)+17+(- 1)^n 9 + 8\omega^n + 8\omega^{2n}}{72}$

We are interested in a_{20} which comes out as 44 since $\omega^{20} + \omega^{40} = \omega^2 + \omega =$
$= - 1$. So there are 44 ways to put a total postage of 2 rupees from
stamps of 10 paise, 20 paise and 30 paise. This is exactly the same answer
obtained earlier in Chapter 2, Section 3 but this time the method works
for other denominations as well.

It is tempting to try to apply the method of solution above to the

expression $\dfrac{1}{(1 - x)(1 - x^2)...(1 - x^m)}$ and thereby get a formula for the

number of partitions of an integer n into at most m parts (or equivalently,
number of partitions into parts of size not exceeding m). Unfortunately,

the partial fraction resolution of $\dfrac{1}{(1 - x)(1 - x^2)...(1 - x^m)}$ seems quite

complicated. So, as remarked before, there is still no easy formula for $p(n)$.
The reason for applying generating functions to study partitions is not that
they yield such handy formulas. Their real advantage is that by manipulat-
ing the enumerators algebraically we can sometimes show that two appa-

rently different combinatorial problems have the same enumerator. In such a case, even if we are not able to expand the enumerator, we do get a relationship between the counts associated with the two combinatorial problems which may not be obvious directly. By way of illustration, we prove one such relationship here.

2.9 Proposition: For every positive integer n, the number of partitions of n in which the part sizes are distinct equals the number of partitions in which the part sizes are all odd.

Proof: Let a_n and b_n be the respective numbers of partitions of n with the stated restrictions. (For example, let $n = 5$. Then $a_n = 3$ because (5), (4, 1), (3, 2) are the partitions of 5 in which the part sizes are distinct. Similarly $b_n = 3$, because (5), (3, 1, 1) and (1, 1, 1, 1, 1) are the partitions of 5 in which the parts are of odd size.) We set $a_0 = b_0 = 1$. We have to show $a_n = b_n$ for all n. It would be great if we can compute both a_n and b_n and show directly that they are equal. But this is too ambitious. Nor is it necessary. It would suffice if we can establish a one-to-one correspondence (i.e., a bijection) between the sets of the two kinds of partitions of n. There are many instances where this kind of an approach works (see e.g. Exercise (2.3.36)). But in the present case, it is rather clumsy to convert a partition in which the parts are of distinct size to a partition in which the parts are of odd size or vice versa. Another approach would be to try induction on n. This would require us to express a_n in terms of a_{n-1} (or in terms of a_k for some $k < n$) and similarly for b_n. In the next two sections we shall see numerous examples where this approach works. But, again, in the present case, there seems to be no easy way to implement it.

However, with generating functions, we can get the result elegantly. Let $A(x) = \sum_{n=0}^{\infty} a_n x^n$ and $B(x) = \sum_{n=0}^{\infty} b_n x^n$. In view of Theorem (1.2) part (i), it would suffice to show that $A(x) = B(x)$. Now $A(x)$ is the enumerator for partitions in which the part sizes are distinct. Referring to the proof of Proposition (2.5), this means that from each box we can pick at most one bag. So the enumerator for the ith box is only $1 + x^i$. It follows that

$$A(x) = (1 + x)(1 + x^2)(1 + x^3)\ldots(1 + x^m)\ldots = \prod_{m=1}^{\infty} (1 + x^m).$$

As for $B(x)$, we follow the proof of Theorem (2.7), with the restriction that the part sizes must be odd. This gives

$$B(x) = \frac{1}{1-x} \frac{1}{1-x^3} \frac{1}{1-x^5}\cdots \frac{1}{1-x^{2k-1}} \cdots = \prod_{k=1}^{\infty} \frac{1}{1-x^{2k-1}}.$$

So we are reduced to proving the following identity about infinite products:

$$(*) \quad (1+x)(1+x^2)(1+x^3)\ldots = \frac{1}{1-x} \frac{1}{1-x^3} \frac{1}{1-x^5} \cdots$$

As usual, we assume that infinite products are amenable to the same algebraic manipulations as are finite products. (A rigorous approach would require a justification.) Rewriting $1 + x^m$ as $\dfrac{1 - x^{2m}}{1 - x^m}$, the left hand side of (*) becomes $\dfrac{(1 - x^2)(1 - x^4)(1 - x^6)\ldots(1 - x^{2m})\ldots}{(1 - x)(1 - x^2)(1 - x^3)\ldots(1 - x^m)\ldots}$. We see that all the factors in the numerator cancel with the alternate factors in the denominator. This leaves us precisely with the right hand side of (*). ∎

In the proof above, we proved an algebraic identity (namely (*)) by direct arguments and interpreted its two sides suitably so as to yield a combinatorial result. Sometimes it is easier to go the other way. That is, we find the enumerators of two mutually equivalent combinatorial problems. Equating these two enumerators gives an algebraic identity. Such combinatorial proofs of algebraic identities are often ingeneous and provide a testimony to the fact that the relationship between algebra and combinatorics is not a one way affair. Each branch has enriched the other. By way of illustration, we prove here one such identity.

2.10 Proposition

$$(1 + x)(1 + x^3)(1 + x^5)\ldots = \sum_{k=0}^{\infty} \frac{x^{k^2}}{(1 - x^2)(1 - x^4)(1 - x^6)\ldots(1 - x^{2k})}.$$

Proof: We immediately recognise the left hand side as the enumerator for the number of partitions into odd, distinct, part sizes. What we need is another way of counting such partitions so that the enumerator would come out to be the right hand side. The key to this alternate counting is provided by the Ferrer's graph of a partition, introduced in Chapter 2, Section 3. As an application of this concept, it is easy to show (cf. Exercise (2.3.36)) that the number of partitions of an integer n into parts of odd, distinct sizes equals the number of self-dual partitions of n, i.e., partitions which coincide with their dual partitions. We now count the number of self-dual partitions. Such partitions are characterised by the fact that their

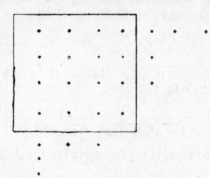

Figure 7.3: Durfee Square of a Partition

Ferrer's graphs are symmetric in terms of rows and columns. Consider the largest square of dots that is contained in the left hand top corner of the Ferrer's graph of a self-dual partition. This square is called the **Durfee square** of that partition. (Actually, this concept makes sense for any partition, but here we need it only for self-dual partitions.) In Figure 7.3 we show the Durfee square of a self-dual partition of 26. It is a 4×4 square.

Conversely suppose we start with a $k \times k$ square of dots. We can then construct a Ferrer's graph of some partition as follows. We add n_i dots to the ith row of the square and an equal number of dots below the ith column. The resulting figure will be the Ferrer's graph of a self-dual partition of the integer $k^2 + 2n_1 + 2n_2 + ... + 2n_k$ provided $n_1 \geqslant n_2 \geqslant ... \geqslant n_k \geqslant 0$. (In Figure 7.3, $n_1 = 3$, $n_2 = 1$, $n_3 = 1$ and $n_4 = 0$). The original $k \times k$ square is the Durfee square of this partition which is uniquely determined by the integers k and $n_1, ..., n_k$. Let $m = 2n_1 + ... + 2n_k$. Then the sequence $(n_1, ..., n_k)$ corresponds to a partition of m into at most k parts of even size. Conversely every such partition of m gives rise to a sequence $(n_1, ..., n_k)$ (we simply divide the part sizes by 2 each). Following an argument analogous to that in Proposition (2.5), the number of partitions of m into at most k parts of even size

is the coefficient of x^m in $\dfrac{1}{(1 - x^2)(1 - x^4)...(1 - x^{2k})}$, which is the same

as the coefficient of x^{m+k^2} in $\dfrac{x^{k^2}}{(1 - x^2)...(1 - x^{2k})}$. It now follows that the

number of self-dual partitions of an integer n is the coefficient of x^n in the

sum $\displaystyle\sum_{k=0}^{\infty} \dfrac{x^{k^2}}{(1 - x^2)(1 - x^4)...(1 - x^{2k})}$. For a fixed k, this sum is really finite

since the terms with $k > \sqrt{n}$ contribute nothing to it. Thus we see that the

enumerator for self-dual partitions is $\displaystyle\sum_{k=0}^{\infty} \dfrac{x^{k^2}}{(1 - x^2)(1 - x^4)...(1 - x^{2k})}$. This

is precisely the right hand side of the identity to be proved. As noted before, this completes the proof. ∎

We now consider an application of generating functions to probability. Many counting problems can be paraphrased as problems in probability because of the definition of probability as the ratio of the number of favourable cases to the total number of cases, assuming the latter are all equally likely. Of course, this simple-minded definition presupposes that the total number of cases is finite. It is possible to extend the theory of probability to the case where the set, say S, of all possible cases is infinite using the concept of what is called a probability measure. The essential idea can be explained as follows. Suppose X is a variable whose values vary randomly over the set S. Such a variable is called a **random** (or a **stochastic**) **variable** (or a **variate**). Now we would like to have a function $\mu: P(S) \to \mathbf{R}$, such that for each $A \subset S$, $\mu(A)$ gives the probability that the value assumed by X lies in the subset A. This function must satisfy certain properties (for example, $0 \leqslant \mu(A) \leqslant 1$ for all $A \subset S$), and when it does, the pair (S, μ) is called a probability measure space. We encountered an example of this in the Continuous House Problem, where the random variable X was the ordered pair (x, y) of two houses on the road, the set S was the unit square in the cartesian plane and for $A \subset S$, $\mu(A)$ was simply the area of A.

Construction of a probability measure on an infinite set S involves certain problems beyond our scope. The situation is somewhat tractable if S is countably infinite. Specifically, let $S = \{s_1, s_2,..., s_n,...\}$. Let X be a random variable taking values in S. For each n, let p_n be the probability that X equals s_n. Obviously $0 \leqslant p_n \leqslant 1$ for all $n \in \mathbb{N}$ and $\sum_{n \in \mathbb{N}} p_n = 1$.

Now for any $A \subset S$, we can define $\mu(A) = \sum_{n : s_n \in A} p_n$. Clearly $\mu(A)$ is the probability that the value of X is in A. In case S is a subset of $\mathbb{R}$, we can speak of the average (or the expected) value of X, because of the algebraic structure on $\mathbb{R}$. Things are especially nice if the set S consists of nonnegative integers, because in this case generating functions can be used handily as we shall soon see. Moreover, this special case has a wide applicability because as the examples below show, there are many real-life situations where the outcome is a non-negative integer.

So, formally we define a **discrete random variable** as a variable X which assumes the value n with probability, say, p_n. We express this by saying that $P(X = n) = p_n$ for $n = 0, 1, 2,...$. Obviously the numbers p_n are non-negative and $\sum_{n=0}^{\infty} p_n = 1$. The sequence $\{p_n\}_{n=0}^{\infty}$ is called the **probability distribution** of X.

Let us consider a few examples of discrete random variables.

(1) Let X denote the figure that appears on top when an ordinary die is rolled. Then X is a discrete random variable for which $p_n = 1/6$ for $1 \leqslant n \leqslant 6$, and $p_n = 0$ for all other n.

(2) Let X denote the score when a pair of ordinary dice is rolled. Here X can assume values from 2 to 12 but not with the same probabilities. Enumerating the various possibilities it is easy to show that X is a discrete random variable with probability distribution $\{p_n\}_{n=0}^{\infty}$ where $p_0 = p_1 = 0$,

$$p_2 = p_{12} = \frac{1}{36}, p_3 = p_{11} = \frac{1}{18}, p_4 = p_{10} = \frac{1}{12},$$

$$p_5 = p_9 = \frac{1}{9}, p_6 = p_8 = \frac{5}{36}, p_7 = \frac{1}{6}$$

and $p_n = 0$ for $n > 12$.

(3) Suppose p is the probability of a head showing up when a coin is tossed, where $0 \leqslant p \leqslant 1$. Let X be the number heads occurring in 10 tosses of the coin (cf. Exercise (1.13)). Then X is a discrete random variable for

which $p_r = \begin{pmatrix} 10 \\ r \end{pmatrix} p^r(1-p)^{10-r}$; $r = 0, 1, 2, \ldots$. (A probability distri-

bution of this kind is called a **binomial disribution**).

(4) In all the three examples above, the discrete random variable assumed only finitely many values and hence the probability p_n was 0 for all sufficiently large n. As an example when $p_n > 0$ for infinitely many n's, suppose an unbiassed coin is tossed until a head shows. It can be shown that the probability of success for this experiment is 1, that is, it is extremely unlikely that we will keep on tossing the coin endlessly and getting a tail every time. A rigorous proof of this statement requires the concept of a measure and is beyond our scope. Hoewver, for every positive integer n, it is easy to calculate the probability that a head will appear for the first time on the nth toss. We identify a head with 1 and a tail with 0. Each outcome of tossing a coin n times corresponds to a binary sequence of length n. There are 2^n such sequences and the assumption that the coin is unbiassed makes all these sequences equally likely. The case where the first head occurs during the nth toss corresponds to the binary sequence $0\ldots01$ which occurs with probability $\frac{1}{2^n}$. So, if X denotes the number of times the coin is tossed till a head shows, then X is a discrete random variable with $p_0 = 0$, and $p_n = \frac{1}{2^n}$ for $n > 0$. Note that

$$\sum_{n=0}^{\infty} p_n = \sum_{n=1}^{\infty} \frac{1}{2^n} = 1.$$

Let us now introduce generating functions into the study of discrete random variables. So far we have been denoting the independent variable in a generating function by x. Since a discrete random variable is itself denoted by symbols like X, Y, it is customary to use the variable t for the generating function instead of x. This is, of course, purely a matter of notation and does not affect the underlying concept.

2.11 Definition: Let X be discrete random variable with probability distribution $\{p_n\}_{n=0}^{\infty}$. Then the **probability generating function** of X, denoted by $p_X(t)$, is defined as $\sum_{n=0}^{\infty} p_n t^n$.

In other words, the probability generating function of a discrete random variable X is nothing but the ordinary generating function of its probability distribution. Because of the uniqueness property of power series expansions (Theorem (1.2) (i)), the random variable X (i.e. its probability distribution) is uniquely determined by its probability generating function. In order to show that the probability generating function is more than just a coding device, we must be able to translate certain concepts about random variables in terms of their probability generating functions. We discuss two such concepts.

The first is about the average or expected value of a discrete random variable. In Exercise (1.10), we defined this concept for a variable which assumed only finitely many real values. The extension to the case of any discrete random variable is straightforward.

2.12 Definition: The **expected value** of a discrete random variable X, with probability distribution $\{p_n\}_{n=0}^{\infty}$ is defined as $\sum\limits_{n=0}^{\infty} np_n$.

$E(X)$ is the standard notation for the expected value of a random variable X. Examples can be given to show that $E(X)$ need not be finite. In the examples given above, in (1), the expected value is

$$\frac{1}{6}(1 + 2 + 3 + 4 + 5 + 6) = \frac{7}{2}.$$

This is also the arithmetic mean of the values assumed by X, because all these values are assumed with equal probability. This is no longer the case in Example (2), where the expected value, $E(X)$ is 7 by a direct calculation. Exercise (1.12) gives the expected value of X in Example (3). In Example (4), to get the expected value we have to sum the series $\sum\limits_{n=0}^{\infty} \frac{n}{2^n}$. This can be done by the methods of the last section. The same methods yield the following alternate description of the expected value.

2.13 Proposition: If $p_X(t)$ is the probability generating function of a discrete random variable X, then $E(X) = p'_X(1)$, where $p'_X \equiv \frac{d}{dt} p_X$.

Proof: By definition, $p_X(t) = \sum\limits_{n=0}^{\infty} p_n t^n$. By (iv) of Theorem (1.2),

$$p_X'(t) = \sum_{n=1}^{\infty} np_n t^{n-1}.$$

Setting $t = 1$, we get

$$p_X'(t) = \sum_{n=1}^{\infty} np_n.$$

which is precisely the expected value of X. ∎

In Example (4) above,

$$p_X(t) = \sum_{n=0}^{\infty} p_n t^n = \sum_{n=1}^{\infty} \frac{t^n}{2^n} = \frac{t/2}{1 - t/2} = \frac{t}{2 - t} = \frac{2}{2 - t} - 1.$$

So

$$p_X'(t) = \frac{2}{(2 - t)^2}.$$

Hence $E(X) = p_X'(1) = 2$. This means that if we keep on tossing a coin then the 'expected' number of times it will have to be tossed before a head shows up is 2. This does not mean that a head will necessarily show in 2 tosses. Indeed, as we saw above, even for a thousand tosses there is always some probability $\left(\text{namely } \dfrac{1}{2^{1000}} \right)$ that no head will show. What, then, is the 'real life' interpretation of the 'expected' value? This is actually a deep question and tends to be more philosophical than mathematical. The mathematical definitions of probability do not really answer what probability is. They merely express one probability in terms of some others. This is analogous to the stand taken in mathematical logic where we never answer what is meant by 'true' or 'false'. Instead our concern is to express the truth or the falsehood of a complicated statement (or, in others words, its truth value) in terms of the truth values of the statements from which it is formed. A vague but intuitively appealing description of expected value is that it is the average over a very large sample. In the present case, suppose one person tosses a coin till a head shows. He may or may not get a head in 2 tosses. But suppose 10000 persons each do the same experiment. Then the average number of tosses will be close to 2. Again this is not a certainty. But the probability that it will differ from 2 by more than, say, .0001 is small, which means that if this experiment in which large numbers of persons flip coins is repeated a large number of times, then in a large majority of them the averges will differ from 2 by at most .0001. But again, we cannot be absolutely sure of this! This only serves to point to the circuitous nature of probability.

We leave this philosophical discussion here and return to showing how generating functions provide a succinct description of some of the concepts associated with discrete random variables. Expected value is one such concept. As another example, we consider a concept which deals with the mutual relationship between two such variables, say, X and Y. Intuitively, we say X and Y are independent, if the values assumed by either one of them are completely independent of each other. In terms of probabilities, we can state this condition as follows.

2.14 Definition: Two discrete random variables X and Y with probability distributions $\{p_n\}_{n=0}^{\infty}$ and $\{q_n\}_{n=0}^{\infty}$ respectively are said to be (mutually) **independent** if for every pair (i, j) of non-negative integers, $P(X = i$ and $Y = j) = P(X = i) \cdot P(Y = j)$.

For example if a pair of dice is rolled and X and Y denote the figures on the faces of the two dice then X and Y are independent. On the other hand suppose we pick two cards, one after the other, randomly from six cards marked 1 to 6. Let X, Y denote the numbers on the first and the second cards respectively. Then X, Y are discrete random variables each having the same probability distribution, namely,

$$\left(0, \frac{1}{6}, \frac{1}{6}, \frac{1}{6}, \frac{1}{6}, \frac{1}{6}, \frac{1}{6}, 0, 0, 0, \dots\right).$$

Here X and Y are not independent, because they can never be equal.

Now let X, Y be any two discrete random variables,. Then their sum, $X + Y$, is also a discrete random variable. Its values are the sums of the various possible values assumed by X and Y. However, very little can be said about its probability distribution unless we know how X and Y are related. In case X and Y are independent, the answer is simple and can be expressed elegantly in terms of probability generating functions as we now show.

2.15 Theorem: If two discrete random variables are independent then $p_{X+Y}(t) = p_X(t)\,p_Y(t)$.

Proof: Let $\{p_n\}$, $\{q_n\}$, $\{r_n\}$ be, respectively, the probability distributions of X, Y and $X + Y$. For a given non-negative integer n, $X + Y$ can assume the value n whenever X and Y assume values i and j for which $i + j = n$. Hence $P(X + Y = n) = \sum_{i=0}^{n} P(X = i \text{ and } Y = n - i)$. Since X, Y are independent, $P(X = i \text{ and } Y = n - i) = p_i q_{n-i}$. So for every $n \geqslant 0$, $r_n = \sum_{i=0}^{n} p_i q_{n-i}$ and hence $\sum_{n=0}^{\infty} r_n t^n = \sum_{n=0}^{\infty} \sum_{i=0}^{n} p_i t^i \, q_{n-i} t^{n-i} = (\sum_{i=0}^{\infty} p_i t^i)(\sum_{j=0}^{\infty} q_j t^j)$. This proves the assertion. (In essence we are applying Theorem (1.2) (iii) here.) ∎

We conclude this section with a discussion of the use of exponential generating functions in problems of enumeration. It was remarked in the last section that theoretically the E.G.G. is not an independent concept inasmuch as the E.G.F. of the sequence $\{a_r\}_{r=0}^{\infty}$ is precisely the O.G.F. of the sequence $\left\{\dfrac{a_r}{r!}\right\}_{r=0}^{\infty}$. However, problems involving permutations can be handled more elegantly with exponential enumerators than with ordinary enumerators. For example, let n be a fixed positive integer. For each r, the number of r-permutations of n objects is $_nP_r = n(n-1)\dots(n-r+1) = \dfrac{n!}{(n-r)!}$.

So the ordinary enumerator for permutations of n objects is $\sum_{r=0}^{\infty} \dfrac{n!}{(n-r)!} x^r$. This is really a finite sum since $_nP_r = 0$ for $r > n$. But there is no closed form expression for it. The exponential enumerator in this case, on the other hand, is $\sum_{r=0}^{\infty} \dfrac{n!}{r!(n-r)!} x^r = \sum_{r=0}^{\infty} \binom{n}{r} x^r$ which is nothing but $(1+x)^n$.

Here the objects are distinct and we are not allowing repetitions in the permutations. If we keep the n objects distinct, but allow unlimited repetitions, then for every r (including $r > n$), the number of permutations is n^r and so the ordinary enumerator for permutations of n objects, with repetitions allowed, is $\sum\limits_{r=0}^{\infty} n^r x^r$, which, now, is an infinite sum whose value is $\dfrac{1}{1-nx}$.

The exponential enumerator in this case is $\sum\limits_{r=0}^{\infty} \dfrac{n^r}{r!} x^r = \sum\limits_{r=0}^{\infty} \dfrac{(nx)^r}{r!} = e^{nx}$. Here although we do have a closed form expression for the ordinary enumerator, the expression $\dfrac{1}{1-nx}$ does not suggest any alternate way to arrive at it. The exponential enumerator, e^{nx}, on the other hand, factors as $e^x \cdot e^x \cdot e^x \cdot \ldots \cdot e^x$ (n times) and suggests that there might be some other method to derive it. Such a method indeed exists and since it has many other applications besides providing an alternate derivation of the formula above, we proceed to study it. As with ordinary enumerators, the crux of this method is the simple formula that $x^i x^j = x^{i+j}$.

Recall that an r-permutation (or a permutation of length r) is an ordered r-tuple of objects*. Equivalently it is an arrangement of r objects in a row, i.e. a way of filling the cells of an $r \times 1$ rectangle. Suppose we are interested in finding, for each value of r, the number of permutations where the objects to be placed in these cells come from two mutually disjoint sources, say, S_1 and S_2. These permutations may be required to obey certain conditions regarding how many times certain objects can appear. We assume that the nature of these conditions is such that the sources S_1 and S_2 are independent in the following sense. Let $(a_1, a_2, \ldots, a_r)$ be a 'permissible' permutation. Out of the r objects, let $a_{i_1}, a_{i_2}, \ldots, a_{i_p}$ (say) come from S_1 and the remaining $r-p$ come from S_2, where $0 \leqslant p \leqslant r$ and $1 \leqslant i_1 < i_2 < \ldots < i_p \leqslant r$. Call the remaining objects $b_1, \ldots, b_q$ where $q = r - p$ and the b's appear in the same order as in $(a_1, \ldots, a_r)$. Then $(a_{i_1}, \ldots, a_{i_p})$ is a permutation of objects from S_1 and $(b_1, \ldots, b_q)$ is a permutation of objects from S_2. We assume that the nature of the conditions on the permutations is such that $(a_1, a_2, \ldots, a_r)$ is a permissible permutation if and only if both $(a_{i_1}, \ldots, a_{i_p})$ and $(b_1, \ldots, b_q)$ are permissible permutations (i.e. satisfy the respective conditions for the sources S_1 and S_2). In other words every permissible permutation is composed by freely merging (and not just concatenating) two permissible 'subpermutations', one coming from the source S_1 and the other from S_2. (By free merging we mean that the entries of one permutation can freely intersperse with those of the other, but the relative order of the objects in each permutation must be maintained. For example, (a, b, a) can be merged with $(1, 2)$ to give $(a, b, a, 1, 2)$, $(a, b, 1, a, 2)$, $(a, b, 1, 2, a)$, $(1, a, 2, b, a)$, $(1, 2, a, b, a)$ etc. but not $(a, a, b, 1, 2)$ nor $(a, 2, b, a, 1)$. It is easy to show that two permutations of length p and q respectively can

---

* In Chapter 2, Section 2 we required that the entries in a permutation be distinct. We remove that restriction here, as in some problems we have to consider permutations with repetitions.

be merged together in $\begin{pmatrix} p+q \\ p \end{pmatrix}$ ways. This fact is the essence of what is to come).

This assumption about the mutual independence of the sources S_1 and S_2 is not very restrictive and in fact, as we shall soon see, holds true in many naturally occurring examples. But first let us see how it helps in identifying the exponential enumerator for the permutations satisfying the given conditions. For $r \geqslant 0$, let a_r be the number of permissible r-permutations. Let $A(x) = \sum\limits_{r=0}^{\infty} \dfrac{a_r}{r!} x^r$. Similarly, let $B(x) = \sum\limits_{r=0}^{\infty} \dfrac{b_r}{r!} x^r$ and $C(x) = \sum\limits_{r=0}^{\infty} \dfrac{c_r}{r!} x^r$ where b_r and c_r are, respectively, the numbers of permissible r-permutations for sources S_1 and S_2. We contend that for every r,

$$a_r = \sum_{p=0}^{r} \begin{pmatrix} r \\ p \end{pmatrix} b_p \, c_{r-p}.$$

To see this, note that every permissible r-permutation arises from merging together a permissible p-permutation from S_1 and a permissible $(r-p)$-permutation from S_2 for some p, $0 \leqslant p \leqslant r$. Every such pair of permutations can be merged together in $\begin{pmatrix} r \\ p \end{pmatrix}$ ways, because out of r cells, we can choose p cells in $\begin{pmatrix} r \\ p \end{pmatrix}$ ways and the cells so chosen are to be used to accomodate the p-permutation from S_1 and the remaining $(r-p)$ cells to accomodate the permutation from S_2. Recalling that $\begin{pmatrix} r \\ p \end{pmatrix} = \dfrac{r!}{p!\,(r-p)!}$, we get that $\dfrac{a_r}{r!} = \sum\limits_{p=0}^{r} \dfrac{b_p}{p!} \dfrac{c_{r-p}}{(r-p)!}$. By Theorem (1.2) part (iii), this amounts to saying that $A(x) = B(x)\,C(x)$. Note that there would have been no such simple relationship, if, instead of exponential enumerators we had taken ordinary enumerators.

The extension to the case of n sources is immediate and we record it as a theorem.

2.16 Theorem: Let $S_1, S_2,..., S_n$ be mutually disjoint sets of objects. For $i = 1,..., n$, let $A_i(x)$ be the exponential enumerator for a certain class of permutations of the objects in the set S_i (to be called permissible permutations hereafter). Then the product, $A_1(x)\,A_2(x)...A_n(x)$ is the exponential enumerator for those permutations of the objects in $S_1 \cup S_2 \cup ... \cup S_n$, which are obtained by freely merging together the permissible permutations from each S_i.

Proof: We proceed by induction on n. For $n = 1$, there is nothing to prove. The argument above proves the case $n = 2$ and also shows how to do the inductive step. ▪

We are now in a position to give alternate derivations of the number of permutations of n distinct objects, say, $a_1, ..., a_n$. Let $S_i = \{a_i\}$. First suppose, no repetitions are allowed. Then for each $i = 1, ..., n$, for $r = 0$, there is the empty permutation, for $r = 1$, there is the lone permutation (a_i) while for $r > 1$, there are no r-permutations of S_i. So each $A_i(x) = 1 + x$. Hence by the theorem above, $A(x) = (1 + x)^n$. On the other hand, suppose unlimited repetitions are allowed. Then for each $i = 1, ..., n$ and $r \geqslant 0$, there is exactly one r-permutation of S_i, namely the r-tuple $(a_i, a_i, ..., a_i)$. So, in this case,

$$A_i(x) = 1 + \frac{x}{1!} + \frac{x^2}{2!} + \frac{x^3}{3!} + ... \quad \text{By } (i) \text{ in Theorem } (1.3), A_i(x) = e^x. \text{ By}$$

the theorem above, $A(x) = (e^x)^n = e^{nx}$.

As another example of the use of exponential enumerators, we give an alternate proof of Theorem (3.2.8) where it was shown that for positive integers k and n, the number of k-ary sequences of length n in which a particular symbol, say 1, appears an even number of times is $\dfrac{k^n + (k-2)^n}{2}$.

The argument given earlier was somewhat tricky, based on classifying all such sequences according to a suitable equivalence relation. With our present technique the problem amounts to counting the permutations of k objects, say $a_1, ..., a_k$, in which one of the objects, say, a_1, appears an even number of times and there is no restriction on the number of occurrences of the other objects. Once again, let $S_i = \{a_i\}$, for $i = 1, ..., k$. Then the desired enumerator, $A(x)$, equals $A_1(x) A_2(x)...A_k(x)$. As before, for $i = 2, ..., k$, $A_i(x) = 1 + \dfrac{x}{1!} + \dfrac{x^2}{2!} + ... + \dfrac{x^r}{r!} + ... = e^x$. However, since a_1 can appear only an even number of times, the enumerator $A_1(x)$ is not the entire sum $1 + \dfrac{x}{1!} + \dfrac{x^2}{2!} + ...$ but only the even powers in it, namely $1 + \dfrac{x^2}{2!} + \dfrac{x^4}{4!} + ...$

It is easily seen that this equals $\dfrac{e^x + e^{-x}}{2}$. So

$$A(x) = \left(\frac{e^x + e^{-x}}{2} \right) (e^x)^{k-1} = \frac{1}{2} [e^{kx} + e^{(k-2)x}].$$

This can be expanded as $\sum\limits_{r=0}^{\infty} \left[\dfrac{k^r + (k-2)^r}{2r!} \right] x^r$. Since this is the exponential and not the ordinary enumerator, the desired number of permutations is n! times the coefficient of x^n, proving the result.

As with ordinary enumerators, it can happen that in some problems it is easy to identify the exponential enumerator but not easy to expand it. Consider, for example the problem of distributing distinct objects into distinct boxes so that no box is empty. In Proposition (2.3.8), we showed that the number of ways to distribute n distinct objects into m non-distinct boxes so that no box is empty is $m! \, S_{n,m}$ where $S_{n,m}$ is Stirling number of second kind, defined as the number of ways to put n distinct objects into m non-distinct boxes with no box empty, or equivalently the number of partitioning a set of cardinality n into m mutually disjoint, non-empty subsets. It was remarked that there was no easy formula for $S_{n,m}$. Note that $S_{n,m}$ is not the same as $P_{n,m}$, which is the number of partitions of the integer n into exactly m parts. Although there is no easy formula for $P_{n,m}$, in Proposition (2.5), we obtained the O.G.F. of the sequence $\{P_{n,m}\}_{n=0}^{\infty}$ for fixed m. Similarly, although there is no formula for the Stirling numbers, $S_{n,m}$, it is easy to obtain the E.G.F. of the sequence $\{S_{n,m}\}_{n=0}^{\infty}$, for a fixed m.

2.17 Proposition: Let m be a positive integer. Then

$$\sum_{n=0}^{\infty} \frac{S_{n,m}}{n!} x^n = \frac{1}{m!} (e^x - 1)^m.$$

Proof: Let us first work with $(e^x - 1)^m$. Consider m distinct boxes, say, $B_1, B_2, \ldots, B_m$ and let us count the permutations of these boxes with repetitions allowed and with the restriction that each box must appear at least once. (Later on we shall put certain objects in these boxes. It is important to note, however, that here we are dealing with the permutations of the boxes and not of the objects that will be put in them. In other words, for the time being, these boxes are our 'objects'.) We repeat the reasoning in the examples following Theorem (2.17). For $i = 1, \ldots, m$, let $S_i = \{B_i\}$. Since the box B_i must appear at least once, we have to exclude the empty permutation (i.e., permutation of length 0) from each $A_i(x)$. Consequently, $A_i(x)$ is not e^x but $e^x - 1$ for each $i = 1, \ldots, m$. By theorem (2.17), $(e^x - 1)^m$ is the exponential enumerator for permutations of the m boxes $B_1, \ldots, B_m$, with repetitions allowed, in which every box appears at least once. We now look at such permutations a little differently. Let n be a fixed integer and let T be a set with n distinct elements, say $T = \{a_1, \ldots, a_n\}$. Let us place these elements (or 'objects') in the boxes $B_1, \ldots, B_m$. Every such placement is equivalent to a permutation $(B_{i_1}, B_{i_2}, \ldots B_{i_n})$ where for $k = 1, \ldots, n$, B_{i_k} is the box in which the object a_k is put. The requirement that every box shall appear at least once corresponds to no box being empty while allowing unlimited repetitions translates into there being no upper bound on how many objects can go into each box. Thus the n-permutations of the boxes with the stipulated restrictions correspond to ways of putting n distinct objects in m distinct boxes with no box empty. By Proposition (2.3.8), this can be done in $m! \, S_{n,m}$ ways. Summing it up, $(e^x - 1)^m$ is the exponential generating function of the sequence $\{m! \, S_{n,m}\}_{n=0}^{\infty}$. To finish the proof we merely divide by $m!$ ∎

Expanding $(e^x - 1)^m$ as a power series in x, we can now show that for each n, $S_{n,m} = \sum_{r=0}^{m-1} \frac{(-1)^r \, (m-r)^{n-1}}{r! \, (m - r - 1)!}$. This is the same as Theorem (2.4.8), which was proved using the principle of inclusion and exclusion. We could have as well started from Theorem (2.4.8) and obtained another proof of Proposition (2.17). Whenever possible, it is generally more desirable to obtain enumerators by combinatorial arguments.

Exercises

2.1 Find the coefficient of x^{10} in the following expressions:

(i) $(1 + x^2 + x^3)(x^4 + x^5)(x + x^3)$

(ii) $(1 + 5x + 10x^2)^6$

(iii) $(1 + x^2)^{20}$

2.2 Devise suitable combinatorial problems which can be solved using the last exercise.

2.3 Find the enumerator for the selections of k kinds of objects, if the selection must not include more than m_i objects of type i, $i = 1$, $2,..., k$.

2.4 Find the enumerator for selections of books from two shelves containing 10 and 7 books respectively if the selection must include an odd number of books from the first shelf.

2.5 Find the enumerator for the total score if n distinct dice are tossed, n being a fixed positive integer.

2.6 A 'crazy' dice is defined as one whose six faces are marked differently than those on an ordinary dice. Prove that if two crazy dice, one marked with figures 0, 1, 2, 3, 4 and 5 and the other with 2, 3, 4, 5, 6 and 7 are rolled together, each possible total score is as likely as that for a pair of ordinary dice.

*2.7 Devise a pair of crazy dice, the markings on whose faces do not form arithmetic progressions, which have the property that when they are rolled together, each possible total score is as likely as that for a pair of ordinary dice. [Hint: Factor $(x + x^2 + x^3 + x^4 + x^5 + x^6)^2$ as a product of unequal factors, say $A(x)$ and $B(x)$ such that $A(1) = B(1) = 6$.]

2.8 Suppose at an international conference, each country sends a team of 4 delegates. There are 5 capitalist countries, 6 communist countries and 7 non-aligned countries at the conference. Let a_n be the number of committees that can be formed in which there are n members and in which the capitalist countries have the same representation as the communist countries. Identify the O.G.F, of $\{a_n\}_{n=0}^{\infty}$. Evaluate a_n for $n \leqslant 10$.

2.9 Generalise Theorem (2.4) to any base b, by first proving the identity

$$(1 + x + ... + x^{b-1})(1 + x^b + x^{2b} + ... + x^{b(b-1)})... = \frac{1}{1 - x}.$$

2.10 Negative coefficients rarely appear in enumerators. (Why?) However, sometimes it is convenient to take the negatives of some coefficients so as to indicate parity. Suppose a box contains infinitely many bags each containing m balls, where m is a positive integer. Prove that $1 - x^m + x^{2m} - x^{3m} + x^{4m} - x^{5m} + \ldots$ still enumerates the number of ways to pick balls from this box (without breaking any bag) and also indicates the parity of the number of bags picked.

2.11 For every positive integer n, consider partitions of n into parts whose sizes are (not necessarily distinct) powers of 2. Among all such partitions, let a_n be the number of those in which the number of parts is even and b_n the number of those in which there is an odd number of parts. Using the last exercise and the identity

$$1 - x = \frac{1}{(1 + x)} \frac{1}{(1 + x^2)} \frac{1}{(1 + x^4)} \cdots \frac{1}{(1 + x^{2^r})} \cdots$$

prove that $a_n = b_n$ for all $n > 1$.

2.12 Prove that the number of partitions of n into exactly m parts equals the number of partitions of $n - m$ into at most m parts and that the number of partitions of n into exactly m parts of distinct sizes equals the number of partitions of $n - \dfrac{m(m-1)}{2}$ into exactly m parts. [Hint: Remove a suitable number of dots from the rows of the Ferrer's graph so as to get direct bijections between the appropriate sets of partitions.]

2.13 Prove the following identity combinatorially:

$$(1 + x)(1 + x^2)(1 + x^3)\ldots(1 + x^n)\ldots = \prod_{n=1}^{\infty} (1 + x^n)$$

$$= 1 + \sum_{m=1}^{\infty} \frac{x^{m(m+1)/2}}{(1 - x)(1 - x^2)\ldots(1 - x^m)}.$$

[Hint: In the Ferrer's graph of a partition consider the largest isosceles triangle of dots at the top left corner.]

2.14 Find the number of ways a 10 rupee note can be exchanged for a collection of coins of denominations 1 rupee, 1/2 rupee and 1/4 rupee.

1.15 What would be the answer to the last exercise if only 10 coins of each denomination are available? What would be the answer if we require further that the change must contain at least 2 half-rupees and at least 4 quarter-rupees?

*2.16 In a triangle ABC, suppose $AB = AC$, D is the midpoint of BC, E is the foot of the perpendicular drawn from D to AC and F is the midpoint of DE. Prove that AF is perpendicular to BE. (Like Exercise (2.3.7), this one too has little to do with combinatorics.

It is meant to illustrate our comment that through a suitable choice of cartesian coordinates, a tricky geometric problem can be transformed into a routine algebraic one.)

2.17 Use Theorem (2.16) to give an alternate proof of Theorem (2.2.17).

2.18 Use Theorem (2.16) to give an alternate solution to Exercises (3.2.17)(b) and (3.2.18).

2.19 Using Proposition (2.17), prove that

$$S_{n,m} = \sum_{r=0}^{m-1} \frac{(-1)^r (m-r)^{n-1}}{r! (m-r-1)!},$$

where m, n are positive integers with $m \leqslant n$, and $S_{n,m}$ is the Stirling number.

2.20 Prove Proposition (2.17) from Theorem (2.4.8).

2.21 Let t_n be the number of triangular partitions of an integer n (cf. Exercise (2.3.35)). Prove that if n is even then t_n equals the number of partitions of $n/2$ into exactly 3 parts (these latter partitions are not required to be triangular).

2.22 Using the last exercise determine t_n if n is even.

2.23 Prove that for all n, $\sum_{m=1}^{n} S_{n,m} x(x-1) \ldots (x-m+1) = x^n$. [Hint: First prove this when x is a positive integer, using Theorem (2.4.7). Then note that both sides of the identity are polynomials in x.]

2.24 The number of all possible partitions of a set of cardinality n is called the nth **Bell number** and denoted by P_n. Show that P_n is also the number of distinct equivalence relations on a set of cardinality n and that $P_n = \sum_{k=0}^{\infty} S_{n,k}$. (This sum is really finite since $S_{n,k} = 0$ for $k > n$.)

*2.25 Combining the last exercise with Exercise (2.19), prove that $P_n = \frac{1}{e} \sum_{k=0}^{\infty} \frac{k^n}{k!}$. [Hint: Interchange the order of a double summation.]

2.26 The **variance** $V(X)$ of a discrete random variable X with probability distribution $\{p_n\}$ is defined as $\sum_{n=0}^{\infty} (n - E(X))^2 p_n$. Prove that

$$V(X) = p_X''(1) + p_X'(1) - p_X'(1)^2.$$

Calculate the variance in Examples (3) and (4).

2.27 Let Y be the discrete random variable denoting the number of times a fair coin is tossed till a head shows for the second time. Prove that $p_Y(t) = [p_X(t)]^2$ where X is the discrete random variable in Example (4). Hence show that $E(Y) = 4$.

2.28 (a) Using the fact that $\sum_{n=1}^{\infty} \frac{1}{n^2}$ is convergent while $\sum_{n=1}^{\infty} \frac{1}{n}$ is not, construct a discrete random variable X for which $E(X)$ is not finite.

(b) What will be $E(X)$ in Example (4) if the probability of a head showing is p where $0 < p < 1$?

2 29   Let Z be the discrete random variable denoting the number of times a fair coin is tossed till two consecutive outcomes are the same. Prove that the probability distribution of Z is $\{p_n\}_{n=0}^{\infty}$ with $p_0 = p_1 = 0$ and $p_n = \dfrac{1}{2^{n-1}}$ for $n \geqslant 2$. Hence show that $E(Z) = 3$.

[Hint: A binary sequence $(x_1, \ldots, x_{n-1}, x_n)$ in which x_{n-1} and x_n is the only pair of consecutive identical digits is uniquely determined by x_n.]

Notes and Guide to Literature

The enumeration techniques discussed here are classic. The identities in Proposition (2.10) and Exercise (2.13) are due to Euler. For more on Bell numbers as well as numerous other identities, see Lovasz [1]. Exercises (2.7) and (2.16) are from Newmann [1] and Larson [1] respectively. All these three books contain an interesting variety of mathematical problems.

A classic treatise on the theory of probability using measures is by Kolmogorov [1]. For a recent treatment, see Tjur [1].

3.   Recurrence Relations

In the last section we studied a few applications of the generating functions. In this section we study what are by far the most important applications of generating functions, namely, towards the solutions of recurrence relations. This is so, because recurrence relations are such a powerful and frequently occurring technique for combinatorial mathematics that any method for solving them is bound to have considerable importance. Recurrence relations are to discrete mathematics what differential equations are to continuous mathematics. This analogy is not based merely on the wide applicability of the two. The relationship between recurrence relations and differential equations is much deeper. Many concepts and results about recurrence relations and their solutions have a striking resemblance with the corresponding concepts and results for differential equations. Recurrence relations are also called **difference equations** because they can be written in terms of the differences between the consecutive terms of a sequence. When so expressed, the analogy between them and the differential equations becomes even more apparent. But we shall not go into that. Instead, in this section we shall study the methods for directly solving recurrence relations and by way of illustration, present the solutions to the Regions Problems, the Shares Problems and the Vendor Problem. The emphasis here, however, will not be on applications. In the next section we shall study various kinds of applications of recurrence relations.

We shall not attempt to define recurrence relations formally. Although such a definition would be necessary for a rigorous development of the theoretical aspects, such as the existence and uniqueness of solutions, it tends to be clumsy. Fortunately, as with differential equations, a formal definition of recurrence relations is not necessary in order to really understand and apply them. The essential idea in a recurrence relation is that it expresses a general term of an (unknown) sequence as a (known) function of its earlier terms. In symbols, if the sequence is $\{x_n\}_{n=0}^{\infty}$, then a recurrence relation for it serves to express x_n in terms of $x_0, x_1, \ldots, x_{n-1}$, for all n after some stage, say, for all $n > r$. This integer r is called the **order** of the recurrence relation. The values of $x_0, x_1, \ldots, x_{r-1}$ have to be given or found from some data. (In some cases we have an option to choose some of these values.) These values are called the **initial conditions**. This is in obvious analogy with differential equations. A differential equation of order r expresses the rth derivative of some function, say $y = f(x)$, in terms of the lower order derivatives; $y, y', y'', \ldots, y^{(r-1)}$ and the initial conditions are usually the values of these derivatives at some point, say x_0, in the domain of f. There is, however, an important difference. For differential equations, the existence of a solution satisfying the given initial conditions is far from obvious. For recurrence relations, on the other hand, this is no problem at all. Suppose we are given the values of $x_0, x_1, \ldots, x_{r-1}$ say $x_i = a_i$ for $i = 0, \ldots, r - 1$. Now, x_r is given as a function of $x_0, x_1, \ldots, x_{r-1}$ and we assume that the initial conditions are such that the values $a_0, \ldots, a_{r-1}$ give us a point in the domain of this function. (If this is not the case, the recurrence relation has no solution. For example the recurrence relation $x_n = \dfrac{1}{x_{n-1} - 1}$, for $n \geqslant 1$ has no solution if the initial condition is $x_0 = 1$.) x_r is uniquely determined, say $x_r = a_r$. Now, x_{r+1} is expressed as a function of $x_0, x_1, \ldots, x_r$. Once again we assume that $(a_0, a_1, \ldots, a_r)$ gives us a point in the domain of this or else there would be no solution. (As an example, consider again, the recurrence relation $x_n = \dfrac{1}{x_{n-1} - 1}$ for $n > 1$. If $x_0 = 2$, we do get $x_1 = 1$, but then x_2 is undefined.) We repeat (or *recur*) this process *ad infinitum*. Either it will stop at some n (in which case there is no solution to the recurrence relation) or else we shall get a sequence $\{a_n\}_{n=0}^{\infty}$ which is a solution of the recurrence relation. By its very construction, this solution is unique, because at every stage the function uniquely determines a_n from $a_0, a_1, \ldots, a_{n-1}$. We record this as a theorem.

3.1 Theorem: A recurrence relation of order r has at most one solution for a given set of initial conditions. ∎

This theorem is too simple to be of any real value. When it comes to compute x_n for large values of n, say $n = 500$, it is hardly practicable to do so by going through the computation of all earlier x_i's. If we want to 'solve' the recurrence relation in the true sense of the term, we must express x_n directly in terms of n, that is, we must have a 'closed form' expression for x_n, like $x_n = n^2 - 2n$, $x_n = \begin{pmatrix} 2n \\ n \end{pmatrix}$ or $x_n = 3^{n+1}$. The trouble is that it is very hard to give a rigorous definition of a 'closed form expression' even though this term is widely used and everybody understands what it means. To illustrate the kind of difficulties that arise in formalising it, consider the sequence $\{x_n\}_{n=0}^{\infty}$ where $x_n = 3^{n+1}$ for all $n \geqslant 0$. It is easily seen that this is the solution of the recurrence relation $x_n = 3x_{n-1}$, with the initial condition $x_0 = 3$. On the face of it, it appears that 3^{n+1} is a closed from expression for x_n. Given any particular value of n, say $n = 500$, all we have to do is to plug 500 for n in it and get $x_{500} = 3^{501}$. But it is not quite as simple as that. What do we really mean by 3^{501}? If we go back to the very definition of 3^{n+1}, we see that it is *defined* as $3^n \cdot 3$ (cf. Chapter 3, Section 4). So by writing $x_n = 3^{n+1}$, we are not really solving the recurrence relation $x_n = 3x_{n-1}$; we are merely writing it differently! Another example would be the relation $y_n = ny_{n-1}$ with $y_0 = 1$. Here $y_n = n!$ because that is the very definition of $n!$.

These example indicate that the concept of a closed form expression is not easy to define. It will be helpful here to recall the comments made after Definition (2.3.7). We often have to enlarge our bag of closed form expressions by including in it certain commonly occurring expressions, even though technically they may have been defined by processes which smack of recursion. For example, we treat 3^n and $n!$ as closed form expressions in n. Methods are availabal to evaluate them, at least approximately, for a large number of values of n. Having let in a few basic expressions like this, we try to express the solutions to other recurrence relations in terms of these 'familiar' expressions. This is the best we can hope to do. But even this may prove to be too ambitious. Sometimes the very nature of the recurrence relation precludes any closed form solution to it. Recall that in a recurrence relation of order r, we merely required that for every $n \geqslant r$, x_n be a function of $x_0, \ldots, x_{n-1}$. But we did not require that it be the *same* function for every n. Indeed, as we shall see below, in many interesting examples, this is not the case. So we have to allow for the possibility that the way x_n depends upon $x_0, \ldots, x_{n-1}$ can itself vary with n. If this variation is of an arbitrary nature, then we can hardly hope to get a closed form expression for x_n. For example let α_n be the digit in the nth place of the decimal expansion of some irrational number, say π. (Since $\pi = 3.141592635 \ldots$, we have $\alpha_1 = 1$, $\alpha_2 = 4$, $\alpha_3 = 1$, $\alpha_4 = 5$, $\alpha_5 = 9$ etc.) Define a recurrence relation by $x_n = x_{n-1} + \alpha_n$ with the initial condition $x_0 = 0$. This is a perfectly well-defined recurrence relation of order 1 and solving it we get $x_1 = 1$, $x_2 = 5$, $x_3 = 6$, $x_4 = 11$, $x_5 = 20$ etc. But a closed form expression for x_n in this case would amount to getting a closed form formula for α_n. So far, no such formula is known. (From the fact that π is irrational, it is easy to show that α_n's cannot repeat periodically. But little positive information is available.)

It follows that if we want a closed form solution to a recurrence relation, then there must be some sort of a uniformity in the way each term depends on the preceding ones. This will be the case in all the examples that we shall consider. But to formalise it and incorporate it as a part of the definition of a recurrence relation is no easy job. That is why a rigorous definition of recurrence relations is beyond our scope. We shall therefore adopt a somewhat naive approach in which, instead of defining recurrence relations in the abstract, we shall concentrate on solving a few kinds of recurrence relations. As foremost examples, we recall three problems discussed in Chapter 1.

(1) In the Regions Problem, we let a_n be the number of regions into which a plane is cut by n lines in general position (i.e., no three of which are concurrent and no two parallel). We got the recurrence relation

$$a_n = a_{n-1} + n \quad \text{for} \quad n \geqslant 1. \tag{1}$$

The initial condition was $a_0 = 1$. In Exercise (3.8), we indicated a method for solving this recurrence relation. Putting $n = 1, 2, ..., k$ (when k is a positive integer) and summing both the sides, we get

$$\sum_{n=1}^{k} a_n = \sum_{n=1}^{k} a_{n-1} + \sum_{n=1}^{k} n.$$

Since every term, except a_k, on the left hand side cancels with some term on the right, we get $a_k = a_0 + (1 + 2 + ... + k)$. By Theorem (1.11),

$$1 + 2 + ... + k = \frac{k(k+1)}{2}. \text{ So } a_k = a_0 + \frac{k(k+1)}{2} = \frac{k^2 + k + 2}{2}.$$

Thus we see that n lines in general position cut the plane into $\dfrac{n^2 + n + 2}{2}$. regions.

(2) In the Shares Problem, b_n is the number of shares had during the nth year. The conditions of the problem give the recurrence relation

$$b_n = b_{n-1} + b_{n-2} \tag{2}$$

which is valid for all $n \geqslant 3$. If we set $b_0 = 0$, (2) becomes valid for $n = 2$ as well. So (2) is a recurrence relation of order 2, with initial conditions $b_0 = 0$ and $b_1 = 1$. Unlike (1), there is no slick way of solving (2). In fact, the solution may come as a surprise to the untutored. Here it is:

$$b_n = \frac{1}{\sqrt{5}} \left[\left(\frac{1 + \sqrt{5}}{2} \right)^n - \left(\frac{1 - \sqrt{5}}{2} \right)^n \right] \text{ for } n = 0, 1, 2,... \tag{3}$$

This is unbelievable! The b_n's, to say the least, are whole numbers (being the numbers of shares during the nth year) and from (3) it does not appear that b_n is even an integer, much less that it satisfies (2). But the surprise wears out if we look at (3) a little closely. First, if we apply the binomial theorem and keep in mind that $(\sqrt{5})^{2k} = 5^k$ and $(\sqrt{2})^{2k+1} = 5^k\sqrt{5}$, we see, after cancellations, that

$$b_n = \sum_{k=0}^{\lfloor (n-1)/2 \rfloor} \binom{n}{2k+1} 5^k/2^{n-1}.$$

This does not quite prove that b_n is an integer, but at least it is a rational number and the initial shock brought about by the presence of the irrational number $\sqrt{5}$ is gone. We still do not know that (3) satisfies (2). This is best done by direct calculation. For each $n \geqslant 2$, we have

$$b_n = \frac{1}{\sqrt{5}}\left[\left(\frac{1+\sqrt{5}}{2}\right)^n - \left(\frac{1-\sqrt{5}}{2}\right)^n\right]$$

$$= \frac{1}{\sqrt{5}}\left[\left(\frac{1+\sqrt{5}}{2}\right)^{n-2}\left(\frac{3+\sqrt{5}}{2}\right) - \left(\frac{1-\sqrt{5}}{2}\right)^{n-2}\left(\frac{3-\sqrt{5}}{2}\right)\right]$$

$$\text{(since } (1 \pm \sqrt{5})^2 = 6 \pm 2\sqrt{5}\text{)}$$

$$= \frac{1}{\sqrt{5}}\left[\left(\frac{1+\sqrt{5}}{2}\right)^{n-2}\left(1 + \frac{1+\sqrt{5}}{2}\right) - \left(\frac{1-\sqrt{5}}{2}\right)^{n-2}\left(1 + \frac{1-\sqrt{5}}{2}\right)\right]$$

$$= \frac{1}{\sqrt{5}}\left[\left(\frac{1+\sqrt{5}}{2}\right)^{n-2} + \left(\frac{1+\sqrt{5}}{2}\right)^{n-1} - \left(\frac{1-\sqrt{5}}{2}\right)^{n-2} - \left(\frac{1-\sqrt{5}}{2}\right)^{n-1}\right]$$

$$= b_{n-2} + b_{n-1}.$$

Thus (3) satisfies (2). Putting $n = 0$ and $n = 1$ in (3), gives $b_0 = 0$ and $b_1 = 1$. So the initial conditions also hold. By Theorem (3.1), (3) is the (unique) solution of (2).

Athough we have now completely solved the recurrence relation (2), the solution seems to have been pulled out of a hat, so to speak. We would naturally like to know if there is a way to *arrive at* (3) rather than merely *verify* it as we just did. As we shall soon see, generating functions provide a way to do just that.

(3) In the Vendor's problem, we saw that the essential part was to count, for every positive integer n, the number of balanced arrangements of n pairs of parentheses. We denoted this number by a_n and got the recurrence relation,

$$a_n = a_{n-1} + a_1 a_{n-2} + a_2 a_{n-3} + \dots + a_{n-3}a_2 + a_{n-2}a_1 + a_{n-1} \quad (4)$$

which is valid for all $n \geqslant 2$. If we set $a_0 = 1$, then (4) can be rewritten as

$$a_n = \sum_{i=0}^{n-1} a_i a_{n-1-i} \quad (5)$$

and is valid for all $n \geqslant 1$. Thus here we have a recurrence relation of order 1 with the initial condition $a_0 = 1$. In this case, we already know the answer by another method. In Chpter 2, Section 3 we counted the number of all unbalanced arrangements of n pairs of parentheses and thereby showed that

$$a_n = \frac{(2n)!}{n!\,(n+1)!} \tag{6}$$

However, the argument given there was a tricky, combinatorial argument. So we look for a more systematic way of solving (5) and as we shall see again, generating functions provide an answer. Unlike in the last examples, in this example, it is not so easy to verify that (6) is a solution of (5).

Let us now see how the method of generating functions provides a systematic solution in all these three examples. The essential idea in this method is to convert the given recurrence relation about a sequence, say $\{a_n\}$, to an algebraic equation for its generating function $A(x)$. Most of the time we take $A(x)$ to be the O.G.F. of $\{a_n\}$, although, with some recurrence relations, the E.G.F. is a more convenient choice. Solving this algebraic equation gives us a formula for $A(x)$. We then expand $A(x)$ as a power series in x. Equating a_n with the coefficient of x^n gives a closed form expression for a_n, thereby solving the recurrence relation.

Let us begin with (1). Multiplying both sides by x^n we get

$$a_n x^n = a_{n-1} x^n + n x^n, \text{ for } n \geqslant 1 \tag{7}$$

We now sum both the sides for $n = 1, 2, \ldots$. Denoting $\sum\limits_{n=0}^{\infty} a_n x^n$ by $A(x)$ and recalling that $a_0 = 1$, the left hand side gives $A(x) - a_0$ i.e. $A(x) - 1$. The first term on the right hand side gives $\sum\limits_{n=1}^{\infty} a_{n-1} x^n$ which is the same as

$$x \sum\limits_{n=1}^{\infty} a_{n-1} x^{n-1}, \text{ i.e., } x \sum\limits_{n=0}^{\infty} a_n x^n, \text{ i.e. } x A(x).$$

The second term yields $\sum\limits_{n=1}^{\infty} n x^n$. In the proof of Theorem (1.11), we already evaluated this sum as $\dfrac{x}{(1-x)^2}$. So, from (7), we get

$$A(x) - 1 = x A(x) + \frac{x}{(1-x)^2} \tag{8}$$

This is an algebraic equation in $A(x)$ and can be easily solved to give

$$A(x) = \frac{1}{1-x}\left[1 + \frac{x}{(1-x)^2}\right] = \frac{1}{1-x} + \frac{x}{(1-x)^3}.$$

We now expand this term by term using Theorems (1.2) and (1.3), (cf Exercise (1.9)). The coefficient of x^n is

$$1 + \binom{n+1}{n-1} = 1 + \binom{n+1}{2} = 1 + \frac{n(n+1)}{2}. \text{ So } a_n = \frac{n^2+n+2}{2},$$

which is the same answer as before. Actually, this solution is not essentially different from the earlier solution because to evaluate the sum $1+2+...+n$ we used generating functions anyway. However, there are slicker methods of evaluating this sum and if we use any one of them, then we do get a solution of (1) without generating functions.

To really drive home the importance of generating functions, let us, therefore, solve (2). Denote $\sum_{n=0}^{\infty} b_n x^n$ by $B(x)$. Multiplying both sides of (2) by x^n and summing over for $n \geqslant 2$ (which is the range of validity of (2)), we get

$$\sum_{n=2}^{\infty} b_n x^n = \sum_{n=2}^{\infty} b_{n-1} x^n + \sum_{n=2}^{\infty} b_{n-2} x^n \tag{9}$$

The sum on the left is $\sum_{n=0}^{\infty} b_n x^n - b_1 x - b_0$ which equals $B(x) - x$ since $b_1 = 1$ and $b_0 = 0$. The first sum on the right equals $x \sum_{n=2}^{\infty} b_{n-1} x^{n-1}$ which reduces to $x \sum_{n=1}^{\infty} b_n x^n$ which equals $x(B(x) - b_0) = xB(x)$. Similarly the second sum reduces to $x^2 B(x)$. Thus (9) gives

$$B(x) - x = x\,B(x) + x^2 B(x) \tag{10}$$

solving which, we get $B(x) = \dfrac{x}{1-x-x^2}$. To expand this, we resolve $\dfrac{x}{1-x-x^2}$ into partial fractions. Suppose $(1 - x - x^2)$ factors as $(1 - \alpha x)$ $(1 - \beta x)$. Then $\alpha + \beta = 1$ and $\alpha\beta = -1$, by comparing coefficients. It follows that α, β are roots of the quadratic $x^2 - x - 1 = 0$. So we let, $\alpha = \dfrac{1 + \sqrt{5}}{2}$ and $\beta = \dfrac{1 - \sqrt{5}}{2}$. [Now we see how on earth these numbers got into (3).] It is easily seen that $\dfrac{x}{1-x-x^2} = \dfrac{1}{\sqrt{5}}\left[\dfrac{\alpha x}{1-\alpha x} + \dfrac{\beta x}{1-\beta x}\right]$.

So, by Theorem (1.3), $b_n = $ coefficient of x^n in $B(x) = \dfrac{1}{\sqrt{5}}(\alpha^n + \beta^n)$. So (3) is the solution of the recurrence relation.

Before tackling the next relation, namely (5), by generating functions, it is perhaps time to comment on the validity of this method. When we recover the nth term of a sequence as the coefficient of x^n in its O.G.F. we are crucially using the uniqueness property of power series expansions (Theorem (1.2) (i)). In essence we are saying that if the power series

$\sum\limits_{n=0}^{\infty} a_n x^n$ and $\sum\limits_{n=0}^{\infty} b_n x^n$ are equal then $a_n = b_n$ for all $n = 0,1,2,\ldots,$. This statement is not quite true as it is. As remarked in the proof of Theorem (1.2), it has to be qualified by the expression 'for all x in the common interval of convergence of the two power series'. Some power series in x may converge only for $x = 0$, that is, their radii of convergence are 0. For example, the power series $\sum\limits_{n=0}^{\infty} n^n x^n$ (where we let $0^0 = 1$) and the power series $\sum\limits_{n=0}^{\infty} n! \, x^n$ converge only for $x = 0$. For such power series, the statement $\sum\limits_{n=0}^{\infty} a_n x^n = \sum\limits_{n=0}^{\infty} b_n x^n$ for all x in the common interval of convergence' is vacuously true and does not really mean much. In particular, it does not guarantee that $a_n = b_n$ for all n. In order to ensure this, we must know beforehand that the two power series have positive radii of convergence. For example, in solving (1), after we get

$$\sum_{n=0}^{\infty} a_n x^n = A(x) = \frac{1}{1-x} + \frac{x}{(1-x)^3},$$

before equating a_n with the coefficient of x^n in the right hand side, we must ensure that both sides have positive radii of convergence. For the right hand side, this is not much of a problem. It is a well-known theorem that the power series expansion of $\dfrac{1}{1-x}$ as well as that of $\dfrac{x}{(1-x)^3}$ are both valid for $|x| < 1$, i.e. the radius of convergence is 1. But when it comes to finding the radius of convergence of $\sum\limits_{n=0}^{\infty} a_n x^n$ we are in a dilemma. As remarked in the proof of Theorem (1.2), the radius of convergence is given by $R = 1/\overline{\lim} \, |a_n|^{1/n}$. But we cannot compute it without knowing a_n. We would know a_n only *after* solving the recurrence relation. But the solution is not justified unless we show $R > 0$. This constitutes a vicious circle. Fortunately, there is a way to break it. We do not really need the exact value of R. It is enough for our purpose to show that $R > 0$, and for this it would suffice if we show that the sequence $|a_n|^{1/n}$ is bounded. For this, any crude upper bound on $|a_n|$ would do. For some recurrence relations, such an upper bound is easy to guess and prove by induction, even before solving them. All we have to ensure is that as n grows, $|a_n|$ grows less rapidly than this upper bound. For example, in (1), a_n exceeds a_{n-1} by n. Since n^2 exceeds over $(n-1)^2$ by $2n + 1$, an upper bound of the form λn^2 for a suitable λ will work. In this case, we guess that $a_n \leqslant n^2$ for all $n \geqslant 2$. For $n = 2$, equality holds. If $a_{n-1} \leqslant (n-1)^2$ then $a_n = a_{n-1} + n \leqslant (n-1)^2 + n = n^2 - n + 1 \leqslant n^2$. So, by induction $a_n \leqslant n^2$ for all $n \geqslant 2$. Hence $|a_n|^{1/n} \leqslant (n^2)^{1/n} = (n^{1/n})^2$. It is well-known that $n^{1/n} \to 1$ as $n \to \infty$. So $|a_n|^{1/n}$ is bounded which means $\overline{\lim\limits_{n \to \infty}} \, |a_n|^{1/n} < \infty$. This is all we need to conclude that $R > 0$. As noted before, we now have an absolutely rigours justification to equate a_n with the coefficient of x^n in the power series expansion of $A(x)$.

Similarly, in solving (2), we got $\sum\limits_{n=0}^{\infty} b_n x^n = B(x) = \dfrac{x}{1-x-x^2}$ Since the power series expansions of $\dfrac{1}{1-\alpha x}$ and $\dfrac{1}{1-\beta x}$ are valid for $|x| < |\alpha|^{-1}$ and $|x| < |\beta|^{-1}$ respectively, we see that the radius of convergence is positive for the power series expansion of $\dfrac{x}{1-x-x^2}$. As for $\sum\limits_{n=0}^{\infty} b_n x^n$, we can easily prove from (2) and induction that $b_n \leqslant 2^n$ for all n. So $\overline{\lim} \, |b_n|^{1/n} \leqslant 2$, showing that $\sum\limits_{n=0}^{\infty} b_n x^n$ also has a positive radius of convergence. We are now justified in equating b_n with the coefficient of x^n in $\dfrac{x}{1-x-x^2}$.

But there are limitations to this approach. For recurrence relations like (5), it is far from easy to obtain an upper bound on $|a_n|$ before solving it. An alternate method proceeds as follows. Instead of appealing to the uniqueness of the power series expansion (Theorem (1.2)), we appeal to the uniqueness of solutions of recurrence relations, satisfying the given initial conditions (Theorem (3.1)). Essentially, this amounts to starting from the O.G.F. and working our way backwards to show that the coefficients of x^n satisfy the given recurrence relation. For example, in the case of (1), let us start from the function $\dfrac{1}{1-x} + \dfrac{x}{(1-x)^3}$. Let the power series expansion of this function be $\sum\limits_{n=0}^{\infty} c_n x^n$. (Note that we have not yet shown that $a_n = c_n$ because we are not in a position to apply Theorem (1.2).) This expansion is valid for $|x| < 1$ and, as we showed above,

$$c_n = \frac{n^n + n + 2}{2}.$$

In particular $c_0 = 1$. If we reverse the steps in arriving at

$$\frac{1}{1-x} + \frac{x}{(1-x)^3},$$

(i.e., going from (7) to (8)), we can show that

$$\sum_{n=1}^{\infty} c_n x^n = \sum_{n=1}^{\infty} (c_{n-1} + n) x^n \qquad \text{(for all } |x| < 1) \tag{*}$$

Now comes the crucial step. The two sides of (*) are power series with known coefficients and their radii of convergence are therefore known to be positive (in this case both equal 1). We are now fully justified in applying Theorem (1.2) (i) to conclude $c_n = c_{n-1} + n$ for all $n \geqslant 1$. So the known sequence $\{c_n\}_{n=0}^{\infty}$ satisfies (1). Since the initial condition (viz. $c_0 = 1$) also holds, it follows from Theorem (3.1) that $\{c_n\}_{n=0}^{\infty}$ is the only solution to (1). So $c_n = a_n$ for all n. As we already know c_n, it now follows that

$$a_n = \frac{n^2 + n + 2}{2}.$$

This approach is somewhat cunning in that we are not answering the question of how we thought of starting from the function

$$\frac{1}{1 - x} + \frac{x}{(1 - x)^3}.$$

It is as if we generated this function behind closed doors and then brought it out in the open and showed that it really worked. Logically, of course, this approach is perfectly unassailable. Nor is this an isolated instance of this kind of reasoning. In mathematics, in many problems we have to 'suspect' the solution first, that is, we have to do reasoning to the effect that if at all the problem has a solution then it has to be this one. Then of course we have to back our suspicion by a rigorous proof that what we have in mind is in fact a solution. If mathematics is viewed as a sterile science then it is only the proof that counts. In many problems, however, the proof is a routine drill and the real art lies in guessing the solution. In a murder trial, as far as the conviction goes, what matters is the hard evidence put forth by the police officer and not what led him to suspect the accused in the first place. In a mystery novel, it is just the other way. (And often, the police officer is not the one who does the brain work!)

Henceforth we shall not worry about the issue of convergence. It will be a good exercise for the fussy minded ones to validate our solutions by either one of the two approaches.

Let us now return to solving recurrence relations. Consider (5). Again let $A(x) = \sum\limits_{n=0}^{\infty} a_n x^n$. The very form of the right hand side of (5) shows that it is the coefficient of x^{n-1} in the product $A(x)A(x)$. So, once again multiplying both sides of (5) by x^n and summing over for $n \geq 1$ (the range of validity of (5)), we get $A(x) - a_0 = x[A(x)]^2$, i.e.,

$$x[A(x)]^2 - A(x) + 1 = 0. \tag{11}$$

This is an algebraic equation for $A(x)$. Unlike the earlier two problems, however, it is a quadratic and so, for each $x \neq 0$, we have, by the quadratic formula, $A(x) = \dfrac{1 \pm \sqrt{1 - 4x}}{2x}$. From this, we must *not* hastily conclude that either $A(x) = \dfrac{1 + \sqrt{1 - 4x}}{2x}$ for all $x \neq 0$ or $A(x) = \dfrac{1 - \sqrt{1-4x}}{2x}$

for all $x \neq 0$. Although for each $x \neq 0$, one of the two possibilities must hold, there is nothing to guarantee that the same possibility must hold for all x. To illustrate the point involved here, it is obviously true that every human being is either a man or a woman. But that does not mean either every human being is a man or every human being is a woman! [cf. Exercise (1.4.9) (viii).] In the present case the choice of sign in the expression

$$A(x) = \frac{1 \pm \sqrt{1 - 4x}}{x}$$

requires considerations from continuous mathematics. Specifically, we need the continuity of the three functions

$$A(x), \frac{1 + \sqrt{1 - 4x}}{2x} \quad \text{and} \quad \frac{1 - \sqrt{1 - 4x}}{2x}$$

in a sufficiently small open interval $(0, R)$. It can then be shown that throughout this interval the same possibility holds, i.e., either

$$A(x) = \frac{1 + \sqrt{1 - 4x}}{2x}$$

for all x or

$$A(x) = \frac{1 - \sqrt{1 - 4x}}{2x}$$

for all x. The details will be indicated in the exercises, because they belong to continuous mathematics and if we take the cunning approach, we can spare them anyway. We still have to decide which of the two possibilities holds. For this, we expand $\sqrt{1 - 4x} = (1 - 4x)^{1/2}$ by the binomial theorem; it comes as $1 + \sum\limits_{r=1}^{\infty} c_r x^r$, where

$$c_r = \frac{(-4)^r \left(\frac{1}{2}\right)\left(-\frac{1}{2}\right)\left(-\frac{3}{2}\right)\cdots\left(\frac{1}{2} - r + 1\right)}{r!}$$

After a little computation, c_r simplifies to

$$-2 \frac{(2r - 2)!}{r!(r - 1)!}.$$

It follows that

$$\frac{1 + \sqrt{1 - 4x}}{2x} = \frac{1}{x} + \frac{1}{2} \sum_{r=1}^{\infty} c_r x^{r-1}$$

for all x in the interval $(0, R)$. But then

$$\frac{1 + \sqrt{1 - 4x}}{2x} \to \infty \quad \text{as } x \to 0^+.$$

whereas $A(x) \to a_0 = 1$ as $x \to 0^+$, $A(x)$ being continuous at 0. So it cannot be the case that $A(x) = \dfrac{1 + \sqrt{1 - 4x}}{2x}$. This forces us to conclude that for all sufficiently small positive values of x,

$$A(x) = \frac{1 - \sqrt{1 - 4x}}{2x} \tag{12}$$

Since we already have the power series expansion of $\sqrt{1 - 4x}$, it follows that

$$A(x) = \frac{1 - \left(1 + \sum\limits_{r=1}^{\infty} c_r x^r\right)}{2x} = \sum_{r=1}^{\infty} -\frac{c_r}{2} x^{r-1}.$$

So $a_n =$ coefficient of x^n in $A(x)$

$$= -\frac{c_{n+1}}{2} = \frac{(2n)!}{(n+1)!n!}.$$

Thus finally we obtain (6) as the solution of (5).

These three examples have hopefully convinced the reader of the power of generating functions. In all these examples, we used the ordinary generating function. Sometimes, it is better to consider the exponential generating function of the given sequence. We illustrate this in the following recurrence relation. In the next section we shall see that this particular recurrence relation provides an alternate way to count the number of derangements of n objects (cf. Theorem (2.4.6)).

3.2 Problem: Solve the recurrence relation

$$d_n = nd_{n-1} + (-1)^n \tag{13}$$

for $n \geqslant 1$, with the initial condition $d_0 = 1$,

Solution: Let us first see what would happen if we try to solve (13) by letting $D(x)$ be the O.G.F. of $\{d_n\}$. As usual we multiply both sides of (13) by x^n and sum over for $n \geqslant 1$. Rewriting nd_{n-1} as $(n-1)d_{n-1} + d_{n-1}$ it is easy to show that we get

$$D(x) - 1 = x^2 D'(x) + xD(x) - \frac{1}{1+x} \tag{14}$$

This is a linear differential equation in x and has for its general solution.

$$D(x) = \frac{1}{xe^{1/x}} \left[c - \int \frac{e^{1/x}\, dx}{x(1+x)} \right] \tag{15}$$

where c is a constant. Unfortunately, the integral in (15) cannot be evaluated in a closed form and so this approach is not workable,

However, if we let $E(x) = \sum\limits_{n=0}^{\infty} \dfrac{d_n}{n!} x^n$ be the E.G.F. of $\{d_n\}_{n=0}^{\infty}$ then (13)

can be solved as follows. We multiply both its sides by $\dfrac{x^n}{n!}$ and sum over

for $n \geqslant 1$. Since $\dfrac{n}{n!} = \dfrac{1}{(n-1)!}$ and $\sum\limits_{n=0}^{\infty} \dfrac{(-1)^n}{n!} x^n = e^{-x}$, we get

$$E(x) - 1 = xE(x) + e^{-x} - 1 \tag{16}$$

Solving which, $E(x) = \dfrac{e^{-x}}{1-x}$. By Proposition (1.11), the coefficient of x^n in

$\dfrac{e^{-x}}{1-x}$ is the sum of the first $n+1$ coefficient of e^{-x}, i.e.,

$$1 - \frac{1}{1!} + \frac{1}{2!} - \frac{1}{3!} + \dots + (-1)^n \frac{1}{n!}.$$

So,

$$d_n = n! \left(\frac{1}{2!} - \frac{1}{3!} + \frac{1}{4!} - \frac{1}{5!} + \dots + (-1)^n \frac{1}{n!} \right). \blacksquare$$

The failure of ordinary generating functions in this problem could have been predicted in advance by consideration of convergence. In the right hand side of (13), the term $(-1)^n$ is insignificant as compared to the other for large n. So

$$d_n \approx n\, d_{n-1} \approx n(n-1)\, d_{n-2} \approx n(n-1)\dots(n-r+1)d_{n-r} \approx \dots \approx \frac{n!}{2}\, d_2.$$

(We do not go further because $d_1 = 0$.) Thus d_n grows at a rate comparable to that of $n!$. Since $(n!)^{1/n} \to \infty$ as $n \to \infty$, the power series $\sum\limits_{n=0}^{\infty} d_n x^n$ will converge only for $x = 0$. As noted before, this will make it impossible to recover d_n from $D(x)$, even if we could obtain a closed form for it. The exponential generating function, $E(x)$, on the other hand has a positive radius of convergence. $n!$ is a very large number and division by it 'brings d_n to proportion' so to speak. Thus the exponential generating functions have a theoretical advantage over the ordinary generating functions. Wherever the O.G.F. converges, so does the E.G.F. But the converse is not true, as shown by this example.

Nevertheless, the attempted solution using $D(x)$ is noteworthy for another reason. We converted the recurrence relation (13) to a differential equation (14). We could just as well have reversed the process and gotten (13) from (14). This is in fact more commonly done. This technique is popularly known as 'series solution of a differential equation'. Although the basic theorems guarantee the existence and uniqueness for a large class of differential equations, they are of little help in actually finding the solution of a differential equation. Various ingeneous methods have to be used for this purpose. Even then, it often happens that the solution cannot be found. In such a case the problem is converted through Taylor series. For convenience, let us suppose that $y = f(x)$ is the solution of the given differential equation and that we are interested in the values of f in a neighboured of the point 0. Let $f(x) = \sum_{n=0}^{\infty} a_n x^n$ be the Taylor series expansion of f near 0. Here the coefficients a_n are related to f by $a_n = \dfrac{f^{(n)}(0)}{n!}$. Solving for f is equivalent to determining a_n for $n = 0, 1, 2, \dots$. The differential equation to be solved can be translated into a recurrence relation about the a_n's. This recurrence relation is usually not easy to solve. Because if it could be solved, then the original differential equation was probably easy to solve directly anyway. However, starting from the initial conditions we can calculate a_n for a fairly large value of n, say $n = N$. Let $g_N(x) = \sum_{n=0}^{N} a_n x^n$. Then, in a small neighbourhood of 0, $g_N(x)$ is very close to $f(x)$ and so $y = g_N(x)$ can be taken as an approximate solution of the differential equation over this neighbourhood.

Although our interest is not in solving differential equations we show by an example how this method works. Suppose we want to solve the differential equation $(x + 2) y'' + y = 0$, in a neighbourhood of 0, with initial condition $y(0) = 1$, $y'(0) = 1$. Let $y = f(x)$ be the solution and $\sum_{n=0}^{\infty} a_n x^n$ its Taylor expansion at 0. Then $y'' = \sum_{n=0}^{\infty} (n + 1)(n + 2) a_{n+2} x^n$. For every n, the coefficient of x^n in $(x + 2) y'' + y$ must be 0. This gives,

$$n(n + 1) a_{n+1} + 2(n + 1)(n + 2) a_{n+2} + a_n = 0 \qquad (17)$$

which can be rewritten, with a shift of indices, as

$$a_n = \frac{-a_{n-2} - (n-2)(n-1) a_{n-1}}{2n(n-1)} \qquad (18)$$

for all $n \geqslant 2$. Starting from $a_0 = 1$ and $a_1 = 1$, we get successively,

$$a_2 = \frac{-a_0 - 0}{4} = -\frac{1}{4}, \ a_3 = \frac{-a_1 - 2a_2}{12} = \frac{-1 + \frac{1}{2}}{12} = -\frac{1}{24} \dots \text{ etc.}$$

These computations can easily be done on a machine. If we compute a_n for $n \leqslant 50$ (say), then $y = \sum_{n=0}^{50} a_n x^n$ will be a very good approximation to the actual solution.

This technique is a classic example of how discrete mathematics is used to approximate continuous mathematics. The other way transition is not very common. That is, differential equations are not generally used to solve recurrence relations. But there are exceptions. Indeed we already saw an instance of this. In Problem (1.12), we found the O.G.F. of the sequence

$\{a_n\}_{n=0}^{\infty}$ where $a_n = \begin{pmatrix} 2n \\ n \end{pmatrix}$. If we analyse the solution, we see that the

crucial step was the equation $\begin{pmatrix} 2n+2 \\ n+1 \end{pmatrix} (n+1) = 2\,(2n+1) \begin{pmatrix} 2n \\ n \end{pmatrix}$.

This can be rewritten as $a_{n+1} = \dfrac{4n+2}{n+1}\,a_n$, or, with a change of indices as

$a_n = \dfrac{4n-2}{n}\,a_{n-1}$ for $n \geqslant 1$. This is a recurrence relation and the first part
of the solution to Problem (1.12) essentially amounts to solving this recurrence relation by converting it to a differential equation.

Among the various recurrence relations we studied so far, (1) and (2) enjoy certain properties not shared by others. They belong to a class of recurrence relations known as linear recurrence relations. As we shall see in the next section, such equations arise in a large number of applications. Consequently, although the method of generating functions is available to solve them, it is worthwhile to have other methods which enable us to write down the solutions more quickly. We study one such method here. First we define linear recurrence relations and a few associated terms.

3.3 Definition: Let r be a positive integer. Then a **linear recurrence relation** (with constant coefficients) of **order** r, is an equation of the form

$$C_r a_n + C_{r-1} a_{n-1} + \ldots + C_1 a_{n-r+1} + C_0 a_{n-r} = f(n)\ (n \geqslant r) \qquad (19)$$

where $C_0, C_1, \ldots, C_r$ are constants (with C_r and C_0 non-zero) and $f(n)$ is a function of n only. If $f(n) \equiv 0$, the recurrence relation is said to be **homogeneous**. Even when $f(n)$ is not identically 0, the equation obtained by replacing the right hand side of (19) by 0, i.e. the equation

$$C_r a_n + C_{r-1} a_{n-1} + \ldots + C_1 a_{n-r+1} + C_0 a_{n-r} = 0\ \ (n \geqslant r) \qquad (20)$$

is called the **associated homogeneous equation** (A.H.E.) of (19).

The assumption that C_0, C_r are non-zero is of a technical nature, intended to ensure that the order r be uniquely defined. Note also that nothing has been said about initial conditions. We shall return to them later.

As examples, a rewriting (1) and (2) we see that (2) is a homogeneous linear recurrence relation of order 2, while (1) is a non-homogeneous linear recurrence relation of order (1). Its A.H.E. is $a_n - a_{n-1} = 0$.

The terminology will undoubtedly remind the reader of the system of linear equations we studied in Chapter 6, Section 4. Those familiar with differential equations will also be reminded of the linear differential equations with constant coefficients and will be in a position to predict what is coming. Actually the resemblance between the two is not coincidental, nor just a matter of form. Using exponential generating functions, it can, in fact be shown that the two are equivalent (see Exercise (3.32)). Consequently the theory of linear recurrence relations can be developed using the corresponding concepts and theorems from linear differential equations. Nevertheless we shall develop this theory directly, because of its simplicity and also because it gives us a chance to use a few basic facts about vector spaces. The other approach is given as an exercise.

Let us first see what is 'linear' about linear recurrence relations. The term comes from linear transformations. We show how the sets of all solutions of (19) and (20) can be interpreted in terms of suitable linear transformations. Although such an interpretation does not, by itself, give a method for solving the recurrence relations, it provides valuable insight into the nature of soultions.

Let V be the set of all infinite sequences, say, $\bar{a} = (a_0, a_1, a_2, ..., a_n, ...)$ of real (or complex) numbers. Every such sequence is a function from the set $\{0, 1, 2, ...\}$ into $\mathbb{R}$ (or $\mathbb{C}$). Consequently, under coordinatewise addition and scalar multiplication, V is a vector space over the field $\mathbb{R}$ (or $\mathbb{C}$) (see Example (4) in Chapter 6, Section 3). Now let r be a positive integer and $C_0, C_1, ..., C_r$ be real (or complex) numbers with C_r, C_0 non-zero. (Some of the other numbers may be 0). Let W be the vector space of all real (or complex) valued functions with domain $\{r, r + 1, ...\} = \{n \in \mathbb{N} : n \geqslant r\}$. Let f be a fixed element of W. We are now ready to look at (19) a little differently.

3.4 Theorem: For a given choice of $r, C_0, ..., C_r$ (with C_0 and $C_r \neq 0$) and f, the solutions to (20) constitute the kernel (or the null-space), say, K, of a linear transformation $T: V \to W$ and those to (19) constitute a translate (or a coset) of the kernel. Further, K is an r-dimensional subspace of V.

Proof: Define $T: V \to W$ by

$$T(\bar{a}) = T((a_0, a_1, a_2, ...)) = T_{\bar{a}}$$

where

$$T_{\bar{a}} : \{r, r + 1, ...\} \to \mathbb{R} \text{ (or } \mathbb{C})$$

is defined by

$$T_{\bar{a}}(n) = C_r a_n + C_{r-1} a_{n-1} + ... + C_0 a_{n-r}$$

for $n \geqslant r$. Then $\bar{a}$ is a solution of (19) iff $T_{\bar{a}} = f$ and it is a solution of (20) iff $T_{\bar{a}} = \bar{0}$ where $\bar{0}$ means the identically zero function; which is also the zero vector in the vector space W. The linearity of T is a routine verification. If $\bar{a}, \bar{b} \in V$, and $\alpha, \beta \in \mathbb{R}$ (or $\mathbb{C}$), then for every $n \geqslant r$, $T_{\alpha\bar{a}+\beta\bar{b}}(n)$

$$= \sum_{i=0}^{r} C_{r-i}(\alpha a_{n-i} + \beta b_{n-i}) = \alpha \sum_{i=0}^{r} C_{r-i} a_{n-i} + \beta \sum_{i=0}^{r} C_{r-i} b_{n-i} = \alpha T_{\bar{a}}(n) + \beta T_{\bar{b}}(n).$$

Let K be the kernel (i.e. the null-space) of T. Then the first assertion follows from the very definition of the kernel.

For the other two assertions, we first show that T is onto, i.e. for every $f \in W$, there exists $\bar{a} \in W$, such that $T_{\bar{a}} = f$. The construction of such $\bar{a}$ is direct and uses the assumption that $C_r \neq 0$ Choose $a_0, a_1, \ldots, a_{r-1}$ arbitrarily. Having chosen them, a_r is uniquely determined by

$$C_r a_r + C_{r-1} a_{r-1} + \ldots + C_0 a_0 = f(r),$$

specifically,

$$a_r = \frac{1}{C_r} \ (f(r) - C_{r-1} a_{r-1} - \ldots - C_0 a_0).$$

Similarly a_{r+1} is uniquely determined by

$$a_{r+1} = \frac{1}{C_r} \ (f(r + 1) - C_{r-1} a_r - \ldots - C_0 a_1),$$

a_{r+2} is uniquely determined by

$$a_{r+2} = \frac{1}{C_r} \ (f(r + 2) - C_{r-1} a_{r+1} - \ldots - C_0 a_2).$$

Continuing in this manner, we determine each a_n for $n \geqslant r$. (Essentially this is a special case of the construction preceding Theorem (3.1).) It now follows that $T_{\bar{a}} = f$ and hence that T is onto. In other words, (19) has at least one solution for a given f. Once we know this, the other solutions of (19) are obtained by adding to this 'particular solution' the elements of K; see Exercise (6.3.8). This proves the second assertion. A popular way of expressing it is that the 'general solution' of a linear recurrence relation is obtained by adding any one particular solution of it to the general solution of its associated homogeneous equation.

The construction above also shows that dim $(K) = r$. If we take $f = \bar{0}$, then the construction shows that every element $\bar{a}$ of K is uniquely determined by $a_0, \ldots, a_{r-1}$ which could be chosen arbitrarily. Thus we have a bijection between K and the r-dimensional vector space $\mathbf{R}^r$ (or $\mathbf{C}^r$). This bijection is easily seen to be a vector space isomorphism. (Under this bijection, every element of K is truncated, by taking only its initial segment of length r.) It follows that dim $(K) = r$. ∎

This theorem shows that the problem of solving a linear recurrence relation can be broken into two parts: (i) finding the general solution of the associated homogeneous equation and (ii) obtaining any one particular solution of the original relation. We tackle these two parts separately. (i) really amounts to finding a basis for the kernel K in the theorem above. One such basis was in fact constructed in the proof. It can be described as follows: For $i = 0, \ldots, r - 1$, let $e_i = (0, \ldots, 0, 1, 0, \ldots, 0)$, i.e. the vector having 1 in the ith place and 0 elsewhere. If we start with $\bar{e}_i$, that is, if we let $a_k = 0$ for $k = 0, \ldots, r-1$, $k \neq i$, and $a_i = 1$ and apply the construction in the proof above, then we get a solution of (20), i.e. a vector, say $\bar{u}_i$, in K which is a continuation of e_i. The vectors $\bar{u}_0, \ldots, \bar{u}_{r-1}$ are linearly independent, because their truncations, $\bar{e}_0, \ldots, \bar{e}_{r-1}$ are so. Since dim $(K) = r$, it follows from Exercise (6.3.11), that $\bar{u}_0, \ldots, \bar{u}_{r-1}$ is a basis for K. Consequently, the general solution of (20) is of the form $c_0\bar{u}_0 + c_1\bar{u}_1 + \ldots + c_{r-1}\bar{u}_{r-1}$ where $c_0, \ldots, c_{r-1}$ are arbitrary constants. However, this form of the general solution is of little practical value, because we have not identified the $\bar{u}_i$'s. The only description we have of them is that they are some sequences constructed recursively so as to satisfy (20). We have no closed form formula for their typical terms.

So we look for a basis for K whose elements can be easily described in terms of the given recurrence relation. The key step in this search is the following. Let α be a non-zero real (or a complex) number. Denote by $\bar{u}_\alpha$ the sequence $\{\alpha^n\}_{n=0}^\infty = (1, \alpha, \alpha^2, \ldots, \alpha^n, \ldots)$. The following proposition tells us when $\bar{u}_\alpha \in K$.

3.5 Proposition: $a_n = \alpha^n$ for $n = 0, 1, 2$, is a solution of (20) if and only if $C_r\alpha^r + C_{r-1}\alpha^{r-1} + \ldots + C_1\alpha + C_0 = 0$. In other words, $\bar{u}_\alpha$ is in K iff α is a root of the polynomial $C_r x^r + C_{r-1}x^{r-1} + \ldots + C_1 x + C_0$.

Proof: Suppose $a_n = \alpha^n$ is a solution of (20). Then for all $n \geqslant r$ we have

$$C_r\alpha^n + C_{r-1}\alpha^{n-1} + \ldots + C_1\alpha^{n-r+1} + C_0\alpha^{n-r} = 0 \qquad (21)$$

for all $n \geqslant r$. In particular, setting $n = r$, we see that α is a root of

$$C_r x^r + C_{r-1}x^{r-1} + \ldots + C_1 x + C_0.$$

So the condition is surely necessary. Conversely if $C_r\alpha^r + C_{r-1}\alpha^{r-1} + \ldots + C_1\alpha + C_0 = 0$ we merely multiply it by α^{n-r} and see that (21) holds for all $n \geqslant r$. This proves sufficiency. ∎

In view of this result, the polynomial $C_r x^r + \ldots + C_1 x + C_0$ and its roots are of tremendous importance in finding the general solution of (20). As we shall soon see, these roots completely characterise all the solutions of (20). This is probably the reason they are given the following name:

3.6 Definition: The polynomial $C_r x^r + \ldots + C_1 x + C_0$ is called the **characteristic polynomial** of (20). Its roots are called **characteristic roots** of (20). The polynomial equation $C_r x^r + \ldots + C_1 x + C_0 = 0$ is called the **characteristic equation** of (20).

Note that because of our assumption that $C_0 \neq 0$, 0 can never be a characteristic root. Proposition (3.5) shows how each characteristic root gives rise to a vector in K. The following proposition goes a little further towards obtaining the general solution.

3.7 Proposition: If $\alpha_1, \alpha_2, \ldots, \alpha_k$ are distinct characteristic roots then the vectors $\bar{u}_{\alpha_1}, \bar{u}_{\alpha_2}, \ldots, \bar{u}_{\alpha_k}$ are linearly independent.

Proof: By Proposition (6.2.25), $k \leqslant r$. Let $\bar{v}_i = (1, \alpha_i, \ldots, \alpha_i^{k-1})$. Then $\bar{v}_i$ is a truncation of $\bar{u}_{\alpha_j}$. In order to show that $\{\bar{u}_{\alpha_1}, ..., \bar{u}_{\alpha_k}\}$ is linearly independent, it obviously suffices to show that the v_i's are linearly independent. By Proposition (6.4.32), the latter is equivalent to showing that det $(A) \neq 0$, where A is the transpose of the $k \times k$ matrix

$$
A = \begin{bmatrix}
1 & 1 & 1 & ... & 1 \\
\alpha_1 & \alpha_2 & \alpha_3 & ... & \alpha_k \\
\alpha_1^2 & \alpha_2^2 & \alpha_3^2 & ... & \alpha_k^2 \\
. & . & . & ... & . \\
. & . & . & ... & . \\
. & . & . & ... & . \\
\alpha_1^{k-1} & \alpha_2^{k-1} & \alpha_3^{k-1} & ... & \alpha_k^{k-1}
\end{bmatrix}.
$$

This is a Vandermonde determinant and by Exercise (6.4.33) it is non-zero since the roots $\alpha_1, ..., \alpha_k$ are distinct. ∎

We are now in a position to completely describe the general solution of a homogeneous linear recurrence relation whose characteristic roots are distinct. We still have to consider the case of multiple roots. However, in the case of most of the linear recurrence relations arising in applications, it turns out that the characteristic roots are distinct. We therefore content ourselves in giving the general solution in this case only, relegating the case of multiple roots to the exercises.

3.8 Theorem: Suppose $\alpha_1, \alpha_2, ..., \alpha_r$ are the distinct characteristic roots of (20). Then the general solution to (20) is

$$a_n = c_1\alpha_1^n + c_2\alpha_2^n + ... + c_n\alpha_r^n, \text{ for all } n \geqslant 0 \tag{22}$$

where $c_1, c_1, ..., c_n$ are arbitrary constants taking values in $\mathbb{R}$ (or $\mathbb{C}$).

Proof: Apply the last proposition with $k = r$. The vectors $\bar{u}_{\alpha_1}, \bar{u}_{\alpha_2}, ..., \bar{u}_{\alpha_r}$ are linearly independent elements of K. But by Theorem (3.4), dim $(K) = r$. So it follows from Exercise (6.3.11), that they span K. This means every element of K, i.e. every solution, say $\bar{a}$, of (20) is of the form $c_1\bar{u}^{\alpha_1} + c_2\bar{u}_{\alpha_2} + ... + c_r\bar{u}_{\alpha_r}$ for some constants $c_1, ..., c_r$. Recalling that the vector space operations in V are termwise, we see that (22) holds. ∎

As a simple application we return to (2). Rewriting it as

$$b_n - b_{n-1} - b_{n-2} = 0$$

we see its characteristic polynomial is $x^2 - x - 1$ and so the characteristic roots are $\dfrac{1 + \sqrt{5}}{2}$ and $\dfrac{1 - \sqrt{5}}{2}$ which are obviously distinct. So its general solution is $b_n = c_1 \left(\dfrac{1 + \sqrt{5}}{2}\right)^n + c_2 \left(\dfrac{1 - \sqrt{5}}{2}\right)^n$ where c_1, c_2 are constants, to be determined from the initial conditions, which in this case are $b_0 = 0$. $b_1 = 1$. Setting $n = 0$ and 1 we get respectively, $b_0 = 0 = c_1 + c_2$ and $b_1 = 1 = c_1 \left(\dfrac{1 + \sqrt{5}}{2}\right) + c_2 \left(\dfrac{1 - \sqrt{5}}{2}\right)$. Solving these two equations simultaneously, we get $c_1 = \dfrac{1}{\sqrt{5}}$ and $c_2 = \dfrac{-1}{\sqrt{5}}$. So the particular solution to (2) is

$$b_n = \frac{1}{\sqrt{5}} \left[\left(\frac{1 + \sqrt{5}}{2}\right)^n - \left(\frac{1 - \sqrt{5}}{2}\right)^n \right],$$

which is the same as (3). Note that the work involved is not qualitatively different, because in the earlier approach, in order to resolve $\dfrac{x}{1 - x - x^2}$ into partial fractions we had to solve the quadratic $x^2 - x - 1$ anyway. The advantage of Theorem (3.8) is that once we obtain the characteristic roots, there is nothing more to do. Note also that whereas in the earlier approach the generating function depended on the initial conditions, in the present approach, the initial conditions figure only at the end. Even when the linear recurrence relation is not homogeneous, we first obtain its general solution and then determine the constants so as to satisfy the given initial conditions.

Although we are omitting discussion of the case of multiple characteristic roots, the case where some of the characteristic roots are complex deserves to be settled here. In our treatment of the linear recurrence relations so far, it made little difference whether the coefficients $C_0, ..., C_r$ in (20) were real or complex. Indeed Theorem (3.8) is equally valid over any field. In most applications, of course, the coefficients are real and in fact integers. The trouble, though, is that even when all C's are real, some of the characteristic roots may be complex, as we see in the following very simple recurrence relation of order 2.

$$a_n + a_{n-2} = 0 \text{ (for } n \geqslant 2) \tag{23}$$

Here the characteristic equation is $x^2 + 1$ and the characteristic roots are i and $-i$. In such a case the solution (22) is valid only if we treat the coefficients of the original recurrence relation as complex numbers. The constants $c_1, ..., c_r$ in (22) must then be allowed to take complex values, if we want a particular solution in which all the a_n's are real. Suppose, for example, that in (23) the initial conditions are $a_0 = 5$ and $a_1 = 6$. By inspection, in this case the solution is $(5, 6, -5, -6, 5, 6, -5, -6, 5, 6, ...)$. Let us try to obtain this from (22). Let $a_n = c_1 i^n + c_2 (-i)^n$. Setting $n = 0$ and 1 we get, from the initial conditions, $a_0 = 5 = c_1 + c_2$ and $a_1 = 6 = c_1 i - c_2 i$. Solving these, we get $c_1 = 5/2 - 3i$ and $c_2 = 5/2 + 3i$. Consequently. $a_n = (5/2 - 3i) i^n + (5/2 + 3i)(- i)^n$. Mathematically, there is nothing wrong with this expression. Its values are real for integer values of n. Still, we would prefer if a_n can be expressed directly in a real form. We now show that such an expression is always possible. That is, as long as the coefficients $C_0, ..., C_r$ in (20) are real, even though some of the characteristic roots are complex we can obtain an alternate basis for the vector space K so that every real solution of (20) can be expressed as a linear combination of these basis elements with real coefficients.

The key idea is that complex roots of real polynomials must always occur in conjugate pairs. Writing complex numbers in the polar form, if $re^{i\theta} = r\cos\theta + ir\sin\theta$ is a characteristic root, so is $re^{-i\theta}$. The vector in K generated by $re^{i\theta}$ is $\bar{u}_{re^{i\theta}} = \{r^n e^{in\theta}\}_{n=0}^\infty$ while that generated by $re^{-i\theta}$ is $\bar{u}_{re^{-i\theta}} = \{r^n e^{-in\theta}\}_{n=0}^\infty$. Now suppose $\bar{a} = \{a_n\}_{n=0}^\infty$ is a real sequence which is a linear combination of $\bar{u}_{re^{i\theta}}$ and $\bar{u}_{re^{-i\theta}}$ with complex coefficients say, c_1 and c_2. That is, $a_n = c_1 r^n e^{in\theta} + c_2 r^n e^{-in\theta}$ for all $n \geqslant 0$. Let $c_1 = \rho_1 e^{i\theta_1}$ and $c_2 = \rho_2 e^{i\theta_2}$. Then a_n must equal the real part of $c_1 r^n e^{in\theta} + c_2 r^n e^{-in\theta}$, which is $r^n \rho_1 \cos(n\theta + \theta_1) + r^n \rho_2 \cos(n\theta + \theta_2)$. Expanding $\cos(n\theta + \theta_1)$ and $\cos(n\theta + \theta_2)$ and rearranging the terms, we get $a_n = d_1 r^n \cos n\theta + d_2 r^n \sin n\theta$ where $d_1 = \rho_1 \cos\theta_1 + \rho_2 \cos\theta_2$ and $d_2 = -\rho_1 \sin\theta_1 - \rho_2 \sin\theta_2$. Note that d_1 and d_2 are real. Thus we have shown that every real linear combination of

$$\{r^n e^{in\theta}\}_{n=0}^\infty \text{ and } \{r^n e^{-in\theta}\}_{n=0}^\infty$$

with complex coefficients can be expressed as a linear combination of

$$\{r^n \cos n\theta\}_{n=0}^\infty \text{ and } \{r^n \sin n\theta\}_{n=0}^\infty$$

with real coefficients. It follows that we can replace the pair of vectors $\bar{u}_{re^{i\theta}}$, $\bar{u}_{re^{-i\theta}}$ in the complex vector space K by the pair of vectors

$$\{r^n \cos n\theta\}_{n=0}^\infty, \{r^n \sin n\theta\}_{n=0}^\infty$$

in the real vector space K. Repeating this procedure for all complex characteristic roots (which we assume to be distinct), we get the following result.

3.9 Theorem: Suppose the coefficients $C_0, C_1, ..., C_r$ in (20) are real and that the characteristic roots are

$$r_1 e^{\pm i\theta_1}, r_2 e^{\pm i\theta_2}, ..., r_k e^{\pm i\theta_k}, \beta_{2k+1}, ..., \beta_r$$

where $\beta_{2k+1}, ..., \beta_r$ are real. Then the general solution of (20) is
$$a_n = d_1 r_1^n \cos n\theta_1 + d_1' r_1^n \sin n\theta_1 + d_2 r_2^n \cos\theta_2 + d_2' r_2^n \sin n_2 + ... +$$

$$+ d_k r_k^n \cos n\theta_k + d_k' r^n \sin n\theta_k + d_{2k+1}\beta_{2k+1}^n + ... + d_r \beta_r^n \quad (24)$$

where $d_1, ..., d_k, d_1', ..., d_k', d_{2k+1}, ..., d_r$ are real consitants. ∎

For example, returning to (23), we have $r = 2, k = 1, r_1 = 1$ and $\theta_1 = \dfrac{\pi}{2}$. So the general solution of (23) is $a_n = d_1 \cos\dfrac{n\pi}{2} + d_1' \sin\dfrac{n\pi}{2}$, where d_1, d_1' are real constants. If the initial conditions are $a_0 = 5$ and

$a_1 = 6$, we get $5 = d_1 \cos 0 + d_1' \sin 0 = d_1$ and $6 = d_1 \cos \frac{\pi}{2} + d_1' \sin$

$\frac{\pi}{2} = d_1'$. So the particular solution is $a_n = 5 \cos \frac{n\pi}{2} + 6 \sin \frac{n\pi}{2}$. This is

exactly the same as $a_n = \left(\frac{5}{2} - 3i\right) i^n + \left(\frac{5}{2} - 3i\right)(-i)^n$, but appears in a

more likable form.

Let us now turn to the second part of obtaining the general solution of (19), namely finding any one particular solution of it. This solution evidently depends on the nature of the function $f(n)$ in (19). Inasmuch as there is no restriction on this function, obviously there can be no golden method which will work in all cases. Nevertheless a few observations are useful. First, suppose $f(n)$ is a linear combination of functions, say, $f(n) = \lambda_1 f_1(n) + ... + \lambda_k f_k(n)$. Because of linearity, it is easily seen that a particular solution to (19) is the corresponding linear combination of the solutions to (19) with $f(n)$ replaced by $f_i(n)$, $i = 1, ..., k$. This reduces the task of finding a particular solution to those cases where $f(n)$ is a 'simple' or 'elementary' function in some sense. In such cases, depending on the nature of the function we try a particular solution of a certain form with some unknown coefficients and then determine these coefficients. The table in Figure 7.4 gives a list of what form of a particular solution works for certain kinds of $f(n)$, assuming that there are no multiple characteristic roots.

Let us illustrate the procedure with the recurrence relation (i). Rewriting it as $a_n - a_{n-1} = n$, we see that 1 is its characteristic root and so by Theorem (3.8) the general solution of the A.H.E, is $a_n = c$ where c is a

No.	Form of $f(n)$	Form of a_n
1	β^n, where β is not a characteristic root	$A\beta^n$, where A is a constant
2	β^n where β is a simple characteristic root	$An\beta^n$ where A is a constant
3	a polynomial in n of degree k	a polynomial in n of degree $k+1$ or k according as 1 is or is not a characteristic root.

Figure 7.4: **Finding a particular Solution of a Linear Recurrence Relation**

constant. As for a particular solution, $f(n) = n$ is a polynomial of degree 1. Here 1 is a characteristic root. So we let a_n be a polynomial in n of degree 2, say $a_n = An^2 + Bn + C$ for a particular solution. Then

$$a_{n-1} = A(n-1)^2 + B(n-1) + C.$$

So $n = a_n - a_{n-1} = 2An + (-A + B)$. If this is to hold for all $n \geqslant 1$, we must have $2A = 1$ and $-A + B = 0$; solving which, $A = B = \frac{1}{2}$. C is undetermined, which happens because 1 is a characteristic root. Setting $C = 0$ arbitrarily, $a_n = \frac{1}{2}n^2 + \frac{1}{2}n$ is a particular solution of (1). By Theorem (3.4), the general solution of (1) is $a_n = c + \frac{1}{2}n^2 + \frac{1}{2}n$. The initial condition $a_0 = 1$ now gives $c = 1$ and this is the same answer as we obtained earlier by the method of generating functions.

Of course we have to verify that the table above really works. This verification, as well as the consideration of the case of multiple roots tends to be messy if done directly. In the exercises we shall indicate how it can be done elegantly using facts about vector spaces.

We have now described how to obtain the general solution of (19). This general solution involves r mutually independent arbitrary constants. To find a particular solution we need to determine these constants. This would require a system of r mutually independent equations in these constants. The initial conditions, i.e., the values of $a_0, a_1, ..., a_{r-1}$ give one such set of equations. This, in fact, is the most common choice of specifying initial conditions, because in applications, the values of a_n can usually be found by inspection for lower values of n. In some problems, however, a_n has no natural meaning for some values of n. For example, let a_n be a number associated with a polygon with n sides. A non-degenerate polygon must have at least three sides. So, for $n < 3a_n$ is undefined. In some problems a_n may be defined for all n, but the recurrence relation may fail for a few small values of n. In such cases, even if we know the values of $a_0, ..., a_{r-1}$, they cannot be used as initial conditions.

In such situations we proceed as follows. Suppose the recurrence relation (19) is valid at least for all $n \geqslant k$ and that the initial conditions specify the values of $a_k, a_{k+1}, ..., a_{k+r-1}$ where k is some integer $\geqslant r$. We can then work backwards and define (or redefine if necessary), a_{k-1}, $a_{k-2}, ..., a_1, a_0$ successively from (19). This determination is always possible because of our assumption that $C_0 \neq 0$. (19) is then valid for all $n \geqslant r$. From $a_0, ..., a_{r-1}$ so defined, we get a particular solution. Alternately, we can substitute the values of $a_k, ..., a_{k+r-1}$ directly in the general solution and solve the resulting system of equations for the arbitrary constants. It is not hard to show that this way the arbitrary constants are uniquely determined. It is necessary, however, that we know a_n for r *consecutive* values of n. For example, in the case of (23), suppose we know, say, that $a_{10} = a$ and $a_{12} = b$. If $a + b \neq 0$, then (23) has no solution with these initial conditions. But if $a + b = 0$, then there are infinitely many such solutions, because these initial conditions do not determine a_n for odd n.

The following numerical problem summarises almost everything we have covered about linear reccurence relations.

3.10 Problem: Solve the recurrence relation

$$2a_n - 5a_{n-1} + 6a_{n-2} - 2a_{n-3} = 2^{-n} + 3^{n+1} - n \qquad (25)$$

subject to $a_4 = 5$, $a_5 = -3$ and $a_6 = 1$.

Solution: The characteristic polynomial is $2x^3 - 5x^2 + 6x - 2$. By the rational root test (Exercise (6.2.34)), the possible rational roots of this polynomial are ± 1, ± 2, $\pm \frac{1}{2}$. By trial we find $\frac{1}{2}$ is a root. So $(x - \frac{1}{2})$ or equivalently, $(2x - 1)$ is a factor. Carrying out the long division we see $2x^3 - 5x^2 + 6x - 2 = (2x - 1)(x^2 - 2x + 2)$. The other characteristic roots are the roots of $x^2 - 2x + 2$, namely $1 \pm i$. These are complex and their polar form is $\sqrt{2}e^{\pm i\pi/4}$. So by Theorem (3.9), the general solution of the A.H.E. is

$$a_n = c_1(\tfrac{1}{2})^n + c_2(\sqrt{2})^n \cos\left(\frac{n\pi}{4}\right) + c_3(\sqrt{2})^n \sin\left(\frac{n\pi}{4}\right) \qquad (26)$$

Now for a particular solution of (25), 2^{-n} equals $(\frac{1}{2})^n$ and since $\frac{1}{2}$ is a characteristic root, we try $An(\tfrac{1}{2})^n$, i.e., $\dfrac{An}{2^n}$. However, $3^{n+1} = 3 \cdot 3^n$ and since 3 is not a characteristic root we try $B3^n$. For the third term, $-n$, since 1 is not a characteristic root, we try $Cn + D$. So let a particular solution be

$$a_n = \frac{An}{2^n} + B3^n + Cn + D \qquad (27)$$

where the constants A, B, C, D are to be determined. Computing a_{n-1}, a_{n-2} and a_{n-3}, we get, after simplification,

$$2a_n - 5a_{n-1} + 6a_{n-2} - 2a_{n-3} = \frac{10A}{2^n} + 25B3^{n-3} + Cn + D - C$$

It follows that if (27) is to be a solution of (25), then we must have $10A = 1$, $25B = 81$, $C = -1$ and $D - C = 0$, which gives $A = \dfrac{1}{10}$, $B = \dfrac{81}{25}$, $C = -1$ and $D = -1$. So $a_n = \dfrac{1}{10 \cdot 2^n} + \dfrac{3^{n+4}}{25} - n - 1$ is particular solution of (25). By Theorem (3.4), the general solution to (25) is

$$a_n = c_1\left(\frac{1}{2}\right)^n + c_2(\sqrt{2})^n \cos\frac{n\pi}{4} + c_3(\sqrt{2})^n \sin\frac{n\pi}{4} + \frac{1}{10 \cdot 2^n} + \frac{3^{n+4}}{25} - n - 1 \qquad (28)$$

To find the particular solution with the given initial conditions, we set $n = 4, 5, 6$ in (28) and get the following system of equations:

(i) $\dfrac{c_1}{16} - 4c_2 = a$ (ii) $\dfrac{c_1}{32} - 4c_2 - 4c_3 = b$ (iii) $\dfrac{c_1}{64} - 8c_3 = c$

where

$$a = 5 - \frac{1}{160} - \frac{6,561}{25} + 5 = \frac{-201,637}{800}, \quad b = -3 - \frac{1}{320} - \frac{19,683}{25} + 6$$

$$= -\frac{1,254,917}{1,600}$$

and

$$c = 1 - \frac{1}{640} - \frac{59049}{25} + 7 = -\frac{7,532,677}{3,200},$$

Solving this system we get

$$c_1 = \frac{64(2a - 2b + c)}{5}, \quad c_2 = \frac{3a - 8b + 4c}{20} \quad \text{and} \quad c_3 = \frac{a - b - 2c}{20}$$

Substituting the valves of a, b, c and then substituting in (28) gives the answer. ▧

We conclude the section with a brief discussion of recurrence relations with two indices. A sequence is a function of a single integer variable. As remarked in Section 1, we often have functions of two (or more) integer variables. Denote such a function by $\{a_{m,n}\}_{\substack{m=0 \\ n=0}}^{\infty}$. A recurrence relation for such functions is an expression which expresses $a_{m,n}$ in terms of $a_{i,j}$'s where $i \leqslant m$, $j \leqslant n$ and at least one inequality is strict. Such recurrence relations can often be written down by combinatorial considerations. For example,

let $a_{m,n} = \begin{pmatrix} m \\ n \end{pmatrix}$, the number of ways to choose n objects from a collection

of m objects. Even if we did not know a formula for $\begin{pmatrix} m \\ n \end{pmatrix}$, we can easily

show by purely combinatorial arguments (cf. Proposition (2.2.19)) that for all $m \geqslant 1$, $n \geqslant 1$,

$$a_{m,n} = a_{m-1,\,n-1} + a_{m-1,\,n} \tag{29}$$

Here the 'boundary conditions' are $a_{m,0} = 1$ for all m and $a_{m,n} = 0$ for $m < n$.

This is a recurrence relation with two indices, m and n. As a matter of

fact, solving (29), we can show $a_{m,n} = \begin{pmatrix} m \\ n \end{pmatrix}$ as follows. Multiply both sides

of (29) by x^n and sum over $n \geqslant 1$. Then we get

$$\sum_{n=1}^{\infty} a_{m,n} x^n = x \sum_{n=1}^{\infty} a_{m-1,\,n-1} x^{n-1} + \sum_{n=1}^{\infty} a_{m-1,\,n} x^n \tag{30}$$

Now, for each i, let $F_i(x) = \sum_{n=0}^{\infty} a_{i,n} x^n$. In other words, $F_i(x)$ is the

O.G.F. of the sequence $\{a_{i,n}\}_{n=0}^{\infty}$ for fixed i. With this notation, and keeping in mind that $a_{i,0} = 1$ for all i, we get

$$F_m(x) - 1 = xF_{m-1}(x) + F_{m-1}(x) - 1 \tag{31}$$

or

$$F_m(x) = (1 + x)F_{m-1}(x) \ (m \geqslant 1) \tag{32}$$

(32) is a recurrence relation for the sequence of generating functions $(F_0(x), F_1(x), F_2(x),..., F_m(x),...)$. Since $a_{0,0} = 1$ and $a_{0,n} = 0$ for $n > 0$, $F_0(x) = \sum_{n=0}^{\infty} a_{0,n}x^n = 1$. So $F_1(x) = 1 + x$, $F_2(x) = (1 + x)^2...$ and in general $F_m(x) = (1 + x)^m$ for all m, as can be easily proved by induction on m. We have thus solved (32). (Of course we could have also solved it by considering $F(x, y) = \sum_{m=0}^{\infty} F_m(x)y^m$, which is the O.G.F. of the sequence $\{F_m(x)\}_{m=0}^{\infty}$ and then deriving a formula for $F(x, y)$ by multiplying both sides of (32) by y^m and summing over $m \geqslant 1$. We would then get $F(x,y) = \dfrac{1}{1 - (1 + x)y}$ and expanding this as a power series in y, we get $F_m(x) =$ coefficient of $y^m = (1 + x)^m$.)

It thus follows that $a_{m,n}$ is the coefficient of x^n in $(1 + x)^m$. (Of course we already know this from the binomial theorem. But the point was to derive it strictly from (29), without knowing how $a_{m,n}$ came about.) Knowing this, it follows that $a_{m,n}$ must equal $\begin{pmatrix} m \\ n \end{pmatrix}$.

It is instructive to look at recurrence relations with two indices geometrically, in the light of the cartesian representation of $a_{m,n}$ (cf. Figure 7.2(a)) introduced at the end of Section 1. Consider the rectangle of dots with sides parallel to the axes, one vertex at $(0, 0)$ and the diagonally opposite vertex at (m, n). A recurrence relation then expresses the value of the function at the corner (m, n) in terms of its values at some of the other points of this rectangle. In general these points need not lie in the same row or column as the point (m, n). (In the case of (29), both $(m-1, n-1)$ and $(m-1, n)$ lie in the column on the left of (m, n).) In fact, if they did, the corresponding recurrence relation would not be very interesting. For example, consider the recurrence relation

$$b_{m,n} = b_{m, n-1} + b_{m, n-2} + n \tag{33}$$

for $n \geqslant 2$.

In this case, if we let $G_m(x) = \sum_{n=0}^{\infty} b_{m,n} x^n$, then for each fixed m, multiplying both sides of (33) by x^n and summing over $n \geqslant 2$, yields a separate equation for each $G_m(x)$. The generating functions $G_m(x)$'s are not interlinked in any way as the $F_m(x)$'s are interlinked by (32). In other words (33) is nothing more than a collection of mutually unrelated recurrence relations of one index, namely, n. In (29), on the other hand, the terms of the mth sequence, $\{a_{m,n}\}_{n=0}^{\infty}$, are linked with those of the $(m-1)$th sequence, $\{a_{m-1,\,n}\}_{n=0}^{\infty}$. That is why solving (29) is more interesting than solving (33).

Such inter-linkage can occur for two ordinary recurrence relations as well. As we shall see in the next section, in some enumeration problems, even though our primary interest may be in a sequence $\{a_n\}$, instead of obtaining a recurrence relation for $\{a_n\}$ directly, it is more convenient to introduce an auxiliary sequence $\{b_n\}$ (often several such sequences) and to inter-relate the terms of the two sequences. This gives rise to what is known as a **system of simultaneous recurrence relations**. The standard method for solving such a system is to translate the inter-relationship between the two sequences into an equation in their generating functions. We illustrate this technique with an example.

3.11 Problem: Solve the system of recurrence relations:

$$a_n = a_{n-1} + b_n \quad (n \geqslant 1) \tag{34}$$

$$b_n = a_{n-1} + b_{n-1} \ (n \geqslant 1) \tag{35}$$

with the initial conditions $a_0 = 1$, $b_0 = 0$,

Solution: As usual we let $A(x) = \sum\limits_{n=0}^{\infty} a_n x^n$ and $B(x) = \sum\limits_{n=0}^{\infty} b_n x^n$. Multiplying throughout by x^n and summing over $n \geqslant 1$, gives

$$A(x) - 1 = xA(x) + B(x) \tag{36}$$

and

$$B(x) = xA(x) + xB(x) \tag{37}$$

Solving these equations simultaneously gives $A(x) = \dfrac{1-x}{1-3x+x^2}$ and $B(x)$

$= \dfrac{x}{1-3x+x^2}$. Expanding these, we get closed form expressions for a_n and b_n. This is left as an exercise.   ▪

In this problem we could have delinked the sequences $\{a_n\}$ and $\{b_n\}$ from each other as follows. Substituting (35) in (34) and then again using (34) with $n-1$ replacing n, we get

$$a_n = 2a_{n-1} + b_{n-1} = 2a_{n-1} + (a_{n-1} - a_{n-2}) = 3a_{n-1} - a_{n-2}.$$

Similarly, $b_n = a_n - b_n + b_{n-1} = b_{n+1} - 2b_n + b_{n-1}$. We can rewrite these as

$$a_n + 3a_{n-1} + a_{n-2} = 0 \quad (n \geqslant 2) \tag{38}$$

and

$$b_n - 3b_{n-1} + b_{n-2} = 0 \quad (n \geqslant 2) \tag{39}$$

These are linear recurrence relations, and can be solved separately. But in some problems such delinking may not be easy and so it is desirable to use the method above.

Exercises

3.1 Let $0.\beta_1\beta_2\beta_3\ldots\beta_n\ldots$ be the decimal expansion of the number $\dfrac{50,482}{99900}$. Show that $\beta_1 = 5$, $\beta_2 = 0$ and for $n \geqslant 3$, β_n depends only on the congruence class of n modulo 3.

3.2 Define a recurrence relation by $x_n = x_{n-1} + \beta_n$ with $x_0 = 0$ where the β's are as the last exercise. Prove that the solution of this recurrence relation is the same as the that of the recurrence relation $x_n - x_{n-3} = 10$ for $n \geqslant 5$ with the initial conditions $x_2 = 5$, $x_3 = 10$ and $x_4 = 13$. Hence obtain a closed form formula for x_n.

3.3 Prove that a real number is rational if and only if its decimal expansion either terminates or repeats periodically after some stage. (From the decimal expansion of any real number we get a recurrence relation. If the number is rational then this recurrence relation can be solved by generalising the method in the last exercise. But in the case of an irrational number like π, there is no known way of solving it as we remarked in the text.)

3.4 Recall from Exercise (1.3.9) that the numbers b_n given by **(3)** are called Fibonacci numbers and are more commonly denoted by F_n. Prove that $\displaystyle\lim_{n\to\infty} \frac{F_n}{F_{n-1}} = \frac{1+\sqrt{5}}{2}$. What is wrong if we try to prove this by the following argument? Let $\displaystyle\lim_{n\to\infty} \frac{E_n}{F_{n-1}} = L$. Then $\displaystyle\lim_{n\to\infty} \frac{F_{n-1}}{F_{n-2}}$ also equals L and consequently $\displaystyle\lim_{n\to\infty} \frac{F_{n-2}}{F_{n-1}} = \frac{1}{L}$. Dividing the recurrence relation $F_n = F_{n-1} + F_{n-2}$ throughout by F_{n-1} and letting $n \to \infty$, we get $L = 1 + \dfrac{1}{L}$ or $L^2 - L - 1 = 0$, solving which $L = \dfrac{1+\sqrt{5}}{2}$, the other root being negative.

3.5 Suppose $ABCD$ is a rectangle with the property that when a square on its shorter side is removed from it, the remaining rectangle is similar to the original rectangle (see Figure 7.5). Prove that the ratio of the length to the width of $ABCD$ is $\dfrac{1+\sqrt{5}}{2}$. Prove that if we go on removing squares on the shorter sides we keep on getting rectangles of the same shape. (For this reason, the ratio $\dfrac{1+\sqrt{5}}{2}$ is called the **golden ratio**. It is believed that among all rectangles

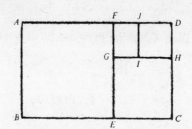

Figure 7.5: The Golden Ratio

which are not squares, the shape of a rectangle with the golden ratio is most pleasing to the eye!. A possible reason might be that according to the last exercise, a rectangle whose sides are the Fibonacci numbers F_n and F_{n-1} for a large n is very close to a rectangle with a golden ratio and, as remarked in Chapter 1, Section 2 Fibonacci numbers arise in nature in biological growth. Maybe we have an innate fascination for anything related to Fibonacci numbers!)

3.6 Using only the recurrence relation (2) for the Fibonacci numbers (and not the formula (3)), show by induction that $|F_n| \leqslant 2^n$ for all n.

*3.7 Let f, g, h be continuous real-valued functions defined on an open interval $(0, R)$, where $R > 0$. Suppose $g(x) \neq h(x)$ for all $x \in (0, R)$ and that for every $x \in (0, R)$ either $f(x) = g(x)$ or $f(x) = h(x)$. Prove that either $f(x) = g(x)$ for all $x \in (0, R)$ or $f(x) = h(x)$ for all $x \in (0, R)$. [Hint: Let $A = \{x \in (0, R) : f(x) = g(x)\}$ and $B = \{x \in (0, R) : f(x) = h(x)\}$. Note that A, B are closed subsets of $(0, R)$. Use connectedness of $(0, R)$.]

3.8 Starting from the power series expansion of $\dfrac{1 - \sqrt{1 - 4x}}{x}$, work backwards to justify that (6) is indeed the solution of (5).

3.9 Solve the following recurrence relations using O.G.F.'s:

(i) $a_n + a_{n-2} = n$ for $n \geqslant 2$ with $a_0 = 1$, $a_1 = -1$.

(ii) $a_n = \dfrac{1}{2n} a_{n-1}$ for $n \geqslant 1$ with $a_0 = 3$ (this can also be solved by inspection).

(iii) $a_n = 3^n - 2^n a_0 - 2^{n-1} a_1 - \ldots - 2^2 a_{n-2} - 2a_{n-1}$, for $n \geqslant 1$, with $a_0 = 1$.

3.10 Solve $a_n = na_{n-1} + 3^n$ with $a_0 = 0$ using exponential generating functions.

3.11 Solve $a_n = 5a_{n-1}$, $a_0 = 2$ by three methods; by inspection, by ordinary generating functions and by exponential generating functions.

3.12 Obtain the first four terms in the series solution of the following differential equations at the point 0.

(i) $y'' + xy' + y = 0$, $y(0) = 2$, $y'(0) = 1$
(ii) $y'' + e^x y' + y = 0$, $y(0) = 1$, $y'(0) = 2$

3.13 Suppose α is a multiple characteristic root of (20). Prove that $a_n = n\alpha^n$ is a solution of (20). [Hint: For each fixed $n \geqslant r$, note that α is a multiple root of $C_r x^n + C_{r-1} x^{n-1} + \cdots + C_0 x^{n-r}$. Apply Exercise (6.2.23).]

3.14 Suppose α is a characteristic root of multiplicity k. Prove that $a_n = \alpha^n$, $a_n = n\alpha^n$, $a_n = n^2\alpha^n, \ldots$, $a_n = n^{k-1}\alpha^n$ are all solutions of (20). Prove further that the vectors $\{\alpha^n\}_{n=0}^{\infty}$, $\{n\alpha^n\}_{n=0}^{\infty}, \ldots$, $\{n^{k-1}\alpha^n\}_{n=0}^{\infty}$ are linearly independent in the subspace K of V. [Hint note that $\alpha \neq 0$. Take suitable truncations of these vectors and a Vandermonde determinant.]

3.15 Generalise Theorem (3.8) to cover the case where some of the characteristic roots are multiple roots.

3.16 (a) Let n, k be positive integers and $A = (a_{rs}) = (x_{rs} + iy_{rs})$ be a $2k \times n$ matrix over $\mathbb{C}$ in which the entries in even numbered rows are the complex conjugates of the corresponding entries in the preceding rows, i.e., $a_{2p,s} = \bar{a}_{2p-1,s}$ for all $p = 1, \ldots, k$; $s = 1, \ldots, n$. Let $B = (b_{rs})$ be the $2k \times n$ matrix over $\mathbb{R}$ obtained by taking the real and the imaginary parts of the entries in A; specifically for $b = 1, \ldots, k$ and $s = 1, \ldots, n$, $b_{2p-1,s} = x_{2p-1,s}$ and $b_{2p,s} = y_{2p,s}$. Prove that A and B have the same ranks. (Hint: Using the formula

$$\begin{pmatrix} x \\ y \end{pmatrix} = \begin{pmatrix} \frac{1}{2} & \frac{1}{2} \\ -\frac{1}{2} & \frac{1}{2} \end{pmatrix} \begin{pmatrix} x + iy \\ x - iy \end{pmatrix}$$

construct a non-singular matrix C such that $B = CA$.)

(b) Suppose the coefficients $C_0, \ldots, C_r$ in (20) are real and $re^{i\theta}$ is a complex characteristic root of multiplicity k. Prove that the vectors $\{r^n \cos n\theta\}_{n=0}^{\infty}$, $\{r^n \sin n\theta\}_{n=0}^{\infty}$, $\{nr^n \cos \alpha\theta\}_{n=0}^{\infty}$, $\{nr^n \sin n\theta\}_{n=0}^{\infty}, \ldots$, $\{n^{k-1} r^n \cos n\theta\}_{n=0}^{\infty}$, $\{n^{k-1} r^n \sin n\theta\}_{n=0}^{\infty}$ are in K and are linearly independent over $\mathbb{R}$. (Hint: First work over $\mathbb{C}$ and then use (a).)

3.17 Generalise Theorem (3.9) to cover the case of multiple complex roots and thereby to give a complete answer to finding the general solution of (20) in all possible cases with real coefficients.

3.18 Let β be a non-zero real (or complex) number and consider the recurrence relation

$$C_r a_n + C_{r-1} a_{n-1} + \ldots + C_r a_{n-r+1} + C_0 a_{n-r} = \beta^n \, (n \geqslant r) \tag{40}$$

(which is, of course, a special case of (19)). Prove that any solution of (40) is also a solution of

$$C_r a_n + (C_{r-1} - \beta C_r) \, a_{n-1}, + \ldots + (C_0 - \beta C_1) a_{n-r} - \beta C_0 a_{n-r-1} = 0$$
$$(n \geqslant r + 1) \tag{41}$$

which is a linear homogeneous recurrence relation of order $r + 1$.

3.19   Continuing the notation of Theorem (3.4), let K, K' and L be the solution spaces of (20), (40) and (41) respectively. Prove that L contains both K and K'. Further suppose $\bar{u} \in L$ and $\bar{u} \notin K$. Then show that there exists a unique $\lambda \in \mathbb{R}$ (or $\mathbb{C}$) such that $\lambda \bar{u} \in K'$. [Hint: Note that K and L are subspaces of V of dimensions r and $r + 1$ respectively and that K' is a coset of K. This exercise shows how we can get a particular solution of (40) which is not a solution of (20).]

3.20   Using the last exercise and Exercise (3.15), show that if β is not a characteristic root of (40), then there exists a unique $\lambda \neq 0$ such that $a_n = \lambda \beta^n$ is a particular solution of (40). [Hint: Note that the characteristic polynomial of (41) is simply $(x - \beta)$ times the characteristic polynomial of (40).]

3.21   With the same reasoning, show that if β is a characteristic root of (40) of multiplicity k $(k \geqslant 1)$, then there exists a unique $\lambda \neq 0$ such that $a_n = \lambda n^k \beta^n$ is a solution of (40). These two exercises show that the first two lines in the table in Figure 7.4 are correct and also take care of the case of multiple characteristic root. Direct computational proofs tend to be clumsy.)

3.22   Suppose $\{b_n\}_{n=0}^{\infty}$ is a sequence which satisfies some homogeneous, linear recurrence relation of order p, say,

$$D_p b_n + D_{p-1} \, b_{n-1} + \ldots + D_1 \, b_{n-p+1} + D_0 b_{n-p} = 0 \, (n \geqslant p) \tag{42}$$

Suppose further that $\{a_n\}_{n=0}^{\infty}$ is a sequence which satisfies a (non-homogeneous) linear recurrence relation of order r, say

$$C_r a_n + C_{r-1} a_{n-1} + \ldots + C_1 a_{n-r+1} + C_0 a_{n-r} = b_n \, (n \geqslant r). \tag{43}$$

(This is the same as (19) if we write $f(n) = b_n$.) Denote by $D(x)$ and $C(x)$ respectively the characteristic polynomials of (42) and (43). By direct calculation, show that $\{a_n\}_{n=0}^{\infty}$ satisfies a homogeneous linear recurrence relation of order $r + p$ whose characteristic polynomial is $D(x) \, C(x)$. (It is a little messy to write down this recurrence relation. But as we shall see, we are more interested in its characteristic polynomial than its individual coefficients. Because of this exercise we can convert a solution of a non-homogeneous relation to a solution of a homogeneous relation if we happen to know a homogeneous relation satisfied by the right hand side. Note that Exercise (3.18) is a special case of this exercise with $D(x) = x - \beta$.)

3.23 Let k be a non-negative integer and let $b_n = n^k$, with the understanding that $0^o = 1$. Prove that $\{b_n\}_{n=0}^\infty$ satisfies a homogeneous, linear recurrence relation whose characteristic polynomial is $(x-1)^{k+1}$. [Hint: Apply Exercise (3.14) with $\alpha = 1$. A direct proof is messy].

3.24 With k as above, consider the linear recurrence relation of order r,

$$C_r a_n + C_{r-1} a_{n-1} + \dots + C_1 a_{n-r+1} + C_0 a_{n-r} = n^k \, (n \geqslant r). \tag{44}$$

Consider the homogeneous linear recurrence relation of order $r + k + 1$,

$$E_{r+k+1}\, a_n + E_{r+k}\, a_{n-1} + \dots + E_1\, a_{n-r-k} + E_0\, a_{n-r-k-1} = 0 \tag{45}$$

whose characteristic polynomial, say $E(x)$, is the product $(x-1)^{k+1}$ $C(x)$ where $C(x)$ is the characteristic polynomial of (44). Let K, K' and L be the solution spaces of (20), (44) and (45) respectively. Prove that L contains both K and its coset K'. [Hint: Apply Exercise (3.22) with $b_n = 0$ and then again with $b_n = n^k$, $D(x)$ being equal to $(x-1)^{k+1}$ in both cases.]

3.25 In (44), suppose 1 is not a characteristic root. Prove that there exist constants $A_0, \dots A_k$ such that $a_n = A_0 + A_1 n + A_2 n^2 + \dots + A_k n^k$ is a particular solution of (44). [Hint: Use the last exercise 1 is a characteristic root of (45) with multiplicity $k + 1$. Apply Exercise (3.15) to (45). For $i = 0, 1, 2, \dots$, let $\bar{v}_i$ be the vector $(0^i, 1^i, 2^i, 3^i, \dots, n^i, \dots)$ in V. Prove that L is spanned by $K \cup \{\bar{v}_0, \bar{v}_1, \dots, \bar{v}_k\}$.]

3.26 (44) suppose 1 is a characteristic root of multiplicity $m \, (m \geqslant 1)$, Prove that there exist constants $A_m, A_{m+1}, \dots, A_{m+k}$ such that $a_n = A_m n^m + A_{m+1} n^{m+1} + \dots + A_{m+k} n^{m+k}$ is a particular solution of (44). [Hint: Same as the last exercise, except that this time 1 is a characteristic root of (45) with multiplicity $m + k + 1$. The vectors $\bar{v}_0, \bar{v}_1 \dots \bar{v}_{m-1}$ are already in K and so L is spanned by $K \cup \{\bar{v}_m, \bar{v}_{m+1}, \dots, \bar{v}_{m+k}\}$. With these exercises we have now completely established the validity of the table in Figure 7.4.]

3.27 Solve the following recurrence relations:

(i) $a_n - 5a_{n-1} + 6a_{n-2} = 2^n + n$

(ii) $a_n - 3a_{n-1} + 3a_{n-2} - a_{n-3} = n^2$ with $a_0 = 1, a_1 = -1, a_2 = \sqrt{2}$

(iii) $6a_n + 7a_{n-1} + 6a_{n-2} - a_{n-4} = 0$

3.28 Show that if the characteristic roots $\alpha_1, \dots, \alpha_r$ of (19) are distinct then for any $k \geqslant 0$ the following system in the unknowns $c_1, \dots, c_r$ has a unique solution:

$$c_1 \alpha_1^k + c_2 \alpha_2^k + \dots + c_r \alpha_r^k = f(k)$$

$$c_1\alpha_1^{k+1} + c_2\alpha_2^{k+1} + \ldots + c_r\alpha_r^{k+1} = f(k+1)$$

$$\vdots \tag{46}$$

$$c_1\alpha_1^{k+r-1} + c_2\alpha_2^{k+r-1} + \ldots + c_r\alpha_r^{k+r-1} = f(k+r-1)$$

Hence show that the values of any r consecutive a_n's determine a particular solution of (19).

[Hint: Vandermonde determinant again.]

3.29   Extend the result of the least exercise to the general case.

3.30   Solve the recurrence relation with two indices:

$$a_{m,n} = \sum_{i=0}^{n} \binom{n}{i} a_{m-1,i} \ (m \geqslant 1, n \geqslant 0) \tag{47}$$

with the boundary conditions $a_{0,0} = 1$ and $a_{0,n} = 0$ for $n > 0$, by considering for each m, the exponential generating function of the sequence $\{a_{m,n}\}_{n=0}^{\infty}$.

3.31   Solve the system of simultaneous recurrence relations:

$$a_n = \frac{(n+1)[1+(-1)^n] - \sum_{i=1}^{n-1} a_i a_{n-i} - \sum_{i=0}^{n} b_i b_{n-i}}{2} \ (n \geqslant 1)$$

and

$$b_n = \frac{1+(-1)^n}{2} - \sum_{i=1}^{n} a_i b_{n-i} \tag{$n \geqslant 1$}$$

with

$$a_0 = 1, a_1 = -1, b_0 = 1.$$

3.32   Let $E(x) = \sum_{n=0}^{\infty} a_n \dfrac{x^n}{n!}$ be the E.G.F. of a sequence $\{a_n\}$. Prove that $\{a_n\}$ is a solution of (20) if and only if $y = E(x)$ is a solution of

$$C_r \frac{d^r y}{dx^r} + C_{r-1} \frac{d^{r-1}y}{dx^{r-1}} + \ldots + C_1 \frac{dy}{dx} + C_0 y = 0$$

which is a homogeneous linear differential equation with constant coefficients. Hence give an alternate proof of Theorem (3.8).

Notes and Guide to Literature

The results in this section are classic. For more on difference equations, see Goldberg [1] or Levy and Lessman [1]. Historically, Fibonacci relation (Equation (5)), was the first recurrence relation, published in 1220. It appears in such diverse contexts ranging from phyllotaxy (arrangement of leaves on a branch) to searching algorithms that its solution by De Moivre (which is essentially the same as given here and appeared in eighteenth century) is probably the strongest testimony of the 'practical' utility of the power series. For more on the golden ratio as well as the real life occurrences of the Fibonacci numbers, see Coxeter [1].

There are numerous treatises on differential equations. See for example, Diwan & Agashe [1] or Piaggio [1]. As an example of an article illustrating the comments about suspecting the solution to a problem, see Rudin [3].

4. Applications of Recurrence Relations

In the last section we studied how to solve recurrence relations and as applications, solved three problems, namely the Regions Problem, the Shares Problem and the Vendor Problem. In this section we present a greater variety of applications of recurrence relations. The principle behind these applications is similar to that of mathematical induction. We take a problem with an integer parameter n. We solve it by inspection for some low values of n. Then we relate the solution of the case where n has value m to the solutions of the cases where n has values lower than m. But there is an important difference. In induction, we have to guess the answer before-hand and we can only verify it. This is indeed a drawback because as in the Shares Problem, sometimes the answer to the general case is far from easy to guess. The method of recurrence relations serves to patch up this gap. That is why it is a powerful tool in enumeration problems, comparable to differential equations in continuous mathematics.

As with differential equations, there are two major steps in applying the technique of recurrence relations to a combinatiorial problem. First, to write down a recurrence relation for the quantity we are looking for and secondly to solve it. The success of this method depends on how easy it is to carry out these two steps in a particular problem. The first step naturally varies considerably from problem to problem and in some problems it cannot be carried out satisfactorily. For the second step too, there is no golden method that will work to solve every possible recurrence relation. Although we discussed the method of generating functions in the last section, as remarked in Section 2 its success hinges upon two factors (i) identifying the generating function in a closed form and (ii) expanding it. As we shall see, in some problems, even if we can write a recurrence relation, there is no way to solve it and so the answer has to be left in a summation form or as the coefficient of a suitable power of x in some function of x. It should also be noted that in some problems even though the method of recurrence relations works, there may be other methods which are more elegant. We already saw an example of this in the Regions Problem and, more strikingly, in the Vendor Problem.

Despite these limitations, the method of recurrence relations is an important one. First of all, whenever it works, it usually works very systematically. Other methods may be more elegant. But they are also often tricky and uncertain, involving ad-hoc reasonings. We recall our earlier analogy that while it may be more artistic to do a problem by the methods of Euclid, doing it with coordinates is often a surer way. Secondly, in some problems, recurrence relations are the only means. An example of this was the Shares Problem.

We first give a few applications where we already had obtained the answer by other methods. For example, in Theorem (3.2.8) we showed that the number of k-ary sequences of length n in which a particular symbol, say 1, appears an even number of times is $\dfrac{k^n + (k-2)^n}{2}$. In Section 2 we rederived this using exponential enumerators. With recurrence relations, the problem almost reduces to a triviality. Let a_n be the desired number. Then $a_0 = 1$. For $n \geqslant 1$, $a_n = |A_n|$, where A_n is the set of all k-ary sequences of length n in which 1 occurs an even number of times. Clearly $A_n = C_n \cup B_n$ where C_n consists of those sequences in A_n which end with 1 and $B_n = A_n - C_n$. If we remove the last 1 appearing in a sequence in C_n, we get a sequence of length $n - 1$ in which 1 appears an odd number of times. There are $k^{n-1} - a_{n-1}$ such sequences (k^{n-1} being the total number of k-ary sequences of length $n - 1$). Conversely every such sequence of length $n - 1$ gives a sequence in C_n if we append a 1 at its end. It follows that $|C_n| = k^{n-1} - a_{n-1}$. As for $|B_n|$, every sequence in B_n must end in one of the remaining $k - 1$ digits other than 1. If we take out the last digit, we get a sequence of length $n - 1$, in which 1 appears an even number of times. By our assumption, there are a_{n-1} such sequences. Since each one of them comes from $k - 1$ sequences in B_n, it follows that $|B_n| = (k - 1)a_{n-1}$. So we have,

$$a_n = |A_n| = |C_n| + |B_n| = (k^{n-1} - a_{n-1}) + (k - 1)a_{n-1}.$$

Or equivalently,

$$a_n - (k - 2)a_{n-1} = k^{n-1} \quad (n \geqslant 1) \tag{1}$$

This is a linear non-homogeneous recurrence relation of order 1. $(k - 2)$ is the only characteristic root. So $a_n = c(k - 2)^n$ is the general solution of the A.H.E. by Theorem (3.8). For a particular solution, we try $a_n = \lambda k^n$ for a constant λ. Substitution in (1) gives $\lambda[k^n - (k - 2)k^{n-1}] = k^{n-1}$ or $\lambda = \frac{1}{2}$. So the general solution of (1) is $a_n = c(k - 2)^n + \frac{1}{2}k^n$. The constant c is determined by the initial condition $a_0 = 1$, giving $c = \frac{1}{2}$. It follows that that for $n \geqslant 0$, $a_n = \dfrac{(k - 2)^n + k^n}{2}$, the same answer as before.

As another example of this kind, which also serves to illustrate how it is sometimes more convenient to consider a system of recurrence relations rather than a single recurrence relation, we count the number of those k-ary sequences of length n in which both 0 and 1 appear even number of times each. (cf. Exercises (3.2.17) (b) and (2.18)).

4.1 Problem: For each n, find the number of k-ary sequences of length n in which two symbols, say 0 and 1, appear an even number of times each.

Solution: For simplicity denote the symbols by 0, 1,..., $k - 1$. Denote the desired number by a_n. Then $a_0 = 1$. However, it is not easy to express a_n directly in terms of a_{n-1}. So we introduce other auxiliary sequences. Let b_n be the number of those k-ary sequences of length n in which 0 appears an even number of times and 1 an odd number of times, c_n the number of those in which 0 appears an odd number of times and 1 an even number of times and d_n the number of those in which both 0 and 1 appear an odd number of times each. There is considerable inter-relationship among $\{a_n\}$, $\{b_n\}$, $\{c_n\}$ and $\{d_n\}$. For example, by symmetry b_n obviously equals c_n. Also $a_n + b_n + c_n + d_n = k^n$, this being the total number of k-ary sequences of length n. Thus, apparently it is wasteful to introduce c_n and d_n when they could have been expressed in terms of a_n and b_n as

$$c_n = b_n \qquad\qquad (n \geqslant 0) \qquad\qquad (2)$$

$$d_n = k^n - a_n - 2b_n \quad (n \geqslant 0) \qquad\qquad (3)$$

But there is a reason for introducing all four sequences. Let A_n, B_n, C_n, D_n denote respectively the sets of sequences whose cardinalities are a_n, b_n, c_n, d_n. Thus $A_n = \{\bar{x} = (x_1,..., x_n): 0 \leqslant x_i \leqslant k - 1$, for all $i = 1,..., n$ and $x_i = 0$ for an even number of i's and $x_j = 1$ for an even number of j's$\}$. Similarly define B_n, C_n, D_n. Now let $\bar{x} = (x_1,..., x_n) \in A_n$. Let $\bar{y} = (x_1,..., x_{n-1})$ be the truncation of $\bar{x}$ obtained by removing the last digit, x_n. Then $\bar{y}$ belongs to C_{n-1}, B_{n-1} or A_{n-1} according as $x_n = 0$, $x_n = 1$ or $x_n = 2,...,k-1$. Conversely starting from $\bar{y}$ in $C_{n-1} \cup B_{n-1} \cup A_{n-1}$ we can get $\bar{x} \in A_n$ by appending a suitable x_n depending upon whether $\bar{y} \in C_{n-1}$, $\bar{y} \in B_{n-1}$ or $\bar{y} \in A_{n-1}$, in the last case there being $k - 2$ possible choices for x_n. It follows that

$$a_n = c_{n-1} + b_{n-1} + (k - 2)a_{n-1} \quad (n \geqslant 1) \qquad\qquad (4)$$

By a similar reasoning, we get

$$b_n = d_{n-1} + a_{n-1} + (k - 2)b_{n-1} \quad (n \geqslant 1) \qquad\qquad (5)$$

$$c_n = a_{n-1} + d_{n-1} + (k - 2)c_{n-1} \quad (n \geqslant 1) \qquad\qquad (6)$$

and

$$d_n = b_{n-1} + c_{n-1} + (k - 2)d_{n-1} \quad (n \geqslant 1) \qquad\qquad (7)$$

It is *now* time to use (2) and (3) to get from (4) and (5),

$$a_n = (k - 2)a_{n-1} + 2b_{n-1} \qquad (n \geqslant 1) \qquad\qquad (8)$$

and

$$b_n = (k - 4)b_{n-1} + k^{n-1} \qquad (n \geqslant 1) \qquad\qquad (9)$$

[Note that (6) and (7) are not used. This is not surprising since, in presence of (2) and (3), they convey no new information.]].

We could solve this system using generating functions as was done in Problem (3.11). But the resulting expressions for $A(x)$ and $B(x)$ come out rather complicated. Fortunately, in the present case there is another way out. Note that (9) is a linear recurrence relation of order 1 and does not involve a_n. Solving it analogously to (1), with the initial condition $b_0 = 0$, we get

$$b_n = \frac{k^n}{4} - \frac{(k-4)^n}{4} \quad (n \geqslant 0) \tag{10}$$

With this, (8) reduces to

$$a_n - (k-2)a_{n-1} = \frac{k^{n-1}}{2} - \frac{(k-4)^{n-1}}{2} \quad (n \geqslant 1) \tag{11}$$

Solving this, with the initial condition $a_0 = 1$, we finally get

$$a_n = \frac{1}{2}(k-2)^n + \frac{k^n}{4} + \frac{(k-4)^n}{4} \quad (n \geqslant 0) \tag{12}$$

as the answer to the problem. █

Notice that although our interest was in finding a_n, along the way we also found b_n (which answers Exercise (3.2.18)). In fact, we *had to* find b_n. Without it, it would have been difficult to find a_n directly. In other words, it was easier to find both a_n and b_n than finding a_n alone!

As an illustration of the use of recurrence relations with two indices, we count, once more, the number of n-selections of m kinds of objects, with unlimited repetitions allowed. Denote this number by $a_{m,n}$ for $m \geqslant 0$, $n \geqslant 0$. A recurrence relation for $a_{m,n}$ is obtained as follows. Suppose $m \geqslant 1$, $n \geqslant 1$. Pick one kind of objects and classify the selections of n objects into two types, those that contain at least one object of this chosen kind and those that do not. In the first case, the remaining $n - 1$ objects can be chosen from the m kinds without restriction. In the second case, all the n objects must be from the remaining $m - 1$ kinds. Consequently,

$$a_{m,n} = a_{m,n-1} + a_{m-1,n} \quad (m \geqslant 1, n \geqslant 1). \tag{12a}$$

The boundary conditions are $a_{m,0} = 1$, for $m \geqslant 0$ and $a_{0,n} = 0$ for all $n > 0$. Let $F_m(x) = \sum_{n=0}^{\infty} a_{m,n}x^n$, $m = 0, 1, \ldots\ldots$ Then from (12) we get, as usual, $F_m(x) - 1 = xF_m(x) + F_{m-1}(x) - 1$, which gives

$$F_m(x) = \frac{1}{1-x} F_{m-1}(x) \quad (m \geqslant 1) \tag{13}$$

The boundary conditions imply $F_0(x) = 1$. Solving (13) as a recurrence relation for the sequence of functions, $\{F_m(x)\}_{m=0}^{\infty}$, we get

$$F_m(x) = (1-x)^{-m} \quad (m \geqslant 0) \tag{14}$$

By Exercise (1.9), $a_{m,n} = $ coefficient of x^n in $(1-x)^{-m} = \begin{pmatrix} m+n-1 \\ n \end{pmatrix}$.

This is the same as Theorem (2.3.12), except for a different notation.

Recurrence relations can also be used to count d_n, the number of derangements of n symbols, which was counted earlier by the principle of inclusion and exclusion (Theorem (2.4.6)). Recall that a derangement of n objects is an n-permutation without a fixed point. (Unlike in Section 2, here by a permutation, we again mean one without repetitions. Equivalently a permutation of a set S is a bijection of S onto itself.) Denote the objects by $1, 2,.... n$ and let D_n be the set of all derangements of these objects. Each such derangement must take 1 to some i for $2 \leqslant i \leqslant n$. For $i = 2,..., n$, let E_i be the set of those derangements in D_n which map 1 to i.

Evidently $|E_2| = |E_3| =...= |E_n|$ and $D_n = \bigcup_{i=2}^{n} E_i$. Further $E_i \cap E_j = \phi$ for $i \neq j$. So $d_n = (n-1)\,|E_2|$. We now classify the derangements in E_2 into two types, those which take 2 to 1 and those which take 2 to some other object. In the case of a derangement of the first type, the symbols 1 and 2 are interchanged and the remaining $n-2$ symbols are permutated among themselves, without a fixed point. It follows that in E_2 there are d_{n-2} derangements of the first type. As for derangements of the second type, every such derangement is a bijection $f: \{2, 3,..., n\} \rightarrow \{1, 3, 4,..., n\}$ with no fixed points, and in which $f(2) \neq 1$. If we call 1 as 2 then this is the same as a bijection $f: \{2, 3,..., n\} \rightarrow \{2, 3,..., n\}$ with no fixed point. There are d_{n-1} bijections like this. Putting together, $|E_2| = d_{n-2} + d_{n-1}$ and hence, finally, from $d_n = (n-1)\,|E_2|$ we get

$$d_n = (n-1)(d_{n-2} + d_{n-1}) \quad (n \geqslant 2) \tag{15}$$

It is a little complicated to solve this recurrence relation. So we rewrite it as

$$d_n - n\,d_{n-1} = -(d_{n-1} - (n-1)\,d_{n-2}) \quad (n \geqslant 2) \tag{16}$$

or as

$$a_n = -a_{n-1} \quad (n \geqslant 2) \tag{17}$$

where $a_n = d_n - n\,d_{n-1}$. The initial conditions are $d_0 = 1$, $d_1 = 0$ and hence $a_1 = -1$. If we define $a_0 = 1$, (17) is valid for all $n \geqslant 1$. It can be solved by inspection to give $a_n = (-1)^n$ for $n \geqslant 0$. So we get,

$$d_n = nd_{n-1} + (-1)^n \quad (n \geqslant 1) \tag{18}$$

with the initial condition $d_0 = 1$. We already solved this recurrence relation in Problem (3.2) to give

$$d_n = n!\left(\frac{1}{2!} - \frac{1}{3!} + \frac{1}{4!} - \frac{1}{5!} +...+ (-1)^n \frac{1}{n!}\right).$$

Note that in this case the answer has to be left in the summation form.

The only non-trivial, non-linear recurrence relation we solved in the last section by the method of ordinary generating functions was (5) of the last section. It arose in connection with the Vendor Problem. There are many apparently unrelated problems in which essentially the same relation holds. Recall from Chapter 1, Section 2 that the essence of the Vendor Problem was to count a_n, the number balanced arrangements of n pairs of parentheses. It turns out that when suitably interpreted, a balanced arrangement of parentheses corresponds to a certain way of doing something in a wide variety of contexts. Exercises (2.3.10) and (3.4.22) already provide two examples of this. Not surprisingly, then, the numbers given by (6) in the last section appear in the solution to a number of problems. Because of their frequent occurrences, these numbers are given a name. Specifically, the nth **Catalan number** is defined as $\dfrac{1}{n}\dbinom{2n-2}{n-1}$ for $n > 0$ (see Exercise (1.16)) and is often denoted by C_n. Here we present one more problem whose solution involves the Catalan numbers.

4.2 Problem: Find the number of ways to divide a convex polygon with n sides into triangles by non-intersecting diagonals.

Solution: Let b_n denote this number. By inspection we see that $b_3 = 1$, $b_4 = 2$ and $b_5 = 5$, b_0, b_1, b_2 have no natural meaning. A recurrence relation for b_n can be obtained directly as follows: Denote the vertices of the polygon by $A_0, A_1, \ldots, A_{n-1}$ in a particular order. Now, in any triangulation of the polygon, $A_0 A_{n-1}$ will form a side of a unique triangle. Let A_r be the third vertex of this triangle (see Figure 7.6), $1 \leqslant r \leqslant n-2$. For a given r in this range, let us count the number of triangulations in which one of the triangles is $A_0 A_r A_{n-1}$. For $2 \leqslant r \leqslant n-3$, any such triangulation amounts to further triangulating the two subpolygons, $A_0 A_1, \ldots, A_r$ and $A_r A_{r+1}, \ldots, A_{n-2} A_{n-1}$. These polygons have $r+1$ and $n-r$ sides respectively and since

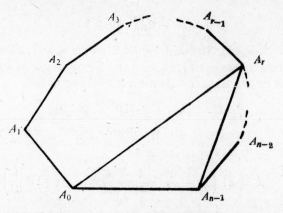

Figure 7.6: **Triangulation of Convex n-gon.**

each can be triangulated independently of the other, it follows that for $2 \leqslant r \leqslant n - 3$, the number of triangulations in which $A_0 A_r A_{n-1}$ is one of the triangles is $b_{r+1} \times b_{n-r}$. For $r = 1$ the first subpolygon is degenerate and the second can be triangulated in b_{n-1} ways. For the sake of uniformity of notation we set $b_2 = 1$ and write b_{n-1} as $b_2 b_{n-1}$. This is also the number of triangulations for the other degenerate case, namely $r = n - 2$. We now have

$$b_n = \sum_{r=1}^{n-2} b_{r+1} b_{n-r} \qquad (n \geqslant 3) \tag{19}$$

We set $b_0 = b_1 = 0$. Then the right hand side of (19) can also be written as $b_0 b_{n+1} + b_1 b_n + b_2 b_{n-1} + \ldots + b_{n-1} b_2 + b_n b_1 + b_{n+1} b_0$, which is precisely the coefficient of x^{n+1} in $[B(x)]^2$ where $B(x)$ is the O.G.F. of $\{b_n\}_{n=0}^{\infty}$. Since $b_0 = b_1 = 0$, the coefficients of x^0, x, x^2 and x^3 in $B(x)B(x)$ are all 0. So, multiplying (19) by x^{n+1} and summing over $n \geqslant 3$ gives

$$x[B(x) - x^2] = B(x)\, B(x) \tag{20}$$

which yields $B(x) = \dfrac{x \pm \sqrt{1 - 4x}}{2}$. $b_1 = 0$ forces the choice of the negative sign for the radical giving $B(x) = \dfrac{x - x\sqrt{1 - 4x}}{2}$. This is the same as $x^2 A(x)$ where $A(x)$ is given by (12) in the last section. Consequently

$$b_n = a_{n-2} = \frac{(2n - 4)!}{(n-1)!\,(n-2)!} = \frac{1}{n-1}\binom{2n-4}{n-2} = C_{n-1}. \quad \blacksquare$$

Instead of working our way through the generating function for $\{b_n\}_{n=0}^{\infty}$, we could have directly related b_n to some other problem. Suppose $A_1, \ldots, A_{n-1}$ are elements of a set on which some non-associative operation $*$ has been defined. Then every triangulation of the polygon corresponds to one possible meaning of the expression $A_1 * A_2 * \ldots * A_{n-1}$. Specifically, a triangulation in which $A_0 A_r A_{n-1}$ is a triangle corresponds to an interpretation of the form $(A_1 * \ldots * A_r) * (A_{r+1} * \ldots * A_{n-1})$ where $1 \leqslant r \leqslant n - 2$. By exercise (3.4.22), the number of all possible interpretations of $A_1 * A_2 * \ldots * A_{n-1}$ is

$$\frac{1}{n-1}\binom{2n-4}{n-2}.$$

In the solution above, b_0, b_1, b_2 had no natural meanings. We set $b_0 = b_1 = 0$ and $b_2 = 1$ so as to simplify our computations. With some other values, we would have ultimately gotten the same answer, but the calculations would have been a little more clumsy. This point should always be kept in mind. In many problems, the first few terms of a sequence are not defined by the conditions of the problem and we are free to choose them. With a judicious choice, the computations become easier.

Counting problems, by far, constitute the vast majority of the kind of problems that are amenable to the method of recurrence relations. Occasionally, however, the recurrence relations can be made to do some other jobs such as proving identities. Suppose the identity to be proved is of the form $a_n = b_n$ where both $\{a_n\}$ and $\{b_n\}$ are sequences. Usually, we can establish the identity by direct verification for lower values of n. Let us suppose that we can identify some recurrence relation satisfied by one of the sides of the identity, say by b_n. If we can then show that the other side, a_n, also satisfies the same recurrence relation, then, in view of the uniqueness of solutions of recurrence relations (Theorem (3.1)) it would follow that $a_n = b_n$ for all n. (An analogous technique is also used in continuous mathematics. If f, g are functions of a real variable x, then in order to show that $f(x) = g(x)$ for all x, it suffices to show that $f(x_0) = g(x_0)$ for some x_0 and $f'(x) = g'(x)$ for all x, or more generally that f and g are solutions of some first order differential equation, satisfying the same initial condition.)

By way of illustration, we prove the identify in Exercise (1.8).

4.3 Proposition: If $n \geqslant 0$, then $\displaystyle\sum_{k=0}^{n} \binom{n+k}{n} \frac{1}{2^k} = 2^n$.

Proof: If $b_n = 2^n$, the right hand satisfies the recurrence relation $b_n = 2b_{n-1}$. Let a_n denote the left hand side. Clearly $a_0 = 1$. Now, for $n \geqslant 1$ we have, by the identity $\dbinom{m}{r} = \dbinom{m}{m-r}$ and Proposition (2.2.19),

$$a_n = \sum_{k=0}^{n} \binom{n+k}{k} \frac{1}{2^k}$$

$$= \sum_{k=0}^{n} \binom{n+k-1}{k} \frac{1}{2^k} + \sum_{k=0}^{n} \binom{n+k-1}{k-1} \frac{1}{2^k}$$

$$= \sum_{k=0}^{n-1} \binom{n-1+k}{k} \frac{1}{2^k} + \binom{2n-1}{n} \frac{1}{2^n} + \frac{1}{2} \sum_{k=1}^{n} \binom{n+k-1}{k-1} \frac{1}{2^{k-1}}$$

$$= a_{n-1} + \frac{1}{2} \left[\binom{2n-1}{n} \frac{1}{2^{n-1}} + \sum_{j=0}^{n-1} \binom{n+j}{j} \frac{1}{2^j} \right]$$

$$= a_{n-1} + \frac{1}{2} \sum_{j=0}^{n} \binom{n+j}{j} \frac{1}{2^j} \qquad \left(\text{since } \binom{2n}{n} = 2 \binom{2n-1}{n} \right)$$

$$= a_{n-1} + \frac{1}{2} a_n.$$

In other words, $a_n = 2a_{n-1}$ for all $n \geqslant 1$. Thus the left hand side satisfies the same recurrence relation as the right hand side. The initial conditions also match because $a_0 = b_0 = 1$. So by Theorem (3.1), $a_n = b_n$ for all n, proving the result. ∎

The argument above is not substantially different from a proof by induction. But there is a subtle difference. If, instead of proving the identity, the problem was to evaluate the sum $\sum_{k=0}^{n} \binom{n+k}{n} \frac{1}{2^k}$, then induction would not have worked unless we were able to guess the answer somehow (which, in the present problem, is not very easy). But the argument above would still allow us to get a recurrence relation for the sum in question and solving it we would have got the answer.

All the applications of recurrence relations considered in this section so far involved problems whose solutions were already known to us by other methods. Let us now turn to new problems where recurrence relations are useful. One very celebrated such problem is popularly called the **Tower of Hanoi problem**. In this problem there are three pegs. On the first one, there is a stack of n discs of different radii (with holes at the centres) arranged in descending order starting from the largest disc at the bottom to the smallest disc at the top (see Figure 7.7). The problem is to move these discs from one peg to another, one at a time and to eventually stack them on another peg in the same order. But the catch is that at no time can a larger disc be placed over a smaller one on the same peg. Obviously, if $n > 1$, the transfer would be impossible without a third peg. For $n = 3$ a solution can be given as follows. Call the pegs A, B, C. Number the discs as 1 to n from the top to the bottom and suppose originally they are stacked on A. For short, denote by (m, X) the movement which consists of moving the mth disc

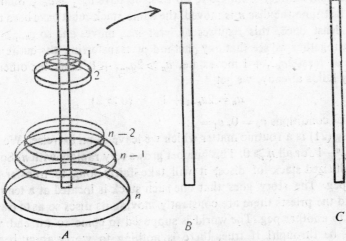

Figure 7·7 : The Tower of Hanoi Problem

(from wherever it is) to the top of the stack on the peg X. Thus $(3, B)$ means 'move the disc 3 to the top of the peg B'. It is easy to see that the sequence of moves $(1, B), (2, C), (1, C), (3, B), (1, A), (2, B), (1, B)$ results in transferring the stack of 3 discs from A to B. Note that this requires 7 moves.

The problem now is to find the minimum number of moves needed to move a stack of n discs from one peg to another. Denote this number by a_n. We have to express a_n as a function of n, or, in other words, find a closed form expression for a_n. By its very nature, the problem is twofold. We have to give some method to effect the transfer in a_n moves and we also have to show that it cannot be done in less than a_n moves. Obviously $a_0 = 0$ and $a_1 = 1$.

A recurrence relation for a_n can be built as follows. Let $n > 1$. Ignore for the moment the disc n at the bottom of the stack on peg A. By our assumption, the stack of the remaining $n - 1$ disc can be moved to the peg C in a_{n-1} moves, without even touching the disc n. Now move n to B(which will be empty at that time). Keep n at the bottom of B and without touching it, move the stack of $n-1$ discs from C to B. This can be done in a_{n-1} moves. So we have shown that the stack of n discs can be moved from one peg to another in $a_{n-1} + 1 + a_{n-1}$, i.e. in $2a_{n-1} + 1$ moves. This proves that $a_n \leqslant 2a_{n-1} + 1$. To show that equality holds, we must prove that *any* method of transferring the stack of n discs would require at least $2a_{n-1} + 1$ moves. In any such method the disc n must move at least once. Let b_{n-1} denote the number of moves (in that method) before the first move of the disc n and let c_{n-1} be the number of moves after the last time disc n is moved. Then the total number of moves in the method is at least $b_{n-1} + 1 + c_{n-1}$. Now, the disc n can only be placed on an empty peg. This means, before it was first moved, the entire stack of $n - 1$ discs must have been moved from one peg to another at least once (without even touching the disc n). Since a_{n-1} is the minimum number of moves to do this, we have $b_{n-1} \geqslant a_{n-1}$. Similarly, after the last time the disc n is moved, the same stack must have been transferred at least once, this requires at least a_{n-1} moves and so $c_{n-1} \geqslant a_{n-1}$. Putting it together we see that any method of transferring the entire stack will need at least $2a_{n-1} + 1$ moves, i.e. $a_n \geqslant 2a_{n-1} + 1$. Since the other way inequality holds already, we get

$$a_n = 2a_{n-1} + 1 \qquad (n \geqslant 2) \qquad (21)$$

with initial conditions $a_0 = 0, a_1 = 1$.

Solving (21) is a routine matter which we leave as an exercise. We then get $a_n = 2^n - 1$ for all $n \geqslant 0$. This number grows very rapidly with n. So even for a small sized stack of discs, it will take fairly long to transfer it to another peg. The story goes that one such stack is located at a temple in Hanoi and the priests there are constantly moving its discs so as to transfer the stack to another peg. The world is supposed to come to an end when they will be through! If true, there is nothing to worry about because assum ng it takes one second to move one disc, a stack with as few as 55 discs would take more than a billion years. And if that is too short, we just have to add one more disc to double the life of the world!

The derivation of the recurrence relation in the Tower of Hanoi Problem was relatively easy because the nature of the problem was such that to solve it for the case of n discs we had to solve the 'sub-problem' for $n-1$ discs and once we did it, the solution to the original problem was very easy to express in terms of the solution to the subproblem. However, things are not always so easy as we saw in Problem (4.1). The following problem is another illustration of this sort of a situation.

4.4 Problem: Find the number of ways to tile a $3 \times n$ rectangle by dominoes (i.e. by rectangle of size 2×1).

Solution: When n is odd, $3 \times n$ is odd and so there is no way to completely cover a $3 \times n$ rectangle with 2×1 rectangles. Let us therefore suppose n is even say $n = 2m$. Let a_m denote the number of ways to cover a $3 \times 2m$ rectangle by dominos. We see by inspection that $a_0 = 1$ and $a_1 = 3$. Now suppose $m > 1$. Let us try to express a_m in terms of a_{m-1}. Every tiling of a $3 \times (2m-2)$ rectangle can be readily extended to a tiling of a $3 \times 2m$ rectangle by covering the remaining 3×2 rectangle in any of the 3 possible ways. This gives $3a_{m-1}$ possible tilings of the $3 \times 2m$ rectangle. But the trouble is that not every tiling of a $3 \times 2m$ rectangle has to be an extension of some tiling of a $3 \times (2m-2)$ rectangle. Let us count those that are not. It is easy to see that every such covering must be either of type A or of type B as shown in Figure 7.8 where squares covered by

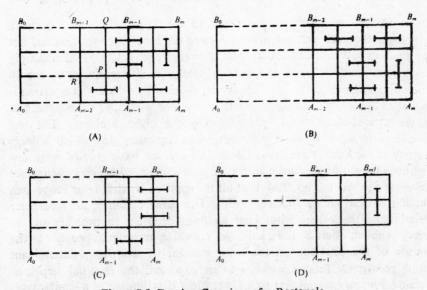

(A)

(B)

(C)

(D)

Figure 7-8: Domino Coverings of a Rectangle

the same domino are joined by a line segment. By symmetry there are as many coverings of type A as there are of type B. So it suffices to count coverings of type A. In Figure 7.8 (A), consider the figure with vertices $A_0, A_{m-2}, R, P, Q, B_{m-2}$ and B_0. It is a $3 \times (2m - 4)$ rectangle to which a 2×1 rectangle has been attached at one end. Denote by b_{m-2} the number of domino coverings of this figure. It is clear that this is also the number of domino coverings of type A. Consequently, we have

$$a_m = 3a_{m-1} + 2b_{m-2} \ (m \geqslant 2) \tag{22}$$

where $b_m =$ number of domino coverings of a $3 \times 2m$ rectangle to which a 2×1 rectangle has been attached. A domino covering of such a figure must end either like (C) or like (D) in Figure 7.8. It is then clear that

$$b_m = b_{m-1} + a_m \ (m \geqslant 1) \tag{23}$$

Substituting for the b's from (22) into (23), we get

$$a_m - 4a_{m-1} + a_{m-2} = 0 \quad (m \geqslant 3) \tag{24}$$

with $a_0 = 1$ and $a_1 = 3$. This is a homogeneous linear relation with characteristic polynomial $x^2 - 4x + 1$. The characteristic roots, therefore, are $2 \pm \sqrt{3}$. By theorem (3.8), the general solution is $a_m = c_1 (2 + \sqrt{3})^m + c_2 (2 - \sqrt{3})^m$ where c_1, c_2 are constants. The initial conditions give $c_1 + c_2 = 1$ and $c_1(2 + \sqrt{3}) + c_2(2 - \sqrt{3}) = 3$. Solving this system gives $c_1 = \dfrac{1 + \sqrt{3}}{2\sqrt{3}}$ and $c_2 = \dfrac{\sqrt{3} - 1}{2\sqrt{3}}$. Thus the number of domino coverings of a $3 \times 2m$ rectangle is

$$a_m = \frac{1}{2\sqrt{3}} \ [(\sqrt{3} + 1)(2 + \sqrt{3})^m + (\sqrt{3} - 1)(2 - \sqrt{3})^m]. \quad \blacksquare \tag{25}$$

It should be clear by now that in the applications of recurrence relations, the real art lies in constructing the recurrence relation and not so much in solving it. In this respect, the recurrence relations differ sharply from their continuous counterparts, namely the differential equations. In the applications of differential equations, writing the differential equation is a matter of recalling the appropriate laws of the particular science (such as physics, economics etc.) which govern the given problem. The real challenge is often in solving the differential equation, for which a large variety of methods is available. Comparatively, we have studied very few methods for solving recurrence relations. Of course, not every recurrence relation can be solved. But if at all the solution is within our scope then finding it is a rather routine matter. The construction of the recurrence relation, on the other hand, changes from problem to problem and so every problem offers a new challenge, requiring techniques peculiar to the nature of the problem. In the next problem we evaluate a determinant using recurrence relations. As is to be expected, the essential step is to relate an $n \times n$ determinant to a lower order determinant of a similar type.

4.5 Problem: Evaluate the determinant of the $n \times n$ matrix A_n which has all 1's on its principal diagonal and the two immediate subdiagonals and 0's everywhere else. (Equivalently $A_n = (a_{ij})$ where $a_{ij} = 1$ if $|i - j| \leqslant 1$ and 0 otherwise.)

Solution: Let $a_n = \det (A_n)$ where A_n is the $n \times n$ matrix,

$$A_n = \begin{bmatrix} 1 & 1 & 0 & 0 & 0 & 0 & ...0 & 0 & 0 & 0 \\ 1 & 1 & 1 & 0 & 0 & 0 & ...0 & 0 & 0 & 0 \\ 0 & 1 & 1 & 1 & 0 & 0 & ...0 & 0 & 0 & 0 \\ \vdots & \vdots & & & & & \vdots & & & \vdots \\ 0 & 0 & 0 & & & & 1 & 1 & 1 & 0 \\ 0 & 0 & 0 & & & & 0 & 1 & 1 & 1 \\ 0 & 0 & 0 & & & & 0 & 0 & 1 & 1 \end{bmatrix} \qquad (27)$$

If we expand $\det (A_n)$ w.r.t. the first row we get

$$a_n = \det (A_n) = \det A_{n-1} - \det (B_{n-1}) \qquad (27)$$

where B_{n-1} is the $(n-1) \times (n-1)$ matrix

$$B_{n-1} = \begin{bmatrix} 1 & 1 & 0 & 0 & 0 & ...0 & 0 & 0 \\ 0 & 1 & 1 & 0 & 0 & ...0 & 0 & 0 \\ 0 & 1 & 1 & 1 & 0 & ...0 & 0 & 0 \\ 0 & 0 & 1 & 1 & 1 & ...0 & 0 & 0 \\ \vdots & & & & & & & \\ 0 & 0 & 0 & 0 & 0 & ...0 & 1 & 1 \end{bmatrix}$$

Expanding $\det (B_{n-1})$ by its first column we see

$$\det (B_{n-1}) = \det (A_{n-2}) \qquad (28)$$

From (27) and (28), we get the recurrence relation

$$a_n - a_{n-1} + a_{n-2} = 0 \quad (n \geqslant 3) \qquad (29)$$

with the initial conditions $a_1 = 1$, $a_2 = 0$. The characteristic roots here are $\frac{1 \pm \sqrt{3}i}{2}$, or in the polar form, $\exp \left(\frac{\pm i\pi}{3} \right)$. By Theorem (3.9), the general solution of (29), is $c_1 \cos \frac{n\pi}{3} + c_2 \sin \frac{n\pi}{3}$ where c_1, c_2 are real constants. The initial conditions imply $c_1 \frac{1}{2} + c_2 \frac{\sqrt{3}}{2} = 1$ and $-c_1 \frac{1}{2} + c_2 \frac{\sqrt{3}}{2} = 0$, giving $c_i = 1$ and $c_2 = \frac{1}{\sqrt{3}}$. So the given determinant equals

$$\cos \frac{n\pi}{3} + \frac{1}{\sqrt{3}} \sin \frac{n\pi}{3}. \quad \blacksquare$$

The rather impressive variety of problems considered so far might lead to the belief that recurrence relations are some kind of a golden tool that can tackle almost any problem which has an integer parameter. Unfortunately this is not the case. Take for example, the problem of determining $p(n)$, the number of partitions of a positive integer n. There is no easy recurrence relation for $p(n)$ and consequently the problem of determining $p(n)$ cannot be tackled by the method of recurrence relations. In fact, as we mentioned earlier, there is no known closed form expression for $p(n)$. Also, in some problems, even if we can write down a recurrence relation, there is no closed form solution to it. This is what happened with (15) which arose in connection with counting the number of derangements of n symbols. Although we could simplify (15) to (18), the solution to the latter had to be left in a summation form. Even for the relatively simple case of a linear recurrence relation, things are not always rosy. Theorem (3.8), along with its extensions, does give a complete answer, at least for the homogeneous case. But those theorems are applicable only after the characteristic roots are found. They do not tell us how to find the characteristic roots. This requires the solution of a polynomial equation. For polynomials of degree 2, there is the well-known 'quardratic formula' for the roots. For polynomials of degree 3 or 4 also there are formulas for roots but they are too complicated to be of much practical use. For polynomials of degree 5 or more, there is no general formula for the roots*. It follows that in general we cannot satisfactorily solve even a linear recurrence relation of order 3. In such cases we may have to be content on a generating function. Finally, there do exist enumeration problems which can be solved quite easily by elementary methods but whose generating functions do not have easy closed form expressions. For such problems, even if a recurrence relation can be written down, it is complicated, if not impossible, to solve it. An example of such a problem will be given in the exercises, e.g. Exercise (4.9) part (iv).

Although there is no recurrence relation for the number of all partitions of an integer n, things may be different if we confine ourselves to partitions of a particular type. For example, Exercise (2.3.35) gives a recurrence relation for the number of triangular partitions of an integer. Solving it, we can count the number of such partitions, or equivalently, the number of mutually incongruent triangles with integer sides having a given perimeter. We leave this determination as an exercise.

To conclude the section, we consider applications of recurrence relations to problems dealing with occurrences of a pattern. Let $\bar{x} = (x_1, x_2,..., x_n)$ be a k-ary sequence of length n. (Usually we shall assume $k = 2$, that is, the case of binary sequences.) Suppose $\bar{a} = (a_1, a_2,..., a_m)$ is a fixed sequence of length m over the same alphabet as $\bar{x}$. If there exists a block of m

---

* See the Epilogue.

consecutive x_i's which equal the a's in the respective order then we say that the **pattern** $\bar{a}$ **occurs** (or **appears**) in x. In other words, the pattern $\bar{a}$ is said to occur in $\bar{x}$ if there exists some r such that $x_{r+i} = a_i$ for all $i = 1,\ldots, m$. (This in particular means $r + m \leqslant n$). For example the pattern 101 occurs in the binary sequences 011011101, 0001101011 and 10010100 but not in the sequence 1110010. Note that in the first two sequences, there are two subsequences each of which matches with 101. But in the case of the first sequence these two subsquences are non-overlapping while in the second one they overlap. In determining the occurrences of a pattern, we always 'scan' the sequence from the left to the right and require distinct occurrences to be non-over-lapping. Each occurrence is denoted by under-lining together the block of consecutive entries which matches with the given pattern. If this block ends with x_j (say), then we say that the given pattern **occurs at** the jth digit. For example, in the sequence 011011101 the pattern 101 occurs twice, namely 011011101. Here the first occurrence is at the 5th digit and the second at the 9th. Note that because of our requirement that distinct occurrence of the pattern be non-overlapping, and because of the convention about scanning from left to right, in the sequence 0001101011 the pattern 101 occurs only once, at the 7th digit (as 0001101011) and not at the 9th digit. This may appear a little strange. But the requirement is designed to meet practical needs. Sometimes a particular pattern is undesirable and a provision is made to erase it as soon as it occurs. A subsequent occurrence of the same pattern must begin fresh. If we let it overlap with an earlier occurrence then the erasing mechanism may be falsely activated.

Because of the requirement that distinct occurrences of a pattern must not overlap, counting problems involving occurrences of patterns require careful handling. Some of the problems done earlier about k-ary sequences of length n can be viewed as problems involving patterns. For example, Problem (4.1) amounts to counting the number of k-ary sequences of length n in which the pattern 0 and the pattern 1 appear an even number of times each. As we saw, this problem can be done in a number of ways including the method of exponential enumerators (Exercise (2.18)). For non-trivial patterns (that is, patterns of length greater than 1), however, the method of exponential enumerators cannot be applied readily because it requires free merging of subsequences and the very idea of a pattern precludes free merging because in a pattern the given sequence of digits must appear consecutively, without anything interspersed. However, the method of recurrence relations goes through as we show.

4.6 Problem: Find the number of binary sequences in which the pattern 101 occurs at the end.

Solution: Denote this number by a_n. For $n \geqslant 3$, there are 2^{n-3} binary sequences of length n which end up with 101. But not all of them will have the pattern 101 at the end. For example, the sequence 0100110101 of length

10 ends with 101. But the pattern 101 already appears at the 8th digit. So the 8th digit cannot be counted in a later occurrence of the pattern. On the other hand, in the sequence 0001010101, the pattern does occur at the end (and also earlier, namely at the $\overline{\text{6th}}$ $\overline{\text{digit}}$). It should now be clear that a sequence of length n ending with 101 will have the pattern 101 appearing at the nth digit or else at the $(n-2)$th digit. As these two possibilities are mutually disjoint, we get the recurrence relation

$$a_n + a_{n-2} = 2^{n-3} \quad (n \geqslant 3) \tag{30}$$

The initial conditions are $a_1 = 0$ and $a_2 = 0$. This is a linear relation with characteristic roots i and $-i$, which, in the complex form are $e^{+i\pi/2}$ and $e^{-i\pi/2}$. By Theorem (3.9), the general solution to the associated homogeneous equation is $a_n = c_1 \cos \dfrac{n\pi}{2} + c_2 \sin \dfrac{n\pi}{2}$. For a particular solution we try $a_n = \lambda 2^n$, since 2 is not a characteristic root. Then from (30), $\lambda(2^n + 2^{n-2}) = 2^{n-3}$, giving $5\lambda = \frac{1}{2}$ or $\lambda = \frac{1}{10}$. So the general solution to (30) is $a_n = c_1 \cos \dfrac{n\pi}{2} + c_2 \sin \dfrac{n\pi}{2} + \dfrac{2^n}{10}$ where c_1, c_2 are real constants. From the initial conditions we get $0 = a_1 = c_2 + \frac{1}{5}$ and $0 = a_2 = -c_1 + \frac{2}{5}$ giving $c_1 = \frac{2}{5}$ and $c_2 = -\frac{1}{5}$. So ultimately, for every $n > 0$, the number of binary sequences of length n having the pattern 101 at the end is

$$\frac{2 \cos \dfrac{n\pi}{2} - \sin \dfrac{n\pi}{2} + 2^{n-1}}{5} . \blacksquare$$

The sequences considered in this problem were allowed to have the pattern 101 occurring several times. Sometimes the occurrence of a particular pattern signifies the termination of some process. In that case obviously it occurs only once. Such situations give arise to problems of the following 'first time occurrence' kind.

4.7   Problem: Find the number of binary sequences of length n in which the pattern 101 appears at the nth digit for the first time.

Solution: This time let b_n be the number of such sequences. Obviously $b_n \leqslant a_n$ where a_n is as in the last problem. We could study b_n independently. But it is instructive to relate it to a_n, which we already know. By definition, $a_n = |A_n|$, where A_n is the set of all sequences of length n in which the pattern 101 appears at the end (and possibly earlier). In any such sequence the first occurrence of the pattern will be at the rth digit for some r, $3 \leqslant r \leqslant n$. If $3 \leqslant r \leqslant n-1$, the number of sequences in A_n in which the first occurrence of the pattern is at the rth digit is evidently $b_r a_{n-r}$, because if we break such a sequence at the rth digit then the first portion can be formed in b_r ways and the second portion is a sequence of length $n-r$ having the pattern 101 at its end. This gives,

$$a_n = \sum_{r=3}^{n-1} b_r a_{n-r} + b_n \quad (n \geqslant 3) \tag{31}$$

We now set $a_0 = 1$. This is not consistent with the formula for a_n in the last problem. But that formula was valid only for $n > 0$. Similarly we set $b_0 = 0$. The advantage of doing this is that we can add dummy terms to the right hand side of (31) and rewrite it as $\sum_{r=0}^{n} a_r b_{n-r}$, which is precisely the coefficient of x^n in the product $A(x) B(x)$ where

$$A(x) = \sum_{n=0}^{\infty} a_n x^n \text{ and } B(x) = \sum_{n=0}^{\infty} b_n x^n.$$

Also with our choices of a_0 and b_0, (31) is valid for all $n > 0$ but not for $n = 0$. So multiplying (31) by x^n and summing over $n \geqslant 1$, we get

$$A(x) - a_0 = A(x) B(x) - a_0 b_0 \tag{32}$$

Since $a_0 = 1$ and $b_0 = 0$, this gives

$$B(x) = 1 - \frac{1}{A(x)} \tag{33}$$

Using

$$a_n = \frac{2 \cos \frac{n\pi}{2} - \sin \frac{n\pi}{2} + 2^{n-1}}{5} \quad (n > 0)$$

or directly from (30), it is not hard to show that

$$A(x) = 1 + \frac{x^3}{1 - 2x + x^2 - 2x^3}.$$

So from (33),

$$B(x) = \frac{x^3}{1 - 2x + x^2 - x^3} \tag{34}$$

If we could resolve $B(x)$ into partial fractions we could expand it and compute b_n as the coefficients of x^n. But this would require us to find the roots of the polynomial $1 - 2x + x^2 - x^3$. It is easy to show that this polynomial has one real and two complex roots. But there is no rational root. So although the roots could be theoretically found by the formula for solving a cubic, it is not practicable to do so. It is best to leave (34) as it is. If we want b_n for a particular value of n we could expand $B(x)$ as

$$x^3 \sum_{r=0}^{\infty} (2x - x^2 + x^3)^r,$$

and take the coefficient of x^{n-3} in

$$\sum_{r=0}^{\infty} x^r (x^2 - x + 2)^r.$$

Although this is an infinite sum, only the terms from $r = 0$ to $r = n - 3$ need be considered because the other terms involve only the higher power of x.

But there is a better way out and it is worth pointing out because of its novelty. So far our approach has been to derive a recurrence relation through a combinatorial argument (or some other consideration peculiar to the problem) and then solve it with generating functions or some other method. For the sequence $\{b_n\}$ we can turn the tables around. From (34) we get

$$B(x) \, (1 - 2x + x^2 - x^3) = x^3 \tag{35}$$

For $n > 3$, the coefficient of x^n in the left hand side is

$$b_n - 2b_{n-1} + b_{n-2} - b_{n-3}$$

while in the right hand side it is 0. So we get

$$b_n = 2b_{n-1} - b_{n-2} + b_{n-3} \quad (n \geqslant 4) \tag{36}$$

with the initial conditions $b_1 = 0$, $b_2 = 0$ and $b_3 = 1$. We can now compute b_n, one be one, giving

$$b_4 = 2, \, b_5 = 3, \, b_6 = 5, \, b_7 = 9, \, b_8 = 16, \, b_9 = 28 \text{ etc,}$$

We could have derived (36) directly. Let B_n be the set of all binary sequences of length n in which the pattern 101 appears at the end for the first time. Then $b_n = |B_n|$, If $\bar{x} = (x_1, x_2, \ldots, x_n)$, denote by $\bar{y}$, $\bar{z}$, the successive truncations of $\bar{x}$ from the left, i.e. $\bar{y} = (x_2, x_3, \ldots, x_n)$ and $\bar{z} = (x_3, x_4, \ldots, x_n)$. Clearly, for $n \geqslant 4$, whenever $\bar{x} \in B_n$, we have $\bar{y} \in B_{n-1}$ and $\bar{z} \in B_{n-2}$. Conversely let $\bar{y} \in B_{n-1}$. If $x_1 = 0$ then $\bar{x} \in B_n$. If $x_1 = 1$, then also $\bar{x} \in B_n$ except when $x_2 = 0$, $x_3 = 1$. So if we let

$$C_n = \{\bar{x} \in B_n : x_1 = 0, \, x_2 = 1\},$$

then we get $b_n = 2b_{n-1} - c_{n-1}$ where $c_n = |C_n|$. To get hold of c_n, note that $\bar{x} \in C_n$ iff $\bar{z} \in B_{n-2} - C_{n-2}$. Thus $c_n = b_{n-2} - c_{n-2}$. We already have $c_n = 2b_n - b_{n+1}$. Substituting for c_n and c_{n-2} in $c_n = b_{n-2} - c_{n-2}$, we get (36).

Despite this, the earlier derivation of (36) from (34) has some desirable features. In some problems, even though the recurrence relation is fairly simple, a direct derivation of it may not be so obvious. In such cases, if we are somehow able to get hold of the generating function, we can convert it to a recurrence relation, the way we get (36) from (34). Note, however, that we cannot solve (36). Solving (36) would amount to solving (31) in the first place. ∎

Although we were unable to get a closed form expression for b_n in this example, (36) gives a reasonable method of calculating it for small values n, which is worthwhile because first time occurrences of patterns are important in many real life problems, where they may signify the termination of an algorithm, the victory in a game etc. To determine the probability of such events we have to know the number of sequences of a given length which end with a first time occurrence of a given pattern. Once we know the generating function $B(x)$ for the number of such sequences, it is easy to get the generating function for the number of sequences in which the pattern occurs exactly twice, exactly thrice etc. (with or without the last occurrence being at the end). Having obtained the generating function, we can write a linear recurrence relation in the same manner as we got (36) from (34). A few problems based on such variations will be given as exercises. One of them will give the answer to the Casino Problem.

Exercises

4.1   For every positive integer n, prove that the number of ways to stack 1 rupee and 2 rupees coins so that the total value of the stack is n rupees is

$$\frac{1}{\sqrt{5}} \left[\left(\frac{1 + \sqrt{5}}{2} \right)^{n+1} - \left(\frac{1 - \sqrt{5}}{2} \right)^{n+1} \right].$$

[Hint: Show that the numbers in question satisfy the same recurrence relation as the Fibonacci numbers but with a different initial condition.]

4.2   Using the least exercise, prove that for every positive integer n,

$$\sum_{k=0}^{\lfloor n/2 \rfloor} \binom{n-k}{k} = \frac{1}{\sqrt{5}} \left[\left(\frac{1 + \sqrt{5}}{2} \right)^{n+1} - \left(\frac{1 - \sqrt{5}}{2} \right)^{n+1} \right].$$

(Essentially, the same sum was evaluated in Exercise (1.18) using generating functions. The present solution is of the same spirit as the proof of Proposition (4.3) because it consists of obtaining a recurrence relation for the sum and solving it.).

4.3   Prove that Fibonacci numbers also arise in finding the number of ways to cover a $2 \times n$ rectangle by dominos.

4.4   Prove that for every positive integer n, the number of binary sequences of length n in which the pattern 11 appears for the first time at the end is

$$\frac{1}{\sqrt{5}} \left[\left(\frac{1 + \sqrt{5}}{2} \right)^{n-1} - \left(\frac{1 - \sqrt{5}}{2} \right)^{n-1} \right].$$

[Hint: Instead of following the method of Problem (4.7), it is much easier to consider the sequence of length $n - 2$ obtained by removing the pattern 11 at the end. This and the last three exercises illustrate the ubiquitous nature of the Fibonacci numbers.]

4.5 Let X be the discrete random variable which denotes the number of times a fair coin has to be tossed before two consecutive heads show up. Prove that the probability generating function of X, $p_X(t)$, is given by

$$p_X(t) = \frac{t}{2\sqrt{5}} \left[\frac{1}{1 - t\left(\dfrac{1 + \sqrt{5}}{4}\right)} - \frac{1}{1 - t\left(\dfrac{1 - \sqrt{5}}{4}\right)} \right].$$

4.6 In a gambling game, the gambler has to pay one rupee for each toss of a fair coin. The game is over and a reward of 5 rupees is given if two consecutive heads show. If you are the gambler, would you play this game? [Hint : Apply Proposition (2.13).]

4.7 Solve the recurrence relation (9) and then (11).

4.8 A restaurant serves three kinds of snacks for tiffin, say A, B, and C costing rupees 1, 2 and 2 respectively. A person has a tiffin allowance of n rupees. If he eats one snack each day till allowance is exhausted, in how many ways can he spend it? (Note that the order of the snacks is important. That is, snack A today and B tomorrow is not the same as B today and A tomorrow.)

4.9 For non-negative integers m, n (with $n \leqslant m$) let $a_{m,n} = \binom{m-n+1}{n}$

and $b_{m,n}$ be the number of ways to select n integers from $\{1, 2,...,m\}$ so that no two consecutive integers are selected. Set $a_{m,n} = 0$ for $m < n$.

(i) Prove that $a_{m,n} = a_{m-2, n-1} + a_{m-1, n}$ $(m \geqslant 2, n \geqslant 1)$

(ii) Prove combinatorially that $b_{m,n} = b_{m-2,n-1} + b_{m-1,n}$

(iii) From (i), (ii) and induction prove that $b_{m,n} = \binom{m-n+1}{n}$.

(The kind of induction that is needed here is called **double induction** because two indices are involved. It proceeds very much like ordinary induction except that in order to prove the truth for some values of m and n, we are allowed to assume the truth for all pairs (i, j) where $i \leqslant m, j \leqslant n$ and at least one inequality is strict.)

(iv) By a direct combinatorial argument show that $b_{m,n} = \binom{m-n+1}{n}$.

[Hint: See Exercise (2.3.5).]

(v) Let $F_m(y)$ be the O.G.F. of the sequence $\{b_{m,n}\}_{n=0}^{\infty}$ for a fixed m. From (ii) show that $\{F_m(y)\}_{m=0}^{\infty}$ satisfies the recurrence relation

$$F_m(y) - F_{m-1}(y) - yF_{m-2}(y) = 0 \quad (m \geqslant 2)$$

with the initial conditions $F_0(y) = 1$ and $F_1(y) = 1 + y$.

(vi) Let $B(x, y) = \sum_{\substack{m=0 \\ n=0}}^{\infty} b_{m,n} x^m y^n$ be the O.G.F. of the doubly infinite

sequence $\{b_{m,n}\}$ (and hence also of $\{a_{m,n}\}$). From (v) show that

$B(x, y) = \dfrac{1 + xy}{1 - x - yx^2}$. (This is the same as Exercise (1.24). But

there we merely verified the answer while now we are arriving

at it through (ii). Expanding $\dfrac{1 + xy}{1 - x - yx^2}$, we can get yet ano-

ther proof of (iv).)

(vii) Using (vi), or directly from (v), show that for all $m \geqslant 0$,

$$\sum_{n=0}^{m} \binom{m-n+1}{n} y^n = \frac{(1+\sqrt{1+4y})^{m+2} - (1-\sqrt{1+4y})^{m+2}}{2^{m+2}\sqrt{1+4y}}$$

(viii) Using (vii), prove that if m, n are positive integers with $m \geqslant n$ then

$$\sum_{k=0}^{\infty} \binom{m+2}{2k+1}\binom{k}{n} = 2^{m-2n+1}\binom{m-n+1}{n}$$

4.10 A stick of n units of length is to be broken into n parts of 1 unit length. In how many ways can this be done if at any stage we simultaneously cut every part of length greater than 1 into two parts? What would be the answer if at every stage only one part can be cut.

4.11 Define f and g by $f(x) = \sum_{n=0}^{\infty} (-1)^n \dfrac{x^{2n+1}}{(2n+1)!}$ and

$$g(x) = \sum_{n=0}^{\infty} (-1)^n \frac{x^{2n}}{(2n)!} \text{ for } x \in \mathbf{R}.$$

Let $h(x) = [f(x)]^2 + [g(x)]^2$. Prove that $f'(x) = g(x)$ and $g'(x) = -f(x)$ for all x. Hence show that $h'(x) = 0$ for all x. Deduce that

$$h(x) = h(0) = 1 \text{ for all } x \in \mathbf{R}.$$

(This is the analytic proof of the well-known trigonometric identity $\sin^2 x + \cos^2 x = 1$. It illustrates the remark preceding Proposition (4.3).)

4·12   Consider the set of all partitions of a set with n elements, partially ordered by the refinement relation. (cf. Example (4) in Chapter 3, Section 3). Find the number of chains of length n in this set.

4.13   Find the number of regions into which a plane is divided by n ellipses in it if every two ellipses intersect at two points and no three of them pass through the same point.

4.14   Prove that for each n, the number of sequences $(a_0, a_1, ..., a_{2n})$ of non-negative integers in which $a_0 = a_{2n} = 0$ and $\mid a_i - a_{i-1} \mid = 1$ for all $i = 1, ..., 2n$ equals the Catalan number $\dfrac{1}{n+1} \dbinom{2n}{n}$. [Hint: Associate a left parenthesis if $a_i - a_{i-1} = 1$ and a right parenthesis if $a_i - a_{i-1} = -1$.]

4.15   Prove that the number of monotonically increasing functions $f:\{1, 2, ..., n\} \to \{1, 2, ..., n\}$ such that $f(x) \geqslant x$ for all $x = 1, ..., n$ equals $\dfrac{1}{n+1} \dbinom{2n}{n}$.

4.16   Prove that the number of monotonically increasing functions $f: \{1, 2, ..., n\} \to \{1, 2, ..., n\}$ such that $f(x) \leqslant x$ for all $x = 1, ..., n$ also equals $\dfrac{1}{n+1} \dbinom{2n}{n}$. [Hint: Consider $g(x) = n+1-f(n+1-x)$.]

4.17   Solve the recurrence relation (21).

*4.18   Design an electrical version of the Tower of Hanoi problem. That is, design a circuit with n switches $x_1, ..., x_n$, n relays $Y_1, Y_2, ..., Y_n$ and a lamp L such that

(i) Y_1 can be operated and released at will by closing or opening the switch x_1 respectively,

(ii)   for $i = 2, ..., n$, Y_i can be operated or released by closing or opening x_i if and only if Y_{i-1} is in the operate state and every Y_j for $j < i - 1$ is in the released state (otherwise operation of x_i has no effect on the state of Y_i) and

(iii)   the lamp L is on only when Y_n is in the operate state and $Y_1, ..., Y_{n-1}$ are in the released state. Find directly how many flips of switches are necessary to light the lamp L. [Hint: Introduce suitable contacts $y_1, y_2, ..., y_n$ on the relays and auxiliary relays. First find the closure functions for their operate and release paths in terms of the Boolean variables $x_1, x_2, ..., x_n, y_1, ..., y_n$. Do the same for the control circuit of the lamp. See Chapter 4 Section 3.]

**4.19   How many moves will be needed in the Tower of Hanoi problem if instead of three pegs, there are four pegs, all other conditions remaining the same?

*4.20 Find the number of ways to cover a $4 \times n$ rectangle by dominos. (These two problems illustrate how even a slight change can considerably increase the complexity of a problem, a typical feature of combinatorics, and more generally, of mathematics.)

4.21 Let r, λ be real numbers with $r \neq \lambda$. Let A_n be the $n \times n$ matrix (a_{ij}) where $a_{ij} = \lambda$ for $i \neq j$ and $a_{ii} = r$ for $1 \leqslant i \leqslant n$. Using recurrence relations prove that $\det (A_n) = (r - \lambda)^{n-1} (\lambda n + r - \lambda)$.

4.22 Generalise Problem (4.5) to the case where in the matrix A_n, $a_{ij} = \alpha$ for $i = j$, $a_{ij} = \beta$ if $|i - j| = 1$ and $a_{ij} = 0$ otherwise, where α, β are some fixed positive real numbers. (Notice that the answer will differ depending upon $\alpha > 2\beta$, $\alpha = 2\beta$ or $\alpha < 2\beta$.)

4.23 Show that the recurrence relation for t_n in Exercise (2.3.35) is equivalent to the linear recurrence relation

$$t_n - t_{n-3} = \frac{n - n(-1)^n - i^{n-1} - (-i)^{n-1}}{8} \quad (n \geqslant 4)$$

4.24 Solve the recurrence relation in the last exercise subject to the initial conditions $t_1 = t_2 = 0$, $t_3 = 1$. Hence determine the number of mutually incongruent triangles of perimeter n whose sides are integers.

4.25 Prove that the number $\dfrac{2 \cos \dfrac{n\pi}{2} - \sin \dfrac{n\pi}{2} + 2^{n-1}}{5}$, obtained in the answer to Problem (4.6) equals the integer closest to the real number $\dfrac{2^{n-1}}{5}$. (This makes it a little easier to calculate it and also gives a rough idea of how rapidly it grows as n increases.)

4.26 Verify that the O.G.F. of the sequence $\{a_n\}_{n=0}^{\infty}$ where $a_0 = 1$ and

$$a_n = \frac{2 \cos \dfrac{n\pi}{2} - \sin \dfrac{n\pi}{2} + 2^{n-1}}{5} \text{ is indeed } 1 + \frac{x^3}{1 - 2x + x^2 - 2x^3} \cdot$$

$\left[\text{Hint: Work with (30). For a direct proof rewrite } \cos \dfrac{n\pi}{2} \text{ and} \right.$

$\left. \sin \dfrac{n\pi}{2} \text{ in terms of powers of } i. \right]$

4.27 Let $\bar{a} = (a_1, a_2, \ldots, a_m)$ be a fixed binary sequence of length m. For each positive integer n, let a_n be the number of those binary sequences of length n which end with the pattern $\bar{a}$ and b_n the number of those which end with the first occurrence of the pattern $\bar{a}$. Set $b_0 = 0$, $a_0 = 1$. For a positive integer k, let $b_{n,k}$ be the number of those binary sequences of length n, in which the pattern $\bar{a}$ occurs exactly k times, the last occurrence being at the end (clearly $b_{n,1} = b_n$), $c_{n,k}$ be the number of binary sequences of length n in which the pattern $\bar{a}$ occurs at least k times (with the last occurrence not necessarily at the end) and $d_{n,k}$ the number of those in which $\bar{a}$ occurs exactly k times (but not necessarily at the end). Let $A(x)$, $B(x)$ be the O.G.F.'s of $\{a_n\}$, $\{b_n\}$ respectively. Prove that:

(i) the O.G.F. of $\{b_{n,k}\}_{n=0}^{\infty}$ is $B^k(x)$ (i.e. $[B(x)]^k$)

(ii) the O.G.F. of $\{c_{n,k}\}_{n=0}^{\infty}$ is $\dfrac{B^k(x)}{1-2x}$

(iii) the O.G.F. of $\{d_{n,k}\}_{n=0}^{\infty}$ is $\dfrac{B^k(x) - B^{k+1}(x)}{1 - 2x}$

(iv) $A(x) = \dfrac{1}{1 - B(x)}$ [Hint: Proceed as in (32) or sum (i) for $k = 1, 2, \dots$.]

4.28 Combining the last exercise with Problems (4.6) and (4.7), obtain linear recurrence relations for the number of binary sequences of length n with one of the following properties:

(i) the pattern 101 occurs at least once in them
(ii) the pattern 101 appears exactly once in them.

4.29 Let b_n be the number of binary sequences of length n in which the pattern 111 appears for the first time at the end. Prove that b_n satisfies the linear recurrence relation

$$b_n = b_{n-1} + b_{n-2} + b_{n-3} \qquad (n \geqslant 4)$$

with $b_1 = b_2 = 0$ and $b_3 = 1$. Count b_n for $n \leqslant 10$.

4.30 In a gambling game, at every round the gambler tosses a fair coin. The game is over when three heads show consecutively (called a win) or at the end of the tenth round whichever is earlier. Prove that the probability of a win is 65/128. [Hint: Even after a win, continue the game with dummy tosses of the coin till 10 rounds are over. Among the 1024 ($= 2^{10}$) equally likely outcomes count those in which there is a win, using the last exercise.]

4.31 Using the last exercise, show that in the Casino Problem, the amount of the reward for a win should be 14 rupees. [Hint: Calculate the total amount of money the machine will get from all possible 2^{10} outcomes corresponding to 10 tosses of the coins keeping in mind that the dummy tosses cost nothing. Note also that as soon as a player is sure to have lost the game (for example if he does not win by the 7th round or earlier and gets a tail on the eighth round), his subsequent tosses must be considered as dummy tosses. Divide this amount equally among all winners.]

Notes and Guide to Literature

The Tower of Hanoi Problem is one of the most well-known problems in mathematics and often features in books on recreational mathematics. The solution to Problem (4.4) is due to Tomescu [1], another problem book, from where Proposition (4.3) is also taken.

The matrix considered in Exercise (4.21) arises naturally in the theory of designs (see the Epilogue).

Epilogue

Perview of 'Applied Discrete Structures'

By now we have (hopefully) taken the reader fairly deep into the spirit of discrete mathematics and have given him a rather thorough acquaintance with its important techniques. It is high time that we apply the tools we have picked. A beginning in this direction was already made in Chapter 4 where we applied Boolean algebras to the problems of designing switching circuits. Also in Chapters 2 and 7 we solved a number of 'real-life' problems (some of them being admittedly somewhat contrived) involving counting techniques.

But there is a lot more to go. In the original plan of this book, the subsequent chapters were devoted to applications of discrete mathematics to a variety of interesting problems. Considerations of space have forced us to carve them out in a separate book titled 'Applied Discrete Structures'. Here we content ourselves by giving a rather detailed perview of that book.

(1) we begin by applications of groups, in a chapter titled 'Group Actions'. Figuratively, group actions are groups put into action! The formal definition is of course different. In essence an action of a group G on a set S is a homomorphism, say θ, from G into P(S), the permutation group of the set S. Quite frequently G itself is a subgroup of P(S) and in that case θ is simply the inclusion function. For x, $y \in S$, we write $x \sim y$ iff there exists some $g \in G$ such that $(\theta(g))(x) = y$. It is easily seen that $\sim$ is an equivalence relation on the set S. The equivalence classes are called the **orbits.** We shall prove theorems due to **Berstein** about the sizes of the orbits and also the number of orbits. (Theorem (5.2.19) will come out as a special case.)

The importance of these theorems stems form the fact that in many counting problems certain apparently distinct objects are to be treated as equivalent to each other and the problem therefore amounts to counting the number of equivalence classes. Such problems can often be reduced to problems of counting the number of orbits under suitable group actions. As a typical example consider the problem of colouring the vertices of a regular pentagon with 3 colours so that 2 vertices are red, 2 are yellow and one is blue. Let $V = \{v_1,\ v_2,\ v_3,\ v_4,\ v_5\}$ be the set of vertices of the pentagon.

Let $R = \{r, y, b\}$ be the set of the three colours. Then a colouring with the given restrictions amounts to a function $f: V \to R$ for which $|f^{-1}(\{r\})| = |f^{-1}(\{y\})| = 2$. Let S be the set of all such functions. We easily calculate $|S|$ as 30. But that is not the answer to the problem. The pentagon being regular, it is impossible to tell apart two colourings if one of them can be obtained from the other by a rotation and/or a flip of the pentagon. Because of this, the problem reduces to counting the number of orbits under the action of G on S where G is the group of isometries of a regular pentagon. (We already identified G in Chapter 5, Section 1.)

For certain applications, we need more information than just the number of orbits. This information can be coded algebraically using polynomials in several variables. This idea is analogous to that behind the enumerators studied in Chapter 7, Section 2 where we saw that an enumerator is nothing but an algebraic coding of a combinatorial problem. The resulting theory, called **Polya's theory of counting,** is a powerful counting technique. (Polya originally developed it for certain enumeration problems in chemistry. But we shall not go into it.)

Finally the theory of group actions can be applied to study groups as well. In Chapter 5, Section 4, we saw that the converse of Lagrange's theorem is not true in general. That is, if G is a group of order n and m is a divisor of n then G need not always contain a subgroup of order m. However, this does hold if m is of the form p^r where p is a prime. If p^r is the highest power of p dividing n, then a subgroup of G of order p^r is called a **Sylow subgroup** of G. We shall prove a number of results about them using suitable group actions. As an application we shall obtain certain invariants which will characterise a finite abelian group upto isomorphism (see the comments following Corollary (5.3.16)).

(2) Next, we move to applications of a much richer algebraic structure, namely fields. In the chapter titled 'Applications of Field Theory' we shall consider two kinds of applications. Some will be purely theoretical while some will be downright practical. The theoretical applications will be of a novel type in that they will be aimed at showing not how something can be done but rather at showing that certain things *cannot* be done. Take the centuries old problem of **trisecting a given angle** using ruler and compass only. In school geometry we learn an elementary construction for bisecting an angle. It is but natural to look for a similar construction for trisecting an angle. People tried this for centuries without success. But, of course, that does not mean it is inherently impossible to trieset an arbitrary angle with ruler and compass only. That it is in fact so can be proved quite rigorously, using only a few simple properties of field extensions.

Another 'impossibility result' we shall prove deals with solving a polynomial equation. In Chapter 7, in the comments following Problem (7.4.5), we remarked that while there are formulas for expressing the roots of a polynomial of degree 4 or less in terms of its coefficients, no such formula is available for polynomials of degree 5 or more. But once again, our own inability to find such a formula does not by itself constitute a proof that no such formula exists. That it is indeed so is a remarkable theorem of **Abel**. But the proof we shall give is even more remarkable. It is based on what is called **Galois theory**, one of the most exciting achievements of human mind. The key idea in it is to assign to a field extension a certain group called its Galois group. The properties of the field extensions are reflected in those of their Galois groups and vice versa. In particular, solving a polynomial equation reduces to the solvability of the corresponding group, as defined in Exercise (5.4.16). The proof ultimately hinges on the result of Exercise (5.4.18) namely that the permutation group S_n is not solvable for $n \geqslant 5$. Incidentally this explains the peculiar name 'solvable group'.

For the 'practical' applications of field theory, we first construct finite fields. Theorem (6.2.26) gives a construction for field extensions starting from an irreducible polynomial over a 'ground' field. (Proposition (6.4.13) gives the same construction in a somewhat 'less abstract' form.) Taking the ground field as $\mathbb{Z}_p$, where p is a prime and applying this construction repeatedly, it is possible to construct a field with p^m elements for any positive integer m, thereby proving a converse to Corollary (6.3.11). The existence of a field structure on a finite set permits certain combinatorial constructions with certain subsets of that set. It can for example be shown that the Religious Conference Problem with n states and n religions has a solution if n is a prime power (cf. Exercise (6.2.33)). Actually a much stronger combinatorial structure, namely a projective plane of order n (see the Notes on Chapter 1, Section 2) exists if n is a prime power. We shall study all this under what is called **design theory**, this peculiar name coming from the design of experiments, where in order to have unbiased samples, it is necessary that the trials of experiments be conducted on batches formed in a way which displays some kind of symmetry. This is an important real-life application, but we shall not go into its details.

We shall, however, consider another very practical application of finite fields, to an area called **coding theory**. When we want to transmit some messages across a communication channel, they first have to be coded as 'strings' or sequences of digits coming from some alphabet S. Such sequences are called codewords. For a given n, the codewords of length n constitute a subset, say C, of S^n (which is the set of all sequences of length n over the alphabet S). The set C is called a code. Because of errors in transmission, the received message may differ from the intended message. The basic problem in coding theory is to design the code C in such a way that even if a certain number of errors occurs, the intended message can still be deciphered unambiguously from the received message. To do this, it is necessary that the Hamming distance (see Theorem (3.1.5) and Exercise (3.1.20)) between every two distinct codewords be sufficiently large. At the same time we do not want the length n to be too large, because this increases the cost of transmission. It is a challenging problem to design a code which meets these conflicting demands. In case the alphabet S can be given a field structure, there is an elegant solution, resulting in what are called **Bose-Choudhuri-Hocquenghem** codes. We shall study these codes and in doing so will use facts about principal ideal domains and matrices.

(3) Having discussed the applications of algebraic structures like groups and fields, we move to a structure which is of a combinatorial nature and is considerably more adaptable. It is called a **graph**. We already mentioned this concept in connection with the Konigsberg Bridge Problem. In the chapter titled 'Graph Theory' we shall formally define graphs and give a solution to this problem. But we shall do a lot more. We shall study the internal structure of a graph. The most crucial concept will be that of a path from a vertex to another. Many real-life problems can be paraphrased in terms of finding such paths in suitably defined graphs. We shall see many instances of this, some of them being solutions to some popular puzzles (typically involving rowing a boat across a river).

Problems of **colouring** the vertices or the edges of a graph subject to certain restrictions are important both theoretically and in applications and will be studied briefly. Another very important problem about graphs is to decide when they are **planar**, i.e. can be drawn in a plane in such a way that the curves representing two distinct edges do not intersect except possibly at the end points. Although we shall not prove a complete characterisation of such graphs, we shall study a number of interesting consequences of planarity.

Trees constitute an especially important class of graphs from the point of view of applications, because they serve to abstract the process of branching, which arises in one form or the other in a variety of contexts. In many problems the solution cannot be obtained by an explicit formula but has to be found out from a set of possible 'candidate solutions' by a search process consisting of performing some tests and eliminating some possibilities depending on the outcome of a test. Such a search can be modelled very conveniently in terms of trees, called **search trees.** Again there are many applications, both of a serious nature and to puzzles (typically involving the detection of a fake coin, given a collection of coins and a balance).

For certain applications, it is necessary to assign certain real numbers (which are called weights, lengths, costs, capacities etc. depending upon the application) to the edges of a graph. The resulting structure is called a **network**. Problems like the Head Office Problem or the Travelling Salesman Problem can be paraphrased as certain minimisation problems in networks and we shall give algorithms for solving them. We shall also discuss the concept of a **flow** in a network and present a well-known algorithm, due to **Ford and Fulkerson,** to obtain the maximum flow.

Although a graph is a purely combinatorial structure, it can be completely represented in terms of certain matrices. Two most important matrices associated with a graph are its **adjacency matrix** and its **incidence matrix**. This association gains strength from the fact that the graph theoretic properties of a graph can often be translated into the algebraic properties (such as eigenvalues or ranks) of the associated matrices. Thereby the machinery of algebra becomes available for studying graphs. We shall prove a few illustrative results of this kind.

(4) A graph being a very flexible structure, it is hardly surprising that graphs can be applied to many problems. Indeed so varied and numerous are these applications that even though we shall devote an entire chapter 'Applications of Graph Theory' to them, we shall barely scratch the surface. We already noted that many combinatorial problems can be reduced to the problems of paths and colouring in suitably defined graphs. We shall present a number of examples of this kind. Among the more profound applications to combinatorial problems, we shall study **matching theory**. Using popular terminology, one of the results we shall prove can be stated as follows: 'Suppose in a town there are n unmarried men and n unmarried women. If each one of them is acquainted with exactly k persons of the opposite sex, then they can all be married to persons of their acquaintance without committing bigamy'. That this can always be done (whenever $k \leqslant n$) is not immediately obvious. But this fact will follow from the Ford-Fulkerson theorem about flows! Of course 'matching' need not always be interpreted as a 'marriage'. Interpreting is suitably, a variety of interesting applications results, one of them being a well-known theorem of **P. Hall** on what are called **'systems of distinct representatives'** and another being **Dilworth's theorem** (See Exercise (3.3.9)).

A graph is conceptually, a very uncomplicated structure. Most of the graph theoretic concepts can be easily related to our everyday experience and some of the theorems are so close to 'common sense' that they are sometimes criticised as lacking in depth. It is remarkable on this background that these very concepts and these 'utterly trivial' results of graph theory, can work wonders when ingeneously applied. For example, certain non-trivial theorems from other branches of mathematics can be proved using some very elementary concepts about graphs. As illustrations of such 'theoretical' applications of graph theory, we shall prove two theorems. One of them is a celebrated theorem of **Nielsen and Schrier**, which asserts, among other things that a subgroup of a free group is free, a result we mentioned without proof in Chapter 5, Section 3. The other result we shall prove using graphs is called **Brouwer's fixed point theorem**. It asserts that if $f: D^n \to D^n$ is a continuous function where D^n is the 'closed unit ball' in the euclidean space $\mathbb{R}^n$, i.e. $D^n = \{(x_1,..., x_n) \in \mathbb{R}^n : x_1^2 + x_2^2 + ... + x_n^2 \leqslant 1\}$, then f has a fixed point, i.e. a point $(a_1,..., a_n) \in D^n$ such that $f(a_1,..., a_n) = (a_1,..., a_n)$. This is one of the most classic theorems of continuous mathematics. To prove it we shall first prove its discrete analogue, popularly called the **Sperner's lemma,** and then apply a limiting process, based on the well-known Bolzano-Weierstrass theorem. The proof, therefore, provides an excellent illustration of our symbolic equation in Chapter 1, Section 1 that continuous mathematics equals discrete mathematics plus the limiting process.

Graphs also have many 'recreational' applications. We already mentioned that many puzzles can be reduced to problems of finding certain paths in suitably constructed graphs. It turns out that certain kinds of games can also be represented in terms of graphs. The problem of finding a winning strategy for such a game reduces to finding a certain set of vertices in its associated graph, which leads to what is called **'game theory'.** We shall take a brief look at it. The techniques are applicable not just to games but in other competitive situations, like planning war strategies, economic policies, etc.

(5) In Chapter 1, Section 2 we remarked that the problems of discrete mathematics are generally finitistic in nature and hence can be done on a computer. The real problem before a programmer is not to devise some algorithm but to devise a good, i.e. an efficient algorithm. We take up this line further in the chapter 'Analysis of Algorithms'. This is a topic which provides some of the most challenging problems of applied discrete structures. The difficulty in fact begins from deciding what constitutes efficiency. Various yardsticks for measuring efficiency will be discussed along with illustrations. An important feature in the analysis of an algorithm is to study what is called its **asymptotic behaviour.** The same algorithm operates on different pieces of data. Let n be an integer variable which measures the 'size' of a given piece of data in some sense. As n increases, so will the time it takes to process the data. Asymptotic behaviour involves the study of how rapid this growth is. Methods from continuous mathematics are needed to estimate the 'order of magnitude' of a function. We shall study one such method and apply it to derive two well-known asymptotic approximations. The first says that for large n, the Harmonic number H_n (see Chapter 7, Section 1) very nearly equals $\ln n + \gamma$ where γ is a fixed constant. The other result is called **Stirling's formula** which asserts that for large n,

$$n! \approx \sqrt{2\pi}\, \frac{n^n \sqrt{n}}{e^n}.$$

The Stirling formula and properties of search trees will be applied in the study of **sorting algorithms.** The sorting problem is to list the elements of a linearly ordered set in an ascending (or descending) order. Numerous algorithms are known for sorting. We shall discuss four of them, **Straight Insertion, Bubble Sort, Merge Sort and Distribution Sort.**

As a crude classification, an algorithm is considered efficient if the time it takes to process a data of size n grows proportionately to some polynomial in n. Such algorithms are called **polynomial time algorithms** (or loosely 'good' algorithms). Problems for which such algorithms are available are called (polynomially) **tractable** and their class is denoted by $\mathscr{P}$. For example the sorting problem is in $\mathscr{P}$. On the other hand it is not known whether the Travelling Salesman Problem is in $\mathscr{P}$ or not. No polynomial time algorithm is known for it; nor is the non-existence of such an algorithm established. This, in fact, is the case with hundreds of problems from highly diverse branches of mathematics. What makes the situation so interesting is that many of these problems are equivalent in some sense, so that the fate of any one of them (namely, knowing whether it is in $\mathscr{P}$ or not) will decide that of all of them. We shall see a few instances of such equivalences.

(6) We frequently come across optimisation problems, i.e. problems where we have to maximise or minimise some real-valued function subject to certain constraints. In notations, we have to maximise (or minimise) a function $f:D \to \mathbf{R}$ where D is a subset of $\mathbf{R}^n$ consisting of all points $(x_1,...,x_n)$ which simultaneously satisfy constraints of the form $g_i(x_1,..., x_n) \leqslant c_i$ for $i = 1,..., k$ (say) where $g_1,..., g_k$ are some real valued functions on $\mathbf{R}^n$ and $c_1,..., c_k$ are some given constants. The simplest case is the one where the functions $f, g_1,..., g_k$ are all linear. Such problems are called **linear programming problems**. The Diet Problem is a typical such problem. In the chapter 'Linear Programming', we shall give a systematic method, called the **simplex method** for solving such problems. We shall also consider briefly variations of linear programming such as integer programming (see the comments on the Cattle Problem).

Although the simplex method works quite well for most linear programming problems, in some pathological cases it is not efficient. It is not a polynomial time algorithm. Recently, a polynomial time algorithm for linear programming has been invented by **Karmarkar.** We shall study it briefly.

References

Ahlfors, Lars
[1] *Complex Analysis*, Second Edition, McGraw-Hill Book Company, New York 1966.

Bender, E.A. and Goldman, J.R.
[2] *On the Application of Mobius Inversion in Combinatorial Analysis*, Amer. Math. Monthly **82** (789–803) 1975.

Birkhoff, Garrett
[1] *Lattice Theory*, AMS Colloquium Publication 1948.

Bishop, D.M.
[1] *Group Theory and Chemistry*, Clarendon Press, Oxford 1973.

Bishop, E.
[1] *Foundations of Constructive Analysis*, McGraw-Hill Book Company, New York 1967.

Bose, R.C. and Shrikhande, S.S.
[1] *On the Falsity of Euler's Conjecture About the Non-existence of Two Orthogonal Latin Squares of Order* $4t + 2$, Proc. Nat. Acad. Sci. USA, **45** (734–737), 1959.

Burnside, W.
[1] *Theory of Groups of Finite Order*, Dover Publications Inc., New York 1955.

Carmichael, Robert
[1] *Introduction to the Theory of Groups of Finite Order*, Dover Publications Inc., New York 1956.

Chandrasekharan, K.
[1] *Introduction to Analytic Number Theory*, Springer Verlag, Berlin 1968.

Chapra, S.C. and Canale, R.P.
[1] *Numerical Methods for Engineers with Personal Computer Applications*, McGraw-Hill, 1985.

Coxeter, H.S.M.
[1] *The Golden Section, Phyllotaxis and Withoff's Game*, Scripta Mathematica, **19** (135–143), 1953.

Crowell, R. and Fox, R.
[1] *Introduction to Knot Theory*, Blaisdell Publishing Co., New York 1963.

Deo, Narasingh
[1] *Graph Theory with Applications to Engineering and Computer Science*, Prentice-Hall of India Pvt. Ltd., New Delhi 1980.

Diwan, G.S. and Agashe, S.S.
[1] *Differential Equations*, Popular Publication, Bombay 1961.

Dornhoff, L.L. and Hohn, F.E.
[1] *Applied Modern Algebra*, The Macmillan Company, New York, 1978.

Gillman L. and Jerison, M.
[1] *Rings of Continuous Functions*, Van Nostrand, Princeton 1960.

Goffman, C.
[1] *Real Functions*, Prindle, Weber and Schmidt, Boston, 1967.

Goldberg, S.
[1] *Introduction to Difference Equations*, John Wiley & Sons, New York 1958.

Gorenstein, Daniel
[1] *Finite Simple Groups: An Introduction to Their Classification*, Plenum Press, New York 1982.

Graham, R.L., Rothschild, H.L. and Spencer, J.H.
[1] *Ramsey Theory*, John Wiley & Sons, New York, 1980.

Greenberg, Marvin
[1] *Note on the Cayley-Hamilton Theorem*, Amer. Math. Monthly, **91** (193–195) 1964.

Hall, G.G.
[1] *Applied Group Theory*, Longmans Publications, London 1967.

Hall, Marshall Jr.
[1] *Combinatorial Theory*, Ginn/Blaisdell, Waltham, Massachusetts 1967.
[2] *Theory of Groups*, The Macmillan Company, New York 1961.

Halmos, P.R.
[1] *Naive Set Theory*, Van Nostrand, Princeton 1960.
[2] *Measure Theory*, Van Nostrand, Princeton, 1950.

Hamersmith, Morton
[1] *Group Theory and Its Applications to Physical Problems*, Addison-Wesley, Reading, Massachusetts 1962.

Harary, F.
[1] *Graph Theory*, Addison-Wesley, Reading, Massachusetts 1969.

Hardy, G.H.
[1] *Course of Pure Mathematics*, 10th Edition Cambridge University Press, 1955.

Herstein, I.
[1] *Topics in Algebra*, Blaisdell Publishing Company, New York, 1964.
[2] *Non-Commutative Rings*, Mathematical Association of America, 1968.

Hoffman, K. and Kunze, R.
[1] *Linear Algebra*, 2nd Edition, Prentice-Hall, Englewood-Cliffs, N.J., 1971.

Hua, L.K.
[1] *Introduction to Number Theory*, Springer-Verlag Berlin, 1982.

Isaacson, E. and Keller, H.B.
[1] *Analysis of Numerical Methods*, John Wiley & Sons, Inc., New York, 1966.

Jacobson, Nathan
[1] *Lectures in Abstract Algebra* (Vols. 1, 2, 3) Van Nostrand, Princeton, N.J., 1951.

Joshi, K.D.
[1] *Introduction to General Topology*, Wiley Eastern Ltd., Publishers, New Delhi, 1983.

Kelley, J.L.
[1] *General Topology*, Van Nostrand, Princeton, N.J., 1955.

Knuth, D.E.
[1] *The Art of Computer Programming*, (Vols. 1, 2, 3), Addison-Wesley Publishing Co. Inc., Reading, Masshchusetts, 1968.

Kolmogorov, A.N.
[1] *Foundations of the Theory of Probability*, 2nd English Edition, Chelsea Publishing Co., New York, 1956.

Kreyszig, E.
[1] *Advanced Engineering Mathematics*, John Wiley & Sons, Inc., 1972.

Krishnamurthy, V.

[1] *Combinatorics, Theory and Applications,* Affiliated East-West Press Pvt. Ltd., New Delhi, 1985.

Kurosh, A.G.

[1] *The Theory of Groups*—2nd Edition, Chelsea Publishing Co., New Delhi, 1960.

Lang, Serge

[1] *Algebra,* Addison-Wesley Publishing Company, Reading, Massachusetts, 1965.

Larsen, M.E.

[1] *Rubik's Revenge: The Group Theoretical Solution,* Amer. Math. Monthly **92** (381–390), 1985.

Larson, Loran C.

[1] *Problem-Solving through Problems,* Springer-Verlag, New York, 1983.

Levy, H. and Lessman, F.

[1] *Finite Difference Equations,* Macmillan, New York, 1961.

Limaye, B.V.

[1] *Functional Analysis,* Wiley Eastern Limited, Publishers, New Delhi, 1981.

Liu, C.L.

[1] *Introduction to Combinatorial Mathematics,* McGraw-Hill Book Company, New York, 1958.

Lovasz, L.

[1] *Combinatorial Problems and Exercises,* North Holland Publishing Company, Amsterdam, 1979.

MacMahon, P.A.

[1] *Combinatory Analysis,* Volumes 1 and 2, Chelsea Publishing Co., New York, 1960.

Mendelson, E.

[1] *Introduction to Mathematical Logic,* Van Nostrand Princeton, 1964.

Meyer, P.L.

[1] *Introductory Probability and Statistical Applications,* Addison-Wesley Publishing Company, Reading, Massachusetts, 1966.

Motzkin, Th.

[1] *The Euclidean Algorithm,* Bull. Amer. Math. Soc. **55** (1142–1146), 1949.

Newmann, D.J.

[1] *Problem Seminar,* Springer-Verlag, New York, 1982.

Niven, I.
[1] *Irrational Numbers*, Mathematical Association of America, Carus Monograph, 1956.

Noble, Ben
[1] *Applied Linear Algebra*, Prentice-Hall, Inc., Englewood-Cliffe, N.J. 1969.

Parzen, Emanuel
[1] *Modern Probability Theory and Its Applications*, John Wiley & Sons Inc., New York, 1960.

Piaggio, H.T.H.
[1] *Elementary Treatise on Differential Equations and Their Applications*, G. Bell and Sons Ltd., London, 1960.

Pontryagin, L.S.
[1] *Topological Groups*, 2nd Edition, Gordon and Breach, New York, 1966.

Riordan, J.
[1] *An Introduction to Combinatorial Analysis*, John Wiley & Sons, Inc., New York, 1958.

Robinson, D.J.S.
[1] *Course in the Theory of Groups*, Springer-Verlag, New York, 1982.

Roth, Leonard
[1] *Old Cambridge Days*, Amer. Math. Monthly, **78** (223–236), 1971.

Royden, H.
[1] *Real Analysis*, The Macmillan Company, New York, 1968.

Rudin, Walter
[1] *Principles of Mathematical Analysis*—Third International Edition, McGraw Hill/Kogakusha, 1976.
[2] *Functional Analysis*, Tata-McGraw Hill, New Delhi, 1974.
[3] *Unique Right Inverses are Two-sided*, Amer. Math. Monthly **92** (489–490), 1985.

Simmons, G.
[1] *Introduction to Topology and Modern Analysis*, McGraw-Hill Book Co., New York, 1963.

Strangio, C.E.
[1] *Digital Electronics*: *Fundamental Concepts and Applications*, Prentice-Hall, Englewood-Cliffs, N.J. 1980.

Thomas, G.B. and Finney, R.L.
[1] *Calculus and Analytic Geometry*, Addison Wesley/Narosa, New Delhi, 1985.

Tjur, Tue
   [1] *Probability Based on Radon Measures,* John Wiley & Sons, Ltd., Chichester, N.Y., 1980.

Tomescu, Ioan
   [1] *Problems in Combinatorics and Graph Theory,* John Wiley & Sons, New York, 1985.

Tremblay, J.P. and Manohar, R.
   [1] *Discrete Mathematical Structures with Applications to Computer Science,* McGraw-Hill Book Company, New York, 1961.

Tucker, Alan
   [1] *Applied Combinatorics*—Second Edition, John Wiley & Sons, New York, 1982.

Vilenkin, N.
   [1] *Combinatorial Mathematics for Recreation,* MIR Publishers, Moscow, 1969.

Whitesitt, J.E.
   [1] *Boolean Algebra and Its Applications,* Addison Wesley, Reading, Masachusetts, 1961.

Zariski, Oscar and Samuel, Pierre
   [1] *Commutative Algebra,* Volumes 1 and 2, Van Nostrand, New York, 1958.

Zassenhaus, Hans
   [1] *The Theory of Groups*—Second Edition, Chelsea Publishing Company, New York, 1958.

Answers to Exercises

These answers are not to be looked at as complete solutions. They are merely meant to help the reader *after* he has honestly tried a problem. If he feels he has a correct solution, he may compare it with the answer here. If the two match, most likely he is right. If they don't, he is advised to check his solution again and/or to discuss it with others. If he still believes the answer given here is incorrect, he is urged to write to the author. In the case of thought-oriented problems, the answers given are often in the form of extended hints. Again, if the reader believes there is a flaw in the answer given here or that he has a more elegant solution, the author would appreciate hearing from him.

In case the reader is unable to do a problem, he is strongly urged to resist the temptation to give up too soon and look up the answer. Keep trying. Do a few special cases of the problem to see if they give any clues. Discuss the problem with others. If nothing comes up, simply leave the problem temporarily. A flash may come after a few days. (In the personal experience of the author, some of the problems took several months.) Look up the solution only when you are convinced it is not worth your time to try further.

No answers are provided in some of the following cases:

(i) Where the problem asks for a straightforward verification of something or for a routine calculation

(ii) where the problem is intentionally vague as it asks the reader to give examples of a general nature or make comments

(iii) where the hint given is either sufficiently detailed or sufficiently incisive to reduce the rest of the work to a routine

(iv) where a reference has been given to the solution in the 'Notes and Guide to Literature'

(v) where, to the author's knowledge, the problem is an unsolved one.

CHAPTER 1

Section 1.1

1. Let the lengths be λ_1, λ_2. If $\lambda_1 = \frac{p}{q} \lambda_2$ then $\lambda_2 = pu$ and $\lambda_2 = qu$

    where $u = \frac{\lambda_2}{q}$ is taken as the unit of length. The converse is clear.

2. This amounts to showing that $\sqrt{2}$ is irrational. If not, let $\sqrt{2} = \frac{p}{q}$

    where p, q are positive integers with no common divisor other than ± 1. Then $p^2 = 2q^2$, whence p^2 and hence p is even, say $p = 2k$. But then $q^2 = 2k^2$ whence q is also even, a contradiction.

3. Depending upon whether n is even or odd there do or do not exist two houses exactly $\frac{1}{2}$ km apart.

4. $2\alpha - \alpha^2$.

5. $\dfrac{2mn + m - m^2}{n(n + 1)}$ where m is the unique integer such that

    $$\frac{m}{n} \leqslant \alpha < \frac{m + 1}{n}.$$

6. $\left(\sum\limits_{i=0}^{10} f_i \right) \Big/ 11$, $\sum\limits_{i=0}^{10} f_i$, jth house where $f_j = \max\limits_{0 \leqslant i \leqslant 10} f_i$

7. $\int\limits_0^1 f(x)dx$, $\int\limits_0^1 f(x)dx$, point in $[0, 1]$ (if any) where f is maximum.

8. Let $f(x, y) =$ population density at (x, y)

    $$= \lim_{\Delta \to 0} \frac{\text{population on } R}{\Delta}$$

    where R is a region containing (x, y) and Δ is its area. The total population now is $\iint\limits_D f(x, y)dA$, the average population density equals

    $$\frac{\iint\limits_D f(x, y)dA}{\text{area of } D}.$$

9. $\frac{2}{5}$. $\left(\text{The distance between two distinct houses can have value } \frac{i}{10}\right.$ with probability $\frac{11-i}{55}$, for $i = 1, 2\cdots, 10.\Big)$

10. $\frac{1}{3}$. (Let α be the distance between two houses. Then α assumes values in the interval $[0, 1]$. For any $r, r + \Delta r$ in $[0, 1]$, the probability that α lies in $[r, r + \Delta r]$ is, by Exercise (1.4) above.

$$2(r + \Delta r) - (r + \Delta r)^2 - 2r + r^2.$$

Dividing by Δr and taking limits as $\Delta r \to 0^+$, the probability distribution of α is $p(r) = 2 - 2r$. The average distance

$$= \int_0^1 r(2 - 2r) \, dr.)$$

11. $\dfrac{10\alpha(2m - 9) + 55 - m(m + 1)}{11}$ where $\dfrac{m}{10} \leqslant \alpha < \dfrac{m+1}{10}$

(m an integer),

$$\left(\sum_{i=0}^{10} f_i \left| \alpha - \frac{i}{10} \right| \right) \Big/ \sum_{i=0}^{10} f_i.$$

12. $\alpha^2 - \alpha + \dfrac{1}{2}$, $\displaystyle\int_0^1 |\alpha - x| f(x)dx \Big/ \int_0^1 f(x)dx$.

13. $\left(1 + \dfrac{1}{10n}\right)^n$.

14. $e^{1/10}$.

Section 1.2

1. Every number except the largest one must appear as the smaller in at least one comparison.

2. Let numbers correspond to players and suppose when two numbers 'play' (i.e. are compared), the larger is the winner.

3. Minimum of x, y is $\frac{1}{2}(x + y - |x - y|)$,

4. Let $u_n = \max \{x_1, \ldots, x_n\}$. Then for $n \geqslant 2$.

$$u_n = \frac{1}{2}(u_{n-1} + x_n + |u_{n-1} - x_n|).$$

5. No, because to find $|x|$ you still have to compare x with 0.

6. Let $a, b, c, \ldots$ denote the states and $1, 2, 3, \ldots$ the religions. Let x_i denote the representative of religion i, from state x. For $n = 3$, a desired arrangement is

$$a_1 \quad b_2 \quad c_3$$

$$b_3 \quad c_1 \quad a_2 \quad .$$

$$c_2 \quad a_3 \quad b_1$$

For $n = 2$, without loss of generality, assume one of the rows of arrangement is (a_1, b_2). Now the first column must be $\begin{pmatrix} a_1 \\ b_2 \end{pmatrix}$. So b_2 appears twice.

7. 4, 5, 7, 8 and 9 are prime powers. For these values of n, there exists a field with n elements (see the comments following corollary (6.3.11)) using which the problem can be solved (cf. Exercise (6.2.33)). The case $n = 6$ was disposed of by exhaustively checking all possibilities by Tarry.

8. Let y be the second largest element. In order to establish that y is not the largest, y must have been 'defeated' sometime in the execution of the algorithm. But the only element which defeats y is the largest element.

9. The third largest element must have been defeated at least twice. Given x, y, z, if we compare x with y and then the smaller of the two with z we know the third largest without knowing the first and the second largest.

10. Determine the largest out of 11 elements in 10 comparisons by the process in the Tournaments Problem. The second largest element must be from among those that have been compared with the largest. There are at most 4 such elements. So with 3 more comparisons, the second largest can be determined.

11. In general the second largest element of n elements can be found with $n - 2 + \lceil lg\ n \rceil$ comparisons where $\lceil lg\ n \rceil$ is the least integer $\geqslant \log_2 n$.

Section 1.3

1. If $n = 2m$, in the first round there are m matches. Out of the m winners, the champion is found by playing $m - 1$ more matches by induction hypothesis. A similar reasoning holds if n is odd.

2. $\frac{1}{2}, \frac{1}{3},$ and $\frac{3}{8}$.

3. Each solution corresponds to an ordered triple (i, j, k) of non-negative integers for which $3i + 2j + k = 20$. k is uniquely determined by i and j. Possible values of i and j are given by the following table:

Value of i	Possible values of j
0	0, 1, 2, 3, 4, 5, 6, 7, 8, 9, 10
1	0, 1, 2, 3, 4, 5, 6, 7, 8
2	0, 1, 2, 3, 4, 5, 6, 7
3	0, 1, 2, 3, 4, 5
4	0, 1, 2, 3, 4
5	0, 1, 2
6	0, 1

4. Out of the 44 solutions above we take only those where each i, j, k lies between 2 and 4. Out of the 9 possible solutions in the table above, the requirement $2 \leqslant k \leqslant 4$ rules out 6 of them. So there are only 3 combinations.

5. If not, let $p_1, ..., p_n$ be the only primes. Let q be a prime dividing $p_1 p_2 \cdots p_n + 1$. Then $q \neq p_i$ for $i = 1, ..., n$ as otherwise p_i would divide 1.

6. If a business has an import license then by (ii) it must employ skilled personnel while by (i) it must not! On the other hand if the business has no import license, then by (iii) it cannot employ local personnel and hence by (ii) it must employ local personnel!

7. Let $s_n = 1 + 2 + ... + (n - 1) + n$
    Then $s_n = n + (n - 1) + ... + 2 + 1$.
    Adding $2s_n = (n + 1) + (n + 1) + ... + (n + 1) + (n + 1)$
    $$[n \text{ times}].$$
    So $s_n = \dfrac{n(n + 1)}{2}$.

8. $\displaystyle\sum_{k=1}^{n} a_k = \sum_{k=1}^{n} a_{k-1} + \sum_{k=1}^{n} k$ gives, after cancellations,
    $$a_n = a_0 + \frac{n(n + 1)}{2} = 1 + \frac{n(n + 1)}{2}.$$

9. (b) Let $c_n = F_{n+1} F_{n-1} - F_n{}^2$. Writing F_{n+1} as $F_n + F_{n-1}$ and $F_n{}^2$ as $(F_{n-1} + F_{n-2}) F_n$ we see that $c_n = - c_{n-1}$ for all $n \geqslant 2$. Since $c_1 = - 1$, the result follows by induction on n.

(c) Let $q = kp$ where k is an integer $\geqslant 2$. Put $n = kp - p$ and $m = p$ in (a).

(d) From (b), $F_n^2 + F_{n+1}^2 = F_{n+2}F_n + F_{n+1}F_{n-1} = F_{2n+1}$ by (a).

10. The desired point P must be within (or on the boundary of) the tringle ABC. For otherwise P is on the 'wrong' side of at least one

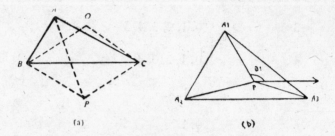

(a)   (b)

line, say, BC. But then its reflection, say, Q, in BC will be a better choice, since $|QB| = |PB|$ and $|QC| = |PC|$ and but $|QA| < |PA|$. Now call A, B, C as A_1, A_2, A_3 respectively. Let $A_i = (x_i, y_i)$. If $P = (x, y)$ is a point in the triangle, we have to minimise $f(x, y) = \sum_{i=1}^{3} [(x - x_i)^2 + (y - y_i)^2]^{1/2}$. Setting the partial derivatives of f equal to 0 gives $\cos \theta_1 + \cos \theta_2 + \cos \theta_3 = 0$ and $\sin \theta_1 + \sin \theta_2 + \sin \theta_3 = 0$ where θ_i is the angle PA_i makes with the x-axis. From this, $1 = \cos^2 \theta_3 + \sin^2 \theta_3 = (\cos \theta_1 + \cos \theta_2)^2 + (\sin \theta_1 + \sin \theta_2)^2$ which ultimately gives $\cos (\theta_2 - \theta_1) = - \frac{1}{2}$ or $\theta_2 - \theta_1 = 120$ degrees. Similarly $\theta_3 - \theta_2 = 120$ degrees. This yields the answer in the text.

12. Let $|BC| = 2x$ and O be the mid-point of BC. Then $|OA| = x$. Let $|OP| = \lambda x$, where $0 < \lambda < 1$. Of the three tours, two have lengths $(1 - \lambda)x + x\sqrt{1 + \lambda^2} + 2x + \sqrt{2}x$ while the third has length

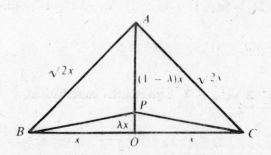

$2x\sqrt{1+\lambda^2}+2\sqrt{2}x$. The three tours are of equal lengths iff $\lambda = \dfrac{9-4\sqrt{2}}{7}$. For $\lambda < \dfrac{9-4\sqrt{2}}{7}$, the third tour is shorter.

13.

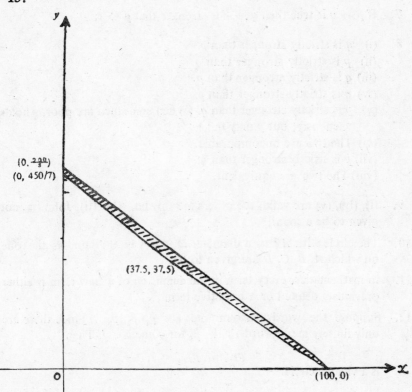

Section 1.4

1. (i) There exists a man for whom there exists no woman who loves him.
   (ii) Gopal is either unintelligent or poor.
   (iii) Gopal is either unintelligent or rich.
   (iv) Gopal is neither intelligent nor rich.
   (v) It rains and the streets do not get wet.
   (vi) There exists $\epsilon > 0$ such that for every $\delta > 0$, there exist x, y in X such that $|x - y| < \delta$ and $|f(x) - f(y)| \geqslant \epsilon$.

2. In general the negation of '$p \to q$' is 'p and $\sim q$', which is quite different from $p \to \sim q$.

4. As a simple example, take the theorem in geometry which says that a quadrilateral is a parallelogram iff its opposite sides are equal.

5. 'If two triangles are congruent, they have equal areas', or 'If a series $\sum_{n=1}^{n} a_n$ is convergent then $a_n \to 0$ as $n \to \infty$.

6. If a triangle ABC is equilateral then $\cos A + \cos B + \cos C = 3/2$.

7. If $p \to q$ is true then $q \to r$ is stronger that $p \to r$.

8.   (i) q is strictly stronger than p.
    (ii) p is strictly stronger than q.
    (iii) q is strictly stronger than p.
    (iv) q is strictly stronger than p.
    (v) p is strictly stronger than q. (When some men are poor, q holds vacuously, but p may fail.)
    (vi) The two are uncomparable.
    (vii) p is strictly stronger than q.
    (viii) The two are equivalent.

9. (i), (iv), (v) are valid, the others are invalid. [In (ii), John is not given to be a man.]

10. The circle with AC as a diameter need not be the same as the one on which A, B, C, D are given to lie.

11. In mathematics, every term in the definition of a new term is either previously defined or a primitive term.

12. Suppose the words encountered are $p_1, p_2, p_3 \ldots$. Since there are only finitely many words, $p_i = p_j$ for some $i < j$. Then

$$p_i, p_{i+1}, \ldots, p_{j-1}$$

is a vicious circle.

CHAPTER 2

Section 2.1

2. See Theorem (4.1.3).

4. (a) $x \in A \Rightarrow x \in A \cup B$. So $A \subset A \cup B$. Similarly $B \subset A \cup B$. So (i) holds. For (ii), let $x \in A \cup B$. Then $x \in A$ or $x \in B$. In either case $x \in C$. So $A \cup B \subset C$.
    (b) $A \cap B$ is the largest subset of X contained in both A and B.

5. If $A_i \subset X_i$ for $i = 1, \ldots, n$ then $A_1 \times A_2 \times \ldots \times A_n \subset X_1 \times \ldots \times X_n$.

6. $(A_1 \times B_1) \cap (A_2 \times B_2) = (A_1 \cap A_2) \times (B_1 \cap B_2)$. $A \times B = (A_1 \times B_1) \cup (A_1 \times B_2) \cup (A_2 \times B_1) \cup (A_2 \times B_2)$.

7. (i), (ii), (iv) (vi) are functions; (ii) is surjective, (iv) is injective. None is bijective.

8. Define $f : (X \times Y) \times Z \to X \times (Y \times Z)$ by $f((x, y), z) = (x, (y, z))$ and $g : X \times Y \to Y \times X$ by $g(x, y) = (y, x)$ for all $x \in X$, $y \in Y$, $z \in Z$.

9. Let $\mathbf{R}^+ = \{x \in \mathbf{R} : x > 0\}$. Let $f : \mathbf{R}^+ \to \mathbf{R}$ be $f(x) = \sqrt{x}$ for $x \in \mathbf{R}^+$ and $g : \mathbf{R} \to \mathbf{R}$ be $g(y) = \sin y$ for $y \in \mathbf{R}$. Then $g \circ f$ is defined but not $f \circ g$.

11. (a) (i) $\Rightarrow$ (ii). We always have $A \subset f^{-1}(f(A))$. Conversely, let $x \in f^{-1}(f(A))$. Then $f(x) \in f(A)$. So there is some $y \in A$ such that $f(x) = f(y)$. Since f is one-to-one, $x = y$. So $x \in A$.
      (ii) $\Rightarrow$ (i). Let $f(x) = f(y)$. Call $\{x\}$ as A. Then $y \in f^{-1}(f(A)) = \{x\}$. So $y = x$. Similarly each statement is proved equivalent to (i). In the implication (i) $\Rightarrow$ (iv), fix some $a \in X$. For $y \in Y$ define $g(y) = x$ if $y = f(x)$ for some $x \in X$ (which is unique by (i)) and $g(y) = a$ otherwise.

   (b) The corresponding statements are:

      (ii) for every $B \subset Y$, $f(f^{-1}(B)) = B$.
      (iii) for every $B_1, B_2 \subset Y$, $B_1 \cap B_2 = f(f^{-1}(B_1) \cap f^{-1}(B_2))$.
      (iv) for any set Z and any two functions $g, h : Y \to Z$, $g \circ f = h \circ f$ implies $g = h$.
      (v) there exists a function $g : Y \to X$ such $f \circ g = id_Y$. (To construct g, for each $y \in Y$ choose some x such that $f(x) = y$. This requires the axiom of choice.)

   (c) Let $X \xrightarrow{f} Y \xrightarrow{g} Z$. Then

      (i) $g \circ f$ is injective iff f is injective and $g/f(X)$ is injective.
      (ii) $g \circ f$ is surjective iff there exists $h : Z \to Y$ such that $g \circ h = id_Y$ and $h(Z) \subset f(X)$.

      (iii) $g \circ f$ is bijective iff f is injective and $g/f(X)$ is bijective.

12. $f(x_0) = x_0 \Rightarrow f(f(x_0)) = f(x_0) = x_0$. Converse is false. Let $X = \mathbf{R} - \{0\}$ and $f : X \to X$ be $f(x) = -x$ for all $x \in X$. Then f has no fixed point. But $f^2 = id_X$ has every $x \in X$ as a fixed point.

13. $f_{A \cap B}(x) = f_A(x) f_B(x)$, $f_{A \cup B}(x) = \max\{f_A(x), f_B(x)\}$, $f_{X-A}(x) = 1 - f_A(x)$ for all $x \in X$. f_ϕ is identically $0 \cdot f_X$ is identically 1 on X.

14. (i) $H \cap R \cap (M - I)$ (ii) $(M - I) \cap (H \cup R)$ (iii) $R \cap (M - H)$ $(= R - H)$ (iv) $(M - R) \cup H$.

15. (i) $f(M)$ (ii) $W - f(M)$ (iii) $R \cap f^{-1}(B)$ (iv) $(W - f(M)) \cup f(I - H - R)$ (v) $H \cap f^{-1}(W - C)$.

16.   See Chapter 4, Section 1.

17.   Once a player loses a match, he never plays again. So the same player cannot be a loser in more than one match. The range is the set of all players except the champion.

18.   (a)   Define $h: X \to X \times \{y\}$ by $h(x) = (x, y)$ for $x \in X$. Then h is a bijection. $f_y = (f/X \times \{y\}) \circ h$. If we identify (x, y) with x (or equivalently treat h as the identity function) then $f_y = f/X \times \{y\}$.

(b) $f(x, y)$ is nothing but $\hat{f}(y)$ evaluated at x.

(c)   Define $\theta: Z^{X \times Y} \to (Z^X)^Y$ by $\theta(f) = \hat{f}$. Then θ is a bijection.

Section 2.2

3.   The pigeonhole principle results by taking each $n_i = 1$.

4.   Consider the two neighbours of 12. At least one of them exceeds 1. Or use the generalized pigeon-hole principle. Think of each adjacent pair as a box. The sum of the contents of the 12 boxes is 156. No two overlapping boxes can have the same sum.

5.   In the clock positions 1 to 12 put, for example, 1, 7, 6, 8, 5, 9, 4, 10, 3, 11, 2, and 12.

6.   Divide the triangle into four mutually congruent subtriangles by lines parallel to sides. At least two points are in the same subtriangle.

7.   Let S be the set of selected integers; $A = \{1, ..., 100\}$ and $B = \{101, ..., 200\}$. The idea is to show that the selection of k integers from A prevents the selection of at least k integers from B, whence $|S|$ cannot exceed 100. Let $A_i = \{x \in A: 2^i x \in B\}$, for $i = 1, ..., 7$. If $x \in A_i$ is in S, it forces $2^i x$ to be out of S. Show that the same $y \in B$ does not get excluded because of two different x's.

8.   Consider the remainder after division by n.

9.   Represent the six persons by six vertices of a hexagon. Colour the edge joining two vertices blue or red according as the corresponding two persons know or do not know each other. The problem is equivalent to showing that there is at least one triangle whose three edges have the same colour. Pick any vertex x. Of the 5 edges meeting at x at least three, say, xa, xb, xc have the same colour. If ab also has this colour, xab is a desired triangle. Similarly for bc and ca. Otherwise abc is a desired triangle.

11. In the pigeonhole principle, $r = 1, k = 2$ and $N = n + 1$. In Exercise (2.9), $n = 2, r = 2, k = 3$ and $N = 6$.

12. k^n.

13. Let $X = \{x_1, \ldots, x_{n-1}, x_n\}$; $Y = \{x_1, \ldots, x_{n-1}\}$. Let $\mathscr{C} = \{A \subset X : x_n \in A\}$ and $\mathscr{D} = \{A \subset X : x_n \notin A\}$. Apply induction hypothesis to $\mathscr{C}, \mathscr{D}$.

14. No, because a binary sequence of length n is nothing but a function from $\{x_1, \ldots, x_n\}$ into $\mathbb{Z}_2$.

15. $_nP_2 = |X \times X - \Delta X|$ where X is any set with n elements and ΔX is the 'diagonal' on X, i.e. $\Delta X = \{(x, x) : x \in X\}$.

16. (i) $\binom{n}{k}$ is the number of k subsets of an n-set.

(ii) Suppose from n objects we choose k and put a white tag on the selected objects. Then out of these k objects we select r and put a black tag on those selected. This is equivalent to first selecting r objects (and putting a white and a black tag on each) and then selecting $k - r$ objects from the remaining $n - r$ putting a white tag on the selected objects.)

(iii) Let $X = Y \cup Z$ where $Y \cap Z = \phi$ and $|Y| = |Z| = n$. If two objects are chosen from X, either both can be from Y or both from Z or one from Y and one from Z.

(iv) Let T be the set of all bijections from $\{1, 2, \ldots, n\}$ to itself. For $i = 1, \ldots, n$, let $T_i = \{f \in T : i$ is the smallest integer such that $f(i) \neq i\}$. Then $|T_i| = (n - i) \times (n - i)!$.

17. (a)

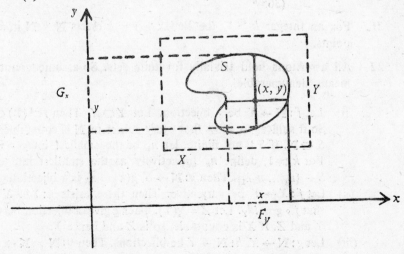

(b) Both sides equal $|S|$.

18. Let X, Y be the sets of boys and girls respectively. Let $S = (x, y) \in X \times Y: x$ and y dance together. In the notation of the last exercise, for each $x \in X$, $|G_x| = 2$. So $\sum\limits_{y \in Y} |F_y| = \sum\limits_{x \in X} |G_x| = 2|X|$, which would be impossible if $|F_y| \leqslant 2$ for all $y \in Y$, since $|Y| < |X|$.

19. (i) Consider the number of arrangements of m objects each of n types and p objects each of q types.

   (ii) Write $\dfrac{(n^2)!}{(n!)^n}$ as $p_1 p_2 \ldots p_n$, where for $i = 1, \ldots, n$,

   $$p_i = \frac{(ni - n + 1)(ni - n + 2) \ldots (ni)}{n!}$$

   Then $p_i = i \, q_i$ where $q_i = \dfrac{(ni - n + 1) \ldots (ni - 1)}{(n-1)!}$

   So $\dfrac{(n^2)!}{(n!)^2} = n! \, q_1 q_2 \ldots q_n$. Each q_i is an integer. For a combinatorial proof, suppose there are n colours and there are n chips of each colour. Count the number of different colour contrast patterns that result by arranging these n^2 chips in a row. Note that two contrast patterns are to be regarded the same if one results from the other by a permutation of the colours.

20. $1 - \dfrac{365 . 364 \ldots (366 - n)}{(365)^n}$.

21. For an integer $k \geqslant 1$, the line $x + y = k$ meets $\mathbb{N} \times \mathbb{N}$ in $k - 1$ points.

22. All assertions hold trivially for finite sets. So assume 'countable' means 'denumerable'.

   (i) Let $f : \mathbb{N} \to X$ be a bijection. Let $Y \subset X$. Then $f^{-1}(Y) \subset \mathbb{N}$. So it suffices to show that every subset of $\mathbb{N}$ is countable. Let $S \subset \mathbb{N}$. If S is not fiinite, let n_1 be the smallest integer in S. For $k > 1$, define n_k inductively as the smallest integer in $S - \{n_1, \ldots, n_{k-1}\}$. Then $g : \mathbb{N} \to S$, $g(k) = n_k$ is a bijection.

   (ii) Let $f : X \to Y$ be surjective. Then there exists $g : Y \to X$ such that $f \circ g = id_Y$. Let $Z = g(Y)$. Then g gives a bijection between Y and Z. If X is countable, so is Z and hence Y.

   (iii) Let $g : \mathbb{N} \to X$, $h : \mathbb{N} \to Y$ be bijections. Then $\theta : \mathbb{N} \times \mathbb{N} \to X \times Y$, defined by $\theta(m, n) = (g(m), h(n))$ is a bijection. Since $\mathbb{N} \times \mathbb{N}$ is countable, so is $X \times Y$.

(iv) Let $g: \mathbb{N} \to X$, $h: \mathbb{N} \to Y$ be bijections. Define $f: \mathbb{N} \times \{1, 2\}$ $\to X \cup Y$ by $f(n, 1) = g(n)$ and $f(n, 2) = h(n)$ for $n \in \mathbb{N}$. Then f is a surjection. Apply (iii) and (ii) to get the countability of $\mathbb{N} \times \{1, 2\}$ and of $X \cup Y$.

(v) Let X be denumberable. For each positive integer r, find an injective function from $P_r(X)$ into X^r. Hence $P_r(X)$ is countable. Let $f_r: P_r(X) \to \mathbb{N}$ be a bijection. Let $F(X) =$ set of all non-empty, finite subsets of X. Define $f: F(X) \to \mathbb{N} \times \mathbb{N}$ by $f(A) = (f_r(A), r)$ where $r = |A|$. Then f is a bijection. So $F(X)$ is countable. By (iv) $F(X) \cup \{\phi\}$ is also countable.

23. $\mathbb{Z} = \mathbb{N} \cup \{0\} \cup (-\mathbb{N})$ where $-\mathbb{N}$ is the set of all negative integers. All three are countable. Similarly write $\mathbb{Q} = \mathbb{Q}^+ \cup \{0\} \cup \mathbb{Q}^-$ where $\mathbb{Q}^+$, $\mathbb{Q}^-$ are sets of positive and negative rationals respectively. Since $f: \mathbb{N} \times \mathbb{N} \to \mathbb{Q}^+$ defined by $f(m, n) = m/n$ is a surjection, $\mathbb{Q}^+$ is countable, which implies $\mathbb{Q}^-$ is also countable.

25. Go on picking one element at a time.

26. Let Z be a denumerable subset of X. Then $Z \cup Y$ is countable. Get a bijection between Z and $Z \cup Y$ and extend it suitably.

27. Let if possible $f: X \to P(X)$ be a bijection. Let $A = \{x \in X : x \notin f(x)\}$. Consider $x^* \in X$ such that $f(x^*) = A$. The contradiction reached is of the same spirit as the Russell's paradox.

28. A bijection need not preserve such attributes as lengths or areas. (As a still more shocking example, it is possible to construct a bijection from a line segment to a solid cube.)

29.

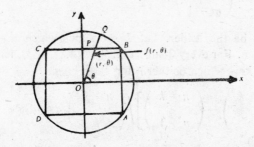

Let S denote the set of points in the square $ABCD$ inscribed on the unit circle as shown in the figure and T denote the set of points in the unit circle. The idea is to define $f: S \to T$ by 'stretching' each segment like OP to the segment OQ. In terms of polar coordinates, (r, θ), f is given explicitly by $f(r, \theta) = (\lambda(\theta)r, \theta)$ where

$$\lambda(\theta) = \left|\begin{array}{l} \sqrt{2}\cos\theta \quad \text{if } \dfrac{-\pi}{4} \leqslant \theta \leqslant \dfrac{\pi}{4} \\[2mm] \sqrt{2}\sin\theta \ \text{if } \dfrac{\pi}{4} \ \leqslant \theta \leqslant \dfrac{3\pi}{4} \\[2mm] -\sqrt{2}\cos\theta \ \text{if } \dfrac{3\pi}{4} \ \leqslant \theta \leqslant \dfrac{5\pi}{4} \\[2mm] -\sqrt{2}\sin\theta \ \text{if } \dfrac{5\pi}{4} \ \leqslant \theta \leqslant \dfrac{7\pi}{4}. \end{array}\right.$$

To get a bijection between any other square and circle compose f with suitable similarity transformations (which are all bijections).

30. Let X be infinite. Let Z be a denumerable subset of X and $f: \mathbb{N} \to Z$ a bijection. Define $g: X \to X$ by $g(x) = x$ if $x \notin Z$ and $g(x) = f(f^{-1}(x) + 1)$ if $x \in Z$. Then g is a bijection from X to $X - \{f(1)\}$. For the converse, let x_1 be any element in $X - Y$. For $n > 1$, define x_n inductively as $f(x_{n-1})$. Then the elements $x_1, x_2, \ldots$ are all distinct.

Section 2.3

1. There are $\dbinom{n}{m-1}$ locks, each lock has $n - m + 1$ keys, each person can open $\dbinom{n-1}{m-1}$ locks.

2. Let P_1 be the leader. Let L_1 be a lock which any of p_2, p_3, p_4, p_5 can open. For every 2-subset $\{i, j\}$ of $\{2, 3, 4, 5\}$, let L_{ij} be a lock with 3 keys which neither i nor j can open.

3. $\left[\dbinom{n+k}{n} - \dbinom{n+k}{n+1}\right] \Big/ \dbinom{n+k}{n}$. (Andre's method goes through.)

5. $\dbinom{n-r+1}{r}$. (Think of a selected integer as a woman and a rejected integer as a man and modify Problem (3.2).)

6. When x_1 is at an end, x_2 has $n - 2$ possible places. When x_1 is not at an end x_2 has $n - 3$ possible places. So x_1 and x_2 can be seated in $2(n-2)+(n-2)(n-3)$ ways in all. For each such seating, the remaining guests can be seated in $(n-2)!$ ways.

7.

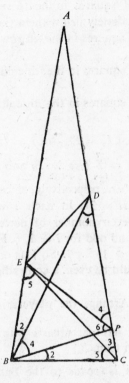

(For angles, x denotes $10\,x$ degrees).

Join *EP*. The triangles *BPC*, *BPD* and *BEC* are isosceles. So $DP = BP = BC = BE$ Since $\angle EBP = 60$ degrees, triangle *EBP* is equilateral. So $\angle EPD = (100-60) = 40$ degrees and $EP = DP$ whence $\angle EDP = \frac{1}{2}(180-40) = 70$ degrees. Since $\angle BDP = 40$ degrees, the result follows.

8. $\dbinom{m}{2}\dbinom{n}{2}$ rectangles; $\dbinom{m+n-2}{m-1}$ paths. (In all $m+n-2$ inter-road distances must be covered of which $m-1$ are North-South and $n-1$ are East-West. Each path corresponds to a binary sequence with $m-1$ 0's and $n-1$ 1's).

9. Without loss of generality let the removed corners be white. In the 62 remaining squares, 32 are black and 30 are white. In any adjacent pair there is one black and one white square. So the desired pairing is impossible.

For a solution without colouring, suppose the 62 squares can be decomposed into 31 pairs of adjacent squares. Let us call the two squares in each pair as 'mates' of each other. Now, for $i = 1, ..., 8$ let

$T_i =$ the number of squares in the ith row (from the top) whose mates lie immediately below them (i.e. in the $i + 1$th row).

$B_i =$ the number of squares in the ith row whose mates are above them.

$R_i =$ the number of squares in the ith column whose mates are to their left.

$L_i =$ the number of squares in the ith column whose mates are to their right.

Clearly $T_i = B_{i+1}$ and $L_i = R_{i+1}$ for $i = 1, ..., 7$. Also $B_1 = T_8 = R_1 = L_8 = 0$. Let $V = \sum_{i=1}^{7} T_i = \sum_{i=2}^{8} B_i$ and $H = \sum_{i=1}^{7} R_i = \sum_{i=1}^{8} L$. Then V and H are the numbers, respectively, of 'vertical' and 'horizontal' pairs, so that $V + H = 31$. In each row, the horizontal pairs (if any) occupy an even number of places. So we get $T_i + B_i$ is even for $2 \leqslant i \leqslant 7$ and odd for $i = 1, 8$, From this it follows that all T_i's are odd for $i = 1, ..., 7$. So V is odd. Similarly H is odd. But then $V + H$ would be even, a contradiction.

10. $\dbinom{20}{10} - \dbinom{20}{11}$. (Arrange the players in descending order of heights. Associate a left parenthesis with those in team A and a right one for those in team B.)

11. At a junction which is i roads to the East and $j - i$ roads to the north of the starting point, there will be $2^{k-j} \dbinom{j}{i}$ persons, $0 \leqslant i \leqslant j$. (Because of the binomial coefficients, such a distribution of persons is called a binomial distribution.)

12. 0 if $t + k$ is odd; $\dbinom{t}{\dfrac{t+k}{2}} \Big/ 2^t$ if $t + k$ is even.

13. $\displaystyle\sum_{k=\lfloor t/2 \rfloor - 5}^{k=\lfloor t/2 \rfloor + 5} \dbinom{t}{k} \Big/ 2^t$. (Apply Exercise (3.4) with $n + k = t$ and $q = 10$).

14. 2^6.

15. The answer for a convex $n -$ gon is $2 \dbinom{n}{4} + \dfrac{n(n-3)}{2}$. (There

are $\dfrac{n(n-3)}{2}$ diagonals and $\begin{pmatrix} n \\ 4 \end{pmatrix}$ points of intersections. If k of them lie on a diagonal, it is cut into $k+1$ segments.)

16. $m(m-1)$ if $n=2m$ and m^2 if $n=2m+1$. (Consider the middle term of the subprogression.)

17. $\begin{pmatrix} n \\ 4 \end{pmatrix} + \begin{pmatrix} n-1 \\ 2 \end{pmatrix}$. (Let r_n be the number of regions. Get a

recurrence relation $r_n = r_{n-1} + 1 + \begin{pmatrix} n-1 \\ 3 \end{pmatrix} + n - 3$ and solve

it by summing, using formulas for $\sum\limits_{k=1}^{n} k$, $\sum\limits_{k=1}^{n} k^2$ and $\sum\limits_{k=1}^{n} k^3$. A some-

what easier method for summing $\sum\limits_{k=1}^{n} \begin{pmatrix} k \\ 3 \end{pmatrix}$ will be given later, in

problem (7.1.5). A much slicker argument to do the present exercise is to suppose that the diagonals are drawn one by one in some fixed but arbitrary order. Every time a new diagonal is drawn, the number of regions goes up by $k+1$ where k is the number of previous diagonals which interest the new diagonal.)

18. The first sum equals the number binary sequences of length n in which 0 occurs an even number of times.

19. k^m if $n=2m$; k^{m+1} if $n=2m+1$.

20. $k(k-1)^n$.

21. 0 if $n \leqslant 2$, $2 \cdot 3^{n-3}$ if $3 \leqslant n \leqslant 5$, and $2 \cdot 3^{n-3} - 3^{n-6}$ if $n \geqslant 6$.

23. $3 \cdot 4 \cdot 5 \cdot \ldots \cdot 101 \cdot 102$.

24. $\begin{pmatrix} n \\ 2 \end{pmatrix}$.

25. Both identities can be proved by induction. However, combinatorial proofs are more interesting. Let $X = \{x_1, \ldots, x_n, x_{n+1}\}$ be a set with $n+1$ elements. Let B be the box in which x_{n+1} is put. Let k be the number of elements in the remaining m boxes. Then $k \geqslant m$. These k elements can be chosen in $\begin{pmatrix} n \\ k \end{pmatrix}$ ways and distributed in $S_{k,m}$ ways into m boxes. For the second formula, suppose the objects are put into boxes one by one, in the order $x_1, x_2, \ldots, x_{n+1}$. Let x_{k+1} be the last object to fall into a box which is empty previously. Then

DISCRETE MATHEMATICS

the first k objects occupy the remaining m non-distinct boxes and the last $n - k$ objects can be put into any of the $m + 1$ boxes (which are now distinguished by their contents.)

26. $\binom{6}{3} \binom{7}{3} \binom{5}{3}$.

27. $\binom{103}{2} - 3 \binom{52}{2}$.

28. First put one object in each box. Now put the remaining $r - n$ objects without any restriction.

29. $45 \left(= \binom{12}{2} - \binom{7}{2} \right)$.

30. Label the types $1, 2, ..., n$. A selection in which k_i objects of type i are selected for $i = 1, ... n$ can be represented by a sequence whose first k_1 terms are 1, next k_2 terms are 2, ... and so on.

32. Suppose $a_1, a_2, ..., a_m$ is a partition of n with $a_1 \geqslant ... \geqslant a_m$. If $a_m > 1$ then $a_1 - 1, a_2 - 1, ..., a_m - 1$ is a partition of $n - m$. If $a_m = 1$, then $a_1, ..., a_{m-1}$ is a partition of $n - 1$. For the second assertion, given a partition of $2n$ into n parts, remove 1 from each part to get a partition of n (into possibly less than n parts).

34. Given a partition of X into m parts, if under the reshuffle of the indices, the x's in each part get permuted among themselves, it is the same partition of X. So, it is not true that $n!$ distinct partitions of X give rise to the same partition of n.

35. A triangular partition of n correspond uniquely to an ordered triple of integers (a_1, a_2, a_3) in which $a_1 \geqslant a_2 \geqslant a_3 \geqslant 1$, $a_1 + a_2 + a_3 = n$ and $a_2 + a_3 > a_1$. If (a_1, a_2, a_3) is a triangular partition of n, then $(a_1 - 1, a_2 - 1, a_3 - 1)$ is a triangular partition of $n - 3$ except when $a_3 = 1$ or when $a_2 + a_3 = a_1 + 1$. In either case n is odd and so of the form $4k \pm 1$. In each case show that the number of exceptions comes to k.

36. Given a self-dual partition of n, consider a new partition of n whose parts are the 'L-shaped' subsets of the Ferrer's graph consisting of the first row and the first column, the second row and the second column and so on.

37. $2^9/10!$. $\Big($ Suppose the rth person leaves last, $1 \leqslant r \leqslant 10$. Then for $i = 1,..., r - 1$, i must leave before $i + 1$ and for $j = r,...,9$, j must leave after $j + 1$. So the number of favourable cases in which the rth person leaves last is $\dbinom{9}{r-1}$. $\Big)$

Section 2.4

1. $18 (= 81 - 4 \times 27 + 6 \times 9 - 4 \times 3 + 3)$.

2. $12 (= 27 - 2 \times 9 + 3)$.

3. $\dfrac{1}{2!} - \dfrac{1}{3!} + \dfrac{1}{4!} - ... + (-1)^n \dfrac{1}{n!}$.

4. $8! - 4 \times 7! + 6 \times 6! - 4 \times 5! + 4!$

5. $\displaystyle\sum_{j=0}^{n-1} (-1)^j \binom{n-1}{j} (n-j)!$

6. The order of the products on the two machines determines a pair (f, g) of permutations of $(1, 2,..., n\}$ such that $f(i) \neq g(i)$ for each $i = 1,..., n$. This condition is equivalent to $f^{-1} \circ g$ being a derangement.

7. This reduces to counting ordered triple (f, g, h) of permutations of $1,..., n$ for which $f^{-1} \circ g$, $f^{-1} \circ h$ and $g^{-1} \circ h$ are all derangements. This comes out to be $n! \, E_n$ where $E_n = |\{(\theta, \psi): \theta, \psi$ and $\theta \circ \psi$ are all derangements of $\{1,...,n)\}|$. But there seems no easy way to compute E_n.

8. $(2n)! - n(2n-1)! + \dbinom{n}{2} (2n-2)! + ... + (-1)^n n!$. (For each colour, consider two objects, one a triplet and the other a duplex of balls of that colour. Arrangements in which these two objects are together get counted again.) With 4 balls of each colour the answer is simply $\dfrac{(2n)!}{2^n}$.

9. $(12)! - \dbinom{6}{1} 12 \cdot (10)! + \dbinom{6}{2} 12 \cdot 10 \cdot 8! - \dbinom{6}{3} 12 \cdot 10 \cdot 8 \cdot 6! +$

$\qquad + \dbinom{6}{4} 12 \cdot 10 \cdot 8 \cdot 6 \cdot 4! - \dbinom{6}{5} 12 \cdot 10 \cdot 8 \cdot 6 \cdot 2 \cdot 2! + 12 \cdot 10 \cdot 8 \cdot 6 \cdot 4 \cdot 2.$

10. 40 and 15. For sharpness see, for example, the Venn diagrams below where the figures represent the cardinalities of appropriate subsets of the set of all casualities whose cardinality is assumed to be 100.

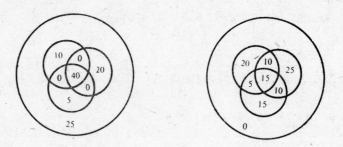

11. $\displaystyle\sum_{k=0}^{5} (-1)^k \dbinom{5}{k} \frac{(20-k)!}{3!} \frac{10!}{(10-k)!}.$

12. A solution using the principle of inclusion and exclusion is clumsy. Instead, first serve the three vegetarians in $\dbinom{10}{2} \dbinom{8}{2} \dbinom{6}{2}$ ways. For each such serving, the remaining 14 packets can be distributed in $\dfrac{(14)!}{2^7}$ ways.

13. $\displaystyle\sum_{j=0}^{k-m} (-1)^j \dbinom{m+j}{m} \dbinom{k}{m+j} = \sum_{j=0}^{k-m} (-1)^j \dbinom{k}{m} \dbinom{k-m}{j}$

$\qquad\qquad\qquad\qquad = \dbinom{k}{m} \displaystyle\sum_{j=0}^{k-m} (-1)^j \dbinom{k-m}{j}$

$\qquad\qquad\qquad\qquad = 0$ by Exercise (3.18).

15. (i) 80, 80 and 408.

(ii) x is relatively prime to p^r iff p does not divide x. There are p^{r-1} positive integers $\leqslant p^r$ which are divisible by p.

(iii) For $i = 1,..., k$ let $S_i = \{x : 1 \leqslant x \leqslant n,\ p_i$ divides $x\}$. Then $\phi(n)$ is simply $\bigcap_{i=1}^{k} S_i'$ which equals $\sum_{j=0} (-1)^j s_j$ in the usual notation. Here s_j is the sum of the cardinalities of sets of the form $S_{i_1} \cap ... \cap S_{i_j}$ for $1 \leqslant i_1 < i_2 < ... < i_j \leqslant k$. It is easily seen than $S_{i_1} \cap ... \cap S_{i_j}$ consists of multiples of $p_{i_1} p_{i_2} ... p_{i_j}$. Consequently $|S_{i_1} \cap ... \cap S_{i_j}| = p_1^{m_1} p_2^{m_2} ... p_k^{m_k}$ where $m_u = r_u - 1$ if $u = i_1,..., i_j$ and $m_u = r_u$ otherwise. But these are also precisely the terms in the expansion of $(p_1^{r_1} - p_1^{r_1-1})...(p_k^{r_k} - p_k^{r_k-1})$.

(iv) Since m, n are relatively prime, the primes appearing in their prime factorisations are all different. The result follows from (iii).

CHAPTER 3

Section 3.1

2. For (b).

3. The simplest example is a graph consisting of just two vertices and an edge joining them.

4. C_1 and C_3 are in the same class. C_2 is not.

5. None. Simple counter-examples exist to show that equality of lengths is neither necessary nor sufficient for C_1 and C_2 to be in the same class. It is also possible to construct C_1, C_2, C_3 such that the area between C_1 and C_2 = the area between C_1 and C_3 but C_1, C_2 are in the same class while C_3 is not. As for curvature, consider mirror images.

6. Let z_0 be a point in the pond. Then it can be shown that C_1 and C_2 are in the same class iff

$$\int_{C_1} \frac{dz}{z - z_0} = \int_{C_2} \frac{dz}{z - z_0}.$$

8. Positivity and symmetry are immediate from the definition. The triangle inequality amounts to proving that for any $x_1, y_1, x_2, y_2, x_3, y_3 \in \mathbf{R}$,

$$(a + b)^2 + (c + d)^2 \leqslant a^2 + b^2 + c^2 + d^2 + 2 \sqrt{a^2 + b^2} \sqrt{c^2 + d^2}$$

where $a = x_1 - x_2,\ b = x_2 - x_3,\ c = y_1 - y_2,\ d = y_2 - y_3$. This further reduces to proving $(ab + cd)^2 \leqslant (a^2 + b^2)(c^3 + d^2)$ which follows from the fact that $0 \leqslant (ac - bd)^2$. For the mid-point property, set $M = \left(\dfrac{x_1 + x_2}{2}, \dfrac{y_1 + y_2}{2} \right)$. For uniqueness of M, shift the origin to $\left(\dfrac{x_1 + x_2}{2}, \dfrac{y_1 + y_2}{2} \right)$. Then P, Q take the form $P = (h, k)$, $Q = (-h, -k)$. If $M = (x, y)$ satisfies $d(M, P) = d(M, Q) = \frac{1}{2}d \ (P, Q)$ then $(x - h)^2 + (y - k)^2 = (x + h)^2 + (y + k)^2 = h^2 + k^2$ which gives $xh + yk = 0$ and $x^2 + y^2 - 2xh - 2yk = 0$ implying $x = y = 0$.

9. For d_1 the balls are squares with sides having slopes ± 1; for d_2 they are squares with sides parallel to the axes. In both cases the points on the sides are included in the closed balls but not in the corresponding open balls.

10. The side of each square should equal the minimum of $d_2(P_i, P_j)$, $i \neq j$ where d_2 is as in the last exercise, with the axes going east-west and north-south.

11. $\min \{ d_1(P_i, P_j) : i \neq j \}$ where d_1 is as above.
The diagonal of each square should equal.

12. If, for example, all P_i's are on the x-axis, then the vertical side of the rectangles can be any positive number.

14. Let the distinct elements of X be $x_1, \ldots, x_n$. Then
$$\lambda = \min \{ d(x_i, x_j) : i \neq j, 1 \leqslant i \leqslant n, 1 \leqslant j \leqslant n \}.$$

15. The arguments are analogous to those in Exercise (1.8) and (1.9). The name Pythagorean comes from the well-known Pythagorean theorem. If d_1 (and also d_2) is the usual metric for $\mathbf{R}$, then by Pythagorean theorem, the usual metric on $\mathbf{R}^2$ is precisely that given by (i).

16. In $\mathbf{Z}_2$ let $d_1(x, y) = 0$ if $x = y$ and 1 if $x \neq y$. Then d_1 is the discrete metric on $\mathbf{Z}_2$. If $\bar{x}$, $\bar{y}$ are two binary sequences of length n, then $\sum\limits_{i=1}^{n} d_1(x_i, y_i)$ is simply the number of i's for which $x_i \neq y_i$.

17. Let, if possible, $\bar{x}, \bar{y}, \bar{z}$ be three such sequences. Let $\bar{w}$ be the 'complementary' sequence of $\bar{z}$, i.e. $\bar{w} = (w_1 \ldots, w_{10})$ where for $i = 1, \ldots, 10, w_i = 1 - z_i$. It is easy to check that for any $\bar{u}, d(\bar{u}, \bar{z}) + d(\bar{u}, \bar{w}) = 10$. So we have $d(\bar{x}, \bar{w}) = 10 - d(\bar{x}, \bar{z}) \leqslant 3$ and similarly $d(\bar{y}, \bar{w}) \leqslant 3$. But then $d(\bar{x}, \bar{y}) \leqslant 3 + 3 = 6$, a contradiction.

18. Let X be the set of all binary sequences of length n. Start with any $\bar{x} \in X$. If $C(\bar{x}, t - 1) \subsetneq X$, pick $\bar{y} \in X - C(\bar{x}, t - 1)$. If
$$C(\bar{x}, t - 1) \cup C(\bar{y}, t - 1)$$

does not exhaust X, pick $\bar{z}$ outside it. Continuing, we are sure to get at least M distinct points, every two of which are, by very construction, at least t apart.

19. The Gilbert bound here is 2; but we can easily find at least 4 elements, for example, 0000000000, 1111110000, 1110000111 and 0001110111.

20. In Theorem (1.7), the conclusion would be

$$m \leqslant \left\lfloor \frac{k^n}{\displaystyle\sum_{j=0}^{r} \binom{n}{j} (j-1)^{n-j}} \right\rfloor$$

21. Let (A, f), (B, g) be multisets. Then their union is $(A \cup B, h)$ while their intersection is $(A \cap B, k)$ where the functions h, k are defined by

$$h(x) = \begin{cases} f(x) & \text{if } x \in A, x \notin B \\ g(x) & \text{if } x \in B, x \notin A \\ \max \{f(x), g(x)\} & \text{if } x \in A \cup B \end{cases}$$

and $k(x) = \min \{f(x), g(x)\}$ for all $x \in A \cap B$.

22. As the simplest counter example, let $X = \{x\}$ and $g(x) = 2$. If $A = X$ and $f(x) = 1$, then $B = A$ and $h = f$. Thus (A, f) equals its own complement.

24. In the case of a set S with n elements $f(x) = 1$, for all $x \in X$. So a permutation of the multiset (S, f) is simply a bijection

$$\theta: \{1, \ldots, n\} \to S.$$

25. This is essentially a duplication of Exercise (2.3.9) except that instead of an 8×8 chess-board, we have a 10×10 chess-board.

26. Birbal would have been guilty of over-generalisation and consequent loss of depth. Anything will go wrong, if you fail to treat it properly. There is no big deal in saying it.

Section 3.2

1. Reflexive: (i), (vii)
   Symmetric: (i), (ii), (vii)
   Transitive: (i), (vii), (ix).

3. R^{-1} is precisely the 'reflection' of R in the diagonal ΔX. Other assertions are trivial.

4. $2^{n^2} - 2^{n^2-n} - 2^{n(n+1)/2} + 2^{n(n-1)/2}; \sum_{m=1}^{n} S_{n,m}.$

6. As the simplest counter-example, let $X = \{1, 2, 3\}$ and $R = \{(1, 2),$ $(1, 3)\}$. Then $S = X \times X$. But if $Y = \{2, 3\}$, then $S/Y = Y \times Y$ while $R/Y = \phi$ and so the equivalence relation generated by R/Y is merely ΔY. (The key idea is that although both 2, 3 are in Y, they are related not directly but through 1, which is not an element of Y.)

7. $\{X\}$ and $\{\{x\}: x \in X\}$ respectively. The corresponding equivalence relations are $X \times X$ and ΔX.

8. Let S be the relation of congruency modulo n on $\mathbb{Z}$ and T be the equivalence relation on $\mathbb{Z}$ generated by R. Then $R \subset S$ and so $T \subset S$. For the other way inclusion, let $(x, y) \in S$. Then $y - x = kn$ for some integer k. If $k = 0$, then $y = x$ and $(x, y) \in T$ by reflexivity of T. If $k > 0$, apply induction on k to show that $(x, y) \in T$. If $k < 0$, consider (y, x) and apply symmetry of T to show that $(x, y) \in T$. So $S \subset T$, and hence $S = T$.

10. As the simplest counter-example, let $X = \{1, 2, 3\}$, $R_1 = \Delta X \cup \{(1, 2),$ $(2, 1)\}$ and $R_2 = \Delta X \cup \{(2, 3), (3, 2)\}$.

11. Let $\mathscr{F}$ be the collection of all equivalence relations on X which contain T. $\mathscr{F}$ is non-empty since $X \times X \in \mathscr{F}$. Let $V = \bigcap_{T \in \mathscr{F}} T$. (If X is finite, so is $\mathscr{F}$. If X is infinite $\mathscr{F}$ may be infinite. Still $\bigcap_{T \in \mathscr{F}} T$ makes sense. It is defined as the set $\{(x, y) \in X \times X: (x, y) \in T$ for every $T \in \mathscr{F}\}$. Generalising the last exercise, V is an equivalence relation on X. Also V contains R. So V contains S. On the other hand, $S \in \mathscr{F}$ and so $V \subset S$.

12. Let $\mathscr{D}, \mathscr{E}, \mathscr{F}$ be the decompositions of X corresponding to R_1, R_2 and $R_1 \cap R_2$ respectively. Then members of $\mathscr{F}$ are of the form $D \cap E$ for some $D \in \mathscr{D}, E \in \mathscr{E}$. Conversely every intersection of this form is either empty or a member of $\mathscr{F}$.

13. Let R, S be the equivalence relations corresponding to $\mathscr{D}, \mathscr{E}$ respectively. Then $\mathscr{D}$ and $\mathscr{E}$ are mutually orthogonal iff $R \cap S = \Delta X$.

14. If xRy, then yRx by symmetry and hence xRx by transitivity.

15. Let $Y = \{x \in X: \text{there exists some } y \in X \text{ such that } xRy\}$.

16. Example (8) results as a special case if T is the identity relation on Y, i.e. $T = \Delta Y$.

17. (a) Follows from Theorem (2.8), since the sum on the left is the number of k-ary sequences of length n in which the symbol 1 occurs an even number of times.

(b) Let 1 occur $2r$ times in a k-ary sequence of length n. Then the remaining entries constitute a $(k - 1)$-ary sequence of length $n - 2r$. If 0 is to occur an even number of times in them, then by Theorem (2.8), there are $\dfrac{(k-1)^{n-2r} + (k-3)^{n-2r}}{2}$ such sequences. The result follows by applying (a) twice (once for k and then for $k - 2$) and adding.

18. $\dfrac{1}{4}[k^n - (k - 4)^n]$.

(Find $\displaystyle\sum_{r=0}^{\infty} \binom{n}{2r}\left[(k - 1)^{n-2r} - \dfrac{(k - 1)^{n-2r} + (k - 3)^{n-2r}}{2}\right]$.)

19. Let D be an equivalence class under S. Let $x \in D$. Then $g(D)$ has to be defined as $[f(x)]$, i.e., the equivalence class of $f(x)$ under T, g is well defined, because if $y \in D$, then xSy and hence $f(x)Tf(y)$ so that $[f(x)] = [f(y)]$. The following diagram is then commutative.

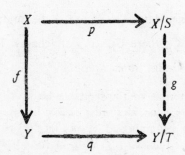

If $T = \Delta Y$, Y/T may be identified with Y and we get Proposition (2.14).

20. Each equivalence class under T is a 'box' whose sides are equivalence classes under R and S.

21. An equivalence class under R is a set of the form $f^{-1}(B)$ where B is an equivalence class under T.

Section 3.3

3. The poset consists of $\{0\}$ and all sets of the form $\{n, - n\}$ where n is a positive integer. Clearly this poset may be identified with $\mathbf{N}_0$ (the set of non-negative integers) with the partial order '$x \leqslant y$' iff x divides y.

5.   (i)   If $S \subset T$ and $T - S$ has at least two distinct elements say, a, b x then $S \cup \{a\}$ lies properly between S and T.

       (ii)   Let the distinct elements of B be $b_1, \ldots, b_n$. Then $\{\phi, \{b_1\}, \{b_1, b_2\}, \ldots, B\}$ is a chain of length $n + 1$. Since no two distinct members of a chain can have the same cardinality, no chain can be of length exceeding $n + 1$.

       (iii)   Each chain of length $n + 1$ corresponds to an arrangement of the elements of B.

6.   In $P(B)$, where $B = \{1, 2\}$, $\{\phi, \{1\}\}$ and $\{\phi, \{2\}\}$ are chains, but their union, $\{\phi, \{1\}, \{2\}\}$ is not.

7.   (i)   Let $B = \{1. \ 2, \ 3\}$, $X = \{\phi, \ \{1\}, \ \{2\}, \{1, 3\}\}$ and $\leqslant$ be defined by set inclusion. Then $\{\phi, \{2\}\}$ is a maximal chain in X. But the longest chain has length 3.

       (ii)   Apply Theorem (3.7) to the poset $(\mathscr{C}, \subset)$. Given a chain $C \in \mathscr{C}$. let $\mathscr{B}$ be the set of all chains which contain C. A maximal element of $\mathscr{B}$ is also a maximal chain.

       (iii)   Apply (ii), to singleton chains, i.e. chains of the form $\{\{x\}\}$ for $x \in X$.

8.   Let the elements $x_1, \ldots, x_m$ of X be mutually incomparable. Let $X = \bigcup_{i=1}^{n} C_i$ where $C_1, \ldots, C_n$ are chains. Then no two x's can be in the same chain. So $n \geqslant m$.

10.   Let the terms of the sequence be $x_1, x_2, \ldots, x_r$ where $r = mn + 1$. For each k, consider all possible monotonically increasing subsequences beginning at x_k and let $f(k)$ be the length of the longest such subsequence. If $f(k) \leqslant m$ for all k, then by the pigeonhole principle, f assumes the same value at at least $n + 1$ distinct points, say $k_1, \ldots, k_{n+1}$ where $1 \leqslant k_1 < k_2 < \ldots < k_{n+1} \leqslant r$. But then we must have $x_{k_1} > x_{k_2} > \ldots > x_{k_{n+1}}$; for if, say, $x_{k_i} \leqslant x_{k_{i+1}}$, then any monotonically increasing subsequence starting from $x_{k_{i+1}}$ can be elongated to a similar subsequence starting from x_{k_i} which would mean $f(k_i) \geqslant f(k_{i+1}) + 1$.

11.   Suppose there are infinitely many boys b_1, b_2, $b_3, \ldots, b_n, \ldots$ and infinitely many girls g_1, $g_2, \ldots, g_n, \ldots$. Consider the case where the only dancing pairs are (b_i, g_j) for which $j \geqslant i$.

       If it is known only that B is finite, say $B = \{b_1, \ldots, b_n\}$, let G_i be the set of girls who danced with b_i, $i = 1, \ldots, n$. If $\{G_1, \ldots, G_n\}$ is a chain then let G_r be the smallest element of it. Then a girl in G_r dances with all boys.

12. Proceed by induction on the cardinality. Let $(X, \leqslant)$, (Y, α) be linearly ordered with $|X| = |Y| = n > 1$. Let x, y be their smallest elements. Apply induction hypothesis to $X - \{x\}$ and $Y - \{y\}$ each with the restriction of the corresponding order.

 If $f: \mathbb{N} \to \mathbb{Q}$ were an order isomorphism, then $f(1)$ would have to be the smallest element of $\mathbb{Q}$. But $\mathbb{Q}$ has no smallest element.

13. Let $f: X \to Y$ be a bijection. Define $\theta: P(X) \to P(X)$ by $\theta(A) = f(A)$ for $A \subset X$. Then θ is an order equivalence.

14. Let $Y = \{2, 3, 5\}$. Define $g: Y \to X$ by $g(A)$ to be the product of the elements in A with the understanding that $g(\phi) = 1$. Then g is an order equivalence.

 If $\leqslant$ is the usual order on X ($= \{1, 2, 3, 5, 6, 10, 15, 30\}$), then $(X, \leqslant)$ is a chain but $(P(Y), \subset)$ can never be a chain if $|Y| \geqslant 2$.

15. In view of Exercise (3.13), it suffices to show that $(P(\mathbb{Q}), \subset)$ contains an uncountable chain. For every real number α, let $S_\alpha = \{x \in \mathbb{Q} : x < \alpha)\}$. Then for $\alpha < \beta$, $S_\alpha \underset{\neq}{\subset} S_\beta$ as there are rationals in the interval (α, β). So $\{S_\alpha : \alpha \in \mathbb{R}\}$ is a chain in $P(\mathbb{Q})$. This chain is uncountable since the set $\mathbb{R}$ is uncountable.

16. (i) If X is finite, let x be the smallest element of X and y be the smallest element of $X - \{x\}$. Then there is no z such that $x < z < y$.

 (ii) Enumerate $\mathbb{Q}$ and X as $q_1, q_2, \ldots, q_n, \ldots$ and $x_1, x_2, \ldots, x_n, \ldots$ respectively. Construct an order-preserving bijection $f: \mathbb{Q} \to X$ by defining $f(q_n)$ inductively. Let $f(q_1) = x_1$. Let $n > 1$ and suppose $f(q_1), \ldots, f(q_{n-1})$ have been defined. Suppose $q_{i_1} < q_{i_2} < \ldots < q_{i_{n-1}}$, where $i_1 i_2 \ldots i_{n-1}$ is a permutation of $\{1, 2, \ldots, n-1\}$. Then $f(q_{i_1}) < f(q_{i_2}) < \ldots < f(q_{i_{n-1}})$. Now q_n is precisely in one of the intervals $(-\infty, q_{i^1}), (q_{i_1}, q_{i_2}), \ldots, (q_{i_{n-1}}, \infty)$. [By $(-\infty, q_{i_1})$ we understand $\{q \in \mathbb{Q} : q < q_{i_1}\}$, similarly for (q_{i_1}, ∞).] Consider the corresponding interval in X (which is nonempty) and let r be the smallest integer such that x_r is in that interval. Define $f(q_n) = x_r$.

17. Let $\mathscr{D}$, $\mathscr{E}$ be two decompositions of a set X. Let R, S be the corresponding equivalence relations on X. Then $\mathscr{D} \wedge \mathscr{E}$ is the decomposition induced by $R \cap S$ which is an equivalence relation (cf. Exercise (2.12)). Let T be the equivalence relation generated by $R \cup S$. Then the decomposition induced by T is $\mathscr{D} \vee \mathscr{E}$.

18. Apply induction on the cardinality.

20.

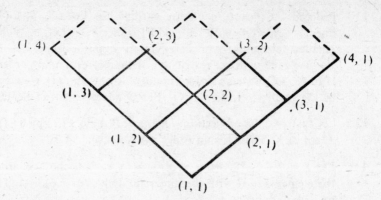

21. (i)   For x, $y \in X$, let z be the least element of $\{x, y\}$. If $z = x$, then $x \leqslant y$. If $z = y$, then $y \leqslant x$.

   (ii) Neither $\mathbb{R}$ nor $\mathbb{Z}$ has a least element. (Note that the set of non-negative reals is also not well-ordered even though it has a least element. The subset $(0, 1)$ has no least element).

   (iii) Let $A \subset \mathbb{N}$ and $n \in \mathbb{N}$. If $k \in A$ for some $k < n$ apply induction hypothesis. Otherwise n is the least element of A. (Actually, the fact that $\mathbb{N}$ is well-ordered is equivalent to the second principle of mathematical induction which can be formally stated as 'If $A \subset \mathbb{N}$, $1 \in A$ and for every $n > 1$, $(k \in A$ for all $k < n) \Rightarrow n \in A$, then $A = \mathbb{N}$. In a rigorous development of number system, positive integers are first so defined that they form a well-ordered set. Then the principle of induction is proved as a theorem.)

   (iv) Let $A \subset X^*$ and $A \neq \phi$. If $X \cap A \neq \phi$, the least element of $X \cap A$ is also the least element of A. Otherwise $*$ is the least element of A.

   (v) Let $x \in X$. Let $A = \{y \in X : x < y\}$. If $A \neq \phi$, the least element of A covers x.

22. With the notation in the text, let θ and σ denote respectively the old and the new permutations. To show that σ is the successor of θ, we have to show two things: (i) $\theta < \sigma$ and (ii) if $\theta < \tau$ then $\sigma \leqslant \tau$. (i) follows easily from the fact that the first $j - 1$ entries of θ and σ match while the jth entry of θ is less than that of σ. For (ii), let T be the set of those permutations of $\{1, 2, \dots, n\}$ whose first $j - 1$ entries are a_1, $a_2, \dots, a_{j-1}$ respectively. Let $M = \{1, 2, \dots, n\} - \{a_1, \dots, a_{j-1}\}$. For any $\lambda \in T$, the jth entry of λ must be from M. For each $x \in M$, let $T_x = \{\lambda \in T :$ the jth entry of λ is $x\}$. Clearly $T = \bigcup_{x \in M} T_x$. Then for $x, y \in M$, $x < y$ implies that every element of T_x is less than every element of T_y. Note that θ is the largest element of T_{a_j} while σ is the smallest element of T_{a_r}. Now suppose $\theta < \tau$. Let i be the smallest integer such that the ith entry of θ is less than that of τ. If $i \leqslant j - 1$, then certainly $\sigma < \tau$. If $i > j - 1$, then $\tau \in T$. So $\tau \in T_x$ for some $x \in M$. Since $\theta < \tau$ we must have $a_j < x$. Now note that in M, a_r is the immediate successor of a_j. So $a_r \leqslant x$. If $a_r \lessdot x$, then again $\sigma < \tau$. If $a_r = x$, then $\sigma \leqslant \tau$ since σ is the least element of T_{a_r}.

23. Given an r-subset $(a_1, a_2, ..., a_r)$ (with $a_1 < a_2 < ... < a_r$) of $\{1, 2, ..., n\}$, first locate the largest j such that $a_j < n - r + j$. Then $(a_1, a_2, ..., a_{j-1}, a_j + 1, a_j + 2, ..., a_j + r - j + 1)$ is the successor. To get the index of $(a_1, a_2, ..., a_r)$, note that the number of r-subsets whose smallest element is less than a_1 is $\binom{n}{r} - \binom{n - a_1 + 1}{r}$. So

the index of $(a_1, a_2, ..., a_r)$ equals $\binom{n}{r} - \binom{n - a_1 + 1}{r} +$ the index

of $(b_1, b_2, ..., b_{r-1})$ where $b_i = a_{i+1} - a_1$ for $i = 1, ..., r - 1$ and $(b_1, ..., b_{r-1})$ is an $(r - 1)$-subset of $\{1, 2, ..., n - a_1\}$. Working inductively, the index of $(a_1, a_2, ..., a_r)$ comes out as

$$\binom{n}{r} - \binom{n - a_1}{r} - \binom{n - a_2}{r - 1} - \binom{n - a_3}{r - 2} - ... - \binom{n - a_r}{1}.$$

$$= \binom{n}{r} - \sum_{k=1}^{r} \binom{n - a_k}{r - k + 1}.$$

24. For $i \leqslant r \leqslant n$, the index of $(a_1, a_2, ..., a_r)$ is

$$\sum_{k=0}^{r} \binom{n}{k} - \sum_{k=1}^{r} \binom{n - a_k}{r - k + 1}.$$

25. Use uniqueness of binary expansion of an integer. If A, $B \subset X$ and $g(A) = g(B)$, then $f_A(i) = f_B(i)$ for all $i = 1, ..., n$ whence $f_A = f_B$ and hence $A = B$. So g is one-to-one. Since $|P(X)| = |Y|$, by the pigeonhole principle g is also onto.

26. Let if possible $(a_1, a_2, ..., a_k)$ be a cycle. Then there is an arrow from a_i to a_{i+1} for $i = 1, ..., k - 1$ and also an arrow from a_k to a_i. But then $a_1 < a_2 < ... < a_k$ and $a_k < a_1$, which is a contradiction in view of transitivity and asymmetry of $<$. (In topological terminology, the absence of cycles is called acyclicity.)

27. Follow the hint. If the assertion in it does not hold then at every stage x_i is the minimum element of X_i. So $x_1 < x_2 < ... < x_n$ in the original order.

28. Both.

29. Let $(X, \leqslant)$ be a poset and let $y_1,..., y_n$ be any listing of elements of X consistent with $\leqslant$, i.e. if $y_i \leqslant y_j$ then $i \leqslant j$. Then for each i, y_i is a minimal element of $\{y_i, y_{i+1},..., y_n\}$. So there is a topological sorting of X in which the elements output successively are $y_1,..., y_n$.

30. (i) Let $\leqslant$ be complete and $B \subset X$, $B \neq \phi$ and b be a lower bound for B. Let A be the set of all lower bounds of B in X. Then $b \in A$ and so $A \neq \phi$. Also A is bounded above, since any element of B is an upper bound for A. The proof is completed by showing that the supremum of A in X is also the infimum of B. The converse is proved by a dual argument.

(ii) Every non-empty subset has a least element, which is also an infimum. Apply (i).

(iii) Apply (ii) to $\mathbb{N}$. That $\mathbb{Q}$ is not complete is seen from the set $A = \{x \in \mathbb{Q} : x^2 < 2\}$ which is bounded above (by 3 for example), but has no supremum. (If $y = p/q$ is the supremum of A in $\mathbb{Q}$, then $y \neq \sqrt{2}$, since $\sqrt{2}$ is not rational. If $y < \sqrt{2}$, there exist $x \in A$ such that $y < x < \sqrt{2}$ contradicting that y is an upper bound for A. If $y > \sqrt{2}$, there exists $b \in \mathbb{Q}$ such that $\sqrt{2} < b < y$. b is then an upper bound for A, contradicting that y is the *least* upper bound for A.

The completeness of $\mathbb{R}$ under usual order is a matter of how the real numbers and their order are constructed. One approach is to *define* a real number as a subset, say, S of $\mathbb{Q}$ having the following properties:

(a) S is non-empty, bounded above but has no greatest element

(b) for all $x, y \in \mathbb{Q}$ if $x < y$ and $y \in S$ then $x \in S$.

Given two real numbers S and T we say $S \leqslant T$ if $S \subset T$ (as subsets of $\mathbb{Q}$). With this approach, if A is a non-empty subset of real numbers which is bounded above then $\bigcup_{S \in A} S$ becomes the supremum of A.

Section 3.4

1 $n^{(n^2)}$; $n^{n(n+1)/2}$; $n^{(n-1)^2+1}$.

3. If $f : A \to B$ and $g : C \to D$ are two functions, let $f \times g : A \times C \to B \times D$ be the function $(f \times g)(a, c) = (f(a), g(c))$ for $a \in A$, $c \in C$. Then associativity of a binary operation $*$ on a set X is equivalent

to the commutativity of the diagram

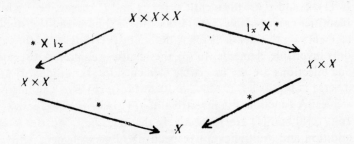

Left distributivity of $*$ over $+$ is equivalent to the commutativity of

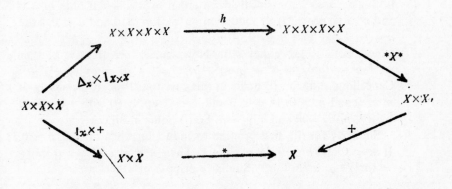

where $\Delta_X : X \to X \times X$ is defined as $\Delta_X(x) = (x, x)$ for $x \in X$ and $h: X \times X \times X \times X \to X \times X \times X \times X$ is defined by $h(a, b, c, d) = (a, c, b, d)$ (i.e. h interchanges the second and the third coordinates.)

For $a \in X_i$ let a also denote the constant function with value a. Then a is a right identity for $*$ iff the diagram

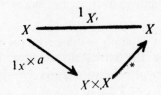

commutes.

4. $(a, x, y) \in g^{-1}(\Delta X) \Leftrightarrow g(a, x, y) \in \Delta X \Leftrightarrow a * x = a * y.$

$(a, x, y) \in X \times \Delta X \Leftrightarrow (x, y) \in \Delta X \Leftrightarrow x = y.$

Since left cancellation law holds iff $[x = y \Leftrightarrow a * x = a * y]$ the results follows.

5. In (1), ϕ and S are the identities for $\cup$ and $\cap$ respectively. Only the identities are invertible. In a lattice the maximum and the minimum elements are the identities for $\wedge$ and $\vee$ respectively. Also they are the only invertible elements. In (3), the identity function is the identity and bijections are the invertible elements. In (4), the identity and the inverse is pointwise. (5) has no identity. In (6) $\circledast$ has an identity iff $*$ does. A function f is invertible iff $f(x)$ is invertible for all $x \in X$ In (7), [0] and [1] are respectively the identities for the modulo n addition and multiplication respectively. Every element is invertible w.r.t. addition. For multiplication modulo n, $[m]$ is invertible iff m is relatively prime to n (This fact is proved in Chapter 6.) In (8), the null sequence is the identity and the only invertible element.

6. Both assertions hold trivially when either m or n is 0. If both $m > 0$ and $n > 0$, prove (i) by induction on n. If $m > 0$ and $n < 0$, write $n = -k$ so that $k > 0$. If $m \geqslant k$, then $a^{(m-k)} a^k = a^m (a^{-1})^k a^k$ (since both sides equal a^m). Now cancel a^k. If $m < k$, then $a^{m-k} a^k = (a^{-1})^{k-m} a^k = (a^{-1})^{k-m} a^{k-m} a^m = a^m = (a^m)(a^{-1})^k a^k$. Cancelling a^k again, (i) holds in this case too. Proceed similarly if $m < 0$ and $n > 0$. If $m < 0$ and $n < 0$, apply (i) with a replaced by a^{-1}, m by $-m$ and n by $-n$. So (i) holds in all cases.

    Similarly for (ii), first proceed by induction when $m > 0$, $n > 0$. If $m < 0$ and $n > 0$, let $m = -k$. Then $(a^m)^n = (a^{-k})^n = ((a^{-1})^k)^n = (a^{-1})^{kn} = a^{-kn} = a^{mn}$. Similarly dispose of other cases.

7. Suppose $\wedge$ is distributive over $\vee$. Let $x, y, z \in X$. Then

$$(x \vee y) \wedge (x \vee z)$$

$$= [(x \vee y) \wedge x] \vee [(x \vee y) \wedge z]$$

$$= x \vee [(x \wedge z) \vee (y \wedge z)] = [x \vee (x \wedge z)] \vee (y \wedge z)$$

$$= x \vee (y \wedge z).$$

So $\vee$ is distributive over $\wedge$. For the converse interchange $\wedge$ and $\vee$.

8. The argument is purely set theoretic for the power set lattice. The argument for the lattice of statements is equally straightforward. For the lattice of positive integers, consider prime factor decomposition. If $a, b, c \in \mathbb{N}$, find primes $p_1, ..., p_r$ so that

$$a = p_1^{\alpha_1} p_2^{\alpha_2} ... p_r^{\alpha_r}; \quad b = p_1^{\beta_1} p_2^{\beta_2} ... p_r^{\beta_r} \text{ and } c = p_1^{\gamma_1} p_2^{\gamma_2} ... p_r^{\gamma_r}$$

where α's, β's, γ's are non-negative integers Use the fact that $a \wedge b = p_1^{\delta_1} p_2^{\delta_2} ... p_r^{\delta_r}$ and $a \vee b = p_1^{\epsilon_1} p_2^{\epsilon_2} ... p_r^{\epsilon_r}$ where for $i = 1, ..., r$, $\delta_i = \min \{\alpha_i, \beta_i\}$ and $\epsilon_i = \max \{\alpha_i, \beta_i\}$. Similarly for $a \wedge c$ etc.

9. Let x_1, x_2, x_3 be three mutually incomparable elements of a lattice such that for $i \neq j$, $x_i \wedge x_j = a$ (say) and $x_i \vee x_j = b$. Then $x_1 \wedge (x_2 \vee x_3) = a \wedge b = a$. But $(x_1 \vee x_2) \wedge (x_1 \vee x_3) = b \wedge b = b$.

10. This follows straight from definitions. However, it is more interesting to prove such results in the context of particular algebraic structures to be encountered in subsequent chapters.

11. That f is a monoid homomorphism follows from the properties of exponential function, namely $e^{x+y} = e^x e^y$ and $e^0 = 1$. f is not a bijection and hence not an isomorphism. However, if $(\mathbf{R}^+, .)$ is the submonoid of $(\mathbf{R}, .)$, consisting of positive real numbers then $f: \mathbf{R} \to \mathbf{R}^+$ is a monoid isomorphism.

12. For example, let $X = \mathbf{N}$, $* =$ usual multiplication. Then $(\mathbf{N}, *)$ is a commutative monoid in which cancellation law holds. Let R be the equivalence relation of congruency modulo 10. The quotient structure X/R is a commutative monoid but does not satisfy cancellation law.

13. $a * c \leqslant a * d \leqslant b * d$. On $\mathbf{R}$, we have $2 < 3$ but $(-1) 2 > (-1) 3$. So $<$ is not compatible with multiplication.

14. For all x, y, $z \in \mathbf{R}$, $d(x + z, y + z) = |(x + z) - (y + z)| = |x - y| = d(x, y)$. Similarly for (x_1, y_1), (x_2, y_2) and (x_3, y_3) in $\mathbf{R}^2$, $d((x_1, y_1) + (x_3, y_3), (x_2, y_2) + (x_3, y_3)) = d((x_1 + x_3, y_1 + y_3), (x_2 + x_3, y_2 + y_3)) = \sqrt{(x_1 + x_3 - x_2 - x_3)^2 + (y_1 + y_3 - y_2 - y_3)^2} = \sqrt{(x_1 - x_2)^2 + (y_1 - y_2)^2} = d((x_1, y_1), (x_2, y_2))$. If $d(x, y) = |x^3 - y^3|$ for $x, y \in \mathbf{R}$ then $d(0, 2) = 8$ but $d(0 + 1, 2 + 1) = d(1, 3) = 26$.

15 If $*$ is commutative or associative, $\circledast$ has the corresponding properties. If e is a right/left identity for $*$, then $\{e\}$ is a right/left identity for $\circledast$. However, even if every element of X is invertible, the same is not always true for $P(X)$. Similarly even if cancellation law holds for $*$, it need not hold for $\circledast$ (e.g. let $X = \mathbf{Z}$, $* = +$. Then $\mathbf{Z} \oplus \{1\} = \mathbf{Z} \oplus \{2\}$ even though $\{1\} \neq \{2\}$.)

16. Let $a_0, a_1, ..., a_k$ be the digits in the decimal expansion of an integer n. Then $n = \sum_{i=1}^{k} a_i 10^i$. Since $10 \equiv 1 \pmod 3$, $100 \equiv 10^2 \equiv 1^2 \pmod 3$ and in general $10^i \equiv 1 \pmod 3$ for all $i = 0, 1, 2, ...,$. So $n \equiv \sum_{i=0}^{k} a_i \pmod 3$. In particular, 3 divides n iff 3 divides $\sum_{i=0}^{k} a_i$.

17. Since $10 \equiv 1 \pmod 9$, a similar criterion holds for divisibility by 9 as for divisibility by 3. In case of 11, $10 \equiv -1 \pmod{11}$. So

$10^i \equiv (-1)^i$ (mod 11). So $n = \sum\limits_{i=0}^{k} a_i 10^i$ is divisible by 11 iff $a_0 - a_1 + a_2 - a_3 + \ldots + (-1)^k a_k$ is divisible by 11.

18. Represent the days of the week by residue classes modulo 7, with Sunday for [0]. Monday for [1] etc. Suppose 13th January falls on [x]. Then 13th February comes after 31 days and since $31 \equiv 3$ (mod 7), it will fall on $[x + 3]$. In an ordinary year the subsequent 13ths will fall on $[x + 3]$, $[x + 6]$, $[x + 1]$, $[x + 4]$, $[x + 6]$, $[x + 2]$, $[x + 5]$, $[x]$, $[x + 3]$ and $[x + 5]$. Regardless of what x is, all residue classes appear in this list. So every ordinary year has at least one Friday the thirteenth. Also at most three of these classes are equal. $[x + 3]$ occurs thrice in the list, namely February, March and November. So the thirteenths of these months fall on the same days, which is a Friday iff $[x + 3] = [5]$, i.e. iff $[x] = [2]$, i.e. iff x is Tuesday (or equivalently the year begins on Thursday).

For a leap year simply add [1] to the classes for the months from March to December. Then January, April and July have their 13ths falling on the same day.

19. $\dfrac{\dbinom{34}{3} + 2\dbinom{33}{3} + 34 \times 33 \times 33}{\dbinom{100}{3}}$. (In $\mathbb{Z}_3$, $a + b + c + 0$ holds iff a, b, c are distinct or all equal).

20. For $i = 1, \ldots, 35$, let x_i be the number of lectures on the ith day. Let $y_i = x_1 + \ldots + x_i$. Then $1 \leqslant y_1 < y_2 < \ldots < y_{35} = 60$. Let $Y = \{y_1, y_2, \ldots, y_{35}\}$. We have to show that for some $i < j$, $y_j - y_i = 13$. For $k = 1, \ldots, 13$ let $S_k = \{n : 1 \leqslant n \leqslant 60, n \equiv k \pmod{13}\}$. Then $|S_k| = 5$ for $1 \leqslant k \leqslant 8$ and $|S_k| = 4$ for $9 \leqslant k \leqslant 13$. Also $Y = \bigcup\limits_{k=1}^{13} (Y \cap S_k)$. Now if the assertion fails then $|Y \cap S_k|$ can contain at most 3 elements for $1 \leqslant k \leqslant 8$ and at most 2 elements for $9 \leqslant k \leqslant 13$. But then $|Y| \leqslant 34$, a contradiction. However, a set Y with 34 elements can be constructed so that no two of its elements differ by 13, In fact let $Y = \{1, \ldots, 13\} \cup \{27, \ldots, 39\} \cup \{53, \ldots, 60\}$. The corresponding x_i's give the schedule of the seminar over 34 days.

21. Let x be a positive integer. Write x as $4m + k$ where m, k are integers and $k = 0, 1, 2,$ or 3. Then modulo 8, x^2 equals k^2 which is 0, 1, 4 or 1. The result follows from the fact in $\mathbb{Z}_8$, [7] cannot be expressed as the sum of three elements from [0], [1], [4], with repetitions allowed.

22. The hint gives one solution. The problem can also be done by recurrence relations. Let b_n be the number of possible interpretations of $a_1 * a_2 * \ldots * a_n$. In any such interpretation, the last application of $*$ must be of the form $x * y$ where for some r, $1 \leqslant r \leqslant n - 1$, x is an interpretation of $a_1 * \ldots * a_r$ and y is an intepretation of $a_{r+1} * \ldots * a_n$. So we get $b_n = \sum\limits_{r=1}^{n-1} b_r b_{n-r}$, for $n \geqslant 2$. This is analogous to the recurrence relation for the Vendor Problem obtained in Chapter 1, Section 3. It can be solved by the methods in Chapter 7.

23. Since X is finite, in the sequence x, x^2, x^4, $x^8, \ldots, x^{2^n}, \ldots$, we must have $x^{2^m} = x^{2^n}$ for some $m < n$. But then $x^{2^n - 2^m}$ is an idempotent.

24. (i) Suppose m, n are relatively prime. Then mn is divisible by a prime square iff at least one of m and n is. When this is not the case, number of prime factors of mn is the sum of the numbers of prime factors of m and n. This proves multiplicativity cf μ. Others are trivially multiplicative.

(ii) For commutativity note that for $n \in \mathbb{N}$, both $(f * g)\,(n)$ and $(g * f)\,(n)$ equal $\sum\limits_{(a,\,b)} f(a) g(b)$ where the sum ranges over all ordered pairs $(a,\ b)$ of positive integers for which $ab = n$. Similarly both $((f * g) * h)(n)$ and $(f * (g * h))(n)$ equal $\sum\limits_{(a,\,b,\,c)}$ $f(a)g(b)h(c)$ where a, b, c are positive integers with $abc = n$.

(iii) Prove the hint using prime factorisations of m, n. With the notation of the hint, if $k = uv$ then $\dfrac{mn}{k} = u'v'$ where $u' = \dfrac{m}{u}$, $v' = \dfrac{n}{v}$. Note that u, v are relatively prime and so are u', v'. So $f(uv) = f(u)f(v)$ etc. Then $(f * g)\,(mn)$ comes out to be $\Sigma f(u)$ $f(v)g(u')g(v')$ where the sum ranges over all the quadruples $(u,\ v,\ u',\ v')$ for which $uu' = m$ and $vv' = n$. This factors as

$$\Big(\sum\limits_{(u,\,u')} f(u)g(u') \Big) \Big(\sum\limits_{(v,\,v')} f(v)g(v') \Big)$$

which is precisely $(f * g)(m)$ times $(f * g)(n)$.

(iv) If $f * g = \delta$ for some $g \in S$, then $(f * g)(1) = 1 = f(1)g(1)$. So $f(1) \neq 0$. For the converse follow the hint. Define $g(1) = \dfrac{1}{f(1)}$. For $n > 1$, $g(n)$ has to be defined as $\dfrac{-1}{f(1)} \sum\limits_{(a,\,b)} f(a)$ $g(b)$ where (a, b) ranges over all ordered pairs (a, b) of positive integers for which $ab = n$ and $b < n$. Then $(f * g)(n) = 0$ for $n > 1$.

(v) By (i) and (iii) $\mu * c$ is multiplicative. So it suffices to show that $(\mu * c)(p^r) = 0$, for a prime p and a positive integer r. The only divisors of p^r are p^k, $0 \leqslant k \leqslant r$. Since $\mu(p^0) = 1$, $\mu(p) = -1$ and $\mu(p^k) = 0$ for $k > 1$, the result follows.

(vi) and (vii) are immediate from definitions.

(viii) For $k = 1, ..., n$, let $G_k = \{x: 1 \leqslant x \leqslant n$, g.c.d. of x and n is $k\}$. Clearly $G_k = \phi$ if k does not divide n. If k divides n and the g.c.d. of x and n is k, we can write $x = kr$ and $n = ks$. Then $r \leqslant s$. Also r is relatively prime to s for any common factor of them will give a larger common divisor of x and n than k. Conversely if $x = kr$ and r is relatively prime to s, then the g.c.d. of x and n is k. So $|G_k| = \phi(s) = \phi\left(\dfrac{n}{k}\right)$. Hence

$$n = \sum_{k \mid n} |G_k| = \sum_{k \mid n} \phi\left(\frac{n}{k}\right) c(k), \text{ showing that } i(n) = (\phi * c)(n)$$

for all n.

(ix) Multiply both sides of $\phi * c = i$ by μ and use $c * \mu = \delta$.

(x) Use (i), (iii) and (ix).

25. The arguments are analogous to that of the last exercise except that some care is necessary since the convolution is not necessarily commutative. If $f(x, x) \neq 0$ for all $x \in X$, define the inverse, say g, of f on (x, y) by induction on the number of elements z which satisfy $x \leqslant z \leqslant y$. (First we construct a right inverse and a left inverse for f separately. Then apply Proposition (4.7).)

CHAPTER 4

Section 4.1

2. Define $f: P(S) \times P(T) \to P(S \cup T)$ by $f(A, B) = A \cup B$.

3. This is a generalisation of Exercise (3.3.14). If n is not square-free and p is a prime such that $p^2 \mid n$, then p and n/p are not complements of each other. Also the number of positive divisors of n is not a power of 2 in this case.

4. From the tables, if at all X is a Boolean algebra then $a = 0$ and $d = 1$. So b and c must be complements of each other. This defines $'$. The axioms can then be verified. Alternatively, let

$$Y = P(\{1, 2\}).$$

Define $f(a) = \phi$, $f(b) = \{1\}$, $f(c) = \{2\}$ and $f(d) = \{1, 2\}$.

5. Observe that a subset A of S is in Y iff it is the union of some members of $\mathcal{D}$, i.e. iff there is some subcollection $\mathcal{C}$ of $\mathcal{D}$ such that $A = \bigcup_{C \in C} C$. (The collection $\mathcal{C}$ need not be finite. Even then the union of its members is well defined. It consists of all elements of S which belong to at least one member of $\mathscr{C}$.) An isomorphism between Y and $P(\mathscr{D})$ is obtained by associating A with $\mathscr{C}$.

6. (a) If such a proof were possible, it would apply equally well to the algebraic structure $(\mathbf{R}, +, \cdot)$. But tautology does not hold in $\mathbf{R}$ either for $+$ or for $\cdot$.

 (b) In the second step, associative law has been tacitly assumed. Although associativity does hold, its proof requires part (ii) of Theorem (1.3).

7. Denote the smallest and largest elements by 0 and 1, suppose for some x, y, z we have $x \vee y = x \vee z = 1$ and $x \wedge y = x \wedge z = 0$. Then $y = y \wedge 1 = y \wedge (x \vee z) = (y \wedge x) \vee (y \wedge z) = 0 \wedge (y \vee z) = y \wedge z$. Similarly $z = y \wedge z$. So $y = z$.

8. $y = y(x + y) = y(x + z) = yx + yz = zx + yz = z(x + y)$
 $= z(x + z) = z.$

9. By absorption $x_i = x_i(x_1 + \dots + x_n)$. If $x_1 + x_2 + \dots + x_n = 0$, then $x_i = x_i(x_1 + \dots + x_n) = x_i \cdot 0 = 0$ for $i = 1, \dots, n$. Converse is trivial. The other assertion follows by duality.

10. If $xy' + x'y = 0$, then by the last exercise $xy' = 0$ and $yx' = 0$. So $x = x(y + y') = xy + xy' = xy$. Similarly $y = xy$. So $x = y$. Converse is trivial.

11. Let H, I, R, M denote appropriate subsets of the set of all men, Then (i) says $HIR' = 0$ while (ii) says $HR'I' = 0$. Adding, $HR' = 0$. Similarly (iii) gives $RH'M' = 0$ and (iv) says $MRH' = 0$. So $RH' = 0$. By the last exercise, $R = H$.

12. Let S, H, C denote appropriate subsets. (ii) says $SHC' = 0$ while (iii) says $H'SC' = 0$. Adding $SC' = 0$, i.e. $S \subset C$. But this contradicts (i).

13. Rule (iii) is redundant. The simplified system of rules is 'Every student must register for course B and at least one more course' and 'No student can register both for A and C'.

14. The first assertion follows from the fact that $x \leqslant y$ iff $y' \leqslant x'$. For the second assertion apply Proposition (1.10) to the complement of the given element.

15. (i) Let $S - A$, $S - B$ be finite. Then $S - (A \cup B)$ is a subset of $S - A$ and hence is finite. $S - (A \cap B)$ equals $(S - A) \cup (S - B)$ and so is finite.

 (ii) If A and $S - A$ are both finite, S would be finite. If $S = \mathbf{R}$

and $A = \mathbb{N}$ then A is an infinite but not a cofinite subset of S.

(iii) Let $\mathscr{C}$, $\mathscr{F}$ be the collections respectively of all cofinite subsets and all finite subsets of S. Define $\theta : \mathscr{C} \to \mathscr{F}$ by $\theta(A) = S - A$ for $A \in \mathscr{C}$, Then θ is a bijection. By Exercise (2.2.22) (v), $\mathscr{F}$ is countable.

16. Clearly Y is closed under complementation. Let $A, B \in Y$. If at least one of them is finite, so is $A \cap B$. If both are cofinite, then so is $A \cap B$ by the last exercise. In either case $A \cap B \in Y$. Similarly $A \cup B \in Y$.

17. Obviously Y is infinite. However, with the notation of Exercise (1.15) (iii) above, $Y = \mathscr{C} \cup \mathscr{F}$ and so Y is countable by Exercise (2.2.22) (iv). Hence $|Y| = \aleph_0$. Now if S is finite so is $P(S)$. If S is infinite, then by Exercise (2.2.25), $|S| \geq |\mathbb{N}|$, from which it follows that $|P(S)| \geq |P(\mathbb{N})|$. But by Exercise (2.2.27),

$$|P(\mathbb{N})| \neq \aleph_0.$$

18. No element equals its own complement. So every pair contains exactly two distinct elements.

19. (i) Commutativity of Δ is obvious. For associativity, $(x \Delta y)\Delta z$ equals $(xy' + x'y)z' + (xy' + x'y)'z$ which simplifies to $xyz + x'y'z + x'yz' + xy'z'$. By symmetry, $x \Delta (y \Delta z)$ also reduces to the same.

(iv) $(xy) \Delta (xz) = xy(x' + z') + (x' + y') xz$
$$= x(yz' + y'z) = x(y \Delta z).$$

(vi) Putting $z = xy$ in (i) above, $x \Delta y \Delta xy$ equals

$$xyxy + x'y'xy + x'y(x' + y') + xy'(x' + y')$$

which reduces to $xy + x'y + xy'$ and further to $x + y$.

20. The operations in Y are pointwise. In particular,

$$fg' = 0 \Leftrightarrow fg'(s) = 0$$

for all $s \in S \Leftrightarrow f(s) \leq g(s)$ for all $s \in S$. For $s_0 \in S$, the function f defined is non-zero. If $g \leq f$ then $g(s) = 0$ for all $s \neq s_0$, while $g(s_0) \leq f(s_0)$ implies $g(s_0) = 0$ or $f(s_0)$ since $f(s_0)$ is an atom of X. So either $g = 0$ or $g = f$.

21. Let X, Y be Boolean algebras. Denote the operations on them by the same symbols. Then $(a, b) \leq (c, d)$ iff $a \leq c$ in X and $b \leq d$ in Y. (x, y) is an atom of $X \times Y$ iff one of x and y is 0 and the other is an atom.

Section 4.2

1. The disjunctive and conjunctive normal forms are:

   (i) $x_1'x_2x_3' + x_1'x_2x_3$ and $(x_1 + x_2 + x_3)(x_1' + x_2 + x_3') \times$
   $(x_1' + x_2 + x_3)(x_1' + x_2 + x_3')(x_1' + x_2' + x_3)(x_1' + x_2' + x_3')$.

   (ii) $abc + abc' + ab'c + ab'c' + a'bc' + a'bc + a'b'c'$ and $(a + b + c')$.

   (iii) 0 and the complete C.N.F., i.e.

   $$(x + y)(x + y')(x' + y)(x' + y').$$

2. If $g = f + h$, then for every $\bar{y}$, $f(\bar{y}) \leqslant g(\bar{y})$. So $f(\bar{y}) = 1 \Rightarrow g(\bar{y}) = 1$. For the converse, we can take h to be g itself. (Other choices are also possible. But note that there is no such thing as $g - f$ in a Boolean algebra.)

3. f is a factor of g iff g vanishes wherever f does.

4. Whenever $f = gh$, the C.N.F. for f is obtained by multiplying the C.N.F.'s of g and h. (In view of tautology, the common factors need not be repeated.)

5. Introduce a new Boolean variable y which is set equal to 1 if the veto power is not exercised and 0 if it is. Then b, i.e. the state of the box, comes out as the product of 11 factors, namely y and 10 factors of the form $(x_i + x_j + x_k)$. So the box has 11 locks, ten as before but the eleventh lock has only one key which lies with p_1.

6. Let $a, b, \ldots, g$ be Boolean variables representing passage or failure in $A, B, \ldots, G$ respectively. Set $s = 1$ or 0 according as the candidate is or is not from a scheduled caste. Let p denote passage in the examination. Then $p = p_1p_2 + p_1p_3 + p_2p_4 + p_3p_4$, where p_1 is a sum of 21 terms of the form $abcde$, $p_2 = a(bc + cf + bf)$, $p_3 = cdg$ and $p_4 = s$.

7. Let a, b, c be Boolean variables representing the sizes, the locations and the personnel respectively. Let $f = 1$ or 0 according as the industry gets a license or not. The condtions of the problem imply $f \geqslant abc + abc' + a'b'c'$ and $f' \geqslant a'b'c + a b'c'$. This does not determine f uniquely. The remaining three terms of the complete D.N.F., namely $ab'c$, $a'bc$ and $a'bc'$ can be put at will either in f or in f', independently of each other. (Such terms are sometimes called **neutral** or **'don't care'** conditions). They are to be used for simplifying f. In the present case if we put $a'bc$ and $a'bc'$ in f, f simplifies to $b + a'c'$. So one possible system of rules is that 'All urban industries will get licenses' and 'All small scale industries employing unskilled personnel will get licenses'.

8. (i) and (ii) are symmetric with characteristic numbers 1, 2, 3 and 3 respectively.

9. g is the minimum of all symmetric functions k such $f \leqslant k$. To construct g explicitly from f, suppose a term like $x_1^{\varepsilon_1} x_2^{\varepsilon_2} \ldots x_n^{\varepsilon_n}$ appears in the D.N.F. of f where, say, r of the ε_i's are 1 and the remaining $n - r$ ε's are 0. Then in g, put $\binom{n}{r}$ terms corresponding to all possible binary sequences of length n in which exactly r terms are 1. Do this for every term in the D.N.F. of f. To construct h, take only those terms in the D.N.F. of f, for which all possible $\binom{n}{r}$ terms are also in the D.N.F. of f. (g may be called the symmetric closure of f).

10. Apply the definition repeatedly.

11. The n-tuple $(x_2, \ldots, x_n, x_1)$ can be obtained from $(x_1, x_2, \ldots, x_n)$ by a series of interchanges of variables; for example, going through $(x_2, x_1, x_3, \ldots, x_n)$, $(x_2, x_3, x_1, x_4, \ldots, x_n), \ldots, (x_2, x_3, \ldots, x_{n-1}, x_1, x_n)$. A symmetric function must assume the same value at all these points. A counter-example for the converse is given by (iii) in Exercise (2.8).

12. The result is trivial for $n = 1, 2$. For $n = 3$, classify the terms in the complete D.N.F. with 3 variables, into four classes depending upon the number of dashes. Thus, $A_0 = \{x_1 x_2 x_3\}$, $A_1 = \{x_1 x_2 x_3', x_1 x_2' x_3, x_1' x_2 x_3\}$, $A_2 = \{x_1' x_2' x_3, x_1' x_2 x_3', x_1 x_2' x_3'\}$ and $A_3 = \{x_1' x_2' x_3'\}$. Then $f(x_1, x_2, x_3)$ is symmetric iff it has the property that whenever its D.N.F' contains any one term from A_i, it contains all terms of A_i, $i = 0, \ldots, 3$. This is always true for $i = 0$ and 3. For $i = 1, 2$ it follows from cyclical symmetry of f.

13. For example, let $f(x, y, z) = x e^y \sin z + y e^z \sin x + z e^x \sin y$.

14. Classify the atoms as in the solution to Exercise (3.12) above, into $n + 1$ classes $A_0, A_1, \ldots, A_n$. Then for $1 \leqslant k \leqslant n - 1$ if f contains an atom in A_k, it must also contain all atoms obtained by cyclical permutations. The point is that if n is a prime, the atoms so obtained are all distinct. So there are $n - 1$ of them besides the original one. This need not be so if n is not a prime e.g. if $n = 4$ then the atom $x_1 x_2' x_3 x_4'$ gives only one more atom, $x_2 x_3' x_4 x_1'$. (Results like this are special cases of a theorem on group actions. See the Epilogue.)

15. Continuing the notation above, each class $A_k, k = 1, \ldots, n - 1$ gets divided into subclasses. Each subclass has n distinct atoms. A cyclically symmetric function must contain either none or all n of these atoms. Thus in all the 2^n atoms get classified into classes,

two of which are singleton and the remaining have n elements each.
The number of classes is $2 + \dfrac{2^n - 2}{n}$. Any cyclically symmetric
function of n variables is determined uniquely by which of these
classes it contains.

16. With the notation in the text, let $m_1 > 1$. Then $m_i \geqslant 2$ for all
$i = 1, 2, 3, 4$. If any p_i is not used or is in the left pan then it is
impossible to weight more than 38 kg. So all p_i's must be in the
right pan. But then the weight is 40 kg. So it is impossible to
weight 39 kgs.

17. Start from $0 \in A_0$. By (i), $1 \in A_1$ and $-1 \in A_{-1}$. If $2 \in A_0$ then
by (i), $1 \in A_{-1}$, a contradiction. Similarly if $2 \in A_1$ then by (ii)
$1 \in A_0$, also a contradiction. So $2 \in A_{-1}$, whence by (iii), $3 \in A_0$
and hence by (i) again $4 \in A_1$. Consider 5. Continue like this.
Similarly work with negative integers. (Actually it is not necessary
to assume beforehand that $|A_i| = 27$ for $i = -1, 0, 1$.)

18. Show that the sets A_{-1}, A_0, A_1 in the hint satisfy the conditions in
the last exercise. For example, A_0 is precisely the set of weights
that can be weighted without using p_1. If m is such a weight, then
putting p_1 in the right pan we can weigh $m + 1$. But this means
$m + 1 \in A_1$, because A_1 consists of precisely the weights that can
be weighed with p_1 in the right pan.

19. More generally, for every positive integer n, a stone weighing
$\dfrac{3^n - 1}{2}$ kgs can be uniquely cut into n parts so that every integral
weight upto $\dfrac{3^n - 1}{2}$ kgs can be weighed with only one use of the
balance. The parts must weigh $1, 3, \ldots, 3^{n-1}$ kgs.

20. In the D.N.F. of f, combine all terms containing x_n as well as all
those containing x_n'. The significance is that a Boolean function of
n variables can be looked upon as a combination of two functions
of $n - 1$ variables, because each of the functions $f(x_1, \ldots, x_{n-1}, 1)$
and $f(x_1, \ldots, x_{n-1}, 0)$ is essentially a function of the $n - 1$ variables,
$x_1, \ldots, x_{n-1}$. Note that such a result is possible only because a
Boolean variable assumes finitely many (in fact only two) values.

21. In $\mathbb{Z}_2$, $0 \oplus 0 = 1 \oplus 1 = 0$ and $1 \oplus 0 = 0 \oplus 1 = 1$. This proves
the assertion for $n = 2$. For the general case apply induction.

Section 4.3

1. (a) $x_2 + x_1'$
   (b) a

(c) $a(bc + bqz + pyz + pyqc) + x(yz + pbc + yqc + pbqz)$.

Only (a) and (b) can be simplified.

2. $(ab + xy)(cd + zw)$

3. Let the switches be x, y, z and f denote the state of the lamp. Without loss of generality we may suppose $f(1, 1, 1) = 1$. This gives

$$f(0, 1, 1) = f(1, 0, 1) = f(1, 1, 0) = 0;$$

$$f(0, 0, 1) = f(0, 1, 0) = f(1, 0, 0) = 1$$

and

$$f(0, 0, 0) = 0.$$

So

$$f = xyz + x'y'z + x'yz' + xy'z'.$$

4. Let the switches be $x_1,..., x_n$. If we assume $f(0, 0,..., 0) = 1$ then $f(x_1,..., x_n) = 1$ iff the number of x_i's having values 1 is even. So f is a symmetric function with characteristic numbers $0, 2, 4, 6,...$. If $f(0, 0,..., 0) = 0$, then also f is symmetric, but its characteristic numbers are 1, 3, 5, 7.

5. Each of the 2^n possible combinations of the states of the switches provides one path for the current starting at the initial terminal. This path will terminate at the lower or the upper level depending upon whether the number of x_i's equal to 1 is even or odd.

6. Let n_i be the number of persons at the ith table and x_i be the corresponding switch. Σn_i is odd iff an odd number of the n_i's are odd, i.e. iff an odd number of x_i's are equal to 1. So the closure function of the lamp is a symmetric function with characteristic numbers 1, 3, 5, 7,.... Use the last exercise.

8. Let x, y be the switches and f, g be the closure functions of the two machines. One choice is to let $f = xy + x'y'$ and $g = xy' + x'y$. Using Exercise (3.5) again, a possible circuit is,

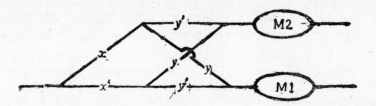

9. Let switches a, b, c indicate whether A, B, C have tests respectively. Let d be the switch which closes on odd numbered days. Then a possible circuit is

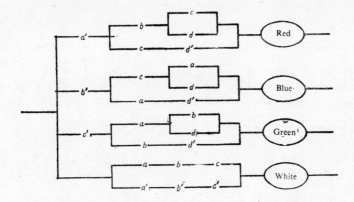

10. Attach lamps of different colours to the terminals in the circuit of Fig. 4.11.

12. Introduce some shorthand notation. Let X_5, Y_5 denote replicas of the circuit in Fig. 4.10 with $n = 5$. In X_5 the switches are $x_1,\ldots,x_5$; in Y_5 they are $y_1,\ldots,y_5$. Also let $Y_{5,2}$ be the portion of Y_5 upto and including level 2 (i.e. terminals T_0, T_1, T_2). Then a desired circuit is

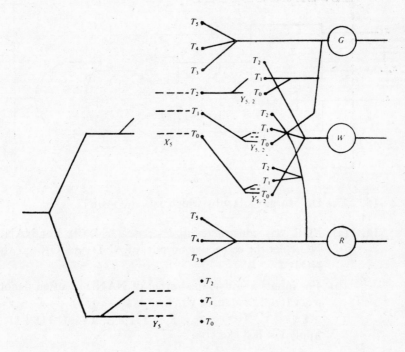

13. $\lceil \log_2 n \rceil$.

14. (i)

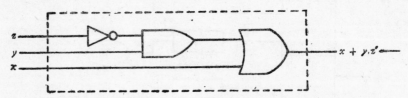

(ii)

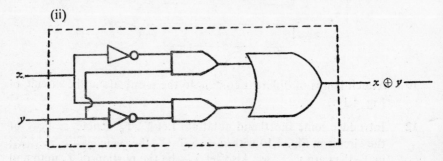

(iii) Let B_n be a black-box representation of $x_1 \oplus - \ldots - \oplus x_n$. An inductive construction for B_n is

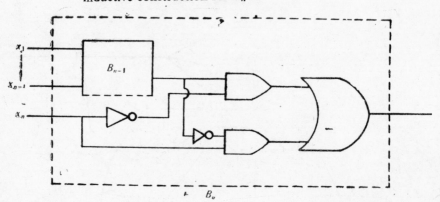

15. Use De Morgan's law by which $xy = (x' + y')'$.

16. (a) XOR was constructed in Exercise (3.14) (ii). For NAND and NOR, let the respective outputs of AND and OR go through NOT.

(b) For inputs x and y, the output of NAND is often denoted by $x \downarrow y$ (read 'x dagger y'), i.e., $x \downarrow y = (xy)' = x' + y'$. Now $x \downarrow x = x'$. Hence $x + y = x' \downarrow y' = (x \downarrow x) \downarrow (y \downarrow y)$. Now apply the last exercise.

(c) dual of (b).

17. Between the terminals T_1 and T_2 the closure function is $a + bc$ i.e. $(a+b)(a+c)$. Similarly for other pairs. The equivalent wye circuit is

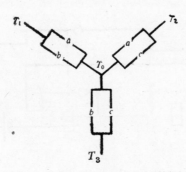

18. After a is pressed contact x is made and the control path of X remains complete through x even after a is released. If now b is pressed and released, the control path of Y remains complete through x and y. If b is pressed while x is not made, Y does not operate.

19. The tricky part is to construct a device which will be activated when C is pressed twice in a row but not when pressed only once. This can be done through a relay which is activated when C is released.

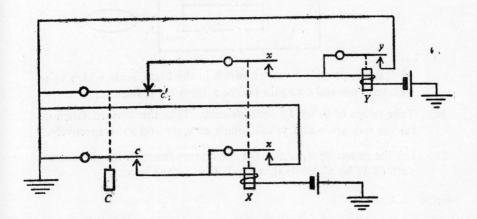

The relay Y 'remembers' that C had been pressed and released. Similarly let Z be a relay which remembers that A had been pressed and released after C.

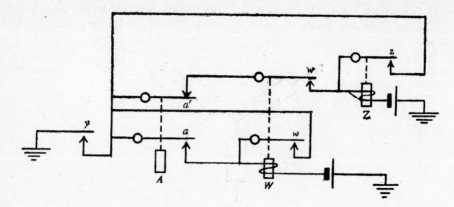

Finally, let the circuits for the bell and the red lamp be respectively, and

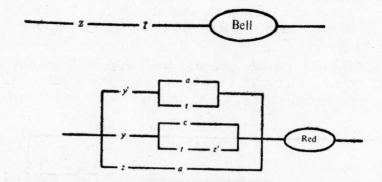

(If desired, provide a master switch in the first circuit which is to be turned off and on again before a fresh try is started.)

20. Take relays of delay 15 seconds each. Take the closure functions for the red, green and yellow lamps as x, xy and xy' respectively.

21. Let the relays be $X_1,...,X_n$. Let the closure function for the control path of X_1 be x'_n while that for X_i ($2 \leqslant i \leqslant n$) be x_{i-1}.

Section 4.4

1. All except (4) and (11) are statements. (4) is not a sentence. (11) is a sentence but cannot be assigned any truth value.

2. In essence we have here a pair of sentences p and q. If p says 'q is false' and q says 'p is false', then we can have p true and q false (or vice versa). So both are statements. However, if p says 'q is true' and q says 'p is false' then they cannot be assigned truth values and hence are not statements.

3. Let V be the set of villagers, b the barber and R the binary relation on V defined by xRy iff x shaves y. Then the announcement says 'For every $x \in V$, bRx iff $\sim (xRx)$. This is a statement which has to be false, as otherwise bRb iff $\sim (bRb)$. (This example is popularly called the **barber's paradox**, and structurally resembles the Russel's paradox in Chapter 2, Section 1.) If the announcement were 'for every $x \in V$, $x \neq b$, xRb iff $\sim (xRx)$, then it could be either true or false.

4. The verb 'to learn' is used to convey metalearning, i.e. learning about learning.

5. No. The given statement is thus false. Here again doing a thing about which it is impossible to do something is 'metadoing'. God could possibly make a stone which nobody other than Himself can lift.

6. There is nothing wrong in the reasoning. It sounds paradoxical because of the definition of a surprise test, which involves meta-knowledge. In practice the instructor could either announce, 'There *may be* a surprise test' or announce 'There will be a surprise test' but change the definition of the surprise test to mean that on no day *except possibly Thursday* can the class know definitely that there would be a test the next day.

7. The first question should be 'If my next question is going to be 'Are you guilty', will your answer to it be the same as your answer to this question?' This is, of course, a metaquestion.

8. Let $s = x + y$ and $p = xy$. To start with, A knows s but not p and B knows p but not s. Now p cannot be a prime since neither of the numbers is 1. If p were a product of two primes then B would know the two numbers. More generally, if p is such that p can be factored in only one way as product of two integers in the range 2 to 99, then also B would know the two numbers (e.g. if $p = 730$, then the numbers have to be 10 and 73). Let q be the largest prime divisor of p. Then p must have some factor $k \geqslant 2$ such that $kq \leqslant 99$, for otherwise the two numbers are uniquely determined as q and p/q. In particular we must have $q \leqslant 47$. This much we conclude from B's first statement.

Now A knows only s. But he knows for sure that B is unable to determine the two numbers uniquely. That means the integer s is such that no matter how it is split into two parts, it is impossible to determine the two parts uniquely from their product. For example s cannot be 52 because 52 can be split as $37 + 15$ which would give p as 555 from which B could have determined the numbers uniquely as 37 and 15. Let us call such a splitting as a 'good' splitting of s. ($47 + 5$ is another good splitting of 52). Now

$4 \leqslant s \leqslant 198$. Let $S = \{s : 4 \leqslant s \leqslant 198, s$ admits no good splitting$\}$. Clearly no integer greater than 54 is in S (since it has a good splitting in which one part is 53). Similarly if s is a sum of two primes it has a good splitting. This leaves $S = \{11, 17, 23, 27, 29, 35, 37. 41, 47, 51, 53\}$.

When B makes his second statement, he knows p and he also knows $x + y \in S$. For each $s \in S$, let A_s be the set of products of the parts in the various possible splitting of s. For example, if $s = 11$, the possible splitting of s are $2 + 9$, $3 + 8$, $4 + 7$ and $5 + 6$ and the corresponding products of parts are 18, 24, 28 and 30. So $A_{11} = \{18, 24, 28, 30\}$. Similarly $A_{17} = \{30, 42, 52, 60, 66, 70, 72\}$. B knows p. If $p \in A_s$ and A_t for two distinct $s, t \in S$, then $x + y$ could be s or t and B would not be able to determine the numbers uniquely. So from B's second statement we conclude $p \in A_s$ for a unique $s \in S$. For each $s \in S$, let B_s be the subset of A_s consisting of those elements which do not belong to any A_t for $t \neq s$, i.e.,

$B_s = \bigcap_{t \in S-\{s\}} A_s - A_t$. For example, $B_{11} = \{18, 24, 28\}$, $B_{17} = \{52\}$

etc. From B's second statement we know $p \in B_s$ for some $s \in S$.

Now consider A's position when he makes his second statement. He already knows s. He also knows that $p \in B_s$. If B_s had more than one element, then A would not know p uniquely and hence would not be able to determine the two numbers. The fact that he could do so means B_s is a singleton. Calculating B_s for all $s \in S$, we see that B_{17} is the only singleton set among them. So $s = 17$ and $p = 52$. Solving $x + y = 17$ and $xy = 52$ we get the two numbers and 4 as 13.

It might appear at first sight that B's two statements contradict each other and hence both cannot be true. But actually they are not the negations of each other. The real meaning of 'to know' in these statements is 'to be able to find out using the available information'. The available information is different at the time of the two statements made by B.

10. Let a be the statement that a particular student x registers for course A. Similarly define b, c, d. The system of five rules is equivalent to $f = 1$ where $f = f_1 f_2 f_3 f_4 f_5$ and where $f_1 = ab + ac + ad + bc + bd + cd$, $f_2 = a' + bc' + b'c$, $f_3 = c' + b + d$, $f_4 = a' + c'$ and $f_5 = c' + d' + b$. $f_4 f_5 = c' + a'd' + a'b$ whence $f_2 f_4 f_5 = a'c' + a'd' + a'b + bc'$ and thus $f_1 f_2 f_4 f_5$ comes out as $b(a' + c')(a + c + d)$ after simplification. Now the C.N.F. of f_3 is $(a + c' + b + d)(a' + c' + b + d)$. Since both these factors occur in the C.N.F. of b, f_3 divides b and hence f is the same as

$f_1 f_2 f_4 f_5$. So the third rule is redundant. The equation $b(a' + c')$ $(a + c + d) = 1$ also gives the simplified system of rules.

11. Letting h, r, i and m stand for appropriate statements, the system is given by $f = 1$ where f is the product of $(r+h'+i')$, $(i+r+h')$, $(m + h + r')$ and $(h + m' + r')$. This reduces to $(r+h')(r'+h)=1$, i.e. $r + h' = 1$ and $r' + h = 1$. So $h \to r$ and $r \leftrightarrow h$. Hence $r \leftrightarrow h$.

12. (i) {Gopal} (ii) {Ram, Rahim, Gopal, Goliath}

(iii) {Rahim, Robert, Gopal} (iv) {Ram, Gopal, Goliath}.

13. In modus ponens we have to show that $(p' + q) p \to q$ is true or equivalently that $[(p' + q) p]' + q = 1$. But $[(p' + q)p]' + q = (pq)' + q = p' + q' + q = p' + 1 = 1$. So modus ponens is valid. For validity of chain rule apply induction.

14. In (v) for example, we have to show $(p' + q) (s' + r) \to (ps)' + qr$ is true, or equivalently that $[(p' + q) (s' + r)]' + (ps)' + qr = 1$. But $[(p' + q) (s' + r)]' + (ps)' + qr = (p' + q)' + (s' + r)' + (p' + s') + qr = pq' + sr' + p' + s' + qr = (pq' + p') + (sr' + s') + qr = pq' + p' + r' + s' + qr = q' + qr + p' + r' + s' = q' + r + p' + r' + s' = 1 + q' + p' + s' = 1$. Similarly the validity of other arguments is proved.

15. With the notation in the text the argument proceeds as follows;

(1) $p + q' \to rs'$ (premise)

(2) $p + r' \to sq'$ (premise)

(3) $p' \to r'$ (premise)

(4) $p + p' \to p + r'$ (from (vi) in the last exercise)

(5) $p + p' \to sq'$ ((4), (2) and (i) above)

(6) $sq' \to s$ ((ii) above)

(7) $p + p' \to s$ ((5), (6) and (i))

(8) $s \to s + r'$ ((iii) above)

(9) $p + p' \to s + r'$ ((7), (8) and (i))

(10) $s + r' \to p'q$ (contrapositive of (1))

(11) $p + p' \to p'q$ ((9), (10) and (i))

(12) $p'q \to q$ ((ii) above)

(13) $p + p' \to q$ ((11), (12) and (i))

(14) $sq' \to q'$ ((ii), above)

(15) $p + p' \to q'$ ((5), (14) and (i))

(16) $(p + p')(p + p') \to qq'$ ((13), (15) and (v) above)

(17) $1 \to 0$ (since $p + p' = 1$ and $qq' = 0$ in (16)).

16.   No. (ii) and (iii) imply that all smokers have cancer. This contradicts (i) only because of numerical data.

17.   With appropriate symbols, the premises are $r \to m$ and m and the conclusion is r. For the argument to be valid $((r' + m)m)' + r$ would have to be 1. But $((r' + m)m)' + r = m' + r$ which is 0 if $m = 1$ and $r = 0$. So there is at least one instance (namely, when missiles are costly and it does not rain) when all premises hold but the conclusion does not. Hence the argument is invalid. However, the question of corelation between rain and the cost of missiles is irrelevant here. In mathematical logic, we are not concerned with the truth or falsehood of statements *per se* but only with how the truth of some statements leads to the truth or falsehood of some other statements.

18. $((r' + m)m)' + r' = m' + r'$ is 0 if $m = 1$ and $r = 1$. There is no contradiction here because in testing the validity of the arguments we are concerned with the statement $((r \to m) \wedge m) \to r$ (which reduces to $m' + r$ as shown above) and the statement $((r \to m) \wedge m) \to r'$ (which reduces to $m' + r'$). These two statements are not the negations of each other, even though r and r' are negations of each other.

CHAPTER 5

Section 5.1

1.   Let I denote the set of all invertible elements of a monoid $(M, *)$. I is closed under $*$ by Proposition (3.4.7). Clearly the identity element is in I. Finally $a \in I$ implies $a^{-1} \in I$ since a^{-1} is invertible with inverse a.

2. For example let $X = \{e, a, b\}$ and $*$ be given by

$*$	e	a	b
e	e	a	b
a	a	e	b
b	b	b	e

3. For each x, choose some $f(x)$ such that $x \cdot f(x) = e$. Now, $ax = bx \Rightarrow ax$ $f(x) = bxf(x) \Rightarrow a = b$. For $a \in G$, $af(a) = e = ee = eaf(a)$, so by right cancellation, $a = ea$. Finally, for any $a \in G$, $f(a) af(a) = f(a)e = ef(a)$ so again by cancellation, $f(a)a = e$. Thus every element has a two-sided inverse.

4. For example let $X = \{e, a\}$ and $ae = ee = e$, $aa = a = ea$.

5. $e = (xy)^{-1} xy = x^{-1} y^{-1} xy$. So $yx = yxe = yxx^{-1}y^{-1} xy = xy$. More generally, let $(xy)^n = x^n y^n$ for $n = k, k+1, k+2$. Then $x^{k+1} y^{k+1} = (xy)^{k+1} = (xy)^k xy = x^k y^k xy$. So by cancellation, $xy^k = y^k x$. Similarly, $xy^{k+1} = y^{k+1}x$. So $y^k xy = y^k yx$. Cancel y^k.

6. Clearly every subgroup H has the property. For the converse, let $a \in H$. Then $a \; a^{-1} \in H$, i.e. $e \in H$. Hence for any $y \in H$, $e \cdot y^{-1}$ $\in H$, i.e. $y^{-1} \in H$. Finally, for x, $y \in H$, $y^{-1} \in H$ and so $x(y^{-1})^{-1} \in H$, i.e. $xy \in H$.

7. They are generated by [2] and [3] respectively.

8. This is a consequence of properties of finite fields (see the Epilogue). The crucial step in the argument is that for every m dividing $p-1$, there are precisely m distinct residue classes p whose mth power is [1]. See Exercise (6.2.38).

9. $x^n = e = y^n$ implies $(x^{-1})^n = e$ in any group and $(xy)^n = e$ in an abelian group. For a counter-example in non-abelian groups. Let G be the group of isometries of an equilateral triangle and $n = 2$.

10. They are $y^d, y^{2d}, ..., y^{md}$ where y is a generator for G and $d = n/m$. If y^k is also a solution, find q, r such that $k = qd + r$ with $0 \leqslant r < d$. Then y^r would also be a solution, a contradiction.

11. Let $G = (x)$. If H is a non-trivial subgroup of G, show that $H = (x^k)$ where k is the least positive integer such that $x^k \in H$.

12. f, g have order 2 each. Let $h = f \circ g$. Then by induction, for every positive integer n, $h^n(1) = 2n$. So h^n cannot be the identity mapping for any n. Similarly $g \circ f$ is of infinite order.

13. Let m, n be the orders of x, y respectively. Then $(xy)^{mn} = x^{mn} y^{nm} = (e)^n e^m = e$. So xy has order $\leqslant mn$. A better upper bound is the l.c.m. of x and y. For sharpness, consider [2] and [3] in $\mathbb{Z}_6$.

14. From the hint, if yx has order n, then $(xy)^{n+1} = xy$ and hence $(xy)^n = e$. So order $(xy) \leqslant$ order (yx). By symmetry the other inequality also holds.

15. Define $f: \{1, 2, ..., n\} \to G$ by $f(k) = x_1 x_2 ... x_k$. Either f is onto or there exists p, q with $1 \leqslant p < q \leqslant n$ such that $f(p) = f(q)$. In the latter case let $i = p + 1, j = q$.

16. Decompose G into sets of the form $\{x, x^{-1}\}$ for $x \in G$. Each such set contains 1 or 2 elements. Since $o(G)$ is even, the number of such sets having only 1 element each is even. But $\{e, e\}$ is one such set. So there is at least one more, say $\{y, y^{-1}\}$ where $y^{-1} = y$. But then $y^2 = e$, $y \neq e$.

17. Let f, $g: \mathbb{R} \to \mathbb{R}$ be given by $f(x) = ax + b$, $g(x) = cx + d$ for $x \in \mathbb{R}$. Then $g \circ f(x) = g(ax + b) = cax + bc + d$ which is linear and non-singular since $ca \neq 0$. f^{-1} is given by $f^{-1}(x) = \dfrac{1}{a} x - \dfrac{b}{a}$ which is also non-singular, linear. Hence such functions form a subgroup of $S(\mathbb{R})$. It is not a normal subgroup, e.g. let $f(x) = x + 1$, $g(x) = x^3$. Then $gfg^{-1}(x) = (x^{1/3} + 1)^3$ which is not linear.

18. Analogous to the last exercise.

19. Consider (r) where r is a rotation through an angle of $\dfrac{360}{n}$ degrees. r and any flip generate D_n.

20. Follows from the comments regarding when the composite of two function preserves/reverses orientation.

22. Let HK be a subgroup of G. Certainly $H = H\{e\} \subset HK$. Let h, $k \in G$. Then $kh = h^{-1}(hk)h$ is in HK. So $kh \in HK$, i.e. $KH \subset HK$. For the other way inclusion, first $(hk)^{-1} \in HK$. So $k^{-1}h^{-1} = h_1 k_1$ for some $h_1 \in H$, $k_1 \in K$. But then $hk = k_1^{-1} h_1^{-1} \in KH$. Conversely suppose $HK = KH$. Then for h_1, $h_2 \in H$, k_1, $k_2 \in K$, $h_1 k_1 h_2 k_2 = h_1 (h_3 k_3) k_2$ for some $h_3 \in H$, $k_3 \in K$. So HK is closed under multiplication. As for inversion, $(h_1 k_1)^{-1} = k_1^{-1} h_1^{-1} \in KH = HK$. So HK is a subgroup of G.

If, say, H is normal, then for $h \in H$, $k \in K$, $hk = k(k^{-1}hk) \in KH$ and $kh = (khk^{-1}) k \in HK$. So $HK = KH$. If both H, K are normal in G, then for any $h \in H, k \in K, g \in G, ghkg^{-1} = (ghg^{-1})$ $(gkg^{-1}) \in HK$.

24. Let $h = x_1^{n_1} \dots x_k^{n_k}$ be a typical element of H. Then for $g \in G$, $ghg^{-1} = (gx_1g^{-1})^{n_1} \dots (gx_kg^{-1})^{n_k} \in H$.

25. In Exercise (1.14), let $x = ab$ and $y = a^{-1}$.

28. (iii) If (g, h) is a generator for $G \times H$, then for every $x \in G$, there is some n such that $(g^n, h^n) = (x, e)$. In particular $g^n = x$. So $G = (g)$. Similarly $H = (h)$. (iv) $\mathbb{Z}_2 \times \mathbb{Z}_2$ is not cyclic.

30. If $A \neq \phi$ and $| X - A | \geqslant 2$, there exist $a \in A, x, y \in X - A$, and $x \neq y$. Let $g: X \to X$ interchange x and y, leaving all other elements fixed. Let $h: X \to X$ interchange a and x, leaving all other elements fixed. Then $g \in F_A$. But $hgh^{-1} \notin G_A$.

31. The elements of finite order in the circle group form an infinite group.

32. Let $G = A \cup B$. If A, B are proper subgroups, there exist $x, y \in G$ such that $x \notin A, y \notin B$. But then $y \in A, x \in B$ and so $xy \notin A$ and $xy \notin B$. For the quaternion group $Q, Q = (i) \cup (j) \cup (k)$.

Section 5.2

1. The cosets of H in G are planes parallel to H.

2. The cosets of K are straight lines parallel to the line K.

3. For $w \in \mathbb{C} - \{0\}$, S^1w is the set of all complex numbers of absolute value $| w |$, or equivalently $S^1w = \{(r, \theta): r = | w |\}$, which is a circle of radius $| w |$ centred at 0. If S^1w_1, S^1w_2 are two such circles, their product is a circle whose radius is the product of their radii, i.e., $| w_1 | | w_2 |$.

4. If $z \in xH \cap yK$ show that $xH \cap yK = z(H \cap K)$. Note that $xH = zH$ and $yK = zK$.

5. Let $x_1H, \dots, x_mH$ and $y_1K, \dots, y_nK$ be the distinct left cosets of H and K respectively. Then $H \cap K$ has at most mn left cosets, they are of the form $x_iH \cap y_jK, i = 1, \dots, m; j = 1, \dots, n$.

7. Let $x_1H, \dots, x_mH$ be the distinct left cosets of H in G and $y_1L, \dots, y_nL$ be those of L in H. Consider $\{x_iy_jL: 1 \leqslant i \leqslant m, 1 \leqslant j \leqslant n\}$. Letting $L = H \cap K$ gives Exercise (2.5).

8. Consider S_3.

9. The left coset multiplication is well-defined iff H has the property that for all $x, y, z, w \in G$, $x^{-1}z \in H$ and $y^{-1}w \in H$ imply that $y^{-1}x^{-1}zw \in H$. If H is normal in G then this is the case because $y^{-1}x^{-1}zw = [y^{-1}(x^{-1}z)y] (y^{-1}w)$. Conversely if the condition holds

and $g \in G$, $h \in H$, put $x = e$, $y = w = g^{-1}$ and $z = h$ to get $ghg^{-1} \in H$.

10. Let $\mathscr{L}$ be the set of all cosets of H in G. Then $\mathscr{L} \subset P(G)$ and $\mathscr{L}$ is closed under $\odot$. So $\odot$ induces a binary operation, the coset multiplication on $\mathscr{L}$. $\{e\}$ is the identity for $\odot$ on the larger set $P(G)$ while H is the identity for the induced binary operation on the subset $\mathscr{L}$. They need not be equal. Proposition (3.4.5) is not violated since it deals with the identities of the same operation on the same set (see the comments following that proposition.)

11. $(x + 1)^p = x^p + 1 + \sum\limits_{k=1}^{p-1} \binom{p}{k} x^k$. Now p divides $\binom{p}{k}$ for $1 \leqslant k < p$ (p being a prime), and $x^p = x$ by induction hypothesis.

12. Apply induction on m and use Theorem (2.16).

13. For all x, $y \in G$, $(xy)^{-1}yx = y^{-1}x^{-1}yx \in C(G)$ and so xy and yx are in the same coset of $C(G)$. For the second part, $(xN)(yN) = (yN)(xN)$ implies $(xy)N = (yx)N$, i.e. $y^{-1}x^{-1}yx \in N$. Since such elements generate $C(G)$, $C(G) \subset N$.

14. For the hint, $hx \notin H$ and so $hxhx = e$ whence $hxh = x$ since $x^2 = e$. Now for h_1, $h_2 \in H$, $h_2x \in G-H$ and so $h_1(h_2x)h_1 = h_2x = h_2h_1xh_1$. So $h_1h_2 = h_2h_1$. For the example, take $G = D_n$, the dihedral group.

15. Suppose $y = g \, x g^{-1}$. Then $h \in N_y$ if $g^{-1}hg \in N_x$.

16. Let Z be the centre of G where $o(G) = p^3$. Then $o(Z) = 1$, p, p^2 or p^3. Since G is non-abelian, $Z \neq G$. So $o(Z) \neq p^3$. Also by Theorem (2.22), $o(Z) > 1$. If $o(Z) = p^2$ then G/Z would be cyclic by Proposition (2.23) and G would be abelian by Proposition (2.24). This leaves $o(Z) = p$.

17. Let x, $h \in H$ and y, $k \in K$. If $f(x, y) = hk$, then $h^{-1}x = ky^{-1} = u$ (say). Then $u \in H \cap K$ and $x = hu$, $y = u^{-1}k$. Conversely for every $u \in H \cap K$, $f(hu, u^{-1}k) = hk$.

18. Let K be a subgroup of order p. Then $H \cap K$ is a subgroup of H. So $o(H \cap K) = 1$ or p. In the first case $|HK| = p^2 > |G|$, a contradiction. So $o(H \cap K) = p$, whence $K = H$.

19. $o(Z) = 1$, p, q, pq. If $o(Z) \neq 1$, then G/Z is cyclic.

21. Let $S_i = x_i^{-1}S$ for $i = 1, 2, 3, 4$. Then $|S_i| = |S|$. So

$$|G - \bigcap_{i=1}^{4} S_i| = |\bigcup_{i=1}^{4} (G - S_i)| \leqslant \sum_{i=1}^{4} |G - S_i| = 4|G| - 4|S| < |G|.$$

So $\bigcap\limits_{i=1}^{4} S_i \neq \phi$.

22. Let $S = \{x \in G : x^2 = e\}$. Then S is closed under inversion. Also let $x, y \in S$. By the last exercise there exists $g \in G$ such that eg, xg, yg and xyg are all in S. So, $x^2 = y^2 = g^2 = (xg)^2 = (yg)^2 = (xyg)^2 = e$. But then $gxg^{-1}x^{-1} = gxgx = e$, showing x commutes with g. Similarly, y commutes with g. So xy commutes with g. This gives $xyg = (xyg)^{-1} = y^{-1}x^{-1}g^{-1} = (gxy)^{-1} = yxg$. So $xy = yx$. Hence $(xy)^2 = x^2y^2 = e$, i.e. $xy \in S$. Thus S is closed under multiplication. Hence S is a subgroup of G. Since $o(S) > \frac{3}{4} o(G)$, $S = G$ by Lagrange's theorem.

25. Without loss of generality, let $x_1 = e$, $x_2 = y$. For any other $x \in G$, x and x^{-1} are inverses of each other and are distinct. So we may suppose for $k > 1$, $x_{2k} = x_{2k-1}^{-1}$.

26. Let $[y] \in \mathbb{Z}_p - \{[0]\}$ where $1 \leqslant y \leqslant p - 1$. If $[y]^2 = e$ in $\mathbb{Z}_p - \{[0]\}$, then p divides $y^2 - 1$, whence either p divides $y - 1$ or p divides $y + 1$. The first possibility gives $[y] = e$ while the second gives $y = p - 1$. So $[p - 1]$ is the only element of order 2 in $\mathbb{Z}_p - \{[0]\}$.

27. Let $x \in G$, $x \neq e$. Then $G = (x)$. If G is infinite, it has proper subgroups, e.g. (x^2). Finally, apply Theorem (2.16).

28. Certainly, $(x) \subset N_x$. So $o(N_x) \geqslant n$. If $o(N_x) > n$, then $N_x = G$ and x is in the centre.

Section 5.3

2. Let R, S be respectively the subgroups (x) and $(f(x))$. Then $S = f(R)$ and so S is isomorphic to a quotient group of R. Hence $o(S)$ divides $o(R)$. But $o(S) =$ order of $f(x)$ while $o(R) = n$. $_1K$, Let y $y_2K, ..., y_mK$ be the distinct left cosets of K in H. For each $i = 1, ... m$ fix $x_i \in f^{-1}(\{y_i\})$. Then $x_1 f^{-1}(K), ..., x_m f^{-1}(K)$ are the distinct left cosets of $f^{-1}(K)$ in G.

3. $R = G/K$ where K is the kernel of f. So $o(R) = o(G/K) = o(G)/o(K)$. Hence $o(R)$ divides $o(G)$. (This fact was also used in the last exercise with different notations.)

4. The composite of two isomorphisms as well as the inverse of an isomorphism are isomorphisms.

5. If $f: G \to H$ is an isomorphism, then for $\theta \in A(G)$, $f\theta f^{-1} \in A(H)$. This correspondence gives an isomorphism from $A(G)$ onto $A(H)$. For the counter-example, take $G = \mathbb{Z}_3$ and $H = \mathbb{Z}_6$. Both $A(G)$ and $A(H)$ have two element each.

6. $f \in A(\mathbb{Z}_p)$ is uniquely determined by $f([1])$, which can be any element of $\mathbb{Z}_p - \{[0]\}$. $A(\mathbb{Z})$ has 2 elements, [the identity function and the function $f: \mathbb{Z} \to \mathbb{Z}$ defined by $f(x) = -x$ for all $x \in \mathbb{Z}$.

7. For $g \in G$, let $T_g: G \to G$ be defined by $T_g(x) = gxg^{-1}$ for $x \in G$. Then $T_g \in I(G)$. The function $\theta: G \to I(G)$ defined by $\theta(g) = T_g$ is a homomorphism with kernel Z.

8. The inversion function is always a bijection. It is a homomorphism iff for all x, y in the group, $(xy)^{-1} = x^{-1} y^{-1}$. Apply Exercise (1.5).

9. If G is non-abelian some inner automorphism is non-trivial. If G is abelian, inversion is a non-trivial automorphism except when $x^2 = e$ for all $x \in G$. In this case G can be considered as a vector space over the field $\mathbb{Z}_2$ (see Exercise (6.3.27)) and from a basis for the vector space, an automorphism for G can be constructed.

10. Let $S = \{x \in G : xf(x) = e\}$.

11. Define $f: H \to HK/K$ by $f(h) = hK$ for $h \in H$. Prove that f is a homomorphism with kernel $H \cap K$.

14. Define $f : \mathbb{Z}_2 \times \mathbb{Z}_2 \to K$ by $f(1, 0) = a, f(0, 1) = b, f(1, 1) = c$ and $f(0, 0) = e$.

15. Write G additively. For any $x \in G$, $p(h(x)) = p(x) - p(j(p(x)))$ $= p(x) - p(x)$ (since $pj = id_{G/H}$). So $h(x) \in \operatorname{Ker} p = H$. Conversely if $y \in H$, then $h(y) = y - j(p(y)) = y$. So h has range H. Define $f: G \to (G/H) \times H$ by $f(x) = (p(x), h(x))$. Show that f is an isomorphism.

16. Define $\psi(xK) = f(x)L$. ψ is well-defined since

$$xK = yK \Rightarrow x^{-1}y \in K \Rightarrow f(x^{-1}y) \in L \Rightarrow f(x)^{-1}f(y) \in L.$$

17. A homomorphism takes a commutator to a commutator. In an abelian group every commutator equals the identity. To get the second assertion of Exercise (2.13), let $H = G/H$ and f the quotient homomorphism.

18. Let $L = \{x \in G : f(x) = g(x)\}$. L is easily seen to be a subgroup of G. Also $S \subset L$. Hence $K \subset L$. An alternate solution is to write a typical element of K in the form

$$x_1^{n_1}, \dots, x_k^{n_k} \text{ for } x_1, \dots, x_k \in S, n_1, \dots, n_k \in \mathbb{Z}.$$

19. Analogous to the second part of Proposition (3.13).

20. Let $o(G) = p^2$. If G has no element of order p^2, let $x \in G, x \neq e$ and let $y \in G - (x)$. Define

$$f: \mathbb{Z} \times \mathbb{Z} \to G \text{ by } f(m, n) = x^m y^n.$$

Show that

$$\operatorname{Ker} f = \{ (m, n) \in \mathbb{Z} \times \mathbb{Z} : p \mid m \text{ and } p \mid n\}.$$

Hence, get an isomorphism from $\mathbb{Z}_p \times \mathbb{Z}_p$ onto G.

21. Let G be a non-abelian group with order 8 and centre Z. Then $o(Z) = 2$. So G/Z is isomorphic to the Klein group by Propositions (3.13) and (2.24). Elements of G may be denoted by $\pm 1, \pm a$, $\pm b, \pm c$ where 1 is the identity, -1 is the other element of Z and $(-1)x$ is written as $-x$. G must have an element of order 4, for otherwise every element except 1 would be of order 2 and G would be abelian by Exercise (1.5). Let a be of order 4. Then $a^2 = -1$ (because in G/Z, $(aZ)^2 = a^2 Z$, but $a^2 \neq 1$). Without loss of generality, let $ab = c$ (the other possibility being $ab = -c$). Then $ac = -b$. Also $ba = -c$ (otherwise $ba = c$, putting b in N_a which would mean $o(N_a) \geqslant 5$ and hence $N_a = G$, i.e., $a \in Z$). Similarly $ca = b$. This determines the multiplication on G except for bc which must equal a or $-a$. Depending upon which possibility holds, G is isomorphic to Q or to D_4.

22. Let G be an abelian group of order 8. If every element of G (except e) is of order 2, choose $x \in G - \{e\}$, $y \in G - (x)$ and $z \in G - H$ where H is the subgroup generated by $\{x, y\}$. Define $f: \mathbb{Z} \times \mathbb{Z} \times \mathbb{Z} \to G$ by $f(i, j, k) = x^i y^j z^k$ and work as in Exercise (3.20). If G contains an element of order 8, G is isomorphic to $\mathbb{Z}_8$. The only case left is where G has an element, say x, of order 4 but no element of order 8. Let $H = (x)$. At least one element of $G - H$ must be of order 2. For otherwise, let $y \in G - H$. Then $xy \in G - H$. y^2 and $(xy)^2$ are of order 2 and hence equal x^2, the only element of order 2 in G. But this gives $x^2 = e$, a contradiction, So some $y \notin H$ has order 2. Now, $G/H = \{H, yH\}$. Define $j: G/H \to G$ by $j(H) = e$ and $j(yH) = x^2$. Exercise (3.15) is now applicable.

23. If T_g is the right translation by g, then for $g, h \in G$, T_{gh} equals $T_h \circ T_g$ and not $T_g \circ T_h$. So the function f would not be a group homomorphism. However, if $f(g) = T_{g^{-1}}$ then $f(gh) = T_{(gh)^{-1}}$ $= T_{h^{-1}g^{-1}} = T_{g^{-1}}T_{h^{-1}} = f(g)f(h)$. So an alternate proof is possible.

24. Any group G of order n, induces a group structure on the set $S = \{1, 2, ..., n\}$, through any bijection from G to S. Isomorphic groups induce isomorphic group structures. But on S there are only finitely many binary operations and only a few of them make S into a group.

26. Let $f: S \to T$ be a bijection with $g: T \to S$ as its inverse. Think of S, T as subsets of $F(S), F(T)$, respectively. Regarding f as a function from S to $F(T)$ let $F: F(S) \to F(T)$ be the unique homomorphism which extends f. Similarly let $G: F(T) \to F(S)$ extend g. Then $G \circ F$ is a homomorphism which extends the inclusion function from S to $F(S)$. Since the identity function $1_{F(S)}$ also does the same, $G \circ F = 1_{F(S)}$. Similarly $F \circ G = 1_{F(T)}$. So F, G are isomorphisms.

27. Define $\theta : F(\{a, b\}) \to \mathbb{Z}_{10}$ by $\theta(a) = [2]$, $\theta(b) = [5]$. Show that the kernel of θ is the smallest normal subgroup of $F(\{a, b\})$ containing a^5, b^2 $aba^{-1}b^{-1}$. The argument is analogous to that in the text for D_5. More generally for any positive integer n, the relations $a^n = b^2 = abab = e$ given D_n while the relations $a^n = b^2 = aba^{-1}b^{-1} = e$ given $\mathbb{Z}_{2n}$.

28. The problem reduces to proving $b^j a^r = a^{rk^j} b^j$. For $j = 1$, apply induction on r. Then apply induction on j.

29. Let k be a positive integer. Define a binary operation $*$ on the set $\mathbb{Z}_7 \times \mathbb{Z}_3$ by $([i], [j]) * ([r], [s]) = ([i + rk^j], [j + s])$. Then $*$ is well-defined iff $i \equiv i'$ (mod 7), $r \equiv r'$ (mod 7), $j \equiv j'$ (mod 3) and $s \equiv r'$ (mod 3) imply that $i + rk^j \equiv i' + r'k^{j'}$ (mod 7) and $j + s \equiv j' + s'$ (mod 3). This finally reduces to $k^3 \equiv 1$ (mod 7), giving $k = 1, 2$ or 4 (mod 7). For these values of k, $*$ does make $\mathbb{Z}_7 \times \mathbb{Z}_3$ into a group. On the other hand if $k = 3$, then the relation $b^j a^r = a^{rk^j} b^j$ gives, for $j = 3$ and $r = 1$, $a = a^{27}$. So $a^{26} = e$, i.e., $a^2 = e$ and hence $a = e$ (since $a^7 = e$). But then G is generated by the single element b with relation $b^3 = e$.

30. This time start with the homomorphism $\theta : F(\{a, b\}) \to Q$ which takes a to i and b to j.

31. First extend $f : S \to G$ to a group homomorphism $\theta : F(S) \to G$. Now apply Exercises (3.17) and (3.16) (with appropriate changes of notation). For the second assertion, take $S = G$.

32. Let $f, g \in B(S)$. Suppose f vanishes outside a finite subset C of S and g vanishes outside a finite subset D of S. Then $f + g$ vanishes outside the finite set $C \cup D$ and so $f + g \in B(S)$. Similarly $-f$ vanishes outside C and hence is in $B(S)$. So $B(S)$ is a subgroup of $\mathbb{Z}^S$. Following the hint, let $\theta : F(S) \to B(S)$ be the unique homomorphism which extends f. Let $K =$ kernel of θ. By Exercise (3.17), $C(S) \subset K$. Conversely let $x \in K$. Write $x = s_1^{n_1} s_2^{n_2} \dots s_r^{n_r}$ where $r \in \mathbb{N}$, $s_1, \dots, s_r \in S$, $n_1, \dots, n_r \in \mathbb{Z} - \{0\}$ and the adjacent s_i's are unequal. To show $x \in C(S)$, apply induction on r. Note that $\theta(x)$ is the function $g : S \to \mathbb{Z}$ which equals $n_1 f_{s_1} + n_2 f_{s_2} + \dots + n_r f_{s_r}$. In particular $g(s_1) = 0$. But $g(s_1) = n_{i_1} + n_{i_2} + \dots + n_{i_p}$ where $1 \leqslant i_1, \dots, i_p \leqslant r$ are the indices for which $s_{i_1} = s_{i_2} = \dots = s_{i_p} = s_1$ and we let $i_1 = 1$. Since $n_1 \neq 0$, $p \geqslant 2$. Thus there is some $k > 0$ such that $s_{k+1} = s_1$, (Actually $k > 1$, since the adjacent s_i's are unequal.) Now write $x = yz$ where

$$y = s_1^{n_1}(s_2^{n_2} \dots s_k^{nk}) s_1^{-n_1} (s_k^{-nk} \dots s_2^{-n_2})$$

and

$$z = s_2^{n_2} \dots s_k^{nk} s_1^{n_1 + n_{k+1}} s_{k+1}^{nk+1} \dots s_r^{nr}.$$

Then $y \in C(S) \subset K$. So $z \in K$. Since length of z is $r - 1$ (or even less if $n_1 + n_{k+1} = 0$), induction hypothesis applies.

33. Given $f \in B(S)$, let $s_1, \ldots, s_r$ be the points of S where f has non-zero values and let $n_1, \ldots, n_r$ be respectively these values. Then $f = \sum_{i=1}^{r} n_i s_i$. For uniquenss suppose f also equals $\sum_{j=1}^{k} m_j t_j$. Then $\{s_1, \ldots, s_r\}$ and $\{t_1, \ldots, t_k\}$ are the same sets, since both consist of points where f does not vanish. Now note that $f(s_i) = n_i$ while $f(t_j) = m_j$.

34. Let $G = A(S)$, $H = A(T)$, where the sets S, T are disjoint. Every pair of functions $(f_1, f_2) \in B(S) \times B(T)$ determines $f \colon S \cup T \to \mathbb{Z}$, defined by $f(x) = f_1(x)$ if $x \in S$ and $f(x) = f_2(x)$ if $x \in T$. This gives an isomorphism from $B(S) \times B(T)$ onto $B(S \cup T)$ and hence from $A(S) \times A(T)$ onto $A(S \cup T)$. If $S \cap T \neq \phi$, first disjunctify them by replacing them with disjoint sets, $S \times \{1\}$ and $T \times \{2\}$.

35. (i) can be proved directly or as a consequence of (iii).

    (ii) let $x \in H$. Write x as $x_1^{n_1} x_2^{n_2} \ldots x_r^{n_r}$. The x_i's are alternately a's and b's. Apply induction on r. Consider various cases depending upon whether $x_1 = a$ or b, and $n_1 > 0$ or $n_1 < 0$. If for example, $x_1 = a$ and $n_1 > 0$, write $x = yz$ where $y = (a^{n_1} b^{n_1})$ and $z = b^{n_2 - n_1} x_3^{n_3} \ldots x_r^{n_r}$. Note that $z \in H$ and has length less than r. For (iv), suppose S is a finite subset of H which generates H. By (ii), write each element of S as a finite product of elements of the form $a^n b^n$, $n \in \mathbb{Z}$ (or their inverses). Since S is finite only finitely many elements of this form are involved. So there exist $n_1, \ldots, n_k \in \mathbb{Z}$ such that the set $T = \{a^{n_i} b^{n_i} : i = 1, \ldots, k\}$ also generates H. Now let m be an integer greater than all $|n_i|$, $i = 1, \ldots, k$. Then $a^m b^m \in H$ but cannot be expressed as a product of elements in T (or their inverses).

Section 5.4

1. Let $\sigma \in H$. Then $\sigma(\{1, 2, \ldots, n - 1\}) = \sigma(\{1, \ldots, n\} - \{n\}) = \{1, \ldots, n\} - \sigma(\{n\})$, σ being a bijection. So σ induces $\theta \colon \{1, 2, \ldots, n-1\} \to \{1, 2, \ldots, n - 1\}$ defined by $\theta(x) = \sigma(x)$ for $x = 1, \ldots, n - 1$. (θ is often called the restriction of σ to $\{1, 2, \ldots, n - 1\}$, although strictly speaking when we take the restriction of a function, it is only the domain and not the codomain that gets reduced.) The correspondence $\sigma \leftrightarrow \theta$ gives the desired isomorphism. The second assertion follows from the fact that σ and θ have the same parity. S_{n-1} is not normal in S_n for $n > 2$ (consider conjugation by the transposition $(n - 1, n)$.)

2. Let $X =$ be the set of ordered r-tuples with distinct entries from $1,..., n$; and Y the set of all r-cycles in S_n. Define $f: X \to Y$ by $f(x_1,..., x_r)$ to be the cycle $(x_1 x_2 ... x_r)$. Each $y \in Y$ has r preimages.

4. Let $\tau = \sigma \theta \sigma^{-1}$. Express θ as a product of disjoint cycles say $\theta = c_1 c_2 ... c_k$. Then $\tau = (\sigma c_1 \sigma^{-1}) (\sigma c_2 \sigma^{-1})...(\sigma c_k \sigma^{-1})$. Apply the last exercise.

5. Let $\theta = (i_1 i_2 ... i_{r_1}) (i_{r_1+1} ... i_{r_1+r_2}) ... (i_{r_1+...r_{k-1}+1} ... i_{r_1+...+r_k})$ and $\tau = (j_1 ... j_{r_1}) (j_{r_1+1} ... j_{r_1+r_2}) ... (j_{r_1+...+r_{k-1}+1} ... j_{r_1+...+r_k})$ where $r_1 \geqslant r_2 \geqslant ... \geqslant r_k$ and $r_1 + r_2 + ... + r_k = n$. Let $\sigma (i_p) = j_p$ for $p = 1,..., n$.

6. With the notation in the solution to the last exercise, the conjugacy class of θ corresponds to the partition $r_1 + r_2 + ... + r_k$ of n.

7. The conjugacy class of θ has $(n - 1)$! elements and so N_θ, the normaliser of θ, has order n. Since the elements $1, \theta, \theta^2,..., \theta^{n-1}$ are distinct, they exhaust N_θ.

8. By the last exercise any element in the centre is a power of every n-cycle. For $n > 2$, the only common power of $(123... n)$ and $(213... n)$ is the identity.

9. For $i \geqslant 2$, $j \geqslant 2$ and $i \neq j$, $(ij) = (1j) (1i)(1j)$. So S_n is generated by $\{(1i): 2 \leqslant i \leqslant n\}$. Let $\theta \in A_n$ Write $\theta = (1i_1) (1i_2)...(1i_{k-1})(1i_k)$ where $i_1, i_2,..., i_k \geqslant 2$. Then k is even. If for any $r = 1,..., \frac{k}{2}$, $i_{2r-1} = i_{2r}$ then the product $(1 i_{2r-1}) (1 i_{2r})$ equals identity and may be deleted from the expression for θ. If $i_{2r-1} \neq i_{2r}$, then $(1 i_{2r-1}) (1 i_{2r}) = (1 i_{2r} i_{2r-1})$. So θ is a product of 3-cycles, [Actually, it suffices to take 3-cycles of the form $(1 r s)$.]

10. For $2 < r < s$, $(1 r s) = (1s) (1r) = (1 s) (1 2) (1 2) (1 r) = (1 2 s) (1 r 2) = (1 2 s) [(1 2 r)]^{-1}$.

11. Let $\theta = (1 2)$ and $\sigma = (1 2 3...n - 1 n)$. Let G be the subgroup of S_n generated by θ and σ. Using Exercise (4.3) and induction on i, it follows that $\sigma^i \theta \sigma^{-i} = (i + 1, i + 2)$ for $i = 0, 1,..., n - 1$ (with the understanding that $n + 1 = 1$). Now for $i = 2,... n-1$, $(1, i) (i, i + 1) (1, i) = (1, i + 1)$ and so it follows again by induction that $(1, i) \in G$ for $i = 2,..., n$. But transpositions of this form generate S_n as noted in Exercise (4.9) above.

12. Let N be a normal subgroup of A_4. If $N \neq \{e\}$ and $N \neq K$, then N contains some 3-cycle σ. But then $o(N)$ would have to be 12, 6 or 3. As shown in the text, A_4 contains no subgroup of order 6. So if $A_4 \neq N$, then N would have to be (σ) which is not normal in A_4.

13.

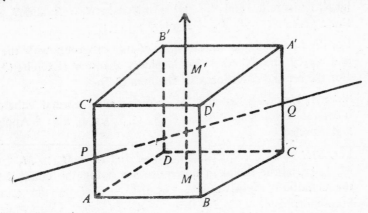

Let the cube be as shown and G the group of its orientation preserving isometries. Let $\theta \in G$. $\theta(A)$ can be any of the 8 vertices. $\theta(B)$ must be adjacent to $\theta(A)$. So for a given $\theta(A)$, $\theta(B)$ has 3 choices. Once $\theta(A)$, $\theta(B)$ are fixed $\theta(C)$ is uniquely determined because of preservation of distance and orientation. Similary θ is determined at all other vertices. It follows that $o(G) \leqslant 24$. That equality holds is shown by enumerating 24 orientation preserving isometrtis. They are (i) identity (ii) 3 rotations each about axes like MM', joining midpoints of opposite faces; one of which has cycle structure $(ABCD)(A'B'C'D')$ (iii) six 180° rotations about axes like PQ joining mid-points of opposite sides; one of them has cycle structure $(A'C)$ $(C'A)$ (BB') (DD') and (iv) two 120° rotations about each of the four diagonals, one of them has cycle structure $(A)(A')(C'BD)$ $(D'C\,B')$.

14. A pair of diagonal vertices is characterised as being $\sqrt{3}\,a$ units away where a is the side of the cube. So every isometry of the cube induces a permutation of the four diagonals. Indeed with the answer to the last exercise. isometries of type (iii) correspond to transpositions in S_4, those of type (iv) correspond to 3-cycles. As for (ii), the 90° rotations give 4-cycles while 180° rotations give elements of $K \subset A_4$.

15. By Exercise (3.11), $N \cap A_5$ is a normal subgroup of A_5. If $N \cap A_5 = A_5$, then $N = A_5$ if $o(N) = 60$, while $N = S_5$ if $o(N) >$ 60. If $N \cap A_5 = \{e\}$ then $N = \{e\}$, for otherwise every permutation in N except e is odd. If θ, σ are two such permutations then $\theta\sigma$, θ^2 are even and must equal e. So N has exactly one element other than e. But then N is not normal in S_5.

16. If G is abelian, take $N_0 = G$, $N_1 = \{e\}$. For D_n, take $N_1 =$ the subgroup of order n consisting of rotations and $N_2 = \{e\}$. For Q, let N_1 be its centre and $N_2 = \{1\}$. For S_3 take $N_1 = A_3$ and $N_2 = \{e\}$. For S_4 take $N_1 = A_4$, $N_2 = K$ and $N_3 = \{e\}$.

17. For the first part take intersections of the subgroup with the N_i's and apply the Noether isomorphism theorem (Exercise (3.11)). For the second assertion, use Theorem (3.7).

18. Since A_5 is non-abelian and has no non-trivial normal subgroups, it cannot be solvable. Also $A_5 \subset A_n \subset S_n$ for all $n \geqslant 5$. Apply the last exercise.

19. Let $G/H = N_0 \supset N_1 \supset ... \supset N_r = H$ and $H = M_0 \supset M_1 \supset ... \supset M_s = \{e\}$ be the chains of subgroups satisfying the conditions in the definition of solvability. Let $f: G \to G/H$ be the quotient homomorphism. For $i = 1,..., r$, let $K_i = f^{-1}(N_i)$, while for $i = r + 1,..., r + s$, let $K_i = M_{i-r}$. Then $G = K_0 \supset K_1 \supset ... K_r \supset K_{r+1} \supset ... \supset K_{r+s} = \{e\}$ is a chain of subgroups showing G is solvable.

20. Let $z \neq e$ be in the centre Z of G. Then $o(z) = p^r$ for some $r \geqslant 1$. If $r = 1$, let $H = Z$. If $r > 1$, let $H = (z^{p^{r-1}})$. In either case $o(H) = p$. So $o(G/H) = p^{m-1}$. Apply induction hypothesis and then Theorem (3.7).

21. Under a suitable bijection between $\{1, 2,..., 9\}$ and $\mathbb{Z}_3 \times \mathbb{Z}_3$, it is easy to see that θ and σ are given respectively by, $\theta(x, y) = (x, y+1)$ and $\sigma(x, y) = (x + y, y)$ for $x, y \in \mathbb{Z}_3$. Since $\theta^3 = \sigma^3 = e$, the various elements of G are functions $f: \mathbb{Z}_3 \times \mathbb{Z}_3 \to \mathbb{Z}_3 \times \mathbb{Z}_3$ of the form $f(x, y) = (x + ay + b, y + c)$ for some $a, b, c, \in \mathbb{Z}_3$. So $o(G) \leqslant 27$. That equality holds can be seen by showing that all 27 functions of this form (resulting from various choices of a, b, c) are in fact the composites of suitable sequences of θ and σ. (Actually it sffices to do this verification for only ten of them by Lagrange's theorem.) It is clear that every function of this form has order 3, except when a, b, c are all [0] which, corresponds to the identity function.

22. Elements of $\mathbb{Z}_3 \times \mathbb{Z}_3 \times \mathbb{Z}_3$ are of the form (x, y, z) for $x, y, z \in \mathbb{Z}_3$. Clearly $3(x, y, z) = (3x, 3y, 3z) = (0, 0, 0)$. $\mathbb{Z}_3 \times \mathbb{Z}_3 \times \mathbb{Z}_3$ is abelian while the group in the last exercise is not.

23. A_5 is simple by Theorem (4.13). Let $n > 5$ and suppose $N \neq \{e\}$ is a normal subgroup of A_n. It suffices to show that N contains a permutation say ϕ other than identity which has at least one fixed point. For in that case we may think of ϕ as an element of A_{n-1}, whence by induction hypothesis $N \cap A_{n-1}$ would be the entire group A_{n-1}. But then N contains a 3-cycle in $A_{n-1} \subset A_n$, from which it can be shown that N contains all 3-cycles in A_n (as in the proof of Theorem (4.13)) and finally that $N = A_n$ (using Exercise (4.9)).

To show that N contains a permutation with a fixed point, let $\theta \in N$, $\theta \neq e$. If θ has a cycle of length 1, θ itself has a fixed point. Otherwise let $\theta = \theta_1\theta_2...\theta_r$ be the cyclic decomposition of θ. If $r=1$, then θ is an n-cycle which may be taken to be $(1\ 2\ 3...(n-1)n)$. Let $\sigma = (1\ 2)(3\ 4)$. Then $\sigma \in A_n$ and $\sigma \theta \sigma^{-1} = (2\ 1\ 4\ 3\ 5...(n-1)n)$. Now $\theta \sigma \theta \sigma^{-1}$ is in N and has 2 as a fixed point. Next suppose $r \geqslant 2$ and at least one cycle, say θ_1, has length > 2. Let $\theta_1 = (i_1, i_2,..., i_p)$ and $\theta_2 = (j_1, j_2,..., j_q)$ with $p > 2$. Again let $\sigma = (i_1, i_2)(j_1, j_2)$. Then $\theta \sigma \theta \sigma^{-1} \in N$ and has i_2 and j_2 as fixed points. Also $\theta \sigma \theta \sigma^{-1} \neq e$ since $p > 2$. The only case left is when each θ_i is a 2-cycle. In that case $n = 4k$ for some integer $k > 1$ and we may suppose θ is $(1\ 2)(3\ 4)(5\ 6)...(4k - 1, 4k)$. In this case let $\sigma = (1\ 3\ 5)$. Then $\theta \sigma \theta \sigma^{-1} = (1\ 5\ 3)(2\ 4\ 6)(7\ 8)...(4k-1, 4k)$ is in N and the earlier argument (where $p > 2$) can be applied to it.

24. Let $\theta \in S_n$ and suppose $\theta = \theta_1\theta_2,..., \theta_k$ is the cycle decomposition of θ. Let n_i be the length of θ_i. Then n_i is also order of θ_i. Let m be the least common multiple of $n_1,..., n_k$. Then since the θ_i's commute with each other, we certainly have $\theta^m = e$. On the other hand suppose $\theta^r = e$. For each i, let x be any element in the cycle θ_i. Then $\theta_i{}^r(x) = \theta^r(x) = x$ gives n_i divides r. So m divides r. Hence m is the order of θ.

CHAPTER 6

Section 6.1

2. 0 is always an idempotent. Suppose x is a non-zero idempotent of an integral domain R. Suppose first R has an identity. Then $x^2 = x = x \cdot 1$ implies $x = 1$. If R has no identity let F be the field of quotients of R. Then x as an element of R corresponds to the element $[x^2, x]$ of F. But this is the same as $[x, x]$ which is precisely the identity element of F. So x is an identity for R, a contradiction. In a Boolean ring every element is an idempotent.

3. The centre of Q consists of quaternions of the form $a_0 + 0i + 0j + 0k$, which is clearly isomorphic to $\mathbf{R}$. The centre of $M_2(\mathbf{R})$ consists of all matrices of the form $\begin{pmatrix} a & 0 \\ 0 & a \end{pmatrix}$ for $a \in \mathbf{R}$. It is also isomorphic to $\mathbf{R}$.

4. Let $u = xy^2 - y^2x$. Then $yuy = yxy^3 - y^3xy = yxy - yxy = 0$. So $yu = (yu)^3 = yuyuyu = 0$, giving $yxy^2 = yx$ and hence $y^2x =$

y^2xy^2. Similarly $uy = 0$ gives $xy^2 = y^2xy^2$. This proves the hint. Now for any x, y; x^2, y^2 and $(yx)^2$ commute with all elements. So $xy = (xy)^3 = x(yx)^2y = (yx)^2xy = yxyx^2y = yx^3y^2 = y^3x^3 = yx$.

5. The first assertion is a consequence of a well-known result, called **Cauchy-Schwarz inequality**, which says that for any real numbers

   $a_1,..., a_n, b_1,..., b_n$, $B^2 \leqslant AC$ where $B = \sum\limits_{i=1}^{n} a_ib_i$, $A = \sum\limits_{i=1}^{n} a_i^2$ and

   $C = \sum\limits_{i=1}^{n} b_i^2$. To prove it observe that the quadratic $f(x) = Ax^2$

   $+ 2Bx + C = \sum\limits_{i=1}^{n} (a_ix + b_i)^2$ is always non-negative and hence its

   discriminant cannot be positive.

   This second assertion can be proved directly by a lengthy computation. Easier proofs are available using properties of vector spaces. One such proof will be indicated in Exercise (3.26).

6. $\{a_0 + 0i + 0j + 0k: a_0 \in \mathbb{R}\}$ and $\{a_0 + a_1i + 0j + 0k: a_0, a_1 \in \mathbb{R}\}$ respectively.

7. Define $f: Q \to M_4(\mathbb{R})$ by $f(a_0 + a_1i + a_2j + a_3k)$ to be

$$\begin{pmatrix} a_0 & -a_1 & -a_2 & -a_3 \\ a_1 & a_0 & -a_3 & a_2 \\ a_2 & a_3 & a_0 & -a_1 \\ a_3 & -a_2 & a_1 & a_0 \end{pmatrix}$$

   Then f is a one-to-one ring homomorphism. Note that since matrix multiplication is associative, this exercise gives an easier proof of the associativity of the quaternionic multiplication.

8. To avoid confusion between complex numbers and quaternions denote by θ the complex number $\sqrt{-1}$ (which is commonly denoted by i). Every complex number z then has the form $x + \theta y$ with $x, y \in \mathbb{R}$ and its complex conjugate is $\bar{z} = x - \theta y$. Now define

$$g: Q \to M_2(\mathbb{C}) \text{ by } g(a_0 + a_1i + a_2j + a_3k) = \begin{pmatrix} z & -\bar{w} \\ w & \bar{z} \end{pmatrix}$$

   where

$$z = a_0 + \theta a_2 \text{ and } w = a_1 + \theta a_3.$$

10. Let $(F, +, \cdot)$ be a field. Suppose $x \in F$ generates the additive group $(F, +)$. Clearly $x \neq 0$. First we claim that F must be of a prime

characteristic. For otherwise, let $u = 1 + 1 \in F$. Then $u \neq 0$. So u^{-1} exists in F. Let $y = u^{-1}x$. Then it is easily seen that $x = y + y = 2y$. Now $y = mx$ for some $m \in \mathbb{Z}$. So $2mx = x$ or $(2m-1)x = 0$, whence x is of finite order. By Corollary (1.11), F has characteristic p for some prime p. But then $px = 0$ and so F consists of the p elements $0, x, 2x, \ldots, (p-1)x$.

11. Let $R_1 = R_2 = R$. Then $\Delta R = \{(x, x): x \in R\}$ is a subring of $R \times R$, but is not of the form $S_1 \times S_2$.

12. For example, in $M_2(\mathbb{R})$, let

$$R_1 = \left\{ \begin{pmatrix} a_{11} & a_{12} \\ 0 & 0 \end{pmatrix} : a_{11}, a_{12} \in \mathbb{R} \right\}$$

and

$$R_2 = \left\{ \begin{pmatrix} a_{11} & 0 \\ a_{21} & 0 \end{pmatrix} : a_{11}, a_{21} \in \mathbb{R} \right\}.$$

Then R_1 is a right ideal and R_2 is a left ideal of $M_2(\mathbb{R})$. But $R_1 \cap R_2$ is neither.

13. No.

14. If $A = \{x_0\}$ and J is an ideal of R such that $M_A \underset{\neq}{\subset} J$, let $f \in J$, $f \notin M_A$. Let $g : [0, 1] \to \mathbb{R}$ be $g(x) = (x - x_0)^2$ and $h(x) = (f(x))^2 + g(x)$. Then $g \in M_A \subset J$ and so $h \in J$. h is never 0 and hence is an invertible element of R. So $J = R$. Conversely if A contains two distinct elements say a, b, then the function $g(x) = (x - a)^2$ is in $M_{\{a\}}$ but not in M_A.

15. For $f \in R$, let $Z(f) = \{x \in [0, 1]: f(x) = 0\}$. Each $Z(f)$ is a closed subset of $[0, 1]$. Let M be a proper ideal of I and let $Z = \underset{f \in M}{\cap} Z(f)$. We claim $Z \neq \phi$, for otherwise by compactness of $[0, 1]$, there exist $f_1, \ldots, f_n \in M$ such that $\overset{n}{\underset{i=1}{\cap}} Z(f_i) = \phi$. Now the function $\overset{n}{\underset{i=0}{\Sigma}} f_i^2$ is in M and never vanishes. So M contains an invertible element, contradicting $M \underset{\neq}{\subset} R$. Let $x_0 \in Z$. Then $M \subset M_{\{x_0\}}$.

16. That R is a subring follows from the fact that the sum, the difference and the product of two analytic functions are analytic. To show R contains no zero divisors let $f, g \in R$. Then $Z(fg) = Z(f) \cup Z(g)$ (where $Z(f) = \{z \in \mathbb{C}: f(z) = 0\}$ etc.). Since $Z(f), Z(g)$ are discrete, so is $Z(fg)$. So fg cannot vanish identically.

17. In general, for any positive integer m, $\{a + \sqrt{m}b: a, b \in \mathbb{Z}\}$ and

$\{a + i\sqrt{mb} : a, b \in \mathbb{Z}\}$ are subrings of $\mathbb{R}$, $\mathbb{C}$ respectively. Unless m is a perfect square, the expressions of the elements are unique, i.e. $a + \sqrt{mb} = c + \sqrt{md}$ iff $a = c$ and $b = d$ etc. More generally for any positive integers n and k, let $\alpha = (n)^{1/k} e^{2\pi i/k}$. Then complex numbers of the form $\sum_{j=0}^{n-1} a_j \alpha^j$ where a_j's $\in \mathbb{Z}$ form a subring of $\mathbb{C}$. All these subrings are integral domains, but their properties depend considerably on the integers n and k.

18. (a) This is a special case of Exercise (3.4.25) since the poset $(\mathbb{N}_0, \leqslant)$, where $\leqslant$ is the usual order, is locally finite. Given $f: \mathbb{N}_0 \to R$, where R is a ring, we define $f: \leqslant \to R$ as $f(x, y) = f(y - x)$. In Exercise (3.4.25), $\mathbb{R}$ could be replaced by any ring R in the proof of associativity of convolution.

    (b) An element of $M_n(R[x])$ is an $n \times n$ matrix say $F(x) = (f_{ij}(x))$ where each $f_{ij}(x)$ is a polynomial over R. These n^2 polynomials may have different degrees. Let r be the maximum of their degrees. Then we may identify $F(x)$ with the polynomial $A_0 + A_1 x + \ldots + A_r x^r$ where for $k = 0, \ldots, r$, A_k is an $n \times n$ matrix whose (i, j)th entry is the coefficient of x^k in the polynomial $f_{ij}(x)$. Then $A_k \in M_n(R)$. This correspondence is a ring isomorphism.

19. This is also a special case of Exercise (3.4.25).

20. The first assertion is straightforward. For the second let $(fg)(x) = \sum_{n=0}^{\infty} c_n x^n$ where for each $n \geqslant 0$ $c_n = \sum_{j=0}^{n} a_j b_{n-j}$. The problem now amounts to showing that for each $n \geqslant 0$, $(n+1)c_{n+1} = \sum_{r=0}^{n} a_r (n - r + 1)b_{n-r+1} + \sum_{r=0}^{n} (r + 1)a_{r+1}b_{n-r}$. This follows since for each j, $0 \leqslant j \leqslant n + 1$, the coefficient of $a_j b_{n+1-j}$ in the first sum is $(n-j+1)$ while that in the second sum is j.

21. The ideal generated by a non-zero element equals the whole ring iff that element is invertible.

22. Define the isomorphism $\theta: R/K \to T$ by $\theta(x + K) = f(x)$ for $x \in R$. Since $x + K = y + K \Leftrightarrow x - y \in K \Leftrightarrow f(x) = f(y)$, θ is a well-defined and one-to-one. That it preserves the binary operations is routine to verify. If U is a subring or an ideal of T, $f^{-1}(U)$ is a subring (respectively an ideal) of R. This gives a one-to-one correspondence between subrings of T and subrings of R containing K etc.

23. Let $f : R \to R/I$ be the quotient homomorphism. By the last exercise I is a maximal ideal of R iff the ring R/I has no proper ideals. Now apply Exercise (1.21) above.

24. Let K be the kernel of f. If f is not identically 0, then $K \neq R$ and so $K = \{0\}$, the only other ideal of R. But then f is one-to-one.

25. For $x, y \in R$, with $y \neq 0$, define $g[x, y] = f(x)[f(y)]^{-1}$.

26. If $x > 0$, then $x \cdot x > 0 \cdot 0 = 0$. If $x < 0$, then $-x > 0$ for otherwise we would have $x + (-x) < 0 + 0$, i.e. $0 < 0$ a contradiction. So $(-x)^2 > 0$, i.e. $x^2 > 0$. In particular $1 = 1^2 > 0$. If F has characteristic $p \neq 0$, then $1 + 1 + \ldots + 1$ (p times) is > 0; a contradiction.

27. For example, in Q, the ring of quaternions, if $(i + j)(i - j)$ equals $i^2 - j^2$, then either $i + j$ or $i - j$ would have to vanish, there being no zero-divisors in Q. For binomial theorem, apply induction on n along with Proposition (2.2.19).

Section 6.2

1. Apply the usual euclidean algorithm. If $\frac{1}{2}|b| < r < |b|$, increase/decrease q by 1 depending upon whether $b > 0$ or $b < 0$ and change r accordingly.

2. The first assertion is a standard property of complex numbers (called 'multiplicativity of absolute value'). For the second assertion, following the hint, $d(r + is) = r^2 + s^2 \leqslant \frac{1}{4}n^2 + \frac{1}{4}n^2 < n^2$.

3. Let $n = a^2 + b^2 = (a + ib)(a - ib)$. Applying the last exercise to $ax + yb + i(ay - bx)$ [which equals $(x + iy)(a - ib)$] find $p, q, r, s \in \mathbb{Z}$ such that $r^2 + s^2 < n^2$ and $(x + iy)(a - ib) = n(p + iq) + (r + is)$. Dividing by $a - ib$, we see, $x + iy = (ap - bq) + i(bp + aq) + \dfrac{r + is}{a - ib}$. This means $\dfrac{r + is}{a - ib} \in G$. Also $d\left(\dfrac{r + is}{a - ib}\right) = \dfrac{d(r + is)}{d(a - ib)} = \dfrac{r^2 + s^2}{a^2 + b^2} < \dfrac{\frac{1}{2}n^2}{n} = \dfrac{1}{2}n$.

4. $x^2 + y^2$ factors into $(x + iy)(x - iy)$. Neither factor is a unit.

5. If p is a prime in G, then either $p \mid x + iy$ or $p \mid x - iy$ in G. In the first case $x + iy = (a + ib)p$ for some $a, b \in \mathbb{Z}$. But then $x = pa$ and $y = pb$, contradicting that x, y are relatively prime in $\mathbb{Z}$. Similarly $p \mid x - iy$ gives a contradiction.

6. In general, in G, $(x + iy) \mid (a + ib) \Rightarrow (x - iy) \mid (a - ib)$ because of complex conjugation. So $a + ib$ is a prime in G iff $a - ib$ is a prime in G. Moreover, these two primes are not associates of each other in G unless either a or b is 0. If $a + ib$ is a prime divisor of

p in G then certainly neither a nor b is 0 (or else p would not be a prime in $\mathbb{Z}$). So the prime factorisation of p in G contains both $a + ib$ and $a - ib$. Hence their product, $(a^2 + b^2)$, divides p in G and hence in $\mathbb{Z}$. But then $p = a^2 + b^2$.

7. For each $r = 1, 2, ..., q$, $[p - r] = [r]$ in the group $\mathbb{Z}_p - \{[0]\}$ under modulo p multiplication. So modulo p, y^2 equals the product $1 \cdot 2 \cdot ... \cdot q \cdot (q + 1) \cdot ... \cdot (p - 1)$ which equals -1 (mod p.). Let $1 \leqslant x \leqslant p-1$ be such that $y \equiv x$ (modulo p). Then $x^2 \equiv y^2 \equiv -1$ (modulo p).

8. For uniqueness, if $x + iy$ has a proper divisor say $u + iv$ in G, then $d(u + iv)$ would be a proper divisor of $d(x + iy)$ in $\mathbb{Z}$ by Exercise (2.2) contradicting that $x^2 + y^2$ is a prime in $\mathbb{Z}$. So $p = (x + iy)(x - iy)$ is a factorisation of p in G into primes. If $p = (a + ib)(a - ib)$ is another such factorisation then by Theorem (2.28), $a + ib$ is an associate of $x + iy$ or of $x - iy$. In either case $\{a, b\} = \{x, y\}$.

9. (i) follow hint and get a contradiction since 4 does not divide $4n + 2$, which is the order of the group $\mathbb{Z}_p - \{[0]\}$.

    (ii) Let $p = x^2 + y^2$. Obviously $0 < y < p$. So there exists u such that $uy \equiv 1$ (mod p), giving $(xu)^2 + 1 \equiv 0$ (mod p), contradicting (i).

    (iii) Argue as in Exercise (2.6) to get a contradiction to (ii).

10. Let $S = \{n \in \mathbb{N} : n$ can be expressed as a sum of squares of two integers$\}$. If $x, y \in S$ then $xy \in S$. (For an easier proof of this, if $x = (a + ib)(a - ib)$ and $y = (c + id)(c - id)$ then $xy = (u + iv)(u - iv)$ and where $u + iv = (a + ib)(c + id)$.) Now 2, p_1, ..., p_r and q_1^2, ... $q_s^2 \in S$. So the condition is sufficient. For necessity, let $n = (x + iy)(x - iy)$. Since q_j is a prime in G, q_j divides either $x + iy$ or $x - iy$. Obviously the same power of q_j divides both $x + iy$ and $x - iy$. So n_j is even.

11. (i) an l.c.m. of a and b (ii) a g.c.d. of a and b.

12. If (x) and (y) are relatively prime then $\lambda x + \mu y = 1$ for some $\lambda, \mu \in R$. So x, y are relatively prime in R. For the converse, reverse the steps, using Theorem (2.14).

13. Since the m_i's are pairwise relatively prime, it follows that m is their l.c.m. So for any $n \in \mathbb{Z}$, $m \mid n$ iff $m_i \mid n$ for every $i = 1, ..., k$. Consequently for $x, y \in \mathbb{Z}$, $x \equiv y$ (mod m) iff $x \equiv y$ (mod m_i) for every $i = 1, ..., k$. This proves that θ is well-defined and one-to-one. Also its domain and codomain have the same cardinality.

14. Let $a = 5832$ and $b = 6639$. Then $a = 0$ (mod 8); so $ab = 0$

(mod 8). Similarly $ab \equiv 0$ (mod 3) since $b \equiv 0$ (mod 3). Since $a \equiv 2$ (mod 5) and $b \equiv 4$ (mod 5), $ab \equiv 3$ (mod 5). Finally, $ab \equiv 1.3 \equiv 3$ (mod 7). So in the notation of the hint, $x_1 = x_2 = 0$, and $x_3 = x_4 = 3$. The first two conditions imply x is a multiple of 24 and this considerably reduces the search for x in the present problem. Let $x = 24y$. Then $x \equiv -y$ (mod 5), giving $y \equiv 2$ (mod 5). Similarly $y \equiv 1$ (mod 7). By trial y is of the form $22 + 35z$. Since $24y < 840$, y has to be 22. So $x = 528$.

15. It suffices to show that there exist $y_1, \ldots, y_k \in R$ such that for each $i = 1, \ldots, k$, $y_i \equiv 1$ (mod A_i) (i.e. $y_i - 1 \in A_i$) and $y_i \equiv 0$ (mod A_j) for $j \neq i$. Once such y's are found, we need only set

$$x = \sum_{i=1}^{k} x_i y_i.$$

To get such y's, let B be the ideal generated by all products of the form $a_2 a_3 \ldots a_k$ with $a_j \in A_j$ for $j = 2, \ldots, k$. (B is often denoted by $A_2 A_3 \ldots A_k$.) Show that A_1 and B are relatively prime to each other. Hence there exists $y_1 \in B$ such that $y_1 - 1 \in A_1$. But then $y_1 \in A_j$ for all $j = 2, \ldots, n$. Similarly find $y_2, \ldots, y_k$.

16. $(u_1, \ldots, u_k)$ is a unit of $R_1 \times R_2 \times \ldots \times R_k$. iff u_i is a unit of R_i for every $i = 1, \ldots, k$. The second assertion follows from the fact for every $m \in \mathbb{N}$, $\phi(m)$ is simply the number of units of the ring $\mathbb{Z}_m$.

17. Every homomorphism $f: \mathbb{Z}_m \to \mathbb{Z}_m$ is uniquely determined by $f([1])$. Let $f([1]) = [k]$. Then f is an automorphism iff $[k]$ generates $\mathbb{Z}_m$, which is the case iff k is relatively prime to m. Let G be the group of units of the ring $\mathbb{Z}_m$. The function $\theta : A(\mathbb{Z}_m) \to G$ defined by $\theta(f) = [f(1)]$ is an isomorphism.

18. Find $a, b \in \mathbb{Z}$ such that $am + bn = 1$. Then $x = x^{am+bn} = x^{am} x^{bn} = (x^m)^a (x^n)^b = e^a e^b = e$.

19. If $y = 0$, then $x = 0$. If $y \neq 0$, then apply the last exercise to the element xy^{-1} of the multiplicative group of the field F. If F is an integral domain, pass over to its field of quotients. For a counter-example, take $x = [3]$, $y = [0]$ in $\mathbb{Z}_9$ $m = 2, n = 3$.

20. The proof of Proposition (2.23) goes through since all that really matters is that a commute with elements of R.

22. For the first part write $f(x) = (x - \alpha_1)^{m_1} g(x)$ and apply induction hypothesis to $g(x)$. For the second part write $f(x) = a_n \prod_{i=1}^{k} (x - \alpha_i)^{m_i}$ and compare coefficients of like powers.

23. If α is a multiple root of $f(x)$ then $f(x) = (x - \alpha)^2 g(x)$ for some $g(x) \in F[x]$. It is easily seen that the formal derivative of $(x - \alpha)^2$ is $2(x - \alpha)$. By exercise (1.20), $f'(x) = (x - \alpha)^2 g'(x) = 2(x - \alpha)g(x)$. So α is a root of $f'(x)$. Conversely, if α is a common root of $f(x)$ and $f'(x)$, write $f(x) = (x - \alpha)^m g(x)$ where $m \geqslant 1$, and $g(\alpha) \neq 0$. If $m = 1$, then $f'(x) = (x - \alpha)g'(x) + g(x)$ which does not have α as a root, contradiction. So $m \geqslant 2$.

24. Let $a = up_1^{m_1} \ldots p_k^{m_k}$ and $b = vq_1^{n_1} \ldots q_r^{n_r}$ with the usual notation. Then a divides b iff for every $i = 1, \ldots, k$, there exists j such that q_j is an associate of p_i and $n_j \geqslant m_i$.

25. With the notation of the last answer, suppose without loss of generality that s is the integer $\geqslant 0$ such that for $i = 1, \ldots, s$, $p_i = q_i$ and for $i > s$, p_i is not an associate of any q_j nor q_i of any p_j. For $i = 1, \ldots, s$, let $\lambda_i = \min \{m_i, n_i\}$ and $\mu_i = \max \{m_i, n_i\}$. Then a g.c.d. of a and b is $p_1^{\lambda_1} \ldots p_s^{\lambda_s}$ while an l.c.m. is $p_1^{\mu_1} \ldots p_s^{\mu_s} p_{s+1}^{\mu_{s+1}} \ldots \ldots p_k^{m_k} q_{s+1}^{n_{s+1}} \ldots q_r^{n_r}$. If a, b are relatively prime then $s = 0$. In that case if $a \mid bc$, the primes $p_1, \ldots, p_k$ must appear in bc and hence in c with powers at least $m_1, \ldots, m_k$ respectively. So $a \mid c$.

26. Both 2 and x are prime elements of $\mathbb{Z}[x]$. Their g.c.d. is 1. Let if possible, $1 = 2f(x) + xg(x)$ where $f(x) = a_0 + a_1 x + \ldots + a_m x^m$ and $g(x) = b_0 + b_1 x + \ldots + b_n x^n \in \mathbb{Z}[x]$, with $a_m \neq 0$, $b_n \neq 0$. This gives $1 = 2a_0$ which is impossible.

27. Clearly a unit cannot vanish anywhere. Conversely if $f : \mathbb{C} \to \mathbb{C}$ is analytic and never vanishes, then $g(z) = \dfrac{1}{f(z)}$ is analytic and is an inverse of f. If f has a single zero, say α of multiplicity 1 then $f(z) = (z - \alpha) g(z)$ where $g(z)$ is analytic and never vanishes. So f is an associate of the polynomial $z - \alpha$ which is clearly irreducible. If, on the other hand, f has two distinct zeros say α_1, α_2 or a single multiple zero, α, then $f(z) = (z - \alpha_1) [(z - \alpha_2)g(z)]$ or $f(z) = (z - \alpha) [(z - \alpha)g(z)]$ gives a non-trivial factorization of f. So primes in R have the desired form. Any finite product of primes must therefore have only finitely many zeros.

28. $4 = 2.2 = (1 + i\sqrt{3})(1 - i\sqrt{3})$. Here 2 is a prime. For $2 = (a + i\sqrt{3}b)(c + i\sqrt{3}d)$ would give $4 = (a^2 + 3b^2)(c^2 + 3d^2)$, the only possible solutions of which are $a = \pm 1, b = \pm 1, c = \pm 1, d = 0$ or $a = \pm 1, b = 0, c = \pm 1, d = \pm 1$. Similarly, $1 + i\sqrt{3}$, $1 - i\sqrt{3}$ are primes. Further if $u + i\sqrt{3} v$ is a unit, then $u^2 + 3v^2$ will be a factor of 1 in $\mathbb{Z}$. So ± 1 are the only units in the ring $\mathbb{Z} + i\sqrt{3}\mathbb{Z}$. Hence, neither $1 + i\sqrt{3}$ nor $1 - i\sqrt{3}$ is an associate of 2.

29. Let $f(x) = a_0 + \ldots + a_n x^n$, $g(x) = b_0 + b_1 x + \ldots + b_m x^n \in \mathbb{Z}[x]$

be primitive. Let $h(x) = f(x)\,g(x) = c_0 + c_1 x + ... + c_{m+n}x^{m+n}$. Given a prime p, let i be the smallest integer such that $p \nmid a_i$ and j be the smallest integer such that $p \nmid b_j$. Let $k = i + j$. Then $c_k = \sum_{r=0}^{k} a_r\, b_{k-r}$. For $r < i$, $p \mid a_r b_{k-r}$ while for $r > i$, $k - r < j$ and so $p \mid a_r b_{k-r}$. However, $p \nmid a_i\, b_{k-i}$. So $p \nmid c_k$. Hence there is no prime which divides all the c's, i.e. $h(x)$ is primitive.

30. Let $g(x) = \dfrac{p_0}{q_0} + \dfrac{p_1}{q_1} x + ... + \dfrac{p_m}{q_m}$ where $p_0, ..., p_m, q_0, ..., q_m \in \mathbb{Z}$. Let $q = q_0 q_1 ... q_m$. Then $g(x) = \dfrac{1}{q} k(x)$ where $k(x) \in \mathbb{Z}[x]$. Let p be the g.c.d. of the coefficients of $k(x)$. Then $k(x) = p g_1(x)$ where $g_1(x) \in \mathbb{Z}[x]$ is primitive. Similarly let $h(x) = \dfrac{r}{s}\, h_1(x)$ with $r, s \in \mathbb{Z}$ and $h_1(x) \in \mathbb{Z}[x]$ is primitive. Then $qs\, f(x) = pr\, g_1(x)\, h_1(x)$. Call this common polynomial $\psi(x)$. Then the g.c.d. of the coefficients of $\psi(x)$ is qs since $f(x)$ is primitive and is also equal to pr since $g_1(x)\, h_1(x)$ is primitive. So $qs = pr$ and we get $f(x) = g_1(x)\, h_1(x)$ in $\mathbb{Z}[x]$.

31. Let $f(x) \in \mathbb{Z}[x]$, $f(x) \neq 0$. In view of the last exercise $f(x)$ is a prime element of $\mathbb{Z}[x]$ iff one of the two possibilities holds: (i) $\deg f(x) = 0$ and $f(x)$ is a prime element of $\mathbb{Z}$ or (ii) $\deg f(x) > 0$, $f(x)$ is primitive and irreducible as an element of $\mathbb{Q}[x]$. Given $f(x) \in \mathbb{Z}[x]$ write $f(x) = c f_1(x)$ where c is the g.c.d. of the coefficients of $f(x)$ and $f_1(x)$ is primitive. Factor c as an element of $\mathbb{Z}$ and $f_1(x)$ as an element of $\mathbb{Q}[x]$, which is a u.f.d.

32. For any $\alpha \in \mathbb{R}$, $\alpha^2 + \alpha + 1 = \alpha^2 + \alpha + \frac{1}{4} + \frac{3}{4} = (\alpha + \frac{1}{2})^2 + \frac{3}{4} > 0$. So $f(x)$ has no root in $\mathbb{R}$. In $\mathbb{Z}_2$ there are only two elements 0 and 1, neither of which is a root of $f(x)$. By Exercise (2.21) $f(x)$ is irreducible over $\mathbb{R}$ and $\mathbb{Z}_2$. In $\mathbb{Z}_3[x]$, however, $f(x) = (x + 2)^2$. If $F = \mathbb{R}$, define $h: F[x] \to \mathbb{C}$ by $h(g(x)) = g(\omega)$ where $\omega = e^{2\pi i/3} = -\frac{1}{2} + i\,\sqrt{3}/2$. Show that h is a ring homomorphism with kernel I. To show h is onto, show that for every complex number z, there exist $\alpha, \beta \in \mathbb{R}$ such that $z = \alpha + \beta\omega$.

If $F = \mathbb{Z}_2$, call $F[x]/I$ as K. The elements of K are cosets of the form $g(x) + I$ for $g(x) \in F[x]$. From the fact that $(x^2 + x + 1) \in F[x]$ it follows that $x^2 + I$ is the same coset as $(x + 1) + I$. (Note that $-1 = 1$ in $\mathbb{Z}_2$.) Similarly $x^3 + I = (x^2 + I)(x + I) = ((x + 1) + I)(x + I) = (x^2 + x) + I = (x + x + 1) + I = 1 + I$. Repeating this argument, it follows that K has only 4 distinct elements, namely, I, $1 + I$, $x + I$ and $(x + 1) + I$. It is customary to denote these elements by 0, 1, ω and $\omega + 1$. (This ω bears only a formal similarity to the complex number ω above since both are roots of $f(x)$.)

33. Let K be a field with 4 elements and denote them by $\alpha_1,\ \alpha_2,\ \alpha_3,\ \alpha_4$, with $\alpha_1 = 0$. Use these same symbols to denote the four religions and the four states. Now form a 4×4 square so that the delegate in the ith row and jth column has religion $\alpha_2\alpha_i + \alpha_j$ and state $\alpha_3\alpha_i + \alpha_j$. Since $\alpha_2 \neq 0$, every religion appears exactly once in every row and in every column. Similarly for the states. Also given any religion r and state s, (i.e., given $r,\ s \in K$) the system of equations $r = \alpha_2\alpha_i + \alpha_j$ and $s = \alpha_3\alpha_i + \alpha_j$ has a unique solution for α_i and α_j, namely $\alpha_i = \dfrac{r-s}{\alpha_2-\alpha_3}$ and $\alpha_j = \dfrac{s\alpha_2 - r\alpha_3}{\alpha_2 - \alpha_3}$. In other words, for every delegate, there is a unique place to go. Hence, an arrangement of 16 delegates is possible.

34. We get $a_0q^n + a_1pq^{n-1} + a_2p^2q^{n-2} + \dots + a_{n-1}p^{n-1}q + a_np^n = 0$. So p divides a_0q^n. By Proposition (2.17), $p \mid a_0$. Similarly $q \mid a_n$.

Section 6.3

1. For $u,\ v \in V$, expand $(1 + 1)\cdot(u + v)$ in two ways; once as $(u + v) + (u + v)$ and then as $(u + u) + (v + v)$.

2. When $\mathbb{R}^3$ is regarded as a vector space over $\mathbb{R}$, only (i) is a subspace. When considered as a vector space over $\mathbb{Q}$, both (i) and (iv) are subspaces.

4. (b) The problem is equivalent to showing that if W is a vector opace over an infinite field F and $W_1, \dots, W_n$ are proper subspaces of W then $\bigcup\limits_{l=1}^{n} W_i$ cannot equal W. Apply induction on n. For each i, there exists $w_i \in W_i$ such that $w_i \notin W_j$ for all $j \neq i$; for otherwise $\bigcup\limits_{k=1}^{n} W_k$ is the same as $\bigcup\limits_{k \neq i} W_k$ and the induction hypothesis applies. Now $\alpha w_1 + w_2 \notin W_1$ for any $\alpha \in F$ (or else $w_2 \in W_1$). For $i = 2, \dots, n$, there is at most one $\alpha \in F$ such that $\alpha w_1 + w_2 \in W_i$. For if $\alpha w_1 + w_2 \in W_i$ and $\beta w_1 + w_2 \in W_i$ and $\alpha \neq \beta$ then $w_1 = \dfrac{1}{\alpha - \beta}(\alpha w_1 + w_2 - \beta w_1 - w_2)$ is in W_i. Since F is infinite there exists $\alpha \in F$ such that $\alpha w_1 + w_2 \notin W_i$ for all $i = 1, \dots, n$.

6. $\{0\} \times V_2$ and $V_1 \times \{0\}$ are the kernels of $\pi_1,\ \pi_2$ respectively. Both are onto.

7. Define $\theta \colon \operatorname{Hom}_F(V, W_1 \times W_2) \to \operatorname{Hom}_F(V, W_2) \times \operatorname{Hom}_F(V, W_2)$ by $\theta(f) = (\pi_1 \circ f,\ \pi_2 \circ f)$. Show that θ is an isomorphism.

8. As in the case of group homomorphisms, define $\theta \colon V/K \to R$ by $\theta(v + K) = f(v)$. Show that θ is a vector space isomorphism.

9. Let T_{i-1} be the set obtained just before v_i is picked. ($T_0 = \phi$.) By induction on i, show that for every $i = 1,\dots, k$, T_i is linearly independent and has the same span as $\{v_1,\dots, v_i\}$. The set T equals T_k.

10. One basis is $\left\{\left(1, 0, -\dfrac{2}{\pi}, 0, -\dfrac{2}{\pi}\right), \left(0, 1, \dfrac{1}{\pi}, 0, \dfrac{\pi+1}{\pi}\right)(0, 0, 0, 1-1)\right\}.$

11. If (i) and (ii) hold but (iii) does not, then some proper subset of S would span V and hence dim $(V) < n$, a contradiction. If (ii) and (iii) hold then S is a basis for V and Theorem (3.13) applies. If (i) and (iii) hold and $L(S) \underset{\neq}{\subset} V$, then for any $v \in V - L(S)$, $S \cup \{v\}$ is a linearly independent set of cardinality $n + 1$ in an n-dimensional vector space, contradicting Proposition (3.12).

12. Evidently, (iii) $\Rightarrow$ (ii) $\Rightarrow$ (i). For (iii), linear dependence of $\{1, \sqrt{2}, \sqrt{3}, \sqrt{6}\}$ over $\mathbb{Q}$ implies a relationship of the form $a + b\sqrt{2} = \sqrt{3}(c + d\sqrt{2})$ for some $a, b, c, d \in \mathbb{Q}$, not all 0. Squaring and simplifying, $2\sqrt{2}(ab - 3cd) = 3(c^2 + 2d^2) - a^2 - 2b^2$ which implies either $\sqrt{2}$ is rational or else $ab = 3cd$ and $3(c^2 + 2d^2) = a^2 + 2b^2$ which further implies $x^2 + y^2 = 3(u^2 + v^2)$ where $x = a + b, y=b$, $u = c + d, v = d$. Here x, y, u, v are rationals. However, writing them with a common denominator we may suppose they are, in fact, integers. Let $m = u^2 + v^2$ and $n = x^2 + y^2$. Then both m, n are expressible as sums of squares of integers. Since $n = 3m$ and 3 is a prime of the form $4k + 3$, this contradicts the result of Exercise (2.10). (iii) also implies (iv). Linear dependence of $\{1, \sqrt{2}, \sqrt{3}, \sqrt{5}\}$ over $\mathbb{Q}$ would imply $\sqrt{5} = a + b\sqrt{2} + c\sqrt{3}$ for some $a, b, c \in \mathbb{Q}$. Squaring we get a relation contradicting (iii).

14. Apply Exercise (3.13) with $X = V \times \{0\}$ and $Y = \{0\} \times W$.

15. $\alpha_1 v_1 + \dots + \alpha_k v_k = 0$ implies $\alpha_1 v_1 + \dots + \alpha_k v_k \in W$, i.e., $\alpha_1 (v_1 + W) + \dots + \alpha_k(v_k + W)$ equals the zero coset W. For the second assertion, combine $\{v_1,\dots, v_k\}$ with a basis for W to get a basis for V. Another approach is to note that V is always isomorphic, as a vector space, to $W \times (V/W)$. For this, first one proves the analogue of Exercise (5.3.15). Given a basis $\{v_1 + W,\dots, v_k + W\}$ for V/W there exists a unique linear transformation $j: V/W \to V$ satisfying $j(v_i + W) = v_i$ for $i = 1,\dots, k$, and hence $j(v + W) \in v + W$ for all $v \in V$. This follows as a consequence of Theorem (4.5)

16. For the first assertion, apply the last exercise after noting that R is isomorphic to V/K. As for the inequalities note that range of $T_2 \circ T_1 \subset$ range of T_2, which implies $r(T_2 \circ T_1) \leqslant r(T_2)$. Also kernel of $T_1 \subset$ kernel of $T_2 \circ T_1$. So $n(T_1) \leqslant n(T_2 \circ T_1)$. Since $n(T_1) = \dim (V) - r(T_1)$ and $n(T_2 \circ T_1) = \dim (V) - r(T_2 \circ T_1)$, the second inequality follows.

17. If T is not one-to-one then there exists $v \neq 0$ such that $T(v) = 0$. So T takes the linearly independent set $\{v\}$ to $\{0\}$ which is linearly dependent. On the other hand, if T is one-to-one and $\alpha_1 T(v_1) + \ldots + \alpha_k T(v_k) = 0$ then $T(\alpha_1 v_1 + \ldots + \alpha_k v_k) = 0$ and so $\alpha_1 v_1 + \ldots + \alpha_k v_k = 0$.

18. T is one-to-one $\Leftrightarrow n(T) = 0 \Leftrightarrow r(T) = \dim V$ (by Exercise (3.16)) $\Leftrightarrow r(T) = \dim W \Leftrightarrow T$ is onto.

19. Get a basis A for X and extend it to a basis B for V. Let Y be the subspace generated by $B - A$. For the second assertion, apply Exercise (3.13) to calculate $\dim (X + Y)$ and conclude $X + Y = V$. Since $X \cap Y = \{0\}$, every $v \in V$ can be uniquely expressed as $x + y$ with $x \in X$, $y \in Y$. This gives the desired isomorphism.

20. For $n \in \mathbb{N}$, $g \in G$, define $n \cdot g$ to be $g + g + \ldots + g$ (n times) if $n > 0$, 0 if $n = 0$ and $(-n)(-g)$ if $n < 0$.

21. For $r \in R$, $m \in M$, $rm \in M$. Define $r \cdot m$ to be rm. For the example, take any ring with identity containing a zero divisor.

22. Let I be the ideal of $F[x]$ generated by $p(x)$. Then $F[x]/I$ is a field. Define $\theta \colon F[x] \to K$ by $\theta(f(x)) = f(\alpha)$. Then θ is a ring homomorphism with kernel I and range $F[\alpha]$. Hence $F[\alpha]$ is a field (isomorphic to the field $F[x]/I$). Hence Theorem (3.20) can be applied to the chain of fields $F \subset F[\alpha] \subset K$. From Theorem (2.26) and the proof of Theorem (3.17), it is clear that $[F[\alpha] : F]$ equals the degree of $p(x)$.

23. Every polynomial in $\mathbb{Q}[x]$ is uniquely determined by a finite sequence of rationals. Arguing as in Exercise (2.2.22) (v), the set of such sequences is countable.

24. Let A denote the set of real numbers which are algebraic over $\mathbb{Q}$. Then A is countable by the last exercise. If $\mathbb{R} - A$ were also countable then $\mathbb{R} = A \cup (\mathbb{R} - A)$ would be countable.

25. $f(u_1 + v_1, u_2 + v_2, \ldots, u_k + v_k) = \Sigma f(w_1, w_2, \ldots, w_k)$ where the sum extends over all 2^k binary sequences $(x_1, x_2, \ldots, x_k)$, and where $w_i = u_i$ or v_i according as x_i is 0 or 1. (For example, $f(u_1 + v_1, u_2 + v_2) = f(u_1, u_2) + f(u_1, v_2) + f(v_1, u_2) + f(v_1, v_2)$.)

26. Associativity of quaternionic multiplication amounts to showing that for any $(a_0, \mathbf{a})$, $(b_0, \mathbf{b})$, $(c_0, \mathbf{c})$, $u_0 = v_0$ and $\mathbf{u} = \mathbf{v}$ where,

$u_0 = a_0 b_0 c_0 - \mathbf{a} \cdot \mathbf{b}\, c_0 - a_0 \mathbf{b} \cdot \mathbf{c} - b_0 \mathbf{a} \cdot \mathbf{c} - (\mathbf{a} \times \mathbf{b}) \cdot \mathbf{c}$

$v_0 = a_0 b_0 c_0 - a_0 \mathbf{b} \cdot \mathbf{c} - b_0 \mathbf{a} \cdot \mathbf{c} - c_0 \mathbf{a} \cdot \mathbf{b} - \mathbf{a} \cdot (\mathbf{b} \times \mathbf{c})$

$\mathbf{u} = a_0 b_0 \mathbf{c} - (\mathbf{a} \cdot \mathbf{b})\mathbf{c} + a_0 c_0 \mathbf{b} + b_0 c_0 \mathbf{a} + c_0 \mathbf{a} \times \mathbf{b} + a_0 \mathbf{b} \times \mathbf{c} + b_0 \mathbf{a} \times \mathbf{c}$
$$+ (\mathbf{a} \times \mathbf{b}) \times \mathbf{c}$$

and

$\mathbf{v} = a_0 b_0 \mathbf{c} + a_0 c_0 \mathbf{b} + a_0 \mathbf{b} \times \mathbf{c} + b_0 c_0 \mathbf{a} - (\mathbf{b} \cdot \mathbf{c})\mathbf{a} + b_0 \mathbf{a} \times \mathbf{c} + c_0 \mathbf{a}$
$$\times \mathbf{b} + \mathbf{a} \times (\mathbf{b} \times \mathbf{c}).$$

Equality of u_0 with v_0 follows from the identity $\mathbf{a} \cdot (\mathbf{b} \times \mathbf{c}) =$

$\mathbf{c} \cdot (\mathbf{a} \times \mathbf{b}) = (\mathbf{a} \times \mathbf{b}) \cdot \mathbf{c}$. For $\mathbf{u} = \mathbf{v}$, use the fact $\mathbf{a} \times (\mathbf{b} \times \mathbf{c}) = (\mathbf{a} \cdot \mathbf{c})\mathbf{b} -$ $(\mathbf{a} \cdot \mathbf{b})\,\mathbf{c}$ and $(\mathbf{a} \times \mathbf{b}) \times \mathbf{c} = -\mathbf{c} \times (\mathbf{a} \times \mathbf{b}) = \mathbf{c} \times (\mathbf{b} \times \mathbf{a}) = (\mathbf{c} \cdot \mathbf{a})\mathbf{b} -$ $(\mathbf{c} \cdot \mathbf{b})\,\mathbf{a}$. All other properties follow routinely from corresponding properties of the vector operations. (For example, for distributivity, use $\mathbf{a} \cdot (\mathbf{b} + \mathbf{c}) = \mathbf{a} \cdot \mathbf{b} + \mathbf{a} \cdot \mathbf{c}$ and $\mathbf{a} \times (\mathbf{b} + \mathbf{c}) = \mathbf{a} \times \mathbf{b} + \mathbf{a} \times \mathbf{c}$ etc.) For multiplicativity of norm, note first that for any vector $\mathbf{a}$, $|\mathbf{a}|^2 = \mathbf{a} \cdot \mathbf{a}$ and use the facts that $\mathbf{a} \cdot (\mathbf{a} \times \mathbf{b}) = \mathbf{b} \cdot (\mathbf{a} \times \mathbf{b}) = 0$ and $(\mathbf{a} \cdot \mathbf{b})^2 + |\mathbf{a} \times \mathbf{b}|^2 = |\mathbf{a}|^2 |\mathbf{b}|^2$.

27. For any $x \in G$, $0 \cdot x$ and $1 \cdot x$ have to be defined as 0 and x respectively. This defines the scalar multiplication and makes G a vector space over $\mathbb{Z}_2$.

28. For any $x \in G$, $nx \in G$ and so in the quotient group G/nG, $n(x + nG) = 0$.

29. $G/2G$ is vector space over $\mathbb{Z}_2$ by the last two exercises. A basis for it consists of elements of the form $s + 2G$ for $s \in S$.

30. If $f: G \to H$ is an isomorphism then $f(2G) = 2H$ and so there is an induced vector space isomorphism $\psi: G/2G \to H/2H$, (cf. Exercise (5.3.16)). So $G/2G$ and $H/2H$ have equal dimensions. Now apply the last exercise. The converse is similar to Exercise (5.3.26).

31. An isomorphism $f: G \to H$ must map $C(G)$ onto $C(H)$ and hence induces an isomorphism from $G/C(G)$ onto $H/C(H)$. Apply the last exercise. The converse was already proved in Exercise (5.3.26).

Section 6.4

1. The (i, j)th entry of $(AC)'$ equals the (j, i)th entry of AC which by definition is $\sum\limits_{k=1}^{n} a_{jk}c_{ki}$. The (i, j)th entry of $C'A'$, on the other hand, equals $\sum\limits_{k=1}^{n} a_i'_k \, a_k'_j$. Since $c_i'_k = c_{ki}$ and $a_k'_j = a_{jk}$ and the ring R is commutative, the result follows. The rest is routine and does not require commutativity of R.

2. For the second assertion instead of a direct argument note that by Proposition (4.4), $(A + A')' = A' + (A')' = A + A'$ proving $A + A'$ to be symmetric. Similarly $(A - A')' = A' - A = -(A - A')$, whence $A - A'$ is skew-symmetric.

3. $A = \frac{1}{2}[(A + A') + (A - A')]$, where $\frac{1}{2}$ is the multiplicative inverse of the field element $1 + 1$.

4. First B is linearly independent. For otherwise some $b_1 \in B$ is a linear combination of the remaining elements of B. Any linear transformation which vanishes on $B - \{b_1\}$ must also vanish on b_1. It follows that if we take $W = V$ and define $f: B \to W$ by $f(b) = 0$

for all $b \in B-\{b_1\}$ and $f(b_1)=$any non-zero vector in W, then f has no linear extension. Secondly B spans V. If not, enlarge B to a basis say C for V. Then the inclusion function $f: B \to V$ can be extended to a linear transformation $T: V \to V$ in at least two ways, because for $v \in C - B$, $T(v)$ may be defined arbitrarily.

5. Let $f: V \to W$ be a vector space isomorphism. Then f must take a basis of V to a basis of W. So dim $(V) =$ dim (W). Converse follows from Theorem (3.13). The second assertion now follows in view of Corollary (4.7). The point to note is that the isomorphism between V and V^* depends on the choice of the basis for V.

6. $e_v(T_1 + T_2) = (T_1 + T_2)(v) = T_1(v) + T_2(v) = e_v(T_1) + e_v(T_2)$. Similarly $e_v(\alpha T) = (\alpha T)(v) = \alpha T(v) = \alpha e_v(T)$. So $e_v \in T^{**}$. Linearity of θ amounts to proving that for all $v_1, v_2 \in V$, $e_{v_1+v_2} = = e_{v_1} + e_{v_2}$ and for all $\lambda \in F$, $v \in V$, $e_{\lambda v} = \lambda e_v$. For any $T \in V^*$, $e_{v_1+v_2}(T) = T(v_1 + v_2) = T(v_1) + T(v_2) = e_{v_1}(T) + e_{v_2}(T) = (e_{v_1} + + e_{v_2})(T)$. Similarly verify $e_{\lambda v} = \lambda e_v$. Now, $v \in$ Ker $\theta \Rightarrow e_v(T) = 0$ for all $T \in V^*$. If $v \neq 0$, extend $\{v\}$ to a basis B for V. Then there exists $T \in V^*$ with $T(b) = 1$ for all $b \in B$. In particular, $T(v) \neq 0$ and so $e_v(T) \neq 0$. So θ is one-to-one. Since dim $(V^{**}) =$ dim $(V^*) = =$ dim (V), θ is onto.

7. $\dfrac{1}{7}\begin{pmatrix} 5 & 24 & 19 \\ 3 & 13 & 3 \\ 3 & -29 & -4 \end{pmatrix}$. T has rank 3.

8. Matrix of D w.r.t. the ordered basis $(1, x, x^2, x^3, x^4)$ is

$$\begin{pmatrix} 0 & 1 & 0 & 0 & 0 \\ 0 & 0 & 2 & 0 & 0 \\ 0 & 0 & 0 & 3 & 0 \\ 0 & 0 & 0 & 0 & 4 \\ 0 & 0 & 0 & 0 & 0 \end{pmatrix}.$$

9. Call $\begin{pmatrix} \alpha & \beta \\ \gamma & \delta \end{pmatrix}$ as B. $T(\alpha_1 A_1 + \alpha_2 A_2) = B(\alpha_1 A_1 + \alpha_2 A_2) + \alpha_1 B A_1 + \alpha_2 B A_2 = \alpha_1 T(A_1) + \alpha_2 T(A_2)$, proving linearity of T. The matrix of T w.r.t. the given basis is

$$\begin{pmatrix} \alpha & 0 & \beta & 0 \\ 0 & \alpha & 0 & \beta \\ \gamma & 0 & \delta & 0 \\ 0 & \gamma & 0 & \delta \end{pmatrix}$$

10. P_σ has a 1 in $(i, \sigma(i))$th place and 0 elsewhere for $i = 1, ..., n$. These entries uniquely determine σ. In det (P_σ) the only non-zero term is 1 and its sign is $+$ or $-$ according as σ is even or odd.

12. By Theorem (5.3.15), G is isomorphic to a subgroup of S_n for some integer n. Now apply (iii) of the last exercise. (An important problem in group representations is to find for a given G, as small n as possible so that G can be embedded in $GL(n, F)$.)

13. If $\sigma \in S_n$ and σ is not the identity permutation, then there exists i such that $\sigma(i) < i$. (Consider any cycle of length $\geqslant 2$ of σ and take the largest symbol in it). So every term except one in det(A) vanishes. An alternate proof can be given by induction.

14. Let the rows of A be $u_1, u_2, ..., u_r, ..., u_s, ..., u_n$ and those of B be $u_1, ..., u_r, ..., u_{s-1}, u_s + \lambda u_r, u_{s+1}, ..., u_n$. Then det $B = \det (u_1, ..., u_{s-1}, u_s, u_{s+1}, ..., u_n) + \lambda \det (u_1, ..., u_r, ..., u_{s-1}, u_r, u_{s+1}, ..., u_n) = \det A + \lambda 0 = \det A$.

15. Using the same notation as in the proof of Theorem (4.26), det $(AB) = \Sigma a_{1j_1} a_{2j_2} ... a_{mj_m} \det (v_{j_1}, v_{j_2}, ..., v_{j_m})$ where the sum ranges over all possible m-tuples $(j_1, ..., j_m)$ where the j's range from 1 to n independently of each other. Again out of the n^m possible terms we need consider only those of the form where the j's are all distinct. Each such term determines an m-subset $J = \{j_1, ..., j_m\}$ of $\{1, ..., n\}$. Each such m-subset J corresponds to $m!$ terms obtained by various permutations of the indices in J.

Assume that $J = \{j_1, ..., j_m\}$ with $1 \leqslant j_1 < j_2 < ... < j_m \leqslant n$. Then the $m!$ terms corresponding to a fixed J are of the form

$$a_{1j_{\sigma(1)}} a_{2j_{\sigma(2)}} ... a_{mj_{\sigma(m)}} \det (v_{j_{\sigma(1)}}, v_{j_{\sigma(2)}}, ..., v_{j_{\sigma(m)}})$$

where σ is a permutation of $\{1, 2, ..., m\}$. By Proposition (4.23), this term equals $(-1)^\sigma a_{1j_{\sigma(1)}} ... a_{mj_{\sigma(m)}} \det (B_{j_1, j_2, ..., jm})$ and so their sum equals det $(A_{j_1, ..., jm}) \det (B_{j_1, ..., jm})$. Thus det (AB) is a sum of $\binom{n}{m}$ terms of this form, each corresponding to a different m-subset of $\{1, ..., n\}$.

16. $Ax = b$ gives $x = A^{-1}Ax = A^{-1}b$. So the solution exists and is unique. By Theorem (4.31), $A^{-1} = \dfrac{1}{\det A}(d_{ij})$ where $d_{ij} = c_{ji}$. So $x_i = \dfrac{1}{\det A}\sum_{j=1}^{n} d_{ij}b_j = \dfrac{1}{\det A}\sum_{j=1}^{n} b_j c_{ji}$ which by Theorem (4.30), equals $\dfrac{1}{\det A}\det A_i$.

19. Let $v_1,\ldots, v_n$ be eigenvectors with corresponding eigenvalues $\lambda_1,\ldots, \lambda_n$ If they form a basis for T then the matrix of T w.r.t. the ordered basis $(v_1,\ldots, v_n)$ is (a_{ij}) where $a_{ij} = 0$ for $i \neq j$ and $a_{ii} = \lambda_i$ for $i = 1,\ldots, n$.

20. The system $2x_1 + x_2 = \lambda x_1$ and $3x_1 = \lambda x_2$ yields $\lambda^2 - 2\lambda - 3 = 0$, i.e. $\lambda = 3$ or $\lambda = -1$. That these are in fact eigenvalues is shown by actually solving the systems for these values of λ. The eigenspace for $\lambda = 3$ is $\{(x, x): x \in \mathbf{R}\}$ while that for $\lambda = -1$ is $\{(x, -3x): x \in \mathbf{R}\}$.

21. f is an eigenvector for $\lambda \Leftrightarrow f'(x) = \lambda f(x)$ for all $x \in \mathbf{R} \Leftrightarrow f(x) = ke^{\lambda x}$ for all $x \in \mathbf{R}$, where k is an arbitrary constant.

22. $A - \lambda$ is the matrix of $T - \lambda I$ which is singular iff $(T - \lambda I)v = 0$ for some non-zero $v \in V$. This is equivalent to saying that λ is an eigenvalue of T.

23. Call $\det(x I_n - A)$ as $f(x)$. In the expansion of the determinant, only one term gives a polynomial of degree n, namely $(x - a_{11})(x - a_{22})\ldots(x - a_{nn})$. For the second assertion, $f(\lambda) = 0 \Leftrightarrow \det(\lambda I_n - A) = 0 \Leftrightarrow \lambda I_n - A$ is singular. Apply the last exercise.

24. Combine last exercise with Proposition (2.25).

25. The problem amounts to showing that for any $a_0,\ldots, a_{n-1} \in F$, $a_0 + a_1 x + \ldots + a_{n-1}x^{n-1} + x^n$ equals the determinant of the matrix

$$
\begin{pmatrix}
x & 0 & 0 & \ldots & 0 & 0 & a_0 \\
-1 & x & 0 & \ldots & 0 & 0 & a_1 \\
0 & -1 & x & \ldots & 0 & 0 & a_2 \\
\cdots & \cdots & \cdots & \ldots & \cdots & \cdots & \cdots \\
0 & 0 & 0 & \ldots & -1 & x & a_{n-2} \\
0 & 0 & 0 & \ldots & 0 & -1 & x + a_{n-1}
\end{pmatrix}
$$

Expand the determinant w.r.t. the first row. Apply induction hypothesis and Exercise (4.13) to the two minors.

26. Let $f(x) = \det(x I_n - A)$ where $A \in M_n(F)$, F being a field. It is tempting to set $x = A$ and conclude $f(A) = \det(AI_n - A) = \det(A - A) = \det(0) = 0$. But the catch is that in this calculation det $(AI_n - A)$ equals the determinant of an $n \times n$ matrix whose entries are themselves $n \times n$ matricts, namely

$$\begin{pmatrix} A - a_{11}I_n & -a_{22}I_n & \cdots & -a_{1n}I_n \\ -a_{21}I_n & A - a_{22}I_n \cdots & & -a_{nn}I_n \\ \cdots & \cdots & \cdots & \cdots \\ -a_{n1}I_n & -a_{n2}I_n & \cdots & A - a_{nn}I_n \end{pmatrix}$$

It is far from obvious why the determinant of this matrix should equal the zero matrix. For a valid proof, regard $xI_n - A$ as a matrix over $F[x]$, which is a commutative ring with identity and let $adj\,(xI_n - A)$ be its adjoint (see the comments after Theorem (4.31)). Then we have

$$f(x)I_n = (xI_n - A)\, adj\,(xI_n - A). \tag{*}$$

This is an equality of two elements of the ring $M_n(F(x))$. By Exercise (1.18) (b), we identify $M_n(F(x))$ with the polynomial ring $(M_n(F))\,[x]$ (except that here we write the powers of the indeterminate x to the left of the coefficients rather than to the right). So (*) may be regarded as an equality of two polynomials over the ring $M_n(F)$. Let $f(x) = a_0 + a_1 x + \ldots + a_n x^n$ for some $a_0, a_1, \ldots, a_n \in F$. (Actually $a_n = 1$). Similarly $adj(xI_n - A)$ equals a polynomial of the form $B_0 + xB_1 + x^2B_2 + \ldots + B_{n-1}x^{n-1}$ for some $B_0, B_1, \ldots, B_{n-1} \in M_n(F)$. Comparing the coefficients of the like powers of x on both sides of (*) we get $(n + 1)$ equations, namely $a_0 I_n = -AB_0$, $a_1 I_n = B_0 - AB_1$, $a_2 I_n = B_1 - AB_2, \ldots, a_{n-1} I_n = B_{n-2} - AB_{n-1}$ and $a_n I_n = B_{n-1}$. Multiply these equations on the left by $I_n, A, A^2, \ldots, A^{n-1}, A^n$ respectively and add to get the result.

27. If $A, B, C \in M_n(F)$ and $B = CAC^{-1}$ then C is also invertible when regarded as an element of the ring $M_n(F(x))$ (which contains $M_n(F)$ as a subring) and has the same inverse, C^{-1}. Also xI_n commutes with C. So $xI_n - B = C(xI_n - A)C^{-1}$. Now apply Corollary (4.28).

28. Let $W = L(\{v_1, \ldots, v_{n-1}\})$. Clearly $T(W) \subset W$. By induction hypothesis $\{v_1, \ldots, v_{n-1}\}$ is linearly independent. If $v_n = \sum_{i=1}^{n-1} \alpha_i v_i$ for some $\alpha_1, \ldots, \alpha_{n-1} \in F$, then applying T we get $\lambda_n v_n = \sum_{i=1}^{n-1} \alpha_i \lambda_i v_i$ and hence $\sum_{i=1}^{n-1} \alpha_i (\lambda_i - \lambda_n) v_i = 0$ which forces each $\alpha_i = 0$ since $\lambda_i \neq \lambda_n$. The three assertions follow by considering the matrix of T w.r.t. the ordered basis $(v_1, \ldots, v_n)$.

29. Let $S: F^n \to F^m$ and $T: F^p \to F^n$ be the linear transformations defined by A, B respectively. Then $S \circ T$ is the linear transformation

defined by AB. Now apply Proposition (4.2) and Exercise (3.16).

30.   (1) $\Rightarrow$ (2) follows from the last exercise. Conversely, if (2) holds, choose $p = n$ and $B = I_n$. Then rank $(A) = n$ whence A is invertible. Similarly prove (1) $\Leftrightarrow$ (3).

31.   Each row operation on a matrix amounts to multiplying it on the left by a non-singular matrix.

33.   Following the hint, the given determinant equals that of the matrix

$$\begin{pmatrix} 1 & 1 & 1 & \ldots & 1 \\ 0 & a_2 - a_1 & a_3 - a_1 & \ldots & a_n - a_1 \\ 0 & a_2^2 - a_2 a_1 & a_3^2 - a_3 a_1 & \ldots & a_n^2 - a_n a_1 \\ \vdots & & & & \\ 0 & a_2^{n-1} - a_2^{n-2} a_1 & a_3^{n-1} - a_3^{n-2} a_1 & \ldots & a_n^{n-1} - a_n^{n-2} a_1 \end{pmatrix}$$

which, after subtracting the first column from all others and taking common factors from them equals $\prod\limits_{j=2}^{n} (a_j - a_1)$ times the determinant of the matrix

$$\begin{pmatrix} 1 & 0 & 0 & \ldots & 0 \\ 0 & 1 & 1 & \ldots & 1 \\ 0 & a_2 & a_3 & \ldots & a_n \\ \vdots & & & & \\ 0 & a_2^{n-2} & a_3^{n-2} & \ldots & a_n^{n-2} \end{pmatrix}$$

Apply induction on n.

34.   Let $X = (x_{ij})$ be the inverse matrix. For $n = 3$, performing suitable row operations we get:

$$\begin{pmatrix} 1 & \frac{1}{2} & \frac{1}{3} \\ \frac{1}{2} & \frac{1}{3} & \frac{1}{4} \\ \frac{1}{3} & \frac{1}{4} & \frac{1}{5} \end{pmatrix} \quad X = \begin{pmatrix} 1 & 0 & 0 \\ 0 & 1 & 0 \\ 0 & 0 & 1 \end{pmatrix}$$

$$\begin{pmatrix} 1 & \frac{1}{2} & \frac{1}{3} \\ 0 & 1 & 1 \\ 0 & 1 & \frac{16}{15} \end{pmatrix} \quad X = \begin{pmatrix} 1 & 0 & 0 \\ -6 & 12 & 0 \\ -4 & 0 & 12 \end{pmatrix}$$

$$\begin{pmatrix} 6 & 3 & 2 \\ 0 & 1 & 1 \\ 0 & 0 & 1 \end{pmatrix} \quad X = \begin{pmatrix} 6 & 0 & 0 \\ -6 & 12 & 0 \\ 30 & -180 & 180 \end{pmatrix}$$

Solving three systems of 3 equations each, we get X as

$$\begin{pmatrix} 9 & -36 & 30 \\ -36 & 192 & -180 \\ 30 & -180 & 180 \end{pmatrix}.$$

For $n = 4$, a similar computation gives the inverse as

$$\begin{pmatrix} 16 & -120 & 240 & -140 \\ -120 & 1200 & -2700 & 1680 \\ 240 & -2700 & 6480 & -4200 \\ -140 & 1680 & -4200 & 2800 \end{pmatrix}$$

However, it is far from easy to guess a formula for x_{ij} in the general case. The answer is,

$$x_{ij} = (-1)^{i+j}\, j \binom{i+j-2}{i-1} \binom{i+n-1}{i-1} \binom{j+n-1}{n-i} \binom{n}{j}!$$

(The last ! is exclamatory and not mathematical !) It is obtained by finding the inverse of a more general matrix called Cauchy's matrix.

35. Let $f: V \times V \to F$ be a bilinear form. Let A and B the matrices of f w.r.t. the ordered bases $(v_1,\ldots, v_n)$ and $(w_1,\ldots, w_n)$ respectively. Then $A' = C'BC$ where C is the matrix of change of basis from $(v_1,\ldots, v_n)$ to $(w_1,\ldots, w_n)$.

36. Let, if possible, $f(x) = g(x)h(x)$ where $g(x) = b_0 + \ldots + b_m x^m$ and $h(x) = c_0 + c_1 x + \ldots + c_r x^r$ are in $\mathbb{Z}[x]$. Then $a_0 = b_0 c_0$. Since $p \mid a_0$ but $p^2 \nmid a_0$, p divides b_0 or c_0 but not both. Without loss of generality, suppose $p \mid b_0$ and $p \nmid c_0$. Now $a_1 = b_0 c_1 + b_1 c_0$. Since $p \mid a_1$, and $p \nmid c_0$ we get $p \mid b_1$. From $a_2 = b_0 c_2 + b_1 c_1 + b_2 c_0$ and $p \mid a_2$ we get $p \mid b_2$. Continuing like this, from $a_m = b_0 c_m + b_1 c_{m-1} + \ldots + b_{m-1} c_1 + b_m c_0$, we get $p \mid b_m$. (We assume here that $m < n$, for otherwise $h(x)$ will have degree 0 which would contradict the primitivity of $f(x)$.) But then $p \mid a_n$ since $a_n = b_m c_r$.

39. Reduce B, C separately to echelon forms by row operations. These two sets of row operators are completely independent of each other. When performed together they reduce A to an echelon form, proving (i). For (ii), note that a term in det (A) corresponding to

a permutation $\sigma \in S_n$ will be non-zero only if $\sigma(i) \leqslant p$ for all $i = 1,..., p$ and $\sigma(i) > p$ for $i = p + 1,..., n$. Every such σ corresponds uniquely to an ordered pair (θ, τ) where θ is a permutation of $\{1,..., p\}$ and τ is a permutation of $\{p + 1,..., n\}$. The number of inversions in σ clearly is the sum of the numbers of inversions in θ and τ. So $(-1)^\sigma = (-1)^\theta(-1)^\tau$. Hence every term in det (A) factors as the product of a term in det (B) and a term in det (C).

CHAPTER 7

Section 7.1

1. (i) $\dfrac{1}{1 + x}$ (ii) $\dfrac{1}{1 - x^2}$ (iii) $\dfrac{1}{1 - x^3}$ (iv) $\dfrac{1 - x - x^2}{1 - x^3}$

   (v) $\dfrac{1}{(1 - x)^2}$.

2. (i) e^{2x} (ii) e^{-x} (iii) $\dfrac{e^x + e^{-x}}{2}$ $(= \cosh x)$ (iv) $\cos x$

   (v) $\sin x$ (vi) $(1 + x)^n$.

3. (i) $a_0, 0, a_2, 0, a_4, 0, a_6....$ (ii) $0, a_1, 0, a_3, 0, a_5, 0,...$ (iii) $0, a_0,$ $\dfrac{a_1}{2}, \dfrac{a_2}{3},...., \dfrac{a_{n-1}}{n},....$ The O.G.F. of $1, \dfrac{1}{2}, \dfrac{1}{3}, \dfrac{1}{4},...$ is $\dfrac{-1}{x} \ln (1 - x)$.

4. $z^m - 1 = (z - 1)(z^{m-1} + x^{m-2} + ... + z + 1)$. Since ξ is a primitive mth root of 1, for $m \nmid n$, $\xi^n \neq 1$ and so ξ^n is a root of $z^{m-1} + z^{m-2} + ... + z + 1$. Thus we get $\sum\limits_{j=1}^{m} (\xi^n)^j = m$ if $m \mid n$ and $= 0$ if $m \nmid n$. Now $\sum\limits_{j=1}^{m} A(\xi^j x) = \sum\limits_{j=1}^{m} \sum\limits_{n=0}^{\infty} a_n \xi^{jn} x^n = \sum\limits_{n=0}^{\infty} a_n x^n \sum\limits_{j=1}^{m} (\xi^n)^j = \sum\limits_{\substack{n=0 \\ m \mid n}}^{\infty} m a_n x^n$.

5. (i) Differentiate both sides of $\sum\limits_{k=0}^{n} \binom{n}{k} x^k = (1 + x)^n$ w.r.t. x and set $x = 1$.

   (ii) Differentiate again and add (i).

   (iii) Integrate both sides of $\dfrac{1 - x^n}{1 - x} = 1 + x + 1... + x^{n-1}$ w.r.t. x

over the interval $[0, 1]$. To evaluate $\int_0^1 \dfrac{1-x^n}{1-x}\,dx$ put $u = 1-x$

and use binomial theorem.

(iv) Let a_n and b_n denote the sums in the two brackets. Then $a_n + ib_n = (1 + i)^n$. So $a_n^2 + b_n^2 = |(1 + i)^n|^2 = |(1 + i)|^{2n} = (\sqrt{2})^{2n} = 2^n$.

(v) Use $(1 + x)^p (1 + x)^q = (1 + x)^{p+q}$.

(vi) $\displaystyle\sum_{j=0}^{\infty} \binom{p}{j}\binom{q}{n+j} = \sum_{j=0}^{\infty} \binom{p}{j}\binom{q}{q-n-j} = \binom{p+q}{q-n}$

$$= \binom{p+q}{p+n}.$$

6. (i) 6^n (ii) $n\dbinom{2n-1}{n-1}$ $\left[\text{Multiply } \displaystyle\sum_{k=0}^{n} k\binom{n}{k} x^k = nx\,(1 + x)^{n-1}\right.$

and $\displaystyle\sum_{k=0}^{n}\binom{n}{n-k} x^k = (1 + x)^n.\Big]$

(iii) $2\left[\text{Differentiate } \dfrac{1}{1 - \frac{1}{2}x} = 1 + \frac{1}{2}x + \frac{1}{4}x^2 + \dots\dots \text{ and set}\right.$

$\left. x = 1. \vphantom{\dfrac{1}{1}}\right]$

7. $\dfrac{(1 + x)^n}{1 - x} = (1 + x)^n (1 + x + x^2 + \dots)$. The coefficient of x^n in

this equals $\displaystyle\sum_{k=0}^{n}\binom{n}{k}$ which is 2^n by Exercise (2.2.16) (i). Similarly

the coefficient, say α, of x^n in $\dfrac{(1 + x)^{2n+1}}{1-x}$ equals $\displaystyle\sum_{k=0}^{n}\binom{2n+1}{k}$. Let

$\beta = \displaystyle\sum_{r=n+1}^{2n+1}\binom{2n+1}{r}$. Then $\alpha+\beta=2^{2n+1}$ by Exercise (2.2.16) (i) again.

Also $\alpha = \beta$ since $\dbinom{2n+1}{k} = \dbinom{2n+1}{2n+1-k}$ for $k = 0,\dots, n$. So

$\alpha = 2^{2n} = 4^n$. Finally, the given sum, say S, equals the coefficient

of x^n in $2^n\displaystyle\sum_{k=0}^{n}\left(x + \frac{1}{2}\right)^{2n-k}$ which, being a geometric series, simplifies to $\dfrac{2(1 + 2x)^n - \dfrac{1}{2^n}(1 + 2x)^{2n+1}}{1 - 2x}$. Putting $2x = y$ and using

the earlier parts, $S = 2 \cdot 2^n \cdot 2^n - \dfrac{1}{2^n} \cdot 2^n \cdot 4^n = 4^n$.

8. The sum equals the coefficients of x^n in $\displaystyle\sum_{k=0}^{n} \left(x + \dfrac{1}{2}\right)^{n+k}$. Now proceed as in the last exercise.

9. Applying Theorem (1.3) (iii), $(1 - x)^{-n} = \displaystyle\sum_{r=0}^{\infty} (-1)^r \binom{-n}{r} x^r$, where

$$\binom{-n}{r} = \dfrac{(-n)(-n-1)\ldots(-n-r+1)}{r!} = (-1)^r \binom{n + r - 1}{r}.$$

11. $\dfrac{4}{5}$

13. Use Exercise (1.12).

14. Putting $k = 1$ and $n = 0$, in Problem (1.5), $1 + 2 + \ldots + r$

$= \binom{r+1}{2} = \dfrac{r(r + 1)}{2}$. Similarly $\displaystyle\sum_{i=1}^{n} i^2 = \sum_{i=1}^{n} 2 \binom{i}{2} + \binom{i}{1}$

$= 2 \binom{n+1}{3} + \binom{n+1}{2} = \dfrac{n(n + 1)(2n + 1)}{6}$. To get the

result from Proposition (1.10), start from $\dfrac{x}{(1-x)^2} = \displaystyle\sum_{n=1}^{\infty} nx^n$.

Differentiating, $\dfrac{x^2 + x}{(1-x)^3} = \displaystyle\sum_{n=1}^{\infty} n^2 x^n$ and so $\displaystyle\sum_{r=1}^{n} r^2$ is the coefficient

of x^n in $\dfrac{x^2 + x}{(1-x)^4}$. Applying binomial theorem and noting that

$$\binom{-4}{n} = (-1)^n \binom{n+3}{3},$$

$$\sum_{r=1}^{n} r^2 = \binom{n+1}{3} + \binom{n+2}{3} = \dfrac{n(n + 1)(2n + 1)}{6}.$$

15. By binomial theorem,

$$(1 - 4x)^{-1/2} = \sum_{r=0}^{\infty} a_r x^r$$

where

$$a_r = (-4)^r \binom{-1/2}{r}$$

$$= (-1)^r 4^r \dfrac{\left(-\dfrac{1}{2}\right)\left(-\dfrac{3}{2}\right) \cdots \left(-\dfrac{(2r - 1)}{2}\right)}{r!}$$

$$= 2^r \frac{1.3.5 \dots (2r-1)}{r!}$$

$$= \frac{1.3.5. \dots (2r-1).2.4. \dots, 2r}{r!} = \frac{(2r)!}{r! \; r!} = \binom{2r}{r}.$$

For the second assertion, $_{2r}P_r = r! \binom{2r}{r}$.

16. From the last exercise, $(1-4t)^{-1/2} = \sum\limits_{r=0}^{\infty} \binom{2r}{r} t^r$. Integrating

both sides from 0 to x and putting $n = r+1$, $\frac{1}{2} - \frac{1}{2}(1-4x)^{1/2}$

$= \sum\limits_{n=1}^{\infty} \frac{1}{n} \binom{2n-2}{n-1} x^n$. Since $a_0 = 0$, the O.G.F. of $\{a_n\}$ is $A(x) =$

$\dfrac{1 - \sqrt{1-4x}}{2}$. From this we get, $[A(x)]^2 = A(x) - x$. For $n > 1$,

the coefficient of x^n in $[A(x)]^2$ is $\sum\limits_{k=1}^{n-1} a_k a_{n-k}$ (since $a_0 = 0$) while on

the R.H.S. it is a_n.

18. Since $\binom{m}{n-m} = 0$ for $m < \lceil n/2 \rceil$, the sum might as well be taken

as $\sum\limits_{m=0}^{n} \binom{m}{n-m}$ which equals the coefficient of x^n in

$$\sum\limits_{m=0}^{n} (1+x)^m x^m = \frac{1 - x^{n+1}(1+x)^{n+1}}{1-x-x^2}.$$

Now $\dfrac{1}{1-x-x^2}$ resolves into partial fractions as

$$\frac{1}{\alpha - \beta} \left(\frac{\alpha}{1 - \alpha x} - \frac{\beta}{1 - \beta x} \right)$$

where $\alpha = \dfrac{1+\sqrt{5}}{2}$ and $\beta = \dfrac{1-\sqrt{5}}{2}$. Since x^n cannot appear in

$x^{n+1}(1+x)^{n+1}$, the answer is $\dfrac{1}{\alpha - \beta}(\alpha^{n+1} - \beta^{n+1})$.

19. $f\left(\dfrac{1}{x}\right) = a_0 + \dfrac{a_1}{x} + \dfrac{a_2}{x^2} + \dots + \dfrac{a_n}{x^n}$. Also $f\left(\dfrac{1}{x}\right) = a_n \left(\dfrac{1}{x} - \alpha_1\right) \dots \left(\dfrac{1}{x} - \right.$

$\left. - \alpha_n\right) = \dfrac{a_n}{x^n}(1 - \alpha_1 x) \dots (1 - \alpha_n x)$. So $a_0 x^n + a_1 x^{n-1} + \dots + a_n$

$= x^n f\left(\dfrac{1}{x}\right) = a_n(1 - \alpha_1 x) \dots (1 - \alpha_n x)$. The roots of this polynomial

are $\dfrac{1}{\alpha_1}, \dots, \dfrac{1}{\alpha_n}$. So their sum equals $-\dfrac{a_1}{a_0}$, as we see by comparing

the coefficients of x^{n-1} in the equation

$$a_0 x^n + a_1 x^{n-1} + \ldots + a_n = a_0 \left(x - \frac{1}{\alpha_1} \right) \left(x - \frac{1}{\alpha_2} \right) \ldots \left(x - \frac{1}{\alpha_n} \right).$$

20. By Exercise $(1.2)(v)$, for $y \neq 0$, $\dfrac{\sin y}{y} = 1 - \dfrac{y^2}{3!} + \dfrac{y^4}{5!} - \dfrac{y^6}{7!} + \ldots$.

21. For $f(x)$ in Exercise (1.20), $a_0 = 1$, $a_1 = -\dfrac{1}{6}$ and the roots are $(n\pi)^2$, $n = 1, 2, 3, \ldots$. So

$$\sum_{n=1}^{\infty} \frac{1}{(n\pi)^2} = \frac{1}{6}; \text{ i.e. } \sum_{n=1}^{\infty} \frac{1}{n^2} = \frac{\pi^2}{6}.$$

22. $\displaystyle \sum_{k=0}^{p} \binom{p}{k}\binom{q}{k}\binom{n+k}{p+q} = \sum_{k=0}^{p} \binom{p}{k}\binom{q}{k} \sum_{j=0}^{\infty} \binom{k}{j}\binom{n}{p+q-j}$

$$= \sum_{j=0}^{\infty} \binom{n}{p+q-j} \sum_{k=0}^{p} \binom{p}{k}\binom{q}{k}\binom{k}{j}$$

$$= \sum_{j=0}^{\infty} \binom{n}{p+q-j} \sum_{k=0}^{p} \binom{p}{k}\binom{q}{j}\binom{q-j}{q-k}$$

$$= \sum_{j=0}^{\infty} \binom{n}{p+q-j}\binom{p+q-j}{q}\binom{q}{j}$$

$$= \sum_{j=0}^{\infty} \binom{n}{q}\binom{n-q}{p-j}\binom{q}{j} = \binom{n}{q}\binom{n}{p}.$$

23. In the last exercise, both sides are polynomials in n (of drgree $p + q$). Since equality holds for infinitely many values of n (namely $n = 1, 2, 3, \ldots$), it holds if n is replaced by any real number. Replace n by $-n-1$. For any positive integer r,

$$\binom{-n-1}{r} = (-1)^r \binom{n+r}{r}.$$

Use this with $r = p$, $r = q$ and $r = p + q$ to get the result.

24. $\dfrac{1+xy}{1-x-yx^2} = \dfrac{1+xy}{1-x} \cdot \dfrac{1}{1 - \left(\dfrac{x^2}{1-x}\right)y} = \dfrac{1+xy}{1-x} \sum_{n=0}^{\infty} \left(\dfrac{x^2}{1-x}\right)^n y^n$

which reduces to $\dfrac{1}{1-x} + \displaystyle\sum_{n=1}^{\infty} \left[\dfrac{x^{2n}}{(1-x)^{n+1}} + \dfrac{x^{2n-1}}{(1-x)^n} \right] y^n$, i.e. to

$\dfrac{1}{1-x} + \sum\limits_{n=1}^{\infty} \dfrac{x^{2n-1}}{(1-x)^{n+1}} y^n$. For $n \geqslant 1$, the coefficient of x^m in $\dfrac{x^{2n-1}}{(1-x)^{n+1}}$ is the same as that of x^{m-2n+1} in $\dfrac{1}{(1-x)^{n+1}}$, which by Exercise (1.9) is $\dbinom{n+1+m-2n+1-1}{m-2n+1}$, i.e. $\dbinom{m-n+1}{m-2n+1}$ or $\dbinom{m-n+1}{n}$. But by Exercise (2.3.5), this is exactly $a_{m,n}$. For a solution without Exercise (1.9),

$$\frac{1+xy}{1-x-yx^2} = \frac{1+xy}{1-x(1+xy)} = \sum_{r=0}^{\infty} x^r (1+xy)^{r+1}$$
$$= \sum_{r=0}^{\infty} \sum_{k=0}^{\infty} \binom{r+1}{k} x^{r+k} y^k.$$

To get the coefficient of $x^m y^n$ set $k = n$ and $r = m - n$.

25. Each path of length m corresponds to a binary sequence of length m, $(a_1, a_2, \ldots, a_m)$ where for $j = 1, \ldots, m$, $a_j = 1$ or 0 depending upon whether in the jth time unit, you go forward or backward. For any k with $0 \leqslant k \leqslant m$, this path terminates at the kth point in the mth row iff, exactly k of the terms in the binary sequence are 1. So there are $\dbinom{m}{k}$ paths terminating at it. The identity follows since there are 2^m paths of length m.

26. Let $P_{m,n}$ denote the nth point in the mth row $(m \geqslant 0, 0 \leqslant n \leqslant m)$. Then (i) says that to reach $P_{m,n}$ you must go either through $P_{m-1,n}$ or through $P_{m-1,n-1}$. For (ii), note that starting with any point in the Pascal triangle there is a replica of Pascal triangle (ignoring the entries) with that point as the apex. Now any path from the apex $P_{0,0}$ to the point $P_{p+q,n}$ will cross the pth row at a point $P_{p,k}$ for a unique k, $0 \leqslant k \leqslant n$. The remainder of the path can be thought of as the replica of a path from $P_{0,0}$ to $P_{q,n-k}$. For (iii), divide the paths arriving at $P_{m,n}$ into two kinds; let a_n of them have their last segment going forward and b_n going backward. $\Big($ Clearly $a_n = \dbinom{m-1}{n-1}$ and $b_n = \dbinom{m-1}{n}$, but that is not very important here. $\Big)$ Then $a_0 = 0 = b_m$ and $a_n = b_{n-1}$ for $n = 1, \ldots, m$. Also $a_n + b_n = \dbinom{m}{n}$. We then get the first sum as $a_0 + a_1 + a_2 + a_3 + \ldots + a_m$ and the second as $b_0 + b_1 + b_2 + \ldots + b_m$.

27.   This is a restatement of the fact that every positive integer can be uniquely expressed as a product of prime powers.

Section 7.2

1.   (i) 2.   (ii) $\sum\limits_{k=0}^{6} \begin{pmatrix} 6 \\ k \end{pmatrix} 5^k 2^{10-k} \begin{pmatrix} k \\ 10-k \end{pmatrix}$. The only non-zero terms arise from $k = 5$ and $k = 6$. So the answer is 4,350,000.   (iii) $\begin{pmatrix} 20 \\ 5 \end{pmatrix}$. (Put $y = x^2$.)

3. $\prod\limits_{i=1}^{k} A_i(x)$ where $A_i(x) = 1 + x + x^2 + \ldots + x^{m_i}$.

4. $\left[x + \begin{pmatrix} 10 \\ 3 \end{pmatrix} x^3 + \begin{pmatrix} 10 \\ 5 \end{pmatrix} x^5 + \begin{pmatrix} 10 \\ 7 \end{pmatrix} x^7 + \begin{pmatrix} 10 \\ 9 \end{pmatrix} x^9 \right] (1 + x)^7$.

5. $x^n (1 + x + x^2 + x^3 + x^4 + x^5)^n$.

6.   The enumerator for total score is

$$(1 + x + x^2 + x^3 + x^4 + x^5) (x^2 + x^3 + x^4 + x^5 + x^6 + x^7)$$

which is the same as that for a pair of ordinary dice.

7.   Take $A(x) = x(1 + x)(1 + x + x^2)$ and $B(x) = x(1 + x^3)(1 + x^2 + x^4)$. The corresponding markings on the faces of the first die are 1, 2, 2, 3, 3 and 4 and those on the second are 1, 3, 4, 5, 6 and 8.

8.   The enumerator is not the entire product $(1 + 4x + 6x^2 + 4x^3 + x^4)^{18}$ but only $(1 + x)^{28} \left(\sum\limits_{r=0}^{\infty} b_r c_r x^{2r} \right)$ where $\sum\limits_{r=0}^{\infty} b_r x^r = (1 + 4x + 6x^2 + 4x^3 + x^4)^5$ and $\sum\limits_{r=0}^{20} c_r x^r$ is the part of $(1 + 4x + 6x^2 + 4x^3 + x^4)^6$ consisting of terms upto x^{20}. Then $b_r = \begin{pmatrix} 20 \\ r \end{pmatrix}$ and $c_r = \begin{pmatrix} 24 \\ r \end{pmatrix}$ for $0 \leqslant r \leqslant 20$. So

$$a_n = \sum\limits_{r=0}^{\lceil n/2 \rceil} b_r c_r \begin{pmatrix} 28 \\ n-2r \end{pmatrix}$$

$$= \sum\limits_{r=0}^{\lceil n/2 \rceil} \begin{pmatrix} 20 \\ r \end{pmatrix} \begin{pmatrix} 24 \\ r \end{pmatrix} \begin{pmatrix} 28 \\ n-2r \end{pmatrix}.$$

9. $1 + x + \ldots + x^{b-1} = \dfrac{1-x^b}{1-x}$.   Similarly, $1 + x^b + \ldots + x^{b(b-1)}$

$= \dfrac{1-x^{b^2}}{1-x^b}$ etc. Multiplying infinitely many factors of this form gives the result.

10. In enumerators the coefficients generally represent the number of ways to do something and hence cannot be negative. In the example, the number of ways to pick n balls is 1 or 0 according as n is or is not a multiple of m. For every $n \geqslant 0$, the absolute value of the coefficient of x^n is the number of ways to pick n balls.

11. For $r \geqslant 0$, let $A_r(x) = 1 - x^{2^r} + x^{2.2^r} - x^{3.2^r} + \ldots + (-1)^k x^{k.2^r} + \ldots$. Let $A(x) = \prod\limits_{r=0}^{\infty} A_r(x)$. Now consider a partition of n in which k_i parts are of size 2^{r_i} each for $i = 1, \ldots, m$ (say) where $0 \leqslant r_1 < r_2 < \ldots < r_m$ and $k_1, \ldots, k_m$ are positive integers. Such a partition corresponds to the term $x^{k_1 2^{r_1}} \cdot x^{k_2 2^{r_2}} \cdot \ldots x^{k_m 2^{r_m}}$ in $A(x)$. The sign of this term is 1 or -1 according as $\sum\limits_{i=1}^{m} k_i$, i.e. the total number of parts, is even or odd. It follows that in $A(x)$, the coefficient of x^n is $a_n - b_n$ for all $n \geqslant 1$. From the hint, $A(x)$ is simply $1 - x$. So $a_1 - b_1 = -1$, while $a_n - b_n = 0$ for $n > 1$.

12. Let $n = n_1 + n_2 + \ldots + n_m$ where $n_1 \geqslant n_2 \geqslant \ldots \geqslant n_m > 0$. Then $n - m = (n_1 - 1) + (n_2 - 1) + \ldots + (n_m - 1)$ is a portition of $n - m$ into at most m parts. If $n_1 > n_2 > \ldots > n_m$, then $[n_1 - (m-1)] + [n_2 - (m-2)] + \ldots + [n_{m-1} - 1] + [n_m - 0]$ gives a partion of $n - \dfrac{m(m-1)}{2}$ into exactly m parts.

13. The coefficient of x^n on the L.H.S. is the number of partitions of n into parts of distinct size. If $n = n_1 + \ldots + n_m$ is any such partition with $n_1 > n_2 > \ldots > n_m > 0$, then $[n_1 - m] + [n_2 - (m-1)] + \ldots + [n_{m-1} - 2] + [n_m - 1]$ is partition of $n - \dfrac{m(m+1)}{2}$ into at most m parts, whose dual is a partition in which the part size is at most m.

14. Taking $\frac{1}{4}$ rupee as a unit the problem asks to find a_{40} where a_n is the number of partitions of n in parts of size 1, 2 or 4. In general a_n equals $\dfrac{(n+2)(n+1)}{16} + \dfrac{n+1}{4} + \dfrac{9}{32} + (-1)^n \dfrac{n+1}{16} + (-1)^n \dfrac{5}{32} + (-i)^n \dfrac{1+i}{16} + i^n \dfrac{1-i}{16}$.

15. Let a_n be as in the answer to the last exercise. Then the answer to the first question is $a_{40} - a_{29} - a_{18} + a_7$ while to the second it is $a_{32} - a_{25} - a_{14} + a_7$.

16. For a direct solution, let M be the midpoint of BD. Then it suffices

to show that $AF \perp MF$. Draw MN and BH perpendicular to AC. Join DH and FN. Since $BH \parallel MN \parallel DE$ and $BM = MD = \frac{1}{2} DC$, we get $HN = NE = \frac{1}{2} EC$, whence it follows that $FN = \frac{1}{2} DH = \frac{1}{2} DC = MD$. So $MDFN$ is an isosceles trapezium and is therefore cyclic. Thus the points M, D, F, N are concyclic. But since $\angle ADM = 90° = \angle ANM$ the points A, M, D, N are concyclic. So A, M, D, F are concyclic, proving that $\angle AFM = \angle ADM = 90°$.

For a solution using coordinates, choose axes so that $D = (0, 0)$, $C = (a, 0)$, $B = (-a, 0)$, $A = (0, b)$. Let $F = (h, k)$. Then

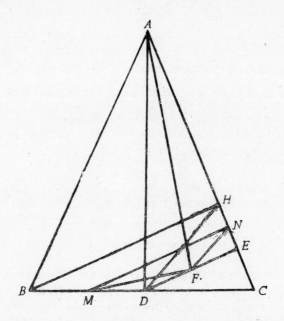

$E = (2h, 2k)$ and $\dfrac{2k - b}{2h} = -\dfrac{b}{a}$. But since $DE \perp AC$, we also have $\dfrac{b}{a} = \dfrac{h}{k}$. A little calculation gives $\dfrac{k - b}{h} \dfrac{2k}{2h + a} = -1$, which means $AF \perp BE$.

17. Let S_i consist of the ith object. Then $A_i(x) = \dfrac{x^{n_i}}{n_i!}$ since the only permissible permutation is of length n_i. So $A(x) = \dfrac{x^n}{n_1! \ldots n_k!}$ where $n = n_1 + \ldots + n_k$. The number of permutations in which ith symbol occurs n_i times is $n!$ times the coefficient of x^n in $A(x)$, i.e.

$$\frac{n!}{n_1! \ldots n_k!}.$$

18. For $i = 1,..., k$, let $S_i = \{i - 1\}$. For exercise (3.2.17) (b), $A_1(x) = A_2(x) = \dfrac{e^x + e^{-x}}{2}$, while for $i = 3,..., k$, $A_i(x) = e^x$. By Theorem

(2.16), $A(x) = \left(\dfrac{e^x + e^{-x}}{2}\right)^2 e^{(k-2)x} = \tfrac{1}{4}[e^{kx} + 2e^{(k-2)x} + e^{(k-4)x}]$. The

coefficient of x^n in $A(x)$ is $\dfrac{1}{4n!}[k^n + 2(k - 2)^n + (k - 4)^n]$. Multi-

plying it by $n!$ gives the result. For Exercise (3.2.18),

$$A_1(x) = \sum_{r \text{ odd}} \frac{x^r}{r!} = \frac{e^x - e^{-x}}{2} \quad \text{(cf. Exercise (1.3) (iii)).}$$

The other $A_i(x)$'s remain unchanged. So

$$A(x) = \frac{e^x - e^x}{2}\,\frac{e^x + e^x}{2} \exp((k-2)x) = \frac{e^{2x} - e^{-2x}}{4} \exp((k-2)x)$$

$$= \tfrac{1}{4}[\exp(kx) - \exp((k - 4)x)]. \text{ So the answer is } \frac{k^n + (k-4)^n}{4}.$$

19. $\displaystyle\sum_{n=0}^{\infty} \frac{S_{n,m}}{n!} x^n = \frac{1}{m!}(e^x - 1)^m = \frac{1}{m!} \sum_{r=0}^{m} (-1)^r \exp((m-r)x)\binom{m}{r}$

$$= \sum_{r=0}^{m} \frac{(-1)^r}{(m-r)!\,r!} \sum_{n=0}^{\infty} \frac{(m-r)^n}{n!} x^n. \text{ Comparing the coefficients of } x^n$$
gives the result.

20. Multiply both sides of $S_{n,m} = \displaystyle\sum_{r=0}^{m-1} \frac{(-1)^r (m - r)^{n-1}}{r!\,(m - r - 1)!}$ by $\dfrac{x^n}{n!}$ and sum
over from $n = 0$ to ∞. The result follows by reversing the steps
in the solution to the last exercise.

21. Let $n = 2m = a + b + c$ where $a + b - c > 0$, $b + c - a > 0$
and $c + a - b > 0$. Note that $a + b - c$ has the same parity as
$a + b + c$, which is even. So $a + b - c = 2x$ for some positive
integer x. Similarly letting $b + c - a = 2y$ and $c + a - b = 2z$,
$x + y + z$ is a partition of m into exactly 3 parts. Moreover, a, b.
c can be uniquely recovered from x, y, z.

22. Let $n = 2m$. Then $t_n = b_m = $ number of partitions of m into exactly
3 parts. b_m is the same as the number of partitions of m in which
the largest part size is 3. So $b_m = $ coefficient of x^m in

$$\frac{x^3}{(1 - x)(1 - x^2)(1 - x^3)}.$$

From the solution to the Postage Problem, $b_m = a_{m-3}$

$$= \frac{6(m - 1)(m - 2) + 18(m - 2) + 17 - (-1)^m 9 + 8\omega^m + 8\omega^{2m}}{72}$$

23.   Let X, Y be sets with x and n elements respectively and let F be the set of all functions from Y to X. Then $| F | = x^n$. Classify the functions in F according to their ranges which can be any non-empty subsets of X. For $1 \leqslant m \leqslant n$, the number of m-subsets of X is $\dfrac{x(x-1)...(x-m+1)}{m!}$. For any one such m-subset, say, R, the number of functions from Y to X with R as their range is $m! \, S_{n,m}$.

24.   Apply Theorem (3.2.7) for the first assertion. For the second, classify the partitions according to the number of parts in them.

25.   The result of Exercise (2.19) can be expressed as

$$S_{n,m} = \frac{1}{m!} \sum_{j=1}^{\infty} (-1)^{m-j} \binom{m}{j} j^n,$$

since $\binom{m}{j} = 0$ for $j > m$. So,

$$P_n = \sum_{k=1}^{n} S_{n,k} = \sum_{k=1}^{\infty} \frac{1}{k!} \sum_{j=1}^{\infty} (-1)^{k-j} \binom{k}{j} j^n$$

$$= \sum_{j=1}^{\infty} j^n \sum_{k=1}^{\infty} (-1)^{k-j} \frac{1}{k!} \binom{k}{j}$$

$$= \sum_{j=1}^{\infty} j^n \sum_{k=j}^{\infty} (-1)^{k-j} \frac{1}{k!} \binom{k}{j} \left(\text{since } \binom{k}{j} = 0 \text{ for } k < j \right)$$

$$= \sum_{j=1}^{\infty} \frac{j^n}{j!} \sum_{k=j}^{\infty} (-1)^{k-j} \frac{1}{(k-j)!} = \frac{1}{e} \sum_{j=1}^{\infty} \frac{j^n}{j!}$$

26.   From $p_X'(t) = \sum_{n=1}^{\infty} n p_n t^{n-1}$ we get $t p_X'(t) = \sum_{n=1}^{\infty} n p_n t^n$ and, after differentiation, $t p_X''(t) + p_X'(t) = \sum_{n=1}^{\infty} n^2 p_n t^{n-1}$. So $V(X) = \sum_{n=0}^{\infty} (n^2 - 2E(X)n + [E(X)]^2) \, p_n = p_X''(1) + p_X'(1) - 2E(X) \sum_{n=0}^{\infty} n p_n + [E(X)]^2 \sum_{n=0}^{\infty} p_n$. Since $\sum_{n=0}^{\infty} n p_n = E(X) = p_X'(1)$ and $\sum_{n=0}^{\infty} p_n = 1$, the result follows. The variances in Examples (3) and (4) are, respectively, $10p(1-p)$ and 2 respectively.

27.   For every $n \geqslant 2$, the probability, say q_n, that a head will occur for the second time at the nth toss is $\frac{n-1}{2^n}$. So $p_Y(t) = \sum\limits_{n=2}^{\infty} \frac{n-1}{2^n} t^n$. In

$[p_X(t)]^2$, the coefficient of t^n is $\sum\limits_{r=1}^{n-1} \frac{1}{2^r} \frac{1}{2^{n-r}}$. So $p_Y(t)=[p_X(t)]^2$. Alter-

natively, tossing a coin till a head shows for the second time is
equivalent to tossing two coins separately till a head shows in each.
Since the two tosses are independent and have the same probability
generating function, namely $p_X(t)$, the result follows. For $E(Y)$,

$\frac{d}{dt}(p_X(t)^2) = 2p_X(t)\, p_X'(1)$. Since $p_X(1) = 1$ and $p_X'(1) = E(X) = 2$,

we get $E(Y) = 4$.

28. (a) Let $\alpha = \sum\limits_{n=1}^{\infty} \frac{1}{n^2}$. (By Exercise (1.21), $\alpha = \frac{\pi^2}{6}$ but that is not very

important here.) Let $p_0 = 0$ and $p_n = \frac{1}{\alpha n^2}$ for $n \geqslant 1$. Then

$\{p_n\}_{n=0}^{\infty}$ is the probability distribution of a discrete random

variable for which $E(X) = \sum\limits_{n=1}^{\infty} \frac{n}{\alpha n^2} = \frac{1}{\alpha} \sum\limits_{n=1}^{\infty} \frac{1}{n}$ is infinite.

(b) For $n \geqslant 1$, p_n would be $q^{n-1} p$, where $q = 1 - p$. Hence $p_X(t)$

would come out to be $\frac{pt}{1-qt}$ differentiating which, $E(X) = \frac{1}{p}$.

Section 7.3

1. $\frac{50,482}{99900} = 0.50 + \frac{532}{99900}$. In the decimal expansion of $\frac{532}{999}$ after first

three stages the remainder is again 532, whence the expansion

must recur with period 3. $\left(\text{(In fact } \frac{50,482}{99,900} = 0.50532532532...\right)$

2. $x_n = \dfrac{-2 - 4\omega + \omega^n (2\omega + 3) + \omega^{2n} (2\omega - 1)}{3(1 + 2\omega)} + \dfrac{10n}{3}$ for $n \geqslant 2$,

where $\omega = -\frac{1}{2} + i \frac{\sqrt{3}}{2}$ is a primitive cuberoot of 1.

3. Let α be a real number. If $\alpha = \frac{p}{q}$ where p, q are integers and $q > 0$,

then in the decimal expansion of α, at any stage the remainder is
some integer from 0 to $q - 1$. So after the digit in the unit's place
of p, there can be at most q distinct remainders. Whenever the
same remainder appears, the digits in the decimal expansion will
recur periodically. (Finding the length of this period is an interesting number theoretic problem.) Conversely if α has a recurring
decimal expansion of period r, say, then for suitable integers m and

n, $10^m \alpha = n + \beta$ where $\beta = 0.\, \alpha_1\alpha_2...\alpha_r\alpha_1\alpha_2...\alpha_r\alpha_1\alpha_2...\alpha_r.....$ Since $(10^r - 1)\beta$ is an integer, β is rational, whence α is rational too.

4. From $F_n = \dfrac{1}{\sqrt{5}}\,(\alpha^n - \beta^n)$, where $\alpha = \dfrac{1 + 5}{2}$ and $\beta = \dfrac{1 - \sqrt{5}}{2}$ we

have $\dfrac{F_n}{F_{n-1}} = \dfrac{\alpha^n - \beta^n}{\alpha^{n-1} - \beta^{n-1}} = \dfrac{(1 - (\beta/\alpha)^n}{1 - (\beta/\alpha)^{n-1}}\,\alpha$. Since $\left|\dfrac{\beta}{\alpha}\right| = \left|\dfrac{1 - \sqrt{5}}{1 + \sqrt{5}}\right|$
< 1, $(\beta/\alpha)^n \to 0$ as $n \to \infty$. The attempted argument is valid to evaluate L in case it is known to exist. But it does not show that L exists and hence is not complete.

5. Let the lengths AD and AB be λx and x respectively. Then FD
$= (\lambda - 1)x$. Since $FDCE$ is similar to $ABCD$, $\lambda = \dfrac{1}{\lambda - 1}$. Solving

$\lambda = \dfrac{1 + \sqrt{5}}{2}$. The property in question depends only on the shape

of the rectangle. So the rectangles $FDCE$, $FDHG$, $FJIG$, ... all have it.

6. Since $F_1 = 1$, $|F_1| = 1 \leqslant 2$. Also $|F_{n-1}| \leqslant 2^{n-1}$ and $|F_{n-2}|$
$\leqslant 2^{n-2}$ give, $|F_n| = |F_{n-1} + F_{n-2}| \leqslant |F_{n-1}| + |F_{n-2}| \leqslant 2^{n-1}$
$+ 2^{n-2} < 2^{n-1} + 2^{n-1} = 2^n$ for $n \geqslant 2$.

7. By assumption, $A \cup B = (0, R)$. Also $A \cap B = \phi$ since $f(x)$ cannot simultaneously equal $g(x)$ and $h(x)$ for any $x \in (0, R)$. So the interval $(0, R)$ has been expressed as the union of two mutually disjoint, closed subsets By connectedness of $(0, R)$, either $A = \phi$ or $B = \phi$. In the first case, $B = (0, R)$, i.e. $f(x) = h(x)$ for all $x \in (0, R)$. In the second case $f(x) = g(x)$ for all $x \in (0, R)$.

For those not familiar with connectedness, but familiar with other versions of completeness of the real line, an alternate (but essentially equivalent) argument can be given as follows. If either $A = \phi$ or $B = \phi$ we are done. If not pick $a_1 \in A$, $b_1 \in B$ and suppose without loss of generality that $a_1 < b_1$. If $\dfrac{a_1 + b_1}{2} \in A$, call

it a_2 and b_1 as b_2. If $\dfrac{a_1 + b_1}{2} \in B$, call it b_2 and a_1 as a_2. In either case, $a_2 \in A$, $b_2 \in B$, $[a_2, b_2] \subset [a, b_1]$ and $b_2 - a_2 = \frac{1}{2}(b_1 - a_1)$. Repeat the argument with $[a_2, b_2]$ to get $[a_3, b_3] \subset [a_2, b_2]$ with $a_3 \in A$, $b_3 \in B$ and $b_3 - a_3 = \frac{1}{4}(b_1 - a_1)$. Continuing, we get sequences $\{a_n\}$, $\{b_n\}$ with $a_n \in A$, $b_n \in B$, $[a_{n+1}, b_{n+1}] \subset [a_n, b_n]$ and $b_n - a_n = \dfrac{1}{2^{n-1}}(b_1 - a_1)$ for all $n \geqslant 1$. By completeness of $\mathbf{R}$, these sequences converge to a common point, say c, of $(0, R)$. But then on one hand, $f(c) = \lim_{n \to \infty} f(a_n) = \lim_{n \to \infty} g(a_n) = g(c)$ while on the other, $f(c) = h(c)$, a contradiction.

8. This is essentially same as Exercise (1.16), except that $a_0 = 1$.

9. (i) $a_n = \dfrac{n+1}{2} + i^{n-1}\left(\dfrac{1}{4}i - 1\right) + (-i)^{n-1}\left(-\dfrac{i}{4} - 1\right)$

$\left(A(x) \text{ comes out as } \dfrac{x}{(1-x)^2(1+x^2)} + \dfrac{1-2x}{1+x^2}\right)$

   (ii) $a_n = \dfrac{3}{2^n n!} \cdot \left(A(x) \text{ satisfies the differential equation } \dfrac{dy}{dx} = \dfrac{1}{2}y.\right)$

   (iii) $a_n = 3^{n-1}$ for $n \geqslant 1$. (Rewriting the relation as $3^n = \sum\limits_{r=0}^{n} a_r 2^{n-r}$

   for $n \geqslant 1$, $C(x) = A(x)B(x)$ where $A(x)$, $B(x)$, $C(x)$ are the
   O.G.F.'s of $\{a_n\}$, $\{2^n\}$ and $\{3^n\}$ respectively.)

10. $a_n = n!\left(\dfrac{3}{1!} + \dfrac{3^2}{2!} + \dfrac{3^3}{3!} + \cdots + \dfrac{3^n}{n!}\right)$.

11. $a_n = 2.5^{n-1}$. (For a solution using E.G.F., write

$$n\,\frac{a_n}{n!}\,x^{n-1} = 5\frac{a_{n-1}}{(n-1)!}\,x^{n-1},$$

summing which the E.G.F. of $\{a_n\}$ satisfies the differential equation

$\dfrac{dy}{dx} = 5y.$)

12. (i) $y = 2 + x - x^2 - \dfrac{1}{3}x^3 + \dfrac{1}{4}x^4 + \cdots$

   (ii) $y = 1 + 2x - \dfrac{3}{2}x^2 - \dfrac{1}{6}x^3 + \dfrac{1}{3}x^4 + \cdots$

13. Let $f(x) = C_r x^r + C_{r-1}x^{r-1} + \cdots + C_1 x + C_0$ and $g(x) = x^{n-r}f(x)$.
Then α is a multiple root of $g(x)$. By Exercise (6.2.23), $g'(\alpha) = 0$
and hence $\alpha g'(\alpha) = 0$. Since $g(x) = C_r x^n + \cdots + C_0 x^{n-r}$, $xg'(x)$
$= C_r n x^n + \cdots + C_0(n-r)x^{n-r}$ and the result follows.

14. Follow the notation of the answer to the last exercise. Let $g_0(x)$
$= g(x)$. For $i = 1, \ldots, k-1$, let $g_i(x) = x g'_{i-1}(x)$. It is easily seen
that α is a root of $g_i(x)$ for $i = 0, \ldots, k-1$. Expanding the polynomials $g_i(x)$, gives the first assertion. For the second assertion it
suffices to show that the matrix

$$\begin{pmatrix} 1 & \alpha & \alpha^2 & \cdots\cdots & \alpha^{k-1} \\ 0 & \alpha & 2\alpha^2 & \cdots\cdots & (k-1)\alpha^{k-1} \\ 0 & \alpha & 2^2\alpha^2 & \cdots\cdots & (k-1)^2\alpha^{k-1} \\ \vdots & \vdots & \vdots & & \vdots \\ 0 & \alpha & 2^{k-1}\alpha^2 & \cdots\cdots & (k-1)^{k-1}\alpha^{k-1} \end{pmatrix}$$

is non-singular. The determinant of this matrix equals $\alpha^{k(k-1)/2}$ times that of the matrix

$$\begin{pmatrix} 1 & 1 & 1 & \ldots\ldots & 1 \\ 0 & 1 & 2 & \ldots\ldots & k-1 \\ 0 & 1 & 2^2 & \ldots\ldots & (k-1)^2 \\ \vdots & \vdots & \vdots & & \vdots \\ 0 & 1 & 2^{k-1} & \ldots..(k-1)^{k-1} \end{pmatrix}$$

which by Exercise (6.4.33) is non-zero.

15. Let $\alpha_1, \ldots, \alpha_p$ be the distinct, (possibly complex) characteristic roots with multiplicities $k_1, \ldots, k_p$ respectively so that

$$\sum_{i=1}^{p} k_i = r.$$

Then the general solution to (20) is

$$a_n = a_{n,1} + a_{n,2} + \ldots + a_{n,p}$$

where for each $i = 1, \ldots, p$, $a_{nsi} = c_{i,1}\alpha_i^n + c_{i,2}n\alpha_i^n + \ldots + c_{i,k_i}n^{k_i-1}\alpha_i^n$, $c_{i,1}, \ldots, c_{i,k_i}$ being arbitrary constants. In proving this, for $i = 1, \ldots, p$ and for $j = 0, \ldots, k_i - 1$, let $\bar{w}_{i,j}$ be the vector $\{n^j\alpha_i^n\}_{n=0}^{\infty}$ in V with the understanding that $0^0 = 1$. We already know these vectors are all in K. It is a little clumsy to prove their linear independence directly using determinants. A different argument runs as follows. Suppose there exist $\lambda_{i,j} \in \mathbb{C}$ not all 0 such that $\sum_{i=0}^{p} \sum_{j=0}^{k_i-1} \lambda_{i,j}\bar{w}_{i,j} = \bar{0}$. Pick any q for which there is some j such that $\lambda_{q,j} \neq 0$ and among all such j's, let s be the largest. Then $\lambda_{q,s} \neq 0$ and $\bar{w}_{q,s}$ is therefore in the linear span of the set S consisting of the vectors $\bar{w}_{i,j}$ for $i \neq q$ and the vectors $\bar{w}_{q,j}$ for $0 \leqslant j < s$. To get a contradiction it suffices to construct a homogeneous linear recurrence relation, such that every member of S is a solution of it but $\bar{w}_{q,s}$ is not. The linear homogeneous relation whose characteristic polynomial is $(x-\alpha_q)^s \prod_{i \neq q} (x-\alpha_i)^{k_i}$ has this property.

16. (a) $C = \frac{1}{2}D$ where $D = (d_{ij})$ is a $2k \times 2k$ matrix over $\mathbb{C}$ in which for $p = 1, \ldots, k$, $d_{2p-1,\,2p-1} = d_{2p-1,\,2p} = 1$, $d_{2p,\,2p} = i$, $d_{2p,\,2p-1} = -i$ and all other entries are 0. Generalising Exercise (6.4.39), $\det(D) = (2i)^n$, whence D and hence C is non-singular. Finally, apply Exercise (6.4.29).

(b) Since the coefficients are real, $re^{-i\theta}$ is also a (complex) root of multiplicity k. For each $0 \leqslant j \leqslant k-1$, express $\{n^j r^n \cos n\theta\}_{n=0}^{\infty}$ and $\{n^j r^n \sin n\theta\}_{n=0}^{\infty}$ as linear combinations (over $\mathbb{C}$) of the vectors $n^j r^n c^{in\theta}$ and $\{n^j r^n e^{-in\theta}\}_{n=0}^{\infty}$, both of which are in K

by Exercise (3.14) above. As for their linear independence, let B be the $2k \times 2k$ matrix whose rows are truncations of the vectors $\{n^j r^n \cos n\theta\}$ and $\{n^j r^n \sin n\theta\}$ and A be the $2k \times 2k$ matrix whose rows are truncations of the vectors $\{n^j r^n e^{in\theta}\}$ and $\{n^j r^n e^{-in\theta}\}$. By Exericse (3.15), A has rank $2k$ over $\mathbb{C}$. So by (a). B has rank $2k$ over $\mathbb{C}$, i.e. its rows are linearly independent over $\mathbb{C}$ and *a fortiori* over $\mathbb{R}$.

17. Let $\alpha_1, \ldots, \alpha_p$ be real characteristic roots of multiplicities $k_1, \ldots, k_p$ respectively and $r_1 e^{\pm i\theta_1}, \ldots, r_q e^{\pm i\theta_q}$ be complex characteristic roots of multiplicities $m_1, m_1, m_2, m_2, \ldots, m_q, m_q$ (so that $\sum_{j=1}^{p} k_i + 2 \sum_{i=1}^{q} m_j = r$). Then a basis for K consists of the vectors of the form $\{n^j \alpha_i^n\}_{n=0}^{\infty}$ for $i = 1, \ldots, p, 0 \leqslant j \leqslant k_i - 1$ and vectors of the form $\{n^j r_i^n \cos n\theta_i\}_{n=0}^{\infty}$ and $\{n^j r_i^n \text{ si } n\theta_i\}_{n=0}^{\infty}$ for $i = 1, \ldots, q$ and $0 \leqslant j \leqslant m_i - 1$.

18. Since $C_r a_n + C_{r-1} a_{n-1} + \ldots + C_1 a_{n-r+1} + C_0 a_{n-r} = \beta^n$ and $C_r a_{n-1} + C_{r-1} a_{n-2} + \ldots + C_1 a_{n-r} + C_0 a_{n-r-1} = \beta^{n-1}$, the result follows by subtracting β times the second equation from the first.

19. That $K \subset L$ follows from subtracting β times the equation $C_r a_{n-1} + C_{r-1} a_{n-2} + \ldots + C_1 a_{n-r} + C_0 a_{n-r-1} = 0$ from the equation $C_r a_n + C_{r-1} a_{n-1} + \ldots + C_1 a_{n-r+1} + C_0 a_{n-r} = 0$. The last exercise shows $K' \subset L$. For the second assertion, K' is of the form $K + \bar{v}$ for some $\bar{v} \in L$. Also $K \cup \{\bar{u}\}$ spans L. So $\bar{v}$ is of the form $\bar{x} + \lambda \bar{u}$ for some λ and some $\bar{x} \in K$. But then $\lambda \bar{u} = (-\bar{x}) + \bar{v} \in K'$. If $\mu \bar{u}$ is also in K' for $\mu \neq \lambda$ then $(\mu - \lambda)\bar{u} \in K$ which would imply $\frac{1}{\mu - \lambda}(\mu - \lambda)\bar{u} \in K$, i.e. $\bar{u} \in K$, a contradiction. So λ is unique.

20. Let $\bar{u} = \{\beta^n\}_{n=0}^{\infty}$. Continuing the notation of the last exercise and following the hint, $\bar{u} \in L$. But $\bar{u} \notin K$. The result now follows from the last exercise.

21. As a characteristic root of (41), β has multiplicity $k + 1$. So the vector $\bar{u} = \{n^k \beta^n\}_{n=0}^{\infty}$ is in L but not in K.

22. This follows directly by substituting in (42), the values of $b_n, b_{n-1}, \ldots, b_{n-p}$ from (43) and regrouping the terms.

24. In (42), let $p = k + 1$ and $D_{p-i} = (-1)^i \binom{k+1}{i}$ for $i = 0,$ $\ldots, p$. Then it becomes a linear homogeneous recurrence relation whose characteristic polynomial is $(x-1)^{k+1}$. Both the null vector and the vector $\{n^k\}_{n=0}^{\infty}$ are solutions of this relation. The former

implies $K \subset L$ while the latter implies $K' \subset L$, both in view of Exercise (3.22).

25. By Exercise (3.15), applied to (45), L has a basis of the form $S \cup \{\bar{v}_0, \ldots, \bar{v}_k\}$ where S consists of vectors corresponding to characteristic roots of (45), other than 1. These are precisely the characteristic roots of (44). So S is a basis for K. Hence $K \cup \{\bar{v}_0, \cdots, \bar{v}_k\}$ spans L. Now K' is of the form $K + v$ for some $\bar{v} \in L$ (since $K' \subset L$). Write $\bar{v}$ as $\bar{x} + A_0\bar{v}_0 + \ldots + A_k\bar{v}_k$ for some $\bar{x} \in K$ and some constants $A_0, \ldots, A_k$. Then $A_0\bar{v}_0 + \ldots + A_k\bar{v}_k \in K'$, implying the result.

27. (i) $a_n = c_1 2^n + c_2 3^n - n2^{n+1} + \dfrac{n}{2} + \dfrac{7}{4}$, where c_1, c_2 are arbitrary

constants. (ii) $a_n = 1 - \left(\dfrac{1}{\sqrt{2}} + \dfrac{37}{20}\right) n + \left(\dfrac{1}{\sqrt{2}} - \dfrac{5}{8}\right) n^2 + \dfrac{1}{3} n^3 + \dfrac{1}{8}$

$n^4 + \dfrac{1}{60} n^5$. (iii) $a_n = c_1 \left(-\dfrac{1}{2}\right)^n + c_2 \left(\dfrac{1}{3}\right)^n + c_3 \cos \dfrac{2\pi n}{3} + c_4 \sin \dfrac{2\pi n}{3}$

where c_1, c_2, c_3, c_4 are arbitrary constants. (Use Exercise (6.2.34) to detect rational characteristic roots. By trial they are $-\frac{1}{2}$ and $\frac{1}{3}$. The

remaining characteristic roots are $\dfrac{-1 \pm \sqrt{3}}{2}$.)

28. Use Cramer's rule (Exercise (6.4.16)) along with Exercise (6.4.33) and the fact that $\alpha_1, \ldots, \alpha_r$ are all non-zero.

29. Let K be the solution space of (20). We know $\dim(K) = r$. Let $\{\bar{w}_1, \ldots, \bar{w}_r\}$ be a basis for K, where $\bar{w}_i = \{w_{i,n}\}_{n=0}^{\infty}$ for $i = 1, \ldots, r$. Let $k \geqslant 0$. For $i = 1, \ldots, r$, let $\bar{v}_i = (w_{i,k}, w_{i,k+1}, \ldots, w_{i,k+r-1})$ and let A be the matrix with rows $\bar{v}_1, \ldots, \bar{v}_r$. It suffices to show that A is non-singular. However, as observed in the answer to Exercise (3.15), it is a little clumsy to do this by computing $\det(A)$. Instead, observe that if A is singular then there exist $\lambda_1, \ldots, \lambda_r$ not all 0 such that $\sum\limits_{i=1}^{r} \lambda_i w_{i,j} = 0$ for every $j = k, k+1, \ldots, k+r-1$. Now, for $i = 1, \ldots, r$, $\bar{w}_i$ is a solution of (20). Since $C_r \neq 0$, each $w_{i,j}$ is a linear combination of the preceeding r terms. So the equation $\sum\limits_{i=1}^{r} \lambda_i w_{i,j} = 0$ holds, successively, for $j = k + r, k + r + 1, \ldots$. Similarly since $C_0 \neq 0$, it holds for $j = k - 1$, $k - 2, \ldots, 0$. So it holds for all $j = 0, 1, \ldots$. But that means the vectors $\bar{w}_1, \ldots, \bar{w}_r$ are linearly dependent, a contradiction.

30. If $F_{m,n}(x) = \sum\limits_{n=0}^{\infty} \dfrac{a_{m,n} x^n}{n!}$, then $F_0(x) = 1$ and for $m \geqslant 1$, $F_m(x) = e^x F_{m-1}(x)$. So, $F_m(x) = e^{mx}$ by induction. Hence $a_{m,n} = m^n$.

31. Let $A(x)$ and $B(x)$ be the O.G.F.'s of $\{a_n\}$ and $\{b_n\}$ respectively. Then the two equations give $[A(x)]^2 + [B(x)]^2 = \dfrac{1}{(1+x)^2} + \dfrac{1}{(1-x)^2}$ and $A(x) B(x) = \dfrac{1}{1-x^2}$. Solving, $A(x) = \dfrac{1}{1+x}$ and $B(x) = \dfrac{1}{1-x}$. (Other possibilities are ruled out by initial conditions.) So $a_n = (-1)^n$ and $b_n = 1$ for all n.

32. If $E(x) = \sum\limits_{n=0}^{\infty} \dfrac{a_n}{n!}$ then $E'(x) = \sum\limits_{n=1}^{\infty} \dfrac{a_n}{(n-1)!} x^{n-1} = \sum\limits_{n=0}^{\infty} \dfrac{a_{n+1}}{n!} x^n$, $E''(x)$ $= \sum\limits_{n=0}^{\infty} \dfrac{a_{n+2}}{n!} x^n, \ldots, E^{(r)}(x) = \sum\limits_{n=0}^{\infty} \dfrac{a_{n+r}}{n!} x^n$. The result follows by direct substitution. It is well-known that $e^{\alpha x}$ is a solution of the differential equation iff α is a characteristic root and that if $\alpha_1, \ldots, \alpha_r$ are distinct characteristic roots, then the solutions $e^{\alpha_1 x}, e^{\alpha_2 x}, \ldots, e^{\alpha_r x}$ constitute a basis for the solution space. From this and the fact that $e^{\alpha x}$ is the E.G.F. of $\{\alpha^n\}_{n=0}^{\infty}$, we get an alternate proof of Theorem (3.8).

Section 7.4

1. Let a_n denote the number of ways to form a stack of value n rupees. Depending upon whether the coin at the top has value 1 or 2 rupees, the rest of the stack can be built in a_{n-1} or in a_{n-2} ways. So $a_n = a_{n-1} + a_{n-2}$ for $n \geqslant 2$. Also $a_0 = 1$ (the empty stack) and $a_1 = 1$.

2. In a stack worth n rupees, let the number of 2-rupee coins be k. Then the total number of coins in it is $n - k$. Now $\dbinom{n-k}{k}$ is the number of stacks that can be formed with k 2-rupee coins and $n-2k$ 1-rupee coins.

3. Let a_n be the number of ways to cover a $2 \times n$, rectangle with dominos. Any such covering must be one of the following forms, depending on how the last 2×1 column is covered. So clearly $a_n = a_{n-1} + a_{n-2}$.

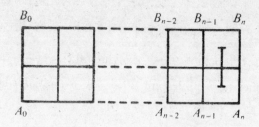

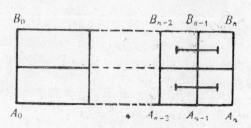

4.   Let b_n be the number of sequences of the desired type. The first $n-2$ entries of any such sequence constitute a binary sequence of length $n-2$ in which every 1 is immediately followed by a 0. Treating 10 as a single symbol which occupies two spaces, the number of such sequences is precisely a_{n-2}, where a_n is as in the answer to Exercise (4.1).

5.   Clearly $p_0 = 0$ while, by the last exercise, for $n \geqslant 1$,

$$p_n = \frac{\left(\dfrac{1+\sqrt{5}}{2}\right)^{n-1} - \left(\dfrac{1-\sqrt{5}}{2}\right)^{n-1}}{\sqrt{5}\,2^n}.$$

The result follows from by summing $\sum\limits_{n=1}^{\infty} p_n t^n$.

6.   The choice is, of course, yours! But monetarily it will be a good idea to play the game only if the expected number of tosses before winning the game does not exceed 5. By Proposition (2.13), this number equals $p_X'(1)$ where $p_\lambda(t)$ is as in the last exercise. A direct computation gives $p_X'(1) = 6$. So if you are 'averagely' lucky it would cost you 6 rupees to win 5 rupees.

8.   The problem is to count b_n, the number of sequences of A, B, C whose total cost is n. As in Exercise (4.1), we classify such sequences depending upon their last entries and get

$$b_n = b_{n-1} + 2b_{n-2} \ (n \geqslant 2) \text{ with } b_0 = 1, b_1 = 1.$$

Solving, $b_n = \dfrac{2^{n+1} + (-1)^n}{3}$.

9. (i) is straight computation. For (ii) classify the selections into two types: those containing m and those not containing m. (v) follows by multiplying both sides of (ii) by y^n and summing over from $n = 0$ to ∞. $F_0(y)$ and $F_1(y)$ are obtained from the boundary conditions, $b_{0,0} = 1$, $b_{0,n} = 0$ for $n > 0$, $b_{1,0} = b_{1,1} = 1$, $b_{1,n} = 0$ for $n > 1$. Multiplying both sides of (v) by x^m and summing over from $m = 2$ to ∞ gives

$$B(x, y) - 1 - x(1 + y) - x(B(x, y) - 1) - yx^2 B(x, y) = 0$$

which gives (vi). Thinking of $B(x, y)$ as a function of x and resolving into partial fractions and expanding again gives (vii). Alternatively (v) can be thought of as a homogeneous, linear recurrence relation for the sequence $\{F_m(y)\}_{m=0}^{\infty}$ whose characteristic polynomial is $x^2 - x - y$. The characteristic roots are $\dfrac{1 \pm \sqrt{1 + 4y}}{2}$ and the initial conditions give (vii) as the solution. The R.H.S. of (vii) equals $\dfrac{1}{2^{m+1}} \sum_{k=1}^{\infty} \binom{m+2}{2k+1} (1 + 4y)^k$ and (viii) follows by expanding $(1 + 4y)^k$.

10. $\dfrac{1}{n} \binom{2n-2}{n-1}$ and $(n-1)!$ respectively. (For the second question, the numbers, say b_n, satisfy the recurrence relation $b_n = (n - 1)b_{n-1}$ for $n \geqslant 2$. To see this, note that at the last stage a piece of length 2 must have been cut into two parts. This piece could be chosen in $(n - 1)$ ways and thought of as a single piece throughout except at the last cut. A solution without recurrence relations is also possible. Each cut must occur at one of the $n - 1$ 'nodes' on the stick and the order of these nodes determines the way the stick is cut.)

11. The key step in the argument is term-by-term differentiation, which is valid for all x because both the series have $(-\infty, \infty)$ as their interval of convergence.

12. $\dfrac{n! \, (n-1)!}{2^{n-1}}$. (Similar to the second part of Exercise (4.10), except that this time the recurrence relation is $b_n = \binom{n}{2} b_{n-1}$.

13. $n(n - 1) + 2$.

14. Follow the hint and apply the solution to the Vendor Problem.

15. Given any balanced arrangement of n pairs of parentheses, define $f: \{1,..., n\} \to \{1, 2,..., n\}$ by $f(x) =$ number of left parentheses occurring before the xth right parenthesis. Then f is monotonically increasing and $f(x) \geqslant x$ for all x. Also the original arrangement can be recovered from f.

16. The function g in the hint is from $\{1,..., n\}$ into itself, monotonically increasing and satisfies $g(x) \geqslant x$ for $x = 1,..., n$. Apply the last exercise.

17. 2 is the only characteristic root. It has multiplicity 1. For a particular solution, try $a_n = A$. Then $A = 2A + 1$ gives $A = -1$. So the general solution to (21) is $a_n = c\, 2^n - 1$ where c is an arbitrary constant. $a_0 = 0$ gives $c = 1$.

18. The crucial part of the construction is to design, for a given ordered pair, say (a, b), of Boolean variables, a Boolean device $D(a, b)$ which can be controlled by b if $a = 1$, but on which b has no control if $a = 0$. ($D(a, b)$ cannot be expressed merely as a Boolean function of a and b, because in that case it will not be able to distinguish between the orders in which a and b are set equal to 1.) One such device is a relay D, shown below (where X is an auxiliary relay).

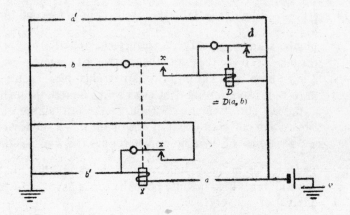

Here a and b may themselves be Boolean functions of some other Boolean variables. Now the desired circuit can be built as follows. Let the control circuit of Y_1 be as shown below.

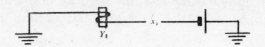

For $i=2,..., n$, let $Y_i=D(y_1'\ y_2'...y_{i-2}'y_{i-1}, x_i)$. Finally, let the closure function of the lamp be $y_1'y_2'...y_{n-1}\ y_n$. To count the number of flips directly, let a_r be the number of flips needed to activate the relay Y_r starting and ending in the position where $Y_1,..., Y_{r-1}$ are all released. Then a_r is also the number of flips needed to release Y_r starting and ending in this position. This gives $a_{r+1} = 2a_r + 1$, with $a_1 = 1$, which is exactly the same recurrence relation as in the Tower of Hanoi problem.

20. Let a_n, b_n, c_n denote respectively the number of ways to cover the following three figures.

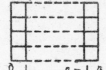  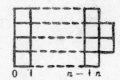

Then we get

$$a_n = a_{n-1} + a_{n-2} + 2b_{n-2} + c_{n-2} \quad (n \geqslant 2)$$
$$b_n = a_n + b_{n-1} \qquad\qquad\qquad (n \geqslant 1)$$
$$c_n = a_n + c_{n-2} \qquad\qquad\qquad (n \geqslant 2).$$

Converting these into equations about their O.G.F.'s and solving, we get $A(x) = \sum_{n=0}^{\infty} a_n x^n = \dfrac{1 - x^2}{1 - x - 5x^2 - x^3 + x^4}$. (equivalently, $a_n = a_{n-1} + 5a_{n-2} + a_{n-3} - a_{n-4}$). But the roots of the polynomial $x^4 - x^3 - 5x^2 - x + 1$ are not easy to find and so a closed form expression for a_n seems difficult.

21. Write the first row as the sum of $(r-\lambda, 0, 0,..., 0)$ and $(\lambda, \lambda, \lambda,..., \lambda)$. Then by Proposition (6.4.25), $\det (A_n) = (r - \lambda) \det (A_{n-1}) + f(n)$ where $f(n)$ is the determinant of the $n \times n$ matrix.

$$\begin{bmatrix} \lambda & \lambda & \lambda & ...... & \lambda \\ \lambda & r & \lambda & & \lambda \\ \lambda & \lambda & r & & \lambda \\ \vdots & & & & \vdots \\ \lambda & \lambda & \lambda & ...... & r \end{bmatrix}$$

Substracting first row from all others, by Exercise (6.4.13) $f(n) = \lambda(r - \lambda)^{n-1}$. So $\det(A_n)$ satisfies the linear recurrence relation

$$b_n - (r - \lambda)\, b_{n-1} = \lambda(r - \lambda)^{n-1} \text{ with } b_1 = r.$$

Solving this gives the result. (Since $r - \lambda$ is a characteristic root, for a particular solution try $b_n = An(r - \lambda)^n$.)

A solution without recurrence relations is also possible. Subtract the first column of A_n from every other column and then add to the first row every other row.

22. $\det(A_n)$ satisfies the homogeneous linear recurrence relation

$$a_n - \alpha\, a_{n-1} + \beta^2 a_{n-2} = 0 \text{ with } a_1 = \alpha, \; a_2 = \alpha^2 - \beta^2.$$

The characteristic roots are real and distinct if $\alpha > 2\beta$, real and coincident if $\alpha = 2\beta$ and complex if $\alpha < 2\beta$. With suitable substitutions the answer in the three cases out as

$$\det(A_n) = \begin{cases} \dfrac{(\sec\theta + \tan\theta)^{n+1} - (\sec\theta - \tan\theta)^{n+1}}{2\tan\theta}\,\beta^n & \text{where } \sec\theta = \dfrac{\alpha}{2\beta} \\[2em] (n+1)\beta^n & \text{if } \alpha = 2\beta \\[1.5em] \dfrac{\sin(n+1)\theta}{\sin\theta}\,\beta^n & \text{where } \cos\theta = \dfrac{\alpha}{2\beta} \end{cases}$$

It is interesting to note that the case $\alpha = 2\beta$ results from either of the other two cases by taking limit as $\theta \to 0$.

23.   Call $\dfrac{n - n(-1)^n - i^{n-1} - (-i)^{n-1}}{8}$ as $f(n)$. If n is even then $n = n(-1)^n$ and $(-i)^{n-1} = -i^{n-1}$ and so $f(n) = 0$. If $n = 4k + 1$ then $f(n) = \dfrac{4k + 1 + (4k + 1) - 1 - 1}{8} = k$. Similarly verify the case $n = 4k - 1$.

24. $t_n = -\dfrac{1}{288} + \dfrac{\omega^n}{9} + \dfrac{\omega^{2n}}{9} + \dfrac{n^2}{48} + \dfrac{n}{16} + (-1)^{n+1}\left(\dfrac{2n + 3}{32}\right)$
$- \dfrac{i^n(1 + (-1)^n)}{16} + \dfrac{i^{n+1}(1 - (-1)^n)}{16},$

where, as usual $\omega = -\dfrac{1}{2} + i\,\dfrac{\sqrt{3}}{2}$. Since ω^n, ω^{2n} depend only on the residue class of n modulo 3 while i^n depends on the residue class of n modulo 4, the formula for t_n can be replaced by 12 separate, simpler formulas, depending upon the congruence class of n modulo 12. For example, $t_n = \dfrac{n^2}{48} + \dfrac{n}{8} + \dfrac{5}{48}$ if $n \equiv 7 \pmod{12}$. See also Exercise (2.22) for n even.

25. Let x_n be the integer closest to $2^{n-1}/5$. Then

$$
x_n = \begin{cases}
\dfrac{2^{n-1}}{5} - \dfrac{1}{5} & \text{if } 2^{n-1} \equiv 1 \pmod{5} \\[2ex]
\dfrac{2^{n-1}}{5} - \dfrac{2}{5} & \text{if } 2^{n-1} \equiv 2 \pmod{5} \\[2ex]
\dfrac{2^{n-1}}{5} + \dfrac{2}{5} & \text{if } 2^{n-1} \equiv 3 \pmod{5} \\[2ex]
\dfrac{2^{n-1}}{5} + \dfrac{1}{5} & \text{if } 2^{n-1} \equiv 4 \pmod{5}
\end{cases}
$$

If $m \equiv n \pmod{4}$, it is easily seen that $2^{m-1} \equiv 2^{n-1} \pmod{5}$. Since the values of $\sin n\pi/2$ and $\cos n\pi/2$ also depend only on the residue class of n modulo 4, it suffices to prove the assertion only for $n = 1, 2, 3$ and 4. This can be done by direct verification.

26. Multiplying both sides of (30) by x^n and summing over from $n = 3$ to ∞, gives $A(x) - 1 + x^2(A(x) - 1) = \dfrac{x^3}{1 - 2x}$ from which the assertion follows.

27. (i) Apply induction on k. Let $B_{n,k}$ be the set of sequences of length n in which the pattern $\bar{a}$ occurs exactly k times, with the last occurrence at the end. Every member say $\bar{x}$ of $B_{n,k}$ can be thought of as a concatenation of a member of $B_{i,1}$ and a member of $B_{n-1,\,k-1}$ for a unique i. So the O.G.F. of $\{b_{n,k}\}$ is the product of the O.G.F. of $\{b_{n,1}\}$ and that of $\{b_{n,k-1}\}$.

(ii) Define the set $C_{n,k}$ so that $|C_{n,k}| = c_{n,k}$. An element of $c_{n,k}$ is the concatenation of an element of $B_{i,k}$ and any binary sequence of length $n - i$ where i is such that the kth occurrence of $\bar{a}$ ends at the ith digit.

(iii) This follows from (ii) since $d_{n,k} = c_{n,k} - c_{n,k+1}$.

(iv) For $n \geq 1$, $a_n = \sum\limits_{k=1}^{\infty} b_{n,k}$. (For a fixed n, the sum is actually finite since $b_{n,k} = 0$ for $k > \lceil n/m \rceil$.) Since $a_0 = 1$, and $b_0 = b_{0,k} = 0$ for $k \geq 1$, we get

$$
A(x) - 1 = B(x) + B^2(x) + \ldots + B^k(x) + \ldots = \frac{B(x)}{1 - B(x)}.
$$

28. (i) $c_n = 4c_{n-1} - 5c_{n-2} + 3c_{n-3} - 2c_{n-4}$. (The O.G.F. is $B(x)/1-2x$ where $B(x)$ is given by (34).)

(ii) $d_n = 6d_{n-1} - 14d_{n-2} + 18d_{n-3} - 17d_{n-4} + 12d_{n-5} - 5d_{n-6} + 2d_{n-7}$.

29. Let $\bar{x} = (x_1, \ldots, x_n)$ be a binary sequence in which the pattern 111 appears for the first time at the end. Then for $n > 3$, $\bar{x}$ must begin with 0 or with 10 or with 110. Adding the three possibilities gives the recurrence relation. The values of $b_1, b_2, \ldots, b_{10}$ are, respectively, 0, 0, 1, 1, 2, 4, 7, 13, 24 and 44.

30. The number of favourable cases is $\sum_{n=3}^{10} b_n 2^{10-n} = 128 + 64 + 64 + 64 + 56 + 52 + 48 + 44 = 520$.

31. Suppose there are 1024 players, corresponding to the 2^{10} binary sequences of length 10. Out of these, 520 are winners as calculated in the last exercise. In all there are 10,240 tosses. The number of dummy tosses from the winners is $\sum_{n=1}^{10} b_n 2^{10-n} (10 - n) = 2,176$. To count the dummy tosses from the losers, note that the number of players who have won by the 7th round or earlier is 376. Out of the 648 remaining players, 324 will get a tail on the eighth round and stop playing, increasing the number of dummy tosses by 648 ($= 2 \times 324$). Among the 324 players who get a head on the eighth round, 52 win the game on the eighth round and have been already accounted for, 272 players remain. Of them, 136 get a tail on the nineth round and hence contribute 136 dummy tosses. The remaining 136 players make no new contribution to the dummy tosses, because the winners among them have already been counted and the losers go for the tenth round anyway. The number of dummy tosses is thus $2,176 + 648 + 136 = 2,960$. Hence the total 'revenue' got from all 1024 players is $10240 - 2960 = 7280$. If this is divided equally among the 520 winners, we get 14 rupees as the amount of the reward for a win.

Index

Errata

Page No.	Location	Correction
4	line 3 from bottom	read 'n, p_n' instead of '$n \cdot p_n$'
35	Exercise (3.12)	read 'Figure 1.7' instead of 'Figure 1.8'.
48	line 13 from bottom	read 'of' instead of 'or'
59	Figure caption	read 'Figure 2.1' instead of 'Figure 21'.
88	Exercise (2.3)	read 'say' instead of 'says'
89	Exercise (2.16)	read 'Proposition' instead of 'Corollary'
93	Figure (2.7)	move the second arrow slightly to the left
103	formula for $|T|$	read '$(n+1)$' instead of '$(n+1$'
106	Figure 2.10	add one more dot to the first row and make the dot in the last row smaller
108	Exercise (3.11)	read 'eastward' instead of 'eastaward'
122	Theorem (4.8)	read '$m-1$' instead of '$m=1$' on top of Σ
145	topmost expression	read '$\binom{n}{k}$' instead of '$\begin{bmatrix} n \\ k \end{bmatrix}$'
171	Exercise (2.3)	read '$R \cup R^{-1}$ and $R \cap R^{-1}$' instead of '$R \cup R^{-1}$ and $R \cup R^{-1}$'
194	Exercise (3.5)(iii)	read 'chains' instead of 'chainr'
219	Exercise (4.24)(v)	read 'μ' instead of 'u'
231	line 16	read 'elements' instead of 'element'
263	line 10	read '$|\blacksquare$' instead of '$\blacksquare$'
263	Figure 4.4(b)	read '$\searrow y$—' instead of '$\searrow$—'
276	line 14	let the second arrow touch the line above
280	Figure 4.20(a)	the labels of switches are x_1, x_2 in line 1, x_1', x_2' in line 2 and x_1, x_2' in line 3
280	Figure 4.20(b)	the labels of switches on wires leading to the vertical wire are a and b'
281	Figure 4.22	replace 'X' by 'x' in the labels
281	line 9 from bottom	delete 'as'
313	line 23	read 'None' instead of 'Note'
343	line 4 from bottom	read 'said to be' instead of 'said to'
368	line 4 from bottom	read '$\pm 3\}$' instead of '± 3'
395	line 22	read 'of F' instead of 'of R'
404	line 13	read '$[x, bx]$' instead of '$[b, bx]$'
404	line 13	read '$[bx, x]$' instead of '$[bx, b]$'

Page No.	Location	Correction
404	line 14	read '$[bx, x]$' instead of '$[bx, b]$'
404	line 15	read '$[ax^2, bx^2]$ and hence '$[a, b]$' instead of '$[a, b]$'
413	line 5 from bottom	read '$\mathbf{Q}$' instead of 'Q'
413	line 5 from bottom	read 'of $\mathbb{Z}$' instead of 'of e'
423	line 12	read '$g(\alpha) = 0$' instead of '$g(x) = 0$'
427	Definition (2.29)	add '(A u.f.d. is also called a factorial ring.)'
430	Exercise (2.15)	add 'Let $A = A_1 \cap \ldots \cap A_k$.' after the second sentence.
459	Exercise (3.26)	read 'Q' instead of '$\mathbf{Q}$'
571	line 4	read 'neighbourhood' instead of 'neighboured'
587	Exercise 3.6(a)	read 'for $p = 1$' instead of 'for $b = 1$'
587	Exercise 3.16(a)	read $\begin{pmatrix} \frac{1}{2} & \frac{1}{2} \\ -\frac{i}{2} & \frac{i}{2} \end{pmatrix}$ instead of $\begin{pmatrix} \frac{1}{2} & \frac{1}{2} \\ -\frac{1}{2} & \frac{1}{2} \end{pmatrix}$
616	line 11 from bottom	read 'Burnside' instead of 'Berstein'
669	line 1 from bottom	read '(a) $x_1 + x_2'$; (b) a'''
732	line 13	read 'comes out' instead of 'out'